THE BIOSYNTHESIS
OF AROMATIC
COMPOUNDS

THE BIOSYNTHESIS OF AROMATIC COMPOUNDS

ULRICH WEISS
National Institutes of Health
Bethesda, Maryland

J. MICHAEL EDWARDS
The University of Connecticut, Storrs

A Wiley-Interscience Publication
JOHN WILEY & SONS, New York • Chichester • Brisbane • Toronto

Library of Congress Cataloging in Publication Data:

Weiss, Ulrich, 1908–
   The biosynthesis of aromatic compounds.
   "A Wiley-Interscience publication."
   Includes bibliographical references and index.
   1. Aromatic compounds—Synthesis.   I. Edwards,
J. Michael, joint author.   II. Title.

QD335.W42   574.1'929   78-1496
ISBN 0-471-92690-6

Printed in the United States of America

10 9 8 7 6 5 4 3 2 1

Preface

It is the aim of this book to give a comprehensive review of the chemical processes and substances involved in the formation of benzenoid rings by living organisms. The book has been written by organic chemists and addresses itself primarily to organic chemists interested in the biosynthetic processes to which the naturally occurring substances that they are studying owe their existence. The emphasis is, therefore, on the chemical nature of the compounds of low molecular weight that form the substrates of the biosynthetic processes, rather than on the enzymatic catalysts which usually promote these reactions. Consequently, the descriptions of these catalysts from the enzymological viewpoint are kept short, and the same is true for the fascinating biochemical and genetic mechanisms that regulate the biosyntheses. We have tried, however, to include enough discussion of these more biological aspects to make possible—we hope—at least a basic understanding of these questions; we have also endeavored to provide a bibliography which could serve as a guide to more detailed study.

Excellent reviews exist for many parts of the field of the biosynthesis of aromatic compounds. The several pathways which are responsible for the formation of the majority of naturally occurring benzenoid substances have been well summarized, and the same is true for some more specialized topics such as the biosynthesis of the oestrogens. On the other hand, no reviews dealing specifically with the aromatic representatives of certain classes of natural compounds (e.g., terpenoids, carotenoids, and polyacetylenes) have appeared; we have attempted to provide these. No comprehensive account of the biosynthesis of aromatic compounds in its entirely has thus been available up to now. The present volume is intended to fill this gap. It is evidently impossible to do justice to every aspect of the widely ramified problems or to mention every single pertinent finding; we have at least endeavored to collect and correlate all information that seemed significant, and we hope that we have not overlooked any facts, theories, or concepts of major importance. The fact that no humanly feasible effort can ensure complete freedom from errors and omissions is clear to everyone who has tried to deal with the recent literature in any scientific field.

Through circumstances beyond our control, the completion of the manuscript took much longer than we had anticipated. We have made every effort to bring the text up to date by insertions; in several cases this was done by providing an addendum to a particular chapter. The literature has been covered through 1972 as far as possible in the main text, through 1973 in the additions.

In principle this book is restricted to those parts of the various biosynthetic pathways which lead from generally available building blocks (e.g., sugars, acetic acid, and mevalonic acid) to the first fully aromatic compound in a given sequence. The further fate of the aromatic ring is not considered to be of primary concern. However, it is obviously impossible to observe this limitation too strictly. It is, for instance, not always known whether a particular structural feature of a given natural compound was introduced before or after the actual formation of the aromatic ring. Examples include, among others, the addition and especially the removal of oxygenated functions and the introduction of methyl groups and isoprenoid residues. When such events take place prior to the actual aromatization, they amount to a modification of the particular biosynthetic sequence leading to the aromatic ring and must thus be included in the discussion of this pathway. In many instances, however, these reactions are evidently modifications of preexisting aromatic structures, whereas in other cases—probably the great majority—the exact stage at which they occur remains unknown. In view of this uncertainty a general discussion of these processes was needed. In other cases the really important product of a sequence is not the earliest aromatic one; an account of the biosynthesis of tryptophan, for instance, would be manifestly incomplete if it were to stop at anthranilic acid, which happens to be the first aromatic compound in this process. Finally, it was absolutely necessary to discuss briefly the significance of the several pathways as sources of the endless variety of aromatic substances that have been found to occur in living organisms, or at least to be obtainable from natural sources. (Actually, many so-called "natural products" must be artefacts formed through postmortem changes or as a consequence of isolation procedures used; several cases of this kind are mentioned in subsequent pages.) A review of the chemical aspects of the biosynthesis of aromatic substances would be inadequate if it did not at least try to provide an insight into the marvellous web of interacting reaction sequences by which organisms build the substances which they require for their existence. Hence it was essential to provide a brief general account of our knowledge of the events that lie beyond the first aromatic compound.

Quinonoid compounds are included, whereas nonbenzenoid rings with aromatic character, such as pyridines, thiophenes, and tropolones, are considered outside the scope of this review, as are the pathways of catabolic breakdown of aromatic compounds.

We wish to give thankful acknowledgment to those colleagues and friends who have helped us through critical evaluation of parts of the manuscript; their criticisms and corrections were invaluable. We want to mention in particular Drs. Marjorie Anchel, A. G. Andrewes, the late Thomas J. Batterham, F. Bohlmann, John W. Daly, Donald M. Jerina, Ellis S. Kempner, Synnøve Liaaen-Jensen, Judith G. Levin, Robert G. Martin, Trevor C. McMorris, John W. Rowe, and D. F. Zinkel. Dr.

Tima N. Johnson has given helpful assistance with the manuscript of the chapters on terpenoids and carotenoids. Our special thanks are due to the staff of the Medical Arts and Photography Branch of the National Institutes of Health, who provided the drawings and photographs for the innumerable formulas and diagrams.

ULRICH WEISS
J. MICHAEL EDWARDS

Bethesda, Maryland
Storrs, Connecticut
March 1980

Contents

PART 2   BIOSYNTHESIS OF AROMATIC
COMPOUNDS FROM ACETIC ACID

PART 3   AROMATIC COMPOUNDS DERIVED
FROM MEVALONIC ACID

PART 4   ADDITIONAL SEQUENCES

THE BIOSYNTHESIS
OF AROMATIC
COMPOUNDS

CHAPTER 1

Introduction

1-1. HISTORY OF THE PROBLEM; MAIN PATHWAYS OF BIOSYNTHESIS OF AROMATIC COMPOUNDS.

The immense variety of aromatic natural compounds and the biological and pharmacological importance of many of them have stimulated research in this field from the earliest times of organic chemistry. Indeed, the work of Liebig and Wöhler (1) in 1832 on "The Radical of Benzoic Acid" marks the very beginning of structural organic chemistry, showing that certain groups of atoms can remain unchanged through a substantial series of chemical transformations. Subsequent elucidation of the structures of an ever-increasing number of natural aromatic substances led by the turn of the century to a sufficiently broad knowledge of the types of such plant products to make possible the earliest attempts at an understanding of their biosynthesis.

The reference to "plant products" requires some comments. The vast majority of known aromatic compounds is produced by plants or bacteria, a fact that led to the conclusion that the formation of aromatic rings from nonbenzenoid precursors is restricted to these organisms. This concept is in evident agreement with the well-known fact that animals must depend upon their food for supplies of the aromatic amino acids, and of a number of vitamin factors (vitamins E, K, p-aminobenzoic acid, etc.). There is, however, one well-authenticated exception: the biosynthesis of the oestrogens from nonaromatic steroids (see Section 24-1), which shows that it is not aromatization per se which the animal organism is incapable of performing. In a few other instances, the formation of aromatic metabolites in animals has been observed. It does not seem, however, that in these cases biosynthesis of aromatic rings by the metazoan organisms themselves has been established, since participation of symbiotic or contaminating microorganisms has not been excluded.

It has been observed, for example, that glucose is converted to protocatechuic acid, presumably by way of tyrosine, in the cockroach, *Periplaneta americana* (2); similarly, the

1

methyl- and ethyl-p-benzoquinones occurring in the defense se-
cretion of the beetle *Eleodes longicollis* are synthesized (3)
in the insect from labeled acetate, or acetate and labeled
malonate, with a distribution of label strongly suggestive of
operation of the normal acetate pathway (see Chapters 17 and
18). (Surprisingly, *p*-benzoquinone itself, which is likewise
present, originates from the preformed aromatic ring of phenylalanine or tyrosine.) In both cases, however, intact insects
were used, and since both species normally harbor symbiotic
microorganisms, the biosynthesis of the aromatic rings may well
be an achievement of the symbionts rather than of the insects
themselves. The same may perhaps be true for the biosynthesis
of the aromatic aphid pigments--typical representatives of the
acetate-derived polyketides (see Chapters 17 and 18)--where
Bowie and Cameron (4) failed to find in the host plant any precursor with the naphthopyran ring system characteristic for
these pigments; presence of symbionts in aphids is well known (5).

   Echinochrome A (2-ethyl-3,5,6,7,8-pentahydroxy-1,4-naphtho-
quinone) is biosynthesized in the sea urchin *Arbacia pustulosa*
from [2-^{14}C] acetate (with strong randomization of label) (6).
Since the animals were kept in running sea water during this
study, intervention of microorganisms can again not be excluded.

   The formation of radioactive tyrosine from labeled acetate,
glucose, or glycine by the nematode *Caenorhabditis briggsae* kept
in axenic culture (7) can hardly be taken as evidence for the
biosynthesis of an aromatic ring in a metazoan organism, since
the localization of the label was not established; it appears
possible, and indeed quite probable, that it was the side-chain
which became labeled.

   There is thus no cogent reason to abandon the concept that
the biosynthesis of aromatic metabolites other than the oestrogens is the monopoly of plants and microorganisms.

   Establishing the biosynthesis of any metabolite would mean
tracing, step by step, the chain of events leading from some
generally available starting material (glucose, acetic acid,
and the like) to the final product. The early hypotheses mentioned before were, of necessity, purely speculative; until
fairly recently, no techniques were available for identifying
the initial building blocks, or for obtaining and recognizing
the various intermediates. From their very function, the
latter will often be highly reactive substances that undergo
further changes as soon as they are formed, without normally
accumulating in quantity. Furthermore, even where such compounds did actually accumulate, and were known and identified
chemically, methods were lacking for recognizing them as

biosynthetic intermediates. Thus, shikimic and methyl-hydroxy-glutaric acids, squalene, lanosterol, sedoheptulose, citrulline, and orotic acid had been isolated and studied long before their importance in biosynthesis could even be suspected, and it is fortunate that their chemical structures should already have been well established prior to the recognition of their exceptional significance.

The early hypotheses had to be made, therefore, by the only approach conceivable at that time: consideration of some aromatization reactions occurring *in vivo*, or *in vitro* under mild conditions, coupled with conclusions drawn from comparison of many aromatic compounds for common structural features. The judicious use of such methods often permitted a remarkably close approach to reality. This is shown by Collie's early (1893, 1907) ideas about the biosynthesis of natural phenolic compounds from acetic acid (8), subsequently vindicated after long neglect (see Chapters 17 through 20), by the work of Winterstein and Trier (9a), Sir Robert Robinson (9b), and Schöpf (10) on the biosynthesis of alkaloids, and by the almost correct hypothesis of H.O.L. Fischer and Dangschat (11), to a certain extent anticipated by Loew (12a) and Hall (12b), which ascribes an essential role to quinic and shikimic acids in biosynthesis, particularly of aromatic acids such as gallic acid (see Chapter 2). Other well-known examples are the isoprene rule (13), and the postulated, and subsequently proven, derivation of the steroids from squalene (14). The discussion in Sir Robert Robinson's Weizmann Lectures of 1955 (14A) gives a fascinating summary of these hypotheses (and of many additional ones), and demonstrates convincingly how often deductions of this kind closely anticipated findings made subsequently when experimental scrutiny became feasible.

The pitfalls of even the most logical speculations in this field may be illustrated by the example of shikimic acid, a key intermediate in the biosynthesis of many aromatic compounds (see Chapter 2). H.O.L. Fischer and Dangschat (15a) had shown in beautiful experiments that the relative and absolute configurations at carbon atoms 3, 4, and 5 of shikimic acid are identical with those of C-3, 4, and 5 of glucose. (In Section 2, the modern numbering for shikimic acid is used, in which the numbers for carbons 3 and 5, and 2 and 6, are interchanged; for the discussion on this page, the older system expresses the situation more clearly, and has been retained for this reason.) The authors' cautious suggestion that this identity is not coincidental but reflects a biosynthetic relationship of the two compounds seemed entirely logical and hence a biosynthesis of shikimic acid from glucose without modification of carbons 3, 4, and 5 quite

plausible. Yet more recent research has shown that the formation
of shikimic acid from glucose is a much more complex process than
could have been anticipated by Fischer and Dangschat. It proceeds through stages in which of the two carbon atoms destined
to become C-4 and C-5 of shikimic acid, the latter definitely,
and the former very probably, are *not* asymmetric.

Again, the ready cleavage of shikimic acid by periodate
into aconitic dialdehyde and formic acid (15b) could perhaps
have suggested, in the absence of abundant evidence to the contrary, that shikimic acid might well be built up from aconitic
acid and a one-carbon fragment by a reversal of this reaction.
Of course, the actual biosynthesis follows an entirely different
path.

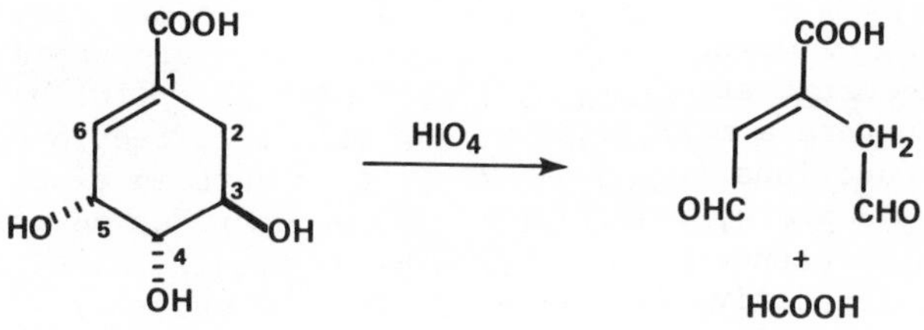

shikimic acid
(older numbering)

During the last three decades, the tools for experimental
investigation of biosynthetic events have become available,
particularly techniques for isotopic labeling, production and
selection of mutants, and detailed study of enzymes. Chiefly
through the use of these techniques, the essential features of
the biosynthesis of aromatic rings have been clarified, and the
existence of three major pathways has emerged. To outline the
entire field of the biosynthesis of aromatic compounds, a preliminary generalized description of these pathways is given here;
their detailed discussion is the topic of Chapters 2 through 25.

One of these sequences, commonly referred to as the "shikimic acid pathway" after its only previously known intermediate,
leads from carbohydrate through a common series of six alicyclic
stages (one of them shikimic acid) to about six or seven primary
aromatic compounds and, beyond these, to the three aromatic
amino acids (phenylalanine, tyrosine, and tryptophan), to the

ubiquinones, to several vitamins (vitamins K, p-aminobenzoic acid), to lignin, and to an immense variety of so called "secondary" metabolites which includes the majority of known alkaloids, natural pigments, etc. This pathway is thus responsible for the biosynthesis of several of the building blocks essential for all life; in the biosynthesis of lignin, which constitutes up to approximately 30% of all wood, it includes a synthetic activity carried out on a truly gigantic scale; from a more anthropocentric viewpoint, we owe to it a great many of our vitally important drugs. The series of common intermediates in this pathway is by now well understood, while the later branches leading individually to the various primary aromatic metabolites still contain some gaps. The regulations that assure the functioning of this sequence in a balanced way to meet the needs for the various metabolites derived from it are rapidly becoming elucidated.

Compounds biosynthesized through this pathway often have certain diagnostic structural features in common. Examples of such features are a marked preference for the occurrence of oxygenated functions (phenolic hydroxyl, methoxyl, etc.) in positions 4, 3 and 4, or 3, 4, and 5 with respect to side-chain, and frequent presence of a three-carbon side-chain, or of structures logically derivable from it. On the other hand, certain groups very commonly encountered in aromatic metabolites originating from the two other major pathways occur only rarely in compounds biosynthesized via shikimic acid. Such groups are methyl and isopentyl (or isopentenyl) side-chains, and modifications of these.

A second pathway starts from acetate and malonate residues, which are aligned in a regular head-to-tail fashion, $-CO-CH_2-CO-CH_2-$; cyclization processes involving the strongly activated methylene groups (mostly aldolization and acylation reactions) lead to a very wide variety of aromatic substances. Most of the aromatic compounds (often termed "polyketides") which arise through this pathway are so-called "secondary metabolites," that is, substances that occur only in a limited number of organisms, rather than universally, and the function of which, where it is known, frequently appears not to be that of an intermediate in metabolism.

It is often considered that these "secondary" metabolites are useless products of a "luxury" metabolism, and hence of little interest. It seems, however, that this concept is based, in part, upon the fact that the function of such substances is usually difficult to ascertain; it may thus be merely unknown to us at present, rather than non-existent. Furthermore, to fulfil

an essential, even vital, role, a substance need not be an intermediate in some metabolic process. Certain secondary metabolites
are known, or at least suspected, to function as execretory or
detoxification products, storage forms of valuable materials,
germination inhibitors in seeds, "chemical warfare" agents in the
fight of plants against competitors or against enemies (molds,
insects, herbivorous mammals) -- activities likely to be vital
for the survival of the particular species. One very evident
type of function of this kind is that of many plant pigments,
for which the anthocyanins and flavonoids may serve as typical
examples: compounds of mixed biosynthetic origins, one part of
their structure being derived from shikimic acid, the other from
acetate (see Chapter 21). As pigments of flowers, they are
essential for fertilization by insects and thus for the per-
petuation of species depending upon this mode of pollination; as
pigments of seeds and berries, they are equally important for the
dispersal and ecological existence of the particular plant.
These pigments thus serve a vital but nonmetabolic role. The
surprising fact about these secondary metabolites is not so much
their existence as their species specificity and immense variety,
both from one species to another and within one species. As an
example of the latter, at least seventy different indole alkaloids are known to occur in one single higher plant, the pink
periwinkle, *Catharanthus roseus (Vinca rosea)*.

In addition, it is becoming evident in an increasing number
of instances that such secondary products can undergo rapid
metabolic changes, a fact which is hardly in agreement with the
concept that they are the final results of blind alleys of metabolism. Interesting discussions of these questions, with many
examples, have been given by Zenk (16a) and by Luckner (16b);
among recent findings, those on rapid metabolic turnover of such
typical "secondary" products as monoterpenes (17) and flavonoids
(18) may be mentioned.

This second major biosynthetic sequence, the acetate or
polyketide pathway, is identical in its early steps with the
well-understood biosynthesis of the fatty acids: initiation by
acetyl-CoA (occasionally replaced as the starter unit by the CoA
derivative of another carboxylic acid: propionic, oleic, benzoic,
cinnamic, in one case nicotinic acid), and stepwise lengthening
of the chain by reaction with malonyl-CoA units. It differs from
the biosynthesis of fatty acids in its later stages; instead of
undergoing the reductive removal of the carbonyl groups, which
yields the fatty acids, the polyketide chain cyclizes to aromatic
(almost always phenolic) structures by intramolecular aldolization or acylation reactions. In contrast to the shikimate

pathway with its many well-investigated intermediates, the acetate pathway seems to produce the aromatic structures in a single concerted process without appearance of defined intermediates. No evidence is as yet available which would provide information on the way in which a polyketide chain consisting of the same number of units, arranged in the same linear head-to-tail fashion, yields a wide diversity of final aromatic compounds.

The structures of aromatic compounds given in Scheme I exemplify some of the characteristic group features of the acetate pathway: preferential occurrence of oxygenated functions in *meta* or *peri* orientation; occasional loss or addition of such functions; retention or loss of the terminal carboxyl (the latter hardly surprising in a biosynthesis which proceeds, at least schematically, through β-keto-acids); changes in oxidation status of this carbon where retained (cf. the carboxyl-derived methyl group of javanicin); tendency for formation of oxygen heterocycles, paralleled in model compounds (see Section 17-2g); appearance of halogen, or isoprenoid residues, on carbon atoms derived from the methyl of acetate.

Two variants of the acetate pathway have been recognized. One of these leads through normal fatty acids to polyacetylenes, some of which have structural features enabling them to undergo very ready cyclization to benzene rings. Processes of this type have been observed so far only in plants belonging to the family Compositae, although polyacetylenes are by no means restricted to this family, occurring also in higher plants of several other families, and in molds. The aromatic acetylenes are discussed in Chapter 19.

The other variant has been found only in the biosynthesis of the o-xylene moieties of riboflavin and the closely related 4,5-dimethylbenzimidazol which forms part of the vitamin B_{12} molecule. Here, the four acetate residues which produce the aromatic ring combine formally in a symmetrical fashion rather than in the non-symmetrical head-to-tail manner found everywhere else. The actual sequence of events which takes place is, however, not yet completely understood. The biosynthesis of these two substances is the topic of Chapter 20.

The third major pathway (see Part 4) starts with mevalonic acid and yields a variety of aromatic mono-, sesqui-, di-, and triterpenoids, carotenoids, and steroids. The actual mechanism of aromatization has been investigated in detail only in one case: the biosynthesis of the estrogens from nonaromatic steroid precursors. Beyond this, the origin of a few of these aromatic compounds from mevalonic acid or from nonaromatic terpenoids has been established experimentally (see Section 22-7), but details

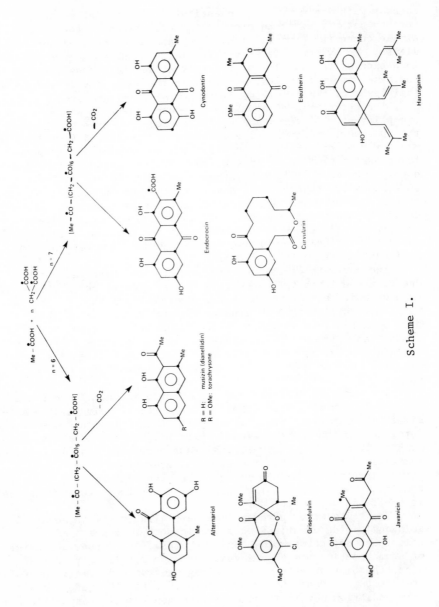

Scheme I.

8

of their formation from these precursors remain unknown. For the majority of the aromatic substances in this group, their typically isoprenoid structures and the frequent co-occurrence with related nonaromatic compounds remain the only indication of their origin.

In a subgroup (Section 22-6), only part of the aromatic ring system is furnished by mevalonic acid, while the remainder is biosynthesized via the shikimate or acetate pathways.

Besides many "secondary" metabolites of limited occurrence and largely unexplored function, the class of mevalonate-derived aromatic substances includes, in the oestrogens, a group of the highest importance. It has already been mentioned that their formation from nonbenzenoid precursors is the only instance, so far, of an unequivocally established biosynthesis of aromatic compounds by the animal organism.

1-2. METHODOLOGY; EXISTENCE AND RECOGNITION OF INTERMEDIATE STEPS IN BIOSYNTHETIC PATHWAYS

Clearly, biosynthesis of even the simplest aromatic structures from generally available starting materials such as glucose, acetic acid, or mevalonic acid, must proceed through many intermediate stages; tracing these sequences of reactions from step to step is one of the major objectives of research into biosynthetic pathways. In most cases, the stages are represented by discrete intermediates capable of being isolated and chemically identified. However, the search for such distinct intermediate stages has not yet been successful in those pathways yielding aromatic rings by head-to-tail linkage of acetic acid groups. Discrete intermediates may not exist here, at least in free form (see Section 17-2f), or may be too elusive for isolation and chemical study (but this latter possibility seems less probable).

In the pathways leading to aromatic structures from carbohydrate via shikimic acid, and from mevalonic acid, the successful isolation of intermediates has led to a more (shikimate) or less (mevalonate) complete understanding of the sequence of individual events. A generalized discussion of the methods used seems needed, especially since some of them differ from those universally employed in chemical and biochemical research. This is particularly true of the method using mutants of microorganisms (bacteria, molds, etc.), which has most commonly permitted the isolation and study of intermediates and their recognition as such; its specific features are not likely to be familiar to chemists.

The following description does by no means intend to give
a complete, up-to-date picture of the present status of mutant
problems and of the very complex questions of biochemistry,
enzymology, microbial genetics, and so on, which enter into it
and are given increasing (and deserved) prominence in the recent
literature. The next few pages may thus perhaps be considered
unsophisticated and obsolete. However, attempts to cover the
aspects that have been mentioned would be outside the stated
scope and limitations of this volume. Rather, the description
intends to provide the background required for an understanding
of mutant methods as they were used for the unraveling of biosynthetic pathways leading to aromatic metabolites, particularly
the shikimate pathway. This research was essentially done between 1950 and 1960, and our account is slanted towards the
mutant methods as used in that work.

Mutant, that is, genetically altered, strains of many organisms can be obtained, more or less readily, by exposure to agents
such as radiation or mutagenic chemicals. For the work pertinent
to this monograph, ultraviolet radiation has been most commonly
employed; this is true particularly for the research on the
shikimate pathway. In more recent literature, a trend towards
increased use of chemical mutagens seems recognizeable.

The action of mutagens results in modifications of the
enzymatic apparatus of the cells; the details of this process,
which operates through primary action upon the genetic material
(DNA), are outside the scope of this monograph. The modifications may consist in the absence of detectable activity of a
given enzyme, in a quantitative alteration of this activity, or
in changes of the properties of the enzyme, such as modification
of its affinity for its substrate, or increased or lessened
sensitivity to inhibitors or to heat inactivation, etc. Of value
for the study of biosynthetic events are primarily those mutations involving an enzyme that is directly concerned with one step
in the biosynthetic chain being studied. Not all mutations, however, fall into this category, and processes other than synthesis
of an intermediate can be affected. As an example pertinent to
our topic, one may mention the mutation observed in the bacterium,
Escherichia coli, which diminishes or prevents the uptake of
exogenously supplied shikimic acid or its immediate precursor,
3-dehydroshikimic acid (19). From the next few paragraphs, it
will be apparent that mutations involving such "permeases" represent one complication of the use of mutants; others will follow.

For a number of years, it was widely held that every genetic
mutation affects one single enzyme ("one-gene-one-enzyme hypothesis"). Modern developments have shown, however, that events

of greater complexity than direct action upon the gene controlling formation and properties of one single enzyme can take
place. In many cases, more than one step in an enzyme-catalyzed
chain of biosynthetic events is promoted by a complex of enzymes,
the individual constitutents of which often do not seem to be
accessible to separate mutation. Some of these problems are
briefly discussed in Chapter 15.

Mutation can thus result in manifold modifications of the
enzymatic equipment of the cell. For work on isolation and
recognition of intermediates in biosynthetic pathways, one will
obviously choose strains with mutations which affect enzymes
directly involved in one step of the pathway, and which completely lack the ability of carrying out this reaction, that is, mutants that have what is called a "total block," rather than incompletely blocked, "leaky" strains.

A total block in the biosynthesis of a vital metabolite will
evidently lead to an organism incapable of existing unless given
an outside supply of that metabolite. It is usually very difficult, if not impossible, to select and maintain such mutants of
higher organisms, while the task is relatively easy for microorganisms. Hence techniques for selection and use of mutants
were developed first for molds, where the classical work of
Beadle and Tatum (20) marks the beginning of the use of mutants
for exploration of biosynthetic pathways, a breakthrough which
first provided the essential tool for the subsequent rapid development of experimental research on problems of biosynthesis.
Subsequently, techniques for obtaining and selecting mutants of
bacteria were worked out by Davis (21); they have formed the
basis for the elucidation of the shikimate pathway by Davis,
Sprinson, and their co-workers, which will be discussed in detail
in Part 2. Because of their preferential use for the study of
the problems which form the topic of this volume, the general
description of mutant methods in the next few pages is slanted
towards those bacterial mutants. Their specific advantages will
be discussed later in this section, as will be the techniques
which permit their isolation. Prior to this, however, a general
discussion of value and limitations of mutant methods for research in the field of biosynthesis seems needed. This discussion will also offer an opportunity to mention certain more
general problems connected with the study of biosynthetic pathways, and some of the difficulties encountered.

It has been stated above that genetic mutation may manifest
itself in alterations of the enzymatic apparatus of the cell. If
the affected enzyme is one participating in the biosynthesis of
an essential metabolite, the mutant will be "auxotrophic" for

this substance; that is, it will require for growth an external supply, either of the metabolite itself, or of an intermediate between the blocked reaction and the final product. Furthermore, the biosynthetic intermediate which serves as the substrate of the enzyme in question will not undergo the reaction normally leading to its transformation into the immediately subsequent stage of the biosynthetic chain. Hence a transient intermediate can survive because of the absence of the catalyst which, in the unaltered ("wild-type" or "prototrophic") strain, is responsible for its rapid disappearance. As a consequence, such intermediates are likely to be accumulated, sometimes in surprisingly large amounts, and often together with smaller quantities of earlier precursors. The great utility of such accumulations for the chemical study of the intermediates will be evident. Actually, the accumulation is due to the absence of the control over early stages of the biosynthetic pathway usually exercised by its end-product(s), which can inhibit the activity, or repress the formation, of enzymes involved in these early steps; these controlling mechanisms are discussed later on, and in Chapter 15.

Use of mutants thus permits interruption of a biosynthetic sequence at a stage that would normally escape detection. Furthermore, genetic blocking of one step should not prevent the functioning of the subsequent reactions of the sequence; intermediates in this part should be convertible to the final metabolite, and should thus enable the mutant to grow normally. Through this growth response, it becomes possible to recognize intermediates, and mutants can serve as very selective and sensitive reagents for them. Consequently, it has often been considered good evidence for the interpretation of a given compound as a biosynthetic intermediate if it is accumulated by one mutant and utilized by another one, and if the enzyme concerned with its transformation to the next following intermediate is present in the wild-type cells but absent from those of the mutant.

The situation can be visualized with the help of the diagram below, considering for the time being only its horizontal part. This diagram indicates the biosynthesis of a metabolite P from some generally available source material A via intermediates B, C, D, E, and others not shown. A genetic block between C and D will prevent growth, unless either P itself, or an intermediate *after* the block, D, E, ..., is supplied, and it will probably lead to the excretion of fairly large amounts of C together with lesser ones of B; the enzyme catalyzing the reaction C→D will be absent.

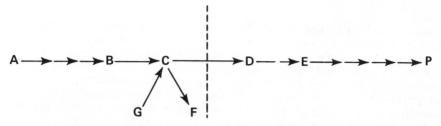

The great value of mutants for research on biosynthesis
thus rests upon their ability to supply substantial quantities
of transient intermediates and to make it possible to recognize
and study their role. Mutant techniques have, however, their
inherent limitations, and the real situation is often more complicated than the necessarily simplified picture just given would
indicate. Especially, the problem of the recognition of intermediates is frequently made difficult by effects that have to be
considered next. More complete discussions of these questions
have been given by Adelberg (22), and by Umbarger and Davis (23);
an excellent brief resumé of most of those problems is due to
Bu'Lock (24).

(1) The methods for the utilization of mutants, as just
outlined, and for their selection (see below) all rest upon their
inability to grow in the absence of the product(s) of the blocked
biosynthetic pathway. It follows that mutant techniques are restricted essentially to the study of the biosynthesis of metabolites necessary for growth, that is, primary ones; they will in
general not be suited for the exploration of pathways leading to
secondary products. Mutants blocked in such pathways do of
course exist, and can be utilized, for instance, where the product of the biosynthesis is pigmented. In other cases, the
techniques for selecting the desired mutants from the large
excess of unaltered organisms tend to become prohibitively
laborious, mutation being a rare event. By and large, the biosynthesis of secondary metabolites is but rarely accessible to
study by mutant methods.

Even in the exploration of the biosynthesis of metabolites
essential for growth, complications occur quite frequently.
Most of them have been encountered in the study of the shikimate
pathway.

(2) A compound isolated from the culture filtrate of a
given mutant may not be the actual biosynthetic intermediate but
the product of some further transformation. An example of this
is the presence of protocatechuic and vanillic acids in filtrates of a mutant of *Neurospora* blocked after dehydroshikimic
acid (see Section 4-4). Only small amounts of the latter compound, which is the true intermediate, are found in the filtrates,
most of it being converted further to the two aromatic acids.

Similarly, 5-enol-pyruvyl-shikimic acid (the so-called "Compound Z_1"), which can be isolated from filtrates of mutants of the bacterium, *Escherichia coli* (see Section 5-3), is demonstrably not an intermediate in the shikimate pathway, but is formed secondarily from its 3-phosphate, the true intermediate, by enzymatic dephosphorylation; the phosphorylated compound can be obtained if the enzyme is inhibited by addition of fluoride. In a late step of the biosynthesis of tyrosine (see Chapter 7), p-hydroxyphenylpyruvic acid, the true intermediate, can undergo secondary reduction in culture filtrates to p-hydroxyphenyllactic acid.

The position of the products obtainable from filtrates in these cases is the one indicated by "F" in the diagram. The situation can be recognized either by the failure of the isolated product (e.g., Z_1) to be converted further by suitable enzymatic action, or sometimes simply by structural criteria. At the time of the discovery of the formation of protocatechuic acid, for example, it was already known that shikimic acid, an alicyclic compound several stages removed from aromaticity, was the next intermediate following dehydroshikimic acid. Aromatic acids such as protocatechuic and vanillic acid could thus hardly be anything but products of a side-reaction.

(3) A compound may have growth-factor activity for appropriate mutants without normally playing a role in the biosynthesis of the metabolite which the auxotroph is unable to produce; it may be converted to a true intermediate by some reaction which is not part of the biosynthetic sequence. The utilization of quinic acid by aromatic auxotrophs of *Aerobacter aerogenes* is a case in point. There is good evidence that in this species, just as in other microorganisms, 3-dehydroquinic acid (see Section 4-1) rather than quinic acid is the true intermediate, and that the growth-factor activity of the latter acid rests upon its enzymatic conversion to the former. Quinic acid can thus be utilized for the biosynthesis of the products of the shikimate pathway, if available, but it is not a normal intermediate in the pathway. The relationship of quinic acid to the biosynthetic sequence is thus the one indicated by "G" in the generalized scheme.

This situation is apt to arise if the compound made available to the organism is chemically close to an actual intermediate. As a consequence of its possible occurrence, mere utilization of a substance *cannot* be accepted as conclusive proof that the particular compound *itself* is an intermediate in the biosynthesis under investigation, although it does suggest that the pathway goes through some substance that is chemically related to the compound supplied. This limitation is a very important one in the evaluation of experiments in which labeled precursors are administered to organisms such as animals or higher plants, where the more selective information obtainable from mutants is not available. As the example of quinic acid shows, however, it cannot be disregarded even in work utilizing mutants.

Detection of cases of this type will in general require a detailed study of the biosynthesis in question. The conclusion, for example, that quinic acid is not directly involved (25) in the shikimate pathway in *A. aerogenes* rests on the finding that no strain of this bacterium accumulates it; that it is neither accumulated nor utilized by any strain of *E. coli;* and that an enzyme converting quinic to 3-dehydroquinic acid occurs (26) in *A. aerogenes* but not in *E. coli* or a variety of other organisms capable of synthesizing their aromatic amino acids. Chemical arguments support the conclusion that quinic acid is not itself an intermediate but owes its nutritional activity for *A. aerogenes* to a conversion, in this organism, to dehydroquinic acid by a reaction outside the normal sequence. In the latter, dehydroquinic acid is known to be transformed to 3-dehydroshikimic acid, shikimic acid, the 3-phosphate of the latter, and so on. Quinic acid could not possibly be an intermediate in any of these transformations. At the time of the discovery of its growth factor activity, it could have been interpreted as a precursor of dehydroquinic acid, but this compound is now known to be formed by an entirely different reaction sequence in which, again, quinic acid has no place.

One criterion that may be helpful for the recognition of cases like the one described is the fact that a product corresponding to G in the diagram will need more than one enzyme for conversion to D, while C, the true intermediate, should require only one enzyme (22, 23).

Cases such as the one of quinic acid have demonstrated the need for a clear definition of the terms "precursor" and "intermediate." Such definitions have been proposed by Davis (27, 23); they are used in the present volume. According to them, a "precursor" is a compound that can be converted by an organism to some further product, while an "intermediate" is a compound that is synthesized by an organism, and is converted by it to the product in question through reactions forming part of a biosynthetic sequence. It can be termed an "obligatory intermediate" if the pathway is the only one by which the final product can be formed from a given initial material. Since the nature of this source material can determine a biosynthetic pathway, the further term "normal intermediate" can be applied to an obligatory intermediate in an organism growing in "minimal medium," that is, one containing only a general carbon source (most commonly glucose), those growth factors which the wild-type is not able to synthesize, and inorganic salts.

(4) A true intermediate may fail to be utilized by intact cells of mutants blocked at an earlier step of a biosynthetic chain. Many instances of this complication have been found during research on the biosynthesis of aromatic compounds. Examples are the complete nutritional unavailability of such well-established obligatory intermediates as 3-deoxy-D-*arabino*-heptulosonic acid 7-phosphate (Chapter 3), shikimic acid 3-phosphate (Section 5-2), 5-enolpyruvyl-shikimic acid 3-phosphate (Section 5-3), chorismic acid (Chapter 6), and prephenic acid (Section 7-1), and the availability of 3-dehydroquinic acid (Section 4-1) to only a limited number of mutants blocked before this compound. Another case is the inability of a number of microorganisms to utilize 6,7-dimethylribityllumazine for the biosynthesis of riboflavin (Chapter 20).

This nutritional unavailability of true intermediates can be explained through the existence of permeability barriers, which prevent the compound from entering the intact cell, or from reaching the intracellular locus of enzymatic transformation.

It has already been mentioned that the uptake of many substances by intact cells is mediated by specific enzymes, the permeases, and that one such enzyme controlling the uptake of shikimic and dehydroshikimic acids in *E. coli* has been studied in detail (19).

It is noteworthy that all the intermediates listed are strongly polar compounds. Inability of such compounds, and in particular of phosphorylated metabolites, to be utilized by intact cells is quite general, and appears to be connected with the well-known fact that most cell membranes are poorly permeable for such hydrophilic substances (28). Surprisingly, the reverse process, *excretion* of highly polar intermediates, seems to proceed smoothly. All the intermediates mentioned above occur in filtrates of appropriate mutants, and were isolated from them; the accumulation of shikimic acid 3-phosphate, for example, reached the remarkably high level of about 1 g/liter (29).

The status of such nutritionally unavailable compounds as true intermediates can of course not be established by the criterion of biosynthetic utilization by intact cells, although it can usually be demonstrated by work with cell-free extracts or purified enzymes.

(5) Still further complications in the exploration of biosynthetic pathways can be caused by the possibility that devia-

tions from the normal sequence of events may be produced by offering nonphysiologically large excesses of normal metabolites, or even by the mere fact that the normal sequence has been interrupted. A few instructive examples of such complications have been observed. The enzymatic conversion of tyrosine to p-coumaric acid in higher plants, for example, is in general irreversible. However, Kindl and Schiefer (30) found in mustard, *Sinapis* (Cruciferae), that labeled p-coumaric acid is incorporated into tyrosine; the authors conclude that this process, normally of little importance, becomes significant if p-coumaric acid is available in large amounts. Another instance has been encountered recently by Hamilton and Cox (31) in work on the biosynthesis of ubiquinone (cf. Chapter 11) in *E. coli;* here, a compound not normally present in detectable amounts is actually formed as a result of mutation. One step of this biosynthesis consists in the assembly of the carbon skeleton of the vitamin from precursors of the aromatic ring and of the isoprenoid sidechain, respectively, which are formed independently through entirely different pathways. In mutations in which the biosynthesis of the aromatic moiety is blocked, that of the side-chain still proceeds as usual; lacking its normal aromatic receptor, the reactive isoprenoid precursor of the side-chain can combine with added p-aminobenzoic acid to give 3-octaprenyl-4-aminobenzoic acid, a compound not found when p-hydroxybenzoic acid, the normal precursor of the aromatic ring of ubiquinone, was present. (We wish to thank Prof. F. Gibson, Canberra, for calling this observation to our attention.)

From what has been said in the preceding paragraphs, it will be clear that identification of a compound as an obligatory intermediate in a given pathway is by no means an easy or straightforward matter; there are cases where a clear-cut decision is hardly possible at all. This is best illustrated by the history of the interpretation of indole as an obligatory intermediate in tryptophan biosynthesis. Indole was early assigned the chemically plausible role as an intermediate in this process on the basis of the observations by Fildes (32a) and Snell (32b) that it can replace the amino acid as a growth factor for certain bacteria, and particularly in the light of the work of Tatum, Bonner, and Beadle (33), who found that some mutants of *Neurospora* excrete indole, while others can synthesize tryptophan from indole and serine (34). Excretion of indole has also been observed in mutants of *E. coli* (findings mentioned in ref. 27). Furthermore, the enzyme responsible for conversion of indole to tryptophan was found to be lacking in appropriate mutants of *Neurospora* (27). The validity of these criteria became doubtful, however, as a result of the studies of Yanofski (35) on the tryptophan synthetase of *E. coli*, the last enzyme in the sequence leading to tryptophan. This enzyme, a homogeneous protein by several criteria, was found to catalyze the reaction of indole glycerol phosphate, the next earlier intermediate, with L–serine to give tryptophan and glyceraldehyde 3–phosphate; it seemed likely that free indole would not appear in such a transfer reaction.

However, many complicating features remained, such as the ability of the enzyme to combine indole with either serine or glyceraldehyde phosphate, and mutational loss of either one or the other of these reactions, or of both. These difficulties were cleared up by the extensive studies of Yanofski, Crawford, and others; a comprehensive account of this work is given by Ginsburg and Stadtman (36), which should be consulted for more detailed information on the rather involved situation, and for references to the very numerous individual papers.

The tryptophan synthetase of *E. coli* was found in these studies to be a complex of two markedly different proteins, designated as α and β_2, respectively, the latter being a dimer; they can be modified independently by mutation. The composition of the complex is $\alpha\beta_2\alpha$. The intact complex is needed for the overall reaction (reaction 3, below) which produces tryptophan from indole glycerol phosphate and serine without the appearance of free indole. However, the individual constituents of the complex, α and β_2, each catalyze one partial reaction, (reactions 1 and 2, respectively), in which free indole does play a role. The catalytic activity of the subunits is rather low, that of the complex very high. The entire set of events can thus be represented schematically as shown on p. 19.

In a case such as this one, the definition of an obligatory intermediate (indole) becomes to a certain extent a question of semantics, and formulation of a detailed biosynthetic pathway somewhat arbitrary.

We have outlined the many problems common to those biosynthetic investigations which are based on the use of auxotrophic mutants; however, difficulties in research aimed at the elucidation of biosynthetic pathways are not restricted to those encountered when such mutants are used. A discussion of these more general problems seems warranted.

Because of rapid turnover of some substance, it can at times even be difficult to decide from which one of the major pathways a given metabolite is derived. Labeled precursors can, for example, be metabolized to CO_2, which can be re-assimilated to sugars, resulting in the labeling of a wide variety of products

$$\text{(1)} \quad \text{indole glycerol phosphate} \underset{}{\overset{\text{protein } \alpha}{\rightleftharpoons}} \text{indole} + \text{glyceraldehyde 3-phosphate}$$

$$\text{(2)} \quad \text{indole} + \text{L-serine} \underset{}{\overset{\text{protein } \beta_2 + \text{pyridoxal phosphate}}{\rightleftharpoons}} \text{L-tryptophan}$$

$$\text{(3)} \quad \text{indole glycerol phosphate} + \text{L-serine} \underset{}{\overset{\text{complex } \alpha_2\beta_2 + \text{pyridoxal phosphate}}{\rightleftharpoons}} \text{L-tryptophan} + \text{glyceraldehyde 3-phosphate}$$

19

not closely related to the initial material. The methyl of acetic acid is known to enter the Krebs cycle (37); as a consequence, label from $2-^{14}C$-acetic acid is apt to be present in all kinds of compounds derived from the cycle. Similarly, the $-CH_2-OH$ group of serine can, through its transformation into the S-methyl of methionine, be distributed among the wide variety of products of the one-carbon metabolism. Since the side-chain of tryptophan may be cleaved off as serine (see reaction 2, above), label from $\beta-^{14}C$-tryptophan can appear in unexpected places.

A particularly striking instance of a difficulty of this type has been mentioned quite recently by Zenk (38): some species of sundew (*Drosera*, Droseraceae) metabolize tyrosine in very high yields to acetic acid, which is then incorporated into the naphthoquinones plumbagin (5-hydroxy-2-methyl-1,4-naphthoquinone) and 7-methyl-juglone (5-hydroxy-7-methyl-1,4-naphthoquinone), thus deceptively suggesting a derivation from the shikimate pathway for two secondary metabolites which are actually known (39) to be biosynthesized from acetate in this genus.

Tyrosine seems especially prone to be broken down into fragments which are subsequently re-utilized for the synthesis of a variety of metabolites. Ibrahim and co-workers, for example, have found (39A) that much of the activity from generally labeled L-tyrosine administered to leaf discs of several higher plants appears in compounds such as sugars, aspartic, glutamic, and malic acids within less than one day in the light; in darkness, very little of this redistribution of label was observed. The breakdown of tyrosine via homogentisic acid, recently shown (38) to take place in higher plants, could well be responsible for these findings.

Similar difficulties can arise from the occurrence of enzymes which cleave the side-chains from amino acids of the general formula $R-CH_2-CHNH_2-COOH$, especially (but not exclusively) those where R represents a cyclic residue, or which exchange one R for another. The enzyme tryptophanase, for example, long known to occur in *E. coli* (40), catalyzes the reversible cleavage of tryptophan into indole, pyruvic acid, and ammonia, but also promotes other related elimination and exchange reactions (41). Similarly, β-tyrosinase or tyrosine phenol lyase from *E. intermedia*, studied in detail by Yamada and co-workers, cleaves tyrosine reversibly into phenol, pyruvate, and ammonia (42). It also synthesizes 3,4- or 2,4-dihydroxy-L-phenylalanine from pyruvate, ammonia, and pyrocatechol or resorcinol, respectively (43), and can produce the former amino acid (L-dopa) from tyrosine and pyrocatechol (44). Even a reaction as strange as formation of 3,4-dihydro-3-amino-7-hydroxycoumarin from S-methyl

cysteine and resorcinol takes place under the influence of this
enzyme (45). Obviously, reactions of this kind could create
great difficulties in interpreting results from studies which
use amino acids labeled specifically in the side-chains.

Other problems might be caused by the common use of D,L-
amino acids in cases where the normal L-antipode is not conveniently available in labeled form. The D-form present in the
racemate is not necessarily an inert diluent; it can exert an
inhibitory action, as has been observed, for example, for D-
serine (46) and D-tyrosine (47). More important is the fact
that the metabolic fate of the D-form is apt to be different
from that of the L-antipode, so that it can be converted to biosynthetic artefacts. A case in point is the formation of N-
malonyl derivatives from several D-amino acids when they are
administered to higher plants (48). D-Phenylalanine, for
example, yields N-malonylphenylalanine when fed to barley seedlings, while the L-antipode gave little, if any, malonyl derivative. Interestingly, α-N-malonyl-D-tryptophan occurs naturally
in the seeds of several plants (49), as do a few acyl derivatives
of other D-amino acids; for references, see (49).

All the problems discussed so far refer to biosynthetic
reactions known, or presumed, to be mediated by enzymes. While
the vast majority of such reactions unquestionably are, the fact
should not be overlooked that this need not always be the case.
Possibilities of spontaneous reactions, or of reactions proceeding under the influence of catalytic agencies other than enzymes,
do exist, even though proof of events of this type may be difficult to obtain. One such possibility in particular, that of
photochemical reactions, has not often been considered, although
it appears plausible enough, and a few examples have actually
been known for a long time, such as the photochemical isomerization of 7-dehydrocholesterol and ergosterol to the precalciferol
precursors of vitamin D, and the formation of melanins from tyrosine and dopa. In addition, however, many natural compounds have
structures commonly encountered only among products of photochemical transformation, and these cases may deserve some discussion. This discussion does not include the reactions in which
the action of light proceeds through such ubiquitous mediators as
the chlorophylls and the phytochrome system. Reactions involving
photosensitized oxidation reactions have likewise not been included; it seems plausible to assume that some, perhaps many, of
the processes known to involve molecular oxygen might proceed
through photochemically generated singlet oxygen. Examples of
the cleavage of aromatic rings by molecular oxygen have been
collected by Thomas (50). Interestingly, the *in vitro* synthesis

of hypericin, the classical example of a naturally occurring photosensitizing pigment, includes a photochemical step (51); it seems quite possible that the laboratory synthesis closely parallels the biogenesis of the pigment.

Among natural products with structures suggestive of a photochemical origin, the long-known isomers α- and β-truxilline [1], [2] (52) belong evidently to the well-established class of cyclobutanes formed by photochemical dimerization of α,β-unsaturated acids and ketones; indeed, a possible monomer, cinnamylcocaine [3], occurs together with the truxillines in coca leaf from *Erythroxylon coca* (Erythroxylaceae). Related more recent examples are provided by the alkaloid thesine [4] from *Thesium minkwitzianum* (Santalaceae) (53), and by the α-pyrone dimers which have been isolated from *Aniba gardneri* (Lauraceae) (54). Of these, dimer A [5] is the dimer of 5,6-dehydrokawain [6], of which both *cis-* and *trans-*isomers occur in the plant; compound B [7] is the co-dimer of [6] with 4-methoxyphenylcoumalin [8], a constituent of other *Aniba* species (55).

The well-investigated photocyclizations of the tropolone ring to derivatives of [3.2.0]-bicycloheptane suggest a similar origin for the α- and β-bourbonenes [9] from Bourbon Geranium oil (56); it is interesting that the oil also contains large amounts of hydrocarbons with guaiane skeleton from which the bourbonenes could be derived by cyclization. The stereoisomeric β- and γ-lumicolchicines [10], which contain the same bicycloheptane system as part of a more complicated structure, are formed photochemically from the tropolone, colchicine, *in vitro*. They have actually been isolated from *Colchicum* and *Merendera* spp. (Liliaceae); a strongly suggestive finding, but hardly proof for photochemical formation in the plant (57).

Again, the well-known photochemical formation of phenanthrenes from stilbenes (58) is responsible for the conversion of the alkaloid septicine [11] (from *Ficus septica,* Moraceae) into tylophorine [12] and tylocrebrine [13] on irradiation *in vitro* (59); since all three alkaloids occur together in this plant, the presumption of an actual photochemical biosynthesis is strong. The conversion of [11] into [12] and [13] involves actual formation of a new aromatic ring (cf. Chapter 25), which could well take place photochemically in the plant.

Even closer to the topic of this volume is the photochemical aromatization, *in vitro* (60), of the natural monoterpene umbellulone [14] (from *Umbellularia californica,* Lauraceae) into thymol [15], which is widely distributed in plants (see Section 22-2). The photolysis to [15] proceeds quantitatively, while pyrolysis of [14] yields [15] together with a small amount of the isomeric

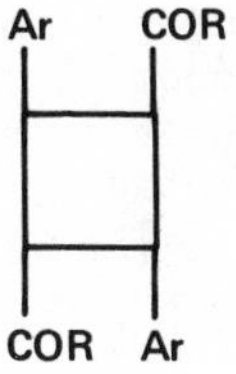

[1] : Ar = C_6H_5

$$R = $$

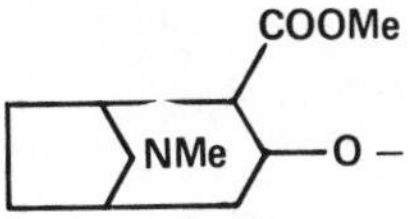

COOMe ... NMe ... O –

[4] : Ar = p – OH – C_6H_4 –

$$R = $$

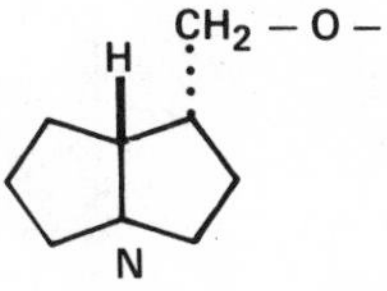

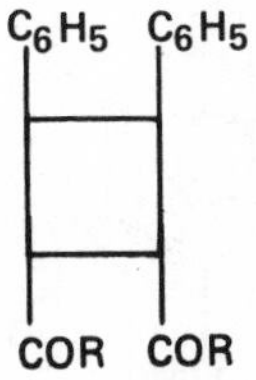

H $CH_2 - O -$

N

[2] : R =

COOMe ... NMe ... O –

COOMe ... NMe – O – CO – CH = CH – C_6H_5

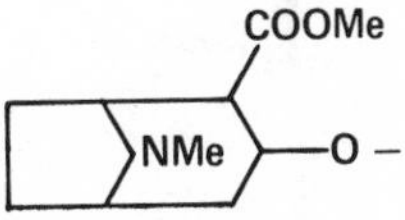

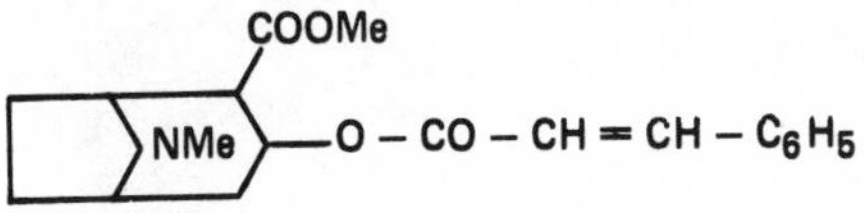

[3]

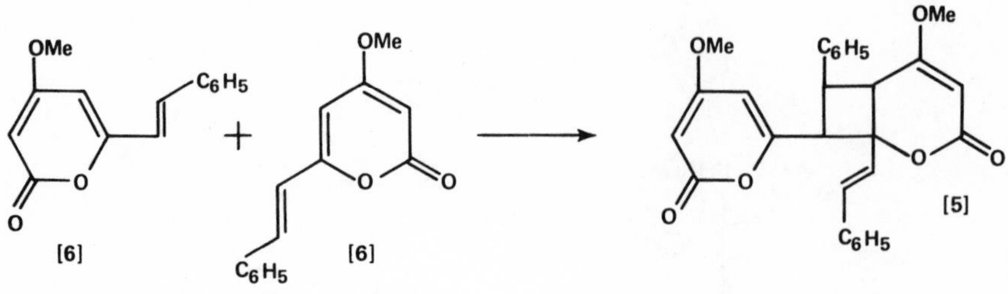

[6] + [6] → [5]

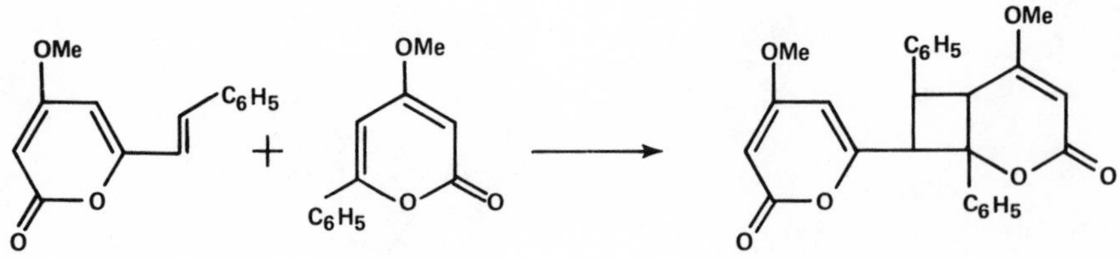

[6] + [8] → [7]

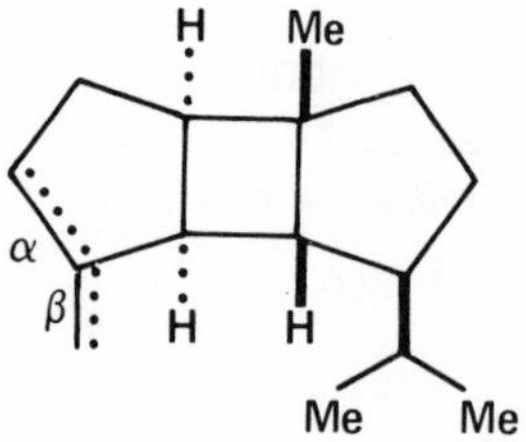

[9]

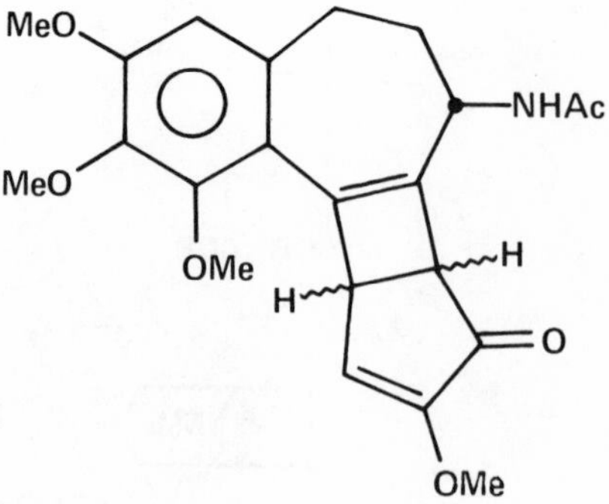

[10]   β = both H's β
        γ = both H's α

24

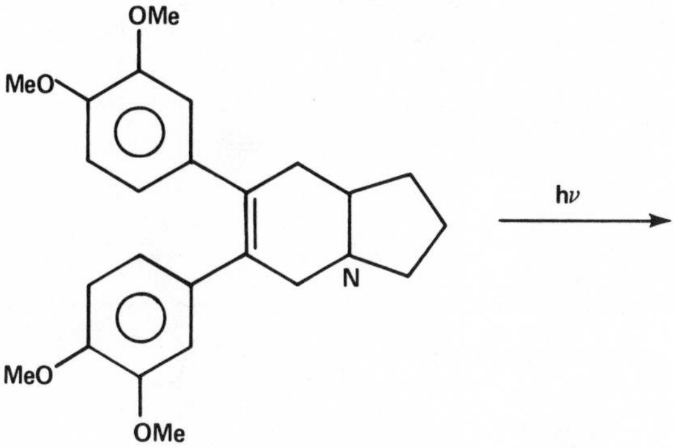

[11]

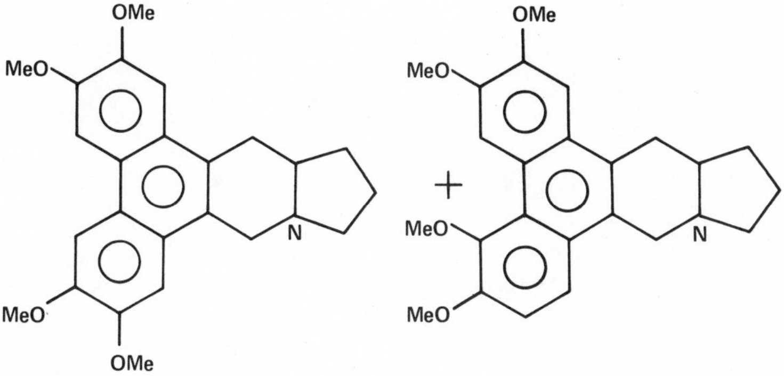

[12]   [13]

3-ethyl-5-isopropylphenol, which is not known as a naturally
occurring compound.

The sponge *Reniera japonica* contains carotenoids with aromatized end-groups of types [16] and [17]. Existence of the former can be explained readily by assuming an origin from a normal
ionone-type ring by an aromatization involving a 1,2-shift of one
of the geminal methyl groups. No such explanation is available
for [17], but its structure can be rationalized by postulating an

isomerization of [16] via an intermediary Ladenburg prism structure (61); there is much precedent for such processes in photochemical transformations of aromatic compounds. However, the
actual biosynthesis of these aromatic carotenoids has not been
investigated. The problem is discussed in detail in Chapter 23.

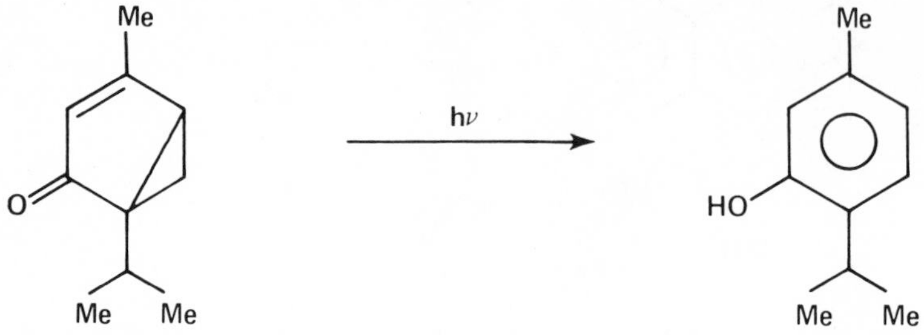

[14]   [15]

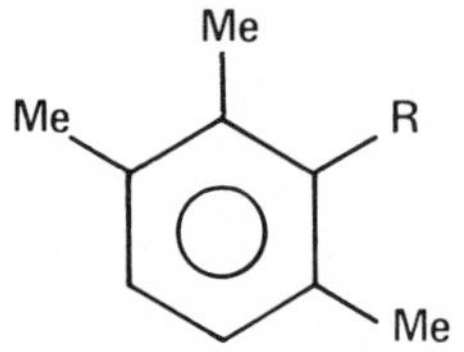

[16] 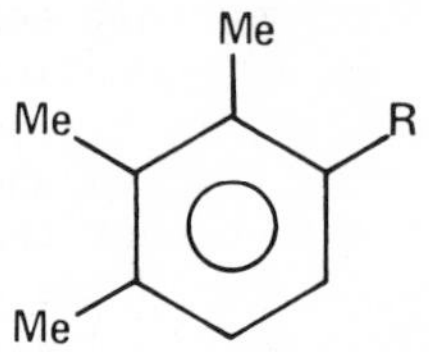   [17]

An observable photochemical reaction may actually take place
in the colorful root system (orange roots, magenta runners) of
Lachnanthes tinctoria (Haemodoraceae) (62). This system contains
the deep red lachnanthofluorone [18] and the orange glycoside
lachnanthoside; the aglycone [19] of the latter, on irradiation
in vitro, very rapidly forms [18]. The runners of the plant have
a magenta outer layer and an orange core; photochemical formation
of [18] in these outer layers seems at least plausible, although
unproven. (The runners grow mostly underneath the soil, but close
enough to the surface to make penetration of light probable; there
is precedent for such penetration.)

All the examples that have been mentioned so far are of

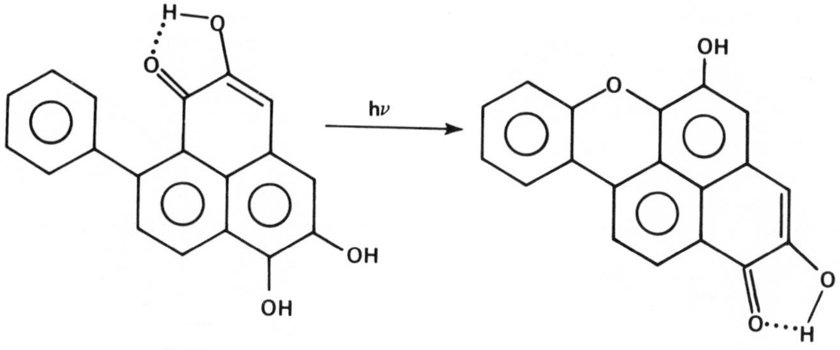

[19]   [18]

course at best only suggestive. There is, however, one type of
reaction, constituting a step in the biosynthesis of a fairly
numerous class of secondary metabolites, which has actually been
proved by experiment to take place nonenzymatically with catalysis by ultraviolet light. The reaction in question: the *trans→*
cis isomerization of the glucosides of *o*-coumaric [20] to those
of *o*-coumarinic acids [21], is a necessary stage in the biosynthesis of coumarins [22]. In the biosynthesis of coumarin itself
[22, R=H] in *Melilotus* (Leguminosae), this step has been shown
(63) to proceed under illumination with light of wavelength below
360 nm, whether the plant material had been previously heated or
not, and at 0° as well as at 29°, while it failed to take place
in the dark, or with light of longer wavelengths. Studies on the
isomerization of [20, R = OMe] during the biosynthesis of hernia-
rin [22, R = OMe] in *Lavandula officinalis* (Labiatae) gave similar results (64).

   In addition to these examples of photochemical reactions
having actual or potential significance for problems of biosynthesis, there exists an extensive literature on light-catalyzed
transformations which could qualify as model reactions for biochemical processes; research on such transformations is particularly active in the field of alkaloids, and of chalcones and
flavonoids. A detailed account of this work would far exceed the
limits of this volume. A few selected examples from the latter
field may, however, find a place here to give an indication of
the existing literature and to provide references. The choice of
this particular area is based on the fact that influence of light
upon the formation of such compounds has often been observed (65).

   Closely related to the case of the glucosides of *o*-coumaric

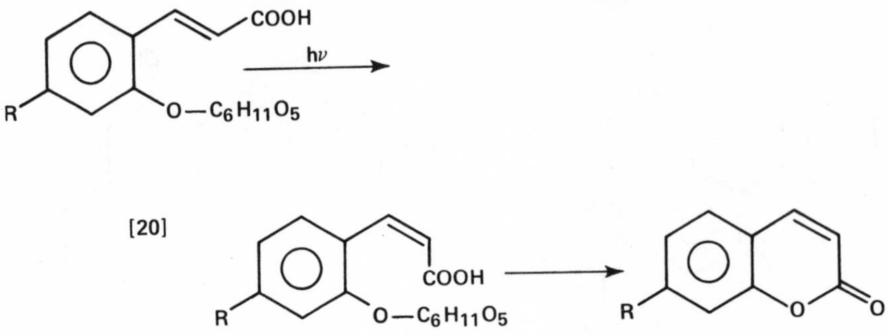

[20]

[21]   [22]

acid just discussed is the transformation of caffeic *(trans*-3,4-
dihydroxycinnamic) acid [23] via the *cis*-isomer [24] into escu-
letin [25] on irradiation in the presence of oxygen (66); both
[23] and [25] are widely distributed plant constituents. The
reaction could thus well be operative in the biosynthesis of some
natural coumarins, and would account for the introduction of the
oxygen *ortho* to the side-chain. 3-Methoxyflavones [26] undergo
photochemical cyclization (67) to products such as [27] with
structures quite analogous to those of certain natural products,
for example, peltogynol [28]. The fact that 5-hydroxylated analogs of [26] are inert under those conditions may perhaps explain
the rarity of compounds of type [28].

       Irradiation of 2-hydroxychalcone [29] in ethanol yields
(68) small amounts of flavone [30], with 2-ethoxy-3-flavene [31]
as the main product; the latter is readily converted to flavylium
salts by acids, so that reactions of this type could well play a
role in the biosynthesis of anthocyanins, which likewise are
flavylium salts.

       After this digression into the field of photochemistry, some
additional aspects of the problem of establishing biosynthetic
pathways need brief mention.

1-3. REGULATION OF BIOSYNTHETIC PATHWAYS

       One of these is the question of the regulation of such se-
quences. Clearly, biosynthetic reactions need to be delicately
controlled if they are to fulfil their purpose of supplying the

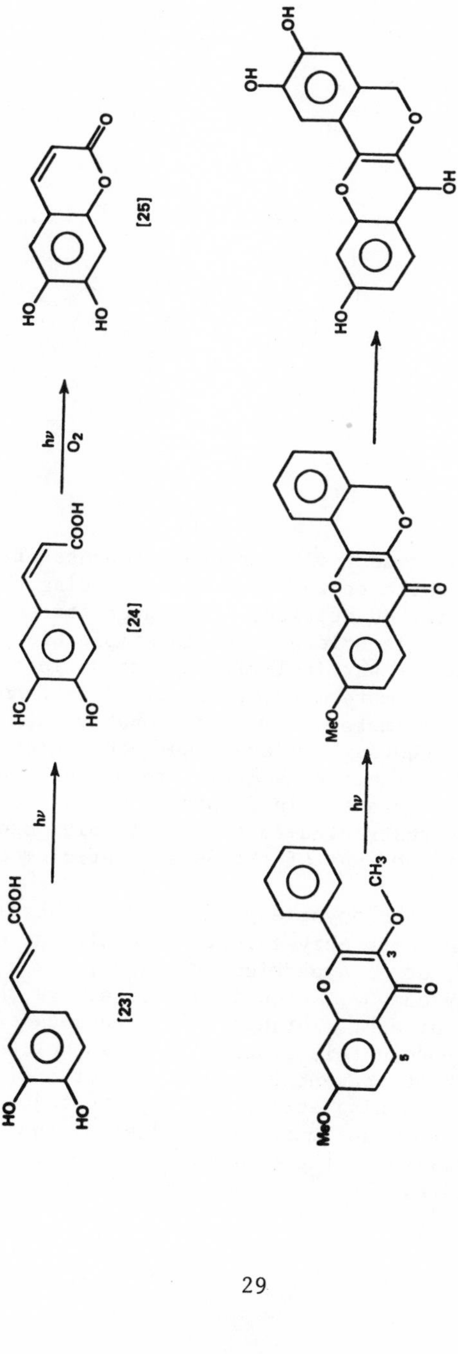
[23]
[24]
[25]
[26]
[27]
[28]
hν
hν
O2
hν
COOH
HO
MeO
CH3
OH

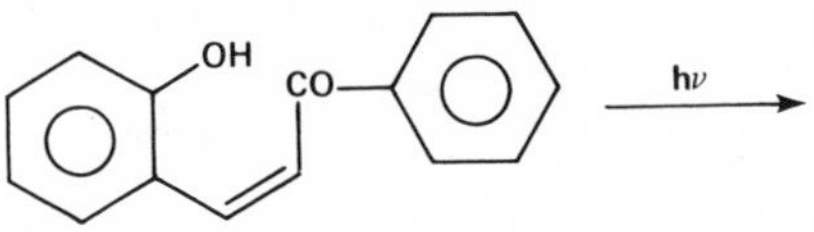

[29]

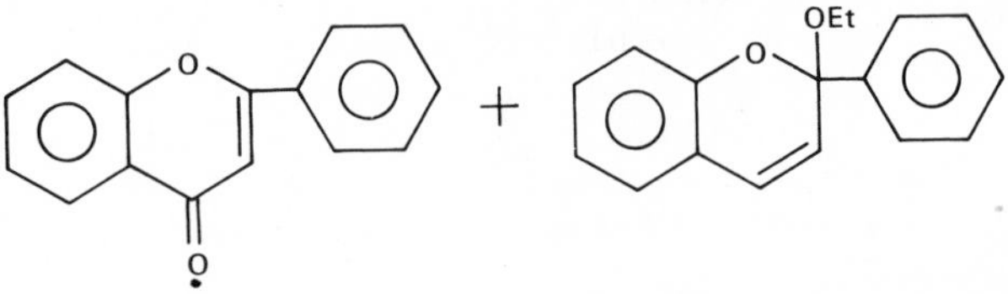

[30]   [31]

organism at the right moment with adequate amounts of the required metabolites, while at the same time avoiding the wasteful
production of excessive quantities. This need for sensitive and
purposeful regulations is evident even in simple biosynthetic
pathways, but it becomes particularly imperative in cases where
the synthesis of two or more metabolites proceeds through several
common steps. The shikimate pathway is probably both the most
complex and the most thoroughly investigated example of such a
branching sequence; for this reason, its regulatory mechanisms
have been discussed separately in Chapter 15. However, the problem is of course not restricted to this particular sequence, and
hence a brief general account of the usual control mechanisms is
needed here.

    In enzyme-promoted biosynthetic pathways, control mechanisms
usually operate upon those enzymes. In general, the end-product
of a given sequence, or of a particular branch, has a restraining
effect upon an early enzyme, which is almost always the first one
in that sequence or branch. Through this feed-back effect,
presence of the end-product in amounts in excess of those required will diminish or prevent the synthesis of additional quantities. For the restraining action itself, two major mechanisms
exist, which are by no means mutually exclusive: the end-product
can inhibit the activity of the enzyme, or it can diminish --
repress--its formation. The duality of mechanisms must clearly

serve some useful purpose, and must reflect multiple needs for control: feedback inhibition of an enzyme will be a rapid, temporary effect, feedback repression a slow, protracted one; also, the former will prevent wasteful synthesis of intermediates, the latter of enzyme protein. In branched pathways, both types of regulation can be achieved in a variety of ways, depending upon the role of the individual end-products (e.g., cumulative, sequential, concerted inhibition). For details on these questions, reviews such as those by Datta (69) and Umbarger (70) should be consulted.

Several alternative mechanisms have been observed occasionally, such as stimulation of the synthesis of one metabolite by the end-product of an alternative branch of the same general pathway, or control of metabolites which are chelators of an inorganic ion by that ion, rather than by the organic chelator, or by the chelate. Examples of such events are discussed in Chapter 15.

One striking aspect of these regulatory schemes is the multiplicity of mechanisms that have been developed by different organisms for the control of the same biosynthetic reaction. One may suspect that this species specificity of regulatory mechanisms has evolved in response to the specific needs of organisms having to adapt to radically different environments.

Another impressive feature of biosynthetic pathways is the frequent occurrence of enzyme complexes, of which one, that of tryptophan synthetase, has already been discussed. Complexes of different enzymes of the same sequence are commonly observed. In some of them, dissociation into their constituents takes place readily, while others have resisted any attempt at resolution, and may be genuine multienzyme particles. On the other hand, some such complexes may be the result of the methods used for isolation and purification. The "native" complexes may be designed to carry out several consecutive transformations on the same substrate molecule, which remains enzyme-bound throughout. Susceptibility of the individual enzymes to mutation can range from the possibility for completely independent mutation to completely concerted reaction.

The existence of such complexes may well be connected with the need for delicate regulation, although this may not be the only purpose; the entire problem of the meaning of these phenomena is far from being completely understood.

In many bacteria, the genes responsible for the production of the consecutive enzymes of one sequence can be combined into one highly organized unit, called an operon. In such a unit, the genes themselves occupy adjacent positions on the chromosome, and are together under the control of an "operator gene," which can turn the entire series on or off. Repression and its opposite, induction, of enzymes are thus actuated in a concerted fashion. Aspects of the function and regulation of operons have been studied in great detail, but need not concern us here, since only one operon is known to play a role in the biosynthesis of aromatic compounds: the tryptophan operon of *E. coli*, and even in this isolated example, only the first enzyme of the operon, anthranilate synthetase, is involved in the biosynthesis of an aromatic metabolite from nonaromatic precursors; the other enzymes all promote the further transformations of this first aromatic material into tryptophan, and are thus outside the stated scope of this volume. In the tryptophan operon, the genes controlling the formation of the several enzymes are not only contiguous, but are also arranged in the sequence in which the various steps take place in the biosynthesis, an arrangement which is by no means always encountered.

The phenomenon of induction of enzymes, which has just been mentioned, merits a brief discussion, since it plays a role in biosynthetic events. The term refers to the specific increase in the amount of an enzyme which can occur in response to the availability of a certain metabolite or nutrient, called the inducer. Induction constitutes formation *de novo* of enzyme protein; it takes place in all cells of a given population, in contrast to mutation which usually affects only a very small minority. Small amounts of inducible enzymes are present even in the absence of the inducer; these amounts can increase by more than two orders of magnitude on induction.

In another class of enzymes, the so-called constitutive ones, the quantity of enzyme present is fairly constant regardless of conditions of growth. It seems plausible to assume that vital biosynthetic processes will usually depend upon constitutive enzymes, and that inducible ones will come into play when the cell has to meet unusual circumstances. However, the distinction is not entirely sharp, since mutation can convert constitutivity into inducibility, and vice versa.

An excellent example of an inducible enzyme is the dehydroshikimate dehydrase of mutants of *Neurospora*, which converts dehydroshikimic acid into protocatechuic acid. In the wild-type, the stationary concentration of dehydroshikimic acid must be quite low, since it is rapidly converted to shikimic acid, the next intermediate in the normal pathway of biosynthesis of aromatic metabolites, and no dehydroshikimate dehydrase was detected (71). In mutants blocked between dehydroshikimic and shikimic acids, however, the former compound accumulates in appreciable amounts and induces the formation of the enzyme which channels it into an alternative pathway.

Other examples of inducible enzymes are the β-galactosidase of *E. coli*, the classical object of studies on induction; the penicillinase of *Bacillus cereus* (interestingly an enzyme inducible by cephalosporin C which is not a substrate of it); and tryptophanase.

1-4. ISOLATION OF BACTERIAL MUTANTS

As has been mentioned already, the mutants first used for the study of biosynthetic pathways were those of the mold *Neurospora* (20, 72). While such mutants have also served in research on the intermediates in the biosynthesis of aromatic compounds, most of this work rests on investigations of uv-induced mutants of bacteria such as *Escherichia coli* and *Aerobacter aerogenes*. Compared to the mutants of *Neurospora*, the ones of the bacteria mentioned offer (73) a number of valuable advantages: ease of cultivation on the fairly large scale needed for study of excreted intermediates, rapid growth, and possibility for investigation of large populations. In addition, they permit observation and assay of growth responses by convenient bacteriological techniques such as turbidimetry, bioautography (74a), and crossfeeding ("syntrophism") (74b): the feeding of one mutant by another one blocked at a different stage of a biosynthetic sequence; this phenomenon is conveniently observed between adjacent streaks on agar plates, and has been very helpful in recognizing the excretion of intermediates. On the other hand, bacterial mutants were less suitable for genetic studies than those of *Neurospora*, at least at the time when the fundamental work on the biosynthesis of aromatic metabolites was done. With the subsequent advances in the investigation of the chromosomal apparatus of bacteria, such studies have become quite feasible; however, they have not played a major role in the elucidation of the pathways leading to aromatic rings, the field of major concern of this monograph. The genetic aspects of these pathways and of their regulations are by now well investigated, but lie outside our topic; as one example, the review by Pittard and Wallace (75) of the genes controlling the shikimate pathway in *E. coli* may be quoted.

While it is easy to isolate mutants that can survive where the wild-type cannot (e.g., drug-resistant ones), special techniques are needed for the isolation of mutant strains such as those which are intended for research on biosynthetic pathways. Such strains, being unable to synthesize some essential metabo-

lite(s) arising from the particular biosynthetic sequence, are
dependent upon an outside supply of these metabolites, or of a
suitable precursor, and are thus not capable of growing on the
minimal medium sufficient for the parent wild-type strain. Methods for isolation of such strains will utilize this inability.

A convenient technique for the isolation of numerous nutritionally deficient bacterial strains, the penicillin method, has
been worked out about simultaneously by Davis (21a) and by Lederberg and Zinder (21b). Prior to its development, techniques were
available (76) for this purpose, but they were too laborious to
permit extensive research to be based on the strains so obtained.
The penicillin method has been extensively utilized by Davis and
co-workers for elucidation of biosynthetic chains leading to many
metabolites, including particularly the pathway that yields aromatic compounds via shikimic acid. This work has been summarized
by Davis (23, 27).

The method rests upon the observation of Hobby, Meyer, and
Chaffee (77) that penicillin attacks only actively dividing cells,
being without action upon resting ones (extensive subsequent research (78) has lead to an understanding of this phenomenon: penicillin exerts its effect by interfering with the formation of the
protective cell-wall). Consequently, exposure to penicillin
after irradiation (or some other mutagenic operation) in minimal
medium kills the wild-type (prototrophic) cells which are able to
multiply under these circumstances (mutation being a rare event,
these cells are present in huge excess); in contrast, the mutants
are protected by their very inability to grow. The penicillin is
subsequently removed by washing the cells or by action of penicillinase, and the desired mutants blocked in a step of the biosynthesis of a given metabolite can then be recognized and isolated by their ability to grow on minimal medium supplemented
with that metabolite, or with an intermediate in the part of the
sequence which follows the block.

Evidently, this efficient method is restricted to penicillin-
sensitive organisms; it can, however, be adapted to other cases
where only actively dividing cells are vulnerable. For mycobacteria, a technique using isoniazid has been developed by
Holland and Ratledge (79). A method that might be generally
applicable, described by Lubin (80), uses ^{3}H-labeled thymidine,
which is incorporated into deoxyribonucleic acid by growing cells
of both plants and animals; subsequent action of the absorbed
radioactive material leads to preferential destruction of the
wild-type cells. In a similar way, auxotrophic mutants of tobacco cells in culture have been isolated by Carlson (81), who
adapted a method initially worked out (82) for mammalian cell

cultures; here, the agent selectively taken up by dividing cells only is 5-bromodeoxyuridine, which is incorporated into deoxyribonucleic acid and renders the cell sensitive to visible light.

1-5. GENERAL ASPECTS OF BIOSYNTHESIS; LITERATURE

It remains for us to consider a few general aspects of biosynthesis, and to quote some of the more important source books. Sir Robert Robinson's *The Structural Relations of Natural Products* (14A), the first synthesis of the field, is still fascinating reading; we shall have opportunity, several times, to give credit to the insight of its author who has correctly anticipated so many later experimental developments. More recently, much of the area of biosynthesis has been reviewed in *Biogenesis of Natural Compounds*, edited by Bernfeld (83), and a great part of it is discussed in Geissman and Crout's *Organic Chemistry of Secondary Plant Metabolism* (84). Among the more specialized works, the following may be mentioned: Richards and Hendrickson, *The Biosynthesis of Steroids, Terpenoids, and Acetogenins* (85); Harborne, *Biochemistry of Phenolic Compounds* (86); *Biosynthesis of Aromatic Compounds*, edited by Billek (87); *Biosynthesis of Alkaloids*, edited by Mothes and Schütte (88); and *Biogenesis of Antibiotic Substances* (89). The smaller volumes by Bu'Lock, *The Biosynthesis of Natural Products* (24) and *Essays in Biosynthesis and Microbial Development* (90), and by Grisebach, *Biosynthetic Patterns in Microorganisms and Higher Plants* (91), give much valuable information and stimulating discussion. Excellent chapters on biosynthesis of particular groups of substances are included in many monographs; the following may be quoted especially: the chapter by Spenser (92) in *Chemistry of the Alkaloids* edited by Pelletier; the one by Goodwin in *The Carotenoids* edited by Isler (93), and the discussion of biosynthesis in Thomson's *Naturally Occurring Quinones* (94). Less generalized review articles are mentioned wherever possible in the individual chapters and sections.

These books, together with others not specifically quoted, and the innumerable individual papers dealing with problems of biosynthesis, permit some generalizations to be made on the status of knowledge which has been gained. The picture that presents itself to the organic chemist is one of a gratifying and increasing unity. As more pathways to natural compounds, both primary and secondary ones, are being explored, it becomes ever clearer that the chemical stages in these pathways tend to be represented by a relatively small number of type reactions, many (by no means all) of them long-established processes well

known to synthetic and mechanistic organic chemistry. Examples
of such reactions are aldol condensations, oxidative coupling of
phenols, Mannich reactions, Wagner-Meerwein rearrangements, and
many others.

Of course, no complete parallelism in the development of
organic chemistry and of knowledge concerning biosynthetic pathways was to be expected during the 100 years--a brief span--since
structural organic chemistry was placed on a firm basis; some
areas (e.g., that of the reactions of phosphate esters) may have
been less extensively developed than might have seemed desirable
in retrospect, after their importance for the understanding of
biochemical processes came to be recognized. In general, however,
it has become increasingly clear that most reactions *in vivo* are
not fundamentally different from those known in the laboratory,
and that both can be interpreted in the light of modern mechanistic concepts in a good percentage of cases. In view of the mar-
velous achievements of enzymes as regards specificity and ability
to make involved reactions proceed at room temperature in aqueous
medium, this may seem surprising, but enzyme catalysts, like catalysts in general, by definition promote reactions which would in
principle be possible without them and, to quote (ref. 14A, p.2):
"Even enzymes are unlikely to disregard stereochemistry or the
mode of electronic displacements in the molecules."

REFERENCES

1. F. Wöhler and J. Liebig, Annalen der Pharmazie 3, 249 (1832).
2. P.C.J. Brunet, Nature 199, 492 (1963).
3. J. Meinwald, K.F. Koch, J.E. Rogers, jun., and T. Eisner,
    J. Amer. Chem. Soc. 88, 1590 (1966).
4. J.H. Bowie and D.W. Cameron, J. Chem. Soc. 5651 (1965).
5. P. Buchner, Endosymbiosis of Animals with Plant Microorgan-
    isms, Interscience Publishers, New York, 1967.
6. A. Salaque, M. Barbier, and E. Lederer, Bull. Soc. Chim.
    Biol. 49, 841 (1967).
7. M. Rothstein and G. Tomlinson, Biochim. Biophys. Acta 63,
    471 (1962).
8. J.N. Collie, J. Chem. Soc. 63, 329 (1893); 91, 1806 (1907).
9. (a) E. Winterstein and G. Trier, Die Alkaloide, Bornträger,
    Berlin, 1910;
    (b) R. Robinson, J. Chem. Soc. 111, 876 (1917).
10. C. Schöpf, Angew. Chem. 50, 779,797 (1937).
11. H.O.L. Fischer and G. Dangschat, Ber. 65, 1009 (1932); Helv.
    Chim. Acta 17, 1200 (1934); 18, 1204 (1935); G. Dangschat
    and H.O.L. Fischer, Biochim. Biophys. Acta 4, 199 (1950).

12. (a) O. Loew, Ber. <u>14</u>, 450 (1881);
    (b) A.J. Hall, Chem. Rev. <u>20</u>, 305 (1937).
13. O. Wallach, Terpene und Campher, Veit and Co., Leipzig, 1909;
    L. Ruzicka, Experientia <u>9</u>, 357 (1953); L. Ruzicka, O. Jeger,
    and D. Arigoni, Helv, Chim. Acta <u>38</u>, 1890 (1955).
14. R. Robinson, Chem. Ind. <u>53</u>, 1062 (1934); R.B. Woodward and
    K. Bloch, J. Amer. Chem. Soc. <u>75</u>, 2023 (1953); R.G. Langdon
    and K. Bloch, J. Biol. Chem. <u>200</u>, 129, 135 (1953).
14A. Sir Robert Robinson, The Structural Relations of Natural
    Products, Clarendon, Oxford, 1955.
15. H.O.L. Fischer and G. Dangschat,
    (a) Helv. Chim. Acta <u>20</u>, 705 (1937);
    (b) Ibid. <u>18</u>, 1204 (1935).
16. (a) M.H. Zenk, Ber. Deut. Botan. Ges. <u>80</u>, 573 (1967);
    (b) M. Luckner, Die Pharm. <u>26</u>, 717 (1971).
17. M.J.O. Francis and M. O'Connell, Phytochemistry <u>8</u>, 1705
    (1969).
18. W. Barz, Z. Naturforsch. <u>24B</u>, 234 (1969).
19. J. Pittard and B.J. Wallace, J. Bact. <u>92</u>, 1070 (1966).
20. G.W. Beadle and E.L. Tatum, Proc. Natl. Acad. Sci. U.S. <u>27</u>
    499 (1941).
21. (a) B.D. Davis, J. Amer. Chem. Soc. <u>70</u>, 4267 (1948); Proc.
    Natl. Acad. Sci. <u>35</u>, 1 (1949);
    (b) J. Lederberg and N. Zinder, J. Amer. Chem. Soc. <u>70</u>, 4267
    (1948).
22. E.A. Adelberg, Bact. Rev. <u>17</u>, 253 (1953).
23. E. Umbarger and B.D. Davis, Pathways in Amino Acid Biosyn-
    thesis, in The Bacteria, Vol. III, I.C. Gunsalus and R.Y.
    Stanier, Eds., Academic, New York and London, 1962, p. 167.
24. J.D. Bu'Lock. The Biosynthesis of Natural Products, McGraw-
    Hill Publishing Company Limited, London, New York, Toronto,
    and Sydney, 1965.
25. B.D. Davis and U. Weiss, Arch. Exper. Pathol. Pharmakol.
    <u>220</u>, 1 (1953).
26. S. Mitsuhashi and B.D. Davis, Biochim. Biophys. Acta <u>15</u>,
    268 (1954).
27. B.D. Davis, Adv. Enzymol. <u>16</u>, 247 (1955).
28. Cf., inter alia, A. Albert, Selective Toxicity, Wiley, New
    York, 2nd ed., 1960, p. 24 ff; L.V. Heilbrunn, An Outline of
    General Physiology, Saunders, Philadelphia and London, 3rd
    ed., 1952, p. 150 ff.
29. U. Weiss and E.S. Mingioli, J. Amer. Chem. Soc. <u>78</u>, 2894
    (1956).
30. H. Kindl and M. Schiefer, Monatsh. <u>100</u>, 1773 (1969).
31. J.A. Hamilton and G.B. Cox, Biochem. J. <u>123</u>, 435 (1971).
32. (a) P. Fildes, Brit. J. Exptl. Pathol. <u>21</u>, 315, (1940);
    (b) E.E. Snell, Arch. Biochem. <u>2</u>, 389 (1943).

33. E.L. Tatum, D. Bonner, and G.W. Beadle, Arch. Biochem. 3,
     477 (1944).
34. E.L. Tatum and D. Bonner, Proc. Natl. Acad. Sci. U.S. 30,
     30 (1944).
35. C. Yanofsky, Bact. Rev. 24, 221 (1960), and refs. there.
36. A. Ginsburg and E.R. Stadtman, Ann. Rev. Biochem. 39, 429
     (1970).
37. S. Aronoff, Techniques of Radiobiology, The Iowa State
     College Press, Ames, Iowa, 1956, p. 91.
38. M.H. Zenk, Z. Physiol. Chem. 353, 123 (1972).
39. R. Durand and M.H. Zenk, Tetrahedron Lett. 3009 (1971).
39A. R.K. Ibrahim, S.G. Lawson, and G.H.N. Towers, Can. J. Bio-
     chem. Physiol. 39, 873 (1961); cf. also V.C. Runeckles, Can.
     J. Botany 41, 823 (1963).
40. W.A. Wood, I.C. Gunsalus, and W.W. Umbreit, J. Biol. Chem.
     170, 313 (1947).
41. T. Watanabe and E.E. Snell, Proc. Natl. Acad. U.S. 69, 1086
     (1972), and refs. quoted there.
42. H. Kumagai, H. Yamada, H. Matsui, H. Ohkishi, and K. Ogata,
     J. Biol. Chem. 245, 1767, 1773 (1970).
43. H. Yamada, H. Kumagai, N. Kashima, H. Torii, H. Enei, and
     S. Okamura, Biochem. Biophys. Res. Commun. 46, 370 (1972).
44. H. Kumagai, H. Matsui, H. Ohgishi, K. Ogata, H. Yamada, T.
     Yamada, T. Ueno, and H. Fukami, Biochem. Biophys. Res.
     Commun. 34, 266 (1969).
45. T. Ueno, H. Fukami, H. Ohkishi, H. Kumagai, and H. Yamada,
     Biochim. Biophys. Acta 206, 476 (1970).
46. B.D. Davis and W.K. Maas, J. Amer. Chem. Soc. 71, 1865,
     (1949).
47. W.S. Champney and R.A. Jensen, J. Bact. 98, 205 (1969).
48. N. Rosa and A.C. Neish, Canad. J. Biochem. 46, 797 (1968).
49. M.H. Zenk and H. Scherf, Biochim. Biophys. Acta 71, 737
     (1963).
50. R. Thomas, in Biogenesis of Antibiotic Substances, Czecho-
     slovak Academy of Sciences, Prague, 1965, p. 155; cf. also,
     inter alia, J.E. Baldwin, H.H. Basson, and H. Krauss, Jr.,
     Chem. Commun. 984 (1968).
51. H. Brockmann, Fortschritte Chem. Org. Naturstoffe 14, 142
     (1957); H. Brockmann and R. Mühlmann, Chem. Ber. 82, 342
     (1949).
52. C. Leibermann and W. Drory, Ber. 22, 680 (1889).
53. A.P. Arendaruk and A.P. Skoldinov, Zhur. Obshcheii Khim. 30,
     484; 489 (1960); Chem. Abstr. 54, 24835 (1960).
54. M.V. von Bülow and O.R. Gottlieb, An. Acad. Brasil. Cienc.
     40, 299 (1968); Chem. Abstr. 71, 46665 (1969).

55. R. Hegnauer, Chemotaxonomie der Pflanzen, Birkhäuser Verlag, Basel and Stuttgart, Vol. 4, 1966, p. 369.
56. J. Křepinský, Z. Samek, F. Šorm, D. Lamparsky, P. Ochsner, and Y.-R. Naves, Tetrahedron. Suppl. 8, Part I. 53 (1966).
57. W.C. Wildman, in Chemistry of the Alkaloids, S.W. Pelletier, Ed., Van Nostrand Reinhold, New York, Cincinnati, Toronto, London, and Melbourne, 1970, p. 199.
58. F. Stermitz, in Organic Photochemistry, O.L. Chapman, Ed., Vol. I, Dekker, New York, 1967, p. 247.
59. J.H. Russel, Naturwiss. 50, 443 (1963).
60. J.W. Wheeler, Jr., and R.E. Eastman, J. Amer. Chem. Soc. 81, 236 (1959).
61. S. Liaaen-Jensen, Pure Appl. Chem. 20, 421 (1969).
62. J.M. Edwards and U. Weiss, Tetrahedron Lett. 4325 (1969).
63. F.A. Haskins, L.G. Williams, and H.J. Gorz, Plant Physiol. 39, 777 (1964); K.G. Edwards and J.R. Stoker, Phytochemistry 6, 655 (1967).
64. K.G. Edwards and J.R. Stoker, Phytochemistry 7, 73 (1968).
65. Cf., e.g., H.W. Siegelman, in Biochemistry of Phenolic Compounds, J.B. Harborne, Ed., Academic, London and New York, 1964, p. 437.
66. J. Kagan, J. Amer. Chem. Soc. 88, 2617 (1966).
67. A.C. Waiss, Jr., R.E. Lundin, A. Lee, and J. Corse, J. Amer. Chem. Soc. 89, 6213 (1967); T. Matsuura and H. Matsushima, Tetrahedron 24, 6615 (1968).
68. D. Dewar and R.G. Sutherland, Chem. Commun. 272 (1970).
69. P. Datta, Science 165, 556 (1969).
70. H.E. Umbarger, Ann. Rev. Biochem. 38, 323 (1969).
71. S.R. Gross, J. Biol. Chem. 233, 1146 (1958).
72. G.W. Beadle, Chem. Rev. 37, 15 (1945).
73. B.D. Davis, Experientia 6, 41 (1950).
74. (a) W.A. Winsten and E. Eigen, J. Biol. Chem. 184, 155 (1950); (b) J. Lederberg, J. Bact. 52, 503 (1946).
75. J. Pittard and B.J. Wallace, J. Bact. 91, 1494 (1966).
76. C.H. Gray and E.L. Tatum, Proc. Natl. Acad. Sci. U.S. 30, 404 (1944); J. Lederberg and E.L. Tatum, J. Biol. Chem. 165, 381 (1946); S. Simmonds, E.L. Tatum, and J.S. Fruton, ibid. 169, 91 (1947).
77. G.L. Hobby, K. Meyer, and E. Chaffee, Proc. Soc. Exp. Biol. Med. 50, 281 (1942).
78. J.L. Strominger, in The Bacteria, I.C. Gunsalus and R.Y. Stanier, Eds., Vol. III, Academic Press, New York and London, 1962, p. 413.
79. K.T. Holland and C. Ratledge, J. Gen. Microbiol. 66, 115 (1971).

80. M. Lubin, Science 129, 838 (1959).
81. P.S. Carlson, Science 168, 487 (1970).
82. T. Puck and F. Kao, Proc. Natl. Acad. Sci. U.S. 58, 1227 (1965).
83. P. Bernfeld, Ed., Biogenesis of Natural Compounds, 2nd ed., Pergamon Press, New York, 1967.
84. T.A. Geissman and D.H.G. Crout, Organic Chemistry of Secondary Plant Metabolism, Freeman, Cooper and Co., San Francisco, Cal., 1969.
85. J.H. Richards and J.B. Hendrickson, The Biosynthesis of Steroids, Terpenes, and Acetogenins, W.A. Benjamin, Inc., New York and Amsterdam, 1964.
86. J.B. Harbone, Biochemistry of Phenolic Compounds, Academic London and New York, 1964.
87. G. Billek, Ed., Biosynthesis of Aromatic Compounds, Pergammon Press, New York, 1966.
88. K. Mothes and H.R. Schütte, Biosynthese der Alkaloide, VEB Deutscher Verlag der Wissenschaften, Berlin, 1969.
89. Z. Vaněk and Z. Hošťálek, Eds., Biogenesis of Antibiotic Substances, Czechoslovak Academy of Sciences, Prague, 1965.
90. J.D. Bu'Lock, Essays in Biosynthesis and Microbial Development, Wiley, New York, 1967.
91. H. Grisebach, Biosynthetic Patterns in Microorganisms and Higher Plants, Wiley, New York, 1967.
92. I.D. Spenser, in Chemistry of the Alkaloids, S.W. Pelletier, Ed., Van Nostrand, New York, 1970.
93. T.W. Goodwin, in Carotenoids, O. Isler, Ed., Birkhäuser, Basel and Stuttgart, 1971.
94. R.H. Thomson, Naturally Occurring Quinones, 2nd ed., Academic Press, London and New York, 1971.

PART 1 BIOSYNTHESIS OF AROMATIC COMPOUNDS
FROM GLUCOSE VIA SHIKIMIC ACID

CHAPTER 2

The General Pathway:
Some Common Features

It has already been mentioned in the Introduction that a
very large number of natural products is formed from glucose by
way of one sequence which proceeds through a series of common
alicyclic intermediates (one of them shikimic acid), to branch
out into a limited number of primary aromatic metabolites. Further transformations of these primary products produce an immense
variety of aromatic compounds ranging from essential constituents
of every living organism to "chemical curiosities" of extremely
limited occurrence and unknown significance.

The present chapter deals with this pathway in some detail.
A description of its pre-aromatic intermediates, and of their biogenetic transformations, is given first (chapters 3-13). In view
of the universal significance of these compounds, and of the absence of a recent review they are treated in greater detail than
are individual substances elsewhere in this volume.[+] A discussion
of the generality of the shikimate pathway, and of its possible
variants, follows in Chapter 14. Chapter 15 deals with the regulatory mechanisms which enable this pathway to function efficiently
in satisfying the various needs for individual aromatic metabolites. Since it is not possible to ignore these questions in the
earlier sections, a few references to recent reviews (la,b) may
find a place here. Finally, Chapter 16 aims at giving information
on the general role of the shikimate pathway for the biosynthesis
of natural substances and on its significance in the overall
scheme of synthetic activities of organisms.

The common origin of all those substances betrays itself by
a number of common structural features. It was recognized quite
early, and well ahead of possibilities for experimental study,
that many of these natural compounds form a coherent group likely
to be joined together by some common biosynthetic background.
The most important one of these shared features is the very frequent occurrence of the grouping C_6-C_3, i.e. an aromatic ring

---

[+] Such a review has appeared since this chapter was written; see
E. Haslam, The Shikimate Pathway, Wiley, 1974.

carrying a straight C_3 side-chain, or of structures readily
derivable from such a chain. Such further elaborations of the
phenylpropanoid unit include shortening to C_6-C_2 and C_6-C_1 (e.g.
C_2 formed through decarboxylation of one of the numerous C_6-C_3
structures with terminal carboxyl); dimerization to lignans; polymerization to lignin; incorporation of oxygen or nitrogen to
yield heterocycles such as coumarins, or isoquinoline alkaloids
(the latter with loss of the terminal carbon atom of the C_3
chain); or combination with units of a different origin, to give,
e.g., the innumerable flavonoids, which are $C_6-C_3-(C_2)_3$ compounds
whereby the C_6-C_3 unit furnishes the three carbons of the heterocyclic ring and the aromatic ring B attached to it, while ring A
is acetate-derived. These substances of mixed origin, and many
related classes of compounds, are discussed in Chapter 21.

   Associated with the occurrence of this C_6-C_3 unit and its
modifications is a distinct selectivity in the distribution of
oxygen-containing substituents on the aromatic ring. Where present, such substituents show a very marked preference for positions 4; 3 + 4; or 3 + 4 + 5 relative to the side-chain; this distribution is in striking contrast to the *meta-* and *peri*-orienta-
tion characteristic of acetate-derived aromatic compounds (see
Chapters 17 and 18). A selection of typical structures is given
in Scheme I; it is instructive to compare the oxygenation patterns
of rings A (acetate-derived) and B (part of the C_6-C_3 unit) in
the formulae of a few $C_6-C_3-(C_2)_3$ compounds given in Section C of
the Scheme. (It should be noted that the isoflavones shown there
belong into this group, despite the fact that their C_6-C_3 unit
corresponds to isopropylbenzene rather than to *n*-propylbenzene as
usual. This structural type is known to be the result of a 1,2-
migration of ring B in some precursor with normal $C_6-C_3-(C_2)_3$
skeleton; see Section 21-2b).

   Although exceptions from the typical location of oxygen-
containing substituents in C_6-C_3 compounds and their relatives
have been encountered in fair numbers, they still form only a
small minority of known structures. Some examples have been
assembled in Scheme II. The question: "how small a minority?"
has been studied by Geissman and Hinreiner (1A) in 1952 in a survey of the pertinent compounds known at that time. The vast numbers of new substances discovered since then may to a certain extent modify the results of this survey, but they are not likely to
invalidate its essential findings: that conformity with the typical pattern is always high (lowest in the coumarins), and reaches
100% in the lignins and anthocyanins. Furthermore, Scheme II
illustrates the fact that most of these atypical compounds have
the normal pattern, modified by the presence of an additional oxygenated substituent *ortho* to the side-chain. This situation is

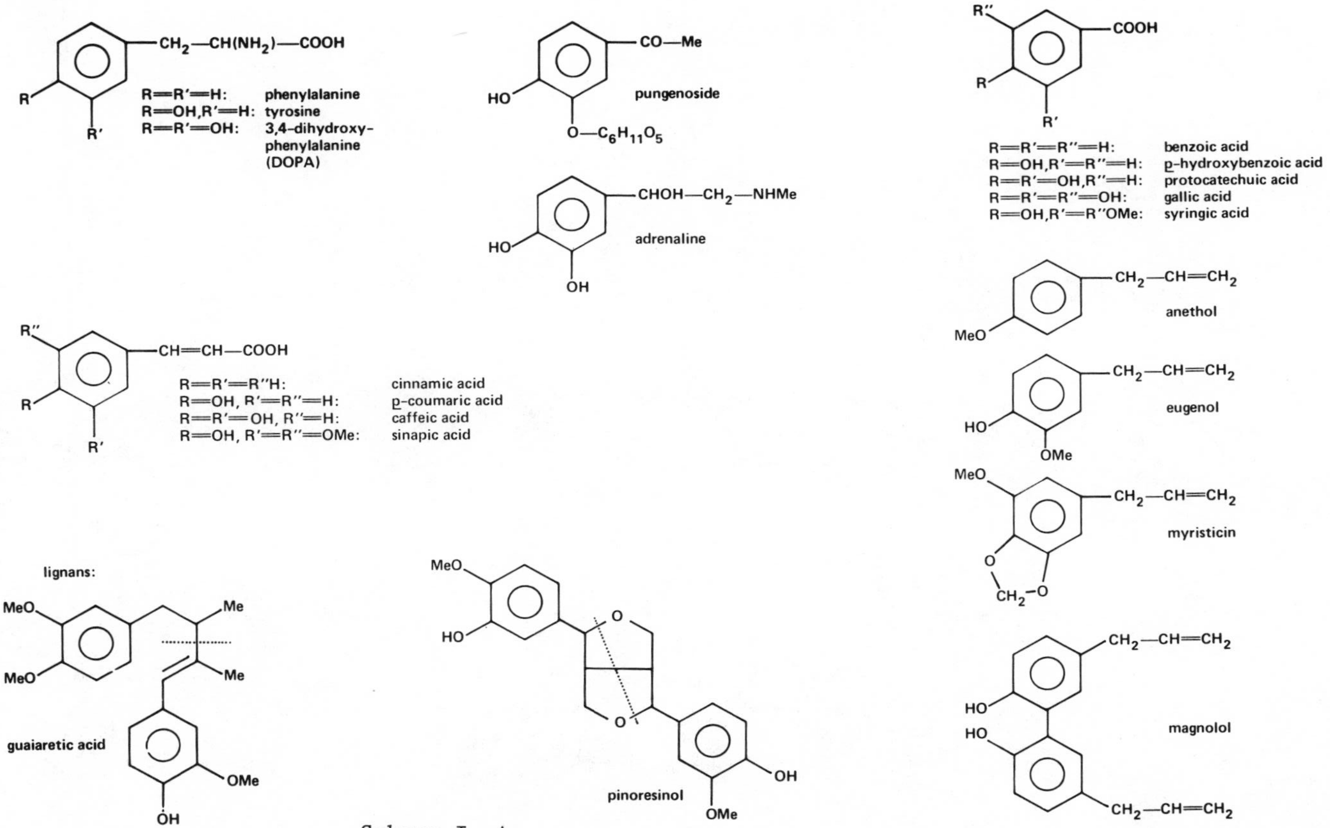
CH2—CH(NH2)—COOH
R═R′═H: phenylalanine
R═OH,R′═H: tyrosine
R═R′═OH: 3,4-dihydroxy-phenylalanine (DOPA)
CO—Me
HO
O—C6H11O5
pungenoside
CHOH—CH2—NHMe
adrenaline
COOH
R═R′═R″═H: benzoic acid
R═OH,R′═R″═H: p-hydroxybenzoic acid
R═R′═OH,R″═H: protocatechuic acid
R═R′═R″═OH: gallic acid
R═OH,R′═R″OMe: syringic acid
CH═CH—COOH
R═R′═R″H: cinnamic acid
R═OH, R′═R″═H: p-coumaric acid
R═R′═OH, R″═H: caffeic acid
R═OH, R′═R″═OMe: sinapic acid
CH2—CH═CH2
anethol
eugenol
myristicin
magnolol
lignans:
guaiaretic acid
pinoresinol
MeO
OMe
OH
Me

Scheme I. A.

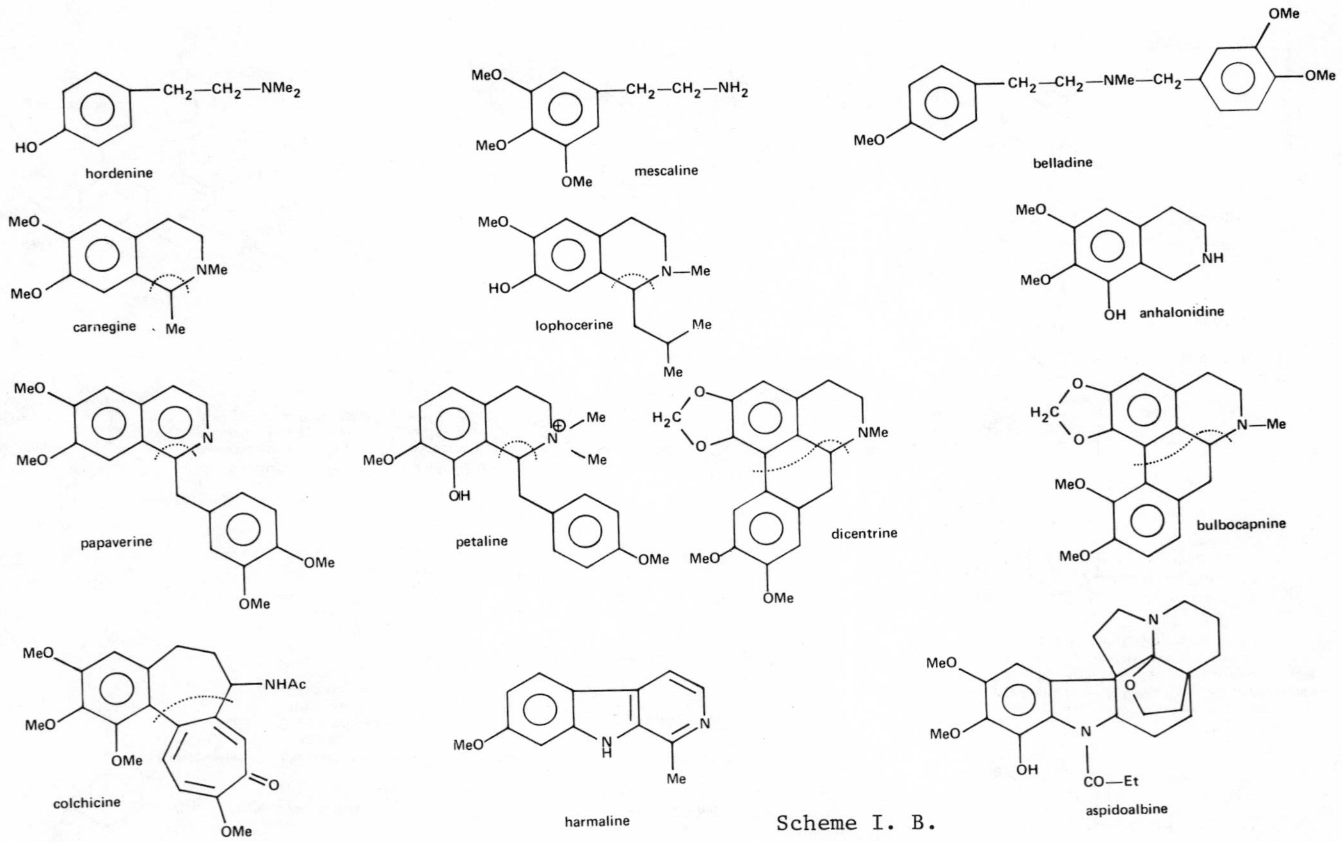

Scheme I. B.

belladine

anhalonidine

bulbocapnine

aspidoalbine

mescaline

lophocerine

dicentrine

petaline

harmaline

hordenine

carnegine

papaverine

colchicine

44

flavonoids

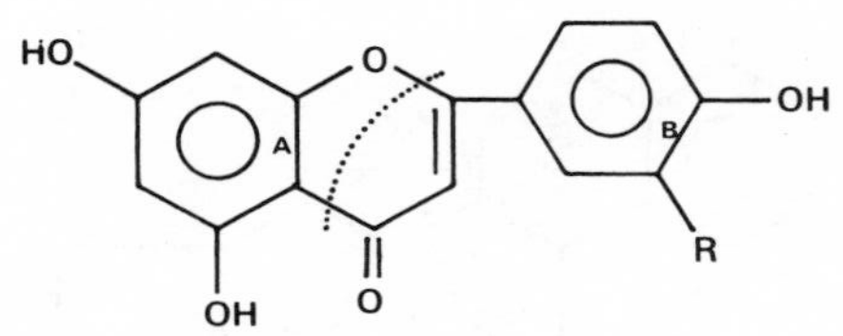

R=H: apigenin
R=OH: luteolin

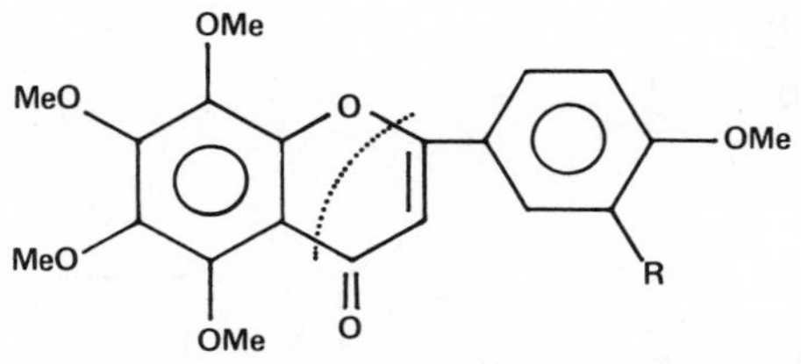

R=H: tangeretin
R=OH: nobiletin

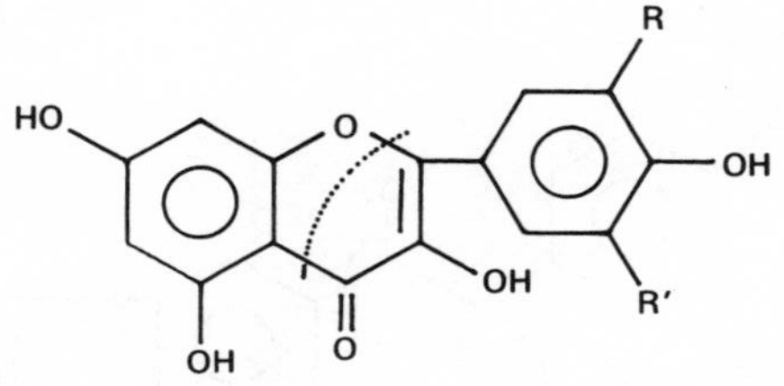

R=R'=H: kaempferol
R=OH, R'=H: quercetin
R=R'=OH: myricetin

isoflavones

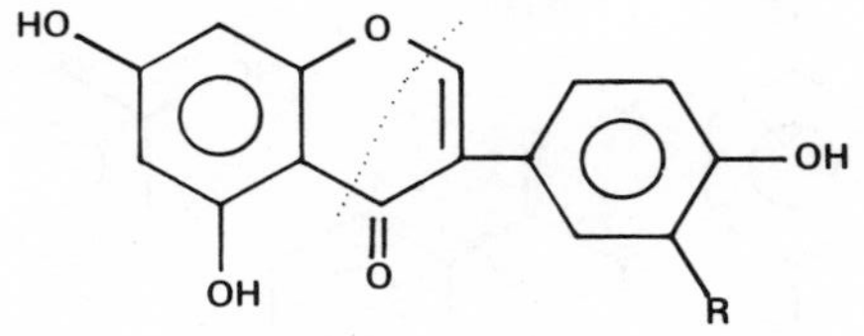

R=H: genistein
R=OMe: biochanin A
R=OH: orobol

anthocyanidins

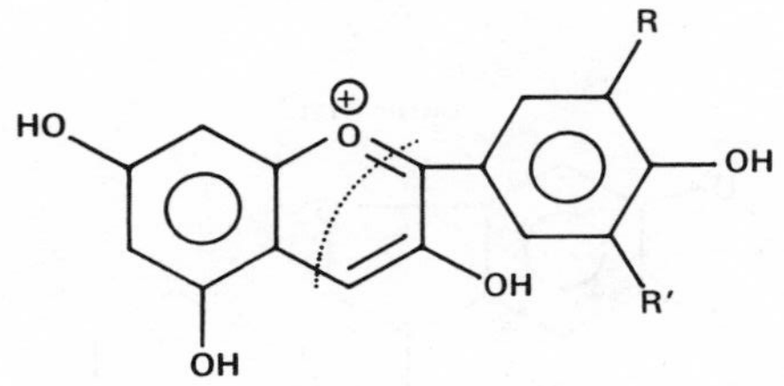

R=R' + H: pelargonidin
R=OH, R'=H: cyanidin
R=R'=OH: delphinidin
R=OMe, R'=H: peonidin
R=OMe, R'=OH: petunidin
R=R'=OMe: malvidin

Scheme I. C.

45

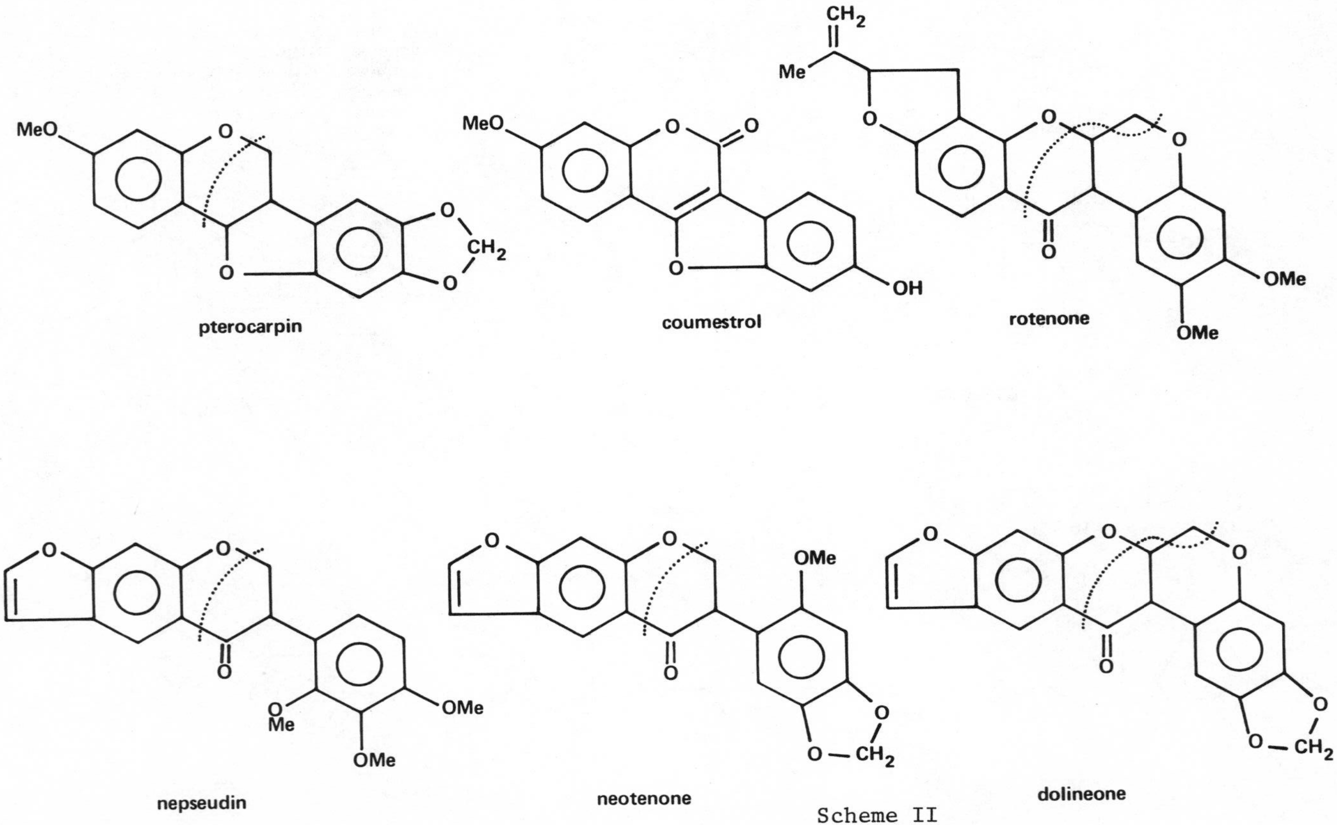

rotenone

coumestrol

pterocarpin

dolineone

neotenone

nepseudin

Scheme II

46

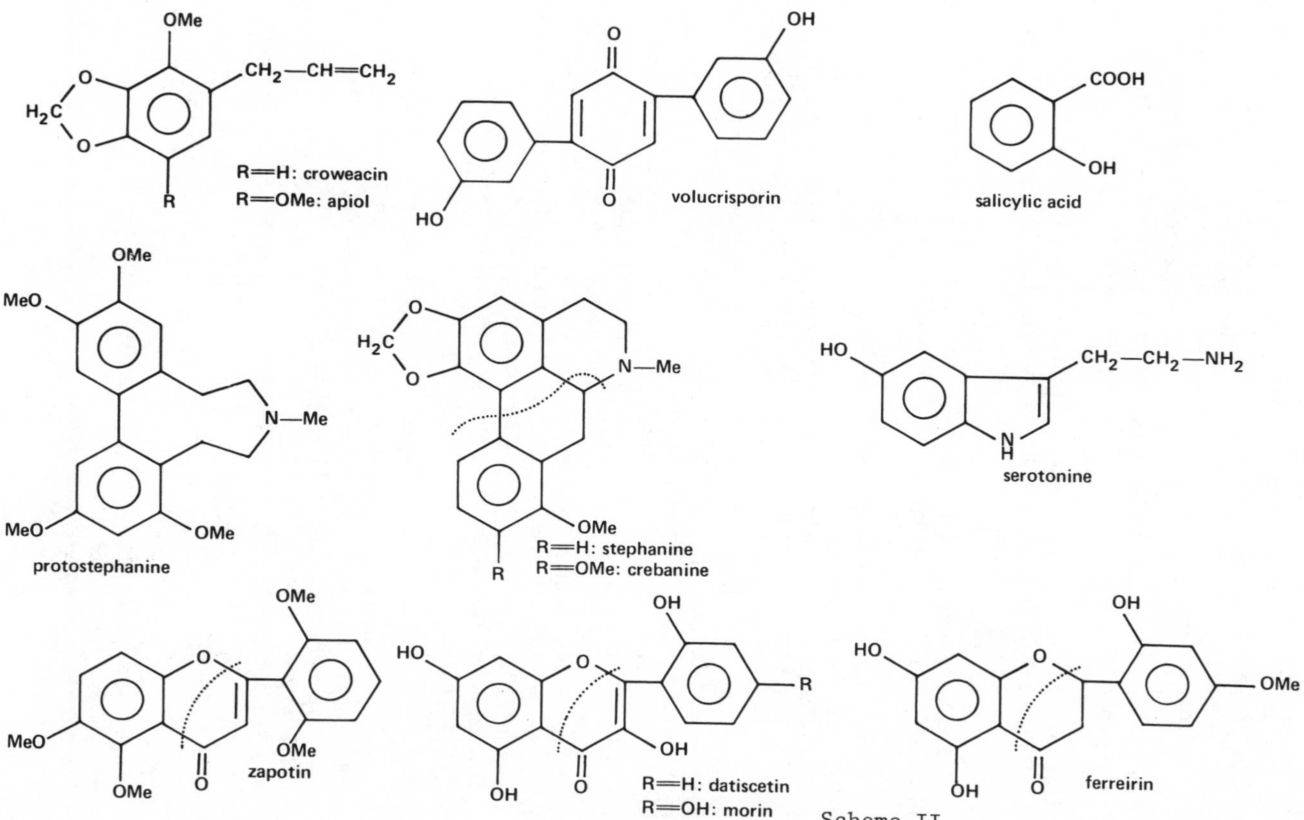

OMe
CH_2—CH=CH_2
H_2C
O
R
R=H: croweacin
R=OMe: apiol
OH
HO
volucrisporin
COOH
OH
salicylic acid
MeO
OMe
N—Me
protostephanine
N—Me
R=H: stephanine
R=OMe: crebanine
HO
CH_2—CH_2—NH_2
N
H
serotonine
MeO
zapotin
R=H: datiscetin
R=OH: morin
ferreirin

Scheme II

particularly common in the isoflavones and in the more complicated structures which contain the isoflavone skeleton; it will be discussed in detail in subsequent parts of this volume, as will be that of the small number of atypical compounds (volucrisporin, protestephanine, stephanine, etc.) that do not show this feature.

Early hypotheses on the biosynthetic background of some of these compounds are due to Winterstein and Trier (2) who recognized, ingeniously and essentially correctly, the probable derivation of the isoquinoline alkaloids from C_6-C_3 amino acids; their formulation of the biosynthesis of laudanosine [1] from two molecules of 3,4-dihydroxyphenylalanine (dopa, [2]) is shown below. Similarly, the significance of the widely distributed cinnamic acids in the derivation of other metabolites containing the C_6-C_3 unit or its variants, and their relationship with the aromatic amino acids, has been recognized quite early.

These interrelations of aromatic metabolites assisted the recognition of many compounds as a unified group. The first fruitful ideas concerning the biosynthetic origin of the aromatic ring itself postulated quinic acid [3] as an intermediate, a role for which its wide distribution in the plant kingdom, its structure, and its ready aromatization *in vitro* and *in vivo* would make it appear very suitable. Its conversion to protocatechuic (3,4-dihydroxybenzoic) acid by bacteria led Loew (3) in 1881, long before its structure was known, to discuss the possibility that [3] might be the precursor of the benzene rings of aromatic compounds in plants. Even earlier, Lücke (4a) and Lautemann (4b) established its aromatization to hippuric acid (N-benzoylglycine) in man; it is also the chief source of this acid in horse urine (5). It has, however, been shown recently (6) that this conversion of [3] (and also of the related shikimic acid, [4]) to hippuric acid is a function of the intestinal microflora, at least in the rat, since it is almost completely suppressed by the antibiotic neomycin.

The correct structure of [3] was conclusively proven by H. O. L. Fischer and Dangschat (7); the location of its three secondary alcohol groups in the ring positions corresponding to those of the phenolic hydroxyls in gallic acid [5] led these authors to postulate (7, 8) a transformation of [3] via [4] to [5], as shown below. Similarly, Hall (7a) ascribed to [3] a central role in various biosynthetic events, and formulated speculative derivations of the cinnamic acids from hypothetical precursors with ring structures resembling that of [3] but carrying C_3 side-chains.

Experimental examination of the problem became possible by the use of labeling and mutant methods; both were made to bear on the question about simultaneously.

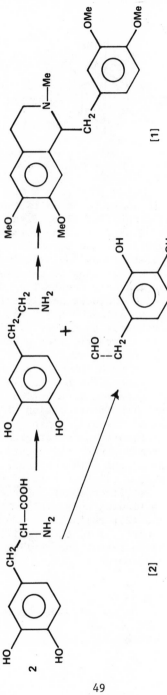

Biosynthesis of Laudanosine according to Winterstein and Trier (2).

49

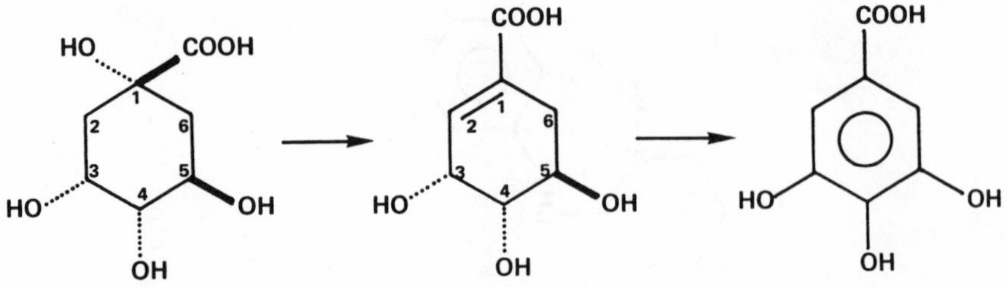

Isotope techniques in non-mutated cells (yeast) were used in 1950-1952 by Gilvarg and Bloch (9a, b) to establich the ultimate origin of two of the most important metabolites now known to come from this pathway, phenylalanine and tyrosine. It was found at first (9a) that labeled acetate was only poorly incorporated into the two amino acids; in yeast growing on glucose as chief carbon source, the incorporation was only about 1% of that found for glutamic acid. This finding excludes acetic acid and metabolites derived from it, such as the acids of the Krebs cycle, pyruvate, or alanine as major contributors to either ring or side-chain, both of which must come more directly from glucose. This conclusion was confirmed (9b) by studies with [1-^{14}C] glucose, from which both phenylalanine and tyrosine incorporated label mainly into atoms 2 and 6 of the ring and β of the side-chain, as shown below. Furthermore, the observed distribution of radioactivity is of course inconsistent with incorporation of the intact glucose chain. The labeling of the side-chain is compatible with its origin from a three-carbon fragment originating from glycolysis and preceding pyruvate; however, the location of label in the ring shows that this part of the structure cannot be formed from two such fragments. The symmetrical nature of the aromatic rings in the amino acids precludes discrimination between C-2 and C-6; with this limitation, the finding agree well with those made subsequently on shikimic acid (20, 28) and on some other products such as lignin (10). In particular, the detailed work on the incorporation of 6-^{14}C glucose into tyrosine (11) and tryptophan (12) bears out the older research.

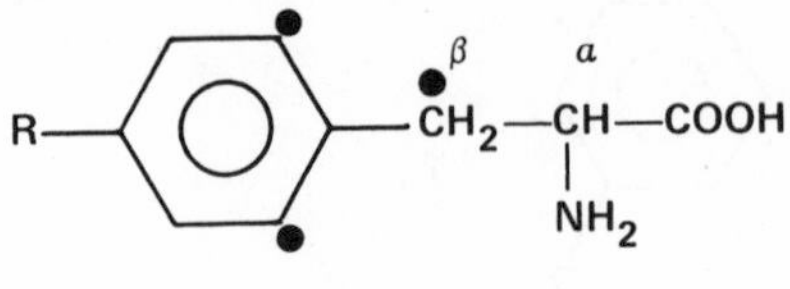

The methods for localization of label in these investigations, and in many others that will have to be mentioned, merit
brief discussion. Many of them are of course the standard procedures of organic chemistry; in the present case (9b), for instance, the β-carbon of the side-chain of phenylalanine was obtained by oxidation to benzoic acid and subsequent decarboxylation with copper oxide and quinoline. However, isolation of individual carbon atoms of an aromatic ring for the purpose of establishing the amount of labeling is a problem encountered almost
exclusively in research on the biosynthesis of such rings, and
the method most commonly used for this purpose needs some comment.
This method consists in nitration of the aromatic ring, followed
by treatment with calcium or barium hypobromite, which yields the
carbon atom bearing the nitro group as bromopicrin, Br_3CNO_2.
Only aromatic rings bearing phenolic OH, or similar activating
groups, undergo this cleavage. In many phenols, particularly pic-
ric and styphnic acids (the trinitro derivatives of phenol and resorcinol, respectively), the carbons adjacent to the nitrated one
appear as CO_2. This cleavage of aromatic rings was discovered
long ago (1854) by Stenhouse (13), who observed the formation of
bromopicrin and CO_2 on distillation of an aqueous mixture of pic-
ric acid, bromine, and lime. The mechanism of the process has
been studied recently by Butler and Wallace (14). The reaction
was introduced for the study of aromatic compounds by Baddiley
and co-workers (15) and has found wide and successful use. It is,
however, reliable only if it is certain that all the carbon of the
bromopicrin comes from the ring position being studied. Serious
errors could result if the carbon under investigation were unlabeled and one of its neighbors strongly active, and if the bromopicrin were contaminated with any C_2 (or C_3, etc.) compound, or
with a C_1 product not originating from the location being studied.
Unfortunately, the bromopicrin as usually obtained is by no means
pure, and is not readily purified in the available micro-quanti-
ties. The chief impurity is CBr_4 (16), and it has been shown by
Wallach (17a) and by Collie (17b) that this compound can form from
a wide variety of organic molecules under the conditions of the
bromopicrin reaction. Even so, serious difficulties seem to have
been encountered only occasionally from use of the initial technique, in which the bromopicrin was subjected to isotope assay
without further purification. However, an improved procedure has
been developed by Birch and co-workers (18), where the bromopicrin
is reduced to methylamine, which is purified as the crystalline N-
methyl-2,4-dinitroaniline. This modified technique enables the
final assay for radioactivity to be carried out on a pure, homogeneous compound.

The work by Gilvarg and Bloch (9a, b) discussed above gave
information on the origin of two of the aromatic end-products
(out of at least six or seven, as was to be found later) from
glucose and to a certain extent on the way in which the carbon
chain of glucose is built into the two amino acids. Nothing
could be said, however, about the substantial number of chemical
processes that must take place between the beginning and end of
the biosynthetic chain, or on the broader validity of the sequence.

Our knowledge of these questions rests essentially on the
mutant work of Davis, Sprinson, and their co-workers, which was
initiated about simultaneously with the research of Gilvarg and
Bloch. Several valuable reviews on the mutant-based investigations exist (19-21), but substantial progress has been made since
appearance of the most recent one of these.

It was found by Davis (22) that the penicillin method yielded
a variety of auxotrophic bacterial strains (mostly of *Escherichia
coli* and *Aerobacter aerogenes)* which required for growth one,
several, or all of four aromatic compounds: phenylalanine, tyrosine, tryptophan, and *p*-aminobenzoic acid, in that order, so that
double auxotrophs required the first two metabolites and the
triple ones the first three. Subsequent research led to the discovery that two more aromatic compounds are essential for these
bacteria, at least under certain conditions. One of these was
identified as *p*-hydroxybenzoic acid (23), while the second one,
referred to as the "Sixth Factor" (24), was not identified. (The
problem of this factor will be discussed in detail in Section
12-2.)

The simultaneous requirement, by certain mutants, for all
these factors suggested that their biosynthesis proceeds through
one or several common intermediates (cf. also ref. 25) and that
the genetic blocks of these strains are located in this part of
the chain of biosynthetic events. These findings opened the
possibility of using bacterial mutants in the search for compounds
capable of replacing the multiple supplement.

Out of approximately 50 compounds, selected as possible precursors of the aromatic ring because of their structures, only
one substance was found to be utilized, very effectively, as a
replacement for the three aromatic amino acids and PABA (22); all
the other compounds were entirely incapable of replacing more than
one aromatic metabolite. Surprisingly, this substance was not the
well-known, widely distributed quinic acid [3], but the closely
related shikimic acid [4], at that time considered a very rare
compound known to occur only in a few species of one single genus
of plants. As found subsequently, [4] also replaces the other

two aromatic metabolites, *p*-hydroxybenzoic acid (23) and the
"Sixth Factor" (24) (see Section 12-2). Utilization of [4] by a
Neurospora mutant with a quadruple aromatic requirement was ob-
served independently by Tatum and co-workers (26).

Shikimic acid [4] had, up to that time, been encountered
only in a few species of the genus *Illicium*. This genus of trees
or shrubs is considered either to belong to the family Magnoliaceae, or to constitute the related separate, unigeneric family
Illiciaceae. The acid had been isolated in 1885 by Eijkman (27)
from the fruits of *I. religiosum* Siebold, the Japanese Star anise
(Japanese name shikimi-no-ki). In this species, it constitutes
up to approximately 20% of the dry matter of the fruits; the
fruits of other species of the genus contain lesser amounts of
the acid. However, after the work of Davis had directed attention to its significance, [4] has been observed in a wide variety
of plants, from one species of moss and several of ferns, throughout the plant kingdom; for an extensive list of plants that have
been examined for [4], see the review by Bohm (21). However, the
heavy accumulation of this compound in *I. religiosum* is still
unique and its significance obscure; concentrations of [4] comparable to those present in this species have not been encountered
elsewhere.

The ability of [4] to replace a plurality of aromatic metabolites supported the idea of a common pathway and suggested, but
of course did not prove, a role as an intermediate in this pathway. That [4] does function as such is further suggested by the
observation that it is accumulated by mutants with later blocks
(22, 24), by its isolation in pure form from culture filtrates of
such mutants (28), and by abundant evidence for its enzymatic
formation and further conversion. Its role as an obligatory intermediate is thus hardly doubtful.

The chemistry and transformations of shikimic acid are discussed in detail in Chapter 5, Section 1.

Evidently, [4] itself must be formed by some specific biosynthetic reaction sequence. Investigation of this problem by
mutant methods led first to the isolation of its immediate precursors, 3-dehydroquinic acid [6] (29) and 3-dehydroshikimic acid
[7] (30).

Use of a traditional numbering system for [3] and [4] with
clockwise numbering of the ring carbon atoms resulted in the
designation of [6] and [7] as 5-dehydro compounds in much of the
published literature. This numbering, however, is contrary to
recent official rules (31) for the numbering of compounds such as
[3], [4], and their derivatives, and has been abandoned in favor
of the correct one in this volume. This change leaves numbers

1 and 4 untouched but necessitates interchange of 2 and 6, and 3 and 5, respectively.

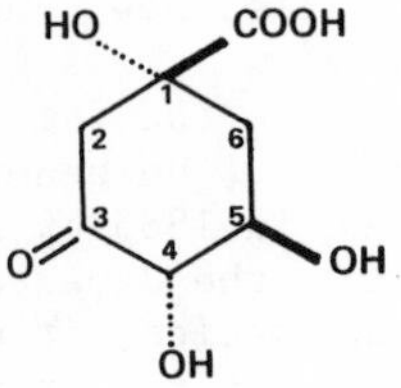

[6] 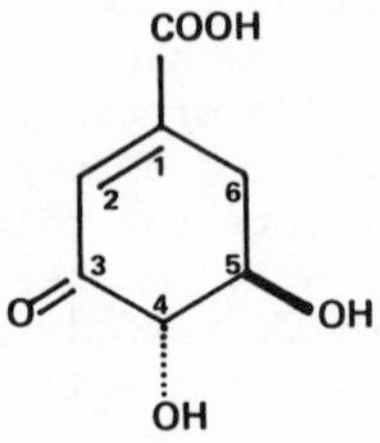   [7]

    The structure of the cyclic aldol [6] suggests strongly (and correctly, as was shown by subsequent findings) that it should be the first *cyclic* intermediate, since it could be, (and actually *is*) formed by aldolization of a suitable straight-chain precursor. The investigation of this earliest, "pre-cyclic" intermediate rests essentially upon interpretation of the incorporation of glucose into shikimic acid, rather than upon mutant methods and isolation of compounds indicated by them. For an understanding of these questions it is thus necessary to discuss these incorporation studies at this point.

    The acyclic precursor of [6] has actually been observed (32) as an excretion product of mutants blocked before [6]; however, because of its nutritional unavailability to mutants with appropriate blocks, it has not played any important role in the elucidation of the early stages of the pathway. It was given the provisional designation "Compound V" and was found to be a phosphorylated, presumably aliphatic α-keto acid; these findings were completely confirmed by later research (see Chapter 3).

    The incorporation of variously labeled glucose into [4] by a mutant of *E. coli* has been studied by Sprinson and co-workers (ref. 20 and papers quoted therein); interpretation of the observed incorporation patterns led to the elucidation of the earliest specific step in the shikimate pathway. The methods used for the degradation of the shikimic acid for the purpose of localizing the label will be discussed in Section 5-1.

    The results of these investigations can be summarized as shown in Scheme III. In this diagram, the contributions of activity from glucose variously labeled with ^{14}C in positions 1; 2; 3 + 4; and 6, to the labeling of the individual carbon atoms of shikimic acid are indicated; numbers in parenthesis give the fraction of this carbon atom derived from the particular carbon of glucose. The observed labeling pattern shows at once that [4]

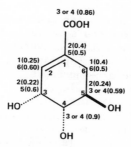

Scheme III

cannot be derived from glucose by any process involving cyclization of the intact C_6 chain; fragmentation and reassembly must
occur. In particular, the fragment COOH-(C-1)-(C-6) has a labeling pattern consistent with a derivation from a C_3 intermediate
of glycolysis, incorporated in such a way that its carbon 3 (i.e.,
C 1 + 6 of glucose) becomes C-6 of [4]. This fragment must come
from a stage of glycolysis preceding pyruvate, since the mutant
does not incorporate pyruvate (or bicarbonate or acetate) to any
significant extent into [4]; the analogy of these findings with
those of Gilvarg and Bloch (9) on the biosynthesis of phenylalanine and tyrosine in yeast is noteworthy.

The rest of the molecule of [4], C-2,3,4 and 5, shows an
isotope distribution which suggests an origin of this C_4 fragment
from a tetrose phosphate formed via the glycolytic and pentose
phosphate pathways, whereby C-1 of the tetrose would furnish C-5
of [4]; for details of this involved derivation, see refs. 20 and
28.

These observations on the labeling patterns of shikimic acid
are in agreement, not only with the findings on biosynthesis of
phenylalanine and tyrosine in yeast (9), but also with those on
biosynthesis of several other products; findings on the biogenesis
of anthranilic acid (and hence tryptophan) (12), lignin (10, 33),
and a number of other plant products are compatible· with the
assumption that [4] or some closely related compound serves as a
biosynthetic intermediate.

The derivation of [4] and its two precursors, [6] and [7],
from carbohydrates was further elucidated by work (34,35) with
cellfree extracts of *E. coli* mutants; intact cells were of little
help in unraveling the earliest stages of the biosynthetic pathway, undoubtedly because permeability barriers prevented the utilization, by the bacteria, of the phosphorylated compounds involved. The mutant chosen, 83-2, was blocked after dehydroshikimic [7], rather than dehydroquinic acid [6]; the observed (36)

high equilibrium ratio (DHS:DHQ = 15) for the conversion of [6]
into[7] should promote the earlier reactions, and the intense u.v.
absorption of [7] (30) provides a convenient analytical tool.

While extracts of the mutant converted a variety of phosphorylated sugars (glucose 1- and 6-phosphate, fructose 1-phosphate
and 1,6-diphosphate, ribose 5-phosphate, and D-*altro*-heptulose
7-phosphate) into [7] in low yield (<5%) (34) very high conversion (∿80%) was achieved (35) with D-*altro*-heptulose 1,7-diphos-
phate [8] (formerly called sedoheptulose diphosphate) in the presence of NAD. Extracts from another mutant, blocked after [4],
produced this compound from [8] labeled biosynthetically in positions 4-7. The resulting [4] was labeled exclusively in the corresponding positions (2-5) with a distribution of label identical
with that found previously for these atoms (see Scheme III).
These results, however, cannot be explained by assuming a simple
cyclization of the C_7 chain. Compound [8] is biosynthesized from
triose phosphate and tetrose phosphate under the influence of
aldolase in such a way that carbons 1, 2, and 3 originate from
(1,6); (2,5); and (3,4) of glucose, respectively. In contrast,
COOH, C-1, and C-6, the corresponding atoms of [4], show
the opposite labeling pattern. The C_7 chain of [8] cannot, therefore, be incorporated as such, but only after fragmentation and
reassembly involving an inversion of the fragment C_1-C_3. Hence
[8] is not an obligatory intermediate but seems to serve as source
of appropriate C_3 and C_4 fragments, which are then recombined to
give the carbon chain of the alicyclic intermediates.

The known cleavage of [8] by the enzyme aldolase into dihy-
droxyacetone phosphate [9] and D-erythrose-4-phosphate [10] (37)
suggested a possible mechanism, if the former compound were converted by well-established enzymatic reactions to enol pyruvate
phosphate [11], and this product condensed with the erythrose phosphate [10] to a new C_7-intermediate [12] capable of being cyclized
to [6] and having the correct labeling pattern. These reactions
are shown in Scheme IV.

This interpretation also agreed with inhibition studies (38).
The conversion of [8] into [6] by cellfree extracts of a suitable
mutant of *E. coli* was completely prevented by fluoride and iodoacetate, which are well-known inhibitors of the enzymes enolase
and D-glycerose 3-phosphate dehydrogenase; these enzymes are involved in the transformation of [9] into [11] (see Scheme IV). The
known NAD-dependence of the latter enzyme would explain the requirement for this cofactor in the synthesis of [6] from [8]. In
agreement with this interpretation, the inhibitions were overcome
specifically by the products of the inhibited reactions: the effect of fluoride by [11], that of iodoacetate by either D-gly-
ceronic acid 3-phosphate [13] or [12].

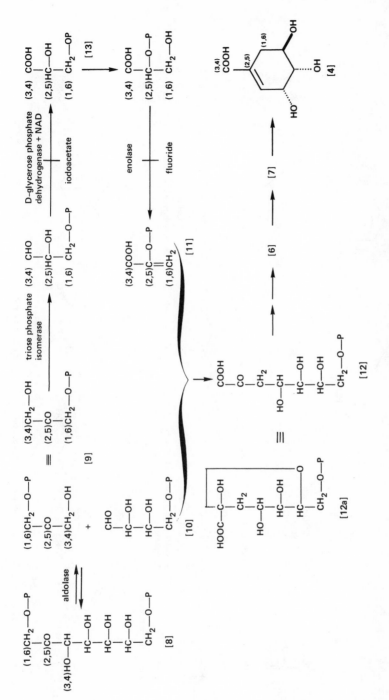

Scheme IV

57

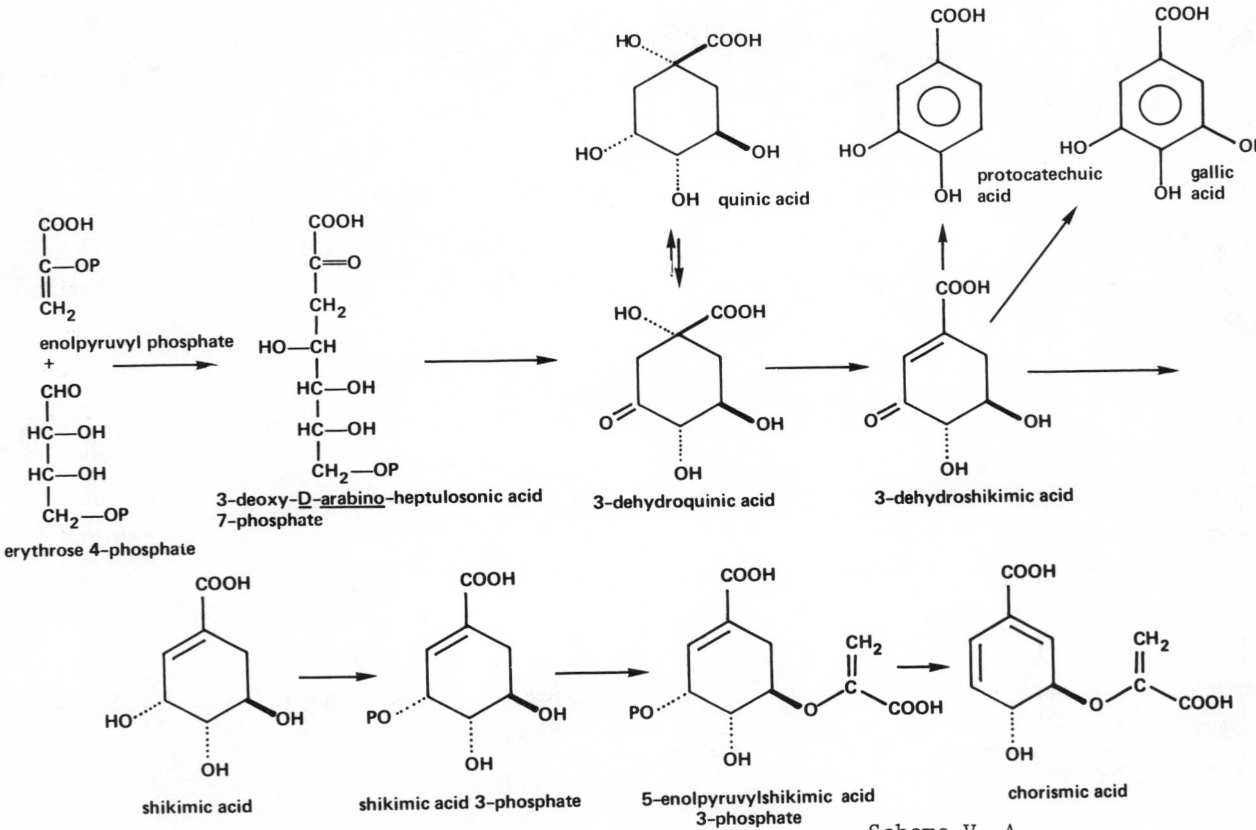
COOH
C—OP
CH2
enolpyruvyl phosphate
+
CHO
HC—OH
HC—OH
CH2—OP
erythrose 4-phosphate
COOH
C=O
CH2
HO—CH
HC—OH
HC—OH
CH2—OP
3-deoxy-D-arabino-heptulosonic acid 7-phosphate
HO
COOH
HO
OH
OH
quinic acid
HO
COOH
O
OH
OH
3-dehydroquinic acid
COOH
O
OH
OH
3-dehydroshikimic acid
COOH
HO
OH
protocatechuic acid
COOH
HO
OH
OH
gallic acid
COOH
HO
OH
OH
shikimic acid
COOH
PO
OH
OH
shikimic acid 3-phosphate
COOH
PO
OH
O
CH2
C
COOH
5-enolpyruvylshikimic acid 3-phosphate
COOH
OH
O
CH2
C
COOH
chorismic acid

Scheme V. A

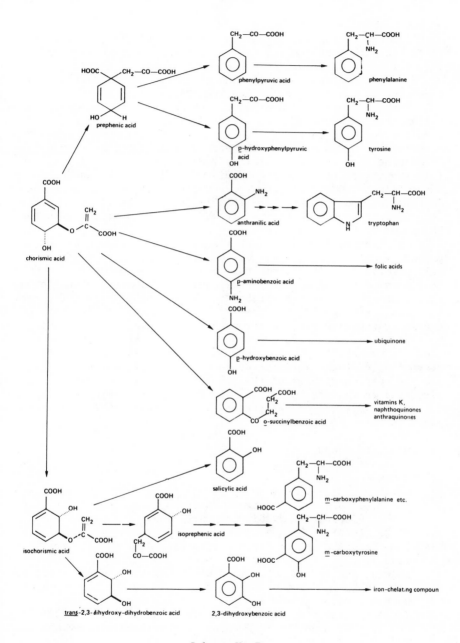

Scheme V. B

59

The direct enzymatic reaction of synthetic [10] (39) with [11] established the correctness of these deductions. Formation of [6] from these products proceeded with almost quantitative yields, faster than the conversion of [8] to [6], and it was insensitive to fluoride or iodoacetate (38).

Both [10] and [11] are general products of carbohydrate metabolism; their interaction to give [12] would then be the first specific reaction in the biosynthesis of aromatic metabolites from carbohydrates. Considering the stereochemistry at C-4 and C-5 (old numbering: C-3) of the alicyclic intermediates [4], [6], and [7], the structure and stereochemistry of [12] given in Scheme IV seemed likely. This compound, correctly named 3-deoxy-D-*arabino*-heptulosonic acid 7-phosphate (DAHP), is thus the first *specific* intermediate in the shikimate pathway. It is identical with the "Compound V" mentioned before (see page 54).

In Chapters 3 and 4 the chemistry of the early intermediates preceding shikimic acid [4] will be discussed; Chapters 5 and 6 will describe [4] itself and the intermediates following it in that part of the shikimate pathway which is common to the biosynthesis of almost all the metabolites known to. be derived from it. Chapters 7 through 13 will then deal with the individual biosynthetic sequences which lead from chorismic acid, the last intermediate of the common part of the pathway, to the various primary aromatic metabolites. The biochemistry of these compounds and the enzymology of their interconversions will be presented briefly, and the complex problem of the regulation of the pathway will be discussed in Chapter 15. Brief accounts of the enzymes involved in the pathway, and of the preparation of their substrates, are given in refs. 40 and 41.

Scheme V shows the entire pathway as far as it is known at present; a few gaps still remain.

REFERENCES

1. (a) F. Gibson and J. Pittard, Bact. Rev. $\underline{32}$, 465 (1968);
    (b) J. Pittard and F. Gibson, Curr. Top. Cell. Regulation $\underline{2}$, 29 (1970).

1A. T.A. Geissman and E. Hinreiner, Botan. Rev. $\underline{18}$, 77, 165 (1952).

2. E. Winterstein and G. Trier, "Die Alkaloide", Bornträger, Berlin, 1910.

3. O. Loew, Ber. $\underline{14}$, 450 (1881).

4. (a) A. Lücke, Virchow's Arch. Pathol. Anat. $\underline{19}$, 196 (1860);
    (b) E. Lautemann, Liebig's Ann. $\underline{125}$, 9 (1863).

5. A.J. Quick, J. Biol. Chem. $\underline{92}$, 65 (1931).

6. A.M. Asatoor, Biochim. Biophys. Acta 100, 290 (1965).
7. H.O.L. Fischer and G. Dangschat, Ber. 65, 1009 (1932).
7A. J.A. Hall, Chem. Rev. 20, 305 (1937).
8. H.O.L. Fischer and G. Dangschat, (a) Helv. Chim. Acta 17, 1200 (1934); (b) ibid. 18, 1206 (1935).
9. C. Gilvarg and K. Bloch, (a) J. Biol. Chem. 193, 339 (1951); (b) ibid. 199, 689 (1952).
10. S.N. Acerbo, W.J. Schubert, and F.F. Nord, J. Amer. Chem. Soc. 82, 735 (1960).
11. M. Sprecher, P.R. Srinivasan, D.B. Sprinson, and B.D. Davis, Biochemistry 4, 2855 (1965).
12. P.R. Srinivasan, Biochemistry 4, 2860 (1965).
13. J. Stenhouse, Ann. 91, 367 (1854).
14. A.R. Butler and H.F. Wallace, J. Chem. Soc. (B) 1758 (1970).
15. J. Baddiley, G. Ehrensvärd, E. Reio, and E. Saluste, J. Biol. Chem. 183, 777 (1950).
16. L. Hunter, J. Chem. Soc. 123, 543 (1923).
17. a. O. Wallach, Ann. 275, 145 (1893);
    b. J.N. Collie, J. Chem. Soc. 65, 262 (1894).
18. A.J. Birch, C.J. Moye, R.W. Rickards, and Z. Vanek, J. Chem. Soc. 3586 (1962).
19. B.D. Davis, (a) Experientia 6, 41 (1950); (b) Adv. Enzymol. 16, 247 (1955); (c) E. Umbarger and B.D. Davis, in "The Bacteria", I.C. Gunsalus and R.Y. Stanier, Eds., Vol. 3, Academic, New York and London, 1967, p. 167.
20. D.B. Sprinson, Adv. Carbohydr. Chem. 15, 235 (1960).
21. B.A. Bohm, Chem. Rev. 65, 435 (1965).
22. B.D. Davis, J. Biol. Chem. 191, 315 (1951).
23. B.D. Davis, Nature 166, 1120 (1950).
24. B.D. Davis, (a) J. Bact. 64, 729 (1952); (b) Bull. Soc. Chim. Biol. 36, 947 (1954).
25. J.F. Nyc, F.A. Haskins, and H.K. Mitchell, Arch. Biochem. 23, 161 (1949); M. Gordon, F.A. Haskins, and H.K. Mitchell, Proc. Natl. Acad. Sci. 36, 427 (1950).
26. E.L. Tatum, in "Plant Growth Substances," F. Skoog, Ed., University of Wisconsin Press, Madison, Wisconsin, 1951, p. 447; E.L. Tatum, S.R. Gross, G. Ehrensvärd, and L. Garnjobst, Proc. Natl. Acad. Sci. U.S. 40, 271 (1954).
27. J.F. Eijkman, Rec. Trav. Chim. Pays-Bas 4, 32 (1885); Ber. 24, 1278 (1891).
28. P.R. Srinivasan, H.T. Shigeura, M. Sprecher, D.B. Sprinson, and B.D. Davis, J. Biol. Chem. 220, 477 (1956).
29. U. Weiss, B.D. Davis, and E.S. Mingioli, J. Amer. Chem. Soc. 75, 5572 (1953).
30. I.I. Salamon and B.D. Davis, J. Amer. Chem. Soc. 75, 5567 (1953).

31. Eur. J. Biochem. 5, 1 (1968); Zeitschr. Physiol. Chem. 350, 523 (1969); Arch. Biochem. Biophys. 128, 269 (1969); Biochim. Biophys. Acta 165, 1 (1968); Biochem. J. 112, 7 (1969); J. Biol. Chem. 243, 5809 (1968).

32. E. Kalan and B.D. Davis, unpublished work mentioned in ref. 19b.

33. K. Kratzl and H. Faigle, Z. Naturforsch. 15b, 4 (1960).

34. E.B. Kalan, B.D. Davis, D.R. Srinivasan, and D.B. Sprinson, J. Biol. Chem. 223, 907 (1956).

35. P.R. Srinivasan, D.B. Sprinson, E.B. Kalan, and B.D. Davis, J. Biol. Chem. 223, 913 (1956).

36. S. Mitsuhashi and B.D. Davis, Biochim. Biophys. Acta 15, 54 (1954).

37. B.L. Horecker, P.Z. Smyrniotis, H.H. Hiatt and P.A. Marks, J. Biol. Chem. 212, 827 (1955).

38. P.R. Srinivasan, M. Katagiri, and D.B. Sprinson, J. Biol. Chem. 234, 713 (1959).

39. C.E. Ballou, H.O.L. Fischer, and D.L. MacDonald, J. Amer. Chem. Soc. 77, 2658, 5967 (1955).

40. B.D. Davis, C. Gilvarg, and S. Mitsuhashi, in "Methods in Enzymology", Vol. II, S.P. Colowick and N.O. Kaplan, Eds., Academic, New York, 1955, p. 300.

41. Various authors, in "Methods in Enzymology", Academic, New York, Vol. VI, S.P. Colowick and N.O. Kaplan, Eds., 1963, p. 493; and Vol. XVIIA, H. Tabor and C.W. Tabor, Eds., 1970, p. 349.

CHAPTER 3

The Acyclic Intermediate: 3-Deoxy-D-*arabino*-heptulosonic Acid 7-Phosphate

The structure and stereochemistry of this intermediate were proven by Sprinson, Rothschild, and Sprecher (1) through total synthesis to correspond to expression [12]. This synthesis proceeded from 2-deoxy-D-*arabino*-hexose (2-deoxyglucose) by way of the cyanohydrin reaction. Introduction of the phosphate ester grouping at C-7, and oxidation of the resulting α-hydroxy acid to the α-keto acid with V_2O_5 in pyridine (2) gave [12], which was isolated as the pure but noncrystalline barium salt; sodium and ammonium salts were crystalline but hygroscopic.

An alternate synthesis of [12] from D-glucose has been described more recently (3).

The synthetic [12] was shown (4) to be identical with the product of the enzymatic reaction of [10] with [11], and also with a sample of Compound V (5). Both the synthetic substance and the product obtained enzymatically showed identical behavior in the further enzymatic conversion to 3-dehydroquinic acid [6] (4). In contrast, the corresponding dephosphorylated compound and other related substances were not attacked by the enzyme.

DAHP [12] actually exists as a pyranoid hemiacetal (no carbonyl stretching band in the ir spectrum of the salts in the solid state, no reaction with semicarbazide in neutral, slow one in acidic medium; uptake of one mole of periodate, while a furanoid form would be resistant).

A substance very probably identical with 3-deoxy-D-*arabino*-heptulosonic acid, the nonphosphorylated parent compound of [12], was observed by Weissbach and Hurwitz (6) to be formed from ribose 5-phosphate by cellfree extracts of wild-type *E. coli*. It was not obtained or characterized in the solid state, but was identified by chromatographic comparison with a product synthesized by methods analogous to those used for the synthesis of [12], and with the product of enzymatic dephosphorylation of the latter. The same extracts also yielded the compound from [10] and [11], strongly suggesting that it arises by such an enzymatic removal of phosphoric acid from [12] initially synthesized. Its formation in high yields from ribose phosphate is readily understandable through conversion of this compound to [10] and [11] by well-known biochemical reactions.

The enzyme, DAHP synthase, which catalyzes the biosynthesis of [12] in *E. coli*, was found (4) to be quite specific; neither [10] nor [11] could be replaced by a variety of related compounds. The enzymatic reaction could not be reversed; it requires Co^{2+} (7).

The detailed mechanism of the biosynthesis of [12] from [10] and [11] is not yet completely understood. Kinetic studies of this reaction by Staub and Dénes (8) in extracts from wild-type *E. coli*, and by Nagano and Zalkin (7) in those from *Salmonella typhi-murium*, suggest that the enzyme interacts first with [11] to give inorganic phosphate and an enzyme-bound intermediate, which then reacts with [10] to form [12]. This process would be an example of a so-called "ping-pong" reaction, where one substrate reacts with the enzyme with release of a product prior to addition of the second substrate. DeLeo and Sprinson (9) and Nagano and Zalkin (7) found that the loss of the phosphate group from [11] takes place by elimination rather than by hydrolysis, since the resulting inorganic phosphate incorporates ^{18}O from [11] labeled in the enolic oxygen, but not from $H_2^{18}O$. On this basis, the former authors (9) interpreted the reaction as shown in Scheme I, involving the intermediate formation of an enolpyruvyl enzyme, represented as a carboxylate ester.

However, this formulation seems in need of revision as a result of recent work by Floss and co-workers (10). These authors prepared the two stereoisomers of $3-^3H-[11]$ from $1-^3H$-glucose and mannose, respectively, by a sequence of enzymatic steps of known stereochemistry. Use of these samples for the biosynthesis of [12], further enzymatic conversion to [4], and degradation of this compound, as shown in Scheme II, proved that C-6 of [4] is stereospecifically tritiated; the sequence of interconversions is given in Scheme II for the (Z)-isomer of $[3-^3H]-[11]$. Even though the stereospecificity was not complete and some scrambling of label occurred, the findings prove retention of the stereochemical identity of the vinylic hydrogens of [11]; such a retention would be unlikely if the biosynthesis of [12] proceeded, as

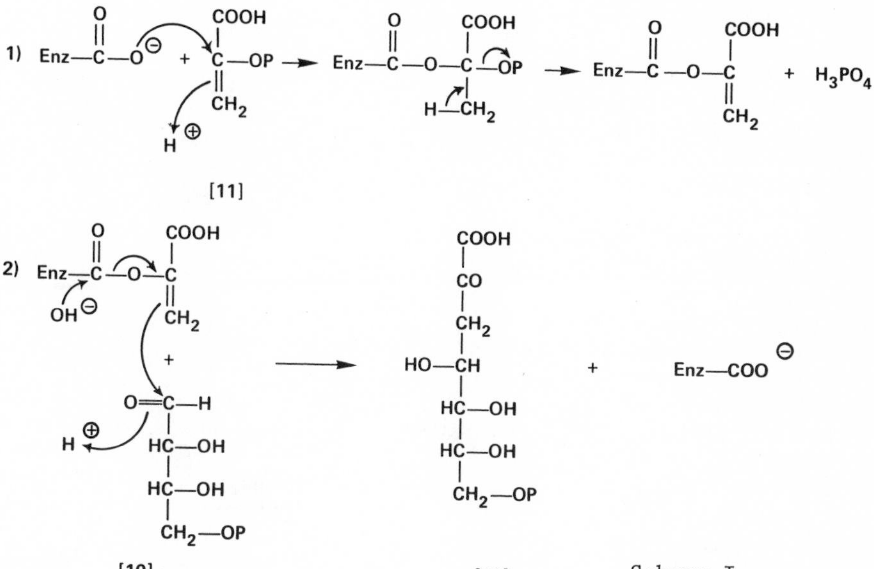

Scheme I

formulated in Scheme I, through a stage where C-3 of [11] is present as a methyl group presumably capable of essentially free rotation. This aspect of the reaction is thus in need of further clarification.

As the earliest specific reaction in a much-branched biosynthetic pathway, the formation of [12] is strongly under the control of end-products and/or later intermediates. These regulation mechanisms have been the object of detailed studies, which have been reviewed by Doy (11). Here, it should suffice to point out that different organisms achieve the necessary regulations in surprisingly diverse manner. In *E. coli*, three isoenzymes are present, which are inhibited and/or repressed by the three major aromatic metabolites, phenylalanine, tyrosine, and tryptophan; additional control is achieved by end-product inhibition in the specific branches leading to those metabolites. In *Bacillus subtilis*, only one DAHP synthase appears to exist, which is subject to feedback inhibition, not by the aromatic amino acids but, to the extent of about 80%, by either prephenic or chorismic acids, of which the former seems to be the physiologically significant inhibitor. Again, in yeast (*Saccharomyces cerevisiae*), two isoenzymes were found, inhibited, but not repressed, by phenylalanine and tyrosine, respectively. In the

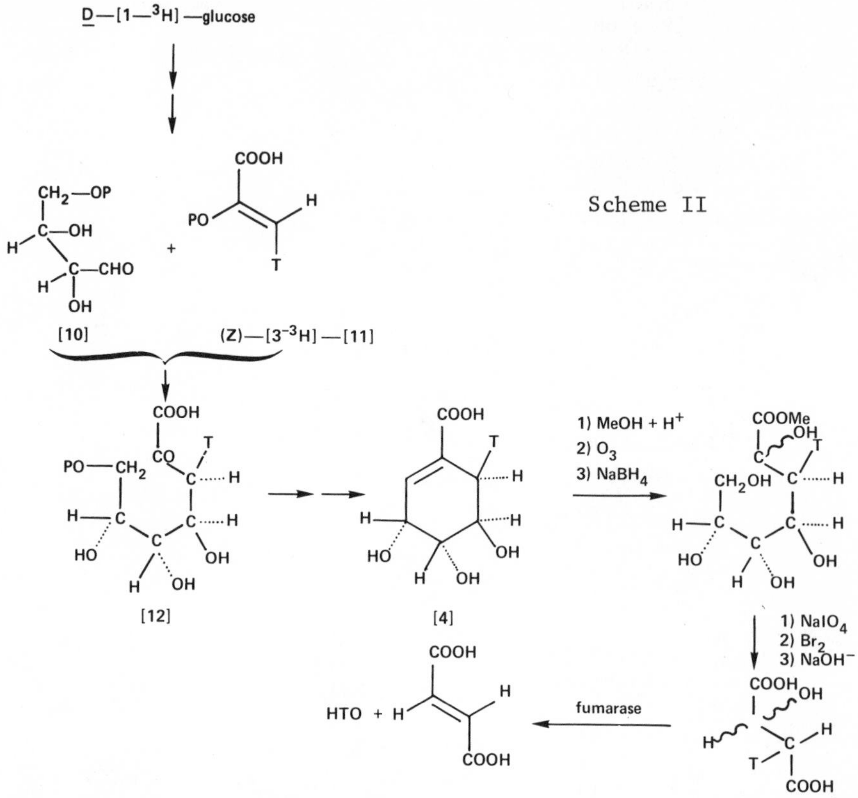

Scheme II

mold, *Neurospora crassa*, three isoenzymes are present, as in *E. coli*, inhibited by phenylalanine, tyrosine, and tryptophan, respectively, but no repression by any of these amino acids seems to exist; additional control is achieved at stages following chorismic acid. Unfortunately, no information on the situation in higher plants seems available, although the enzymatic activity itself has been observed in several species (12).

Compound [12] is the substrate of the enzymatic reaction which leads to the first appearance of the cyclic structure destined to become the aromatic ring. In this reaction, [12] is transformed into 3-dehydroquinic acid [6]; the chemistry of this substance will be discussed in detail in Section 4-1. On purely chemical grounds, the conversion of [12] into [6] must involve several steps; however, no distinct intermediate between these compounds has been obtained.

It has already been pointed out in Chapter 2 (page 54) that [6] is a cyclic aldol; its structure strongly suggests formation through aldol condensation of the acyclic diketone [14]. The ease with which aldolizations occur *in vitro*, and the frequency with which they are encountered in biosynthetic events, make this hypothesis very attractive. The simplest interpretation of the transformation [12]→[6] would then be elimination of phosphoric acid from [12] to give the enol form of [14], followed by aldolization to [6].

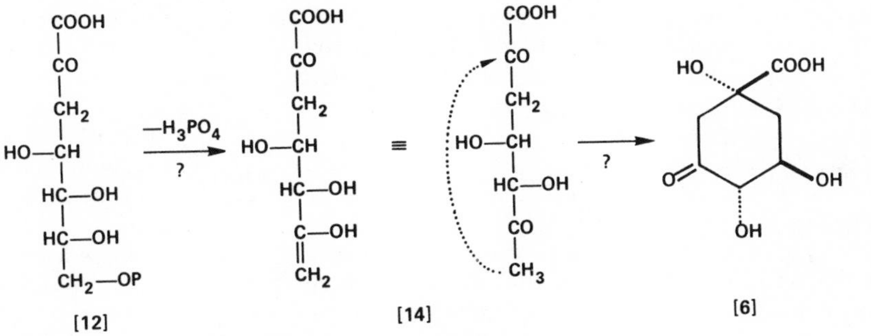

[12]
COOH
|
CO
|
CH$_2$
|
HO—CH
|
HC—OH
|
HC—OH
|
CH$_2$—OP

$\xrightarrow[?]{-H_3PO_4}$

[14]
COOH
|
CO
|
CH$_2$
|
HO—CH
|
HC—OH
|
C—OH
‖
CH$_2$

≡

COOH
|
CO
|
CH$_2$
|
HO—CH
|
HC—OH
|
CO
|
CH$_3$

$\xrightarrow{?}$

[6]

Compound [14], 3,7-dideoxy-D-*threo*-hepto-2,6-diulosonic acid, has been prepared through total synthesis by Adlersberg and Sprinson (13); it was obtained only in solution but was characterized as the crystalline bis-2,4-dinitrophenylhydrazone. As expected, [14] cyclizes very readily to [6] *in vitro*, slowly even at room temperature in neutral solution (5% in 24 hours), more rapidly at 37° and pH 11. However, it is stable in acidic medium. In contrast to this smooth conversion by purely chemical means, the diketone [14] was *not* converted to [6] by extracts of *E. coli* capable of forming this compound from [12]. At least the *free* [14] thus *cannot* be an intermediate in the biosynthetic cyclization. The substance also did not inhibit the normal biosynthetic formation of [6]. Its ready conversion to the latter does suggest, however, that it may, in enzyme-bound form, play a role in the biosynthesis.

In any event, the assumption of a simple, direct elimination of phosphoric acid from [12], as formulated above, is unattractive. Such an elimination reaction from a nonactivated phosphate group would proceed chemically only under forcing conditions, is thus undoubtedly energetically unfavorable, and seems to lack biochemical precedent.

That the conversion of [12] to [6] does indeed follow a more complex course has been demonstrated by the studies of Srinivasan, Rothschild, and Sprinson (5) on the enzyme 3-dehydroquinate synthase, which catalyzes this reaction. With partially purified extracts from a mutant of *E. coli*, conversion of [12] to [6] depended upon the presence of catalytic amounts of both NAD and Co^{2+} which were not replaceable by NADP, or by Zn^{2+}, Fe^{2+}, Mg^{2+}, or Mn^{2+}; addition of NADase, or of EDTA, or dialysis, prevented the reaction, which was restored by Co^{2+}. NADH at low concentrations (10^{-7}–10^{-6} M) proved about as active as NAD, but became inhibitory at higher concentrations (5×10^{-6}–10^{-5} M); this inhibition persisted in the presence of NAD. The requirement for a cofactor such as NAD would be inexplicable if the reaction were a simple elimination of phosphoric acid followed by aldolization of an enzyme-bound form of the resulting diulosonic acid [14], and strongly suggests a reaction sequence of several steps. However, all attempts to separate the individual enzymes by a variety of fractionation procedures were unsuccessful, and kinetic studies showed that the rates of disappearance of [12], of release of inorganic phosphate, and of formation of [6] are practically identical. Furthermore, radioautographic monitoring of the enzymatic conversion of carboxyl-labeled [12] (prepared by use of $K^{14}CN$ in the cyanohydrin reaction which constitutes the first step of the synthesis (42) of [12]) showed very rapid formation of [6] (clearly detectable after 2 minutes) and concomitant decrease of [12] (almost used up after 10, completely consumed after 30 minutes), without any indication for the presence of another labeled substance at any time. The sequence of events between [12] and [6] must thus be completely concerted.

As a hypothetical but very convincing interpretation of the mechanism of the 3-dehydroquinate synthase reaction, the sequence shown in Scheme III was proposed (5) whereby the intermediates between starting material and final product are assumed to be enzyme-bound (and the first two intermediates probably present in a lactol form, like [12]).

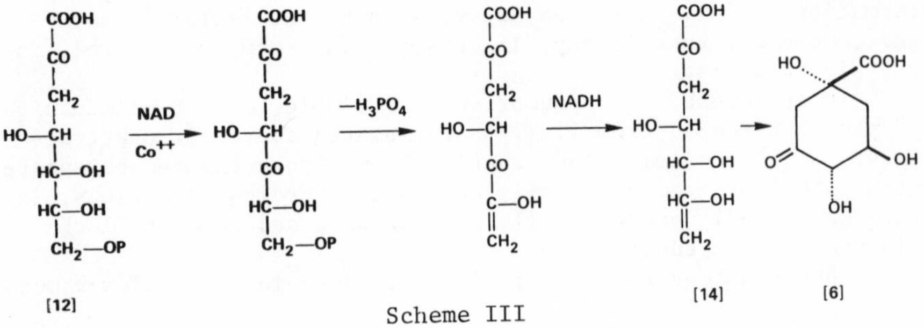

Scheme III

This interpretation assigns to the NAD the role of oxidizing the secondary alcohol grouping at C-5 to a keto group, to be subsequently reduced back, in correct stereochemistry, by the NADH formed. Assumption of intermediary formation of a 5-keto compound has the great advantage of giving a ready explanation for the removal of the phosphate group by an elimination reaction, with formation of the CO-CH$_3$ group required for aldolization to [6]. Elimination reactions of this type in phosphate esters of β-hydroxy carbonyl compounds have often been observed, in purely chemical systems, to proceed very readily; see ref. 14 and earlier investigations quoted there. Well-known biochemical instances of this reaction are the conversion of glyceraldehyde 3-phosphate to pyruvaldehyde (15), and the elimination of phosphate from serine and threonine phosphates, which requires pyridoxal and metal ions (16). This hypothetical scheme thus furnishes an excellent explanation, amply provided with biochemical precedent, for the conversion of [12] to a structure appropriate for cyclization to [6]. The requirement for Co^{2+} in both this and the preceding biosynthetic reaction is quite unusual; the ability of low concentrations of NADH to replace NAD is admittedly not explained.

In Scheme III, the diulosonic acid [14] appears only as the enol form, in contrast to the initial formulation (5) of this reaction sequence. This modification became necessary because of the subsequent findings of Rotenberg and Sprinson (17), which preclude the possibility that C-7 of [12] could at any time be present as a methyl group, as it would in the keto form of [14]. In this work, samples of [12] labeled with ^{14}C in position 1, and with ^{3}H, randomly or stereospecifically, at position 7, were converted enzymatically into [6]. In all three cases, [7-^{3}H] (R,S), [7-^{3}H] (R), and [7-^{3}H] (S), the tritium was completely retained, while no label was present in [6] biosynthesized from unlabeled [12] in tritiated water. If the keto form of [14] were involved as an intermediate, label from the medium would have been incorporated in the latter case, while label would have been lost in the former. Since neither event took place, [14] can be present only as an enzyme-bound enol which is prevented from tautomerizing to the keto form. Furthermore, the aldolization of this intermediate to [6] is stereospecific, since the subsequent conversion of [6] to [7] proceeds with complete retention of ^{3}H in the sample of [6] derived from [7-^{3}H] (7S)-[12], and with complete loss in the one made from [7-^{3}H] (7R)-[12]. The enzyme dehydroquinate dehydratase is known (18) to convert [6] into [7] by a stereospecific cis-elimination.

The enzyme, dehydroquinate synthase, seems influenced but little by feedback inhibition or repression (19), in agreement with the fact that it is not concerned with a reaction involving a branch-point intermediate. In *Neurospora*, this enzyme, and the ones catalyzing the next four steps of the shikimate pathway, are associated as a multienzyme complex (20), which has been obtained in purified form without evidence for separation of the five enzymatic activities during the purification procedure (21). The possible significance of this aggregate will be discussed in Section 4-1 and Chapter 15.

REFERENCES

1. D.B. Sprinson, J. Rothschild, and M. Sprecher, J. Biol. Chem. 238, 3170 (1963).

2. P.P. Regna and B.P. Caldwell, J. Amer. Chem. Soc. 66, 243 (1944).

3. G.B. Paerels and H.W. Geluck, Rec. Trav. Chim. 89, 813 (1970).

4. P.R. Srinivasan and D.B. Sprinson, J. Biol. Chem. 234, 716 (1959).

5. P.R. Srinivasan, J. Rothschild, and D.B. Sprinson, J. Biol. Chem. 238, 3176 (1963).

6. A. Weissbach and J. Hurwitz, J. Biol. Chem. 234, 705 (1959); J. Hurwitz and A. Weissbach, ibid. 710 (1959).

7. H. Nagano and H. Zalkin, Arch. Biochem. Biophys. 138, 58 (1970).

8. M. Staub and G. Dénes, Biochim. Biophys. Acta 132, 528 (1967).

9. A.B. DeLeo and D.B. Sprinson, Biochem. Biophys. Res. Commun. 32, 873 (1968).

10. (a) D.K. Onderka and H.G. Floss, Biochem. Biophys. Res. Commun. 35, 801 (1969); (b) H.G. Floss, D.K. Onderka, and M. Carroll, J. Biol. Chem. 247, 736 (1972).

11. C.H. Doy, Rev. Pure Appl. Chem. 18, 41 (1968).

12. T. Minamikawa, Plant Cell Physiol. (Tokyo) 8, 695 (1967).

13. M. Adlersberg and D.B. Sprinson, Biochemistry 3, 1855 (1964).

14. D.M. Brown, M. Fried, and Sir A.R. Todd, J. Chem. Soc. 2206, (1955).

15. O. Meyerhof and K. Lohmann, Biochem. Zeitschr. 271, 89 (1934); E. Baer and H.O.L. Fischer, J. Biol. Chem. 150, 223 (1943).

16. T.B. Longenecker and E.E. Snell, J. Biol. Chem. 225, 709 (1956).

17. S.L. Rotenberg and D.B. Sprinson, Proc. Natl. Acad. Sci. U.S. 67, 1669 (1970)

18. K.R. Hanson and I.A. Rose, Proc. Natl. Acad. Sci. U.S. 50, 981 (1963).

19. F. Gibson and J. Pittard, Bacteriol. Rev. 32, 465 (1968);

E. Gollub, H. Zalkin, and D.B. Sprinson, J. Biol. Chem. $\underline{242}$, 5323 (1967); K.D. Brown, Genetics $\underline{60}$, 31 (1968).

20. N.H. Giles, M.E. Case, C.W.H. Partridge, and S.I. Ahmed, Proc. Natl. Acad. Sci. U.S. $\underline{58}$, 1453 (1967); M.E. Case and N.H. Giles, ibid. $\underline{68}$, 58 (1971).

21. L. Burgoyne, M.E. Case, and N.H. Giles, Biochim. Biophys. Acta $\underline{191}$, 452 (1969).

CHAPTER 4

Alicyclic Intermediates Prior to Shikimic Acid

4.1 3-DEHYDROQUINIC ACID

3-Dehydroquinic acid [6] (formerly 5-dehydroquinic acid) is the first one of six consecutive alicyclic intermediates in the common part of the shikimate pathway; its enzymatic formation has been discussed in the preceding section. Actually, however, the compound was first observed by Davis (1, 2) in culture filtrates of *E. coli* blocked between it and the next intermediate, 3-dehydroshikimic acid [7]. It was isolated in pure, crystalline form from such filtrates (3).

The recognition of [6] as an intermediate was delayed initially by the fact that available mutants of *E. coli* and *Aerobacter aerogenes* with genetic blocks after it responded poorly to this compound (2). However, secondary mutants (obtained by u.v. irradiation in the case of *E. coli*, appearing spontaneously in that of *A. aerogenes*) were able to utilize [6] smoothly. The poor response of the parent primary mutant of *A. aerogenes* was exhibited only when glucose was used as the carbon source. On succinate or lactate, this strain responded well to [6]; a similar phenomenon has been observed in the case of *p*-hydroxyphenylpyruvic acid (see Chapter 7-2). The growth response on succinic acid was abolished by addition of 0.01% glucose; for another instance of such an effect of the carbon source, see Table 6 of ref. 1.

From culture filtrates of a mutant of *E. coli*, [6] was isolated in pure form (3) by column chromatography on charcoal and elution with aqueous ethanol, a technique worked out previously by Salamon and Davis (4) for the isolation of [7]. The compound was eluted by very dilute ethanol (2.5 - 5%), while [7] requires 60-75%. Precipitation as the ethanol-insoluble brucine salt, removal of the brucine in aqueous medium as the picrate, and evaporation of the filtrate at low temperature gave pure [6] as a glass which crystallized on contact with ether. A simplified method, using ion exchange chromatography instead of charcoal columns, has been described subsequently by Whiting and Coggins (5), who ob-

tained the compound on oxidation of quinic acid [3] by a few
strains of *Acetomonas oxydans* (formerly called *Acetobacter sub-
oxydans*).

Pure [6], $C_7H_{10}O_6$, forms crystals (from acetone-chloroform)
which melt at 140-142° (3). Somewhat above this temperature, the
melt resolidifies, to melt again at about 200°. The product
formed on melting was identified as protocatechuic acid [15] by
m.p., and color reaction with FeCl3; cf. the analogous behavior
of [7], which is undoubtedly an intermediate in this thermal reaction, as it is in the acid-catalyzed conversion (see below).

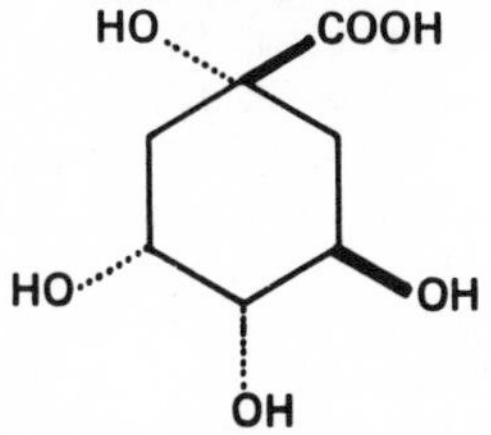

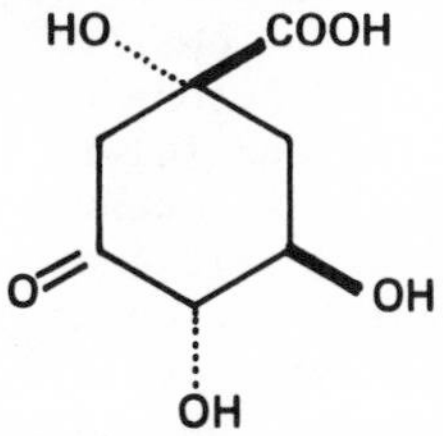

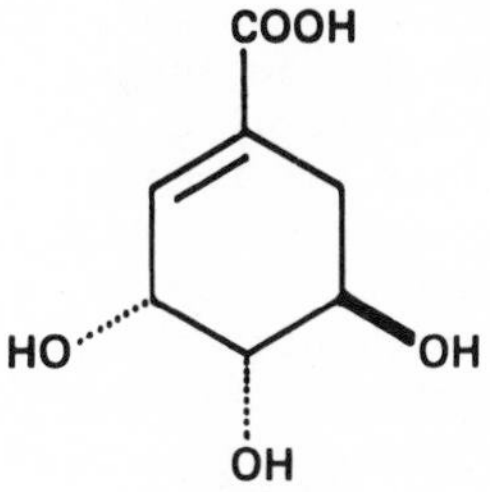

The structure and absolute configuration of [6] follow (3)
from its properties, and from its interrelation with [3] and [7],
whose constitution and absolute stereochemistry had been established previously. Attempts to prepare crystalline derivatives
of [6] were unrewarding. Of the functional groups present, the
carbonyl was indicated by the u.v. absorption of the compound it-
self (λ_{max} 269 nm, ε 97.4 in ethanol) and of its thiosemicarbazone (not isolated, λ_{max} 230 and 266 nm, $\varepsilon \sim$ 10,000 and 17,600,
respectively). The presence of an α-ketol group was suggested by
alkali-sensitivity, orcinol test, and reduction of Fehling's and
Tollens' reagents at room temperature, that of an α-hydroxy acid
grouping by the yellow color with aqueous FeCl3. Brief boiling
with concentrated hydrochloric acid produced [15]. Milder treatment with acid (0.1N HCl at 100°) gave [7] which, however, could
not be isolated; its presence was demonstrated by bioautography

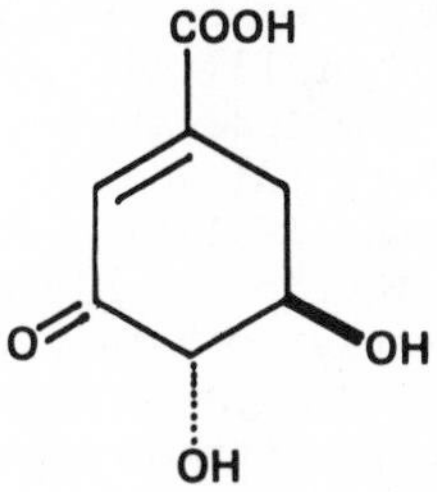

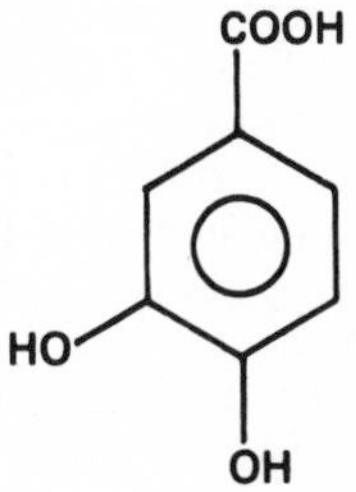

(2, 3) and by reaction with thiosemicarbazide, when the characteristic spectrum (6) of the thiosemicarbazone of an α,β-unsatu-
rated ketone appeared, identical with that obtained similarly
from authentic [7], λ_{max} 236 (ε 15,400) and 314 nm (ε 12,100).
This result of work with the small amounts of material isolated
from bacterial filtrates was subsequently confirmed and extended
by Grewe and Jeschke (7), when larger amounts of [6] became available through HNO_3-oxidation of [3]; the [7] formed on acid-treatment of [6] was isolated and unequivocally identified.

On the basis of the findings discussed so far, [6] could be
the 3- or 5-keto analog of [3] or of its 1-epimer. Position 3
for the carbonyl follows from the fact that [6] remains unchanged
under conditions (acetone with Amberlite IR 120 as catalyst)
which convert [3] and [4] to their isopropylidene derivatives;
[6] thus lacks the adjacent pair of *cis*-hydroxyls present at C-3
and C-4 of [3] and [4], and must, therefore, be the 3-keto compound. The configuration at C-1 is the same as that of [3],
since catalytic reduction over PtO_2 yields this compound, a finding which completes the proof of structure and configuration of
[6]. Microbiological assay of the reduced solution showed, however, that only about 50% of [3] had been formed; the suggestion
(3) that this low yield may indicate a nonstereospecific reduction with formation of both [3] and its 3-epimer has recently
been proven correct through isolation of the latter compound (P.
Dansette, private communication.) Microbiological assay of [3]
is possible because of the growth-factor activity of the compound
for certain mutants of *A. aerogenes* (see below); distinction of
[3] from [6] in this assay rests on the fact that only [6] is de-
stroyed by alkali.

Through an oversight, [6] was written initially (3) with the
antipodal absolute configuration, an error subsequently corrected
(ref. 8, footnote 8); however, it has been pointed out (9) that
this error has entered into a number of publications and reviews
and could thus be misleading. Actually, the absolute configuration of [4], and through it that of [6], rests securely by unequivocal correlation upon that of glucose itself.

Additional proof for the correctness of the formulation of
[6] is available through the observation (7) that catalytic
(PtO_2) reduction in 2N H_2SO_4 yields a new compound identified as
3-deoxyquinic acid [16] by uptake of one mole of periodate (which
excludes the 4-deoxy compound) and ready formation of a lactone,
sterically possible only with the β-hydroxyl on C-5. It is known
(10) that reduction under those conditions in the cyclitol series
specifically removes the carbonyl oxygen.

Those reactions of [3], [4], [6], and [7] which were of

major importance in the elucidation of their structures and configurations are shown in Scheme I.

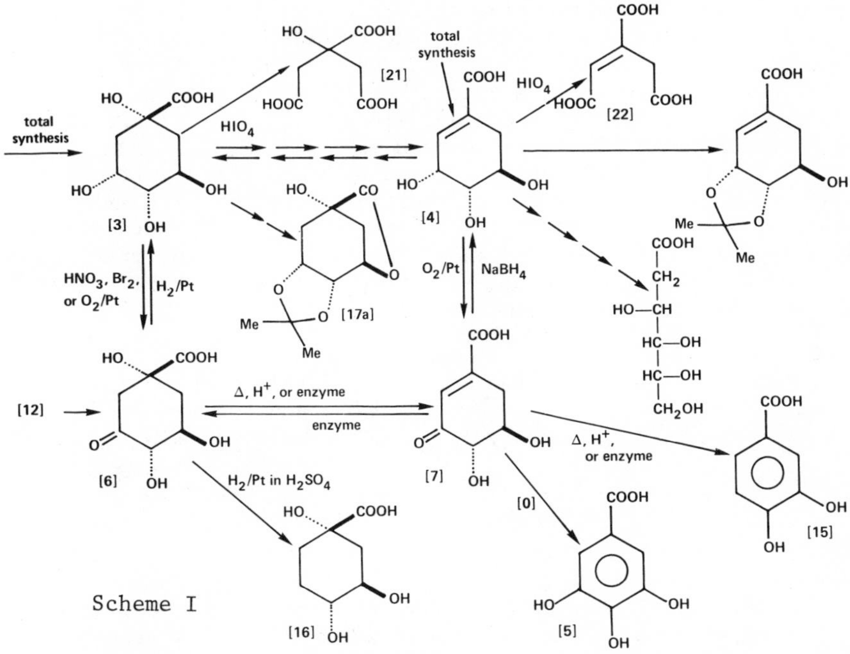

Scheme I

Compound [6] has become readily accessible through direct
chemical oxidation of [3] by several methods. The fact that these
reactions yield [6] quite specifically can be explained by conformational analysis. Investigation of the nmr spectrum of [3]
has shown (11), as expected, that the preferred conformation of
[3] in solution is the one with the bulky carboxyl in equatorial
orientation. Of the three secondary alcohol groups, only the one
at C-3 is axial in this conformer; the specific attack upon this
particular group is thus not surprising. The conversion of [3]
into [6] was first achieved by Grewe and Jeschke (7) in 63% yield
using concentrated nitric acid. Since [3] has been obtained synthetically by Grewe and co-workers (12), its conversion to [6]
constitutes a total synthesis of the latter. Subsequently, Pt-
catalyzed oxidation in aqueous solution with oxygen (13), or oxidation with aqueous bromine (14), gave similar results. Actually,
the formation of [6] under the last-named conditions had been observed at a much earlier time when, however, its isolation or

identification were not yet feasible. While reexamining Hesse's observation (15a) that [3] is converted to [15] by aqueous bromine, Fittig and Macalpine (16) noticed the appearance of an intermediate which is not extractable by ether; the ether-soluble [15] forms only on evaporation of the reaction mixture. Neither Fittig and Macalpine, nor Hesse in a subsequent investigation (15b), were able to isolate this ether-insoluble material; it was obtained in pure form by Grewe and Winter (14) and identified as [6] about 80 years after it had been observed.

Conversion of [3] into its lactone, quinide [17], is sterically possible only from the alternative conformer with axial carboxyl; in this conformer, it is the hydroxyl at position 4 which is axial. As a consequence, catalytic oxidation of [17], followed by opening of the lactone ring with weak alkali, gave the new 4-dehydroquinic acid [18] (17).

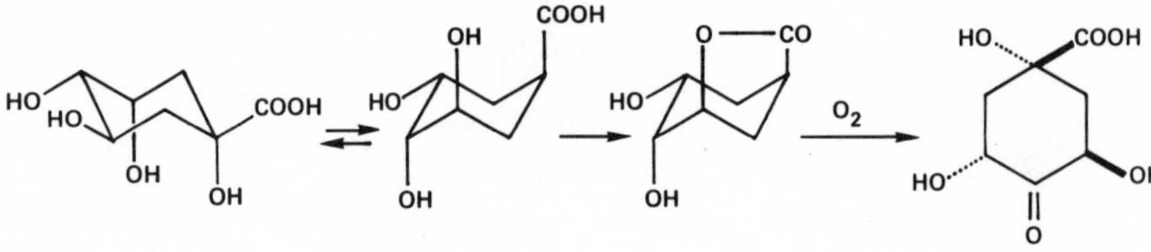

[3]   [17]   [18]

Dehydrogenation of cyclitols by *Acetomonas oxydans* is likewise known (18) to take place specifically at axially oriented secondary hydroxyls; in agreement with this rule, [3] is efficiently converted to [6] by this organism (5).

On paper-chromatograms of [6] (and [7]), spraying with semicarbazide hydrochloride, followed by heating to 100°, produces an intense blue uv fluorescence (3). Several related substances, including [3] and [4], do not show this reaction, which might therefore be valuable for the detection of [6] and [7]. The identity of the fluorescing species is not known.

It is remarkable that structures closely resembling [6], for example [19] and [20], should have been postulated by Robinson (19) as early as 1917 for hypothetical precursors of the benzene rings of certain alkaloids. The insight that led to this postulate: formation of the ring through intramolecular aldolization of an appropriate precursor, has been borne out much later by experimental findings.

Occurrence of [6] is not restricted to *E. coli* and *A. aerogenes*; the compound has been detected, for example, in the leaves of the oak *Quercus pedunculata* (93). The enzyme dehydroquinase,

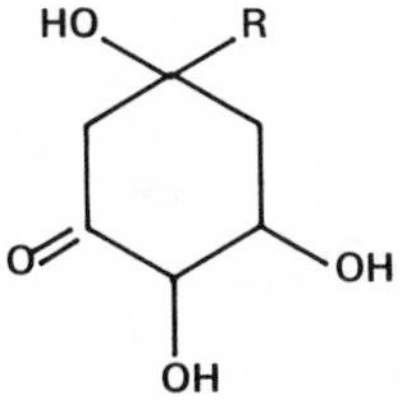

[19] : R = —CH$_2$—CHO
[20] : R = —CH$_2$—NH$_2$

which catalyzes the conversion of [6] to [7], has been found in
many microorganisms and higher plants (see below); these observations make it very probable that [6] functions quite generally
as an intermediate in the biosynthesis of aromatic compounds.

The enzymatic conversion of 3–dehydroquinic acid [6] into 3–
dehydroshikimic acid [7] constitutes the next step in the shikimate pathway. It proceeds in both directions with *cis*-stereo-
chemistry, which is unusual for reactions of this type. This *cis*-
stereochemistry was first demonstrated for the reverse process,
hydration of [7] to [6], by Hanson and Rose (20); it was subsequently established for the forward reaction [6]→[7] by Haslam
and co-workers (21). In the former work, enzymatic hydration of
[7] in tritiated water was used to obtain 2-[3]H-[6], whose stereochemistry was proven by enzymatic conversion to 2-[3]H-[3], oxidation (periodate followed by bromine) to tritiated citric acid
(21) and enzymatic transformation of this by aconitase to *cis*-
aconitic acid (22) with almost complete loss of the label. From
the known *trans*-stereochemistry of the last reaction, it follows
(20) that the citric acid, and consequently the hydration of [7]
to [6], must be formulated as shown in Scheme II.

In the work of Haslam and co-workers (21), equilibration of
[6] at pH 7.0 with D$_2$O was found to give selectively a 2–mono-
deuteriated derivative (at pH 8.5, all three protons in positions
2 and 4 adjacent to the carbonyl were exchanged). Nmr evidence,
confirmed by correlation with the citrate-aconitate system,
showed the deuterium in the mono-deuteriated compound to be in
the axial 2β-position, that is, *trans* to the hydroxyl on C-1 (the
preferred conformation of [6] was shown by nmr spectroscopy to be
the expected one with the carbonyl and the two secondary hydroxyls
in equatorial orientation). The fact that the labeled [6] was
converted via [7] to shikimic acid, [4], with retention of label
proved the *cis*-stereochemistry of the enzymatic elimination reaction.

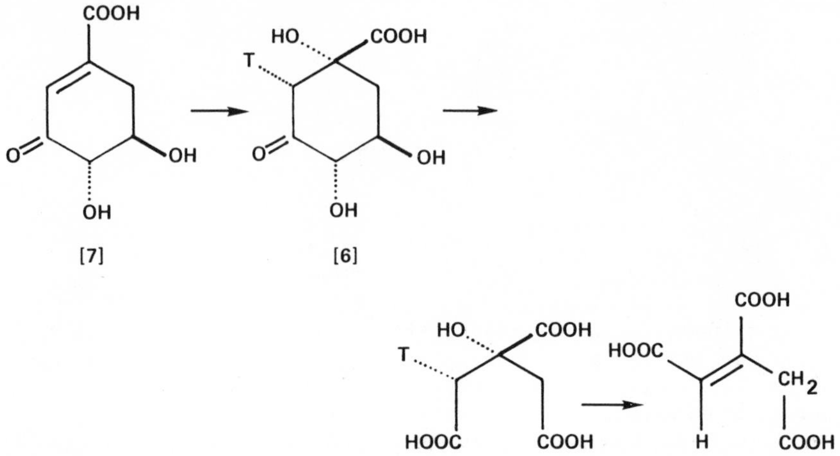

Scheme II

[21]   [22]

The enzyme that catalyzes the conversion of [6] into [7] was first studied in *E. coli* by Mitsuhashi and Davis (22). It is called 3-dehydroquinase (or 3-dehydroquinate dehydratase). It was found to occur in wild-type *E. coli* but to be absent from mutants blocked after [6]. No cofactor requirements were observed. Enzymatic activity was found in other microorganisms such as *A. aerogenes*, yeast, and *Euglena*, and in peas and spinach, but not in guinea pig liver. The enzyme is specific for the interconversion of [6] and [7]; it has no effect on structurally related acids ([3], (21), isocitric, or malic). Reversibility of the enzymatic reaction was demonstrated, but the equilibrium strongly favors [7]. Because of this, and because the pronounced uv absorption of [7] allows convenient spectrophotometric assay, the system has been widely used in enzymatic studies of the two earliest steps of the pathway (see Chapter 3).

Enzymes with very similar properties have been obtained from cauliflower (23) and from cell cultures from a variety of plants (24). In sweet potato tubers, the amount of dehydroquinase increases in response to injury (slicing), as does that of several other enzymes concerned with the biosynthesis of certain phenolic compounds, especially caffeic and chlorogenic acids, which increase after injury or infection with the black-rot fungus, *Ceratocystis fimbriata* (25). In contrast to the very stable enzyme from *E. coli*, the one from cauliflower loses activity very fast.

Like other enzymes not concerned with the transformations of a branch-point intermediate (see Chapter 15), dehydroquinase is only little influenced by the end-products of the shikimate pathway (26).

In *Neurospora crassa*, Giles and co-workers (27) observed the occurrence of two distinct, separable dehydroquinases with very different properties (heat stability, behavior on sedimentation, etc.). One of these enzymes catalyzes a step in an inducible catabolic pathway which exists in *N. crassa* and presumably leads via [7] to protocatechuic acid [15]. This dehydroquinase is induced in wild-type strains by [3] or [6] added to the growth medium; it occurs even in the absence of external [3] in certain mutants blocked in the common sequence of the shikimate pathway; apparently, [6] acts as an internal inducer. The other dehydroquinase of *N. crassa* is constitutive and forms part of the aggregate of the five consecutive enzymes responsible for the second to sixth reaction of the common sequence, which occurs in this species (28) (see Chapters 3 and 15). The biosynthetic and the degradative pathway which coexist in *N. crassa* have the step [6]→ [7] in common; the enzyme complex in this species may have the function of providing a means for keeping these two pathways separate, and of preventing the biosynthetically formed [6] from inducing the catabolic enzyme and being wastefully diverted away from its role in biosynthesis. However, this interpretation seems to be contradicted by the observation (29) that certain other fungi contain enzyme aggregates similar to the one of *N. crassa* but no catabolic dehydroquinase.

An inseparable complex of dehydroquinase with the enzyme responsible for the conversion of [7] into [4] has been studied in extracts from the roots of the oak, *Quercus pedunculata* (Fagaceae) (29a).

4-2. QUINIC ACID [3]

The observation, already mentioned before, that [3] can be used as a growth factor by strains of *A. aerogenes* (2), together with its widespread occurrence in plants and the close chemical connections with [6], make a discussion of the chemistry and biochemistry of [3] desirable. While this acid is not utilized by any aromatic auxotroph of *E. coli* (30), use of such a mutant of the closely related species, *A. aerogenes*, for bioassay during the isolation of [6] led to the surprising observation that this strain, and others of this species, are able to use [3] (2). This finding might suggest that [3] can be an intermediate in the

biosynthesis of aromatic metabolites in *A. aerogenes* but not in
E. coli. However, no mutant of the former species was found to
excrete any detectable trace of [3]; the ability of this organism
to utilize large amounts of [3] is thus probably based upon an
oxidation to [6] which is not an obligatory step in the biosynthetic pathway.

Quinic acid was discovered in *Cinchona* bark by Hofmann (31)
in 1790; it is thus one of the earliest organic plant products
to have been isolated. Its first chemical investigation is due
to Vauquelin (32). It occurs in many other plants, both free and
esterified with a variety of aromatic acids (see below). Its conversion to benzoic acid (excreted as hippuric acid) in man (33)
and in microorganisms (34) were observed early.

The constitution and stereochemistry of [3] follow mostly
from the classical work of H. O. L. Fischer and Dangschat, who
established the correct structure in 1932 (35). Key observations for the constitution are the formation of citric acid [21]
on treatment of methyl quinate with periodate, followed by oxidation of the resulting dialdehyde with bromine, and saponification
(36); for the relative stereochemistry, the fact that [3] readily
forms the lactone, quinide [17], while its 5-monomethyl ether
does not, and that an acetonide [17a] of [17] is easily obtained
from [3] (35) (see Scheme I). The absolute stereochemistry of
[3] follows from multiple interconversions between it and shikimic acid [4], which in its turn has been correlated unequivocally
with glucose (37). Conversion of [3] to [4] was first achieved
by Dangschat and Fischer (38) via a derivative of the nitrile of
[3]. The subsequent work of Grewe and Jeschke (7) resulted in a
transformation of [3] into [4] through oxidation with HNO_3 (see
p. 75) to [6], acid-catalyzed dehydration to [7], and reduction
to [4] with borohydride; this sequence completely parallels the
biosynthetic chain of events and is thus particularly significant.
Quite recently, the conversion of [3] into [4] has been achieved
in a remarkably straightforward manner by Cleophax and co-workers
(39) through treatment of methyl 3,4,5-tri-O-benzoylquinate with
sulfuryl chloride in pyridine, which gave the corresponding derivative of [4] in quantitative yield. It is interesting that the
reaction appears to proceed entirely in the desired direction;
the alternative product with the double bond between carbons 1
and 6, rather than 2, could not be detected. Compare also the
similarly specific conversion of the triacetate of the nitrile of
[3] into the corresponding derivative of [4] (40). The reverse
transformation of [4] into [3] has been carried out by Grewe and
co-workers by way of the corresponding primary alcohols (41), or
of the bromine addition product of [4] (42).

The first total synthesis of [3] was achieved by Grewe, Lorenzen, and Vining (12); it is shown in Scheme III. Here again, it is remarkable that all reactions, with the exception of the first one, proceeded with high stereoselectivity to give good yields of the desired products, although several stereoisomers could have been formed.

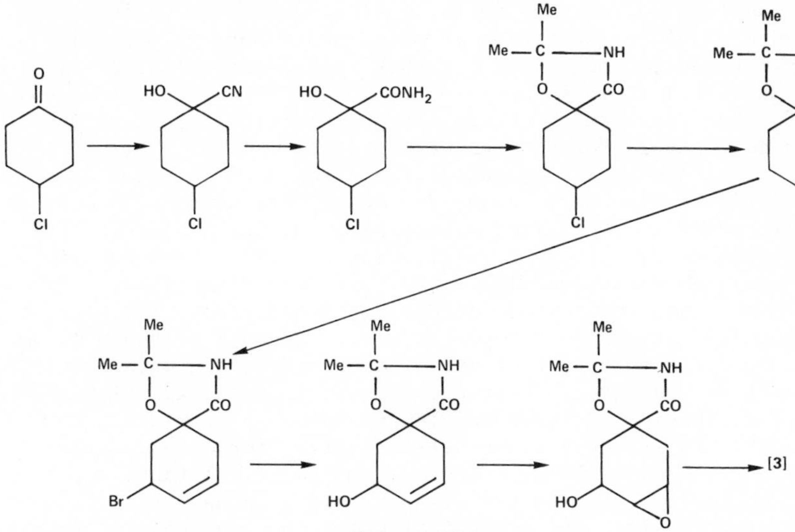

Scheme III

More recently, two related stereospecific syntheses have been worked out by Smissman and Oxman (43a) and by Wolinsky and co-workers (43b); they are intended for the preparation of labeled [3]. Both syntheses start from esters of α-acetoxypyruvic acid, build a cyclohexene structure by Diels-Alder reaction with butadiene or a derivative of it, and hydroxylate the double bond with OsO_4.

An entirely different synthesis, described recently by Bestmann and Heid (44), starts from D-arabinose. Reduction to arabitol and subsequent suitable transformations yielded the tribenzoate of arabitol 1,5-ditosylate, which was converted to a cyclohexane derivative by insertion of one carbon atom through reaction with methylenetriphenylphosphorane. Wittig reaction converted the resulting phosphorus ylide into 3,4,5-tribenzoyloxy-1-methylene-cyclohexane of correct stereochemistry. Replacement of the three benzoyl groups by acetyl, and reaction with $OsO_4/NaIO_4$ yielded the known 3,4,5-triacetoxycyclohexanone, which is also readily

accessible through Hunsdiecker reaction from tetraacetyl-[3] (40).
Conversion of this ketone to [3] by the cyanohydrin reaction completed the synthesis. This last step had been worked out previously by Grewe and Vangermain (40), who found it to be remarkably stereospecific; none of the nitrile epimeric at C-1 was obtained.+ The dehydration of the triacetylated nitrile of [3]
again gave exclusively the corresponding derivative of [4] and
yielded none of the stereoisomer which could form through elimination between carbons 1 and 6 rather than 1 and 2.+ This efficient and specific introduction of the carboxyl of [3] seems suitable for the preparation of the carboxyl-labeled acids, which
should be valuable for biosynthetic studies.

In addition to its ubiquitous occurrence as the free acid in
plants, [3] is also very frequently found esterified with certain
phenolic acids; a variety of esters with gallic, p-coumaric (4-
hydroxycinnamic), caffeic (3,4-dihydroxycinnamic), and ferulic
(3-methoxy-4-hydroxycinnamic) acids have been identified. Of
these compounds, chlorogenic acid [23], 5-0-caffeoyl-[3], (see
below) is by far the most important one. Because of their widespread occurrence throughout the plant kingdom, and of numerous
observations suggesting their possible biological significance,
a brief discussion of these substances seems appropriate. This
account is not intended to be all-inclusive; for more detailed
information, especially on questions of biological function, the
excellent review by Sondheimer (45) should be consulted. Similarities of chemical and analytical properties and ease of intramolecular acyl migrations have caused numerous difficulties in
the earlier work on these esters; however, use of modern analytical methods, and specific total syntheses of many individual compounds, have clarified most of the doubtful points. It should be
noted that almost all publications dealing with these depsides
use the older numbering system for [3].

Only two galloyl esters of [3] seem to be known: the theo-
gallin from tea leaves (*Camellia sinensis,* Theaceae), which has
been identified as 5-0-galloyl-[3] (46), and the main component
of Tara tannin from *Caesalpinia spinosa* (Leguminosae), which is
a 3,4,5-tri-0-trigalloyl-[3] with two or three additional galloyl
residues attached to phenolic hydroxyls (47).

+
Added in proof: However, in very recent work by Rapoport and co-
workers (44a), the cyanohydrin reaction did not exhibit any stereospecificity, and the subsequent dehydration gave almost equal
amounts of the two possible nitriles.

All other compounds of this class are esters of hydroxylated cinnamic acids; the occurrence and biosynthetic role of 5-cinnamoyl-[3] itself are debatable (see below).

5-0-p-Coumaroyl-[3](24) has been observed in several plants (48); its structure was proven by comparison with a specimen prepared by unequivocal synthesis (49). A different mono-p-coumarate of unknown structure has been isolated in crystalline form from a white-flowered mutant of snapdragon, *Antirrhinum majus* (Scrophulariaceae) (50). *Schinopsis* spp. (Anacardiaceae) and the quebracho tannin obtained from them seem to contain two different mono-p-coumaroyl esters of [3] (51). A di-p-coumarate of [3], probably the 1,4-isomer, occurs in the vegetative parts of the pineapple plant, *Ananas comosus* (Bromeliaceae) (52).

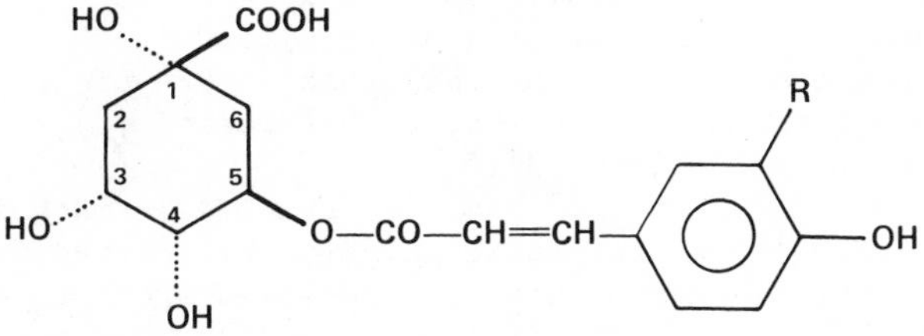

[24] : R=H
[23] : R=OH

The most numerous groups among the esters of [3] are those in which caffeic acid is the esterifying acid. Foremost among these is chlorogenic acid [23], discovered in coffee beans in 1846 (53) and found since then in a vast number of plants (53). It occurs sometimes in surprisingly large concentrations; for example, fresh coffee beans are stated (45) to contain 6.3% [23]. Chlorogenic acid was recognized as a caffeoyl ester of [3] by Gorter (54). It was shown to be 5-0-caffeoyl-[3] by Fischer and Dangschat (55); this structure was proven by total synthesis (56). The numerous claims for biological action and possible physiological function of [23] are discussed in ref. 45.

Two other mono-caffeoylquinic acids have been obtained from several plants. They are 3-0-caffeoyl-[3] or neochlorogenic acid, first isolated from peaches (56), and 4-0-caffeoyl-[3] or cryptochlorogenic acid, discovered in coffee and initially called "Band 510" (57). The structures of these two esters were established by Scarpati and Esposito (58). The remaining isomer, 1-0-caffeoyl-[3], has been detected by paper chromatography in aqueous extracts from artichoke leaves (59). A product initially ob-

tained from coffee or maté (*Ilex paraguayensis*, Aquifoliaceae)
and called "isochlorogenic acid" (60a) was more recently recognized (60b,c) as a mixture of three closely related compounds
which were identified by Scarpati and Guiso (61) as the 3,4-,
4,5-, and 3,5-di-0-caffeoylquinic acids. The 3,4- isomer had
been discovered independently in unroasted coffee beans (62).
Extraction of artichokes with methanol at room temperature yields
1,5-di-0-caffeoyl-[3] (63a). Heating this in phosphate buffer at
pH 7.0-7.2 converts it to the 1,3-isomer, cynarin. The latter is
the active choleretic principle obtained by extraction with hot
water (63b); it had initially been formulated as 1,4-di-0-caf-
feoyl-[3]. Cynarin is thus an isolation artifact; it has also
been obtained from a few other plants of the families Compositae
and Umbelliferae (64).

Esters of [3] with ferulic (3-methoxy-4-hydroxycinnamic)
acid have been found in coffee beans (65a) and in the cambium of
hemlock (*Tsuga canadensis*, Pinaceae) (65b). One of the compounds
from coffee has been shown (66) to be 5-feruloyl-[3], i.e.,
the 3'-monomethyl ether of [23].

The biosynthesis of [23] has been studied in much detail.
It proceeds from [3] and cinnamic acid (itself formed from phenylalanine), but details of the path followed are not yet clear
and need not be the same in different species. The ester could
be formed directly from [3] and caffeic acid, presumably after
suitable activation (e.g., as the CoA derivative), with the caffeic acid being synthesized from cinnamic via p-coumaric acid.
Alternatively, [3] could be esterified with cinnamic or p-coumaric acid, with subsequent introduction of one or two hydroxyls
into the phenylpropanoid moiety. The first possibility, i.e.,
hydroxylation preceding esterification, has been observed by Gamborg (67) in suspension cultures of potato cells. Label from
generally labeled [3] and 3-^{14}C-caffeic acid appeared almost exclusively in the corresponding moieties of [23], showing that the
precursors were incorporated directly rather than by degradation
and subsequent resynthesis. This finding contrasts with those of
other investigators who studied intact plant tissues. Levy and
Zucker (68) and Hanson (69), using discs from potato tubers, concluded that the precursor of [23] is [24]. It seems doubtful
whether 5-cinnamoyl-[3] is the next earlier intermediate. A
cellfree enzyme system capable of converting [24] into [23] was
obtained by Hanson and Zucker (60b). Runeckles (70) found good
incorporation of cinnamic and p-coumaric acids, and of [24], into
[23] in tobacco leaf discs, but caffeic acid was only poorly
utilized, and this utilization proceeded with much randomization
of label, evidently due to breakdown of the caffeic acid. The

results thus support the pathway via [24]. The findings by Steck
(71) with leaf discs of tobacco and other plants suggest that the
sequence cinnamic acid→p-coumaric acid→[24]→[23] constitutes the
major pathway, while hydroxylation of p-coumaric acid to caffeic
acid and combination of this with [3] occurs to a lesser extent.
Surprisingly, the neochlorogenic acid, present in about 1/4 the
amount of [23], was only weakly labeled under conditions that
gave strong incorporation of radioactive precursors into [23].
Similar experiments by Legrand and co-workers with tobacco leaves
(72) likewise showed poor incorporation of label from G-^{14}C-
phenylalanine into neo- and cryptochlorogenic acids, while the
[23] was strongly active.

The questions of the *origin of* [3] and its *possible function
as an intermediate in the biosynthesis of aromatic metabolites*
are still far from clarified.

As mentioned earlier, [3] is utilized by certain mutants of
A. aerogenes but not by any strain of the closely related *E. coli,*
and no strain of either species was found to excrete any detectable amount of it (2). Those observations definitely argue
against the possibility that [3] might be an obligatory biosynthetic intermediate in these bacteria. The utilization of [3] by
mutants of *A. aerogenes* parallels that of [6] and is thus undoubtedly based upon the ability of this species to convert [3]
into [6]. The enzyme responsible for this reaction, quinate dehydrogenase, has been studied by Mitsuhashi and Davis (73). It
requires NAD, and is insensitive to EDTA; its action is reversible. Extracts from *E. coli,* yeast, *Euglena gracilis,* peas, or
spinach failed to show any detectable activity. Several subsequent attempts to find such an activity in a variety of higher
plants were unsuccessful (74 and references therein), but an enzyme resembling the one from *A. aerogenes* was finally discovered
by Gamborg (24, 75) in cell cultures from mung beans (*Phaseolus
aureus,* Leguminosae). No such activity was detectable in similar
cell cultures from several other plants (24).

The enzyme in *Acetomonas oxydans* (5) which dehydrogenates
[3] to [6] (see above) is of a very different type: it is not
influenced by pyridine nucleotides, but requires cytochrome 555
as a cofactor.

Many other microorganisms (bacteria, fungi) are capable of
utilizing [3], which can often serve as the sole carbon source.
This utilization frequently rests upon the ability of these organisms to convert [3], via [6] and [7], to protocatechuic acid
[15], which is the key compound of a well-investigated degradative sequence, the so-called. β-keto-adipate pathway (76). In
this pathway, oxidative cleavage of the aromatic ring of [15]

gives β-carboxy-*cis*,*cis*-muconic acid, which is converted in several steps to ß-keto-adipic acid; this compound undergoes fission into succinic acid and acetic acid (as acetyl CoA). Utilization of these ultimate products for the synthesis of cell constituents presumably explains the fact that compounds convertible into [15] can serve as the only carbon source. The β-keto-adipate pathway is not restricted to microorganisms, since it has also been observed in mung bean seedlings (77). The enzymes catalyzing the various steps of this sequence are inducible. Those enzymes connecting [3] with the β-keto-adipate pathway have been studied in *N. crassa* (78) and particularly in *Acinetobacter calco-aceticus (Moraxella calcoacetica)* (79), in which an enzyme catalyzing the conversion of both [3] and [4] to the corresponding 3-dehydro-compounds, [6] and [7], is induced not only by [3] but also by [15]. This enzyme is independent of pyridine nucleotides. Further conversion of the [6] formed under its influence into [7] provides the connection of [3] with the β-keto-adipate pathway; the aromatization of [7] to [15] is a well-known reaction which will be discussed in Section 4-4.

The processes just described may indicate the metabolic fate of [3] and may have a bearing on its role as a precursor of aromatic compounds (see below). However, the biosynthetic origin of [3] remains something of a puzzle; especially, the wide distribution of [3] is difficult to explain, as are some features of its utilization for the formation of the aromatic amino acids. The simplest assumption would derive [3] from the normal shikimate pathway through reduction of [6] catalyzed by quinate dehydrogenase, but the rare occurrence of this enzyme seems to argue against such a hypothesis. It has been pointed out by Gamborg (24) that enzymes of the type present in *Aerobacter* and in mung beans might be widely distributed but might have escaped detection because of instability; the enzyme from mung beans is quite unstable in solution. Alternatively, enzymes similar to the one in *Acetomonas* might be more common than is known today, and might be responsible for the appearance of [3]. In either case, derivation of [3] from the typical shikimate pathway would become explicable, but alternative modes of formation of [3] can be devised in which [6] is not its direct precursor, and hence no enzyme interconverting these two compounds is needed. For instance, [4] could undergo hydration to [3] in some reaction that would be the reverse of the ready, specific dehydrations of certain derivatives of [3] *in vitro* (cf. refs. 38, 39).

The entire field of biosynthesis and metabolic fate of [3] suffers from the difficulty that studies so far had to be carried out mostly on intact parts of plants, or on intact cells in cul-

ture; the findings made are thus apt to be complicated by problems of permeability and compartmentalization.

Like the biosynthesis, the function of [3] is not clear.
Many observations in higher plants have been interpreted as indicating for [3] a role as an intermediate in a biosynthetic sequence not identical with the shikimate pathway in microorganisms.
These questions are discussed in detail in ref. 74; a few such
findings may be mentioned. Using generally labeled [3] isolated
from rose petals exposed to $^{14}CO_2$ in light (80a), Weinstein, Porter, and Laurencot (80b) found strong labeling of phenylalanine
and tyrosine in many plants. However, the specific activities
of the two amino acids were different; in some plants the former
was more strongly labeled, in others the latter. Similar results
were obtained by Rohringer and co-workers with both [3] and [4]
(81). Both groups of workers also found (80c, 81) that phenylalanine and tyrosine, but not tryptophan, incorporated label from
[3] (or [4]). At first sight, these findings seem difficult to
reconcile with operation of the usual shikimate pathway with its
extensive series of common intermediates. This difficulty is
particularly evident in the case of phenylalanine and tyrosine,
whose established biosynthesis (see Chapter 7) bifurcates
either at the stage creating their immediate precursors, the
corresponding α-keto-acids, or not at all (cf. the cases of formation of tyrosine by direct hydroxylation of phenylalanine discussed in Chapter 7). But here, as elsewhere, interpretation of the results may not be straightforward on account of the
use of intact tissues. In the experiments just quoted (80), for
example, not only phenylalanine and tyrosine were strongly labeled, but also the respiratory CO_2 produced in the dark, and a
number of other metabolites (nonaromatic amino acids, Krebs-cycle
intermediates) contained smaller amounts of ^{14}C.

The results just discussed, and related findings such as differences in the utilization of [3] and [4] (81) have been interpreted as demonstrating the existence of variants of the shikimate pathway (direct utilization of [3], bypassing [4], for the
biosynthesis of phenylalanine; biosynthesis of tryptophan by reactions different from those established in bacteria; etc.). It
seems to us that, on the basis of existing evidence, such assump-
tions are possible but not necessary.

A few additional transformations of [3] by bacteria are
worth mentioning. *Lactobacillus pastorianus* var. *quinicus* (82)
converts [3], and also [4], into a dihydro[4], which is different
from the compound formed on catalytic hydrogenation of [4] (see
Section 5-1), but identical with an acid obtained by Grewe and
Lorenzen (42). This latter compound is known to correspond

stereochemically to [3], while the product from hydrogenation of [4] has been proved to be the l-epimer. The removal of the tertiary hydroxyl of [3] thus proceeds with retention of configuration, if it is a direct process, or with restitution of the initial stereochemistry, if [4] should be an intermediate. Other types of lactobacilli reduce [3], [4], or the dihydro-[4], into all-*cis*-3,4-dihydroxycyclohexane-1-carboxylic acid (83).

Finally, the long-known (33) conversion of [3] into hippuric acid (N-benzoylglycine) in man and various animal species has been shown (84) to be performed essentially by intestinal bacteria; the same is true for the analogous aromatization of [4]. Both [3] and [4] are converted to hippuric acid on oral, but not on intraperitoneal administration, and suppression of the intestinal flora with the antibiotic neomycin decreases or abolishes the formation of hippuric acid from [3] or [4] given *per os*. Actually, the bacteria apparently convert these acids to benzoic acid, which is then conjugated with glycine in the liver or kidney of the host, with subsequent urinary excretion. Unfortunately, the details of this bacterial aromatization are still unexplored; it is not known by what mechanism the intestinal bacteria convert the alicyclic acids to benzoic acid with removal of all hydroxyls. Biosynthesis of phenylalanine or phenylpyruvic acid from [3] or [4] by way of prephenic acid (see Chapter 7), followed by oxidation to benzoic acid, is one possibility, but hardly the only one.

4-3. 3-DEHYDROSHIKIMIC ACID [7]

3-Dehydroshikimic acid (formerly called 5-dehydroshikimic acid) is the intermediate in the common pathway following [6] and preceding [4]; it was the first one of these intermediates to be isolated in pure form (4). The compound was initially observed by Davis (30a), with the help of bioautographic techniques, in culture filtrates from certain mutants of *E. coli*; it was found there as the only active spot in filtrates of strains blocked immediately after it, and as a minor one, besides a large one of [4], in strains blocked after this latter compound. It was distinguished from [4] by lower mobility in n-butanol-acetic acid, sensitivity to autoclaving, and inactivity as a growth factor for mutants blocked between it and [4]. Its occurrence as a minor excretion product besides [4] indicated that it might be the precursor of the latter, an interpretation fully borne out by all later findings.

The compound was subsequently isolated in pure, crystalline form by Salamon and Davis (4) from cultures of mutants of *E. coli* blocked immediately after it. Chromatography on charcoal and elution with 75% ethanol yielded the well-crystallized monohydrate, $C_7H_8O_5 \cdot H_2O$. The behavior of [7] on melting is similar to that of [6]: m.p. 150–152°, resolidification on further heating, second m.p. 201–202° (formation of protocatechuic acid [15]; see below). The intense uv absorption of [7]: λ_{max} 234 nm, ε 12,100 (ethanol) has often been useful for the detection and assay of the compound.

The transformations which prove the structure and stereochemistry of [7] are summarized in Scheme IV; cf. also Scheme I of Section 4–1 and Scheme I of Section 5–1.

The functional groups present in [7] are demonstrated by the following evidence (4): the carboxyl group is shown by the pK_a of 3.2 and the ready esterification with diazomethane, the carbonyl by formation of a remarkably insoluble semicarbazone. Reduction of Fehling's and Tollens' reagents, marked sensitivity to alkali, formation of a phenyl osazone and of a 2,4-dinitrophenylosazone of the methyl ester indicate the presence of an α-ketol unit. Two hydroxyl groups are present, since the methyl ester gives a diacetate. The uv maximum at 234 nm is consistent with the presence of the chromophore –CO–CH=CH–COOH. The conjugated nature of the carbonyl is further confirmed (3) by the absorption spectrum of the thiosemicarbazone (not isolated): λ_{max} 265, 315 nm, as expected (6) for the derivative of an α,β-unsaturated ketone.

The actual structure and stereochemistry of [7] can be deduced from these observations, and from the following reactions (4, 85): the compound readily yields [15] (m.p. 202–204°) on pyrolysis or brief warming with concentrated HCl (it will be remembered that [15] is also formed in exactly the same manner from [6], dehydration to [7] being the first step in this transformation; cf. Section 4–1). Aeration of [7] or, better, action of Fehling's solution at room temperature gave gallic acid [5] (13b). The *enzymatic* formation of [15] and [5] from [7] will be discussed in Section 4–4. These observations strongly suggest a nonaromatic ring-structure closely related to [15] and [5]. Pt-catalyzed reduction of [7] to the known (86) dihydroshikimic acid [25] confirms this point and establishes the orientation of the substituents.

The evidence discussed so far would be compatible with formulae [7] or [26]. The latter structure is made unlikely by the observation (85) that the methyl ester of [7] remains unchanged under conditions (acetone and dry $CuSO_4$ at 37°) where methyl shikimate is rapidly transformed to the 3,4-isopropylidene derivative (86); it is conclusively disproved by conversions of [7] to [4] and its derivatives. Of these, the enzymatic transformation will be discussed below. Chemically, reduction with $NaBH_4$ yielded (85) a mixture of [4] and the heretofore unknown, bio-

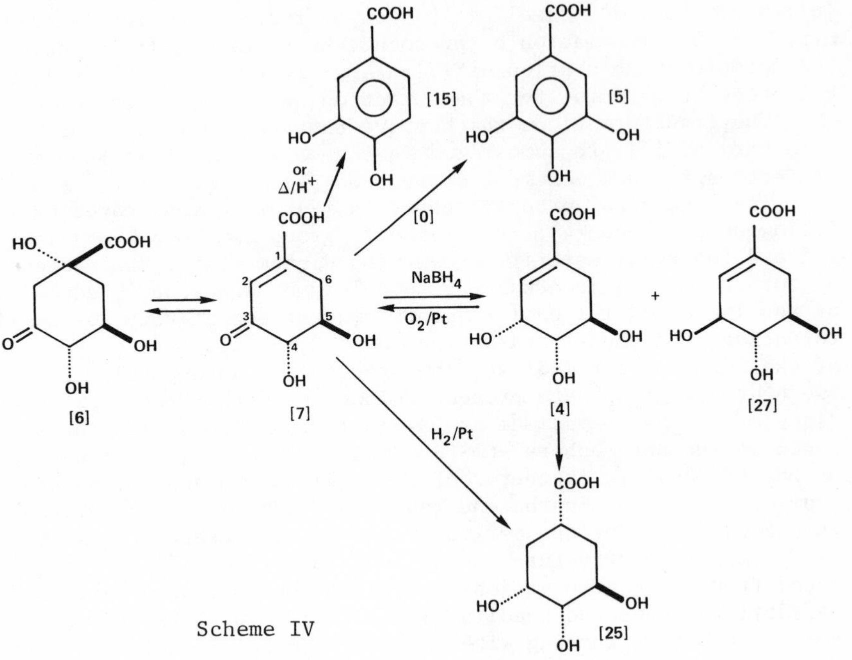

Scheme IV

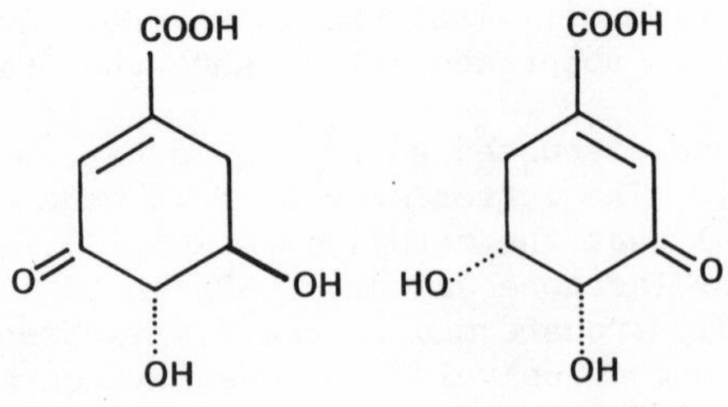

90

logically inactive 3-epishikimic acid [27]. Owing to the small
amounts on hand, isolation of the more soluble [7] in pure form
proved impossible, but its presence was demonstrated by growth
factor activity for appropriate mutants. With the larger quantities of material available through their conversion of [6] (obtained by HNO_3-oxidation of [3]) to [7], Grewe and Jeschke (7)
isolated the methyl esters of both [4] and [27] on borohydride
reduction of the methyl ester of [7], thus completing the proof
of structure and configuration. Since [3] has been synthesized
(12), the work of Grewe and Jeschke constitutes a total synthesis
of [7]. In addition, the connection between [6] and [7] permits
applying to the latter compound the evidence for the location of
the carbonyl in [6], which is firmly established by the failure
of that compound to give an acetonide (2), and by its catalytic
reduction in acidic medium (7) to the 3-deoxy compound [16] (see
Section 4-2).

Platinum-catalyzed oxidation of [4] with oxygen provides
(13a, 87) a convenient laboratory preparation of [7], analogous
to the oxidation of [3] to [6] (13).

Since the initial isolation of [7] (4), the compound has
been encountered repeatedly. Haslam and co-workers (13b) isolated it in crystalline form from cultures of the mold *Phycomyces
blakesleeanus* by ion-exchange chromatography, Whiting and Coggins
(5) from those of *Acetomonas oxydans* grown in presence of [4].
Hattori, Yoshida, and Hasegawa (88) obtained a product with the
properties of [7] from filtrates of *Pseudomonas ovalis* growing on
ammonium shikimate as sole carbon source, and isolated it as the
crystalline semicarbazone; it was similarly identified by Nandy
and Ganguli (89) as the product of the action of the dehydroshikimate reductase from mung bean extracts upon [4]. In addition,
the presence of [7] has been detected repeatedly by chromatographic or bioautographic methods or the like; a few instances may
be mentioned. Ehrensvärd (90) identified traces of [7] in the
seeds of *Illicium religiosum* and *I. verum*, the initial source of
[4]; Tatum and co-workers (91a) and Metzenberg and Mitchell (91b)
observed the accumulation of small amounts of it by mutants of
Neurospora. Simonart and Wiaux (92) found it in *Penicillium griseofulvum;* Boudet and co-workers detected it, together with [6],
in oak leaves *(Quercus pedunculata)* (93).

The enzyme 3-dehydroquinase, which is responsible for the
formation of [7] from [6], has been discussed in Section 4-1.
Its substrate, [7], is capable of being used as a growth factor
by microbial mutants blocked before it. However, in strains of
certain bacteria such as *E. coli, A. aerogenes, Salmonella typhi-
murium, B. subtilis,* this utilization exhibits peculiarities

which were explored by Davis (1, 94). Strains blocked in any
step of the common pathway would be expected *a priori* to require
all five primary aromatic metabolites for growth. However, mutants exist which, from the pattern of intermediates excreted by
them, are evidently blocked between [7] and [4], but which show
growth requirements ranging from double to quintuple. Furthermore, these requirements are not distributed randomly but follow
a very definite pattern. All double mutants need phenylalanine
and tyrosine for growth at normal rates; in triple mutants, a requirement for tryptophan (or anthranilic acid) is added, in the
quadruple ones a further one for *p*-aminobenzoic acid. This phenomenon is interpreted as being based upon incomplete genetic
blocks in the mutants with less than quintuple requirements,
coupled with a preferential conversion of the limited amounts of
the intermediates formed to the various aromatic end-products.
This assumption is supported by the ability of the double and
triple mutants to grow slowly on minimal medium. The very small
amount of such intermediates which can pass the block in the mutants with a quadruple requirement appears to be utilized for the
synthesis of *p*-hydroxybenzoic acid; increasingly incomplete
blocks enable these "leaky" mutants also to synthesize their own
p-aminobenzoic acid, tryptophan, and phenylalanine plus tyrosine,
in that sequence, from which no deviations were observed. Additional support comes from the observation of spontaneous or uv-
induced reversions which, undoubtedly in one step, (multiple mutations being exceedingly rare events), can convert even mutants
with quintuple requirements into prototrophs capable of growing
on minimal medium. In other cases, secondary mutants with diminished growth requirements were observed (e.g., quadruple to
triple), where the disappearance of growth requirements again
followed the same patterns. In addition, work on a double mutant
blocked both before and after [7] has shown that this compound
acts as a competitive inhibitor of its daughter product [4], and
that this inhibition, too, follows the same sequence: increasing
ratios of the inhibitor block successively the utilization of
shikimic acid for the biosynthesis of tyrosine, phenylalanine,
tryptophan, *p*-aminobenzoic acid, *p*-hydroxybenzoic acid, and the
"sixth factor" (94). The sequential order parallels the amounts
of the various aromatic metabolites required for growth: the
quantitative requirements for the two *p*-substituted benzoic acids
are stated to be about 1/1000 of that for tryptophan, and this
latter about 1/4 of that for phenylalanine and tyrosine. For
complications introduced by observed dependence of growth factor
requirements upon carbon source, pH, etc., see refs. 1 and
94.

The inhibition by their own metabolic product, [7], excreted in excess, prevented mutants blocked in its subsequent conversion from normal utilization of [4] as growth factor. No such inability was observed in mutants blocked before [7]. Again, the ability of mutants excreting this compound to utilize [4] for growth fell off in the same familiar sequence with increasing completeness of the genetic block. Similar phenomena, although analyzed in less detail, may play a role in the utilization of intermediates by mutants of *E. coli* with earlier blocks (2). Several of these strains grow very poorly on [4] or [7] as sole supplement, and not at all on [6]. Supplementation with phenylalanine and tyrosine brought the response to the first two intermediates up to normal but left the one to [6] poor. Here, too, spontaneous secondary mutants occurred which grew normally on [4], [7], or [6] alone.

Since structural analogies between consecutive intermediates in biosynthetic chains are quite general, competitive inhibition of the type described ought to be fairly common, and indeed a phenomenon very similar to the preferential utilization of [7] was observed in *S. typhi-murium* by Sprinson and co-workers (26a). Here, however, it is the last common intermediate, chorismic acid (see Chapter 6) which is converted preferentially into *p*-aminobenzoic acid, tryptophan, and phenylalanine plus tyrosine, in that order, by mutants with increasingly incomplete blocks. Again, the needs for the metabolite required in the smallest amount are satisfied first.

No detailed analysis of the competition between [7] and [4], or of the mechanism underlying the preferential synthesis of the aromatic metabolites, seems to have been carried out; in the light of modern insight into genetic and regulatory processes, this is regrettable.

In the absence of a detailed experimental analysis, a speculative attempt at rationalization may be justifiable. The sequential inhibitory action of [7], and the undoubtedly closely related preferential utilization of [4] in mutants with incomplete blocks between these two intermediates, evidently favor the biosynthesis of the metabolites required in the smallest amounts, that is, the aromatic vitamins. It seems possible that the existence of such a mechanism channeling a limited flow of the common intermediates in this direction may be connected with the regulation of 3-deoxy-D-*arabino*-heptulosonic acid 7-phosphate synthetase, the enzyme responsible for the first stage of the shikimate pathway (Chapter 3). Wallace and Pittard (95) found that amounts of the intermediates of the common pathway sufficient for the biosynthesis of the aromatic vitamins are still formed when the three

well-known isoenzymes of the synthetase are fully repressed by
phenylalanine, tyrosine, and tryptophan, respectively; no additional isoenzyme specifically connected with the biosynthesis of
the vitamins seems to exist. The sequential inhibition by [7]
might thus constitute a needed further control which causes the
small amount of available intermediates to be utilized for the
biosynthesis of the required vitamins, rather than being used for
the biosynthesis of further amounts of the aromatic amino acids
already present in adequate quantities. The existence of an
additional control point at the level of chorismic acid, at least
in *Salmonella*, does not necessarily militate against this hypothesis. Dehydroshikimic acid [7] might be a particularly suitable agent for this control, since it is the only one among the
common intermediates which contains an α,β-unsaturated ketonic
carbonyl. This group is notorious for its marked tendency to
react with nucleophiles at its β-position; the hydration of [7]
to [6] (see p. 77) is an obvious example of this reactivity. Since
-SH groups are quite particularly active nucleophiles in reactions
of this type, the graded inhibitory action of [7] might conceivably be based upon a similarly graded ability of the -SH groups of
some enzymes involved in the biosynthesis of the various aromatic
metabolites to react with [7]. The fact that the other common
intermediates (with the exception of chorismic acid) do not show
such an inhibitory action would become intelligible on the basis
of this hypothesis. Admittedly, the similar role of chorismic
acid in *Salmonella* would have to be explained differently, perhaps through the presence of the very reactive diene system only
in this compound, and in isochorismic acid (see Chapters 6 and
12).

   3-Dehydroshikimate reductase (or shikimate dehydrogenase),
the enzyme which interconverts [7] and [4], was first studied by
Yaniv and Gilvarg (96), using *E. coli*. It is dependent upon NADP
as cofactor, which cannot be replaced by NAD. The enzyme was
found to be highly substrate-specific, being without action upon
[3], 3-epi-[4], the dihydro-[4] obtained on catalytic hydrogenation of [4] (86, 97), or the 3- and 5-phosphate of [4] (2). As
expected, dehydroshikimate was absent from mutants of *E. coli*
blocked between [7] and [4]. Enzymatic activity was also observed in other organisms, such as *A. aerogenes*, *Euglena*, yeast,
spinach, and peas.

   Enzymes with similar properties have been found repeatedly
in both microorganisms and higher plants. In *Neurospora*, an enzyme resembling the one from *E. coli* has been observed (98); it
forms part of the aggregate of enzymes 2 - 6 of the common portion of the shikimate pathway which exists in this species (28).

As expected from its position far from any branch-point, it is
hardly subject to feedback control (cf., e.g., ref. 26a). Among
the shikimate reductases from higher plants, those from etiolated
pea epicotyls (99), mung bean (89), the tea plant (100), sliced
sweet potatoes (25), and cell cultures from several species (29)
have been studied. The enzyme from pea epicotyls is inactivated
by the SH-reagent, p-chloromercuribenzoate, and activity is restored by cysteine (99a); this enzyme has recently been purified
over two-hundredfold (99b).

4-4. PROTOCATECHUIC [15] AND GALLIC [5] ACIDS

In addition to its vitally important conversion to [4] in
the common sequence of the shikimate pathway, [7] can also undergo biosynthetically significant direct transformations to [15]
and [5]. These reactions are of particular interest through the
fact that they constitute formation of aromatic metabolites from
a nonaromatic precursor; they are among the most straightforward
examples of biosynthetic aromatization.

Actually, alternative modes of biosynthesis of [15] and [5]
exist; they are discussed more completely in Chapter 16. Both
acids can be formed through further hydroxylation of p-hydroxy-
benzoic acid and [15], respectively, or they can arise through
oxidative shortening of the side-chain of C_6-C_3 precursors. It
seems, however, that both acids usually originate directly from
[7], although the alternatives mentioned do occur.

Chemically speaking, the various alternative possibilities
are distinguished by the origin of the hydroxyl and carboxyl
groups of [15] and [5]. If derived directly from [7], the hydroxyls and carboxyls would all be those of the precursor. In case
of formation by further hydroxylation of p-hydroxybenzoic acid or
[15], the carboxyl and one or two of the hydroxyls, respectively,
would have this origin. Biosynthesis through oxidative degradation of a C_3 side-chain, finally, would proceed with loss of the
original carboxyl during the aromatization of prephenic acid
(Chapter 7), and the carboxyl found in the aromatic C_6-C_1
acids would be derived from C-3 of enolpyruvyl phosphate [11] and
consequently from C-1 and C-6 of glucose, while the carboxyl of
[7] comes from that of [11], and hence from C-3 and C-4 of glucose (101). In principle, therefore, experiments with [7] labeled with ^{18}O in the hydroxyls, or specifically with ^{14}C in the
carboxyl, should yield clear-cut information on the biosynthesis
of [15] or [5]; carboxyl-labeled [7] is potentially available
through Grewe and Vangermain's (40) preparation of the acetylated
nitriles of [3] and [4] from 3,4,5-triacetoxycyclohexanone. No

experiments using ^{18}O seem to have been reported, but precursors with preferential labeling of the carboxyl have been prepared by biochemical methods, and have been utilized to explore the biosynthesis of [5] (see below).

The direct formation of [15] from [7] was first observed by Tatum and co-workers (91a) in a mutant of *Neurospora* blocked between [7] and [4]. This strain accumulated only little [7] but instead produced substantial amounts of [15], undoubtedly by further transformation of [7], a reaction which was interpreted as a detoxification process. The direct derivation of [15] from [7] was demonstrated by Tatum and Gross (102), who found that label from 1- and 6-^{14}C-glucose was incorporated almost exclusively into carbons 2 and 6 of [15], as expected from the known biosynthesis of the alicyclic intermediates of the shikimate pathway from glucose (see Chapters 2 and 5). The alternative modes of biosynthesis mentioned before are excluded in this case already by the location of the genetic block; that from C_6-C_3 precursors is further eliminated by the lack of labeling of the carboxyl of [5], which thus cannot originate from the β-carbon of a C_3 side-chain.

The biosynthetic formation of [15] from [7] is of course chemically analogous to the conversion under the influence of heat or acid (4). Through degradation of a sample of [15] derived from [4] labeled in positions 2, 3, 4, and 5 (from [11] and labeled erythrose 4-phosphate, [10]), Gross (98) showed that it is the hydroxyl on C-5 which is lost, presumably in an elimination reaction from the enol form of [7].

The detailed mechanism of this biosynthetic aromatization has been established recently by Floss and co-workers (103) with the help of [7] labeled stereospecifically with tritium at C-6. It was found that it is the *pro*-R hydrogen (β-oriented) which is lost. It is thus the unusual *cis*-elimination of water which occurs here, as it does in the transformation of [7] into [6] (see Section 4-1). Apparently, so far, biochemical *cis*-dehydration has been observed only in those two reactions. In contrast to the biosynthetic reaction, the conversion by heat or acid is not stereospecific; it proceeds with rapid formation of a very transient intermediate.

The enzyme 3-dehydroshikimate dehydrase of *Neurospora* has been examined by Gross (98). It is very sensitive to heat, metal ions such as Cu^{2+} and Hg^{2+}, and to *p*-chloromercuribenzoate. It is present in aromatic auxotrophs blocked after [7], but not in the wild-type.

Conversion of [7] to [15] is by no means restricted to *Neurospora*. Compound [15] is formed, for example, by *Pseudomonas*

ovalis growing on the ammonium salts of either [7] or [4] (88).
Gibson and co-workers (104) observed the accumulation of [15] in
the culture filtrates of several mutants of *A. aerogenes* blocked
in the biosynthesis of the aromatic amino acids, but not in those
of strains with blocks in the biosynthesis of several other amino
acids; compound [15] should thus be formed from one of the common
intermediates of the shikimate pathway, presumably [7], but the
actual precursor has apparently not been identified experimentally. In higher plants, [15] can be formed either directly from
[7] or via C_6-C_3 compounds; the various, sometimes conflicting
findings have been reviewed by Zenk (105).

Besides [15], certain strains of *Neurospora* blocked in the
biosynthesis of aromatic metabolites also produce vanillic acid,
the 3-methyl ether of [15]; it was isolated in pure, crystalline
form from culture filtrates of such mutants by Metzenberg and
Mitchell (91b). It is not known whether this acid results from
the methylation of [15] itself, or of [7] with subsequent aromatization. Since methylation of aromatic hydroxyls occurs much
more frequently than that of alicyclic ones, the former alternative seems more probable.

Like [15], gallic acid, [5], is very widely distributed in
the plant kingdom. One important class of tanning materials, the
gallotannins, consists of esters of [5] or some closely related
acids; the galloyl esters of [3] have already been discussed (see
Chapter 4-2). Among lower plants, the mold *Phycomyces blakesleea-*
nus is a well-known producer of both [5] and [15] (106).

In principle, [5] could be formed through three different
pathways: (a) direct dehydrogenation of [7]; (b) hydroxylation of
[15]; or (c) oxidative degradation of the side-chain of a C_6-C_3
precursor. Reaction (a) has its chemical precedent in the smooth
oxidation of [7] to [5] by Fehling's solution (13b).

All three processes have been observed at times. The situation is complicated by species differences and the like; it is discussed in ref. 105. In general, pathway (a) seems to be the preferred one; route (b) has been claimed to occur in some higher
plants, but the evidence is in part contradictory; occurrence of
(c) has been established, but it is quantitatively less important
than (a).

In *P. blakesleeanus*, Haslam, Haworth, and Knowles (13b)
found that [6], [7], and [4] stimulate formation of [5], while
[3] does not. In replacement cultures (i.e., cultures in which
the initial glucose medium was replaced by a fresh one, containing probable precursors, after formation of [5] had been established), the synthesis of [5] was resumed immediately when [7]
was offered, but only after a lag period in case of [6] or [4].

Label from [15] was not incorporated. In a recent, detailed study, Dewick and Haslam (107) found path (a) to be the prevalent mode for the biosynthesis of [5] in a number of higher plants, with (c) likewise occurring, but to a much lesser extent. Label from $1\text{-}^{14}C$-glucose was found predominantly in positions 2 and 6 of [5], as expected for its formation from one of the intermediates of the shikimate pathway (cf. Chapters 3 and 5). The carboxyl of [5] bore little label; it would have been strongly active if it had originated in the β-carbon atom of a $C_6\text{-}C_3$ precursor. Generally labeled [4], isolated from shoots of *Gingko biloba* growing in $^{14}CO_2$ in the light (80a), was incorporated with retention of the isotope distribution; interestingly, this distribution was far from uniform, the carboxyl of the [4] containing about 1/4 instead of the expected 1/7 of the total label. With generally labeled phenylalanine as the substrate, both ring and carboxyl of [5] were active, showing the occurrence of pathways (c).

REFERENCES

1. B.D. Davis, J. Bact. 64, 729 (1952).
2. B.D. Davis and U. Weiss, Arch. Exper. Pathol. Pharmakol. 220, 1 (1953).
3. U. Weiss, B.D. Davis, and E.S. Mingioli, J. Amer. Chem. Soc. 75, 5572 (1953).
4. I.I. Salamon and B.D. Davis, J. Amer. Chem. Soc. 75, 5567 (1953).
5. G.C. Whiting and R.A. Coggins, Biochem. J. 102, 283 (1967).
6. L.K. Evans and A.E. Gillam, J. Chem. Soc. 568 (1943).
7. R. Grewe and J.-P. Jeschke, Chem. Ber. 89, 2080 (1956).
8. U. Weiss and E.S. Mingioli, J. Amer. Chem. Soc. 78, 2894 (1956).
9. K.R. Hanson, J. Chem. Educat. 39, 419 (1962).
10. T. Posternak and D. Reymond, Helv. Chim. Acta 38, 195 (1955).
11. (a) J. Corse, R.E. Lundin, E. Sondheimer, and A.C. Waiss, Jr., Phytochemistry 5, 767 (1966);
    (b) W. Gaffield, A.C. Waiss, Jr., and J. Corse, J. Chem. Soc. (C) 1885 (1966); E. Haslam and M.J. Turner, ibid. 1496 (1971).
12. R. Grewe, W. Lorenzen, and L. Vining, Chem. Ber. 87, 793 (1954).
13. (a) K. Heyns and H. Gottschalck, Chem. Ber. 94, 343 (1961);
    (b) E. Haslam, R.D. Haworth, and P.F. Knowles, J. Chem. Soc. 1854 (1961).
14. R. Grewe and G. Winter, Angew. Chem. 71, 163 (1959).

15. O. Hesse, (a) Ann. 112, 52 (1859); (b) ibid. 200, 232 (1880).
16. R. Fittig and F. Macalpine, Ann. 168, 99, 112 (1873).
17. E. Haslam and J.E. Marriott, J. Chem. Soc. 5755 (1965).
18. T. Posternak and D. Reymond, Helv. Chim. Acta 36, 260 (1953); cf. B. Magasanik and E. Chargaff, J. Biol. Chem. 174, 173 (1948).
19. R. Robinson, J. Chem. Soc. 111, 876 (1917).
20. K.R. Hanson and I.A. Rose, Proc. Natl. Acad. Sci. U.S. 50, 981 (1963).
21. B.W. Smith, M.J. Turner, and E. Haslam, Chem. Commun. 842 (1970); E. Haslam, M.J. Turner, D. Sargent, and R.S. Thompson, J. Chem. Soc. (C) 1489 (1971).
22. S. Mitsuhashi and B.D. Davis, Biochim. Biophys. Acta 15, 54 (1954).
23. D. Balinsky and D.D. Davies, Biochem. J. 80, 300 (1961).
24. O.L. Gamborg, Canad. J. Biochem. 44, 791 (1966).
25. T. Minamikawa, M. Kojima, and I. Uritani, Arch. Biochem. Biophys. 117, 194 (1966).
26. (a) E. Gollub, H. Zalkin, and D.B. Sprinson, J. Biol. Chem. 242, 5323 (1967_;
    (b) K.D. Brown, Genetics 60, 31 (1968);
    (c) E.W. Nester, R.A. Jensen, and D.S. Nasser, J. Bact. 97 83 (1969).
27. N.H. Giles, C.W.H. Partridge, S.I. Ahmed, and M.E. Case, Proc. Natl. Acad. Sci. U.S. 58, 1930 (1967); H.W. Rines, M.E. Case, and N.H. Giles, Genetics 61, 789 (1969).
28. N.H. Giles, M.E. Case, C.W.H. Partridge, and S.I. Ahmed, Proc. Natl. Acad. Sci. U.S. 58, 1453 (1967); L. Burgoyne, M.E. Case, and N.H. Giles, Biochim. Biophys. Acta 191, 452 (1969).
29. S.I. Ahmed and N.H. Giles, J. Bact. 99, 231 (1969).
29a. A. Boudet, FEBS Lett. 14, 257 (1971).
30. B.D. Davis, (a) J. Biol. Chem. 191, 315 (1951); (b) Experientia 6, 41 (1950).
31. F.C. Hofmann, Crell's Ann. 1790 II 314.
32. L.N. Vauquelin, Ann. Chim. [1], 59, 162 (1806).
33. E. Lautemann, Ann. 125, 9 (1863).
34. O. Loew, Ber. 14, 450 (1881).
35. H.O.L. Fischer and G. Dangschat, Ber. 65, 1009 (1932).
36. H.O.L. Fischer and G. Dangschat, Helv. Chim. Acta 17, 1196 (1934).
37. H.O.L. Fischer and G. Dangschat, Helv. Chim. Acta 20, 705 (1937).
38. G. Dangschat and H.O.L. Fischer, Biochim. Biophys. Acta 4, 199 (1950).

39. J. Cleophax, D. Mercier, and S.D. Géro, Angew. Chem., Int. Ed. 10, 652 (1971).
40. R. Grewe and E. Vangermain, Chem. Ber. 98, 104 (1965).
41. R. Grewe, H. Büttner, and G. Burmester, Angew. Chem. 69, 61 (1957).
42. R. Grewe and W. Lorenzen, Chem. Ber. 86, 928 (1953).
43. (a) E.E. Smissman and M.A. Oxman, J. Amer. Chem. Soc. 85, 2184 (1963);
    (b) J. Wolinsky, R. Novak, and R. Vasileff, J. Org. Chem. 29, 3596 (1964).
44. H.J. Bestmann and H.A. Heid, Angew. Chem., Int. Ed. 10, 336 (1971).
44a. R.M. Baldwin, C.D. Snyder, and H. Rapoport, J. Amer. Chem. Soc. 95, 726 (1973).
45. E. Sondheimer, Botan. Rev. 30, 667 (1964).
46. R.A. Cartwright and E.A.H. Roberts, Chem. Ind. (London) 230 (1955); E.A.H. Roberts and M. Myers, J. Sci. Food Agri. 9, 701 (1958); G.V. Stagg and D. Swaine, Phytochemistry 10, 1671 (1971).
47. E. Haslam, R.D. Haworth, and P.C. Keen, J. Chem. Soc. 3814 (1962).
48. E.A.H. Williams, Chem. Ind. (london) 1200 (1958).
49. E. Haslam, R.D. Haworth, and G.K. Makinson, J. Chem. Soc. 5153 (1961).
50. H.R. Schütte, W. Langenbeck, and H. Böhme, Naturwissenschaften 44, 63 (1957).
51. H.G.C. King and F. White, Proc. Chem. Soc. 341 (1957).
52. G.K. Sutherland and W.A. Gortner, Austral. J. Chem. 12, 240 (1959).
53. W. Karrer, "Konstitution und Vorkommen Organischer Pflanzenstoffe", Birkhäuser Verlag, Basel and Stuttgart, 1958 p. 395
54. K. Gorter, Ann. 358, 327; 359, 217 (1908).
55. H.O.L. Fischer and G. Dangschat, Ber. 65, 1037 (1932).
56. L. Panizzi, M.L. Scarpati, and G. Oriente, Gazz. Chim. Ital. 86, 913 (1956); J. Corse, Nature 172, 771 (1953).
57. E. Sondheimer, Arch. Biochem. Biophys. 74, 131 (1958).
58. M.L. Scarpati and P. Esposito, Tetrahedron Lett. 1147 (1963); Ann. Chim. 54, 35 (1964).
59. L.I. Dranik and V.T. Chernobai, Rast. Resur. 3, 250 (1967); Chem. Abstr. 68, 910a (1968).
60. (a) H.M. Barnes, J.R. Feldman, and W.V. White, J. Amer. Chem. Soc. 72, 4178 (1950);
    (b) K.R. Hanson and M. Zucker, J. Biol. Chem. 238, 1105 (1963);
    (c) M.L. Scarpati and M. Guiso, Ann. Chim. 53, 1315 (1963).
61. M.L. Scarpati and M. Guiso, Tetrahedron Lett. 2851 (1964).

62. Y. Inoue, S. Aoyagi, and K. Nakanishi, Chem. Pharm. Bull. 13, 100 (1965).
63. L. Panizzi and M.L. Scarpati, (a) Gazz. Chim. Ital. 95, 71 (1965); (b) ibid. 84, 792 (1954).
64. M.L. Scarpati and G. Oriente, Ann. Chim. 47, 155 (1957).
65. (a) G. Pictet and H. Brandenberger, J. Chromatogr. 4, 396 (1960); (b) O. Goldschmid and H.L. Hergert, Tappi 44 858 (1961).
66. J. Corse, E. Sondheimer, and R.E. Lundin, Tetrahedron 18, 1207 (1962).
67. O.L. Gamborg, Can. J. Biochem. 45, 1451 (1967).
68. C.C. Levy and M. Zucker, J. Biol. Chem. 235, 2418 (1960).
69. K.R. Hanson, Phytochemistry 5, 491 (1966).
70. K.R. Hanson and M. Zucker, J. Biol. Chem. 238, 1105 (1963).; V.C. Runeckles, Can. J. Biochem. Physiol. 41, 2249 (1963).
71. W. Steck, Phytochemistry 7, 1711 (1968).
72. M. Legrand, B. Fritig, and L. Hirth, Compt. Rend. 273, 525 (1971).
73. S. Mitsuhashi and B.D. Davis, Biochim. Biophys. Acta 15, 268 (1954).
74. S. Yoshida, Ann. Rev. Plant Physiol. 20, 41 (1969).
75. O.L. Gamborg, Biochim. Biophys. Acta 128, 483 (1966).
76. L.N. Ornston and R.Y. Stanier, J. Biol. Chem. 241, 3776 (1966), and refs. there; L.N. Ornston, ibid. 3787, 3800
77. T.N. Tateoka, Botan. Mag. Tokyo 78, 294 (1965); 81 103 (1968); Chem. Abstr. 68, 112196d (1968).
78. J.A. Valone, Jr., M.E. Case, and N.H. Giles, Proc. Natl. Acad. Sci. U.S. 68, 1555 (1971).
79. M.E.F. Tresguerres, G. de Torrentegui, W.M. Ingledew, and J.L. Cánovas, Eur. J. Biochem. 14, 445 (1970).
80. L.H. Weinstein, C.A. Porter, and H.J. Laurencot, Jr., (a) Contrib. Boyce Thompson Inst. 20, 121 (1959); (b) ibid. 21, 201 (1961); (c) Nature 194, 205 (1962).
81. R. Rohringer, A. Fuchs, J. Lunderstädt, and D.J. Samborski, Can. J. Botan. 45, 863 (1967).
82. J.G. Carr, A. Pollard, G.C. Whiting, and A.H. Williams, Biochem. J. 66, 283 (1957).
83. G.C. Whiting and R.A. Coggins, Biochem. J. 115, No. 5, 60P (1969); J. Sci. Food Agr. 24, 897 (1973).
84. R. Cotran, M.I. Kendrick, and E.H. Kass, Proc. Soc. Exp. Biol. Med. 104, 424 (1960); A.M. Asatoor, Biochim. Biophys. Acta 100, 290 (1965); R.H. Adamson, J.W. Bridges, M.E. Evans, and R.T. Williams, Biochem. J. 116, 437 (1970).
85. I.I. Salamon, unpublished observations; cf. also B.D. Davis, in Symposium sur le Métabolisme Microbien, IIe Congrès Inter-

nat. de Biochimie, Paris, 1952, p. 32.
86. H.O.L. Fischer and G. Dangschat, Helv. Chim. Acta 18, 1206 (1935).
87. E. Haslam, R.D. Haworth, and P.F. Knowles, in: Methods of Enzymology, Vol. VI, S.P. Colowick and N.O. Kaplan, Eds., Academic, New York and London, 1963, p. 498.
88. S. Hattori, S. Yoshida, and M. Hasegawa, Arch. Biochem. Biophys. 74, 480 (1958).
89. M. Nandy and N.C. Ganguli, Arch. Biochem. Biophys. 92, 399 (1961).
90. G. Ehrensvärd, Svensk Kemisk Tidskr. 66, 249 (1954).
91. (a) E.L. Tatum, S.R. Gross, G. Ehrensvärd, and L. Garnjobst, Proc. Natl. Acad. Sci. U.S. 40, 271 (1954);
    (b) R.L. Metzenberg and H.K. Mitchell, Biochem. J. 68, 168 (1958).
92. P. Simonart and A. Wiaux, Nature 186, 78 (1960).
93. A. Boudet, P. Gadal, G. Alibert, and G. Marigo, Compt. Rend. Ser. D 265, 119 (1967).
94. B.D. Davis, J. Bact. 64, 749 (1952).
95. B.J. Wallace and J. Pittard, J. Bact. 99, 707 (1969).
96. H. Yaniv and C. Gilvarg, J. Biol. Chem. 213, 787 (1955).
97. H.O.L. Fischer and G. Dangschat, Helv. Chim. Acta 17, 1200 (1934).
98. S.R. Gross, J. Biol. Chem. 233, 1146 (1958).
99. (a) D. Balinski and D.D. Davies, Biochem. J. 80, 292 (1961);
    (b) D. Balinski, A.W. Dennis, and W.W. Cleland, Biochemistry 10, 1947 (1971).
100. G.W. Sanderson, Biochem. J. 98, 248 (1966).
101. D.B. Sprinson, Adv. Carbohydrate Chem. 15, 235 (1960).
102. E.L. Tatum and S.R. Gross, J. Biol. Chem. 219, 797 (1956).
103. K.H. Scharf, M.H. Zenk, D.K. Onderka, M. Carroll, and H. G. Floss, Chem. Commun. 765 (1971).
104. A.J. Pittard, F. Gibson, and C.H. Doy, Biochim. Biophys. Acta 57, 290 (1962).
105. M.H. Zenk, in Pharmacognosy and Phytochemistry, 1st. Int. Congress, Munich, 1970, H. Wagner and L. Hörhammer, Eds., Springer-Verlag, Berlin, Heidelberg, New York, 1971, p. 314.
106. K. Bernhard and H. Albrecht, Helv. Chim. Acta 300, 627 (1947); H.-B. Schröter, Die Kulturpflanze, Beiheft 1, 49 (1956).
107. P.M. Dewick and E. Haslam, Biochem. J. 113, 537 (1969).

CHAPTER 5

Alicyclic Intermediates: Shikimic Acid and Derivatives

5-1. SHIKIMIC ACID

As the only previously known one among the intermediates in the biosynthesis of aromatic metabolites from glucose, shikimic acid [1] occupies a key position. It has been pointed out earlier (Chapter 2) that it was this compound whose utilization by bacterial mutants was the starting point (1) for the subsequent exploration of the biosynthetic pathway usually named after it. The enzymatic formation of [1] from its immediate precursor 3-dehydroshikimic acid [2] has been discussed in Section 4-3.

An excellent, comprehensive review of the chemistry of [1] has been published by Bohm (2) in 1965; it should be consulted for details that cannot be given here.

Shikimic acid, [1], $C_7H_{10}O_5$, was first isolated in 1885 by Eijkman (3) from the fruits of the Japanese shikimi tree or Japanese (false) star anise, *Illicium religiosum*, where it occurs in surprisingly large amounts, constituting up to about 20% of the dry matter. It was subsequently isolated from the same source by Chen (4), by H.O.L. Fischer and Dangschat (5) and, with improved technique using ion exchange chromatography, by Grewe and Lorenzen (6). Freudenberg and Geiger (7) prepared it from the fruits of the true or Chinese star anise, *I. verum*; however, these fruits contain much less of the compound (3).

For many years following its discovery, [1] had been encountered only in plants of the genus *Illicium*, a genus formerly classed among the Magnoliaceae but at present usually considered to constitute a separate family, Illiciaceae. This seemingly restricted occurrence of [1] led to its being at times classified as a mere "laboratory curiosity"--a striking example of the danger of using this jargon term in the absence of adequate knowledge. After the recognition by Davis (1) of its vital role in

103

the biosynthesis of aromatic metabolites in bacteria had stimulated great interest in the compound, the presence of [1] in a
large number of higher and lower plants was demonstrated. In
1965, a list (2) cited 159 species of vascular plants, out of 278
examined, which were found to contain [1]. Almost all classes
are represented. The distribution of [1] in Gymnosperms has been
studied extensively by Hattori and co-workers (8a) and by Plouvier (8b). However, no plant so far seems to equal *I. religiosum*
in the amounts of [1] present. Instances of isolation of [1] in
pure form from higher plants are too numerous to list.

In microorganisms, [1] is likewise widely distributed. Its
isolation from culture filtrates of *E. coli* in the pure form required for subsequent labeling studies is described by Srinivasan
and co-workers (9); this work was fundamental for the recognition
of the way in which [1] is derived from intermediates of carbohydrate metabolism (see Chapter 2).

The problem of the chemical structure of [1] was attacked
and almost solved by its discoverer only a few years after the
first isolation of the compound. In a remarkable paper, Eykman
(10) described reactions defining the functionality of [1], such
as formation of triacyl derivatives, a dibromide, and a dihydro
derivative; he also reported the acid-catalyzed transformation of
[1] into *p*-hydroxybenzoic acid. These findings were interpreted
correctly as showing that [1] is a trihydroxycyclohexene carboxylic acid related to the long-known quinic acid. An initial choice
of six possible structures was narrowed down to two, one of them
the structure now known to be correct--no mean achievement at that
early date. However, in an attempt to decide between the two remaining structures on the basis of electrical conductivity, Eykman was led to prefer the incorrect alternative; it differs from
the correct one only by placement of one of the hydroxyls in position 6 (new numbering) instead of 5.

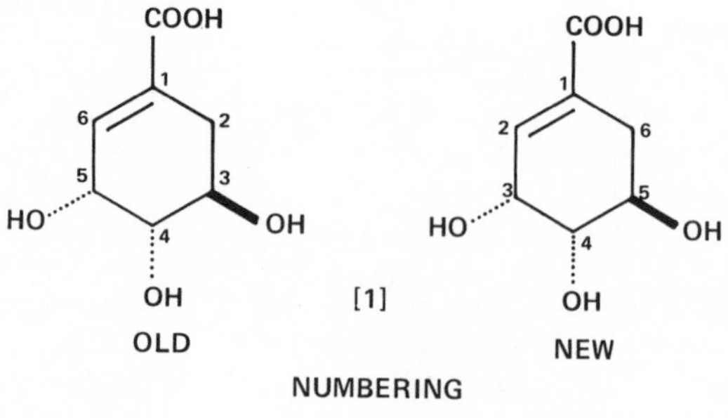

OLD   [1]   NEW

NUMBERING

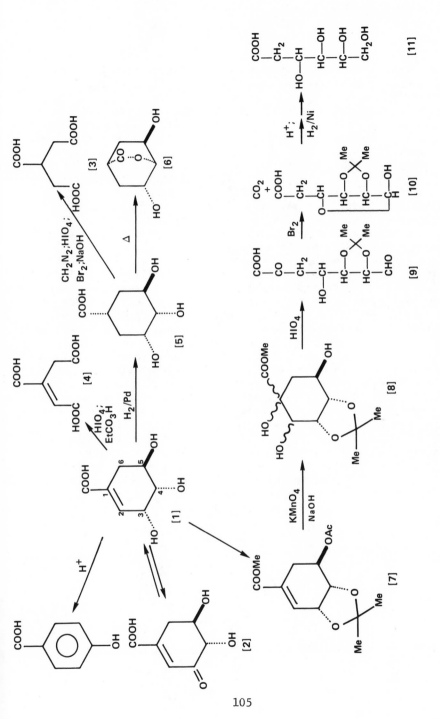

Scheme I

It should be noted that here, as elsewhere, the new, rational numbering scheme for [1] is used, which differs from the traditional one by interchanging positions 2 and 6, and 3 and 4, respectively.

The definitive elucidation of structure and stereochemistry of [1] is due to the classical work of H.O.L. Fischer and Dangschat from 1934 on. The salient features of this work are shown in Scheme I.

At first, the position of the hydroxyls relative to the carboxy group was defined by the periodate cleavage of methyl dihydroshikimate (see below) and of [1]; after oxidation of the resulting dialdehydes and saponification, tricarballylic [3] and *trans*-aconitic acid [4], respectively, were obtained (5, 11). Shikimic acid is thus one of the stereoisomers of 3,4,5-trihydroxy-Δ^1-cyclohexene carboxylic acid.

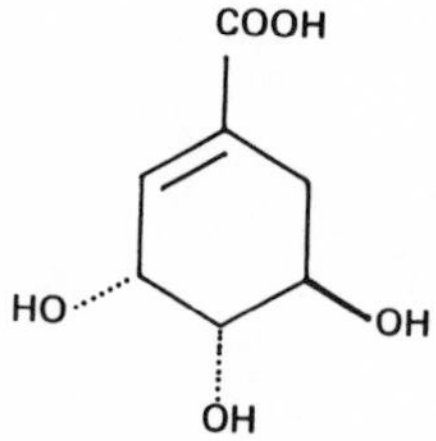

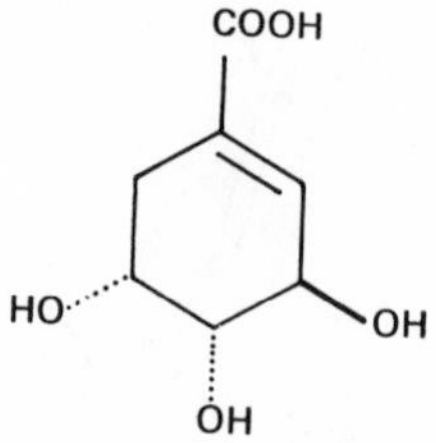

Information on the *relative* configuration of [1] was next derived (12) through investigation of dihydro-[1] = [5]. This acid, first obtained by Eykman (3), is readily prepared by catalytic hydrogenation of [1]. Heating of [5], or distillation of its ester, gave the crystalline, optically active lactone [6], which was resistant to lead tetraacetate. Hence the hydroxyl in position 4 is involved in lactone formation, and the optical activity of the otherwise symmetrical structure proves that its two remaining hydroxyls must have different configurations. That one of them must be *cis* to the hydroxyl at position 4 also follows from the ready formation of cyclic acetals, such as isopropylidene derivatives, by [1] and related compounds under conditions where only *cis*-1,2-cyclohexandiols would react. The resulting relative configuration of the hydroxyl groups is compatible with [1a] and [1b] or their antipodes. The problem of the location of the double bond was solved, and at the same time the absolute stereochemistry of [1] established, through its degradation to a derivative of glucose by Fischer and Dangschat (13). Shikimic acid was converted to the protected derivative [7], whose double bond was hydroxylated with permanganate to give [8]. After alkaline saponification of the ester groups, the cyclohexane ring was cleaved with periodate, one molecule of which was consumed to give the glucoheptonic acid derivative [9], which was not isolated. Action of bromine in acetic acid gave [10] with loss of CO_2. Reduction of the potential aldehyde group of this compound over Raney nickel yielded 2-deoxygluconic acid ("glucodesonic acid"), [11], identical with authentic material from D-glucose. This series of transformations establishes the location of the double bond, and connects [1] in a beautifully direct and unequivocal way with D-glucose, and thus with D-(+)-glyceraldehyde, the primary reference compound. Hence the structure and stereochemistry of [1] are correctly represented by formula [1a]. Fischer and Dangschat assumed, plausibly enough, that the biosynthesis of [1] proceeds with incorporation of the intact carbon chain and hydroxyls of glucose, without change of stereochemistry; this assumption was the basis for their numbering of [1], now abandoned, in which the carbons of [1] corresponding configurationally to C-3, 4, and 5 of glucose were given these numbers.

The proof of stereochemistry of [1] also applies to quinic acid through the multiple interrelations of the two acids (see Section 4-2).

The dihydroshikimic acid [5] is not identical with the compound produced from [1] (or quinic acid; see Section 4-2) by *Lactobacillus pastorianus* var. *quinicus* (14); the latter acid is known from the work of Grewe and Lorenzen (6) to be the 1-epimer of [5].

Through its unequivocal connection with the primary standard of configurational relationships, [5] has played a major role in the elucidation of the absolute stereochemistry of the terpenoids and steriods. Freudenberg and co-workers (7, 15) succeeded in converting [5] into the simplest optically active hydrocarbon. 3-methylhexane (methyl-ethyl-propyl-methane) by methods (periodate cleavage of the 3-acetate of [5], followed by reductive removal of the oxygens) which left the configuration at C-1 untouched. This series of reactions made available a key reference compound of securely established absolute configuration. 3-Methylhexane had been interrelated before in unequivocal ways with optically active amyl alcohol, isoleucine, and numerous terpenes; through one of these, citronellal, the steroids had been attached to this family of compounds. However, it had not been possible to connect this vast group of substances with D-glyceraldehyde. This gap was closed by the work of Freudenberg and his associates on [5].

The conformation of [1] in solution was established by Hall (16) through nmr spectroscopy. As expected, one of two possible half-chairs is the preferred conformation. This work constitutes the first complete nmr analysis of a cyclohexene.

Among numerous chemical transformations of [1] which have been studied, (cf. ref. 2), the work of Plieninger and Schneider (17) on the conjugate addition of ammonia deserves mention; the two stereoisomeric 2-amino-dihydroshikimic acids which were obtained might be of interest in connection with the biosynthesis of anthranilic acid from [1].

Several total syntheses of [1] have been published. Most of them are designed in ways which would make them suitable for the preparation of [1] specifically labeled with ^{14}C; with one exception, they use the Diels-Alder reaction for the construction of the ring.

The first of these synthetic sequences was worked out, about simultaneously and with only minor differences, by Smissman and co-workers (18), and by McCrindle, Overton, and Raphael (19); it is shown in Scheme II. This synthesis is based upon the reaction of trans,trans-1,4-diacetoxybutadiene with acrylic acid (19) or methyl acrylate (18) to give the diacetoxycyclohexane carboxylic acid [1] (or its methyl ester). Hydroxylation with OsO4, methylation with diazomethane (in the case of the free acid), and formation of the acetonide yielded [13]. Elimination of the acetoxyl at C-2 gave the racemate of [7] (see Scheme I), which was converted to D,L-[1]. Resolution of the racemate completed the synthesis. The stereochemistry at C-1 of some of the intermediates in this synthesis had initially been in doubt, but has subsequently been proved by nmr measurements to be the one shown in the scheme (20).

A synthesis explored in its early stages by McCrindle and co-workers (19) and later carried to successful completion by Doshi (21) differs from the preceding one by starting from 3-acetoxy-3, 6-epoxycyclohex-4-ene 1,2-dicarboxylic anhydride, the known Diels-Alder adduct from 2-acetoxyfuran and maleic anhydride; further transformations yielded a methyl tetra-acetoxycyclohexane carboxylate, which on heating in the presence of powdered soft glass gave methyl triacetyl-D,L-shikimate.

Two syntheses by Grewe and co-workers begin with 1,4-cyclohexadiene carboxylic acid [14], the Diels-Alder adduct from propiolic acid and butadiene; they are shown in Scheme III. In the earlier one of these syntheses (22), the methyl ester of [14] was made to react with performic acid, which attacked only the isolated double bond. Acetylation of the resulting trans-diol, and reaction with N-bromosuccinimide followed by silver acetate in presence of one mole of water (conditions known to lead to inversion) gave the

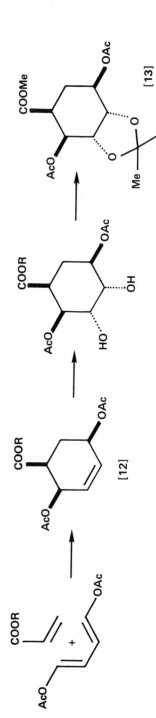

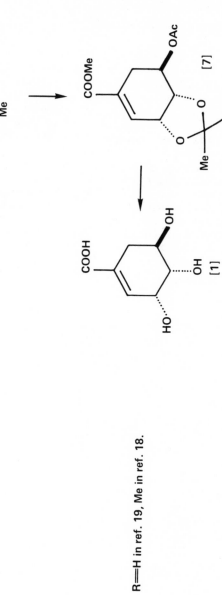

R=H in ref. 19, Me in ref. 18.

Scheme II

109

methyl ester of D,L-[1] in good overall yield. The 3-epimer of
[1] and two stereoisomeric 4,5,6-trihydroxycyclohexene carboxylic
acids were present as minor by-products.
   The second synthesis, worked out by Grewe and Kersten (23),
is of remarkable simplicity, giving high yields of [1] with isolation of only two intermediates, and without the use of chromatography. Here, acetoxylation of [14] with silver acetate and
iodine gave the *cis*-diol diacetate, isolated as the methyl ester
[15]. Reaction with N-bromosuccinimide, followed by potassium
acetate, and removal of the acetyl groups, gave the crystalline
methyl D,L-4-epishikimate [16]. Treatment of its triacetate with
liquid HF led to inversion of the configuration of the central one
of the three adjacent acetoxylated carbon atoms. This general
reaction of vicinal *cis,trans*-triacyloxy systems had been discovered by Hedgley and Fletcher (24), and yielded in the present
case the methyl ester of D,L-[1].

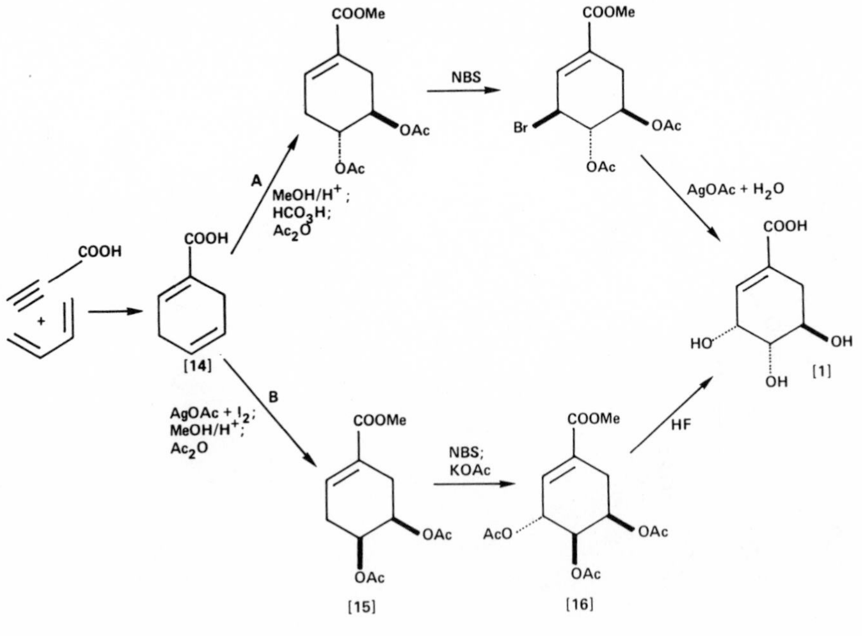

Scheme III

Finally, Bestmann and Heid's synthesis of quinic acid from D-arabinose (see Section 4-2) can be modified in its later stages to give [1] instead, if the intermediate cyanohydrin of 3,4,5-triacetoxycyclohexane is dehydrated with POCl₃, following the method of Grewe and Vangermain (26).[+]

In Chapter 2, the studies of Sprinson and co-workers on the incorporation of specifically labeled glucose into [1] have been discussed (see especially Scheme I there). In view of the great importance of this work for the elucidation of the earliest steps of the shikimate pathway, the very complete degradation scheme developed (9) for the localization of activity in the resulting labeled samples of [1] deserves brief discussion. The reactions used are shown in Scheme IV. It will be seen from it that three of the seven carbons of [1] can be obtained individually: the carboxyl, C-7, by decarboxylation; C-4 by periodate cleavage (11); C-6 as iodoform by degradation (NaIO₄, NaOI) of the product of hydroxylation of [1] with OsO₄. The other carbon atoms are accessible only in various combinations, which will be obvious from Scheme IV. The activity of C-1 of [1], for example, can be determined from the one of the oxalic acid (C-1 + 7) by subtraction of that of C-7 determined individually. The three-carbon tautomerism common in such compounds leads to the indicated equalization of C-2 and 6, and C-3 and 5, in aconitic acid [4] and the fragments derived from it. The actual occurrence of this equalization in the present case was established by separate experiments.

Natural derivatives of [1] have been encountered much less frequently than their analogs derived from quinic acid (cf., Section 4-2); they have apparently not been studied to any great extent. The first indication of the occurrence of such substances was obtained by Goldschmid and Hergert (27), who detected compounds tentatively identified as p-coumaroyl, caffeoyl, and feruloyl shikimates in paper chromatograms of extracts from the cambium of hemlock, *Tsuga canadensis*. Hanson and Zucker (28) observed similar substances in slices of potatoes supplied with phenylalanine and quinic acid. The best-investigated compound of this group is dactylifric acid, isolated in pure form by Maier and co-workers (29) from fresh green dates (fruit of *Phoenix dactylifera*). Hydrolysis of the compound yielded [1] and caffeic (3,4-dihydroxycinnamic) acids in equivalent amounts; enzymatic hydrolysis after treatment with acetone and HCl gave caffeic acid and the 3,4-isopropylidene derivative of [1]. Dactylifric acid is thus

---

[+]
But see the footnote on page 82.

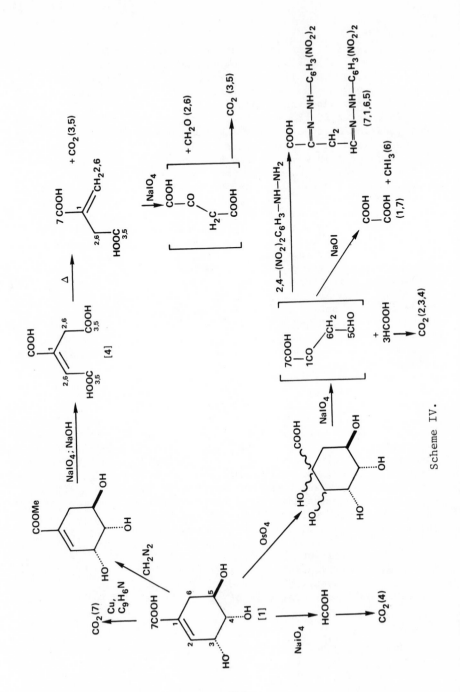

Scheme IV.

112

5-caffeoyl-[1], an interpretation confirmed by total synthesis
from 3,4-isopropylidene-[1] and the diphenylmethylene acetal of
caffeoyl chloride, and removal of the protecting groups. Two more
conjugates of [1] with caffeic acid were likewise isolated from
dates (29). They were named iso- and neodactylifric acids, and
may be 3- and 4-caffeoyl-[1], respectively.

The most important *in vivo* transformation that shikimic acid
[1] undergoes is its phosphorylation to 3-phosphoshikimic acid
[17], the next intermediate in the shikimate pathway. The chemistry of this compound will be discussed in Section 5-2. This process, however, is not the only biochemical transformation of [1].
One other such reaction has already been mentioned: the reduction,
by *Lactobacillus pastorianus* var. *quinicus* (14), to the dihydro-[1]
which corresponds stereochemically to quinic acid. Another
formation of alicyclic compounds from [1] was discovered recently
by de Rosa, Bu'Lock, and co-workers (30a), who found that a thermophilic bacillus closely resembling B. *acidocaldarius* produces
large amounts of two unusual ω-cyclohexyl fatty acids, 11-cyclo-
hexylundecanoic and 13-cyclohexyltridecanoic acid. Uniformly
labeled [1] is efficiently and selectively incorporated into these
acids, while co-occurring acyclic fatty acids contain very little
label (30b). Labeled phenylalanine is hardly utilized for the
biosynthesis of either type. Degradation of the cyclic acids
formed from [1] gives benzoic acid which contains more than 96% of
the label, close to 1/7 of it in the carboxyl. This result shows
that [1] is incorporated as an intact C7 unit with retention of
the carboxyl, a finding that precludes a derivation from the major
branches of the shikimate pathway, where this carboxyl is lost
during the transformation of prephenic acid into phenylalanine and
tyrosine (see Chapter 7), and of anthranilic acid into indole
(Chapter 9). The alicyclic acids might thus be formed through reduction of some aromatic structure biosynthesized from [1] by one
of the minor pathways in which the carboxyl is retained (cf.,
e.g., Chapters 10, 11, 12, 16; etc.); alternatively, they
might be formed more directly by a sequence which does not proceed
through intermediate aromatic stages.

The abundant proof for the role of [1] as an intermediate in
the biosynthesis of the aromatic amino acids and vitamins in microorganisms has already been discussed in Chapter 2. Key pieces of
evidence were, of course, the ability of [1] to substitute for
these essential metabolites in bacterial mutants, and its excretion
by mutants blocked after it. This role has been further confirmed
in a substantial number of cases by proof that [1] itself is specifically incorporated into some aromatic compound, or that known
precursors of [1] are utilized in a manner consistent with the

known pattern of their incorporation into [1]. Only a few selected examples of such observations can be given here.

In microorganisms (yeast, *E. coli*), specifically labeled glucose (1-^{14}C and 6-^{14}C) is incorporated into phenylalanine (31),
tyrosine (9, 31), and anthranilic acid (32) in a way consistent
with the formation of their aromatic rings from [1]. As an example of such a biosynthesis of a secondary mold metabolite, the
incorporation of [1] into the strange fungal pigment volucrisporin,
2,5-bis(*m*-hydroxyphenyl)-1,4-benzoquinone (see Chapter 16) may be
mentioned (33).

Biosynthetic utilization of [1] by higher plants has been established in many instances; several aspects of these questions
have been reviewed by Yoshida (34). The biosynthesis of [1] has
been shown (35) to take place in *Pinus resinosa* from phospho-enolpyruvate and erythrose 4-phosphate, just as it does in microorganisms. Conversion of [1] to phenylalanine and tyrosine has been
demonstrated in *Salvia* (Labiatae) (36), and in wheat and buckwheat,
Fagopyrum (Polygonaceae) (37); tryptophan is biosynthesized from
[1] in barley shoots (38). Particularly detailed evidence for the
biosynthetic role of [1] in a higher plant, *Reseda lutea* (Resedaceae), has been obtained very recently by Olesen Larsen, Onderka,
and Floss (39): feeding of uniformly ^{14}C-labeled [1], also labeled
stereospecifically with ^{3}H at the *pro-6S* and *pro-6R* positions, has
shown that the conversion of [1] into phenylalanine and tyrosine
follows the same chemical course as it does in microorganisms, and
that the stereochemistry of the loss of hydrogen from position 6
is likewise the same as that known (40) to occur in bacteria.
Utilization of [1] with retention of the carboxyl for the biosynthesis of *m*-carboxyaryl amino acids (cf. Section 12-5) was shown
to take place in *Reseda* (39). Several substituted cinnamic acids
(*p*-coumaric, caffeic, ferulic, sinapic acids) are biosynthesized
from [1] in *Salvia* (41). Administration of labeled [1] to *Fagopyrum* (42) gave quercetin (3,3',4',5,7-pentahydroxyflavone) labeled exclusively in ring B, ring A being inactive; this result is
in complete agreement with the known biosynthesis of the flavonoids (see Section 21-2), where ring A originates from acetate,
Ring B and the three carbons of the heterocyclic ring from a shikimate-derived C_6-C_3 unit.

Actually, the first evidence for a biosynthetic role of [1]
in higher plants was apparently provided in 1955 by Brown and
Neish (43), who found that wheat and maple *(Acer negundo* var. *in-
terius)* incorporate labeled [1] and phenylalanine into their lignin. It is appropriate that this earliest indication for the occurrence of the shikimate pathway in higher plants should have
been made through study of a process of such enormous importance.

Among all naturally-occurring aromatic products, lignin is by far the one biosynthesized in the largest amounts. It constitutes up to about 40% of wood (44); it occurs almost universally in vascular plants (ref. 44, p. 198) and has also been observed a few times in lower plants (45). A more complete discussion of lignin is given in Chapter 16; it is certainly a polymer built up from C_6-C_3 units, although details of the chemistry of this rather intractable material are not yet clear. The observed (43) utilization of [1] and phenylalanine thus suggests a biosynthesis of lignin via the shikimate pathway.

Subsequent research has confirmed and extended this interpretation. For instance, Eberhardt and Schubert (46) found that in sugar cane, feeding of 2,6-^{14}C-[1] led to lignin which on oxidative degradation gave vanillin labeled almost exclusively in the corresponding positions, 2 and 6. The ring of [1] had therefore been incorporated as a unit, without any significant rearrangement or fragmentation.

These few instances of experimental proof for the utilization of [1] will demonstrate the universal significance of this compound as an intermediate in the biosynthesis of aromatic metabolites by microorganisms and higher plants.

The labeled [1] used for those experiments, and for many others, has been obtained in a variety of ways. Early work, for example that of Eberhardt and Schubert (46), was carried out with samples of [1] isolated from culture filtrates of mutants of *E. coli* grown on suitably labeled glucose ([6-^{14}C]-glucose in this particular case). Perkins and Aronoff (47a), Millican (47b), and Zaprometov (47c) described methods for isolation of labeled [1] from such filtrates. Higher yields of generally labeled [1] (and also quinic acid) can be obtained by the method of Weinstein, Porter, and Laurencot (48), when shoots of *Gingko biloba* are allowed to photosynthesize for several days in an atmosphere of $^{14}CO_2$, and the acids are isolated and separated by ion exchange chromatography. This method has been widely used.

For recent exploration of the stereochemical course of biosynthetic reactions involving [1] and its relatives, samples of [1] specifically and stereospecifically labeled with isotopic hydrogen in various positions were required. Such samples have been prepared by chemical or enzymatic methods. In the former category, Hill and Newkome (40a) obtained the two epimers of 6-^{2}H-[1] by total synthesis (18) using labeled samples of methyl acrylate. Smith and co-workers (49) found that 3-dehydroquinic acid at pH 7.0 specifically exchanges only the 2(S) hydrogen; the resulting deuteriated species was converted enzymatically to 2-labeled [1]. Floss, Onderka, and Carroll (40b), on the other hand, prepared the two epimers of [6-^{3}H]-[1] entirely by enzymatic methods: glucose and mannose, respectively, labeled with ^{3}H in position 1, were converted enzymatically to either of the stereospecifically tritium-labeled phospho-enolpyruvates, which yielded the desired epimers of 6-tritiated [1] on reaction with erythrose 4-phosphate under the influence of cell-free extracts of a mutant of *E. coli* blocked after [1]. Similar enzymatic methods were used to prepare samples of [1] labeled in other positions; this technique seems to be preferable to other methods because of its simplicity, versatility, and the good yields obtained.

It has already been mentioned briefly that the normal sequence of biosynthetic events leads from [1] to its 3-phosphate [17]. The formation of this compound had first been observed by Davis and Mingioli (50), who designated it as "Compound Z2". The enzyme which catalyzes the formation of [17] from [1] and ATP has been named shikimate kinase (51). The reaction was first observed by Kalan (52); the enzyme responsible for it has been studied in some detail by Fewster (51) in extracts from *E. coli* prepared by ultrasonic disintegration. Shikimate kinase was found to require Mg^{2+} or Mn^{2+}; it was partially inhibited by iodoacetate but was insensitive to phenylmercuric acetate. Enzymatic activity was found in several other microbial species capable of synthesizing their aromatic amino acids; it was absent from others that lack this ability. Shikimate kinase has subsequently been studied in a variety of bacteria and fungi. It has been observed in higher plants only recently by Ahmed and Swain (53), who found it in extracts from seedlings of pea *(Pisum sativum)* and mung bean *(Phaseolus mungo)*.

In microorganisms, the enzymatic reaction leading from [1] to [17] shows several unusual features distinguishing it from most of the other reactions of the common sequence in the shikimate pathway. It had been noted in the initial studies of Davis and Mingioli (50) that none of the available mutants of *E. coli* and related species seemed to be blocked in this reaction. No strain of these organisms was found which would excrete [1] alone, unaccompanied by [17]; only one exception, a strain of *Bacillus subtilis*, is briefly mentioned. On the basis of these observations, it seemed doubtful at first whether [17] is an intermediate in the shikimate pathway at all; however, these doubts were subsequently removed (see Section 5-2); furthermore, mutants of *Neurospora* blocked between [1] and [17] have actually been obtained by Giles and co-workers (54). Yet, detailed studies in the years following the first investigation (50) did reveal further peculiarities in this reaction. The curiously involved situation cannot yet be interpreted in a coherent manner and here, as in other cases (see Chapter 15), different organisms have developed a wide variety of ways for carrying out or controlling a seemingly simple reaction. A brief account of the more significant observations appears warranted; many of them seem to point to some regulatory processes operating on this step of the common sequence. Cases of feedback inhibition and repression, of occurrence of isoenzymes, and of enzyme complexes have been observed.

Feedback inhibition and repression are usually restricted to the first enzyme of a given pathway or branch. Marked effects of this type would thus not be expected for steps 2 to 7 (dehydroquinate synthetase to chorismate synthetase, respectively) of the common sequence of the shikimate pathway. This is indeed so for most of these enzymatic reactions (55a,b), but shikimate kinase, the fifth enzyme of this sequence, is exceptional. The activity of this enzyme in E. coli (51a) and in Salmonella typhi-murium (56) was found to be essentially independent of the aromatic endproducts, as might have been anticipated. However, under certain conditions a significant influence of tyrosine and tryptophan (but not of phenylalanine) upon the levels of enzymatic activity in mutants of E. coli can be observed (55a). In B. subtilis, feedback inhibition by chorismic and prephenic acids (57a) and repression by tyrosine (57a,b) occur; in this organism, however, the situation is involved since its shikimate kinase exists as a complex with two other enzymes (58) (see below).

In E. coli, indications for the existence of two isoenzymes have been obtained (59). The shikimate kinase of S. typhi-murium has actually been separated into two isoenzymes through chromatography on diethylaminoethyl cellulose by Morell and Sprinson (60), who also succeeded in purifying the major, more stable one of these enzymes over a hundredfold. The existence of these isoenzymes could explain the failure (50) to isolate mutants of either species blocked in the conversion of [1] into [17].

The shikimate kinase of Neurospora and certain other fungi is a constituent of the well-known complex of enzymes 2 to 6 of the shikimate pathway (54, 61) which has been mentioned several times in earlier chapters.

In B. subtilis, shikimate kinase forms a complex with two other enzymes, DAHP synthetase and chorismate mutase; this complex has been studied by Nester and co-workers. The three enzymes were initially found (58) to be consistently eluted together from diethylaminoethyl cellulose or Sephadex, and to be inhibited and repressed in the same manner by chorismate or prephenate, and tyrosine, respectively (57). A recent detailed study by Nakatsukasa and Nester (62) has shown that actually DAHP synthetase and chorismate mutase form a very tightly bound complex which can be obtained in highly purified form, while the shikimate kinase is more loosely associated and may be separated from the two other enzymes by chromatographic techniques. The kinase is, however, active only in the presence of the binary complex, of which chorismate mutase seems to be the decisively important constituent; mutants lacking this enzyme invariably also lack the two other enzymatic activities.

Shikimate kinase thus shows a number of peculiarities of the kind usually exhibited only by the first enzyme of a sequence, and assumed to play a role in the control of this sequence. It is difficult to rationalize these observations on the basis of present knowledge, but it may be permissible to speculate that perhaps some as yet unrecognized biosynthetic or degradative pathway branches off from [1]. The recent work on ω-cyclohexyl fatty acids (30), which has been discussed before, may suggest that the search for products of such a pathway need not necessarily be restricted to aromatic compounds.

In the sequence of common intermediates, [1] is followed by two compounds which are derivatives of it. The existence of these conjugates was first observed by Davis and Mingioli (50), who found that autoclaving of culture filtrates of bacterial mutants blocked after [1] often led to a marked increase of growth-factor activity for strains with earlier blocks. Chromatographic and bio-autographic investigation of filtrates of this type from *E. coli, S. typhi-murium,* and *A. aerogenes* revealed the existence of two such conjugates having different stability and chromatographic mobility. Both were inactive as growth factors. Autoclaving at pH 6-7, or treatment with acid, released biologically active material which was identified as [1] by chromatographic techniques and specific mutant response. The less polar, more acid-sensitive compound was provisionally designated as "Compound Z_1", the highly polar, less readily hydrolyzable one as "Compound Z_2". Both compounds were stable to autoclaving at pH 9 or 14 for 30 minutes. It was found that some mutants of the species investigated accumulate Z_2, and small amounts of [1], without any detectable trace of Z_1; in contrast, all those strains which excrete Z_1 also accumulate Z_2 and [1]. These facts indicate that Z_1 must be formed in a later reaction than Z_2.

Subsequent research (see Sections 5-2 and 5-3) has shown that Z_2 is identical with shikimic acid 3-phosphate [17]; it is a true intermediate in the pathway of biosynthesis. In contrast, Z_1 [18] turned out to be a secondary transformation product of the actual intermediate [19], which was discovered only later. The latter compound has been proven to be the 5-enolpyruvyl derivative of [17], while [18] is the corresponding dephosphorylated compound. It will be noted that the correct numbering for [1] and the com-

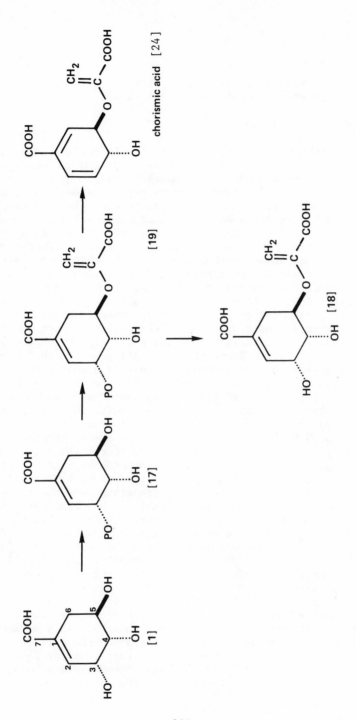

chorismic acid [24]

119

pounds derived from it is used here, while the older, traditional one appears in much of the original literature.

5-2. SHIKIMIC ACID 3-PHOSPHATE

This compound ("Compound Z_2"), initially designated as the 5-phosphate, was isolated in form of the pure, although noncrystalline, K and Ba salts by Weiss and Mingioli (63) from culture filtrates of mutants of *Aerobacter aerogenes*; accumulations of up to 1 g/liter were obtained. From such filtrates, acidified to pH 3, the compound was adsorbed on charcoal columns, which were washed with water, 5%, and 10% ethanol, all acidified with HCl; without this acidification, the substance is eluted even by water alone, a peculiarity not shown by the isomeric 5-phosphate which was synthesized for comparison (see below). Washing with neutral 5% ethanol next eluted the desired compound, which was purified further by precipitation from concentrated alcoholic solution as the brucine salt. Removal of the brucine in aqueous solution as the insoluble picrate, and evaporation of the filtrate gave the free acid as a noncrystallizable glass, which was converted to a solid, noncrystalline K salt. Elementary analysis revealed the then-unexpected presence of phosphorus and gave values in close agreement with those calculated for the mono-hydrate of the mono-potassium salt of a phosphate of [1]. In agreement with this, treatment with potato phosphatase gave equimolecular quantities of inorganic phosphate and free [1], which was isolated in pure form. Hydrolysis with 0.1 N HCl at 120° yielded the same products, but the amount of [1] obtained never exceeded 70% of the theoretical one even on prolonged autoclaving, while that of phosphate reached 90%.

The task of establishing the structure of Z_2 thus resolved itself to that of locating the phosphate group. Of the four possible points of attachment, the carboxyl can be ruled out, since such an acyl phosphate would be much more sensitive than the relatively stable Z_2, and would be resistant to phosphorylase. Uptake of one mole of periodate eliminated the 4-phosphate [20]. Compound Z_2 can thus be the 3- or the 5-phosphate. An authentic sample of the latter [21] was synthesized from the methyl ester of 3,4-isopropylidene-[1], [22] (see Scheme V) by treatment with $POCl_3$ in pyridine, removal of the protecting groups with H^+ and OH', and isolation as the noncrystalline trihydrate of the Ba salt. It proved to be different from the natural phosphate in a variety of chemical tests. Compound Z_2 is thus the 3-phosphate [17]. Most of the tests mentioned were based on the fact that the 5- but not the 3-phosphate contains a pair of vicinal hydroxyls in *cis*-relationship. For steric reasons, and on the basis of much analogy,

only this *cis*-1,2-glycol grouping should readily form cyclic derivatives or metal complexes, while the corresponding *trans*-group-
ing in [17] should be inert. This was indeed found to be the case.
Thus, cupric acetate and NaOH gave a deep blue solution of the
copper complex with [21], and a blue precipitate of cupric hydroxide with [17]. The latter remained unchanged under conditions
where [21] gave an isopropylidene derivative (which was not isolated as such but demonstrated by enzymatic conversion to isopro-
pylidene-[1] and paper-chromatographic identification).

A further test was based on the well-known acid-catalyzed migration of the phosphate residue in mono-phosphorylated 1,2-gly-
cols; it has been demonstrated (64) that this migration proceeds
via a cyclic phosphate which, in a cyclohexene glycol, can be expected to form only between adjacent *cis*-hydroxyls. Consequently,
[21] should remain unchanged on brief refluxing in 80% acetic acid,
a treatment known (64c) to be suitable for phosphate migration;
this was indeed found to be the case. In contrast, [17] with its
hydroxyl at position 4 in *cis*-relation to the phosphoryloxy group
at C-3 was found to be converted to the extent of about 20% to the
4-phosphate [20]. This compound was not isolated, but was shown
to be present by its resistance to periodate: treatment with this
reagent, followed by action of phosphorylase and bioassay with a
suitable mutant, demonstrated the survival of about 20% of [1],
while none survived in the case of [21]. (Phosphate migration is
an equilibrium reaction; it was, however, not determined whether
equilibrium had been reached in the present case.)

Another test rests on the fact that periodate-cleavage of the
5-, but not the 3-phosphate would give an α,β-unsaturated aldehyde,
whose presence would be demonstrable by its uv absorption (see
Scheme V). Action of periodate upon the synthetic acid [21] did
indeed produce the expected chromophore (λ max$^\sim$ 235 nm), while the
analogous reaction product from the natural 3-phosphate [17] showed
only end-absorption, without any maximum down to 230 nm. The compounds resulting from these reactions were very unstable and had
to be examined in the reaction mixtures without isolation. Furthermore, the reaction proceeded much faster in the synthetic acid
[21] than with the natural one [17], consistent with the presence
of a *cis*-1,2-glycol in the former, a *trans*-1,2-glycol in the latter.

All these results are in complete agreement with the formulation of the compound as the 3-phosphate [17] of [1]. They are
summarized in Scheme V.

The demonstration of an acid-catalyzed migration of the phosphate residue from position 3 to position 4 made it appear possible
that the phosphate ester initially present in the culture filtrates

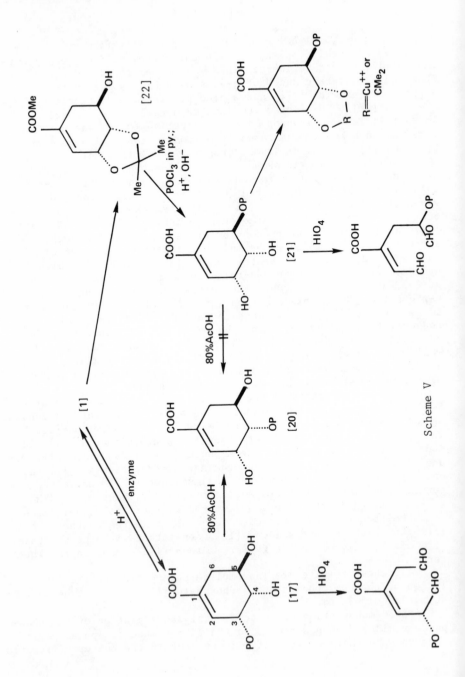

Scheme V

122

is actually [20], which could be converted to [17] during the prolonged exposure to acidic medium in the course of the isolation
procedures. This possibility was, however, eliminated by the
finding that bioassay by mutant response fails to reveal any sur-
viving [1] after treatment of a fresh culture filtrate with periodate followed by enzymatic dephosphorylation.

No synthesis of [17] has been described so far. Unless some
specifically enhanced reactivity of the allylic hydroxyl at position 3 could be exploited, a synthesis from [1] would require selective protection of the pair of *trans*-hydroxyls on C-4 and C-5;
no convenient method for such protection seems to be available.
However, synthesis from dehydroshikimic acid [2] by suitable protection of its hydroxyls, hydride reduction of the carbonyl, phosphorylation of the resulting hydroxyl, and removal of the protecting groups should be feasible.

A modified technique for the isolation of [17] from *A. aerogenes* has been described by Sprinson and co-workers (65). For
isolation from culture filtrates of *E. coli*, Edwards (66a) and
Zaprometov (66b) used chromatography on ion exchange columns rather than charcoal chromatography.

3-Phosphoshikimic acid [17] has apparently not been isolated
in pure form from any organism other than these two bacteria. However, paper chromatographic evidence suggesting its occurrence in
culture filtrates of the Basidiomycete *Lentinus lepideus* has been
obtained by Eberhardt (67), and similar observations made by Hathway (68) with extracts of myrobalans, the fruits of *Terminalia
chebula* (Combretaceae) have been interpreted as indicating its
occurrence in this higher plant. While the general similarities
of the steps of biosynthesis of aromatic metabolites in different
organisms (see Chapter 14) would suggest a widespread occurrence
of [17], experimental evidence for this occurrence is thus still
surprisingly inadequate.

Like many other phosphorylated and hence very polar compounds,
[17] is not utilized by intact cells of any organism tested, undoubtedly because a permeability barrier prevents it from entering
the cell (or perhaps from reaching the intracellular site of biosynthetic transformation). In view of this fact, the exceptionally large accumulation of [17] in some culture filtrates may seem
surprising, since it must involve, very efficiently, the reverse
translocation. Conceivably, different mechanisms might control
the uptake and the excretion of the compound; experimental evidence
suggesting the existence of such differences has been obtained for
a few metabolites (68A). Obviously, the same problem exists, to a
lesser extent, also for other intermediates which are inactive as
growth factors but are excreted into culture filtrates; among

non-phosphorylated intermediates in this category, chorismic acid (Chapter 6) and prephenic acid (Chapter 7), may be mentioned.

Because of the failure of [17] to be utilized, it was not possible to decide by methods based upon growth-factor activity whether [17] is a true biosynthetic intermediate or not. However, use of cell-free extracts provided an affirmative answer. Evidence of this kind was first obtained by Kalan (69), who showed that extracts of mutants blocked after the next intermediate can synthesize 5-enolpyruvyl shikimic acid [18] ("Compound Z_1") from phospho-enolpyruvate [23] and [17], but not from [1]. Such extracts can thus perform the next biosynthetic step, but the complexities of this step were not yet recognized at that time: the fact that not [18] but its 3-phosphate [19] is the actual intermediate. Analogous activity of cell-free extracts was established for the conversion of [17] to intermediates located further along the biosynthetic pathway, or to aromatic metabolites. Clark and Sprinson (70) observed the formation of prephenic acid (see Section 7-1) from [17] and [23] by extracts of *E. coli*. Srinivasan (71, 72) found that such extracts from mutants of *E. coli* were able to convert [17], but not [1], to anthranilic acid in the presence of glutamine; almost quantitative yields were obtained, and [17] could be replaced, although rather inefficiently, by [1] and ATP. Cell-free extracts from yeast converted the same precursors, [17] and glutamine, into *p*-aminobenzoic acid (73); [1] was poorly utilized in this reaction, but addition of ATP increased the yield to about 50% of that obtained from [17]. These findings establish without reasonable doubt that [17] is an obligatory intermediate in the common biosynthetic chain leading to aromatic metabolites.

The reaction that converts [17] to the next intermediate in this sequence was investigated in detail by Levin and Sprinson (74); it was found to consist of the combination of [17] with [23] to give 5-enolpyruvylshikimic acid 3-phosphate, [19]. The compound [18] mentioned before is the product of a secondary enzymatic dephosphorylation of [19]. It is in the formation of [19] that the three-carbon fragment which is destined to become the side-chain of phenylalanine, tyrosine, and the innumerable C_6-C_3 compounds derived from them, first enters the molecule. The fact that [23] is the precursor of this fragment is consistent with the much earlier findings of Gilvarg and Bloch (31) on the derivation of the side-chains of phenylalanine and tyrosine from one of the C_3 fragments preceding pyruvic acid in glycolysis.

In the work (74) leading to the discovery of [19], it was found that extracts of an appropriate mutant of *E. coli* produced [18] when incubated with [17] and [23]. In the presence of fluoride, the amount of [18] was strongly diminished, but the quantity of inorganic phosphate released was much higher than the one expected from the equation: [17] + [23] → [18] + 2 orthophosphate. Also, the acid-sensitive [18] is known to give equivalent amounts of [1] and pyruvic acid on mild hydrolysis, yet such treatment of the fluoride-containing incubation mixtures yielded much more pyruvate than could be explained on the basis of the limited quantity of [18] present. In contrast, the amount of unreacted [17] (measured by growth-factor activity of the [1] released on enzymatic dephosphorylation) was much smaller than that expected from the formation of [18], but action of phosphatase followed by acid hydrolysis gave an additional quantity of [1]. These findings prove the primary formation of a compound, [19], which resembles [18] in giving pyruvic acid on mild acid hydrolysis, but differs from it by yielding [1] on such hydrolysis only after previous enzymatic dephosphorylation. In filtrates or extracts, [19] can be converted further to [18] by a fluoride-sensitive enzymatic dephosphorylation. The isolation of [19] and its identification as 5-enolpyruvylshikimic acid 3-phosphate will be discussed in Section 5-3.

The properties of the enzyme, 5-enolpyruvylshikimate 3-phosphate synthase, of *E. coli* were examined by Levin and Sprinson (74b). It does not seem to require any metal or cofactor; its action is reversible. Enzymatic activity has been established in some higher plants (53). In *Neurospora* (Ascomycetes) (54, 61) and several other fungi belonging to the Asco-, Phyco-, and Basidiomycetes (75), 5-enolpyruvylshikimate 3-phosphate synthase forms part of the complex of enzymes responsible for steps 2 to 6 of the common pathway.

In the enzymatic synthesis of [19], the enolpyruvyl grouping of [23] is transferred unchanged to the acceptor, [17]. This process is quite exceptional; it seems to have only one precedent (65, 74) among the multiple biochemical transformations of [23]. The mechanism of this unusual reaction has been studied in detail by Bondinell and co-workers (65). Use of [23] labeled with [18]O in the enolic oxygen showed that the synthesis of [19] from [23] proceeds with C–O cleavage; all the label from this oxygen was found in the inorganic phosphate released. When the reaction was run in D_2O or tritiated water, the two vinylic hydrogens of the enolpyruvyl side-chain became labeled, while the [17] formed on acid hydrolysis was free of label. These findings support an addition-elimination mechanism first proposed by Levin and Sprinson (74) which assumes protonation at C-3 and nucleophilic attack by the hydroxyl of [17] at C-2 of [23], followed by elimination of phosphoric acid; the process is shown in Scheme VI.

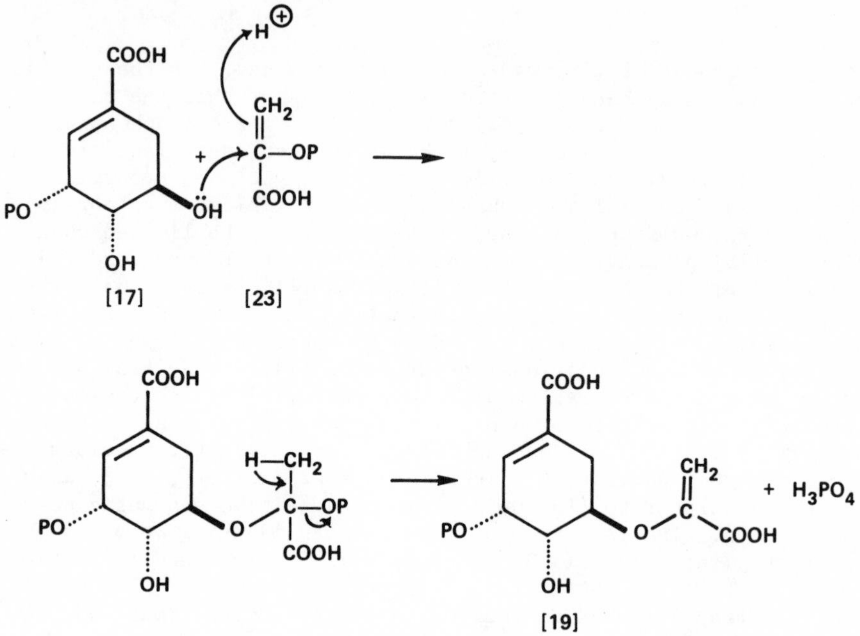

Scheme VI.

5-3. 5-ENOLPYRUVYLSHIKIMIC ACID 3-PHOSPHATE, [19]

As has been mentioned in the preceding section, 5-enolpyruvyl-shikimic acid ("Compound Z_1") [18], an acid-sensitive conjugate of shikimic acid [1], has been observed by Davis and Mingioli (50) in culture filtrates from mutants of *E. coli* and related bacteria. It was obtained as a noncrystalline Ba salt by Gilvarg (76); Edwards (66a) later on isolated it by ion-exchange chromatography. The compound was found (76) to yield equimolecular amounts of [1] and pyruvic acid on acid hydrolysis, and to consume one mole of periodate. On account of these findings, its sensitivity to acid, and its resistance to alkali (50) which precludes an ester linkage, the compound was tentatively formulated as 3- or 5-enolpyruvyl-[1]. The enol ether structure was supported (77, 78) by the ir spectrum, which shows a band at 1220 cm^{-1} typical of vinylic ethers, but lacks any absorption ascribable to an ester carbonyl. The oxidation with periodate was found (77) to yield an unstable compound with an uv absorption very similar to that of the product obtained (63) in the same manner from [21] and caused by the chromophore OHC-C=C-COOH (see Scheme V). Consequently, [18] is the 5-enol ether

of [1]. Its remarkably low polarity in the acidic solvent systems
used for its paper-chromatographic investigation (50) suggests
that [18] may be present as the lactone [18a] under these conditions. The corresponding compound lacking the free hydroxyl in
position 3 has been described (66a).

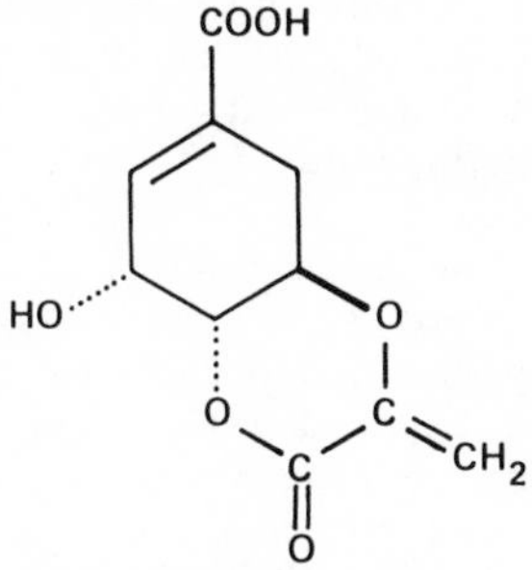

[18a]

Compound [18] is inactive in systems that synthesize anthranilic acid (72) or prephenic acid (74a) from [170; these findings
support the interpretation of [18] as a secondary transformation
product of the true intermediate [19].

The biosynthesis of 5-enolpyruvylshikimic acid 3-phosphate
[19] from [17] and [23] has been discussed in Section 5-2. The
compound was isolated by Levin and Sprinson (74b) from incubation mixtures or mutant filtrates through gradient elution with
LiCl from a column of Dowex 1-X8 (chloride form), conversion to
the Ba salt, and reprecipitation from aqueous solution with
ethanol. The salt ($\sim$94% pure) seems to be a dihydrate. The
assignment of structure [19] to the compound rests essentially
on the following findings: like [18], it is sensitive to acid
but stable to alkali; one mole of pyruvic acid is released by
dilute acid, one mole of orthophosphate by alkaline phosphatase,
while one mole of [1] is liberated on treatment with the enzyme
followed by acid hydrolysis; the ir spectrum, like that of [18],
shows the band at 1220 cm^{-1} caused by the vinyl ether grouping,
and in addition contains bands around 1100 and at 975 cm^{-1} ascribable to the phosphate group and also present in the spectrum of
[17] but not in that of [18]. These findings show that the compound is the phosphate of an enolpyruvyl ether of [1], but they
do not in themselves establish the location of the two groups
attached to the [1]. However, the relationship of the compound
with [18], its biosynthesis from [17], and its further transformation into chorismic acid can only be interpreted in a
simple and rational way if it is assumed that the phosphate and

pyruvyl residues occupy the position they have been shown to have in [17] and [18], respectively. On this basis, formula [19] must be the correct one. Strict chemical proof for [19], for example by isolation and characterization of [17] (perhaps together with [20]) from acidic, and of [18] from enzymatic hydrolyzates, has not been adduced. However, formula [19] is established beyond reasonable doubt by comparison of the nmr spectrum of the Ba salt (65) with that of [1] (16): the signals from the protons on the cyclohexene rings of both compounds are very similar, except for the one assigned by Hall (16) to the proton at C-3, which shows coupling (J = 8.5 Hz) with the nearby phosphorus atom.

3-Enolpyruvylshikimic acid 5-phosphate [19] must be one of the common intermediates in the chain leading to aromatic metabolites, and it cannot be the last one, that is, the branch-point intermediate. The first one of these conclusions is based upon the observation that [19], but not [18], is converted to later intermediates such as prephenic acid (79), anthranilic acid (72, 80), and phenylpyruvic and p-hydroxyphenylpyruvic acids (81); the second conclusion rests upon the fact that mutants which excrete [19] (74b) or its transformation product [18] (50) require the full quintuple aromatic supplement for growth and are thus evidently blocked in a reaction of the common sequence which is later than the formation of [19]; mutants with blocks in one of the individual branches *after* the last common intermediate would normally require only supplementation with those metabolites that are formed through that particular branch. Hence [19] must be an intermediate before the last common one; it was indeed found to be the one immediately preceding it. The need for assuming a specific branch-point intermediate after [19] was recognized already before the actual isolation of this compound (80-84); even the structure [24] of this intermediate, subsequently called chorismic acid, was correctly anticipated (74b, 83) from considerations of the structural requirements for a compound located between [19] and prephenic acid. Details of the chemistry of chorismic acid [24] (see p. 119) will be given in Chapter 6.

The fact that [19] is one of the common intermediates has one unexpected corollary: the three-carbon fragment destined to become the side-chain of phenylalanine, tyrosine, and innumerable other C_6-C_3 compounds is introduced into an intermediate which is also the parent of aromatic compounds that do *not* contain this fragment: anthranilic acid and the host of compounds derived from it, p-aminobenzoic acid, p-hydroxybenzoic acid, and so on. In the biosynthesis of these substances, the C_3-fragment just introduced must therefore be eliminated again. At first sight, this process seems strangely uneconomical; it may conceivably find its explanation in

some particular aptitude of the enolpyruvyl ether residue to function as an efficient leaving group in the elimination reactions
that are involved in the aromatization process.

The enzymatic reaction of which [19] is the substrate leads
to [24]. The enzyme chorismate synthase, which catalyzes this
conjugate 1,4-elimination reaction, has been studied in Sprinson's
laboratory (79) in extracts of *E. coli*. It was purified by fractionation with ammonium sulfate, chromatography on a column of diethylaminoethyl cellulose, and specific precipitation at pH 6.3.
Surprisingly, chorismate synthase requires strongly reducing conditions for activity, although it catalyzes a reaction that does
not involve any change in the oxidation level of the substrate.
The enzyme is inactive under aerobic conditions and needs such reducing agents as reduced flavin adenine nucleotide or riboflavin,
or dithionite, or platinum and hydrogen for activity. Sulfhydryl compounds seem to protect the enzyme but are not specifically
required. Inhibition by specific chelators for Fe^{2+} indicates
that ferrous iron may be a cofactor; this might explain the need
for a reducing environment. However, at higher concentrations,
Fe^{2+} is inhibitory, while Cu^{2+} is not.

Chorismate synthase activity has been encountered in some
other microorganisms (54, 56). No observation has come to our
attention which would suggest any regulatory influence upon this
last enzyme of the common pathway. It seems, however, that it is
not part of the complex of the five preceding enzymes in *Neurospora*, and the gene specifying its synthesis is outside the cluster
of genes for the enzymes of this complex (54).

The 1,4-elimination of phosphoric acid from C-3 and hydrogen
from C-6 of [19] during its enzymatic transformation to [24] could
proceed with *cis* or *trans* stereochemistry, that is, with loss of
the 6α = (6S) or the 6β = (6R) hydrogen. This question has been
studied with concordant results by Hill and Newkome (40a) and Onderka and Floss (40b). In both cases, samples of [1] labeled
stereospecifically with hydrogen at 6α and 6β were prepared and
converted to [24] (40b) or to phenylalanine and tyrosine (40a) by
mutants of *A. aerogenes* or *E. coli*, respectively. (The conversion
of [24] to the aromatic amino acids proceeds without further
change of the hydrogen at position 6, so that retention or loss of
label in these processes also indicate the isotopic status of [24]).
Hill and Newkome (40a) prepared their samples of 6α and 6β-^{2}H-[1]
by the total synthesis of Smissman and co-workers (18), using specifically deuteriated samples of methyl acrylate; Onderka and
Floss (40b) obtained the corresponding tritiated species enzymatically from 1-labeled glucose and mannose, respectively, which
were converted to the specifically tritiated forms of [23] by en-

zymatic transformations of known stereospecificity; biosynthetic combination of those with erythrose 4-phosphate (see Chapter 2) and further conversion via the shikimate pathway gave the desired samples of [1].

It was found in both cases that it is the label from 6β = (6R) which is lost. The overall reaction is thus a 1,4-*trans*-elimination. A concerted E2' reaction of this type would be expected (85) to proceed with *cis*-stereochemistry; the results are therefore consistent with a two-stage process rather than with a concerted one-stage reaction. A possible mechanism or the former type has been proposed (40b). It assumes addition of the enzyme at C-1 with simultaneous formation of the double bond between C-2 and C-3 through elimination of the phosphoric acid by an SN_2' reaction (*cis* stereochemistry) followed by a second stage where the enzyme and the 6β-H are removed in a normal *trans*-elimination.

REFERENCES

1. B.D. Davis, (a) Experientia 6, 41 (1950); (b) J. Biol. Chem. 191, 315 (1951).
2. B. Bohm, Chem. Rev. 65, 435 (1965).
3. J.F. Eijkman, Rec. Trav. Chim. 4, 32 (1885); 5, 299 (1886).
4. S.Y. Chen, Amer. J. Pharm. 101, 550 (1929).
5. H.O.L. Fischer and G. Dangschat, Helv. Chim. Acta 17, 1200 (1934).
6. R. Grewe and W. Lorenzen, Chem. Ber. 86, 928 (1953).
7. K. Freudenberg and J. Geiger, Ann. 575, 145 (1952).
8. (a) S. Hattori, S. Yoshida, and M. Hasegawa, Physiol. Plant. 7, 283 (1954); (b) V. Plouvier, Compt. Rend. 249, 1563 (1959).
9. P.R. Srinivasan, H.T. Shigeura, M. Sprecher, D.B. Sprinson. and B.D. Davis, J. Biol. Chem. 220, 477 (1956).
10. J.F. Eykman, Ber. 24, 1278 (1891).
11. H.O.L. Fischer and G. Dangschat, Helv. Chim. Acta 18, 1204 (1935).
12. H.O.L. Fischer and G. Dangschat, Helv. Chim. Acta 18, 1206 (1935).
13. H.O.L. Fischer and G. Dangschat, Helv. Chim. Acta 20, 705 (1937).
14. J.G. Carr, A. Pollard, G. C. Whiting, and A.H. Williams, Biochem. J. 66, 283 (1957).
15. K. Freudenberg, H. Meisenheimer, J.T. Lane, and E. Plankenhorn, Ann. 543, 162 (1940); K. Freudenberg and W. Hohmann, ibid. 584, 54 (1954); K. Freudenberg and W. Lwowski, ibid. 587, 213 (1954); 594, 76 (1955).

16. L.D. Hall, J. Org. Chem. 29, 297 (1964).
17. H. Plieninger and K. Schneider, Chem. Ber. 92, 1587 (1959).
18. E.E. Smissman, J.T. Suh, M. Oxman, and R. Daniels, J. Amer. Chem. Soc. 81, 2909 (1959); 84, 1040 (1962).
19. R. McCrindle, K.H. Overton, and R.A. Raphael, J. Chem. Soc. 1560 (1960).
20. R. McCrindle, K.H. Overton, and R.A. Raphael, Tetrahedron Lett. 1847 (1968); R.K. Hill and G.R. Newkome, ibid. 1851, (1968); E.E. Smissman and J.P. Li, ibid. 4601 (1968).
21. M.M. Doshi, Dissertation Abstr. 24, 3998 (1964).
22. R. Grewe and I. Hinrichs, Chem. Ber. 27, 443 (1964).
23. R. Grewe and S. Kersten, Chem. Ber. 100, 2546 (1967).
24. E.J. Hedgley and H.G. Fletcher, Jr., J. Amer. Chem. Soc. 84, 3726 (1962).
25. H.J. Bestmann and H.A. Heid, Angew. Chem., Int. Ed. 10, 336 (1971).
26. R. Grewe and E. Vangermain, Chem. Ber. 98, 104 (1965).
27. O. Goldschmid and H.L. Hergert, Tappi 44, 858 (1961).
28. K.R. Hanson and M. Zucker, J. Biol. Chem. 238, 1105 (1963).
29. V.P. Maier, D.M. Metzler, and A.F. Huber, Biochem. Biophys. Res. Commun. 14, 124 (1964).
30. M. de Rosa, A. Gambacorta, L. Minale, and J.D. Bu'Lock, (a) Chem. Commun. 1334 (1971); (b) Biochem. J. 128, 751 (1972).
31. C. Gilvarg and K. Bloch, J. Biol. Chem. 199, 689 (1952).
32. P.R. Srinivasan, Biochemistry 4, 2860 (1965).
33. G. Read and L.C. Vining, Chem. Ind. (London) 1547 (1959).
34. S. Yoshida, Ann. Rev. Plant Physiol. 20, 41 (1969).
35. S. Yoshida and G.H.N. Towers. Can. J. Biochem. Physiol. 41, 579 (1963).
36. D.R. McCalla and A.C. Neish, Can. J. Biochem. Physiol. 37, 531 (1959).
37. O.L. Gamborg and A.C. Neish, Can. J. Biochem. Physiol. 37, 1277 (1959).
38. F. Wightman, M.D. Chisholm, and A.C. Neish, Phytochemistry 1, 30 (1961).
39. P. Olesen Larsen, D.K. Onderka, and H.G. Floss, Chem. Commun. 842 (1972).
40. (a) R.K. Hill and G.R. Newkome, J. Amer. Chem. Soc. 91, 5893 (1969);
    (b) D.K. Onderka and H.G. Floss, ibid. 5894 (1969); H.G. Floss, D.K. Onderka, and M. Carroll, J. Biol. Chem. 247, 736 (1972).
41. D.R. McCalla and A.C. Neish, Can. J. Biochem. Physiol. 37, 537 (1959).

42. E.W. Underhill, J.E. Watkins, and A.C. Neish, Can. J. Biochem. Physiol. 35, 219 (1957).
43. S.A. Brown and A.C. Neish, Nature 175, 688 (1955).
44. R. Hegnauer, Chemotaxonomie der Pflanzen, Vol. 1, Birkhäuser Verlag, Basel and Stuttgart, 1962, p. 202.
45. S.M. Siegel, Amer. J. Botany 56, 175 (1969); J.D. Bu'Lock and H. Smith, Experientia 17, 553 (1961).
46. G. Eberhardt and W.J. Schubert, J. Amer. Chem. Soc. 78, 2835 (1956).
47. (a) H.J. Perkins and S. Aronoff, Can, J. Biochem. Physiol. 37, 149 (1959);
     (b) R.C. Millican, Biochim. Biophys. Acta 57, 407 (1962);
     (c) M.N. Zaprometov, Biokhimiya 26, 597 (1961).
48. L.H. Weinstein, C.A. Porter, and H.J. Laurencot, Jr., Contrib. Boyce Thompson Inst. 21, 439 (1962).
49. B.W. Smith, M.J. Turner, and E. Haslam, Chem. Commun. 842 (1970).
50. B.D. Davis and E.S. Mingioli, J. Bact. 66, 129 (1953).
51. J.A. Fewster, (a) Biochem. J. 73, No. 1, 14P (1959); (b) ibid. 85, 388 (1962).
52. E.B. Kalan, unpublished work mentioned by B.D. Davis, Arch. Biochem. Biophys. 78, 497 (1958).
53. S.I. Ahmed and T. Swain, Phytochemistry 9, 2287 (1970).
54. N.H. Giles, M.E. Case, C.W.H. Partridge, and S.I. Ahmed, Proc. Natl. Acad. Sci. U.S. 58, 1453 (1967).
55. (a) F. Gibson and J. Pittard, Bact. Rev. 32, 465 (1968);
     (b) J. Pittard and F. Gibson, Curr. Top. Cell. Regulation 2, 29 (1970).
56. E. Gollub, H. Zalkin, and D.B. Sprinson, J. Biol. Chem. 242, 5323 (1967).
57. (a) D.S. Nasser, G. Henderson, and E.W. Nester, J. Bact. 98, 44 (1969);
     (b) E.W. Nester, R.A. Jensen, and D.S. Nasser, ibid. 97, 83 (1969).
58. E.W. Nester, J.H. Lorence, and D.S. Nasser, Biochemistry 6, 1553 (1967).
59. M.B. Berlyn and N.H. Giles, J. Bact. 99, 222 (1969).
60. H. Morell and D.B. Sprinson, J. Biol. Chem. 243, 676 (1968).
61. L. Burgoyne, M.E. Case, and N.H. Giles, Biochim. Biophys. Acta 191, 452 (1969); S.I. Ahmed and N.H. Giles, J. Bact. 99, 231 (1969).
62. W.M. Nakatsukasa and E.W. Nester, J. Biol. Chem. 247, 5972 (1972).

63. U. Weiss and E.S. Mingioli, J. Amer. Chem. Soc. 78, 2894 (1956).

64. (a) D.M. Brown and A.R. Todd, J. Chem. Soc. 52 (1952);
     (b) D.M. Brown, D.I. Magrath, and A.R. Todd, ibid. 2708 (1952) and literature quoted there;
     (c) D.M. Brown and A.R. Todd, ibid. 44 (1952).

65. W.E. Bondinell, J. Vnek, P.F. Knowles, M. Sprecher, and D.B. Sprinson, J. Biol. Chem. 246, 6191 (1971).

66. (a) J.M. Edwards, Thesis, University of Melbourne, 1966;
     (b) M.N. Zaprometov, Biokhimiya 26, 597 (1969).

67. G. Eberhardt, J. Amer. Chem. Soc. 78, 2832 (1956).

68. D.E. Hathway, Biochem. J. 63, 380 (1956).

68A. B.L. Horecker, J. Thomas, and J. Monod, J. Biol. Chem. 235, 1586 (1960); Y.S. Halpern, H. Barash, and K. Druck, J. Bact. 113, 51 (1973).

69. E.B. Kalan, unpublished work quoted in refs. 63 and 74.

70. M.J. Clark and D.B. Sprinson, unpublished work quoted in ref. 74a.

71. P.R. Srinivasan, J. Amer. Chem. Soc. 81, 1772 (1959).

72. P.R. Srinivasan and A. Rivera, Jr., Biochemistry 2, 1059 (1963).

73. B. Weiss and P.R. Srinivasan, Proc. Natl. Acad. Sci. U.S. 45, 1491 (1959).

74. J.G. Levin and D.B. Sprinson, (a) Biochem. Biophys. Res. Commun. 3, 157 (1960);
     (b) J. Biol. Chem. 239, 1142 (1964).

75. S.I. Ahmed and N.H. Giles, J. Bact. 99, 231 (1969).

76. C. Gilvarg, unpublished work mentioned by B.D. Davis, The Harvey Lectures, Series L, 230 (1954–1955), and in refs. 74 and 78.

77. D.B. Sprinson and I.J. Borowitz, unpublished work mentioned in ref. 78.

78. D.B. Sprinson, Adv. Carbohydrate Chem. 15, 235 (1960).

79. H. Morell, M.J. Clark, P.F. Knowles, and D.B. Sprinson, J. Biol. Chem. 242, 82 (1967).

80. A. Rivera, Jr., and P.R. Srinivasan, Biochemistry 2, 1063 (1963).

81. C.H. Doy and F. Gibson, Biochim. Biophys. Acta 50, 495 (1961); M.I. Gibson, F. Gibson, C.H. Doy, and P. Morgan, Nature 195, 1173 (1962).

82. E. Umbarger and B.D. Davis, in "The Bacteria", Vol. III, Academic, New York and London, 1962, p. 167 ff., esp. p.225.

83. J.G. Levin, Thesis, Columbia University 1962.

84. M.I. Gibson and F. Gibson, Biochim. Biophys. Acta 65, 160 (1962).

85. K. Fukui, Tetrahedron Lett. 2427 (1965); N.T. Anh, Chem. Commun. 1089 (1968).

CHAPTER 6

Alicyclic Intermediates: Chorismic Acid

Because the mutants that accumulate 5-enolpyruvylshikimic acid 3-phosphate (ES-3P) are multiple auxotrophs, requiring all five aromatic metabolites for growth, and are thus presumably blocked in a single enzymic step after ES-3P in the common pathway, two independent groups of workers (1, 2), suggested that a new "branch-point compound", which should be converted to anthranilic acid in the presence of glutamine (cf. conversion of shikimic acid 3-phosphate to anthranilic acid: Section 5-2) and to prephenic acid in the absence of nitrogen source, would be found after ES-3P.

The successful search (3, 4) for the proposed new intermediate started with a mutant of *A. aerogenes* (T. 17), extracts of which were able to convert shikimic acid to anthranilic acid, or to the phenylpyruvic acids. Since T. 17 was blocked after anthranilic acid, omission of glutamine from the growth medium effectively closed the entire tryptophan pathway. Irradiation of T. 17 gave a secondary mutant (61-3), which required both tryptophan and tyrosine for growth, and which was therefore blocked at the enzymic steps *a* and *b* (see Scheme I). Further irradiation of 61-3 yielded several polyaromatic auxotrophs which could have been blocked at *c* or *d* as well as *a* and *b*, or anywhere on the common pathway. However, mutants unable to carry out reactions *c* and *d* should still be able to accumulate anthranilic acid in the presence of glutamine, and three of the new mutants were able to do this. Of the three, *A. aerogenes* 62-1 was selected; it was found that an extract of cells grown on limiting tryptophan and in the absence of glutamine excreted a substance which was not found if glutamine was present. This substance, which could be extracted into ethyl acetate at pH 2, showed uv absorption in the 270 nm region. Addition of glutamine and an extract of a mutant strain capable of synthesizing anthranilic acid from ES-3P converted the new compound into anthranilic acid.

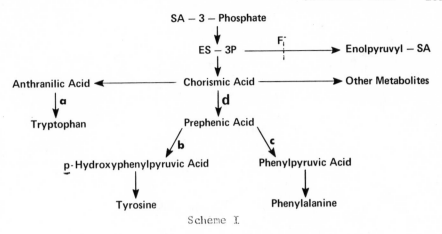

SA − 3 − Phosphate

ES − 3P $\xrightarrow{F_i^-}$ Enolpyruvyl − SA

Anthranilic Acid ◄─────── Chorismic Acid ─────────► Other Metabolites

a

Tryptophan

d

Prephenic Acid

b   **c**

p-Hydroxyphenylpyruvic Acid   Phenylpyruvic Acid

Tyrosine   Phenylalanine

Scheme I

It was also found that washed suspensions of whole cells of *A. aerogenes* 62-1, in the presence of an excess of tryptophan, accumulated the new compound, which was called chorismic acid [1] (derived from χωρ∿σμοο, separation). Tryptophan represses the formation of the enzyme that converts [1] to anthranilic acid. Chorismic acid can be isolated from the culture filtrates by chromatography over an anion exchange resin. The cooled filtrates are adjusted to pH 8 and passed down a column of Dowex 1-Cl⁻ resin; after washing, the acid can be eluted as its ammonium salt by a molar NH₄Cl solution adjusted to pH 8. The eluate is acidified to pH 2 and extracted with large volumes of ether which, after drying and evaporation, yield the crude chorismic acid.

Chorismic acid was initially characterized by Gibson (5) as its noncrystalline barium salt ($C_{10}H_8BaO_6 \cdot 3H_2O$), and by its uv [$\lambda_{max}(H_2O)$ 272 nm (ϵ 2700)], and ir spectrum. The structure of the salt was inferred from an nmr study in D_2O (6), and the stereochemistry assigned on the basis of the conversion of [1] to prephenic acid of known stereochemistry (7, 8).

The structure assigned to chorismic acid [1] is that which would arise if the acid were formed by a concerted 1,4-elimination (E-2') (9) of phosphate from ES-3P to form the cyclohexadiene system. Further investigation has shown, however, that the reaction is a stereospecific *trans*-1,4-elimination; both deuterium (10) and tritium (11) labeling studies have shown that it is the 6β proton which is eliminated in this reaction. Onderka and Floss (11) have suggested an enzyme-bound intermediate [2] in order to explain this unexpected result.

The enzyme (chorismate synthetase) responsible for the conversion of ES-3P to chorismic acid in *E. coli* has been purified and

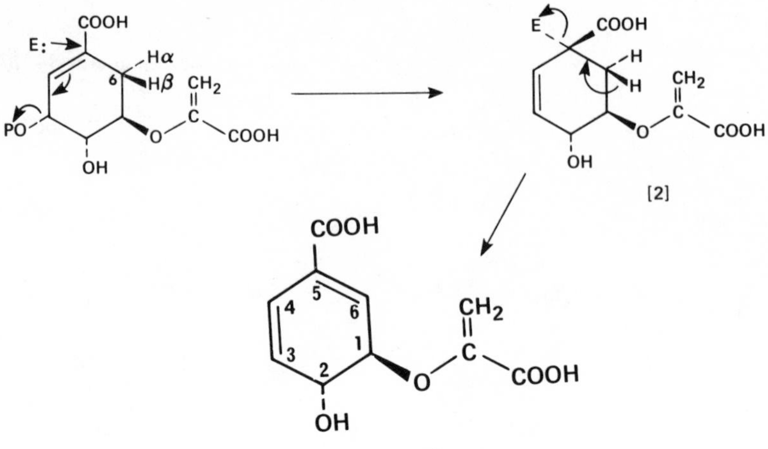

[1]

studied in detail by Morell et al. (12), who found it to be a
single protein; a similar study of the enzyme from *Neurospora
crassa* (50) showed it to have comparable properties (activation
by the substrate, and a requirement for a reduced pyridine nucleotide), but to be resolvable into two proteins. The isolated fractions do not seem to be isoenzymes.

Free chorismic acid has been isolated (13, 14) as a crystalline solid (m.p. 148-149°, from ethyl acetate - petroleum ether).
The acid is freely soluble in polar solvents; its solution in
water shows λ_{max} 275 nm (ε 2630). The ir spectrum exhibits absorption bands characteristic of an enol-ether of pyruvic acid
(1740 and 1620 cm^{-1}) (15, 16). The nmr spectrum confirms the
structural features of [1]: the signals of the side-chain methylene group appear as a pair of doublets at τ 5.46 and 4.79 (J =
2.8 Hz), those of the conjugated diene system as a singlet at τ
3.8, and an AB quartet at τ 3.5 and 3.97 (J = 9.75 Hz), and the
two allylic protons as an AB quartet, with additional long-range
coupling, at τ 5.11 and 5.30 (J = 11.7 Hz). This last coupling
constant, characteristic of a vicinal interaction for which the
dihedral angle is close to 180°, indicates the relative *trans*-
stereochemistry of the allylic oxygen functions.

Chemical proof that the enol-pyruvate side-chain is attached
at C-1 has been provided by Lingens (17), who treated chorismic
acid with 2N HCl at 80° for 10 minutes and isolated α-(3-carboxy-
phenoxy)acrylic acid, identified by hydrolysis to 3-hydroxybenzoic
acid and pyruvic acid.

Chorismic acid readily absorbs one mole of hydrogen to yield
a dihydro acid [3] which can be hydrolyzed to (-)-*trans*-4,5-dihy-

droxy-1-cyclohexene-1-carboxylic acid [4] under mild acidic conditions. Structure [4] has been degraded to (+)-β,β'-dihydroxyadipic
acid of known absolute stereochemistry (18), thus establishing the
absolute configuration indicated for [1] (13). This absolute stereochemistry has been further confirmed (19) by ozonolysis of [1]
to (-)-tartaric acid. See Scheme II.

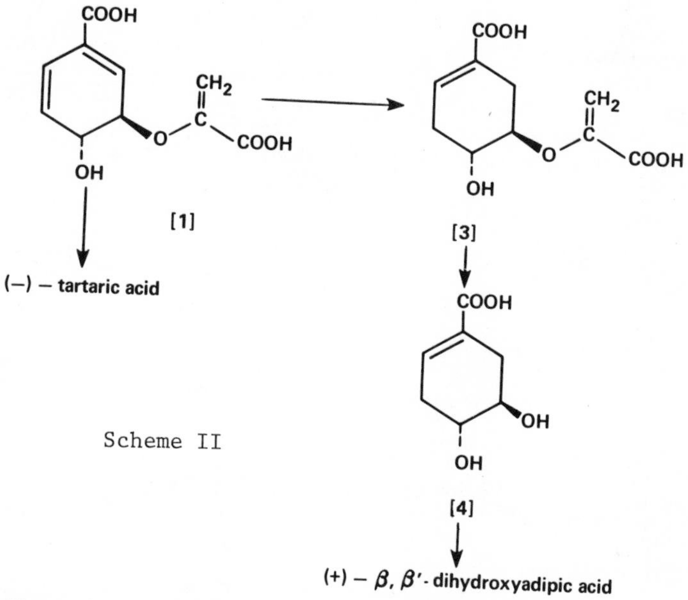

Scheme II

Dreiding models indicate that [1] can exist in two conforma-
tions, [A] and [B]. In view of the remarkable ease of the conversion of chorismic to prephenic acid by an S_{Ni}' rearrangement, [B]
would seem the more likely one. In it the double bond of the side-
chain lies almost parallel to and above the ring double bond between C-2 and C-3, suggesting a possible interaction, and the methylene group of the side-chain is positioned close to C-3, its fu-
ture point of attachment. Conformation [A] would, on the other
hand, require a major change of molecular shape before the conversion to prephenic acid could take place. The relative dihedral
angle between the protons at C-1 and C-6 is about 55° in the case
of the diaxial arrangement of the substituents [B], and approximately 180° in the diequatorial form [A]. The observed vicinal
coupling constant of 11.7 Hz between the C-1 and C-6 protons
clearly indicates the extensive population of the conformation in
which the two substituents are quasi-equatorial. This conformation

is not suitable for the S_Ni' rearrangement to prephenic acid, and
presumably one role of the enzyme (chorismate mutase) which brings
about the isomerization is to invert the conformation of the ring
as well as to orientate the side-chain correctly.

Chorismic acid shows a negative Cotton effect in its ORD spectrum, with a minimum at 283 nm, $[M]_{min}=-12,294°$, and in its CD curve,
$\Delta\varepsilon$ 265-270 nm=-1.8; if the molecule is considered as a simple cyclohexadiene system, these results would indicate (20-23) that the
diaxial conformation (left-handed helix) is the prevailing one.
However, it is unlikely that arguments based upon the helicity of
the diene chromophore alone are valid; the chiroptical results
are open to question because of the cross-conjugated nature of the
chromophore and the two allylic asymmetric centers at C-1 and C-2.

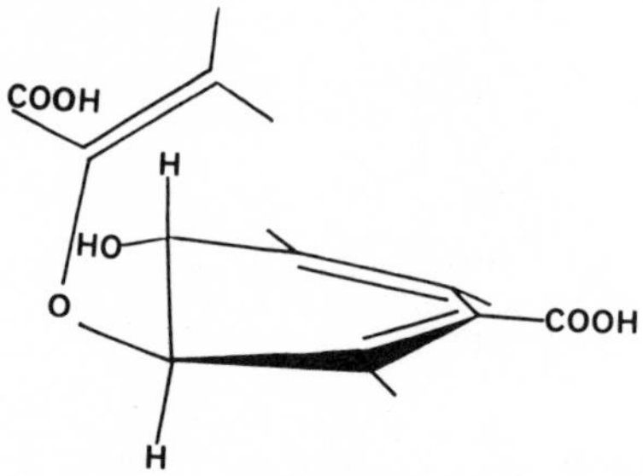

A 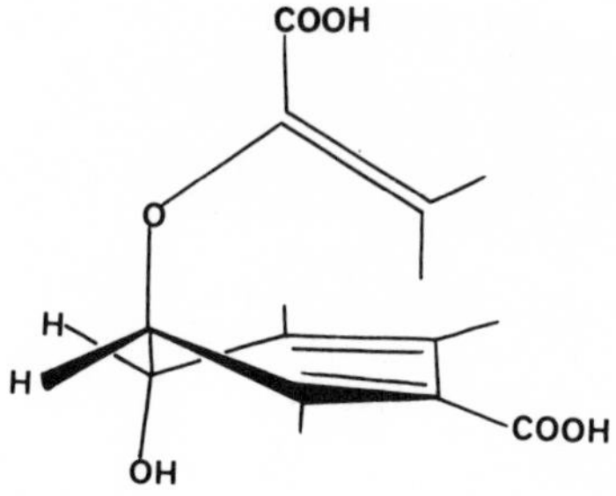   B

Chorismic acid reacts with diazomethane to give a crystalline
product lacking the light absorption of the diene chromophore. The
structure [5] has been assigned on analytical and spectroscopic
grounds (24); this parallels the formation of a similar pyrazoline
from shikimic acid (25).

Chorismic acid is thermally unstable both in solution and in
the solid form. At low temperatures prephenate is formed quite
rapidly; the half-life of [1] is 7 days at 4° for a solution and
19 days at 22° for the solid. If [1] is heated *in vacuo* the expected products, p-hydroxybenzoic acid, pyruvic acid, and phenylpyruvic acid, are formed (13). The decomposition of [1] on heating at various pH's has been studied (26). Heating at 100° in
0.05M HCl produces p-hydroxybenzoic, phenylpyruvic, and 3,4-dihy-

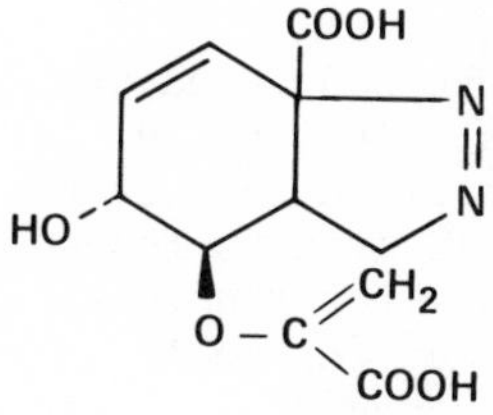

[5]

dro-3,4-dihydroxybenzoic acids.   Similar heating in 1M HCl yields
p-hydroxybenzoic, phenylpyruvic, and *meta*-hydroxybenzoic acids,
and heating in 1M NaOH produces p-hydroxybenzoic, 4-hydroxyphenyl-
lactic, and α-(3-carboxyphenyloxy)acrylic acids.   These thermal
decomposition products are set out in Scheme III; all are quite
compatible with the established structure of [1]. Many of them
are also products of enzyme action, and these are discussed separately later in the chapter.

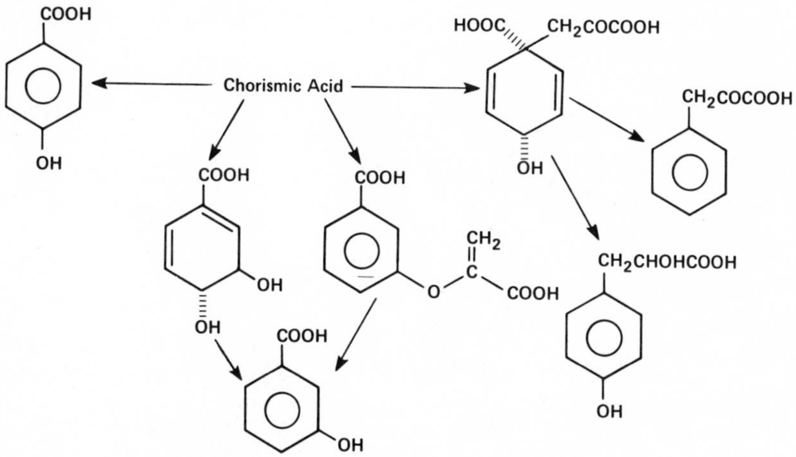

Scheme III. **Thermal Decomposition Products of Chorismic Acid**

Since the publication of the isolation of chorismic acid
from cultures of *A. Aerogens* 62-1, the acid has been detected in
the accumulation media of other, similar, multiple auxotrophs:
N. crassa td 48 R (27), *E. Coli* B-37 (28), and *Saccharomyces
cerevisiae* E. 107, E. 104 (29, 30).

Chemically, [1] can be converted into prephenic acid by warming in aqueous solution at alkaline pH. At pHs below 6, the prephenic acid formed is converted further to phenylpyruvic acid. In both cases *p*-hydroxybenzoic acid is formed.

The proposed mechanism for the conversion of [1] to prephenic acid (6) is an example of an *intra*molecular anionotropic rearrangement which proceeds via a concerted cyclic process. Reactions of this type have been discussed by Ingold (46), who has suggested

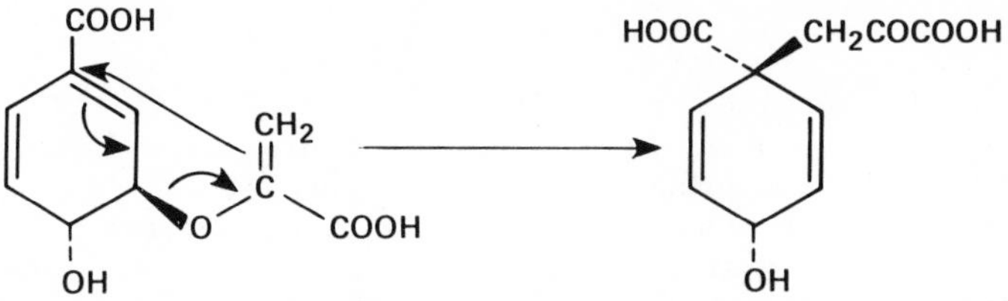

that they be classified as S$_N$i'. Hill and Edwards (47) have pointed out that the rearrangement of vinylcyclopent-2-enyl ether [7] to cyclopent-2-ene-acetaldehyde [6] provides a good stereochemical model for the conversion of [1] to prephenic acid.

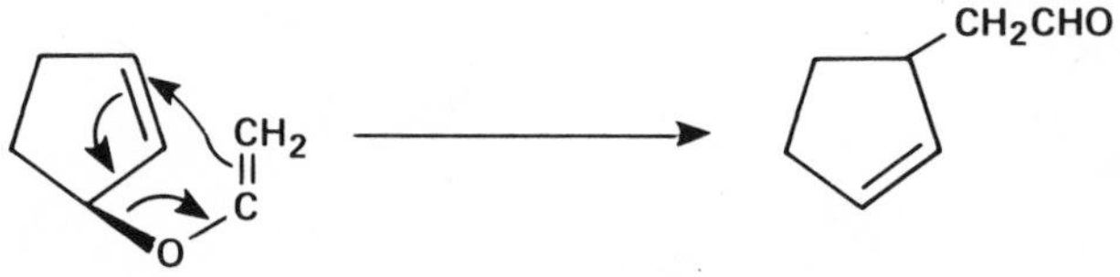

[6]   [7]

The isolation of chorismic acid also allowed direct experiments on the reaction leading to the biosynthesis of prephenic acid, the precursor of phenylalanine and tyrosine. Using purified enzymes from cell-free extracts of mutant strains of *A. aerogens* and *E. coli*, Cotton and Gibson (31) have shown that there are two enzymes, or enzyme complexes, metabolizing [1]; one leads through prephenic acid to phenylpyruvic acid (chorismate mutase P and prephenate dehyratase) and the other through prephenic acid to 4-hydroxyphenylpyruvic acid (chorismate mutase T and prephenate dehydrogenase). The two enzyme systems in *E. coli* are independently subject to end-product inhibition by L-phenylalanine and L-tyrosine, respectively, although studies of chorismate mutase and its related enzymes in other organisms (32-45) have shown considerable variations in inhibitory mechanisms and enzyme patterns (see Chapter 15). For example, the enzyme complex from *Salmonella typhimurium* is quite comparable to those of *E. coli* and *A. aerogenes* (32), whereas the enzyme from cell-free extracts of *Euglena gracilis* is apparently a single species (43). The enzyme is inhibited by phenylalanine and tyrosine, and activated by tryptophan; tryptophan also inhibits anthranilate synthetase, thus effectively diverting the whole shikimate pathway to the production of phenylalanine and tyrosine. The single enzyme from *Neurospora crassa* is, on the other hand, inhibited by phenylalanine and tyrosine, but is hardly sensitive to tryptophan (36). *Saccharomyces cerevisiae* also has a single enzyme, but this is inhibited by tyrosine, activated by tryptophan, and unaffected by phenylalanine (35, 45). Chorismate mutase of *Bacillus subtilis* consists of three isoenzymes, none subject to inhibition by the aromatic amino acids, but two inhibited by prephenic acid (33); interestingly, one of the enzymes is aggregated with DAHP synthetase (see Chapter 15). Two enzymes have been isolated from *Claviceps paspali*; one is activated by tryptophan and inhibited by tyrosine and phenylalanine; the other is activated by tryptophan but shows no inhibition by the other amino acids (41).

The enzyme system has also been studied in higher plants; thus, Cotton and Gibson (40) found a single species of chorismate mutase in pea (*Pisum sativum*), subject to inhibition by phenylalanine and tyrosine, and activated by tryptophan. Mung bean *(Phaseolus aureus)* contains two forms of the enzyme, one inhibited by both phenylalanine and tyrosine and activated by tryptophan, the other unaffected by the amino acids (48). Similarly, enzyme preparations from tissue cultures of tobacco, rice, carrot, and tomato are not repressed by phenylalanine or tyrosine (49).

REFERENCES

1. M.I. Gibson, F. Gibson, C.H. Doy, and P. Morgan, Nature 195, 1173 (1962).
2. A. Rivera and P.R. Srinivasan, Biochemistry 2, 1063 (1963).
3. P.N. Morgan, M.I. Gibson, and F. Gibson, Biochem. J. 89, 229 (1963).
4. M.I. Gibson and F. Gibson, Biochem. J. 90, 248 (1964).
5. F. Gibson, Biochem. J. 90, 256 (1964).
6. F. Gibson and L.M. Jackman, Nature 198, 388 (1963).
7. H. Plieninger and G. Keilich, Z. Naturforsch. 16b, 81 (1961).
8. H. Plieninger, Angew. Chem. (Int. Ed.) 1, 367 (1962).

9. S.J. Cristol, W. Barasch, and C.H. Tieman, J. Amer. Chem. Soc. 77, 583 (1955).

10. R.K. Hill and G.R. Newkome, J. Amer. Chem. Soc. 91, 5894 (1969).

11. D.K. Onderka and H.G. Floss, J. Amer. Chem. Soc. 91, 5894 (1969); J. Biol. Chem. 247, 736 (1972).

12. H. Morell, M.J. Clark, P.F. Knowles, and D.B. Sprinson, J. Biol. Chem. 242, 82 (1967).

13. J.M. Edwards and L.M. Jackman, Aust. J. Chem. 18, 1227 (1965).

14. F. Gibson, Biochem. Preps. 12, 94 (1968).

15. M. Engelhardt, H. Plieninger, and P. Schreiber, Chem. Ber. 97, 1713 (1964).

16. F. Lingens and B. Sprössler, Ann. 709, 173 (1967).

17. F. Lingens and G. Müller, Z. Naturforsch. 54, 492 (1967).

18. Th. Posternak and J. Ph. Susz, Helv. Chim. Acta 39, 2032 (1956).

19. I.G. Young and F. Gibson, Biochim. Biophys. Acta 177, 348 (1969).

20. U. Weiss, H. Ziffer, and E. Charney, Chem. Ind. (London) 1286 (1962).

21. A. Moscowitz, E. Charney, U. Weiss, and H. Ziffer, J. Amer. Chem. Soc. 83, 4661 (1961).

22. A.W. Burgstahler, H. Ziffer, and U. Weiss, J. Amer. Chem. Soc. 83, 4660 (1961).

23. H. Ziffer, E. Charney, and U. Weiss, J. Amer. Chem. Soc. 84, 2961 (1962).

24. J.M. Edwards, Ph.D. Thesis, University of Melbourne (1966).

25. R. Grewe and A. Bokranz, Chem. Ber. 88, 49 (1955).

26. I.G. Young, F. Gibson, and C.G. MacDonald, Biochim. Biophys. Acta 192, 62 (1969).

27. J.A. DeMoss, J. Biol. Chem. 240, 1231 (1965).

28. M.J. Clark, H. Morell, P.F. Knowles, and D.B. Sprinson, Fed. Proc. 23, 313 (1964).

29. F. Lingens and W. Luck, Angew. Chem. 76, 51 (1964).

30. F. Lingens and W. Goebel, Z. Physiol Chem. 342, 1 (1965).

31. R.G.H. Cotton and F. Gibson, Biochim. Biophys. Acta 100, 76 (1965).

32. J.C. Schmit and H. Zalkin, Biochemistry 8, 174 (1969).

33. J.H. Lorence and E.W. Nester, Biochemistry 6, 1541 (1967).

34. E.W. Nester and R.A. Jensen, J. Bact. 91, 1594 (1966).

35. F. Lingens, W. Goebel, and H. Uesseler, Eur. J. Biochem. 1, 363 (1967); Naturwissenschaften 54, 141 (1967); Biochem. Z. 346, 357 (1966).

36. T.I. Baker, Biochemistry 5, 2654 (1966).

37. O.L. Gamborg and F.J. Simpson, Can. J. Biochem. 42, 583 (1964).

38. O.L. Gamborg and F.W. Keeley, Biochim. Biophys. Acta 115, 65 (1966).
39. P. Cerutti and G. Guroff, J. Biol. Chem. 240, 3034 (1965).
40. R.G.H. Cotton and F. Gibson, Biochim. Biophys. Acta 156, 187 (1968).
41. B. Sprössler and F. Lingens, Z. Physiol. Chem. 351, 448, 967 (1970); FEBS Lett. 6, 232 (1970).
42. J.C. Schmit, S.W. Artz, and H. Zalkin, J. Biol. Chem. 245 4019 (1970).
43. H.L. Weber and A. Böck, Eur. J. Biochem. 16, 244 (1970).
44. G.L.E. Koch, D.C. Shaw, and F. Gibson, Biochim. Biophys. Acta 229 795, 805, (1971); 258, 719 (1972).
45. B. Sprössler, U. Lenssen, and F. Lingens, Z. Physiol. Chem. 351, 1178 (1970).
46. C. Ingold, Structure and Mechanism in Organic Chemistry, Cornell University Press, Ithaca, 1953, p. 589.
47. R.K. Hill and A.G. Edwards, Tetrahedron Lett. 3239 (1964).
48. D.G. Gilchrist, T.S. Woodin, M.L. Johnson, and T. Kosuge, Plant Physiol. 49, 52 (1972); ibid. 43, 47 (1968).
49. M. Chu and J.M. Widholm, Physiol. Plant. 26, 24 (1972).
50. F.H. Gaertner and K.W. Cole, J. Biol. Chem. 248, 4602 (1973).

CHAPTER 7

Biosynthetic Conversion of Chorismic Acid to Phenylalanine and Tyrosine

In terms of the number of individual naturally occurring substances arising through it, the importance of many of them, and the huge quantities of others (lignin!), the branch pathway leading from chorismic acid to phenylalanine and tyrosine is unquestionably one of the most significant and interesting areas of biosynthetic events in living nature. It yields, first, a common non-aromatic intermediate, prephenic acid, which, in many organisms, represents a secondary branch-point. Its structure permits extremely ready direct transformation to an aromatic intermediate, phenylpyruvic acid, which is then converted to phenylalanine by transamination. Alternatively, prephenic acid can be converted to p-hydroxyphenylpyruvic acid, the precursor of tyrosine. This, at least, is the series of processes which is found in those instances where phenylalanine and tyrosine are derived from prephenic acid by separate pathways; this is the case in *E. coli* and related bacteria, in *Neurospora*, and often in higher plants.[+] However, tyrosine can also be formed by direct hydroxylation of phenylalanine. This mode of formation is well established in mammals (1), and has been observed in higher plants (2) and in certain microorganisms. In mammals, the conversion of phenylalanine into tyrosine accounts for

---

[+]

Added in proof: It has been found very recently that blue-green bacteria carry out these two reactions in the reverse order, transamination of the side-chain preceding the aromatization of the ring. This variant (which involves pre-tyrosine as a new intermediate) is discussed on page 171.

the fact that the former, but not the latter, is an essential
amino acid (3). As a secondary modification of an existing aromatic ring, the hydroxylation of phenylalanine is, strictly speaking, outside the stated scope of this volume; however, in view of
its great significance, expecially for human nutrition, it merits
a brief discussion.

The reaction seems in general to involve a reduced pteridine
as cofactor. The phenylalanine hydroxylase in rat liver has been
investigated in detail by Kaufman (4). It requires L-*erythro*-di-
hydrobiopterin; the enzyme itself is an iron protein (4b), and no
outside supply of iron is needed. Studies by Guroff (5) on the
formation of tyrosine from phenylalanine in cell-free extracts
of a *Pseudomonas* adapted to growth on phenylalanine as sole carbon
source have shown that this process has much analogy with the one
in mammals; however, the cofactor here is L-*threo*-dihydroneopterin,
and addition of Fe^{2+}, replaceable by Hg^{2+}, Cd^{2+}, Cu^{2+}, or Cu^+, is
needed. In certain higher plants, especially legumes, the occur-
rence of the enzymes which convert prephenic acid to tyrosine (2,
6) demonstrates a biosynthesis analogous to that in *E. coli*, but
the hydroxylation of phenylalanine to tyrosine has been observed
by Nair and Vining (7) in cell-free extracts from spinach leaves;
here, too, a cofactor is involved which seems to be a pteridine.

7-1. PREPHENIC ACID [+]

Research on this intermediate began with an observation by
Simmonds (8) concerning the strange behavior of a phenylalanine
auxotroph of *E. coli*: growth of this strain on a limited supply of
the amino acid ceased when the supplement had been used up, but
was resumed at a rapid rate after a time lag of about 10 hours.
Filtrates from this second growth phase had considerable growth-
factor activity for phenylalanine auxotrophs, and it was subse-
qently shown (9) that at least part of this activity is due to
the presence of phenylalanine itself in these filtrates. In spite
of this, cells isolated during the second phase still fully retained their requirement for phenylalanine (8). The mutant is
thus capable of producing, at the later stages of its growth, an
excess of the very substance it was initially quite unable to

---

[+]
Prephenic acid is discussed in somewhat greater detail than the
other intermediates, since the class of substances to which it belongs present many points of chemical and biosynthetic interest,
and has not yet been reviewed elsewhere. Furthermore, part of
the information has been available so far only in preliminary form.

synthesize. This apparent ability of the strain to bypass its
genetic block, which at first seemed incompatible with the very
nature of genetic mutation, was explained simultaneously by Katagiri and Sato (9a) and by Davis (9b). These authors observed that
filtrates from the initial growth phase contain an excessively
sensitive, nutritionally inactive substance which is converted to
an active growth factor by brief autoclaving (9a) or by exposure
to even slight acidity at room temperature (9b). This latter observation explains the biphasic growth of the mutant: the pH of
the medium changes during the earlier growth phase from its initial value of 7.0 to an acidity (ph $\sim$ 6.5) sufficient for slow
conversion of the inactive substance to the active one. The latter factor is ninhydrin-negative and thus not phenylalanine itself,
from which it also differs on bioautograms. It can be converted
into the amino acid by the cells, a fact which explains the presence of phenylalanine during the second stage of growth. The
active precursor was identified by Davis (9b) as phenylpyruvic
acid [1]; this compound is known (10) to be capable of replacing
phenylalanine for mutants of *E. coli*.

Conversion of the initial, nutritionally inactive material
into [1] is also brought about by extracts of wild-type *E. coli*,
but not by those of the phenylalanine auxotroph (11), a finding
which suggests that this reaction constitutes a step in the biosynthesis of phenylalanine, and that the inactive substance is an
intermediate in this sequence which is devoid of growth-factor
activity because of a permeability barrier. This compound, at one
time provisionally named "Compound X" (9a) or "pre-phenylalanine"
(9b), has been termed prephenic acid (11).

Because of the extreme acid-sensitivity of prephenic acid,
the mutant cultures used for its isolation had to be grown in a
medium strongly buffered to pH 7.5 - 8.0 (9b), the highest value
which the organism will stand; any exposure to a pH below 7.0 during subsequent isolation steps had to be strictly avoided.

From filtrates of cells grown under these conditions, prephenate could be adsorbed on charcoal columns and eluted with
water. These eluates served for identification of the product of
acid-catalyzed transformation of prephenic acid as [1] by uv spectroscopy, characteristic green color of the Fe^{3+} chelate, and isolation of the 2,4-dinitro-phenylhydrazone (9b). The acid-catalyzed
conversion of prephenic to [1] makes possible a convenient spectrophotometric assay of the former, since [1] as the enolate ion shows
a very intense absorption band in alkaline medium: λ_{max} 320 nm.

Addition of barium acetate to the eluates (11) precipitated
impurities, while barium prephenate itself was brought down on
subsequent addition of methanol to the filtrate. Repeated repre-

cipitation from aqueous medium by methanol finally gave the barium salt in pure, crystalline form, albeit with very low recoveries. Improved and simplified methods for isolation of Ba prephenate with the help of column chromatography on ion exchange resins have been worked out subsequently using suitable mutants of *Neurospora* (12, 15). In addition, the conversion of barium chorismate to barium prephenate by gentle warming in aqueous medium at pH 8.0 (16) furnishes an alternative approach to preparation of the latter. A method for isolation of barium prephenate by this reaction has been briefly described by Cotton and Gibson (17). Since chorismic acid is much more stable than prephenic acid (whose preparation as the free acid has never even been tried and may well be impossible), and is thus much more readily handled, this technique may be the method of choice for obtaining barium prephenate. It seems probable (17b) that the *Neurospora* mutant used (12) for isolation of Ba prephenate actually excreted the then-unknown chorismic acid, which was converted to prephenic acid during the isolation procedures. This interpretation would account for the fact that the strain is auxotrophic not only for phenylalanine and tyrosine, but also for tryptophan.

During the isolation work (11) described in the preceding pages, it was observed that the effluents from the charcoal columns showed only end-absorption in the ultraviolet, even when they contained enough prephenate to give intense bands of the enolate of [1] after acid-treatment and addition of NaOH. This finding militated against an initial assumption that prephenic acid might be an aromatic compound related to [1], for example some exceptionally acid-sensitive conjugate of it. Such a substance would almost certainly contain the grouping $C_6H_5-CH_2-CO-COO-$ or its enol form, which is the chromophore responsible for the intense uv bands of [1] in acidic or alkaline medium. (It is true that some structures could be drawn in which the side-chain is modified in such a way that formation of this enolic chromophore $C_6H_5-C=C(OR)-CO$ would be impossible, for example, an acetal $C_6H_5-CH_2-C(OX)_2-CO-$, or an acid $C_6H_5-C(COOH)_2-CO-CO-$ which would lose 2 CO_2 at once on acidification; however, these structures are too implausible from the viewpoint of biosynthesis to deserve serious consideration). Consequently, the absence of specific uv absorption made it most unlikely that prephenic acid should be derived from [1] by modification of the side-chain, and suggested that the difference of the two compounds might reside in the ring, that is, that prephenic acid is a nonaromatic compound and that its facile conversion (*in vitro* and *in vivo*) into [1] constitutes a *de novo* formation of the benzene ring from an appropriate nonaromatic structure. On this basis, formula [2], with X left undetermined, seemed an acceptable expression for prephenic acid, since such a 2,5-cyclohexadienol would not show any selective uv absorption, and would be expected

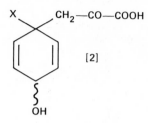

[2]

to undergo very smooth acid-catalyzed aromatization through protonation of the hydroxyl, followed by loss of H_2O and conversion of the resulting cationic species to [1] with simultaneous elimination of X. No cyclohexadienol with secondary hydroxyl was known at the time, and thus no suitable model compound was available; some related compounds with tertiary hydroxyl had been made long before (18, 19) through reaction of the corresponding cyclohexadienones with Grignard reagents and the like; they will be discussed below.

During isolation, prephenic acid was assayed spectrophotometrically after conversion to [1]. The apparent molecular weight of the Ba salt calculated on this basis decreased with progressive purification, to reach constant values of approximately 390–400. If prephenic acid were *mono*basic, this high value (cf. $C_6H_5-CH_2-CO-COOBa/2$ = 223.8) would require an unknown unit of molecular weight of approximately 160–170 to be lost during conversion to [1]. A more biosynthetically meaningful interpretation suggested, however, that Ba prephenate is the salt of a *di*basic acid, with the second carboxyl placed in such a manner that it has to be lost from the molecule by action of acid. This possibility led at once to the expression [2,X = COOH] for prephenic acid, which would be consistent with the known facts and would correspond to a barium salt $C_{10}H_8O_6Ba$ having a molecular weight of 361.5 compatible with the one found spectrophotometrically. This interpretation was confirmed by elementary analysis of the pure Ba salt, which gave values in excellent agreement with $C_{10}H_8O_6Ba \cdot H_2O$; ratio C:Ba 10:1 (in the Ba salt of [1], this ratio is 18:1). The conversion of prephenic acid to [1] would then be an aromatization with simultaneous decarboxylation. At the same time, the resulting formula [2], X = COOH, =[3], for prephenic acid fits very well with the structural features required for a biosynthetic intermediate between shikimic acid and phenylalanine or tyrosine: the results of

tracer experiments on the biosynthesis of these two amino acids
(20, 21) and of shikimic acid (22) in yeast and *E. coli* show that
in the formation of phenylalanine and tyrosine from shikimic acid
the C_3 side-chain attaches itself to position 1 of the ring, with
loss of the carboxyl from the same position. Evidently, then,
prephenic acid is an intermediate in this process in which the
carboxyl is still present, while the side-chain has already been
fixed at its final position, and in which the ring has undergone
changes which permit ready aromatization. Formula [3] is obviously compatible with this interpretation. The fact that this
aromatization must be accompanied by loss of one carbon atom
through decarboxylation is likewise uniquely explained by formula
[3].

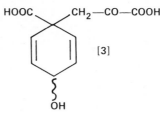

The constitution of prephenic acid [3] so derived was shown
(11) to be completely consistent with the physical and chemical
properties of the substance. As has been mentioned above, analytical results obtained with the pure barium salt were in excellent
agreement with those required for a monohydrate. Confirmation of
this result has been obtained subsequently by others (6a, 12, 14,
15) on samples prepared with the use of barium acetate. However,
Gamborg and Simpson (6a) observed formation of a different salt
when an excess of barium hydroxide was used instead of barium acetate. This well-crystallized hygroscopic salt gave values for C,
H, and Ba which establish its composition as $(C_{10}H_9O_6)_2Ba \cdot H_2O$; its
ir spectrum is quite different from that of the neutral salt (11,
14, 16). This latter spectrum lacks bands ascribable to an aromatic ring (11), in agreement with formula [3]. A carbonyl band
at 1712 cm^{-1} is present; cf. the one of sodium pyruvate at 1715
cm^{-1} (23). The nmr spectrum of Ba prephenate likewise lacks signals in the aromatic region (24). In the uv, only fairly intense
end-absorption occurs above 220 nm, undoubtedly caused by the α-
keto acid grouping in the side-chain (cf. the band of pyruvic acid
at 210 nm (ε 400), and the marked decrease of the end-absorption
on reduction with $NaBH_4$, which is discussed below). Barium prephenate is optically inactive (11, 14), as required by formula [3].

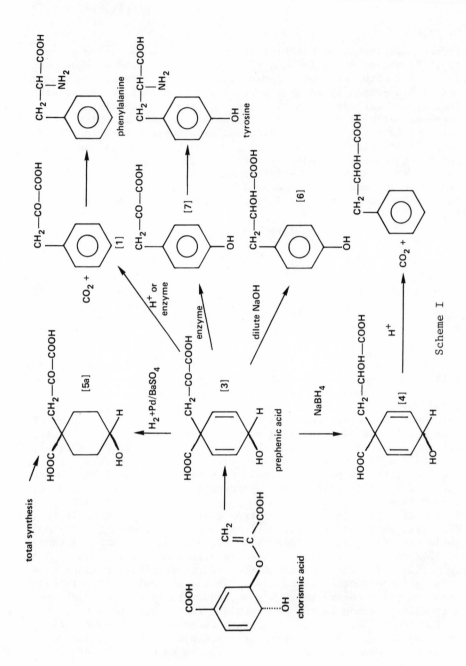

Scheme I

150

It is stable at room temperature, moderately soluble in water, and quite insoluble in organic solvents.

The observed physical properties of its Ba salt are thus thoroughly consistent with structure [3] for prephenic acid. Conclusive proof for this constitution was then obtained (11) through chemical transformations. The evidence for constitution and stereochemistry of [3] is summarized in Scheme I.

As expected, treatment of Ba prephenate with acid produced one mole each of [1] and CO_2; the latter was determined manometrically. Catalytic hydrogenation resulted in uptake of hydrogen well in excess of that compatible with an aromatic structure: with PtO_2 as catalyst, Ba prephenate absorbed between 3 and 4 moles of hydrogen under conditions where [1] took up the expected one molecule. Consumption in excess of the 3 moles required for saturation of the double bonds and reduction of the carbonyl can be explained by hydrogenolytic removal of the doubly allylic hydroxyl in part of the material. Occurrence of this reaction was subsequently established by Plieninger and Keilich (25) in the synthetic model compound 1-methyl-1-dichloromethyl-2,5-cyclohexadiene-4-ol, which was reduced to 1-methyl-1-dichloromethylcyclohexane over Pd/charcoal in methanol with uptake of 3 moles of hydrogen. However, it depends upon the nature of the catalyst and upon the conditions used whether this hydrogenolysis will take place. In both the above model compound and in Ba prephenate, only the double bonds were reduced over $Pd/BaSO_4$, the hydroxyl being retained. Similarly, Metzenberg and Mitchell (12) observed uptake of 3.03 moles of hydrogen on reduction of Ba prephenate in water with a mixed catalyst (Adams' platinum and $Pd/BaCO_3$); nonreactivity of the reduction product towards 2,4-dinitrophenylhydrazine indicated disappearance of the carbonyl, and hence nonoccurrence of hydrogenolysis.

Additional evidence for structure [3] was obtained (11) from the product, [4], of the reduction of Ba prephenate with $NaBH_4$. The marked end-absorption of [3] $\varepsilon_{260}{\sim}210$), which might conceivably have hidden the fairly weak band in this region given by $C_6H_5{-}$, is decreased in [4] to a value ($\varepsilon_{260}{\sim}20$) quite incompatible with the presence of an aromatic ring. The uptake of two molecules of bromine by [4] gave further proof for the two reactive double bonds; the formation of CO_2 and, presumably, phenyllactic acid on treatment of the reduction product with acid showed that the evolution of CO_2 does not depend upon the α-keto-acid grouping in the sidechain. The smooth uptake of the expected amount of bromine is actually somewhat surprising, since several anomalies in the addition of bromine by 1,4-cyclohexadienes have been observed; cf., *inter alia*, the formation of a *di*bromide by 1,4-dihydrobenzoic

acid (26a), and the failure of some of its 1-substituted derivatives to react with bromine at all (26b). However, other synthetic model compounds, to be discussed below, did resemble [3] in
adding the expected 2 moles of bromine. For the total synthesis
of a product [27] with the structure of [4] but undetermined
stereochemical homogeneity and configuration, see below, page
160.

Formula [3] obviously represents two *cis-trans* stereoisomers, [3a] and [3b]. This problem of stereochemistry was
solved by Plieninger and Keilich (25) through catalytic reduction (Pd/BaSO₄) of [3] to a tetrahydroprephenic acid. The 2,4-
dinitrophenylhydrazone of this substance was compared by paper
chromatography with the corresponding derivatives of the two
synthetic stereoisomeric (1-carboxy-4-hydroxycyclohexyl)-
pyruvic acids [5a] and [5b] of unequivocally known structure
and configuration (27a,b). The compound resulting from reduction of [3] proved identical with [5a]. This work provides

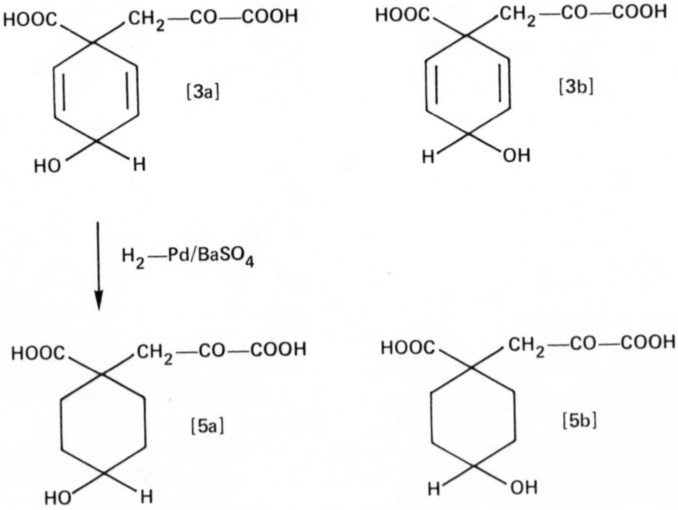

definitive proof for structure [3] by connecting prephenic acid
with a substance prepared by straightforward total synthesis,
and it establishes that it has configuration [3a].

It is interesting that a formula embodying a 2,5-cyclo-
hexadienol ring had been postulated by Sir Robert Robinson (28)
for a hypothetical precursor of ephedrine already before publication of the structure of [3].

From what has been said in the preceding paragraphs, it is
clear that the acid- or enzyme-catalyzed conversion of [3] to [1]
and CO_2 is by far the most important reaction of the compound;
it constitutes the biosynthetic process responsible for the formation of the aromatic ring of phenylalanine and hence of vast
numbers and quantities of further aromatic substances. The
mechanism of this reaction is thus of great interest. That it
takes place with remarkable ease *in vitro* under conditions of
acid catalysis has already been mentioned; it proceeds with
appreciable speed at room temperature already at pH 6 (11) and
is very rapid in stronger acid (9b,11,12,21). The extraordinary
ease of this reaction, which is paralleled in certain synthetic
cyclohexadienols (see below), is evidently due to the driving
force provided by the formation of the stable aromatic ring.
It can hardly be doubted that the acid-catalyzed transformation proceeds in the manner formulated in Scheme II, that is,
by protonation of the hydroxyl and subsequent aromatization with
extrusion of the carboxyl (11,29); it was indeed the recognition
of the possibility of ready acid-promoted aromatization of the
cyclohexadienol ring which led to the initial postulate of the
presence of this unit in [3] and hence to the elucidation of the
structure of the compound. There seems little reason to doubt
that the enzyme-catalyzed biosynthetic reaction proceeds by an
analogous mechanism.

Besides the acid- or enzyme-catalyzed aromatization to [1]
compound [3] can undergo an alternative aromatization on being
warmed with dilute alkali, when *p*-hydroxyphenyllactic acid [6] is
formed (30). This reaction, for which no biological significance
is known so far, is undoubtedly an internal oxido-reduction of
the Cannizzaro type, again proceeding with exceptional ease under
mild conditions because it leads to an aromatic system. The
formulation as an intramolecular hydride-ion shift presented in
Scheme III has not been experimentally proved (the reaction has
never been studied in detail), but seems very attractive in view
of the established stereochemistry of [3]. See Addendum p. 173.

The mechanism shown in Scheme III has now been unequivocally established; see the Addendum, page 173.

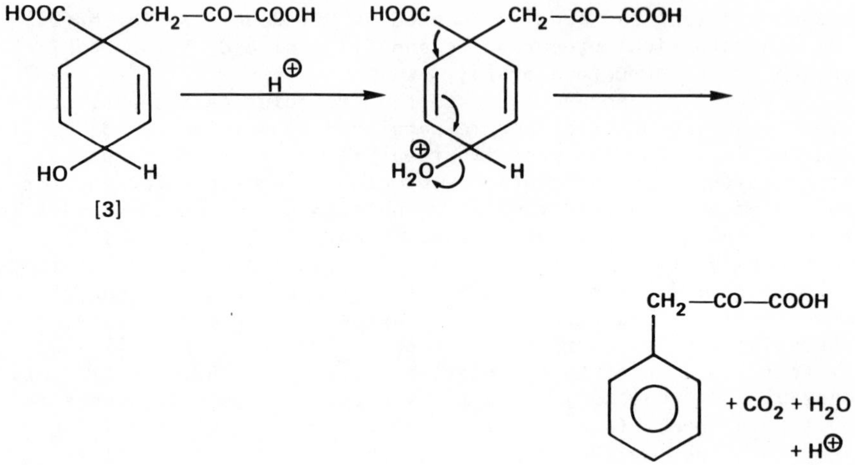

Scheme II

a.   Prephenic Acid as a Source of Tyrosine.

On purely chemical grounds, [3] could be a precursor of tyrosine as well as of phenylalanine; some dehydrogenation reaction might transform it to a species which would retain the oxygen of the hydroxyl during a subsequent aromatization reaction. Such a derivation of tyrosine from [3] (in those cases where the amino acid is not formed by direct hydroxylation of phenylalanine) would be consistent with the close biogenetic relationship of phenylalanine and tyrosine, which has long been recognized. A late common intermediate in their biosynthesis had been postulated by Davis (31) on the basis of the occurrence of one-step mutations resulting in a specific requirement for both amino acids, and of the observation that mutants blocked in the biosynthesis of one of them excrete excesses of the other one. The parallel way in which the production of phenylalanine and tyrosine in E. coli is influenced by incompleteness of a block between dehydroshikimic and shikimic acids, and by the competitive inhibition of the utilization of shikimic acid by dehydroshikimic acid (32) (see Section 4-3), is likewise suggestive of such a particularly close relationship.

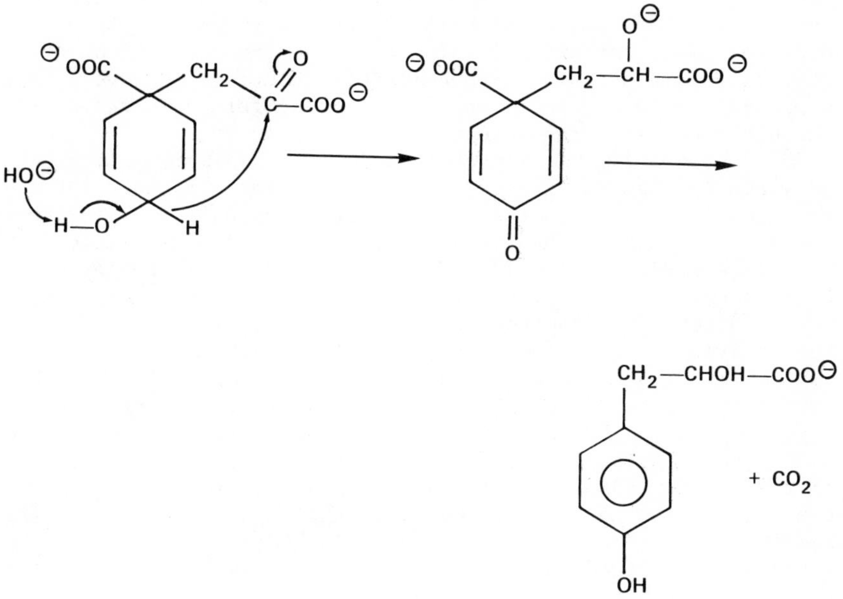

Scheme III   [6]

The postulated dehydrogenation and aromatization would give CO_2 and p-hydroxyphenylpyruvic acid [7]. This acid is able to replace tyrosine in E. *coli* (10a,b) and *Neurospora* (33); it is converted to tyrosine through a transamination reaction which was first studied in E. *coli* by Rudman and Meister (34). The two α-keto acids and their conversion to phenylalanine and tyrosine, respectively, will be discussed in somewhat greater detail in Section 7-2.

Experimental evidence for the assumption that [3] is a precursor of tyrosine as well as of phenylalanine came initially from the observation (35) that certain *tyrosine* auxotrophs of E. *coli* excrete a substance which behaves like [3] on charcoal columns and gives [1] on acidification; this compound was, however, not isolated in pure form. Conclusive proof for this role of [3] was furnished by Schwinck and Adams (17), who obtained from a phenylalanine auxotroph of E. *coli* an enzyme, prephenate dehydrogenase, which transforms [3] into [7] in the presence of NAD^+. Addition of pyridoxal phosphate and glutamate led to tyrosine. Somewhat prior to this work, it had been observed (36) that extracts of a tyrosine auxotroph of *E. coli* convert [3] into *p*-hydroxyphenyllactic acid [6]. However, this compound was subsequently shown (17) to be formed in the extracts through secondary transformation of [7].

Regrettably, the detailed chemistry of the conversion of [3] into [7] has apparently not been studied. The most attractive mode for this process would be dehydrogenation of [3] to the corresponding dienone [8]; as a vinylogous β-keto acid, this intermediate would lose CO_2 to give initially the hypothetical ketonic tautomer of [7]; see Scheme III. This process has been mentioned by Schwinck and Adams (13); it is supported by the behavior of the diethyl ketal of [8] diethyl ester, and of several other 4-carbethoxy-2,5-cyclohexadienones which have been synthesized by Plieninger and co-workers (37). Esters of this type are stable, but alkaline saponification followed by treatment with acid leads at once to decarboxylation with formation of the corresponding phenol, for example [7], in the case of the acetal just mentioned. However, the sequence by way of [8] is not the only one that could be envisaged; the reaction might alternative proceed in one step as an oxidative decarboxylation (13), for which there is much precedent.

The question of the actual occurrence of [8] as a discrete intermediate should be accessible to experimental scrutiny. There seems to be no reason why salts of such a vinylogous β-keto acid should not be capable of existing, even if the free acid is not. It might then be possible to isolate salts of [8] from culture filtrate of suitable tyrosine auxotrophs, just as the Ba salt of [3] has been isolated.

At the time of the proof of the constitution of [3], its 2,5-cyclohexadienol ring was quite unprecedented. A few additional naturally occurring compounds of this type, as well as two 2,5-cyclohexadienylamines, have been discovered since then (see Chapter 8). The novelty of this structural unit, and the interest inherent in [3] as the non-aromatic direct precursor of aromatic metabolites, made the study of suitable <u>model compounds</u> very desirable. However, the only related substances then known were tertiary cyclohexadienols prepared from cyclohexadienones by action of Grignard reagents or, in a few cases, by Reformatzky reaction. The first compounds of this type, [9], had been prepared by von Auwers and Keil (18, 38) through treatment of the dienone [10] with methyl magnesium bromide. In subsequent classical investigations by von Auwers, Zincke, and their co-workers, many additional cyclohexadienols have been

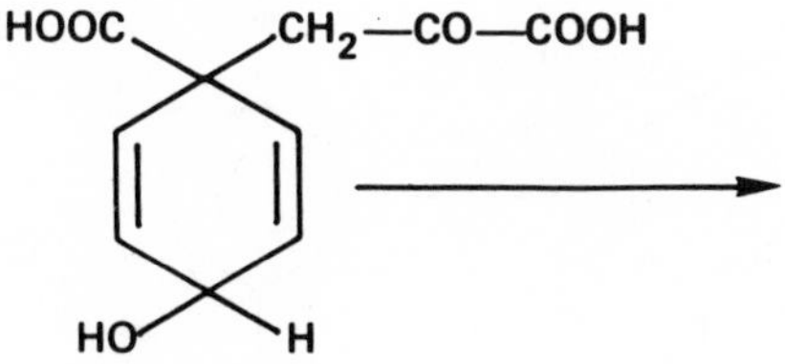

[3]

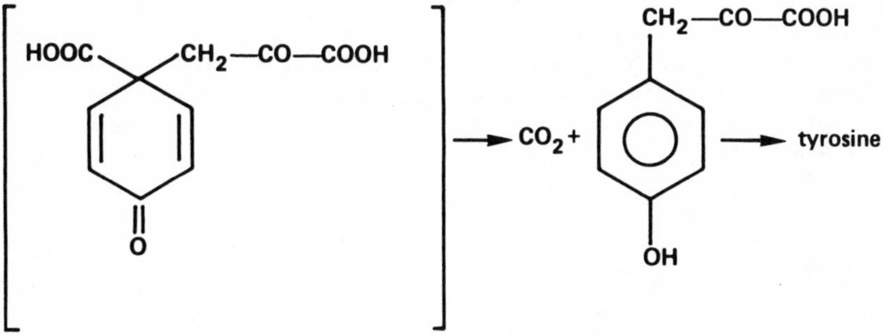

[8]   [7]

prepared; a few leading references (19, 39, 40) may be useful.

All those compounds resemble [3] in undergoing ready acid-catalyzed aromatization; however, most of them differ from it in the mode of this reaction, which usually follows a path not available to cyclohexadienols with secondary hydroxyl: those with tertiary OH, for example [9], undergo extremely ready acid-catalyzed dehydration to so-called "semibenzenes" such as [11] (39), which in turn are converted by acid to aromatic compounds through various migrations of one of the *gem*-substituents in position 4. Biosynthetic aromatizations by way of semibenzenes have apparently not been observed so far; however, brief mention of this reaction sequence seems justified, since such a process may yet be found to play a role in the biosynthesis of aromatic substances.

Because of the intermediacy of semibenzenes, and of the overall aromatization with migration rather than elimination of a group (cf. the carboxyl of [3]), these compounds are less than ideally suited as models. However, one of them, the

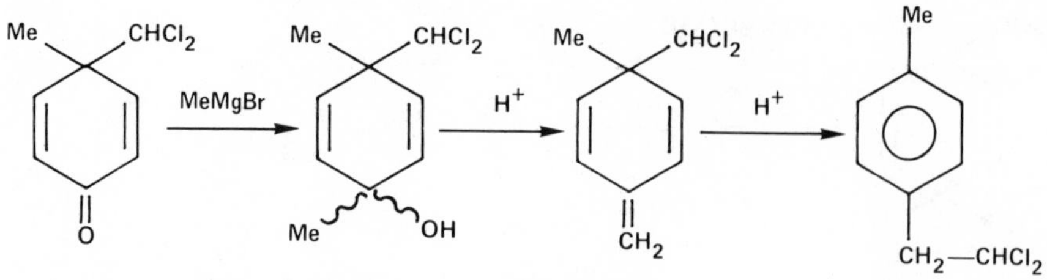

[10]   [9]   [11]

phenylcyclohexadienol [12] of von Auwers and Jülicher (39f)
comes close. This fairly stable substance is obviously unable
to form a semibenzene in the usual fashion, yet it does give
aromatic substances with cold formic acid. This reaction was
shown much later by Newman and co-workers (41) to produce mainly
[13] together with some [14]. While formation of latter takes
place with 1,2-migration of methyl, that of [13] involves loss
of the CCl3, a process fairly analogous to the conversion of
[3] into [1].

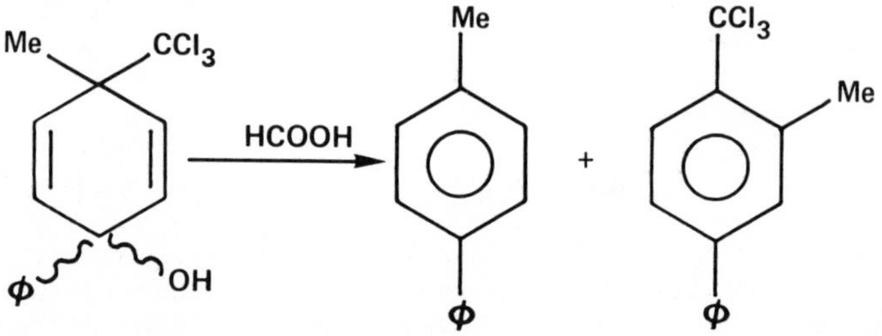

[12]   [13]   [14]

    The first synthetic cyclohexadienols with secondary
hydroxyl, [15] and [16], were synthesized as models for [3] by
Plieninger and Keilich through Meerwein-Ponndorf reduction of the
corresponding cyclohexadienones, for example [10] (42). Later on,
hydride reduction was used (43). In this case, reduction of double
bonds can take place (44), although usually only as a side-
reaction. Cyclohexadienols [15] and [16] were crystalline, relatively stable compounds which survived action of hot alkali or
brief heating above the melting point. Like [3], they lacked
selective uv absorption. Only one of the two possible stereoisomers was observed. Action of acid led to aromatization with
migration of one of the *gem*-substituents, to give [17] and [18],
respectively. The name "dienol-benzene rearrangement" was

introduced (42) for this reaction, in analogy to the well-known
dienone-phenol rearrangement.

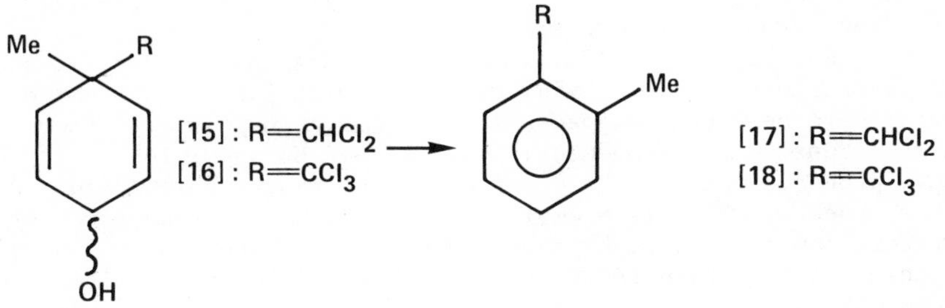

[15] : R$=$CHCl$_2$
[16] : R$=$CCl$_3$

[17] : R$=$CHCl$_2$
[18] : R$=$CCl$_3$

Study of the catalytic hydrogenation of the two dienols
showed (25) strong dependence upon the nature of the catalyst.
The reaction can saturate the double bonds with retention of
hydroxyl group and chlorine atoms (Pd/BaSO$_4$ in presence of a
small amount of triethylamine), or it can in addition also
remove the hydroxyl (Pd/C) or the chlorine atoms (Raney nickel).
In the first two instances, the results are quite analogous to
those obtained with [3].

Properties and behavior of the model cyclohexadienols are
thus very similar to those of [3], and hence give valuable
support to its structure. However, the model compounds again
differ from [3] through the fact that during their aromatization one of the two *gem*-substituents in position 4 migrates,
instead of being lost from the molecule.

Preparation and dienol-benzene rearrangement of many
other 2,5-cyclohexadienols have been studied by Plieninger and
co-workers (37a,b,45) and by other groups, usually with results
similar to the ones just discussed; particularly, the behavior
of steroidal 1,4-dien-3-ols has been examined in detail. In
the work of Gentles and associates (46), for example, reduction
of two 1,4-dien-3-ones of the steroid series with LiAlH4 gave
the corresponding dienols, which underwent dienol-benzene
rearrangement to 4-methylestra-1,3,5(10)-trienes on chromatography over Florisil, or on one minute's heating with dilute
acetic acid, i.e. with an ease strongly reminiscent of the
behavior of [3]. Related reactions have been studied by
Dannenberg (47), who examined the possibility of *in vivo* conversion of steroid ketones to nonphenolic carcinogenic compounds
through reduction to the dienol and dienol-benzene rearrangement. Caspi and co-worker explored similar reactions of

steroidal dienols (48) and established (49) the detailed mechanism by which the resulting 4-methyloestra-1,3,5(10)-trienes are formed; the reaction proceeds as shown in Scheme IV.

Other aspects of the dienol-benzene rearrangements have been studied by Schmid and co-workers (50). Related 2,4-cyclohexadienols have been prepared by them during this work, and by Hart and associates (51). In a few instances, acid-catalyzed aromatizations of cyclohexadienylamines with loss of the amino group have been encountered which are quite analogous to the dienol-benzene rearrangement. Wei-Zan and Huang-Minlon (52) have proposed the name "dienamine-benzene rearrangement" for this process, which is exemplified (52,53) by the conversion of artemisinamine [23] to hypoartemisin [24]. Since two cyclohexadienamines have been found recently as microbial metabolites (see page 187), the process might well have biological significance. Other rearrangements of this type have been observed in the extensive studies of Hey and co-workers (58A).

Esters and salts of 4-carboxy-2,5-cyclohexadienols, close enough to [3] to constitute truly representative model compounds, were finally prepared by Plieninger and his associates (29,37). The compounds in question are the esters [25] (37a,b), [26] (37c) and [27] (37b), and the corresponding Na and Ba salts. Evidently, [26] is the diethyl ester diethyl ketal of [3] and/or its epimer [3b], while [27] is structurally identical with the product [4] of borohydride reduction of [3] which has been mentioned before. Compound [26] has been shown to be a mixture of the two possible epimers; compound [25] seems to be stereochemically homogeneous, but its configuration remains unknown.

These dienols were prepared by borohydride reduction of the corresponding dienones [28] (37a,b) [29] (37c), and [30] (37b); compound [29] is the diethyl ester diethyl ketal of [8], the presumed intermediate between [3] and p-hydroxyphenylpyruvic acid [7]. The dienones themselves are accessible in a variety of ways. Compound [28], for example, can be obtained by alkylation of 1,4-dihydrobenzoic acid, available (26a) through Birch reduction of benzoic acid, with chloroacetic acid and KNH_2 (26b) and oxidation of the resulting 1-carboxy-2,5-cyclohexadieneacetic acid with t-butyl chromate (37a), or through introduction of double bonds into 1-carboxy-4-cyclohexanoneacetic acid (27b; cf. also ref. 54). However, the most versatile approach consists in synthesis of the corresponding cyclohexenone, followed by introduction of the second double bond; it is exemplified for [29] in Scheme V.

On treatment with alkali followed by acidification, the dienones [28] and [29] are converted to p-hydroxyphenylacetic acid and [7], respectively.

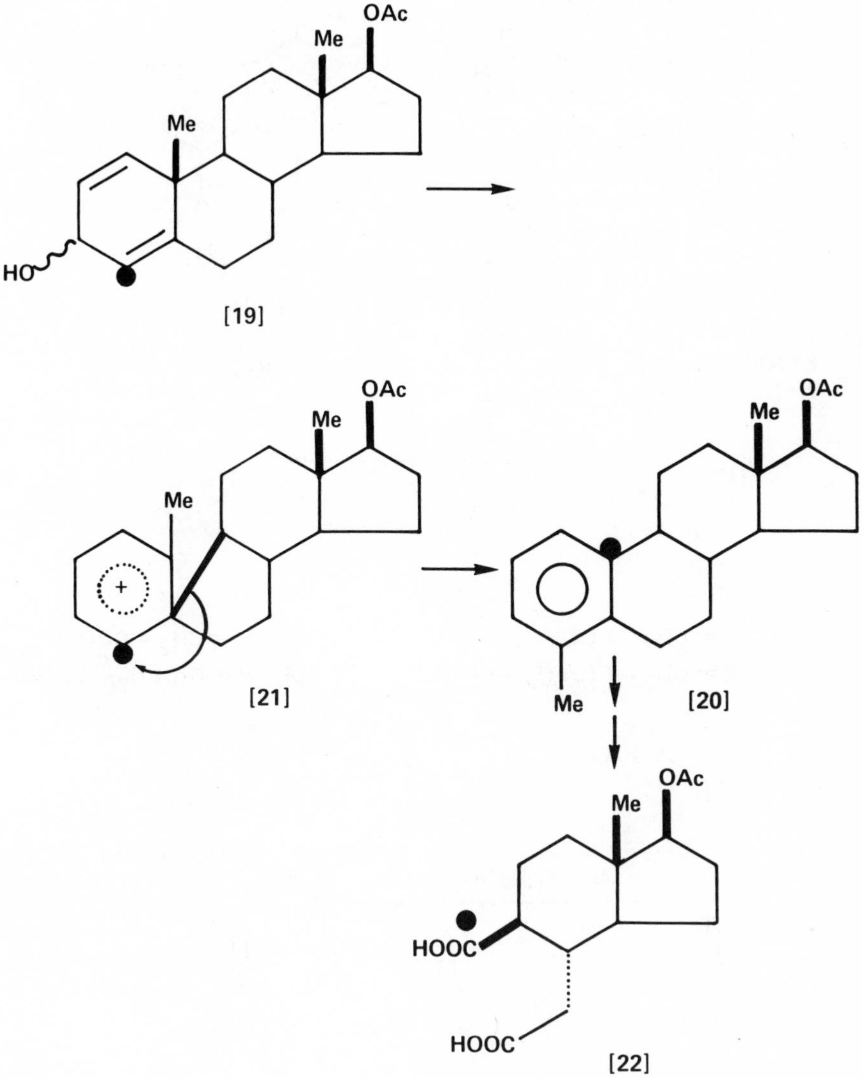

[19]

[21]

[20]

[22]

Scheme IV

161

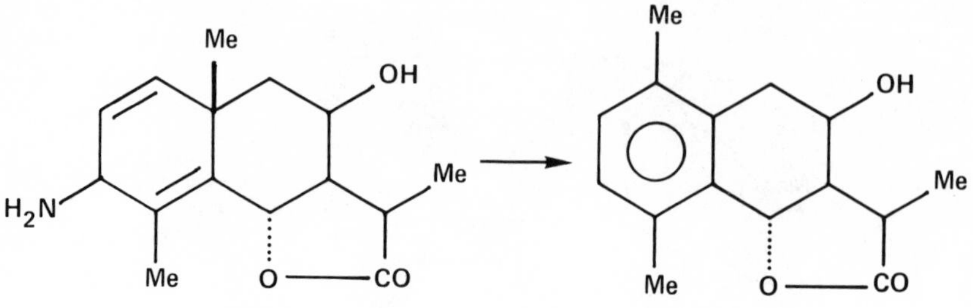

[23]   [24]

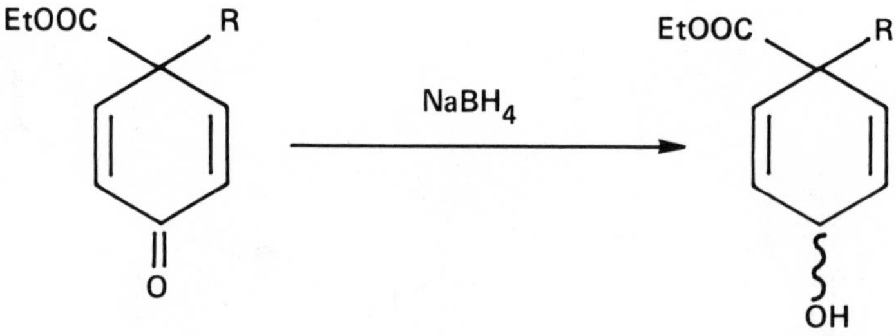

[28] : R=CH$_2$—COOEt   [25] : R=CH$_2$—COOEt
[29] : R=CH$_2$—C(OEt)$_2$—COOEt   [26] : R=C(OEt)$_2$—COOEt

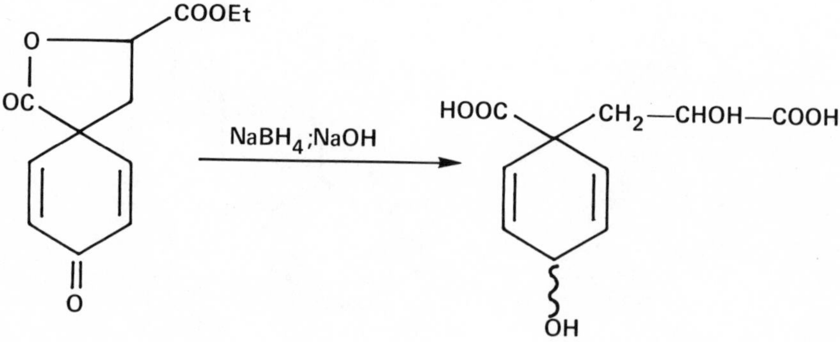

[30]   [27]

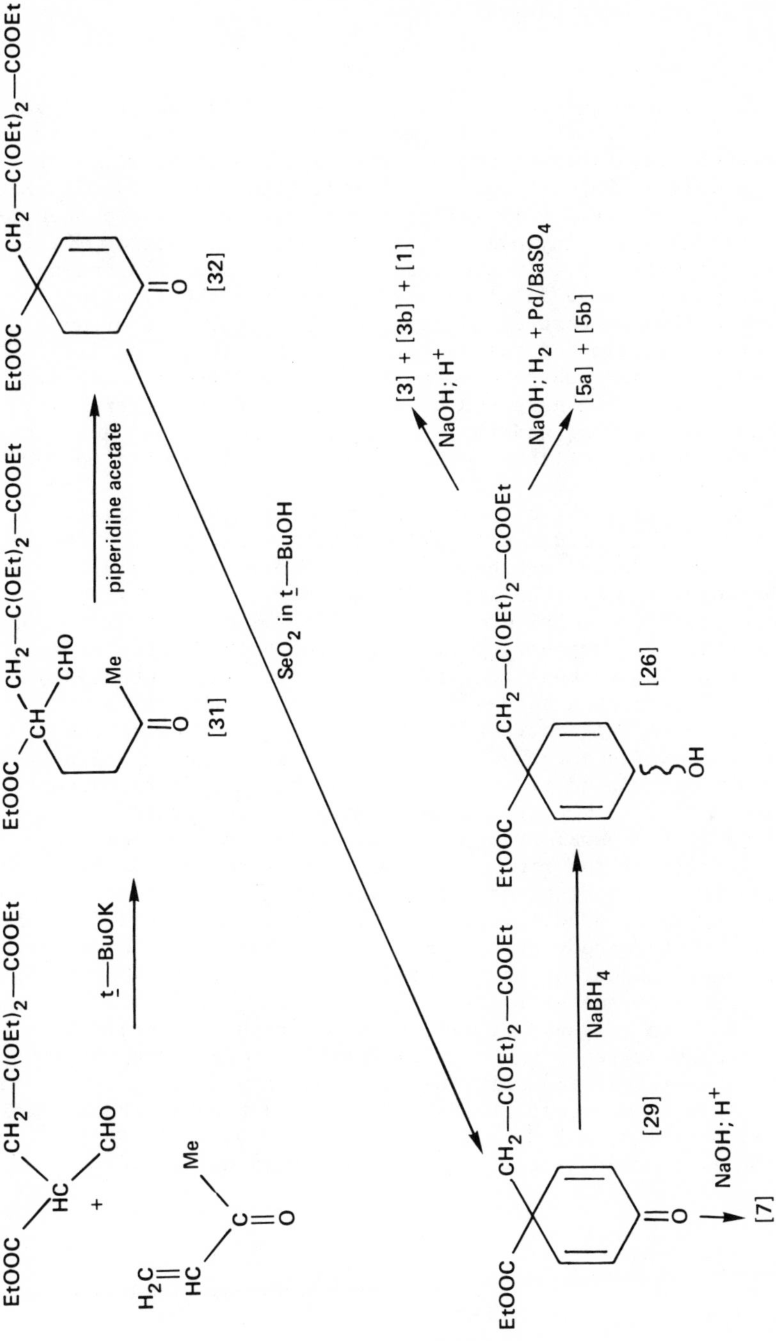

Scheme V

163

The cyclohexadienols [25], [26] and [27] resemble [3] very
closely. Like it, they lack specific absorption down to 240 nm.
The dienol [25] takes up the expected 2 moles of hydrogen or
bromine (37b). Saponification of [25], [26], and [27] with NaOH
at room temperature, followed by acidification, yielded phenylacetic (37a), phenylpyruvic [1] (37c), and phenyllactic (37b)
acids, respectively. In the case of [25], the corresponding
Ba salt was isolated (37b) in solid form from the alkaline
saponification mixture by addition of Ba acetate and ethanol.
Catalytic reduction of the analogous salt derived from [26]
over Pd/BaSO4 and subsequent treatment with 2,4-dinitrophenyl-
hydrazine in 2N HCl gave the hydrazones of the two tetrahydroprephenic acids [5a] and [5b] (27a,b), proving beyond doubt that
[26] is a mixture of the diethyl ketals of [3] and its epimer
[3b] (37c).

The similarities of physical properties and chemical
behavior of [25], [26], and [27] with those of [3] thus furnish
additional evidence for the structure of prephenic acid.
Furthermore, the synthesis of [26] and [29] offers possible
approaches to the total synthesis of [3] itself.[+] Solution of
this problem is made very difficult through the excessive sensitivity of [3] to most reagents. In principle, synthesis of
[3] and [3b] through borohydride reduction of the Na salt
obtained on saponification of [29] should only require the
liberation of the carbonyl in the side-chain; protection of this
group during hydride reduction is obviously needed to prevent
formation of [27]. However, the removal of the protecting ketal
grouping by the usual treatment with acid leads concomitantly to
aromatization of the ring. In experimental tests (29, 37c), the
intermediate appearance of the desired cyclohexadienols was demonstrated by spectroscopic techniques; Ba salts isolated from the
reaction mixture consisted of much Ba phenylpyruvate together
with a small amount of the Ba salts of these cyclohexadienols.
Since action of the extract from wild-type *E. coli* upon this mixture led to an increase of [1], some Ba prephenate seems actually
to have been present, and to this extent a total synthesis of [3]
has been achieved. Ideally, an efficient synthesis of [3] should
be possible from a salt of [8], if a method for the stereospecific
reduction of the carbonyl in the ring could be found, and if
the one in the side-chain could be protected during this

[+]
Added in proof: This synthesis has now been accomplished; see
Addendum on page 173.

reduction by a group that can be removed subsequently without the use of acid, or of other reagents that alter [3], such as warm dilute alkali.

It will be noted that in the 2,5-cyclohexadienols mentioned so far, including [3] (and in all those still to be discussed), C-4 (that is, the atom in *para*-position to the one bearing the hydroxyl) is a quaternary. Related compounds carrying hydrogen in position 4 seem to be as yet unknown, although there is no reason why such substances should not exist. A precedent among naturally occurring compounds has been found in the 2,5-cyclohexadienyl amines stravidin and MSD-235S$_2$, which will be discussed below (see page 187). Chemical synthesis of 2,5-cyclohexadienols with mono-substituted 4-position would, however, require an approach fundamentally different from the one used up to now, in which the cyclohexadienol system has invariably been produced by reduction of the corresponding 2,5-cyclohexadienone. A ketone of this type carrying a hydrogen atom at C-4 would evidently be the hypothetical keto form of a 4-substituted phenol, and would thus be incapable of independent existence. One such alternative approach might be introduction of a hydroxyl into the methylene group of a 1,4-cyclohexadiene, for example, the 1,4-dihydrobenzoic acid obtained on Birch reduction of benzoic acid (see page 160). However, oxidative transformations of this acid led mainly to aromatization (26a).

Prephenic acid [3] is the substrate of two enzymes which catalyze the first steps of its conversion to phenylalanine and tyrosine, respectively. The enzyme that promotes the transformation into phenylpyruvic acid [1] on the path to phenylalanine was initially termed prephenate aromatase but is now known as prephenate dehydratase, while the one concerned with the formation of *p*-hydroxyphenylpyruvic acid [7] as the first step towards tyrosine is called prephenate dehydrogenase (13).

Prephenate dehydratase activity was observed (11) in extracts from wild-type *E. coli*. but was absent from those of mutants blocked between [3] and [1]. The enzyme has been found subsequently in many other microorganisms; the prephenate dehydrogenase of a *Pseudomonas* species has been studied in some detail by Cerutti and Guroff (56). Gamborg and Simpson (6a) found prephenate dehydratase activity in mung bean seedlings (*Phaseolus aureus*).

In the conversion from [3] to [7], the dienone [8] is a plausible, if unproven, intermediate. Prephenate dehydrogenase, the enzyme responsible for this reaction, was first studied in *E. coli* by Schwinck and Adams (13), who found it to be dependent upon NAD$^+$. As expected, it was not detectable in extracts from tyrosine auxotrophs. This enzyme, too, was subsequently observed in many microorganisms. The studies of Gamborg and co-workers have established the occurrence of a prephenate

dehydrogenase in some plants of the family Leguminosae; it
was found in mung bean (*P. aureus*) (6a,c), and in cell cultures
from cotyledons, hypocotyls, and roots of this species, from
roots of wax bean (*P. vulgaris*) and soy bean (*Glycine max*) (6b).
However, similar cultures from plants of other families did not
show any such activity. The enzyme from cotyledons of *P. vulgaris* has been purified about tenfold (6c). The prephenate dehydrogenase from these plants differs from the one of *E. coli* by
requiring $NADP^+$ rather than NAD^+ as cofactor (6a,b,c); it is
inhibited by sulfhydryl reagents, but not by phenylalanine or
tyrosine (6c). The presence of this enzyme suggests that the
major pathway of tyrosine biosynthesis, at least in these
leguminous plants, parallels the one in bacteria; however, the
alternative biosynthesis through hydroxylation of phenylalanine
has been observed (7) in acetone powders from spinach leaves
(see page 145).

Since [3] is the branch point of two highly important biosynthetic pathways, the enzymatic reactions in which it plays
a role are subject to finely balanced control mechanisms which
ensure that the alternative end products phenylalanine and
tyrosine are produced in precisely the required amounts. The
general features of such steering devices have been discussed
in Section 1-3, and those of them which apply particularly
to the shikimate pathway will be brought together in Chapter
15, but a brief account of the controls of formation and transformations of [3] seems required here. As in other cases, the
regulation of the biosynthetic reactions leading to phenylalanine and tyrosine, respectively, is achieved in very different
ways by different organisms, even though the actual chemical
steps are the same ones. A comprehensive discussion of all
those types of control is not our purpose, but a few observations of fairly general interest should be presented here.
The modes of such control in microorganisms are covered more
completely in several reviews (57,59).

Among the cases investigated, that of *E. coli* and other
enteric bacteria has been studied in greatest detail. In *A.
aerogenes* and *E. coli*, Cotton and Gibson (60) observed two
isoenzymes of chorismate mutase, the enzyme responsible for
the formation of [3] from chorismic acid; these enzymes were
called chorismate mutase P and T, respectively. Each of these
is closely associated with one of the enzymes which catalyze
the first step in the further conversion of [3] into phenylalanine and tyrosine: isoenzyme P with prephenate dehydratase,
isoenzyme T with prephenate dehydrogenase. The aggregates were
named proteins P and T. They were not separable into the constituent enzymes by chromatography or other techniques;

presumably, they are both single bifunctional enzyme proteins
(61).   Protein T from *E. Coli* has been purified to homogeneity
(62), the one from *A. aerogenes* nearly so (63).   This extensive
purification did not alter the ratio of the two enzymatic activities, which are also in general affected together by single-
step reversible mutations and tend to react in parallel fashion
to the influence of various agents.

The regulatory role of these proteins is clear from the
action upon them of the two aromatic amino acids.   In *A.
aerogenes*, both enzymatic activities of protein P are inhibited
by phenylalanine (60).   Of the two enzymatic functions of protein T, in both *E. coli* and *A. aerogenes*, only the prephenate
dehydrogenase in inhibited, strongly, by tyrosine, while chorismate mutase T is not (57a,60); however, both activities are
strongly repressed by tyrosine.   In protein P of *E. coli*, only
the prephenate dehydratase is inhibited by phenylalanine (57a,
b).   (In *A. aerogenes*, column chromatography on DEAE cellulose
showed (60) the existence of another species of prephenate
dehydratase, not inhibited by phenylalanine and not associated
with chorismate mutase).

From *Salmonella typhi-murium*, Schmit and Zalkin (64) prepared a similar aggregate of chorismate mutase and prephenate
dehydratase.   Both of these activities are inhibited by phenylalanine, although not to the same extent.   This enzyme protein
was likewise obtained in highly purified form without appreciable
change in the ratio of the two enzymatic activities.   Kinetic
study of the overall reaction from chorismic to phenylpyruvic
acid [1] showed that [3] appears as a transient free intermediate.   A mutant enzyme from a phenylalanine auxotroph
exhibited chorismate mutase activity but lacked the prephenate
dehydratase (65).   This fact, and the finding that the latter
enzyme is selectively inactivated by certain reagents, prove
the existence of different active sites for the two enzymatic
functions.   For protein T of *A. aerogenes*, however, Koch, Shaw,
and Gibson (66) concluded that the active sites appear to be
close to each other or to overlap; no selective inactivation
was obtained and [3] remains bound to the enzyme in this system.

A similar complex of chorismate mutase and prephenate
dehydratase seems to exist in a *Pseudomonas* (56).

Like many other regulatory enzymes (cf., e.g., refs. 61
and 64b, and literature quoted there), the bifunctional enzymes
just discussed are capable of reversible association and dissociation; these changes profoundly influence the enzymatic
activities. Gibson and his co-workers found that protein T of
Aerobacter is split reversibly into two identical or at least
very similar subunits of molecular weight of about 40,000

(62) on dilution (61) or under the influence of acid, sodium
dodecyl sulfate (67), or urea(68). Those subunits are enzymatically inactive (67). The dissociation can be prevented by
[1] or p-hydroxyphenylpyruvic acid [7], but is not influenced
by tyrosine (67). In the case of the aggregate of chorismate
mutase and prephenate dehydratase from *Salmonella*, however,
Schmit and Zalkin (64b) found that the active protein is a
monomer of molecular weight 109,000, which changes to an
inactive dimer under the influence of phenylalanine, the feedback inhibitor of both enzymatic activities. Enzymes made
insensitive to this inhibition by mutation or chemical treat-
meant no longer dimerize. Feedback control by phenylalanine
thus seems in some way connected with the dimerization, but
this relationship is not a simple, straightforward one. For
instance, the enzyme is maximally associated at concentrations
of phenylalanine much smaller than those required for 50%
inhibition of the two enzymatic activities.

Such complexes of chorismate mutase with the two enzymes
which use [3] as their substrate have apparently been found
so far only in enteric bacteria. Occurrence of these enzymes
has been observed in many other organisms, but their regulation is here achieved by different means. A few examples may
be of interest; no complete coverage is intended. In *B. subtilis*, for example, Nester and his co-workers (69) have demonstrated the existence, *in vitro* and probably also *in vivo*, of
the regulatory aggregate of chorismate mutase with DAHP synthetase and shikimate kinase which has already been discussed
(see Chapter 3 and Section 5-2); however, the two prephenate-
based enzymes are not part of this complex. In some strains of
this organism, chorismate mutase exists in two or three
different forms (70).

The common types of regulation by feedback inhibition or
repression have been observed repeatedly for prephenate dehydratase and dehydrogenase. In *B. subtilis*, both enzymes are
inhibited by the corresponding end-products (71). At higher
concentrations, phenylalanine and tyrosine also inhibit the
enzyme of the alternative branch (71a), and some mutants were
found resistant to feedback inhibition by the aromatic amino
acids (72). Feedback control of both enzymes by the end-products of their own branches was also found in yeast (*S. cerevisiae*) (73a) and the ergot fungus *Claviceps paspali* (73b). In
contrast to this, the prephenate dehydrogenase from the cotyledons of wax bean (*Phaseolus vulgaris*) was not inhibited by
tyrosine (6c).

Repression of the formation of prephenate dehydratase and
dehydrogenase by phenylalanine and tyrosine has likewise been

observed.   The prephenate dehydrogenase of *B. subtilis* is repressed by tyrosine, but no evidence for a similar repression of the dehydratase by phenylalanine was found (74).   The situation in *S. cerevisiae* seems the exact opposite: only the dehydratase was repressed by its own end-product phenylalanine (75).

In addition to these controls by the usual mechanisms of feedback inhibition and repression, the portion of the shikimate pathway which contains the adjacent branch-point intermediates chorismic acid and [3] exhibits a surprising wealth of less usual activation and inhibition phenomena.   Some of these can be rationalized fairly readily as having regulatory significance, while the meaning of others seems obscure.   Again, only a few examples can be given.

To the former category belongs the activation of the first enzyme of one branch by the end-product of the other one, which has been observed several times.   The chorismate mutases of *Neurospora* (76), of the chlorophyll-bearing unicellular alga *Euglena gracilis* (77), and of pea cotyledons (*Pisum sativum*) (78) are all under feedback inhibition by phenylalanine and tyrosine, the end-products of the pathway to [3] whose first step is catalyzed by the enzyme.   In addition, however, they are activated by tryptophan, the product of the major alternative branch starting from chorismic acid.   Similar reciprocal activations take place at the two branches that begin with [3]. Cerutti and Guroff (56) found that the prephenate dehydratase of a *Pseudomonas* species is strongly activated by L-tyrosine; the D-antipode has a much smaller effect, as have several other *p*-substituted analogs of phenylalanine.   The counterpart of this process: activation of prephenate dehydratase by phenylalanine, has been found in *Neurospora* by Catcheside (79). Surprisingly, tyrosine in this organism fails to exhibit the activating effect upon prephenate dehydratase which it has in *Pseudomonas*.   In these systems of reciprocal activations, any excess of the end-product of one branch would presumably channel the flow of intermediates into the alternative one by stimulating its first enzyme.

Cases of reciprocal inhibition rather than activation seem more difficult to rationalize.   The inhibition of prephenate dehydratase and dehydrogenase of *B. subtilis* by high concentrations of L-phenylalanine and L-tyrosine, respectively, (71) has already been mentioned.   Both enzymes are also to a certain extent inhibited by tryptophan (71,80).   In addition, the dehydrogenase of this species is inhibited by D-tyrosine and by L-phenylalanine (71b).   Finally, Coats and Nester (72) observed actual activation, rather than inhibition, of prephenate dehydratase by phenylalanine, the end-product of its own branch, in two mutants of *B. subtilis* which were resistant to β-thienylalanine (an inhibitory analog of phenylalanine). The term "regulation reversal mutation" was proposed for this phenonmenon.

In addition to these activations and inhibitions by end-products of parallel branches of biosynthetic pathways, influences by metabolites formed through unrelated sequences have been observed frequently. Pertinent to the present discussion is, for example, the activation of prephenate dehydratase of *B. subtilis* by leucine and methionine, which was discovered by Jensen (81). This author gives detailed discussion of this case, and of other instances of "metabolic interlock."

7-2. PHENYLPYRUVIC AND *p*-HYDROXYPHENYLPYRUVIC ACID

Although these two compounds are, strictly speaking, no longer within the scope of this book as defined initially, their universal importance justifies a brief discussion.

A possible role of phenylpyruvic acid [1] as a precursor of phenylalanine was apparently suggested for the first time by Simmonds, Tatum, and Fruton (10a), who noted its utilization by phenylalanine-auxotrophs of *E. coli* (incidentally, one of the earliest instances of the use of bacterial mutants for the study of biosynthetic pathways). This role of [1] was subsequently confirmed in the work of Davis (10b) with numerous mutants of *E. coli*; his interpretation of the autocatalytic growth of prephenate-excreting mutants, and the identification of the first nutritionally active transformation product as [1], may be considered conclusive proof (9a). The α-keto acid can also satisfy the phenylalanine requirement of the mutant of *Neurospora* used by Metzenberg and Mitchell (12) for the isolation of [3].

Evidence for a similar function of *p*-hydroxyphenylpyruvic acid [7] as a precursor of tyrosine was initially less straightforward. Simmonds and her co-workers found (10a) that [7] was not utilized by their tyrosine-auxotroph of *E. coli*. However, Davis (10b) observed that the compound, while nutritionally inactive for tyrosine-requiring mutants under ordinary conditions, was utilized very efficiently when the pH of the nutrient medium with glucose as the carbon source was lowered for 7.0 to 5.5 or if, at pH 7.0, glucose was replaced by succinate. It is worth noting that such influences of carbon source, pH, etc. upon the nutritional availability of metabolites have also been observed elsewhere in the shikimate pathway. Examples are the utilization of 3-dehydroquinic acid by mutants of *A. aerogenes*

on media with succinate, but not on those with glucose as carbon
source (see Section 4-1), and the dependence of the requirement
for *p*-hydroxybenzoic acid upon pH and carbon source (32) (see
Chapter 11, where these questions are discussed further). Additional evidence for the intermediacy of [7] was obtained by
Schwinck and Adams (13) in their work on its enzymatic formation from [3], where it was observed that [7] is converted
further to tyrosine when the system is supplemented with pyridoxal phosphate and glutamic acid.

Since Doy and Gibson (82) found that both [1] and [7] are
excreted under conditions of nitrogen starvation by wild-type
A. aerogenes, but not by a mutant with quintuple aromatic
requirement, the role of the keto acids as precursors of the
amino acids seems well established.

[*Added in proof*: Work published since completion of this
chapter has shown that [7] does not always function as an
intermediate in the biosynthesis of tyrosine, since a most
interesting variant, in which this compound is by-passed, has
been found (82A) quite recently in *Agmenellum quadruplicatum*
and a few other blue-green bacteria (Chroococcales). *A. quad-
riplicatum* contains several of the enzymes of the shikimate
pathway but lacks prephenate dehydrogenase. However, it contains a transaminase which converts prephenic acid into a
ninhydrin-positive product termed "pretyrosine"; it has been
obtained in partially purified form on enzymatic transamination
of prephenic acid with phenylalanine. Pretyrosine is converted
into phenylalanine on exposure to acidic pH, and into tyrosine
by an NAD^+-dependent enzyme isolated from the organism. This
evidence strongly suggests that pretyrosine is β-(1-carboxy-
4-hydroxy-2,5-cyclohexadienyl)alanine, and that it is an
intermediate in the biosynthesis of tyrosine. In these blue-
green bacteria, transamination of the side-chain of prephenic
acid thus seems to take precedence over the aromatization of
the ring, so that the usual sequence prephenic acid ---→ [7]
──→ tyrosine is replaced by the alternative prephenic acid
──→ pretyrosine ──→ tyrosine.]

The transaminating system which catalyzes the conversion
of [1] and [7] to phenylalanine and tyrosine, respectively, was
first studied by Rudman and Meister (34) in *E. coli*. Two distinct fractions were obtained by chromatography or fractional
precipitation and were named transaminase A and B; a third
one observed by these authors was subsequently named transaminase C (82B). Of these, enzyme A promoted the transamination
of α-ketoglutaric acid with phenylalanine, tyrosine, tryptophan
and, to a lesser extent, leucine and methionine, while transaminases B and C act upon a number of aliphatic *α*-amino acids.·

Later on, Silbert, Jorgenson, and Lin (83) observed that formation of transaminase A is repressed specifically by tyrosine,
but not by phenylalanine. In addition, another enzyme was
detected which acts specifically upon phenylalanine alone; it
constituted about 1/5 of the total activity towards this substrate, was not repressed by it, or by any other individual
amino acid, and was much less heat-sensitive than transaminase
A, which is rapidly inactivated at 60°. In a very recent re-
investigation, Collier and Kohlhaw (84) showed that the activity
towards aspartic acid is actually due to an enzyme different
from transaminase A and separable from it by electrophoresis.
This enzyme was again much more stable to heat than transaminase
A, and was not repressed by tyrosine; a mutant with low activity
to aspartic acid was found to have normal activity towards the
three aromatic acids. Transamination of the latter (and apparently also of leucine and methionine) is thus specifically
catalyzed by transaminase A. Since tryptophan and methionine
are not formed via the corresponding α-keto acids, the transaminase in these two cases seems to be concerned with the further
metabolism of the amino acids rather than with their biosynthesis. The same may be assumed to be true for the tyrosine aminotransferase of mammalian liver, which has been isolated in
highly purified, crystalline form (85).

In higher plants, formation of phenylalanine and tyrosine
from [3] via [1] and [7], respectively, was first established
by Gamborg and Simpson (6a); conversion of [1] to phenylalanine
was also shown to take place in extracts from tissue cultures
of a variety of plants (6b). The aminotransferase responsible
for these transaminations was studied in detail by Gamborg and
co-workers (86,87). In mung bean (*Phaseolus aureus*), one single
enzyme seems to be present which catalyzes the transamination
between α-ketoglutaric or pyruvic acid and a variety of α-amino
acids, including phenylalanine, tyrosine, tryptophan, *m*-tyrosine,
and 3,4-dihydroxyphenylalanine. This enzyme was purified 40- to
60-fold (87).

Through a peculiar variant of the usual pathway, [1] is
formed in certain micro-organisms--most of them strictly anaerobic species--from phenylacetic acid and CO_2 under the influence of reduced ferredoxin; the [1] so formed is converted
further to phenylalanine. This variant was first observed by
Allison (88) in some anaerobic bacteria from sheep rumen;
during growth of these organisms in presence of 1-^{14}C-phenyl-
acetic acid, label was incorporated into phenylalanine but not
into any other amino acid. Allison and Robinson (89) showed
subsequently that the same labeled substrate was specifically
converted to 2-^{14}C-phenylalanine in the strictly anaerobic

bacterium *Chromatium* (Thiorhodaceae). Incorporation into phenylalanine also took place in the facultative anaerobe *Rhodospirillum rubrum* (Athiorhodaceae) growing anaerobically under illumination; no incorporation was found under aerobic conditions in the dark. Hence this reaction seems to be restricted to anaerobic metabolism. That it is likely to proceed through carboxylation of the phenylacetic acid to [1] follows from the frequent occurrence of similar reactions in these organisms (see below), and from the fact that label from $NaH^{14}CO_3$ was likewise incorporated into phenylalanine. Details of the reaction were elucidated through the recent studies of Gehring and Arnon (90) with cell-free extracts of photosynthetic bacteria such as *Chromatium* sp., *Chlorobium thiosulfatophilum*, and *Chloropseudomonas ethylicus*. Such extracts from *Chromatium* converted phenylacetyl-CoA into [1] in presence of bicarbonate and reduced ferredoxin; on addition also of glutamic acid or glutamine, phenylalanine was formed. Similar findings were made with extracts from the two other species. The reaction is quite analogous to the production of other α-keto acids from acyl derivatives of coenzyme A and CO_2 in presence of reduced ferredoxin, which has been observed in photosynthetic bacteria; formation of pyruvate from acetyl CoA, and of α-ketoglutarate from succinyl CoA may serve as examples (for literature, see refs. 89 and 90).

It seems (89) that this formation of phenylalanine from phenylacetic acid is of importance for anaerobic microorganisms in their natural habitat. Phenylacetic acid is available in the rumen (91), apparently as a result of microbial degradation of phenylalanine; information on its occurrence in the environment of the photosynthetic bacteria seems to be lacking, but since these organisms occur as secondary flora growing on metabolic products of other microorganisms, phenylacetic acid may be available to them as well.

ADDENDUM

The problems of the total synthesis of prephenic acid and of the mechanisms of its alkali-catalyzed conversion into *p*-hydroxyphenyllactic acid have recently found their final solution.

Danishefsky and Hirama (92) have synthesized the cyclohexadienone [33], while the diethyl analog of this key intermediate has been prepared independently by Gramlich and Plieninger (93) through a different approach. Reduction of [33] with 9-borabicyclo[3.3.1]nonane gave (92) a separable mixture of the corresponding epimeric cyclohexadienols [34] and [35]. Treatment of these with NaOH in aqueous methanol at room temperature produced in the Na salts, [36] and [370, of prephenic [3a] and epiprephenic [3b] acids, respectively. The salt [36] gave an nmr spectrum identical with the once of a sample of Na prephenate prepared from authentic Ba prephenate, thus establishing identity. The nmr spectrum of [37] was noticeably different.

The availability of the Na salts [36] and [37] enabled Danishefsky and Hirama (94) to prove that the aromatization of prephenate [3a] into p-hydroxyphenyllactic acid [6] by hot dilute alkali is indeed an intramolecular Cannizzaro reaction and that it proceeds with 1,6-shift of hydride, as formulated in Scheme III. The intramolecular nature of the process is shown by the fact that epiprephenic acid [3b] remains unchanged under the conditions which produce [6] from [3a]; this finding also provides a striking confirmation for the configuration of prephenic acid assigned by Plieninger and Keilich (25).

That this alkali-catalyzed aromatization does actually proceed with 1,6-shift as shown in Scheme III follows (94) from the fact that prephenate completely deuteriated in the CH_2 group yields [6] having the same distribution of label; the hydrogen next to the secondary hydroxyl of [6] is thus furnished by the proton at C-4 of [3a], and not by those of the methylene group, as it would be if imaginable alternative mechanisms (94) were valid.

REFERENCES

1. A.R. Moss and R. Schoenheimer, J. Biol. Chem. 135, 415 (1940).
2. O.L. Gamborg and A.C. Neish, Can. J. Biochem. Physiol. 37, 1277 (1959).
3. W. Womack and W.C. Rose, J. Biol. Chem. 107, 449 (1934).
4. (a) S. Kaufman, Adv. Enzymol. 35, 245 (1971).
    (b) D.B. Fisher, R. Kirkwood, and S. Kaufman, J. Biol. Chem. 247, 5161 (1972).
5. B. Guroff and T. Ito, J. Biol. Chem. 240, 1175 (1965); G. Guroff and C.A. Rhoads, ibid. 242, 3641 (1967); 244, 142 (1969).
6. O.L. Gamborg and F.J. Simpson, Can. J. Biochem. 42, 583 (1964); O.L. Gamborg, ibid. 44, 791 (1966); O.L. Gamborg and F.W. Keeley, Biochim. Biophys. Acta 115, 65 (1966).
7. P.M. Nair and L.C. Vining, Phytochemistry 4, 401 (1965).
8. S. Simmonds, J. Biol. Chem. 185, 755 (1950).
9. (a) M. Katagiri and R. Sato, Science 118, 250 (1953);
    (b) B.D. Davis, ibid. 252 (1953).

10. (a) S. Simmonds, E.L. Tatum, and J.S. Fruton, J. Biol. Chem. 169, 91 (1947);

    (b) B.D. Davis, ibid. 191, 315 (1951).

11. U. Weiss, C. Gilvarg, E.S. Mingioli, and B.D. Davis, Science 119, 774 (1954).

12. R.L. Metzenberg and H.K. Mitchell, Arch. Biochem. Biophys. 64, 51 (1956).

13. I. Schwinck and E. Adams, Biochim. Biophys. Acta 36, 102 (1959).

14. R.W. Colburn and E.L. Tatum, Biochim. Biophys. Acta 97, 442 (1965).

15. J. Dayan and D.B. Sprinson, in Advances in Enzymology, Vol. XVII A, H. Tabor and C.W. Tabor, Eds. Academic, New York and London, 1970, p. 559.

16. F. Gibson, Biochem. J. 90, 256 (1964).

17. (a) R.G.H. Cotton and F. Gibson, Biochim. Biophys. Acta 100, 76 (1956).

    (b) F. Gibson and J. Pittard, Bact. Rev. 32, 465 (1968).

18. K. Auwers and G. Keil, Ber. 36, 1861 (1903).

19. Th. Zincke and F. Schwabe, Ber. 41, 897 (1908).

20. C. Gilvarg and K. Bloch, J. Biol. Chem. 199, 689 (1952).

21. M. Sprecher, P. R. Srinivasan, D.B. Sprinson, and B.D. Davis, Biochemistry 4, 2855 (1965).

22. P.R. Srinivasan, H.T. Shigeura, M. Sprecher, D.B. Sprinson, and B.D. Davis, J. Biol. Chem. 220, 477 (1956).

23. W.P. Jencks and J. Carriuolo, Nature 182, 599 (1958).

24. U. Weiss, unpublished observations.

25. H. Plieninger and G. Keilich, Z. Naturforsch. 16b, 81 (1961).

26. H. Plieninger and G. Ege (a) Chem. Ber. 94, 2088 (1961); (b) ibid. 2095.

27. (a) H. Plieninger and H.J. Grasshoff, Chem. Ber. 90, 1973 (1957);

    (b) H. Plieninger and G. Keilich, ibid. 92, 2897 (1959).

28. Sir Robert Robinson, The Structural Relations of Natural Products, Clarendon, Oxford, 1955, p. 54.

29. H. Plieninger, Angew. Chem., Int. Ed. 1, 367 (1962).

30. (a) C. Gilvarg, unpublished observations mentioned in B.D. Davis, The Harvey Lectures, Series L (1954-1955), p. 230, and in ref. 29;

    (b) I.G. Young, F. Gibson, and C.G. Macdonald, Biochim. Biophys. Acta 192, 62 (1969).

31. B.D. Davis, Experientia 6, 41 (1950).

32. B.D. Davis, J. Bact. 64, 729, 749 (1952).

33. A.G. De Busk and R.P. Wagner, J. Amer. Chem. Soc. 75, 5131 (1947).

34. D. Rudman and A. Meister, J. Biol. Chem. 200, 591 (1953).
35. U. Weiss, E.S. Mingioli, and B.D. Davis, unpublished observations mentioned in B.D. Davis, Adv. Enzymol. 16, 247 (1955).
36. J.J. Ghosh, Fed. Proc. 15, 261 (1956).
37. (a) H. Plieninger, G. Ege, H.J. Grasshoff, G. Keilich, and W. Hoffmann, Chem. Ber. 94, 2115 (1961);
    (b) H. Plieninger, L. Arnold, and W. Hoffmann, ibid. 98, 1765 (1965);
    (c) H. Plieninger, L. Arnold, R. Fischer, and W. Hoffmann, ibid. 1774 (1965).
38. K. Auwers and G. Keil, Ber. 36, 3902 (1903).
39. K. Auwers, (a) Ber. 39, 3748 (1906);
    (b) Ann. 352, 219 (1907);
    (c) Ber. 44, 588 (1911);
    (d) K. Auwers and K. Müller, ibid. 1595 (1911);
    (e) K. von Auwers and K. Ziegler, Ann. 425, 217, 280 (1921);
    (f) K. von Auwers and W. Jülicher, Ber. 55, 2167 (1922).
40. Th. Zincke, Ber. 41, 897 (1908).
41. M.S. Newman, J. Eberwein, and L.L. Wood, Jr., J. Amer. Chem. Soc. 81, 6454 (1959).
42. H. Plieninger and G. Keilich, Angew, Chem. 68, 618 (1956).
43. H. Plieninger and G. Keilich, Chem. Ber. 91, 1891 (1958).
44. F. Sondheimer, M. Velasco, E. Batres, and G. Rosenkranz, Chem. Ind. (London) 1482 (1954)
45. H. Pleininger, L. Arnold, and W. Hoffman, Chem. Ber. 101, 981 (1968).
46. M.J. Gentles, J.B. Moss, H.L. Herzog, and E.B. Hershberg, J. Amer. Chem. Soc. 80, 3702 (1958).
47. H. Dannenberg and C.H. Doering, Z. Physiol. Chem. 311, 84 (1958); H. Dannenberg, D. Dannenberg-von Dresler, and T. Köhler, Chem. Ber. 93, 1989 (1960); H. Dannenberg and H.-G. Neumann, Ann. 646, 148 (1961).
48. E. Caspi, P.K. Grover, N. Grover, E.J. Lynde, and Th. Nussbaumer, J. Chem. Soc. 1710 (1962), and subsequent papers.
49. E. Caspi, D.M. Piatak, and P.K. Grover, J. Chem. Soc. (C) 1034 (1966).
50. H.-J. Hansen, B. Sutter, and H. Schmid, Helv. Chim. Acta 51, 828 (1968), and subsequent papers.
51. H. Hart, P.M. Collins, and A.J. Waring, J. Amer. Chem. Soc. 88, 1005 (1966).
52. Ch. Wei-Zan and Huang-Minlon, Acta Chim. Sinica 25, 327 (1959).
53. M. Sumi, W.G. Dauben, and W.K. Hayes, J. Amer. Chem. Soc. 80, 5704 (1958).

54. S. Dorling, J. Harley-Mason, and L.C. Johnson, Chem. Ind. (London) 495 (1960).

55. H. Plieninger, G. Ege, R. Fischer, and W. Hoffmann, Chem. Ber. 94, 2106 (1961).

56. P. Cerutti and G. Guroff, J. Biol. Chem. 240, 3034 (1965).

57. (a) F. Gibson and J. Pittard, Bact. Rev. 32, 465 (1968);
    (b) J. Pittard and F. Gibson, in Current Topics in Cellular Regulation, Vol. 2, Academic, New York, 1970, p. 29.

58. F. Lingens, Angew. Chem., Int. Ed. 7, 350 (1968).

58A. D.H. Hey, G.H. Jones, and M.J. Perkins, J. Chem. Soc. Perkin I 118 (1972), and earlier papers of this series.

59. P. Truffa-Bachi and G.N. Cohen, Ann. Rev. Biochem. 37, 80 (1968).

60. R.G.H. Cotton and F. Gibson, Biochim. Biophys. Acta 100, 76 (1965).

61. R.G.H. Cotton and F. Gibson, Biochim. Biophys. Acta 147, 222 (1967).

62. G.L.E. Koch, D.C. Shaw, and F. Gibson, Biochim. Biophys. Acta 229, 795, 805 (1971).

63. G.L.E. Koch, D.C. Shaw, and F. Gibson, Biochim. Biophys. Acta 212, 375 (1970).

64. J.C. Schmit and H. Zalkin, (a) Biochemistry 8, 174 (1969);
    (b) J. Biol. Chem. 246, 6002 (1971).

65. J.C. Schmit, S.W. Artz, and H. Zalkin, J. Biol. Chem. 245, 4019 (1970).

66. G.L.E. Koch, D.C. Shaw, and F. Gibson, Biochim. Biophys. Acta 258, 719 (1972).

67. R.G.H. Cotton and F. Gibson, Biochim. Biophys. Acta 160, 188 (1968).

68. G.L.E. Koch, D.C. Shaw, and F. Gibson, Biochim. Biophys. Acta 212, 387 (1970).

69. (a) E.W. Nester, J.H. Lorence, and D.S. Nasser, Biochemistry 6, 1553 (1967);
    (b) W.M. Nakatsukasa and E.W. Nester, J. Biol. Chem. 247, 5972 (1972).

70. J.H. Lorence and E.W. Nester, Biochemistry 6, 1541 (1967);

71. (a) E.W. Nester and R.A. Jensen, J. Bact. 91, 1594 (1966);
    (b) W.S. Champney and R.A. Jensen, J. Biol. Chem. 245, 3763 (1970).

72. J.H. Coats and E.W. Nester, J. Biol. Chem. 242, 4948 (1967).

73. F. Lingens, W. Goebel, and H. Uesseler, (a) Biochem. Zeitschr. 346, 357 (1966);
    (b) Eur. J. Biochem. 2, 442 (1967).

74. E.W. Nester, R.A. Jensen, and D.S. Nasser, J. Bact. 97, 83 (1969).

75. F. Lingens, W. Goebel, and H. Uesseler, Eur. J. Biochem. 1, 363 (1967).
76. T. I. Baker, Biochemistry 5, 2654 (1966).
77. H.L. Weber and A. Böck, Europ. J. Biochem. 16, 244 (1970).
78. R.G.H. Cotton and F. Gibson, Biochim. Biophys. Acta 156, 187 (1968).
79. D.E.A. Catcheside, Biochem. Biophys. Res. Commun. 36, 651 (1969).
80. J.L. Rebello and R.A. Jensen, J. Biol. Chem. 245, 3738 (1970).
81. R.A. Jensen, J. Biol. Chem. 244, 2816 (1969).
82. C.H. Doy and F. Gibson, Biochim. Biophys. Acta 50, 495 (1961).
82A. S.L. Stenmark, D.L. Pierson, R.A. Jensen, and G.I. Glover, Nature 247, 290 (1974).
82B. A. Meister, in Biochemistry of the Amino Acids, A. Meister, Ed. 2nd ed., Vol. I, Academic, New York, 1965, p. 362.
83. D.F. Silbert, S.E. Jorgensen, and E.C.C. Lin, Biochim. Biophys. Acta 73, 232 (1963).
84. R.H. Collier and G. Kohlhaw, J. Bact. 112, 365 (1972).
85. S.-I. Hayashi, D.K. Granner, and G.M. Tomkins, J. Biol. Chem. 242, 3998 (1967).
86. O.L. Gamborg and L.R. Wetter, Can. J. Biochem. Physiol. 41, 1733 (1963).
87. O.L. Gamborg, Canad. J. Biochem. 43, 723 (1965).
88. M.J. Allison, Biochim. Biophys. Res. Commun. 18, 30 (1965).
89. M.J. Allison and I.M. Robinson, J. Bact. 93, 1269 (1967).
90. U. Gehring and D.I. Arnon, J. Biol. Chem. 246, 4518 (1971).
91. H. Tappeiner, Z. Biol. 22, 238 (1886); T.W. Scott, P.F.V. Ward, and R.M.C. Dawson, Biochem. J. 90, 12 (1964).
92. S. Danishefsky and M. Hirama, J. Amer. Chem. Soc. 99, 7740 (1977); cf. also S. Danishefsky, R.K. Singh, and T. Harayama, ibid. p. 5810.
93. W. Gramlich and H. Plieninger, Tetrahedron Letters 475 (1978).
94. S. Danishefsky and M. Hirama, Tetrahedron Letters 4565 (1977).
95. K.H. Baggaley, B. Blessington, C.P. Falshaw, W.D. Ollis, L. Chaiet, and F.J. Wolf, Chem. Commun. 101 (1969).
96. T. Yamashita, N. Miyairi, K. Kunigita, K. Shimuzu, and H. Sakai, J. Antibiotics 23, 537 (1970); J.P. Scannell, D.L. Pruess, T.C. Demny, T. Williams, and A. Stempel, ibid. 618 (1970).
97. (a) G.M. Sharma and P.R. Burkholder, Tetrahedron Lett. 4147 (1967);
     (b) G.M. Sharma, B. Vig, and P.R. Burkholder, J. Org. Chem. 35, 2823 (1972).

98. (a) K. Takeda, T. Okanishi, K. Igarachi, and A. Shimaoka, Tetrahedron 15, 183 (1961);
    (b) K. Igarachi, Chem. Pharm. Bull. 9, 722 (1961).
99. M.-M. Janot, Ph. Devissaguet, Q. Khuong-Huu, and R. Goutarel, Ann. Pharm. Fr. 25, 733 (1967).
100. (a) D.H.R. Barton, Chem. Brit. 330 (1967);
     (b) A.R. Battersby, in Oxidative Coupling of Phenols, W.I. Taylor and A.R. Battersby, Eds., Dekker, New York, 1968, Chapter 3;
     (c) T. Kametani and K. Fukumoto, Synthesis 657 (1972).
101. D.H.R. Barton and T. Cohen, Festschrift Prof. Dr. Arthur Stoll, Birkhäuser, Basel, 1957, p. 117.
102. (a) J. Gadamer, Arch. Pharm. 249, 498 (1911);
     (b) C. Schopf and K. Thierfelder, Ann. 497, 22 (1932).
103. M.D. Glick, R.E. Cook, M.P. Cava, M. Srinivasan, J. Kunitomo, and A. I. da Rocha, Chem. Commun. 1217 (1969).
104. (a) L.J. Haynes, K.L. Stuart, D.H.R. Barton, and G.W. Kirby, Proc. Chem. Soc. 208 (1963; 261 (1964); J. Chem. Soc. (C) 1676 (1966);
     (b)K. Bernauer; Helv. Chim. Acta 46, 1783 (1963); 47, 2122 (1964).
105. M.R. Cava, K. Nomura, R.H. Schlessinger, K.T. Buck, B. Douglas, R.F. Raffauf, and J.A. Weissbach, Chem. Ind. 282 (1964); M. Shamma and W.A. Slusarchyk, Chem. Rev. 64, 59 (1964).
106. K. Bernauer and W. Hofheinz, Fortsch. Chem Org. Natur. 26, 245 (1968).
107. L.J. Haynes, K.L. Stuart, D.H.R. Barton, D.S. Bhakuni, and G.W. Kirby, Chem. Commun. 141 (1965); D.H.R. Barton, D.S. Bhakuni, G.M. Chapman, G.W. Kirby, L.J. Haynes, and K.L. Stuart, J. Chem. Soc. (C) 1295 (1967).
108. D.H.R. Barton, D.S. Bhakuni, G.M. Chapman, and G.W. Kirby, J. Chem. Soc. (C) 2134 (1967).
109. (a) A.R. Battersby, T.H. Brown, and J.H. Clements, J. Chem. Soc. 4550 (1965);
     (b) A.R. Battersby, T.J. Brocksom, and R. Ramage, Chem. Commun. 464 (1969);
     (c) A.R. Battersby, R.T. Brown, J.H. Clements, and G.G. Iverach, Chem. Commun. 230 (1965).
110. (a) M. Pailer, Fortschr. Chem. Org. Natur. 18, 55 (1960);
     (b) R. Hegnauer, Chemotaxonomie der Pflanzen, Vol. III Birkhäuser Verlag, Basel and Stuttgart, 1964, p. 191.
111. S.M. Kupchan and J.J. Merianos, J. Org. Chem. 33, 3735 (1968).
112. (a) L.A. Maldonado, J. Herran, and J. Romo, Ciencia (Mex.) 24, 237 (1966); through Chem. Abstr. 65, 15438e (1966);

(b) M. Akasu, H. Itokawa, and M. Fujita, Tetrahedron
Lett. 3609 (1974).

113. H.R. Schütte, U. Orban, and K. Mothes, Eur. J. Biochem.
1, 70 (1967).

114. I.D. Spenser, Lloydia 29, 71 (1966); F. Comer, H.P. Tiwari,
and I.D. Spenser, Can. J. Chem. 47, 481 (1969).

115. L. Kühn and S. Pfeifer, Pharmazie 20, 659 (1965).

116. K.L. Stuart, Chem. Rev. 71, 47 (1971).

117. A.R. Battersby, Proc. Chem. Soc. 189 (1963); cf. also
D. Ginsburg, The Opium Alkaloids, Interscience, New York,
1962.

118. J.W. Fairbairn and S. El-Masry, Phyochemistry 6, 499
(1967).

119. D.H.R. Barton, G.W. Kirby, W. Steglich, G.M. Thomas,
A.R. Battersby, T.A. Dobson, and H. Ramuz, J. Chem. Soc.
2423 (1965); A.R. Battersby and T.H. Brown, Chem. Commun.
170 (1966).

120. D.H.R. Barton, G.W. Kirby, W. Steglich, and G.M. Thomas,
Proc. Chem. Soc. 203 (1963).

121. D.H.R. Barton, D.S. Bhakuni, R. James, and G.W. Kirby,
J. Chem. Soc. (C) 128 (1967); K.W. Bentley, Experientia
12, 251 (1956).

122. T. Kametani, M. Ihara, and T. Honda, Chem. Commun. 1301
(1969).

123. B. Franck, H.J. Lubs, and G. Dunkelmann, Angew. Chem. Int.
Ed. 10, 969, 1075 (1967).

124. K.L. Stuart, V. Teetz, and B. Franck, Chem. Commun. 333
(1969); K.L. Stuart and L. Graham, Phytochemistry 12,
1967 (1973).

125. K. Takeda, Bull. Agr. Chem. Soc. Japan 20, 165 (1956).

126. K.W. Bentley, The Chemistry of the Morphine Alkaloids,
Clarendon, Oxford, 1954.

127. G. Stork, in The Alkaloids, R.H.F. Manske and H.L. Holmes,
Eds., Vol. II, Academic, New York, 1952, p. 189ff.

128. (a) K.W. Bentley and R. Robinson, Experientia 6, 353 (1956);
(b) K.W. Bentley, J. Amer. Chem. Soc. 89, 2464.

129. D.M. Hall and W.W.T. Manser, Chem. Commun. 112 (1967).

130. D.H.R. Barton, Pure Appl. Chem. 9, 35 (1964).

131. A.R. Battersby, A.K. Bhatnagar, P. Hackett, C.W. Thornber,
and J. Staunton, Chem. Commun. 1214 (1968).

132. E. Fattorusso, L. Minale, and G. Sodano, Chem. Commun. 751
(1970); J. Chem. Soc., Perkin Trans. I, 16 (1972).

133. W. Gulmor, G.E. Van Lear, G.O. Morton, and R.D. Mills,
Tetrahedron Lett. 4551 (1970).

134. D.B. Cosulich and F.M. Lovell, Chem. Commun. 397 (1971);
      L. Mazzarella and R. Puliti, Gazz. Chim. Ital. 102, 391
      (1972).
135. L. Minale, G. Sodano, W.R. Chan, and A.M. Chen, Chem.
      Commun. 674, 968 (1972).
136. E. Gattorusso, L. Minale, G. Sodano, K. Moody, and
      R.H. Thomson, Chem. Commun. 752 (1970).
137. K. Moody, R.H. Thomson, E. Fattorusso, L. Minale, and
      G. Sodano, J. Chem. Soc., Perkin Trans. I 16 (1972).
138. E. Fattorusso, L. Minale, K. Moody, G. Sodano, and
      R.H. Thomson, Gazz. Chim. Ital. 101, 61 (1971).
139. J.R.D. McCormick, J. Reichenthal, U. Hirsch, and
      N.O. Sjolander, J. Amer. Chem. Soc. 84, 3711 (1962).
140. D.T. Gibson and D.M. Jerina, private communication.
141. D.T. Gibson, Crit. Rev. Microbiol. 1, 199 (1971).
142. (a) D.T. Gibson, G.E. Cardini, R.C. Maseles, and
      R.E. Kallio, Biochemistry 9, 1631 (1970);
      (b) D.T. Gibson, J.R. Koch, and R.E. Kallio, ibid.
      7, 2653 (1968).
143. D.T. Gibson, M. Hensley, H. Yoshioka and T.J. Mabry,
      Biochemistry 9, 1626 (1970).
143A. D.T. Gibson, J.R. Koch, C.L. Schuld, and R.E. Kallio
      Biochemistry 7, 3795 (1968).
144. N. Walker and G.H. Wiltshire, J. Gen. Microbiol. 8, 273
      (1953).
145. D.M. Jerina, J.W. Daly, A.M. Jeffrey, and D.T. Gibson,
      Arch. Biochem. Biophys. 142, 394 (1971).
146. N. Walker and G.H. Wiltshire, J. Gen. Microbiol. 12, 478
      (1955).
147. L. Canonica, A. Fiecchi, and V. Treccani, Ist. Lombardo,
      Rend. Sci. 91, 119 (1957); through ref. 108.
148. C. Colla, A. Fiecchi, and V. Treccani, Ann. Microbiol.
      Enzimol. 10, 77 (1960); through ref. 141.
149. A.M. Reiner and G.D. Hegeman, Biochemistry 10, 2530 (1971).
150. H. Taniuchi and O. Hayaishi, J. Biol. Chem. 238, 283 (1963).
151. H. Ziffer, D.M. Jerina, D.T. Gibson, and V.M. Kobal,
      J. Amer. Chem. Soc. 95, 4048 (1973).
152. F.A. Catterall, K. Murray, and P.A. Williams, Biochim.
      Biophys. Acta 237, 361 (1971).
153. M. Mori, H. Taniuchi, Y. Kojima, and O. Hayaishi, Biochim.
      Biophys. Ácta 128, 535 (1966).
154. R.Y. Stanier and O. Hayaishi, Science 114, 326 (1951);
      O. Hayaishi, H. Taniuchi, M. Tashiro, and S. Kuno,
      J. Biol. Chem. 236, 2492 (1961).
155. F.A. Catterall and P.A. Williams, J. Gen. Microbiol. 67,
      117 (1971).

156. D.M. Jerina, H. Ziffer, and J.W. Daly, private communication; cf. also ref. 180.

157. M. Reiner, J. Biol. Chem. 247, 4960 (1972).

158. M.R. Bell, J.R. Johnson, B.S. Wildi, and R.B. Woodward, J. Amer. Chem. Soc. 80, 1001 (1958).

159. J. Fridrichsons and A. McL. Mathieson, Acta Cryst. 23, 439 (1967).

160. N. Neuss, R. Nagarajan, B.B. Molloy, and L.L. Huckstep, Tetrahedron Lett. 4467 (1968).

161. J.D. Bu'Lock and A.P. Ryles, Chem. Commun. 1404 (1970).

162. N. Johns and G.W. Kirby, Chem. Commun. 163 (1971).

163. G. Lowe, A. Taylor, and L.C. Vining, J. Chem. Soc. (C) 1799 (1966).

164. M.S. Ali, J.S. Shannon, and A. Taylor, J. Chem. Soc. (C) 2044 (1968).

165. R. Nagarajan, L.L. Huckstep, D.H. Lively, D.C. De Long, M.M. Marsh, and N. Neuss, J. Amer. Chem. Soc. 90, 2980 (1968).

166. R. Nagarajan, N. Neuss, and M.M. Marsh, J. Amer. Chem. Soc. 90, 6518 (1968).

167. D. Hauser, H.P. Weber, and H.P. Sigg, Helv. Chim. Acta 53, 1061 (1970).

168. R. Hodges, J.W. Ronaldson, A. Taylor, and E.P. White, Chem. Ind. (London) 42 (1963); J. Fridrichsons and A. McL. Mathieson, Acta Cryst. 18, 1043 (1965).

169. E. Baumann and C. Preusse, Ber. 12, 806 (1879); M. Jaffe', ibid. 1092.

170 J.W. Daly, D.M. Jerina, and B. Witkop, Experientia 28, 1129 (1972).

171. (a) D.M. Jerina, H. Yagi, and J.W. Daly, Heterocycles 1, 267 (1973);
     (b) D.M. Jerina and J.W. Daly, Science 185, 573 (1974).

172. E. Boyland and A.A. Levi, Biochem. J. 29, 2679 (1935).

173. T. Sato, T. Fukuyama, T. Suzuki, and H. Yoshikawa, J. Biochem. (Tokyo) 53, 23 (1963).

174. (a) J.N. Smith, B. Spencer, and R.T. Williams, Biochem. J. 47, 284 (1950);
     (b) D.M. Jerina, J.W. Daly, and B. Witkop, J. Amer, Chem. Soc. 89, 5488 (1967).

175. (a) L. Young, Biochem. J. 41, 417 (1947);
     (b) J. Booth and E. Boyland, Biochem. J. 44, 361 (1949);
     (c) E.D.S. Corner and L. Young, Biochem. J. 58, 647 (1954).

176. (a) E. Boyland and G. Wolf, Biochem. J. 47, 64 (1950);
     (b) E. Boyland and P. Sims, Biochem. J. 84, 571 (1962).

177. F. Waterfall and P. Sims, Biochem, J. 128, 265 (1972).

178. E. Boyland and P. Sims, Biochem, J. 91, 493 (1964).

179. P. Sims, Biochem. Pharmacol. 19, 795 (1970).

180. D.M. Jerina, H. Ziffer, and J.W. Daly, J. Amer. Chem. Soc. 92, 1056 (1970).

181. J. Holtzman, J.R. Gillette, and G.W.A. Milne, J. Amer. Chem. Soc. 89, 6341 (1967).

182. R. Miura, S. Honmaru, and M. Nakazaki, Tetrahedron Lett. 5271 (1968).

183. P.K. Ayengar, O. Hayaishi, M. Nakajima, and I. Tomida, Biochim. Biophys. Acta 33, 111 (1959).

184. M. Nakajima, I. Tomida, A. Hashizume, and S. Takei, Ber. 89, 2224 (1956).

185. J. Booth and E. Boyland, Biochem. J. 70 681 (1958); cf. also H.S. Posner, C. Mitoma, S. Rothberg, and S. Udenfriend Arch. Biochem. Biophys. 94, 280 (1961).

186. O. Hayaishi, M. Katagiri, and S. Rothberg, J. Amer. Chem. Soc. 77, 5450 (1955); H.S. Mason, W.L. Fowlkes, and E. Peterson, ibid. 2914; H.S. Mason, Ann. Rev. Biochem. 34, 595 (1965); O. Hayaishi, ibid. 39, 21 (1969).

187. H.S. Mason. Adv. Enzvmol. 19, 79 (1957).

188. D.M. Jerina J.W. Daly, B. Witkop, P. Zaltzman-Nirenberg, and S. Udenfriend, J. Amer. Chem. Soc. 90, 5625 (1968); Biochemistry 9, 147 (1970).

189. E. Boyland, Biochem. Soc. Symp. 5, 40 (1950).

190. (a) E. Boyland, P. Sims, and J.B. Solomon, Biochem. J. 66, 41P (1957);
(b) R.H. Knight and L. Young, ibid. 66, 55P (1957); 70, 111(1958).

191. (a) E. Boyland and P. Sims, Biochem. J. 68, 440 (1958);
(b) B. Gillham and L. Young, ibid. 103, 24P (1967).

192. (a) J. Booth, E. Boyland, and P. Sims, Biochem. J. 74, 117 (1960);
(b) E. Boyland and K. Williams, ibid. 94, 190 (1965).

193. E. Boyland and P. Sims, Biochem. J. 77, 175 (1960); E. Boyland, G.S. Ramsay, and P. Sims, ibid. 78, 376 (1961).

194. E. Boyland and P. Sims, Biochem. J. 84, 564 (1962).

195. M.S. Newman and S. Blum, J. Amer. Chem. Soc. 86, 5559 (1964).

196. E. Vogel and H. Günther, Angew. Chem. Int. Ed. 6, 385 (1967).

197. D.M. Jerina, J.W. Daly, and B. Witkop, in Biogenic Amines and Physiological Membranes in Drug Therapy, Part B, J.H. Biel and L.G. Abood, Eds., Dekker, New York, 1971, p. 413.

198. D. Jerina, J. Daly, B. Witkop, P. Zaltzman-Nirenberg, and S. Udenfriend, Arch. Biochem. Biophys. 128, 176 (1968).

199. E. Vogel and R.G. Klärner, Angew. Chem., Int. Ed. 7, 374
      (1968).
200. C. Mitoma, H.S. Posner, H.C. Reitz, and S. Udenfriend
      Arch. Biochem. Biophys. 61, 431 (1956).
201. J. Booth, E. Boyland, and P. Sims, Biochem. J. 79, 516
      (1961).
202. J.K. Selkirk, E. Huberman, and C. Heidelberger, Biochem.
      Biophys. Res. Commun. 43, 1010 (1971); P.L. Grover.
      A. Hewer, and P. Sims, Fed. Eur. Biochem. Soc. Lett. 18,
      76 (1971).
203. D.M. Jerina, J.W. Daly, and B. Witkop, J. Amer. Chem.
      Soc. 90, 6523 (1968).
204. N. Sakabe, H. Harada, Y. Hirata, Y. Tomiie, and I. Nitta
      Tetrahedron Lett. 2523 (1966).
205. G. Guroff, D.M. Jerina, J. Renson, S. Udenfriend, and
      B. Witkop, Science 157, 1524 (1967).
206. D.M. Jerina, Chem. Technol. 4, 120 (1973); and refs. quoted
      there.
207. J. Daly and G. Guroff, Arch. Biochem. Biophys. 125, 136
      (1968).
208. D.M. Jerina, N. Kaubisch, and J.W. Daly, Proc. Natl. Acad.
      Sci. U.S. 68, 2545 (1971); N. Kaubisch, J.W. Daly, and
      D.M. Jerina, Biochemistry 11, 3080 (1972).
209. A.Y.H. Lu, R. Kuntzman, S. West, M. Jacobson, and
      A.H. Conney, J. Biol. Chem. 247, 1727 (1972), and literature
      there.
210. A.Y.H. Lu and M.J. Coon, J. Biol. Chem. 243, 1331 (1968);
      A.Y.H. Lu, K.W. Junk, and M.J. Coon, ibid. 244, 3714
      (1969).
211. H.W. Strobel, A.Y.H. Lu, J. Heidema, and M.J. Coon, J.B.C.
      245, 4851 (1970).
212. F. Oesch, D.M. Jerina, and J. Daly, Biochim. Biophys.
      Acta 227, 685 (1971).
213. F. Oesch and J. Daly, (a) Biochim. Biophys. Acta 227,
      692 (1971);
      (b) Biochem. Biophys. Res. Commun. 46, 1713 (1972).
214. F. Oesch, D.M. Jerina, J.W. Daly, A.Y.H. Lu, R. Kuntzman,
      and A.H. Conney, Arch. Biochem. Biophys. 153, 62 (1972).

CHAPTER 8

Occurrence and Biosynthetic Role of Cyclohexadienols Other than Prephenic Acid

After the recognition of prephenic acid as a cyclohexadienol
had demonstrated the biosynthetic potentialities of this ring
system, it was often postulated that compounds of this kind are
involved in the formation or transformation of aromatic rings;
in many cases, the assumption of such intermediates is almost
inescapable on purely chemical grounds. However, only a small
number of such compounds has actually been isolated and chemically
characterized; in view of the great sensitivity of most cyclohexadienols, this is hardly surprising.

Reactions that have been shown or plausibly assumed to
proceed through cyclohexadienols as intermediates are involved
in many processes described elsewhere in this volume; it seems
appropriate to include here a summary of the occurrence and
function of cyclohexadienols, especially since no review of
this topic exists. Both cyclohexa-2,5-dienols and the related
$\Delta^{2,4}$-compounds with a conjugated diene are included. The latter
show in general the same reactivity as their nonconjugated
analogs; however, synthetic dienols of this type tend to be
more stable than the 2,5-dienols (1). (Some methoxylated cyclohexadienols are capable of reacting in ways not open to their
parent compounds; examples of such reactions will be discussed
below in connection with certain aspects of alkaloid biosynthesis). Likewise included are the two naturally occurring
cyclohexadienylamines which have been found so far.

Because of their close analogy and great biosynthetic significance (exemplified by chorismic and isochorismic acids; (see
Chapters 6 and 12, resp.), the cyclohexadien*di*ols are likewise

185

briefly discussed. Their much-investigated role as intermediates in the transformation of aromatic rings via the arene oxides is pertinent enough for an understanding of certain structural features of naturally occurring aromatic compounds to require some presentation; no exhaustive coverage of this involved topic is intended.

8-1. CYCLOHEXADIENOLS WHICH HAVE BEEN ACTUALLY ISOLATED

The closest known relative of prephenic acid is the corresponding α-amino acid, pretyrosine (cf. page 171), which is formed from it by transamination in certain blue-green bacteria (2). Formation of this compound, and its subsequent dehydrogenative aromatization to tyrosine, constitute an alternative, apparently more primitive biosynthesis of tyrosine; the evolutionary aspects of this duality of pathways have been discussed.

Bohlmann and co-workers (3) have encountered three cyclohexadienols: ferulol [1], the corresponding 2,4-dienol [2], and isoferulol [3], in many plants of the family Umbelliferae, where they occur in free form or esterified with a variety of isoprenoid acids.

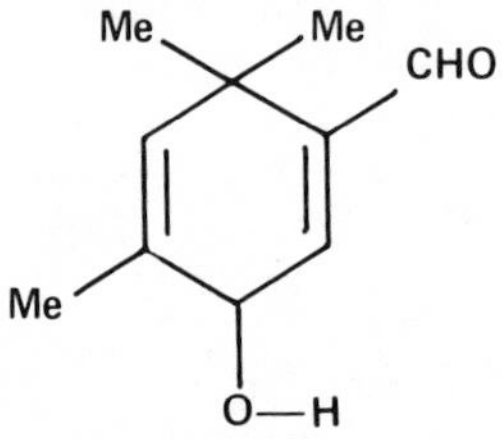

[1] 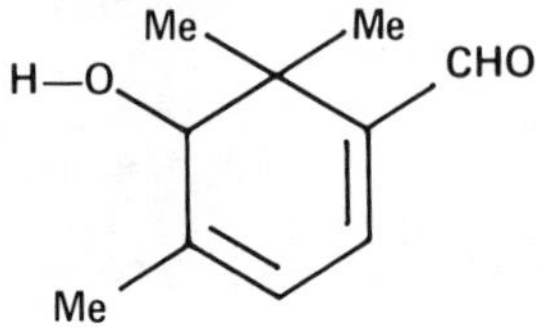   [2] 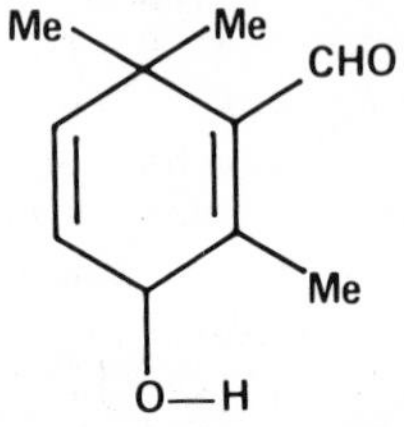   [3]

The structures of [1] and [3] follow from spectroscopic data and from treatment with acid, which yields 2,3,4- and 2,3,6-trimethylbenzaldehyde, respectively, through dienolbenzene rearrangement. The carbon skeletons of [1], [2], and [3] can be dissected into two isoprene units. In [3], these units are joined in the head-to-tail fashion found in most terpenoids. The compound is thus a normal, geraniol-derived monoterpenoid (see Chapter 22) containing the same ring-system as β-ionone [4a] and β-carotene (see Chapter 23); picrocrocin [5] from saffron constitutes a particularly close analog. In contrast, the two units in [1] and [2] are linked in an anomalous fashion which is also found in a few natural terpenoids such as lavandulol [6] from lavender; in the irones from orris root (*Iris* spp.), for example β-irone [4b], the ring system present in [1] carries an additional C^4 side-chain. For the biosynthesis of terpenes with irregular skeleton, including [6], hypothetical schemes have been advanced (4) which derive these compounds from a postulated cyclopropyl precursor analogous to presqualene alcohol (see Sections 22-1 and 22-8) by opening of the three-membered ring in different directions.

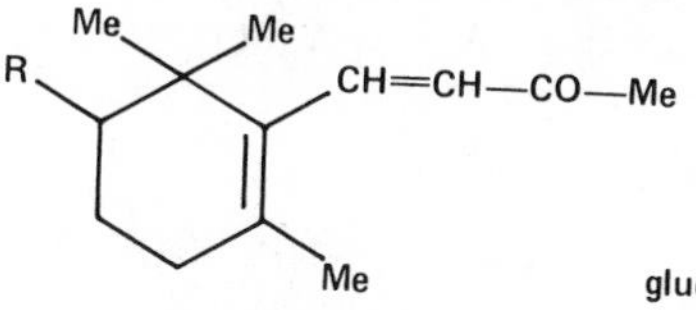

[4]a: R = H
    b: R = Me

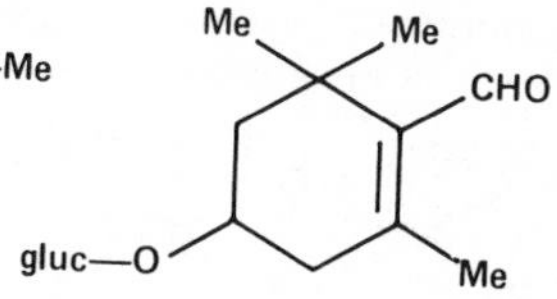

[5]

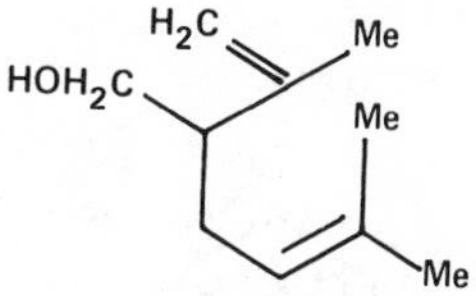

[6]

While cyclohexadienols have often been postulated as precursors of alkaloids (see below), nudaurine [7] from *Papaver nudicaule* (Papaveraceae) is, to the best of our knowledge, the only such base that has actually been isolated and characterized so far (5). No biosynthetic role for this compound has been observed, and the acid-catalyzed aromatization *in vitro* of its cyclohexadienol ring does not proceed through a straightforward dienol-benzene rearrangement.

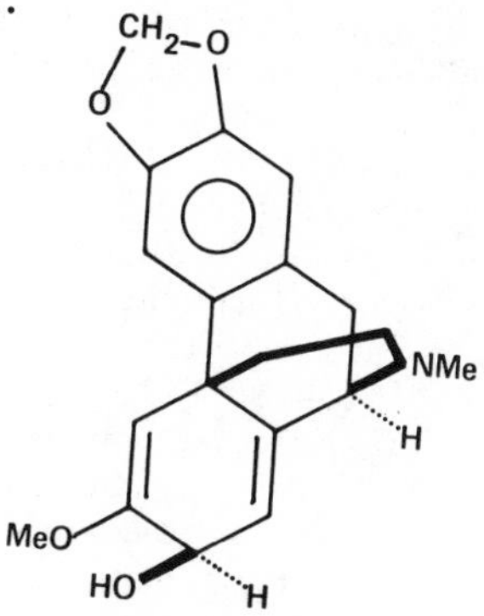

[7]

The analogous cyclohexa-2,5-dienamines are represented by two compounds, stravidin [8a] and MSD-235S$_2$ [8b] which, together with a protein streptavidin, form the antibiotic MSD-235 produced by *Streptomyces avidini* (6). On acid hydrolysis, [8a] gave one mole each of ammonia, (2S)-2-amino-4-phenylbutyric acid, and (2S,3S)-N-methyl-isoleucine; boiling in water alone for one hour transformed the ring of [8a] to C$_6$H$_5$- with liberation of ammonia but without further change of the molecule.

These reactions are entirely analogous to the "dienamine-benzene rearrangements" which have been mentioned before (see page 160). Compounds [8a] and [8b] are the only cyclohexadienamines found in nature so far. In addition, they are unique in having a tertiary, rather than quaternary, carbon atom in position 4 (see p. 165).

In connection with these cyclohexa-2,5-dienamines, the occurrence of L-cyclohexa-1,4-dienylalanine [9] in a streptomycete may be of interest (7).

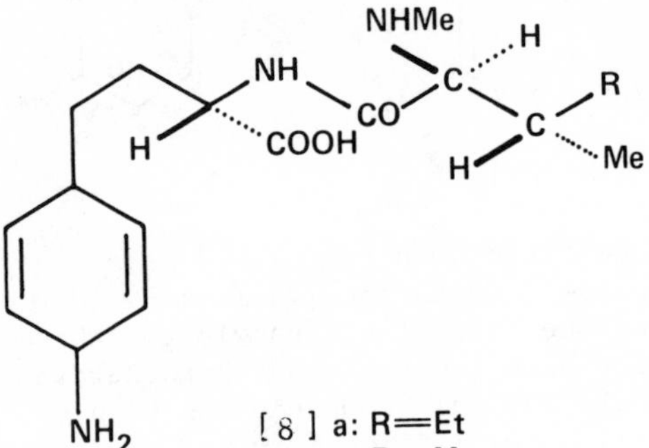

[8] a: R=Et
    b: R=Me

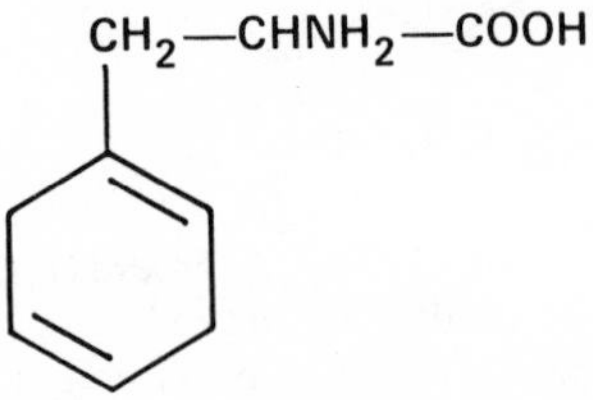

[9]

Certain marine sponges of the genera *Aplysina, Ianthella,* and *Verongia* have yielded a group of closely related bromine-containing antibiotics derived from cyclohexa-2,5-dienol and cyclohexa-2,4-diendiol. The former are represented by [10] (8a) and its dimethyl acetal [11] (8b). The cyclohexadienol rings in these compounds are surprisingly stable to acid: hot dilute sulfuric acid hydrolyzed the acetal and amide groups of [11] but left the ring intact. The related derivatives of cyclohexadiendiol will be discussed below, as will be the probable biosynthesis of all these compounds by way of intermediary arene oxides of type [12]. The way in which such an oxide could give cyclohexa-2,4-diendiols is obvious, but there is laboratory precedent also for conversion of compounds of this type to cyclohexa-2,5-dienols.

8-2.   CYCLOHEXADIENOLS AS POSTULATED INTERMEDIATES

In addition to the cases where cyclohexadienols and their allies have actually been obtained as such, compounds of this type have often, for valid reasons, been postulated to exist but were not isolated. Two such cases have already been mentioned briefly: Sir Robert Robinson's early hypothesis (9) of a

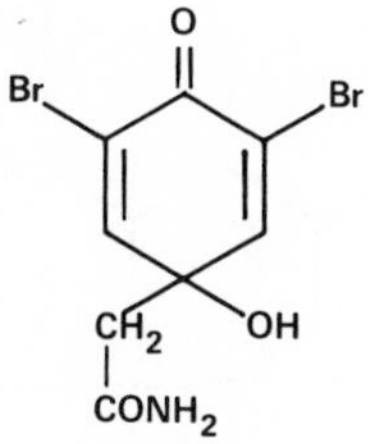

[10] 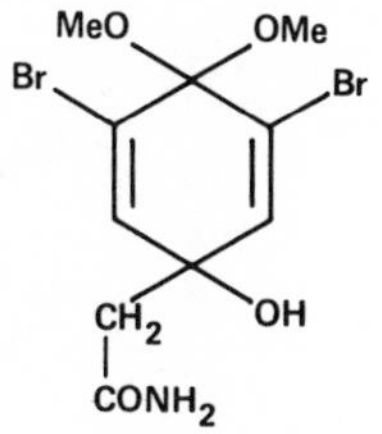   [11] 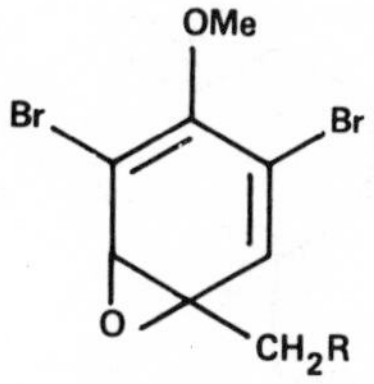   [12]

cyclohexadienol as intermediate in the biosynthesis of ephedrine
(see p. 153), and isoprephenic acid (see Chapter 12). Some
other instances merit discussion.

Three saponins, which must occur in *Metanarthecium luteo-*
viride (Liliaceae) but have not been isolated, are very probably cyclohexadienols. Acidic hydrolysis of methanol extracts
of this plant yields a number of sapogenins, of which several
are normal representatives of their class with saturated ring A
bearing the usual hydroxyl at C-3. However, three of these compounds: luvigenin [13] (10a); meteogenin [14] (10b), and
neometeogenin (10b), the 25-epimer of [14], have aromatic nonphenolic benzene rings instead, although their structures are
otherwise those of typical sapogenins. The constitutions
assigned to these compounds rest on their synthesis from normal
sapogenins (diosgenin, hecogenin) of unequivocally known chemistry through reaction sequences whose final steps involve the
preparation of the corresponding 1,4-diene-3-ones, hydride reduction to the dienols, and dienol-benzene rearrangement. The
rearrangement that gives the 4-methyl compound [13] is completely
analogous to the conversion of authentic steroidal 1,4-diene-
3-ols to 4-methylestra-1,3,5(10)-trienes which has been discussed
before (see page 159).

The vast majority of known steroids have oxygen (in a few
cases nitrogen) at C-3, as would be expected from their formation
through cyclization of 2,3-oxidosqualene (see Chapter 24); saponins characteristically carry sugar residues on the hydroxyl at
C-3. The three aromatic compounds that do not have this group,
and especially [13], which lacks hydroxyl altogether, can therefore hardly be genuine sapogenins, and it is logical to assume
that they are artifacts formed from non-aromatic 3-hydroxylated
saponins by dienol-benzene rearrangements during the treatment

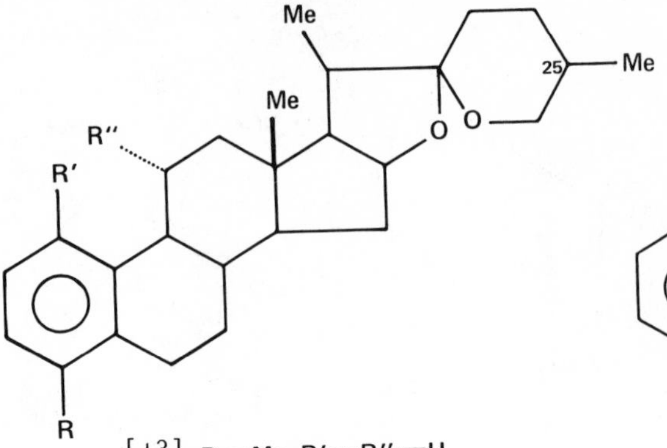

[13]: R=Me, R′=R″=H

[14]: R=H, R′=Me, R″=OH

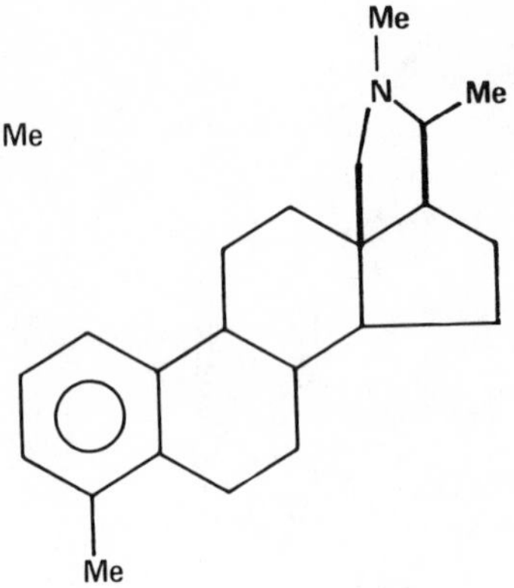

[15]

with acid.   Those precursors are most probably 1,4-cyclohexadien-
3-ols, although alternative structures (e.g. 1,5-cyclohexadien-
3-ol) are possible (10a).*

     A situation perfectly analogous to that of [13] exists in
the alkaloid holaromine [15] which has been obtained from
Holarrhena floribunda (Apocyanceae) but may well be an artefact
formed during isolation (11).   The hypothetical precursor may
again be a cyclohexadienol, and it is suggestive that the corresponding 1,4-cyclohexadien-3-one, holadienine, occurs together
with [15].   Since, however, plants of the genus *Holarrhena* are
well known for the wealth of otherwise rare 3-amino steroids
that they contain, compound [15] might also be formed from a
precursor of this type by a dienamine-benzene rearrangement.

     No information seems to have been obtained which would sug-
gest that the hypothetical cyclohexadienol precursor of [15] or
the alkaloid nudaurine [7] mentioned earlier (see page 187)
function as intermediates in any biosynthetic process.   Abundant
evidence is available, however, for the participation of dienols
of this type in the biosynthesis of a substantial number of
alkaloids, and of some compounds biogenetically related to them,
such as the aristolochic acids.   Even though none of these
cyclohexadienols has been isolated so far, their intervention
has been established beyond reasonable doubt by labeling

---

*Recent evidence indicated that the precursor of [14] is a 2β,
3β-dihydroxy-Δ4-steroid, i.e., a structure formally derivable
from the postulated cyclohexadienol precursor by hydration of
the Δ1 double bond.   Cf. I. Kitagawa, T. Nakanishi, Y. Morii,
and I. Yosioka, Tetrahedron Letters 1885 (1976).

experiments; the beautiful work with multiply labeled precursors
carried out by Barton, Battersby, and their respective co-workers,
must be mentioned in the first place. This research has been
reviewed by these authors (12a,b). A recent review by Kametani
and Fukumoto (12c) contains much information on questions of
biosynthesis.

Occurrence of cyclohexadienols, and their aromatization by
dienol-benzene rearrangement, was first postulated by Barton
and Cohen (13) in their classical discussion of the role of
phenol coupling in biosynthesis. As they pointed out, this
reaction, with its bond formation invariably occurring in
positions *ortho* or *para* to a free phenolic hydroxyl, can account
for the structures of a wide variety of natural compounds.
Actual occurrence of this process was subsequently proved in a
large number of instances. Important for our purposes is the
fact that some of these structures can be explained only if
formation of cyclohexadienols and subsequent re-aromatization
by dienol-benzene rearrangement are postulated as intermediary
steps. This is particularly the case for certain alkaloids.

It had, for example, been realized early by Gadamer (14),
Schöpf and Thierfelder (14), Robinson (9), and others, that
the carbon skeleton [16] of the 1-benzyl-tetrahydroisoquinolines
can yield that [17] of the aporphines through dehydrogenation;
both [16] and [17] represent very large classes of natural
alkaloids. Barton and Cohen (13) called attention to the fact
that phenol coupling, applied to an appropriately substituted
benzyltetrahydroisoquinoline [18], would explain the structures
of two main types of aporphines, those represented by corytuber-
ine [19] and glaucine [20], respectively. The individual alkaloids of these two classes would then arise through unexceptional
peripheral modifications such as O- and N-methylation, and formation of methylenedioxy groups.

However, this direct coupling does not provide an adequate
explanation for the structures of a number of aporphines. For
the biosynthesis of some of these, for example nuciferoline
[21] and stephanine [22], *meta*-coupling would have to be assumed,
a reaction which has never been observed. Others, such as anolo-
bine [23], isothebaine [24], and crebanine [25], could form
through normal *o,o*- or *o,p*-coupling, but the benzyl-tetrahydroisoquinolines analogous to [18] which would be required as precursors
would have to carry oxygenated substituents in *meta*-position
to the carbon side-chain of the benzyl group. This group is
known to be derived from shikimic acid; in view of the strong
preponderance of oxygenation in positions 4, 3+4, or 3+4+5 in
structures having this origin (see Chapter 2), precursors
with the required *meta*-orientation would be unlikely, although

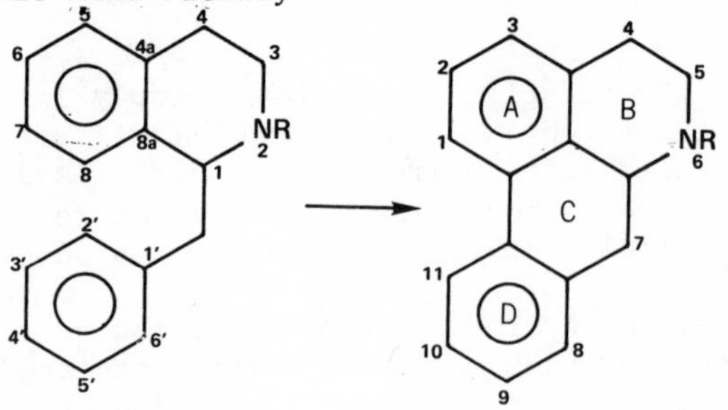

[16]   [17]

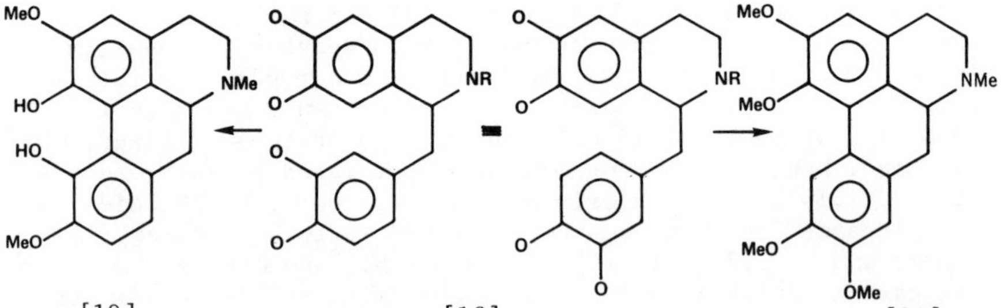

[19]   [18]   [20]

not impossible. Finally, a fairly numerous group of aporphines
lacks any oxygenated substituent on ring D altogether; nuciferine
[26], anonaine [27], roemerine [28], liriodenine [29], and imenine
[30] (15) may serve as examples.

    To account for these structures, Barton and Cohen (13)
postulated the intermediary formation of cyclohexadienones such
as [31] through *o,p*-coupling of benzyl-tetrahydroisoquinolines
of type [32]. These dienones could be transformed to aporphines
through dienone-phenol rearrangement with migration either of
the aromatic ring, or of the –CH$_2$– group C-7 (see routes a and b,
respectively, in Scheme I). A dienone such as [31] would yield
bases represented by [21] and [23]; presence of an additional
oxygenated function in position 9 or 11 could produce types [19],
[20], or [25]. For the biosynthesis of [22], [24], and their
allies, a $\Delta^{2,4}$-dienone might be assumed; however, an alternative
explanation exists, and will be discussed below. Finally, reduction of [31] to the dienol [33], followed by dienolbenzene
rearrangement, would account for the aporphines with unsubstituted ring D ([26]-[30]).

    Experimental support for the postulated biosynthesis via
intermediate cyclohexadienones [31] came soon through the
discovery of the first natural alkaloids of this type: crotono-
sine [34] (=[31], R = H, R' = Me) from *Croton linearis*

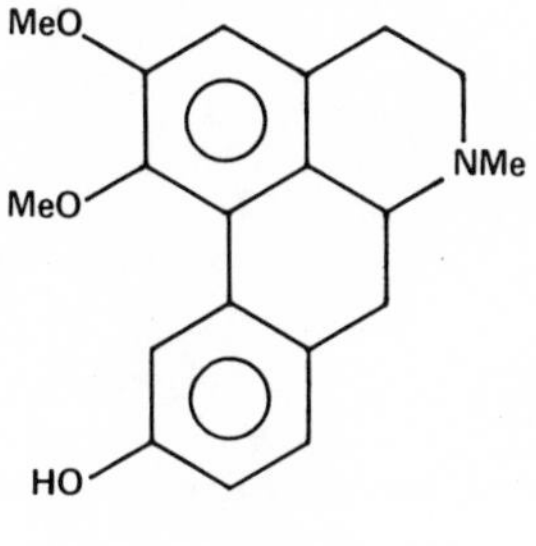

[21]

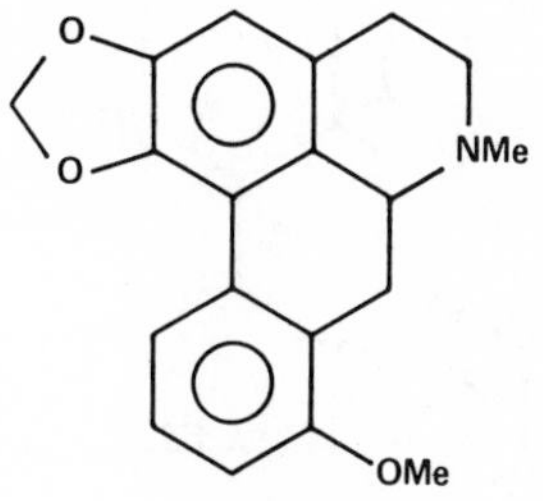

[22]

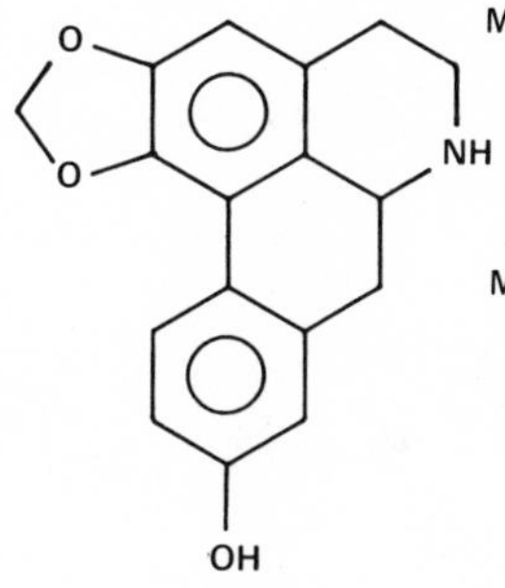

[23]

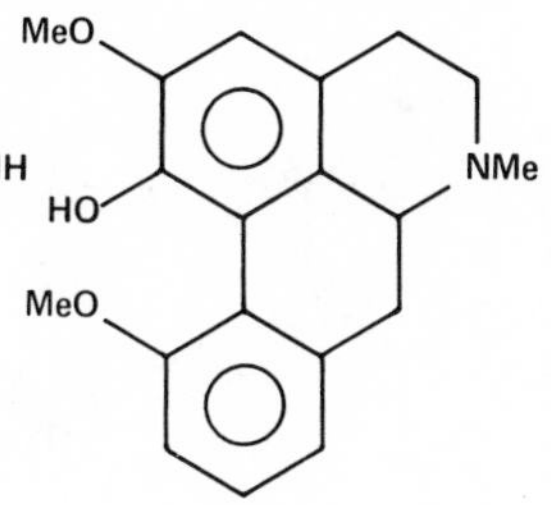

[24]

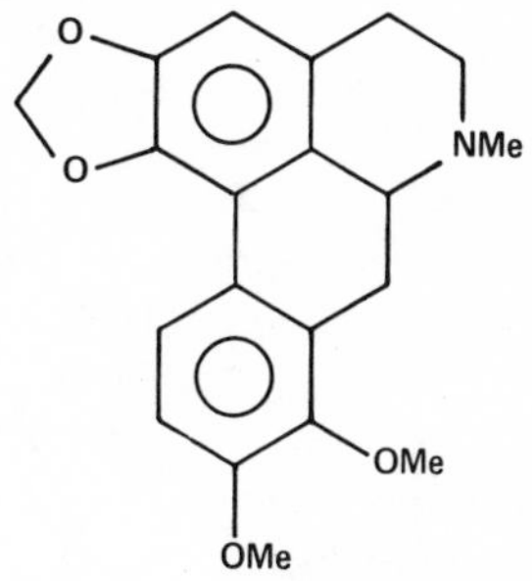

[25]

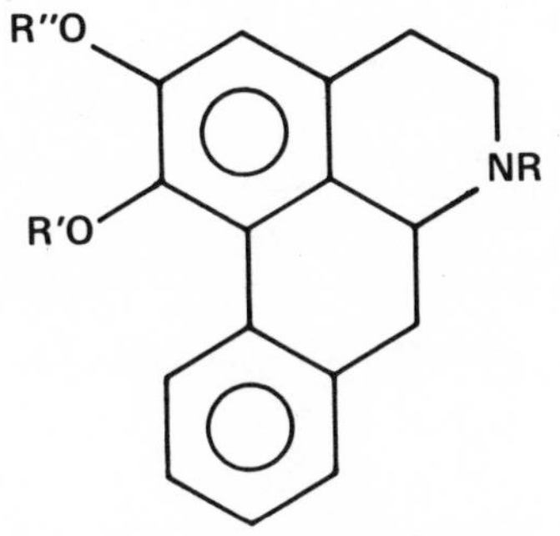

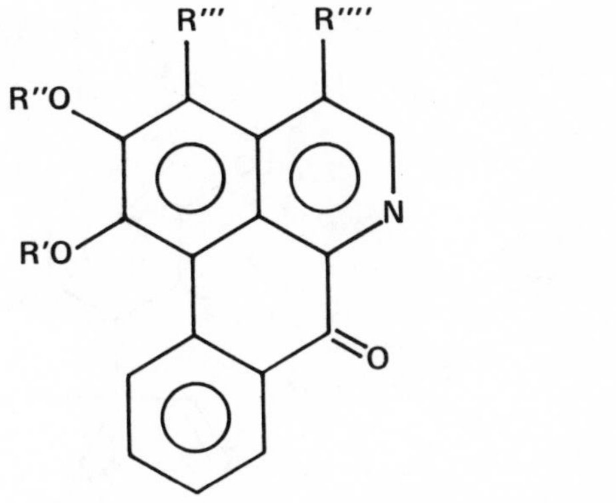

[26] : R=R'=R''=Me
[27] : R=H,R'=R''=—CH₂—
[28] : R=Me, R'=R''—CH₂—

[29]: R'=R''=Me,R'''=R''''=H
[30]: R'=R'=Me,R'''=R''''=OMe

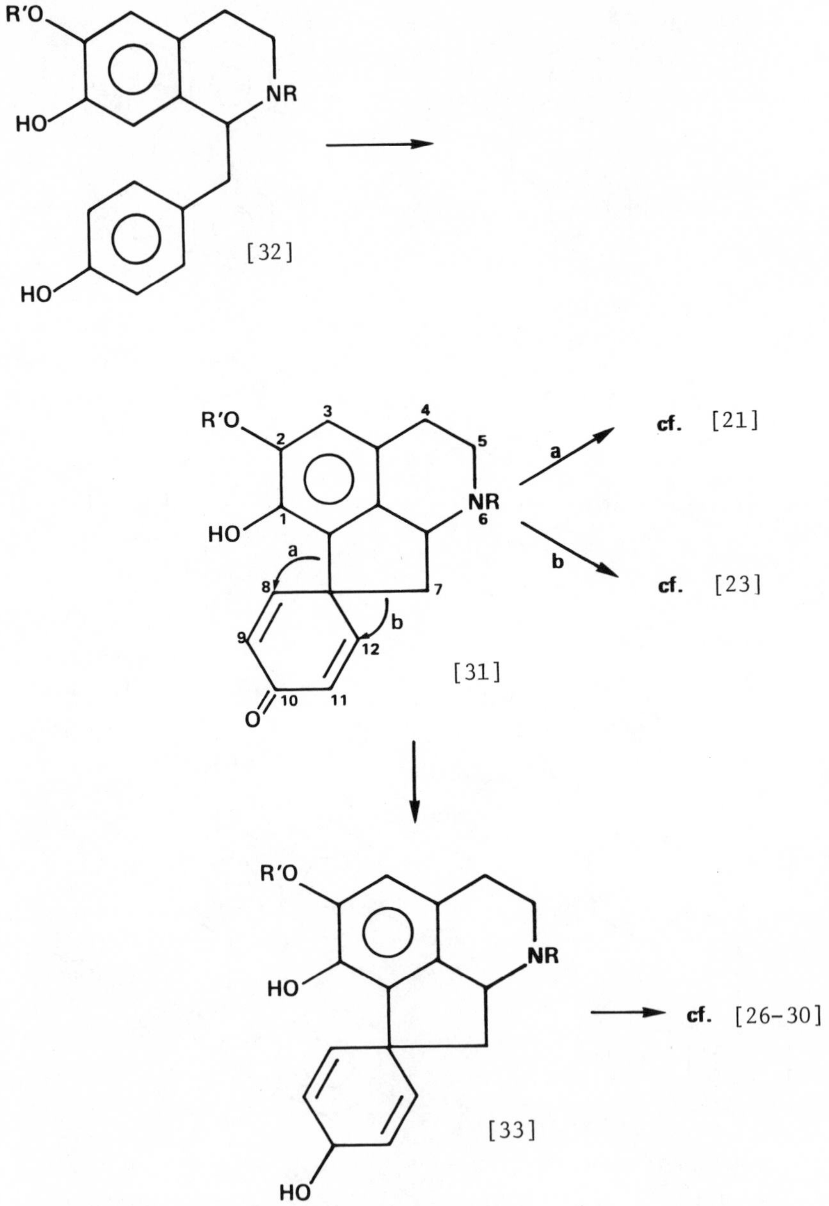

Scheme I

194

(Euphorbiaceae) (16a), and pronuciferine ([31], R = R' = Me) (16b). The latter gave the known alkaloid [21] on acid-cata-lyzed dienone-phenol rearrangement; borohydride reduction yielded the corresponding dienol (the methyl ether of [33], R = R' = Me), which was converted to [26] by acid. The name "proaporphines" was introduced (17) for these cyclohexadienones, of which a fairly large number is now known; they have been reviewed by Bernauer and Hofheinz (18). Convincing proof for their function as intermediates between types [16] and [17] is available; while none of the corresponding cyclohexadienols [33] has been isolated so far from natural sources, their biosynthetic role is likewise unequivocally established.

Evidence for conversion of a precursor of type [16] to a proaporphine [31] was obtained, for example, in *C. linearis* (19): (-)-coclaurine [35] (= [32], R = H, R' = Me) tritiated in posi-tions 8, 3', and 5', gave [34] labeled in the expected positions 9 and 11. The (+)-antipode of [35] was not utilized; neither was the isomeric isococlaurine (the 7-methyl ether of [32], R = H), which lacks the required free hydroxyl in position 7. Both [35] and isococlaurine are naturally occurring alkaloids.

Clear evidence for the intermediacy also of a cyclohexa-dienol of type [33] comes from the work of Barton and co-workers (20) with *Papaver dubium* and *Meconopsis cambrica* (Papaveraceae): labeled (±)-[35] (and its O-nor- and N-methyl derivatives) were incorporated into the proaporphine mecambrine [36], which was converted into mecambroline [37] by dienone-phenol rearrangement in *M. cambrica*, but into roemerine [28] in *P. dubium*, and into [27] (= N-nor-[28]) in *Annona reticulata* (Annonaceae). In all cases, the localization of the label was the expected one.

These findings show that the aporphines with only two oxygen atoms and with unsubstituted ring D are formed from tri-oxygen-ated precursors such as [35] and [36]; the dienol-benzene rear-rangement leads to the loss of one of the three phenolic hydroxyls of [35] and constitutes one of the few biochemical mechanisms for such a removal of phenolic OH; the problem of dehydroxylation of the aromatic ring is discussed further in Chapter 16.

Analogous removal of one oxygen from *tetra*-oxygenated pre-cursors such as [18] can account for the anomalous structures [21] - [24]. Clearly, rearrangement of a pair of dienols epimeric at the spiro carbon atom [38a,b] can explain these structures, as shown in Scheme II.

The work of Battersby and co-workers shows that such a dienol-benzene rearrangement, following route (a) of Scheme II, does indeed take place in the biosynthesis of isothebaine [24] in *Papaver orientale*. This alkaloid can be synthesized in the laboratory (21a,b) from orientaline [39], a base of type [18] and the precursor of many alkaloids, by a reaction sequence (Scheme III) which is entirely analogous to the one shown in Scheme I. In the plant, [3-^{14}C]-[39] gave [24] having 95% of the label in the corresponding position (21b,c). The interme-diary proaporphine orientalinone [40], prepared synthetically by phenol coupling of [38], on reduction with tritiated LiAlH$_4$ gave the two epimeric dienols [41a,b], of which only one was effi-ciently incorporated into [24] with high retention of ^{3}H; the biosynthetic dienol-benzene rearrangement is thus stereospecific. Both [39] and [40] occur in *P. orientale*.

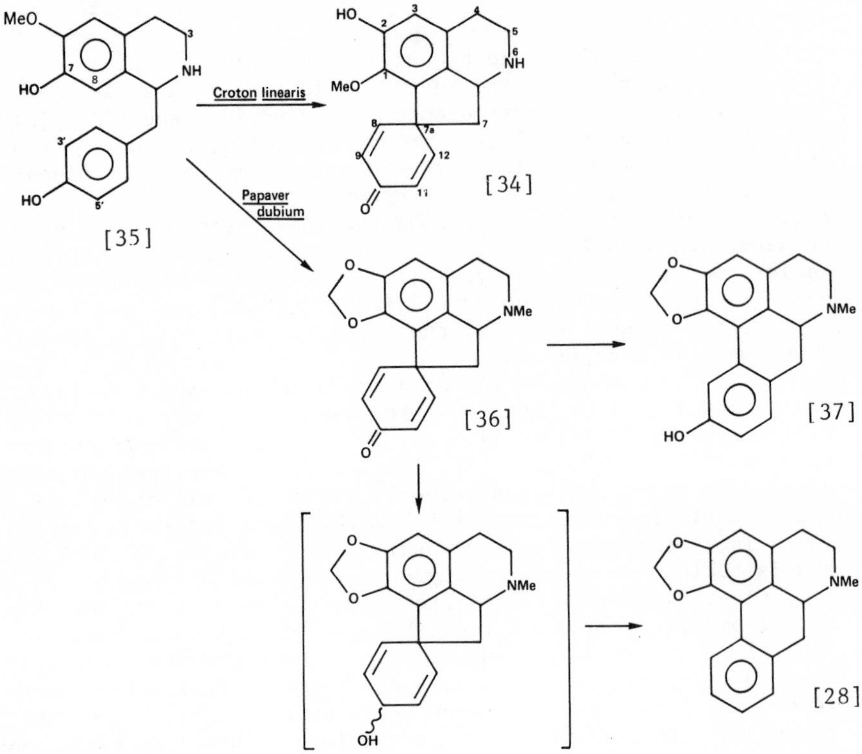

A dienol-benzene rearrangement of type d (see Scheme II) appears to play a role in the biosynthesis of aristolochic acid-I [42], one of a group of closely related nonalkaloidal phenan-threnes occurring in plants of the family Aristolochiaceae; reviews of these compounds are available (22a,b). The group includes, besides [42], the aristolochic acids II [43], C [44], and D [45] (23), debilic acid [46] (22b) (a higher homolog

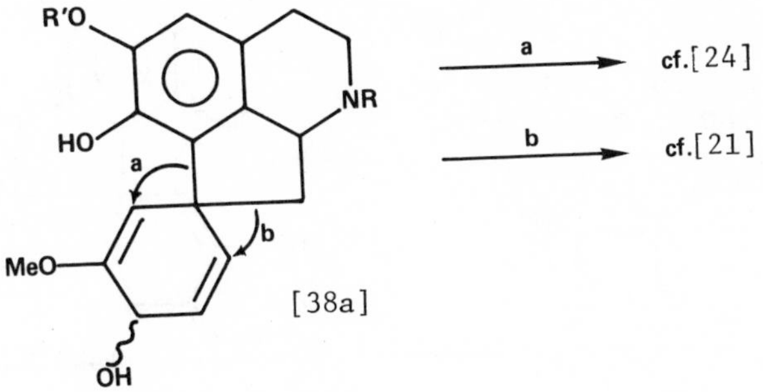

Scheme II

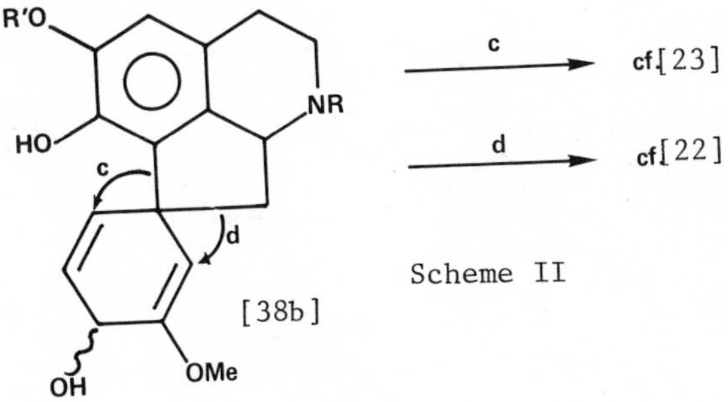

of [42]), and three lactams of the corresponding amino acids: [47], [48], and [49] (22b, 23, 24a).   Several closely related lactams occur in *Stephania cepharantha* (Menispermaceae) (24b). The structures of all these compounds clearly point to a derivation from aporphines through some biosynthetic equivalent of a Hofmann degradation, followed by oxidative transformations (22a); the occurrence of quaternary aporphine bases in certain species of *Aristolochia* is suggestive of such an origin, as is the structure of [46], a possible intermediate in the formation of [42].   Hypothetical aporphine precursors of these phenanthrenes would have the oxygenation pattern of [22] for [42], [45], and [47], of [26] and [27] for [43], and of [21] for [44].   It is of interest that this production of the fully aromatic phenanthrene system from aporphines would involve aromatization of one of the rings; see Chapter 16 for such secondary aromatizations.

Much experimental evidence supports the postulated biosynthesis of [42].   Schütte, Orban, and Mothes (25) demonstrated the specific incorporation of [4-^{14}C]-norlaudanosoline [50], the methyl-free parent compound of [39], almost the entire activity being present in the carboxyl, as expected.   Spenser and co-workers (26) found that tyrosine, dopa, dopamine, and nor-adrenaline are utilized in a way consistent with intermediary

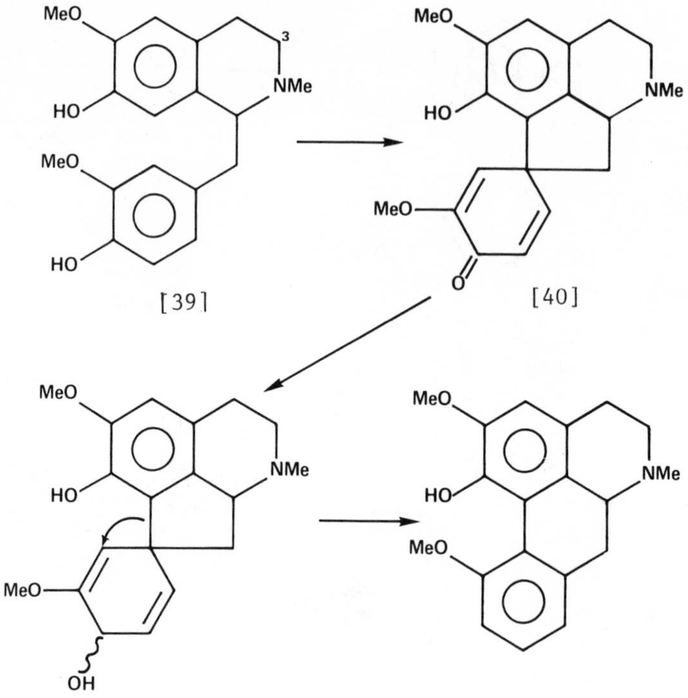

[39]   [40]

[41a and b]   [24]
Scheme III

formation of [50], and that the nitro group of [42] is derived
from the amino group of tyrosine. The incorporation of the
various precursors strongly suggests a biosynthesis of [42] from
[50] via a proaporphine analogous to (or identical with) [40],
reduction to the dienol (cf. [41]), rearrangement by path d to
[22] (or an analog with the same oxygenation type), and subsequent oxidative opening of the heterocyclic ring. While the
individual stages remain conjectural, the postulated dienol-
benzene rearrangement provides the only known reaction of
established biosynthetic occurrence by which a tetraphenolic
precursor such as [50] can yield the tri-oxygenated aromatic
system of [42]. The biosyntheses of the other compounds, [43] -
[49], have not been studied so far; they can be formulated in
analogous fashion.

The orientation of the methoxyls in [45] and [48] is most
unusual for a shikimate-derived structure. No aporphine of this
type seems to be known. The biosynthesis of protostephanine,
which contains a similar pair of *meta*-oriented methoxyls, will

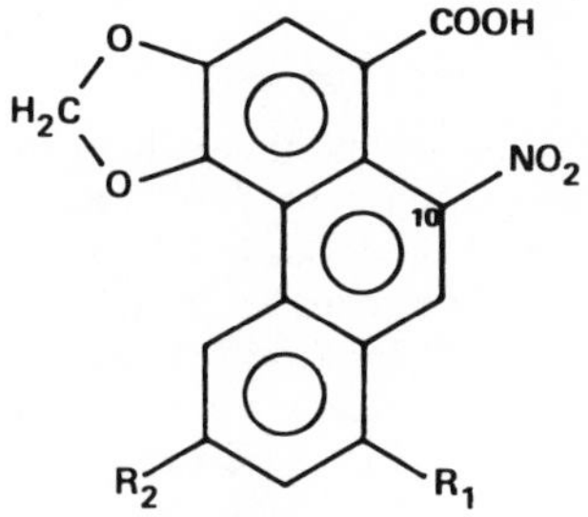

[42]: R₁=OMe, R₂=H
[43]: R₁=R₂=H
[44]: R₁=H, R₂=OH
[45]: R₁=R₂=OMe

[42]: R_1=OMe, R_2=H
[43]: R_1=R_2=H
[44]: R_1=H, R_2=OH
[45]: R_1=R_2=OMe

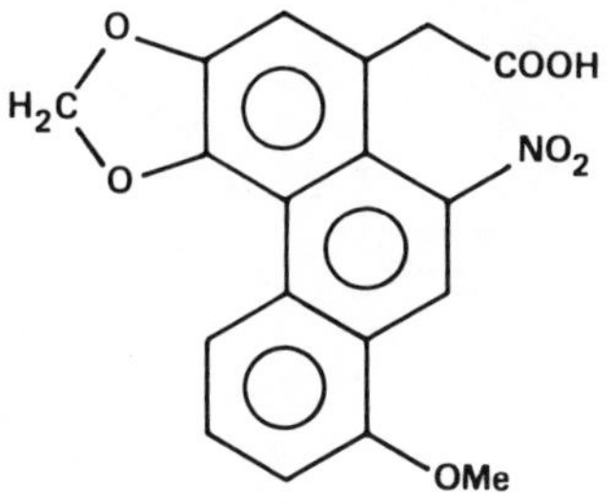

[46]

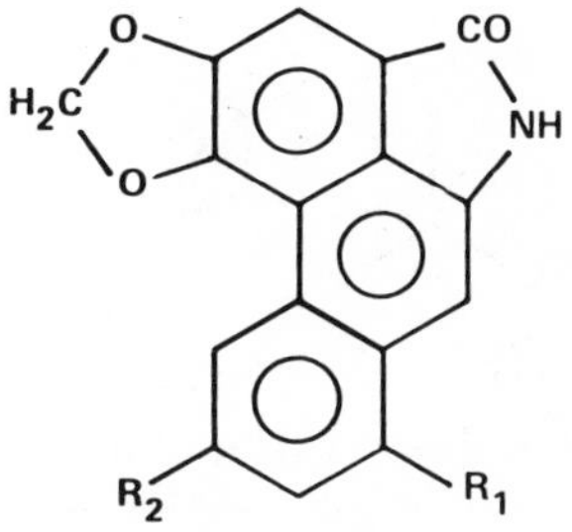

[47] : R_1=OMe, R_2=H
[48] : R_1=R_2=OMe
[49] R_1=OMe, R_2=H,
       (OMe)$_2$ instead of
       —O—CH$_2$—O—

be discussed below; it, too, can only be explained by a dienol-
benzene rearrangement.

In addition to the phenol coupling between positions 8 and
1' which yields proaporphines(cf. [31], [34], [36], [38], [40]),
1-benzyltetrahydroisoquinolines [16] can also undergo such coupling between 2' or 6' of the benzylic ring and either of the two
backbone carbons, 4a or 8a, of the tetrahydroisoquinoline system,
with conversion of the phenolic ring of this system into a cyclohexadienone.   Coupling to 4a or 8a can each give rise to a pair
of such dienones, resulting from involvement of either 2' or 6'.
Since biosynthetic phenol coupling takes place exclusively in

positions *ortho* or *para* to a free phenolic hydroxyl, the methylation pattern of the precursor [16] provides a steering
mechanism (cf. the analogous utilization of [35], but not of its
isomer isococlaurine, for the biosynthesis of [34]).

Coupling to 8a in a 1-benzyl-6-hydroxy-7-methoxytetrahydro-
isoquinoline can give dienones such as [51] and [52]. Compounds
of this type play a role in the biosynthesis of certain aporphines, and of the *Erythrina* alkaloids. No case seems to be
known, however, where a cyclohexadienol would be involved in
these processes, and no further discussion is needed.

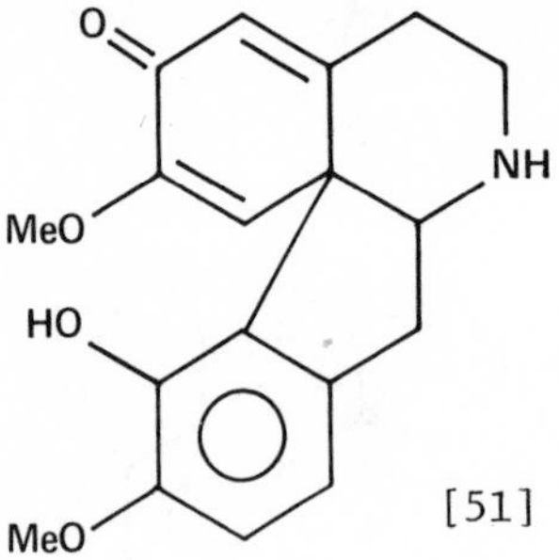

[51]

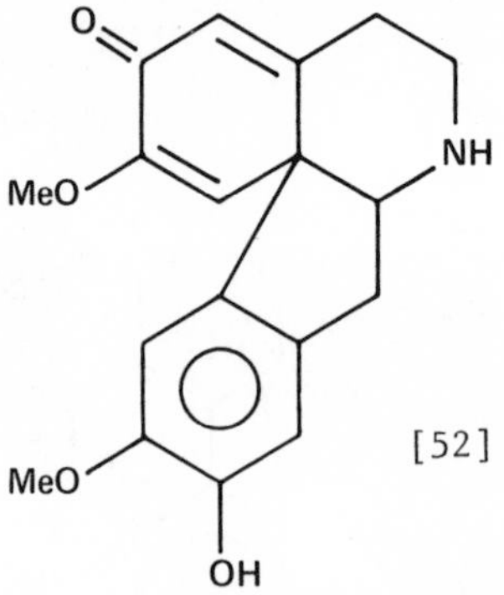

[52]

In 1-benzyltetrahydroisoquinolines with a free hydroxyl in
position 7, for example, reticuline [53], phenol coupling between
carbons 4a and 2' or 6' can take place, to produce dienones
exemplified by [54] and [55]. Cyclohexadienones of both types
are well known as natural alkaloids, and can also be obtained
synthetically by phenol-coupling reactions. Because [54] is
a key intermediate in the biosynthesis of morphine and its
relatives, dienones with the ring system of [54] and [55] are
called morphinadienones (27), and are usually written in the
form shown, which emphasizes the fact that their carbon-nitrogen
skeleton is that of morphine. The morphinadienones and their
relatives have been reviewed by Stuart (28). They occur mostly
in plants of the genera *Papaver* (Papaveraceae), *Croton* (Euphorbiaceae), and *Sinomenium* (Menispermaceae). Interestingly,
Sinomenium produces morphinadienones (and related bases) with an
absolute configuration antipodal to that of the alkaloids of this
type found in *Papaver* and *Corydalis*. *Croton* spp. may contain
representatives of either series; one species, *C. flavens*, yields
bases of both types [54] and [55], of which the latter, flavinan-
tine (as [55] but OH in 3 and OMe in 2) and its N-nor analog fla-
vinine, have the same stereochemistry as the *Papaver* bases, while
the representative of the former, sinoacutine, of structure [54]

has the opposite absolute configuration. Its antipode salutaridine [54] is the key intermediate in the biosynthesis of the
morphine bases.

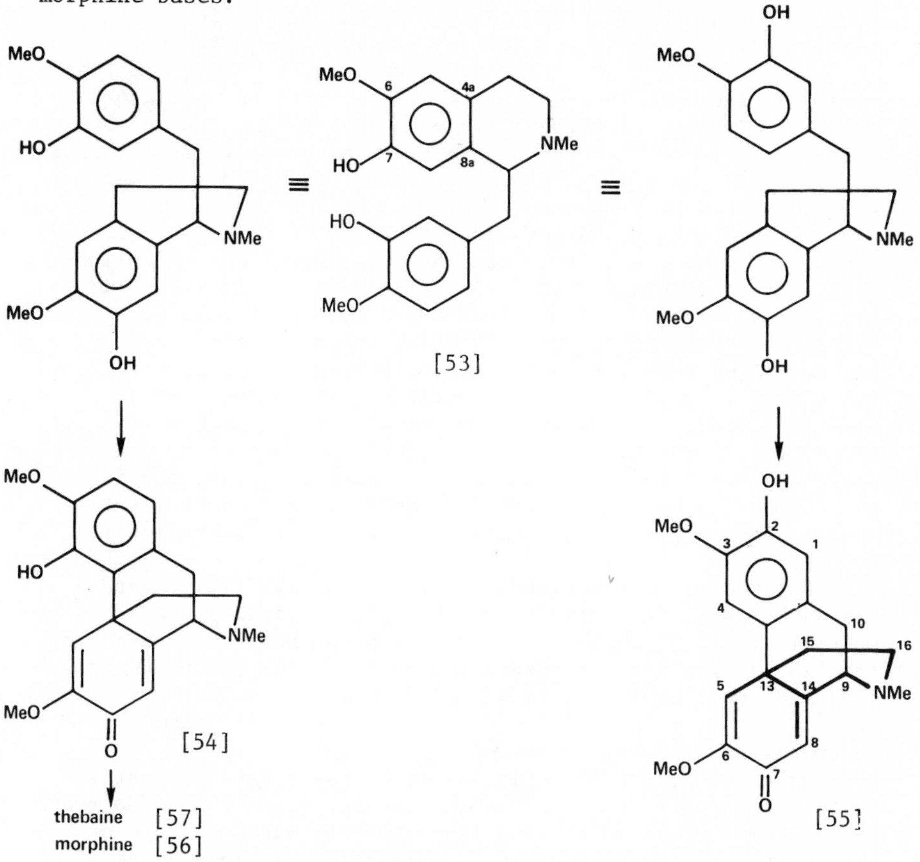

[53]

[54]

thebaine   [57]
morphine [56]

[55]

Morphinadienones of both series have been reduced to the
corresponding cyclohexadienols in the laboratory, and evidence
for the participation of such dienols of both types in biosynthetic sequences is compelling. The only cyclohexadienol alkaloid
actually isolated from a plant so far, nudaurine [7] (see page
187), belongs to the [55] series.

Of the two types of morphinadienones, [54] is the much more
important one because of its role in the biosynthesis of morphine.
Intermediacy of cyclohexadienols is well established and seems
instrumental in the formation of the oxygen bridge, a characteristic structural feature of the morphine group of alkaloids.

Further transformation of these dienols, however, does not proceed by dienol-benzene rearrangement, and does not lead back to aromaticity, as far as is known.

The biosynthesis of morphine [56] and its allies, for example, thebaine [57] and codeine [58], in the opium poppy *Papaver somniferum*, and the closely related *P. bracteatum*, has been elucidated in full detail through studies in the laboratories of Barton, Battersby, Leete, Rapoport, and others; it is one of the best-understood biosynthetic sequences. Its essential features are shown in Scheme IV; for more detailed information, see the review by Battersby (12b).

As predicted by Barton and Cohen (13), the biosynthesis of [56] leads from the 1-benzyltetrahydroisoquinoline reticuline [53] to the morphinadienone salutaridine [54]. In a reaction proposed by Battersby (29), this dienone is next reduced to the cyclohexadienol salutaridinol I [59], which is converted to the methoxy-diene thebaine [57] through the reaction indicated in the Scheme. The further transformations shown there yield codeinone [60], codeine [58], and finally morphine itself, [56] which, incidentally, is rapidly metabolized further to products as yet unidentified (30)--a telling argument against the interpretation of secondary metabolites as inert and biochemically unimportant dead ends of metabolic processes (see Chapter 1, p. 6).

All stages in the biosynthesis of [56] are firmly established by work often involving the highly ingenious use of multiply labeled compounds. Every step has been duplicated in the laboratory, and all intermediates, with the exception of [59] and [60], have been shown to occur in *P. somniferum*, *P. bracteatum*, or *P. orientale*, which produces [57] and its O-demethylated analog oripavine.

Only the stages directly involving the dienol [59] are pertinent to this section. In the laboratory, borohydride reduction of [54] produced (31,32) approximately equal amounts of the two dienols epimeric at C-7, salutaridinols I [59] and II [61]. Both epimers gave [57] on brief exposure to pH 4 at room temperature, [59] reacting faster than [60]. *P. somniferum* converted [59] into [57]. Experiments with [59] doubly labeled with tritium at position 7 and with either ^{14}C at C-16 or ^{3}H at C-1 gave [57] without scrambling of label or significant loss of ^{3}H from C-7. The epimeric [62] was much less efficiently utilized by the plant.

In the transformation of [59] to [57], the oxygen bridge is closed by formal addition of the phenolic hydroxyl to the double bond between C-5 and C-6. Surprisingly, [54] itself does

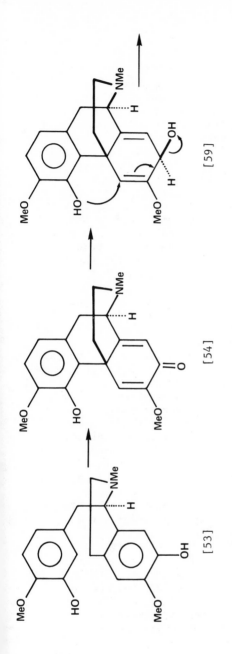

[53] [54] [59]

[57] [60] [58] : R=Me
[56] : R=H

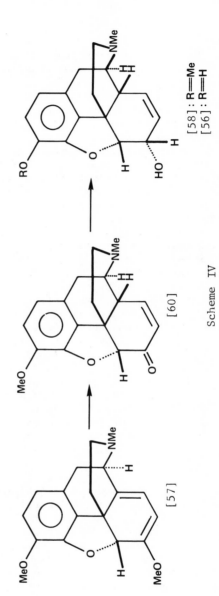

Scheme IV

203

not undergo this reaction, although several analogous dienones do (31). For an understanding of the mechanism of this process, it seems significant that none of the known alkaloids of the morphine group ([56] - [58] and others), which all contain this bridge, carries an oxygen at C-7, while related natural bases without oxygen bridge are oxygenated at C-7. This fact suggested the interpretation (29) that [59] is an intermediate which loses the oxygen at C-7 through an elimination reaction (perhaps as a derivative, e.g., a phosphate ester) with closure of the dihydrofuran ring. The process has certain analogies with the conversion of enolpyruvylshikimate phosphate via chorismic acid to prephenic acid. The biosynthesis of [57] from [59] can, however, not be a concerted one-step process of the S_N2' type, which would require a cis-relationship of the leaving group at C-7 and the phenolic hydroxyl, while these two groups are $trans$ in [59]. Closure of the oxygen bridge and elimination of the hydroxyl from C-7 should thus constitute two distinct steps, for which two possibilities have been discussed (33): inversion at C-7 by a reactive group on an enzyme, to give an enzyme-bound species with the correct stereochemistry for ring-closure by an S_N2' reaction; or allylic shift to give a dienol with conjugated double bonds and a β-oriented hydroxyl at C-5 suitable for S_N2 displacement by the phenolic OH. The actual mechanism of the process has not been studied. It is interesting that both $in\ vivo$ and $in\ vitro$, formation of [57] takes precedence over the alternative possibility of a dienol-benzene rearrangement.

The alkaloids of the morphine group, and quite particularly [57], are notorious for the ease with which they undergo a wide variety of rearrangements leading to aromatization of ring C. No such process has so far been found to occur in the $Papaver$ species that contain those alkaloids. One reaction of this kind, however, must be assumed to take place in the biosynthesis of protostephanine (see below), a base derived from intermediates analogous to [54] and [59].

The class of alkaloids related to the alternative morphinadienone [55] is represented (28) by (-)-[55] itself, pallidine (isosalutaridine), which occurs in $Corydalis\ pallida$ (Fumariaceae) (34), and a few related bases including amurine [62] and the corresponding dienol nudaurine [7] from $Papaver\ nudicaule$ Stereochemically, [7] corresponds to [59], from which it differs only in nature and position of the substituents on the aromatic ring. Both in the laboratory and $in\ vivo$, bases of type [55] have been obtained by phenol coupling of appropriate 1-benzyltetrahydroisoquinolines (35,36,37).

The coupling mode that yields [55] is also known to operate in the biosynthesis of protostephanine [63], an alkaloid from

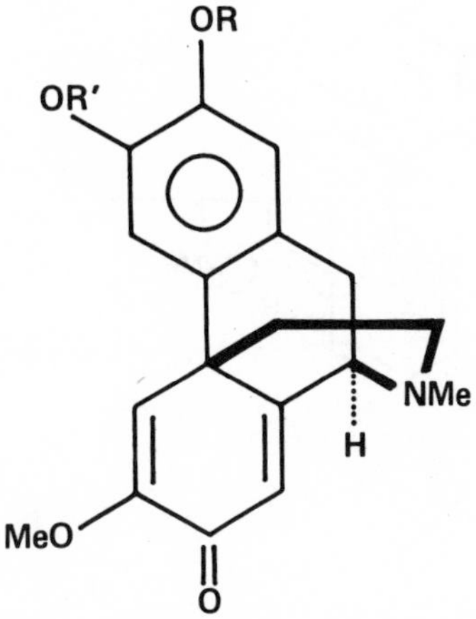

[55]: R=H, R'=Me

[62]: R=R'=—CH₂—

Stephania japonica (Menispermaceae), (37), which has been mentioned before (page 198). The structure of this base is
extraordinary: the *meta*-orientation of the methoxyls on an
evidently shikimate-derived benzene ring has precedent only in
the phenanthrenes [45] and [48] from *Aristolochia*, and the ring
system is apparently unique among natural alkaloids. However,
substances derived from this system have long been known as
transformation products of [57], which yields compounds [64] with
Grignard reagents (38). This reaction has been formulated by
Stork (39) as proceeding through the iminium salt [65] in the way
shown in Scheme V. The hydriodide of [65] was subsequently
obtained from [57] and MgI₂ (40) and shown to give [64] with
Grignard reagents (40a), while hydride reduction (40b, 41) gave
the parent compound [66], a perfect analog of [63].

For the biosynthesis of [63], Barton (42) proposed the
pathway shown in Scheme VI. Phenol-coupling of the 1-benzyl-
tetrahydroisoquinoline [67] is assumed to yield the morphinadienone [68], which would be reduced to the dienol [69].
Conversion of this to the iminium ion [70] (cf. [65]) follows
Stork's interpretation. Finally, reduction would yield [63].

Results obtained by Battersby and co-workers (43) make it
highly probable that Scheme VI represents the actual biosynthesis
of [63]. Oxidation of synthetic [67] with ferricyanide gave
[68]. This dienone was found to be efficiently incorporated
into [63] by *S. japonica*, and it was the key intermediate in a
total synthesis which closely followed the postulated biosynthetic sequence to yield [63] identical with the natural alkaloid.
On the basis of these findings it can hardly be doubtful that
[63] is formed in the plant by a pathway essentially identical
with the one of Scheme VI, and that a dienol such as [69] is an

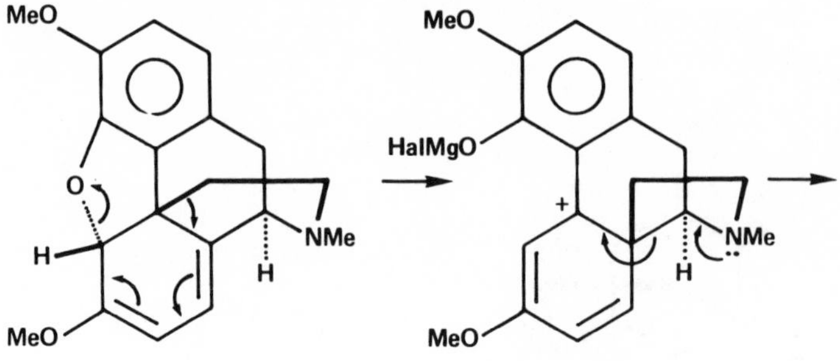

[57]

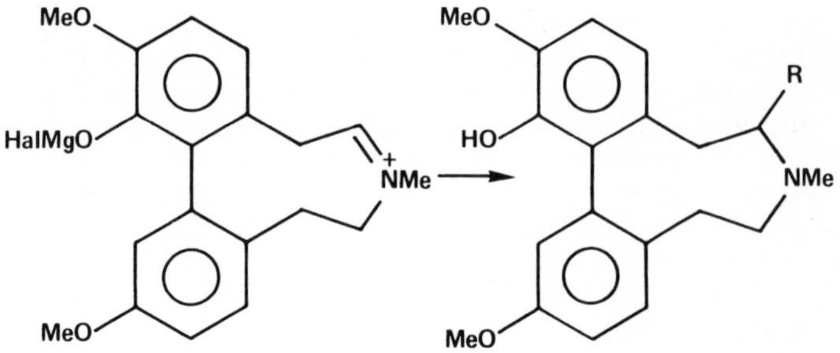

Scheme V

[65]

[64] :R=Ph or Me
[66] :R=H.

intermediate whose further transformations can account for the unusual resorcinol ring of [63].

The most important naturally occurring cyclohexadienediols are chorismic, isochorismic, and *trans*-2,3-dihydroxy-2,3-dihydro-benzoic acid; they are discussed in Chapters 6 and 12, respectively. A few other such diols, however, have been encountered as secondary metabolites, and compounds of this type play a most important role in biological transformations of aromatic rings. Further metabolism of these diols very often leads back to aromatic structures which are different from the initial ones in particular, cyclohexadienediols are frequent intermediates in the introduction of phenolic hydroxyl. They are thus *bona fide* precursors of aromatic rings, and as such have to be included in this account, the more so since the patterns of phenolic hydroxyl provide important clues as to the biosynthetic origin of aromatic structures (cf. the Introduction, pages 5,7,8). Diols derived from

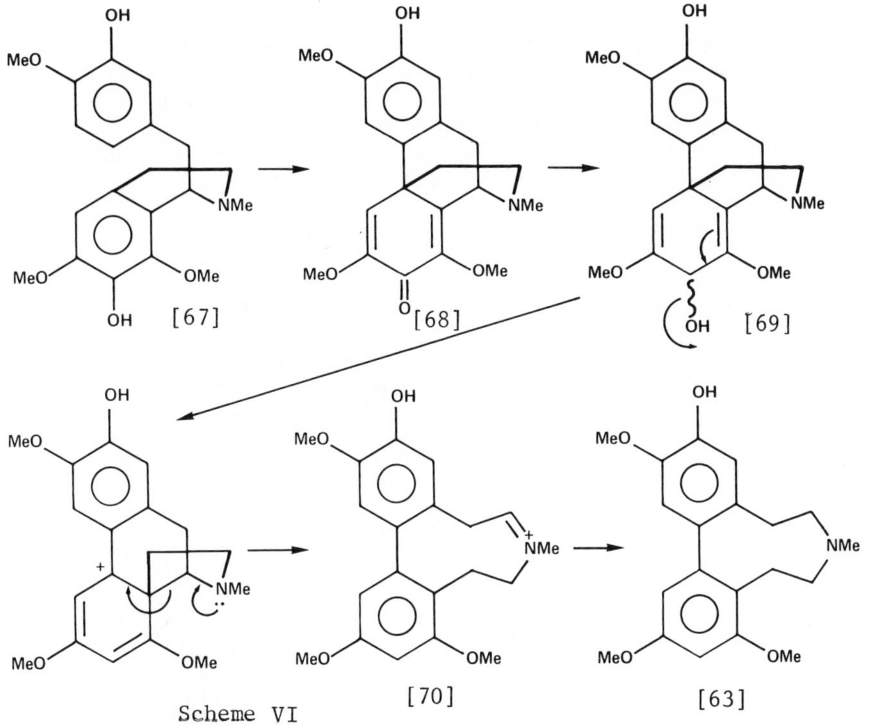

Scheme VI

[67] [68] [69] [70] [63]

either cyclohex-1,3-and 1,4-diene occur, the former being much the more important class.

Representatives of both classes occur among the bromine-containing metabolites of the sponge genera which also produce the cyclohexadienols [10] and [11] (see page 188). *Aplysina aerophoba* yields aeroplysinin-I [71] (44), which is aromatized to 2-hydroxy-4-methoxy-3,5-dibromophenylacetonitrile by dilute alkali, and to the corresponding nitrile by concentrated sulfuric acid at 100°. Some samples of another sponge, *Ianthella ardis,* likewise contain [71], while a different sample of this species gave the optic antipode (45). Occurrence of both antipodes in the same species is surprising, as is the fact that both the (+)- and (-)-forms show the same antibacterial potency. The absolute configurations of both antipodes have been established independently by X-ray crystallography (46). The same genera also contain another brominated compound, aeroplysinin-2 [72] (47); *A. aerophoba* yielded the (+)-form, *Ianthella* sp. the racemate.   Treatment of the acetate of [72] with cold dilute alkali gave 2-hydroxy-4-methoxy-3,5-dibromophenylacetic acid. Finally, two bis-cyclohexadienols: aerothionin [73] (48,49) and homoaerothionin [74] (49,50) occur in *Verongia aerophoba* and *V. thiona*, [73] being the major bromo compound in these two sponges.   Both compounds are aromatized by alkali in a way analogous to the corresponding reaction of [71].

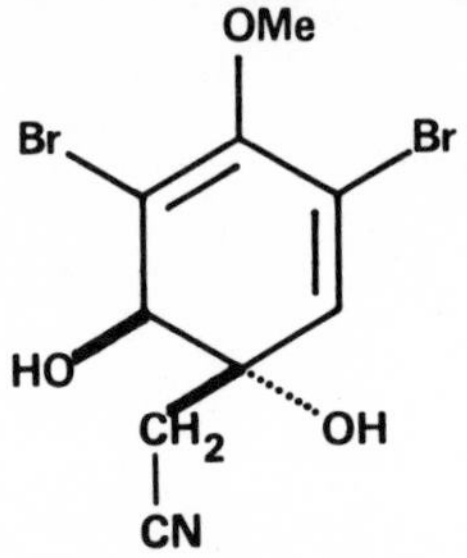

OMe
Br   Br
HO
CH$_2$ OH
CN
(+) — [71]

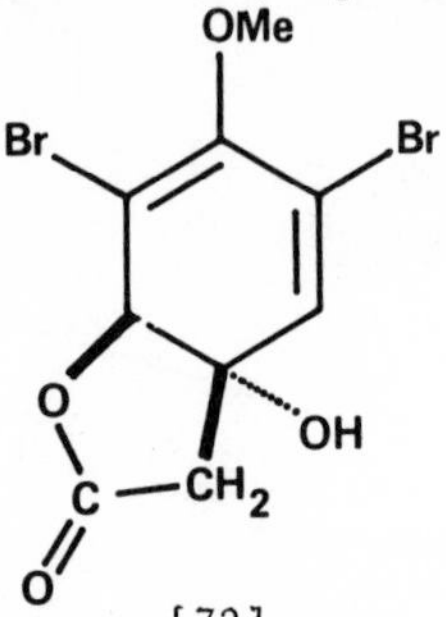

OMe
Br   Br
O
C — CH$_2$ OH
O
[72]

OMe
Br   Br
HO
O   CH$_2$

OMe
Br   Br
H$_2$C   O   OH

N $\equiv$ C—CO—(CH$_2$)$_n$—CO—C $\equiv$ N

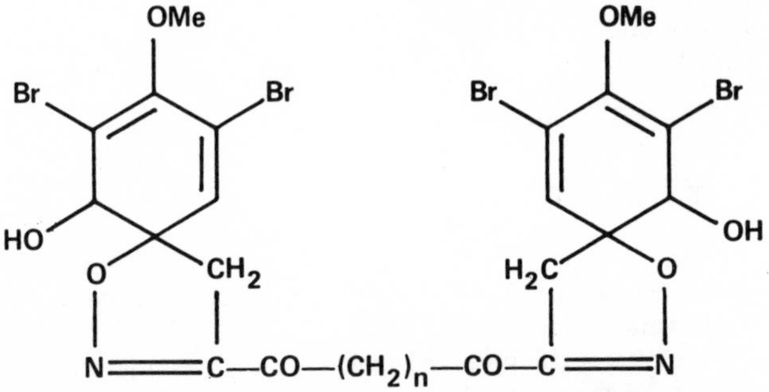

[73] n=4   [74] n =5

Like [10] and [11], the diols [71] - [73] are assumed (44, 47-49) to be derived biosynthetically from arene oxides analogous to [12].

An amino analog of 2,3-dihydroxy-2,3-dihydrobenzoic acid, (+)-*trans*-2,3-dihydro-3-hydroxyanthranilic acid [75], has been isolated by McCormick and co-workers (51) from culture filtrates of *Streptomyces aureofaciens*.   It yields anthranilic acid on treatment with concentrated hydrochloric acid at 60°, and *m*-hydroxybenzoic acid with 25% sodium hydroxide at 93°; action of Pd/C catalyst in boiling water leads to disproportionation into 3-hydroxyanthranilic acid and its hexahydro derivative. Since generally labeled [75] was not incorporated into the

7-chlorotetracycline produced by *S. aureofaciens*, the compound
does not seem to be involved in the biosynthesis of the antibiotic (for the biosynthesis of the tetracyclines, see Section
18-12).

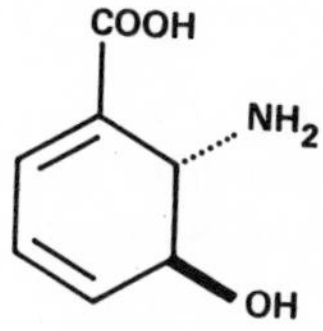

[75]

It has been known for a number of years that many aromatic
compounds are metabolized through sequences that amount to their
transformation into phenolic compounds, often followed in
mammals by urinary excretion in the form of various conjugates.
These catabolic reactions--essentially detoxification processes--
would not *per se* qualify for inclusion in this account of
formation, rather than transformation, of benzenoid structures.
It has been observed repeatedly, however, that vicinal cyclohexadienols are involved as intermediates, and that reactions of this
kind, while most commonly encountered in the metabolism of
extraneous aromatic compounds, are by no means restricted to
these substances. An example for the involvement of such a
dihydrodiol in the catabolic transformation of a normal cell
constituent is provided by the appearance of 7,8-dihydroxy-7,8-
dihydrokynurenic acid in the metabolism of tryptophan by a
Pseudomonas sp., and by its subsequent enzymatic dehydrogenation
to 7,8-dihydroxykynurenic acid (4,7,8-dihydroxyquinoline 2-
carboxylic acid). Formation of N-analogs of such diols in a
reaction which can properly be called biosynthetic is exemplified
by gliotoxin and certain related microbial metabolites. These
cases will be discussed below.

Transformation of aromatic compounds to their dihydroxy-
dihydro derivatives has been found so far in mammals and microorganisms. However, the diols produced by these two classes of
organisms differ significantly in their stereochemistry, and in
the source of the oxygens of their hydroxyl groups. Those of
mammalian origin are *trans*-diols and contain one atom of oxygen
furnished by O_2; in contrast, the diols from microorganisms--
at least those derived from simple aromatic compounds and from
naphthalene (52)--have been shown to have *cis*-stereochemistry
in all cases that have been studied recently by modern methods,
and both oxygen atoms are derived from atmospheric O_2. Theoretically, the *cis*- and *trans*-diols should, therefore, be formed

by fundamentally different mechanisms, and there is abundant evidence that this is indeed the case; consequently, these two types should be treated separately. In the biosynthesis of the *trans*-diols, arene oxides (cf. compound [12], above) have been established as intermediates. The diols from microorganisms shall be discussed first.

Cyclohexadiene-1,2-diols have been encountered as intermediates in the microbial transformation of several aromatic hydrocarbons, some of their halogenated derivatives, and of two carboxylic acids. The oxidative catabolism of aromatic hydrocarbons has been reviewed recently by Gibson (53). Such diols have, to the best of our knowledge, been isolated in pure form from the following aromatic substances: benzene (54), toluene (55), *p*-chlorotoluene (56), ethylbenzene (53), naphthalene (57,58),1-chloronaphthalene (59), 2-methylnaphthalene (60), phenanthrene (61), anthracene (61), and benzoic acid (62). The 7,8-dihydroxy-7,8-dihydrodiol from kynurenic acid (itself a metabolite of tryptophan) has not been obtained entirely pure (63). It is interesting to note that the dihydro-diol from phenanthrene is the 3,4-compound (61); it will be shown below that the analogous compound formed in the mammalian organism is mainly 9,10-dihydroxy-9,10-dihydrophenanthrene, accompanied by lesser amounts of the 1,2-diol and little of the 3,4-compound (all three diols being the *trans*-isomers).

The earliest of these compounds to be obtained in pure form was 1,2-dihydroxy-1,2-dihydronaphthalene [76], produced from naphthalene by a soil bacterium; it was isolated in 1953 by Walker and Wiltshire (57). The diols from unsubstituted aromatic hydrocarbons are generally difficult to obtain because of the speed with which they undergo further enzymatic oxidative transformation to catechols; they are more readily accessible with the help of mutants, particularly of *Pseudomonas* spp., which are blocked in these reactions. However, introduction of halogen into aromatic compounds generally seems to render them more resistant to metabolic degradation, and diol [77], obtained fairly readily through action of wild-type *P. putida* upon *p*-chlorotoluene, was the first one of these compounds for which the *cis*-stereochemistry was proved by Gibson and co-workers (56) through nmr methods. The same relative configuration had been indicated previously (54b) for the diol [78] from benzene by radiochemical trapping experiments with the synthetic *cis*- and *trans*-diols, and was subsequently proved (54a) by isolation of *cis*- [78] in crystalline form; *cis*-configuration was also established for the diols [79], [80], [76], and [81] from toluene (55), ethylbenzene (53), naphthalene (48,65), and benzoic acid (62), respectively. For [76], the *absolute* configuration shown was proved (58) through catalytic reduction of the double bond and hydrogenolytic removal of the α-hydroxyl to give (-)-2(S)-hydroxy-1,2,3,4-tetrahydronaphthalene of known absolute stereochemistry; that of [79] was deduced quite recently (64) from a study of its circular dichroism and by oxidation to (-)-2-methyladipic acid, whose absolute configuration is known. The remaining compounds are formulated in the way shown below, rather than as the antipodes, by analogy with [76j and [79].

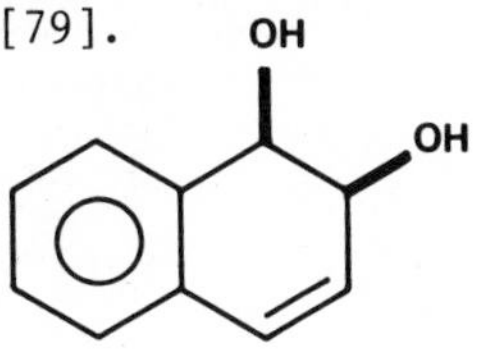

[76] 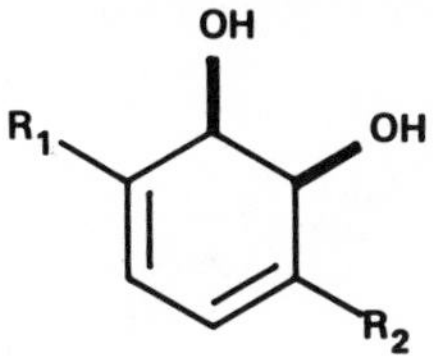 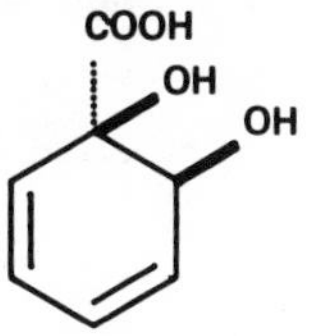   [81]

[77] : R_1= Me, R_2= Cl
[78] : R_1 = R_2 = H
[79] : R_1 = Me, R_2 = H
[80] : R_1 = Et, R_2 = H

Compounds [76] to [80] were all obtained from *Pseudomonas* spp., usually *P. putida*; the acid [81] was isolated from culture filtrates of the bacterium *Alcaligenes eutrophus*. Formation by this organism of dienes analogous to [81] from *m*-methyl-, *m*-chloro, *m*-fluoro-, and *p*-fluorobenzoic acids was detected by appearance of the characteristic uv spectrum (62).

In these *cis*-1,2-diols, the hydroxyls and the hydrogen atoms on the adjacent saturated carbon are of necessity in the *trans*-relationship favorable for elimination reactions. In fact, the compounds undergo very facile acid-catalyzed dehydration to monophenols; the diol [76] was found to be converted to α-naphthol by dilute HCl at room temperature 44 times faster than the *trans*-epimer (58). Where two different phenols can form, one product is usually much favored; diol [79] seems to give *o*-cresol exclusively (55), and [76] yields α-napthol together with less than 5% β-naphthol (58). This preponderance of the α-compound does not seem to operate in the case of the 8-chloro derivative of [76] which is formed from 1-chloronaphthalene (59); it is stated to give mostly 8-chloro-β-naphthol together with only a small amount of 8-chloro-α-naphthol. The diol [81] from benzoic acid is converted to salicylic acid and phenol (62).

The dihydrodiols [76] - [81] have all been isolated in pure form and their *cis*-stereochemistry has been established experimentally. Neither is true for one more compound of this class which has already been mentioned briefly: the dihydrodiol [82] formed from kynurenic acid [83] by cell-free extracts of *Pseudomonas fluorescens*. This diol has been observed by Hayaishi and co-workers (63, 66), who made a detailed study of its formation from [83] and of its further conversion into 7, 8-dihydroxykynurenic acid [84]. These reactions constitute steps in the metabolism of tryptophan by *Pseudomonas*, which leads via kynurenine [85], [83], [82], [84], and further intermediates to glutamic acid, alanine, and acetic acid (67) (Scheme VII).

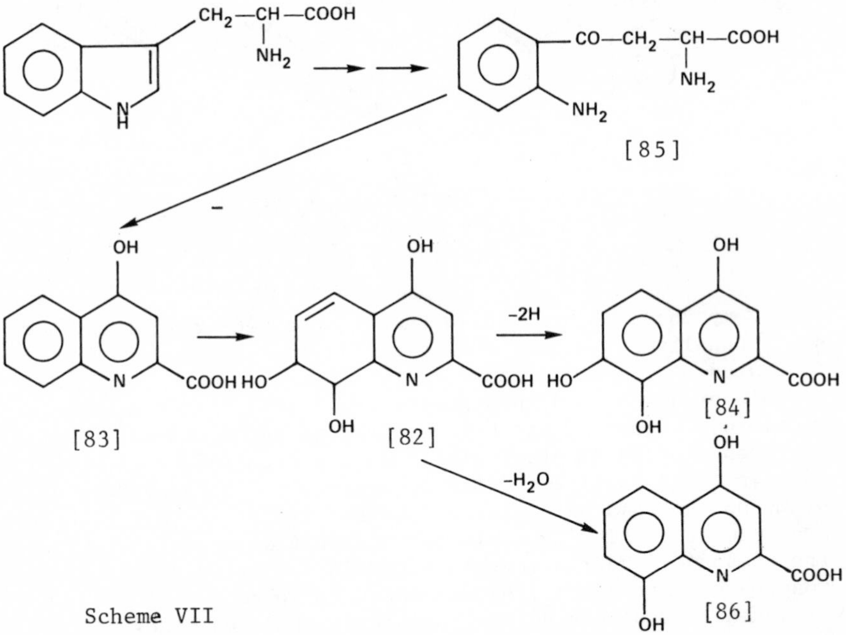

Scheme VII

The dihydrodiol [82] could not be isolated in crystalline form because of the ease with which it loses water to give 8-hydroxykynurenic acid [86]; its structure was not actually proved but can hardly be doubtful on the basis of this conversion into [86] and of the enzymatic dehydrogenation to [84]. The diol was initially assumed to be the *trans*-compound by analogy with the dihydrodiols occurring in mammalian metabolism, the *cis*-stereochemistry of their microbial counterparts being still unrecognized at the time; it was suggested that the compound might be formed via a 7,8-oxide or an 8-hydroperoxide. However, the extremely ready dehydration of [82] to [86] on mere drying strongly suggests *cis*-configuration; furthermore, the properties of the enzymes which catalyze both the formation of [82] from [83] and its further transformation into [84] are quite similar to those involved in the microbial metabolism of aromatic compounds via the *cis*-diols and differ from those concerned with the *trans*-diols in mammals (see below). It is thus probable that [82], like the other microbial cyclohexa-dienediols, is the *cis*-compound (cf. ref. 68).

It may be permissible to speculate that the strikingly frequent occurrence of oxygenated substituents in positions 8 and 7 + 8 in quinoline alkaloids, and in the corresponding positions 7 and 6 + 7 of indoles, might find its explanation in reactions analogous to the formation of [82] and its conversion into [84] and [86]. In these monophenols (OH in 8 or 7, respectively), the hydroxyl occurs in the position *meta* to the one initially occupied by the carboxyl of shikimic acid, which is the precursor of the aromatic ring of indoles (cf. Chapters 9 and 16); this orientation is otherwise quite uncommon (cf. Chapter 16).

The enzymatic formation of these dihydrodiols is of partic- ular interest, since it constitutes the first attack upon the aromatic ring. Its mechanism has been clarified by studies of the incorporation of ^{18}O into diols [78] (54a) and [81] (62). In both cases, the oxygens of *both* hydroxyls were found to be labeled. This result stands in striking contrast to those made on the *trans*-diols from mammals, where only one of the two oxygens is derived from ^{18}O (see below). The findings on the formation of the *cis*-diols suggest that the reaction proceeds through an intermediary cyclic peroxide (1,2-dioxetane) [87], as shown below; opening of the heterocyclic ring would necessar- ily produce the observed *cis* configuration.

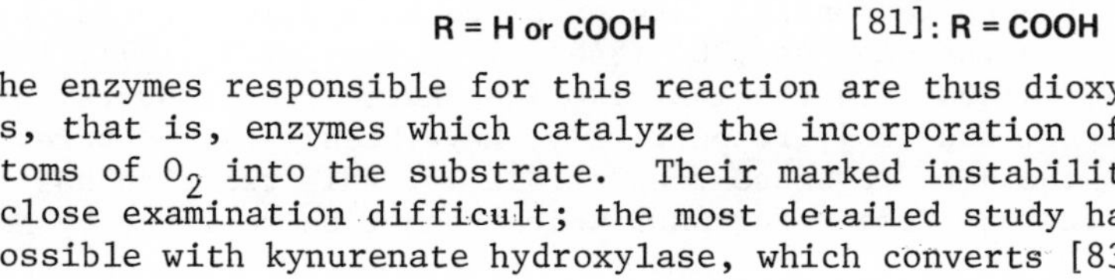

[87]

R = H or COOH

[78]: R = H
[81]: R = COOH

The enzymes responsible for this reaction are thus dioxy- genases, that is, enzymes which catalyze the incorporation of both atoms of O_2 into the substrate. Their marked instability makes close examination difficult; the most detailed study has been possible with kynurenate hydroxylase, which converts [83] into [82]. This enzyme has been obtained in significantly purified form by Hayaishi and co-workers (63,66). It requires NADH as cofactor, NADPH being much less active. The enzyme is a flavoprotein, and precipitation with ammonium sulfate at pH 4.2 yields an inactive apoprotein which can be re-activated by FMN or, less effectively, by FAD, but not by riboflavin. Kynurenate hydroxylase is stimulated by Fe^{2+} and strongly inhibited by chelators such as o-phenanthroline or α,α'-dipyridyl; this inhibition is specifically overcome by Fe^{2+}. Like several other oxygenases, kynurenate hydroxylase is readily inactivated by air; it is protected by cysteine, and the activity of the inactivated enzyme is almost completely restored by NaBH4 or cysteine.

Where it has been possible to study them, the properties of the enzymes catalyzing the formation of the other cis-dihydro-diols were found to resemble those of kynurenate hydroxylase, although certain differences have been observed. Requirement for NADH, replaceable with lowered efficiency by NADPH, is always found, and Fe^{2+} generally plays a role (54b, 57, 68); on the other hand, dependence upon flavines has apparently not been observed for any of these enzymes. Naphthalene oxygenase (68), the enzyme that catalyzes the formation of [76], resembles kynurenate hydroxylase in its sensitivity to air; here, too, activity can be restored by NaBH4 or by SH compounds. However, the enzyme is also inactivated very readily on mere dilution; this feature precluded purification by chromatography which was feasible to a certain extent with the enzymes acting upon kynurenic acid (66) and benzene (54b).

In the oxidative metabolism of aromatic compounds by micro-organisms, the step following formation of the dihydrodiols is their dehydrogenation to o-diphenols (catechols): cf. [82]$\longrightarrow$ [84]. Since this reaction constitutes a return to aromaticity, it is evidently pertinent to our topic.

Catechols have repeatedly been obtained during action of microorganisms on aromatic compounds (cf. (53) and references there). In the case of p-chlorotoluene, both the diol [77] and its dehydrogenation product 3-chloro-6-methylcatechol were found together in the culture filtrates, and were both isolated in pure form (57). The enzymatic transformation of isolated diols into the corresponding catechols has been demonstrated, *inter alia*, in the case of [78] (54b), [79] (55), [77] (57), [80] (53), and [82] (63); in the last two instances, the products, 3-ethyl-catechol and [84], respectively, have been obtained pure.

The dehydrogenase responsible for this reaction has a strict requirement for NAD^+ which cannot be replaced by $NADP^+$

NADH, or NADPH (54b, 55, 63).   It acts specifically upon *cis*-
1,2-diols; diol [78], for example, is rapidly dehydrogenated,
while its *trans*-epimer is entirely resistant to the action of a
partially purified enzyme; however, it is slowly attacked by
washed cell suspensions (54b), and *trans*-1,2-dihydroxy-1,2-
dihydronaphthalene, epimeric with [76], is metabolized by cell-
free extracts of a *Pseudomonas*, although much less rapidly than
[76] itself (65).   The reaction does not require oxygen; conversion of [79] (55), [80] (53), and [82] (63) to the corresponding
o-diphenols proceeds anaerobically, and in the case of [80],
3-ethylcatechol was actually obtained only under those conditions,
while a product of further transformation (probably 2-hydroxy-6-
oxo-2,4-octadienoic acid) was produced in the presence of air.

   With respect to both substrate specificity and cofactor
requirements, the dehydrogenase from *Pseudomonas* spp. differs
strikingly from its counterpart in mammals.   The latter dehydrogenates both *cis*- and *trans*-1,2-diols (69) and depends upon
NADP+ as cofactor.

   The transformation of the diol [81] from benzoic acid
presents a special case, since its dehydrogenation is accompanied
by decarboxylation, the product being catechol.   It has been
shown by Reiner (70) that the entire process is carried out by
one single, NAD+-dependent enzyme protein, which has been
obtained in highly purified form.   The reaction is plausibly
assumed to proceed through initial dehydrogenation of the
secondary hydroxyl at C-2; the resulting β-keto acid would of
course lose CO_2 to give the unstable keto form of catechol.

   The further microbial degradation of aromatic compounds
proceeds by oxidative cleavage of the ring and is thus outside
the scope of this volume.

   The preceding discussion can be summarized briefly by
stating that microorganisms, particularly pseudomonads, metabolize
many aromatic compounds by first transforming them to the *cis*-
dihydrodiols, presumably via a cyclic peroxide.   In contrast,
the mammalian metabolism initially produces arene oxides, which
undergo stereospecific opening reactions of the epoxide ring to
give *trans*-dihydrodiols and their analogs; these processes will
be discussed below.

   There are, however, indications for the occurrence of this
latter mechanism also in microorganisms.   The so-called NIH
shift, diagnostic of intermediary occurrence of an arene oxide,
has been observed during hydroxylation of some aromatic compounds
by microbial enzymes; furthermore, arene oxides [88] and their
oxepin tautomers [89] (see below) play a role in the biosynthesis of one class of microbial metabolites: the group of

[88]   [89]

sulfur-containing fungal antibiotics represented by gliotoxin
[90] (71). This compound, which has been isolated from a variety of fungi, is evidently an N-analog of a cyclohexa-3,5-diene-
1,2-diol. Since X-ray crystallography has established (72) the
trans-relationship of its OH and N, [90] is related to the
mammalian rather than the other microbial diols, and studies
on its biosynthesis make it indeed almost certain that it is
formed from phenylalanine or some derivative of it through
formation of the 2,3-epoxide, which is then attacked by the
N-atom of the side-chain. Scheme VIII shows this reaction
sequence, which was first proposed by Neuss and co-workers (73).
Their interpretation finds powerful support in experimental
results. Bu'Lock and Ryles (74) proved that phenylalanine
completely deuterated in the aromatic ring is incorporated into
[90] by *Trichoderma viride* with complete retention of the label;
dehydrogenation to dehydrogliotoxin [91] proceeded with the
expected loss of two atoms of deuterium. Johns and Kirby (75)
found similarly that [3-^{3}H]-phenylalanine yields [90] without
loss or migration of label; formation of [91] in this case
resulted in retention of 52% of the tritium, since the label
should appear on both positions marked by asterisks in formula
[90]. These findings show that no substitution reaction can
be involved in the biosynthesis of [90] from phenylalanine;
the observed *trans*-stereochemistry is the one to be expected
from the proposed sequence, which seems established beyond
reasonable doubt by the results quoted.

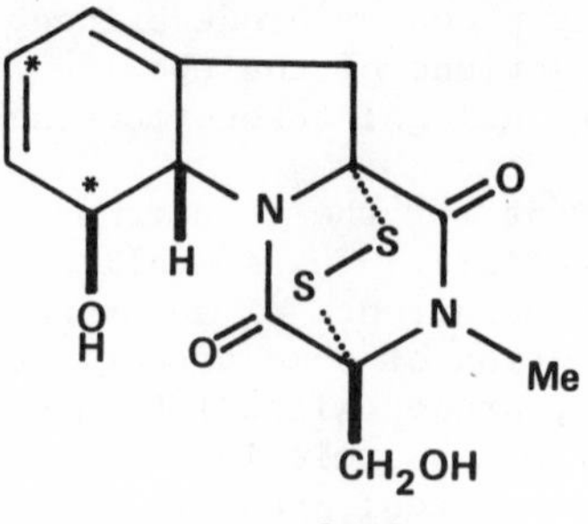

[90] 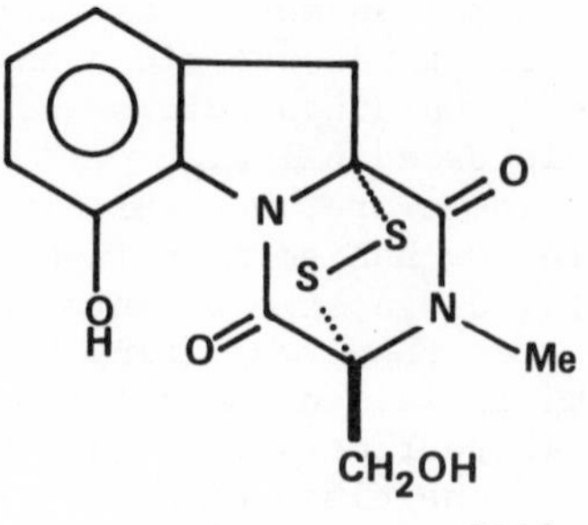   [91]

It is interesting that [91] itself is a natural compound; it has been isolated from *Penicillium terlikowskii* (76), which also contains a number of related sulfur-free metabolites (77).

Involvement of epoxides and oxepins (cf. [88] and [89]) in the biosynthesis of antibiotics related to [90] is quite manifest from the structures of some metabolites of the fungus *Arachniotus aureus*. Representative members of this group are apoaranotin [92] (73), [93] (73), and aranotin [94] (78); of these [92] and [94] have been shown (79) to have the same absolute configuration as [90]. Several other antibiotics have closely related structures in which the rings are aromatic (cf. [91]); chaetocin [95] (80) and sporidesmin [96] (81) may be mentioned. The biosynthesis of [95] and its relatives has been formultaed by Neuss (73) in the way shown in Scheme VIII.

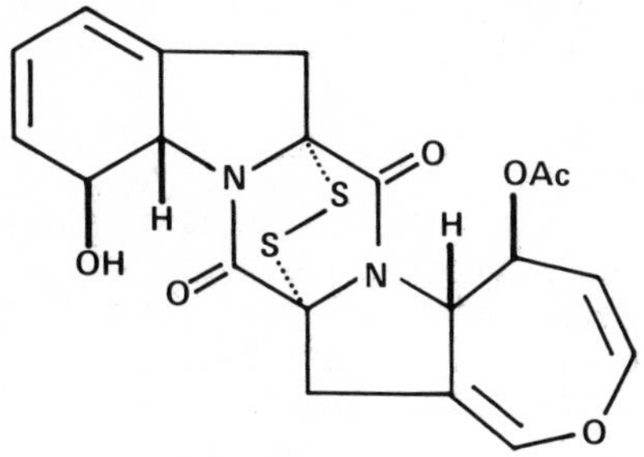

[92]:2 –SMe instead of ...S—S...,
[93] —OAc instead of —OH

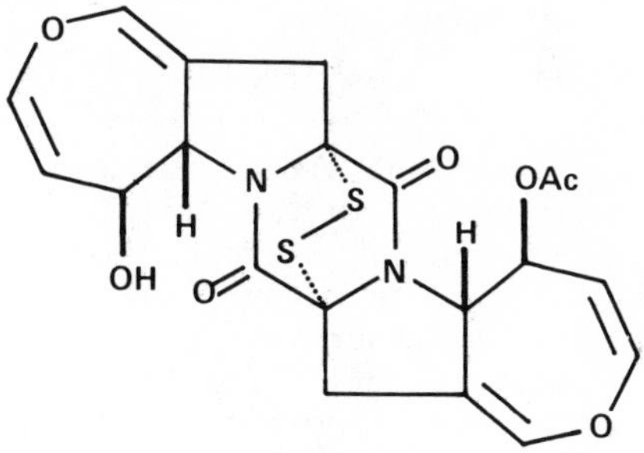

[94]

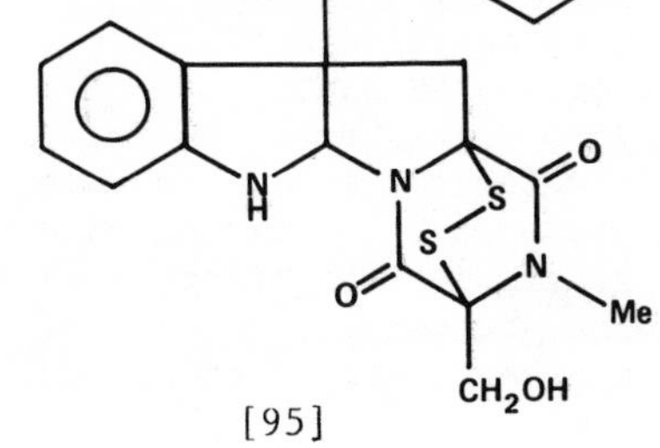

[95]

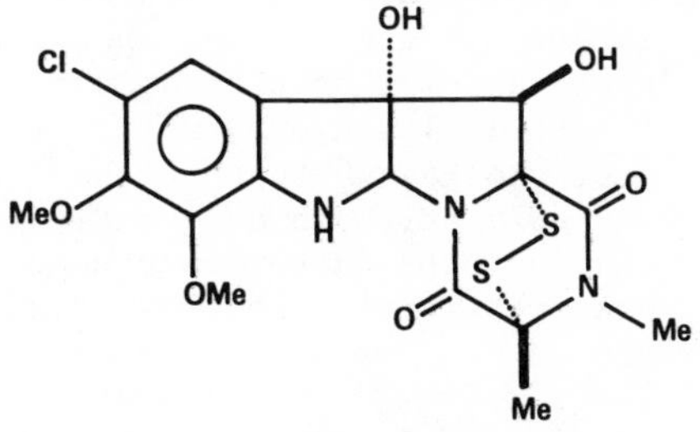

[96]

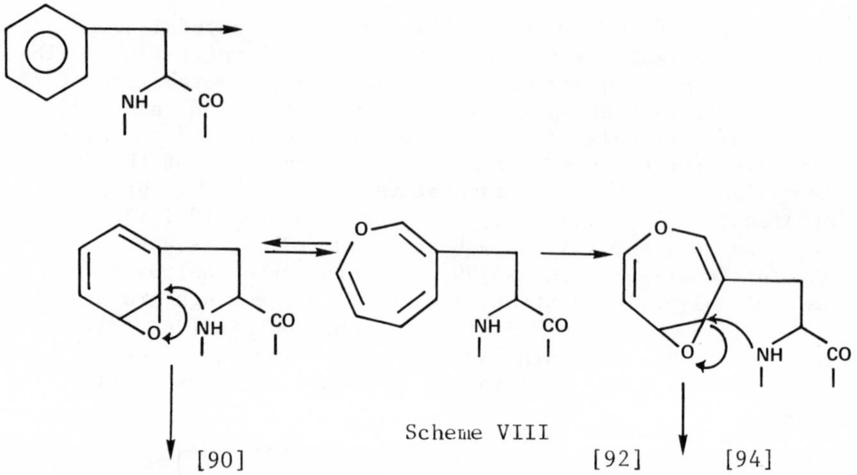

Scheme VIII

[90]   [92]   [94]

As has already been mentioned, the *metabolism* of *aromatic substances* in the *mammalian organism* shows curious discrepancies from that in microorganisms. While cyclohexadienediols are involved in either case, those from mammals are invariably *trans-* rather than *cis-*diols, and the enzymatic apparatus for their formation and further metabolic conversion is radically different from that in microbes.

It has been known for a long time that many aromatic compounds are metabolized by mammals, including man, through conversion into monophenols or *o*-diphenols (catechols), which are usually excreted in the urine as conjugates with sulfuric or glucuronic acid. In a variant first observed in 1879 (82), sulfur analogs of the monophenols, derived from N-acetylcysteine, can be isolated from the urine. It was, however, appreciated already at the time of the discovery of these so-called mercapturic acids that they do not pre-exist in the urine, since they can be isolated only after pretreatment with acid. Subsequent research has shown that the *o*-diphenols are formed *in vivo* through the enzymatic dehydrogenation of cyclohexadienediol precursors, and that the parent compounds of the mercapturic acids, termed premercapturic acids, are derived from mono-S-analogs of these diols. Discussion of these two types of cyclohexadiene derivatives, of their origin, and of their metabolic fate, is thus required. As will become clear later, both these types, as well as the monophenols, arise through transformations of arene oxides (cf. [12], [88]). The entire field has recently been reviewed in detail, with particular attention given to the arene oxides and their transformations (83, 84).

The first of these diols, *trans*-1,2-dihydroxy-dihydroanthra-cene [97], was found in 1935 by Boyland and Levi (85) in the urine of rabbits and rats to which anthracene had been fed. Largely through the work of Boyland and his co-workers, many more such compounds have been isolated since then, too many for complete enumeration.   Extensive lists are given in refs. 83 and 84b.   A few of these diols, however, deserve special mention. The prototype, [98], was obtained by Sato and co-workers (86) in minute amounts from the urine of rabbits after administration of benzene.   The 4-chloro derivative, [99], is formed similarly from chlorobenzene (87).   Naphthalene yields the much-investi-gated *trans*-1,2-dihydroxy-1,2-dihydronaphthalene [100] (88). The urine of rats and rabbits dosed with phenanthrene contains mainly *trans*-9,10-dihydroxy-9,10-dihydrophenanthrene [101] together with smaller amounts of the *trans*-1,2- and *trans*-3,4-isomers (88a, 89a,b); all three diols occur both free and as the glucuronides (89b).   Among the polycyclic carcinogenic hydrocarbons, 3,4-benzopyrene yields the diol [102] and some of its 9,10-isomer (90), and benzo-[a]-anthracene gives three diols analogous to the compounds from phenanthrene (91). Numerous other carcinogenic hydrocarbons are likewise converted into dihydrodiols (84b, 92).

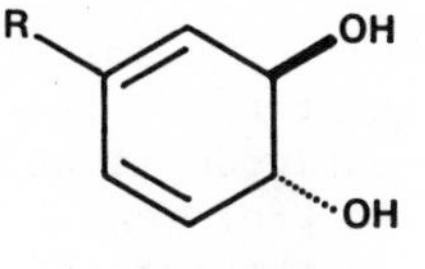

[98] R=H
[99] R=Cl

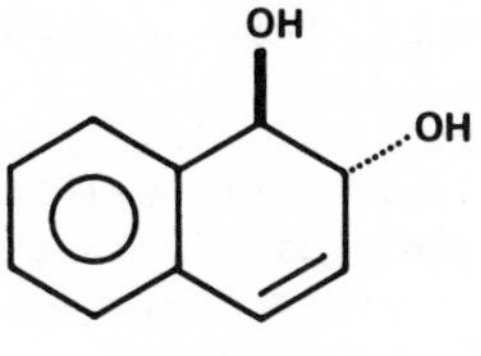

[100]

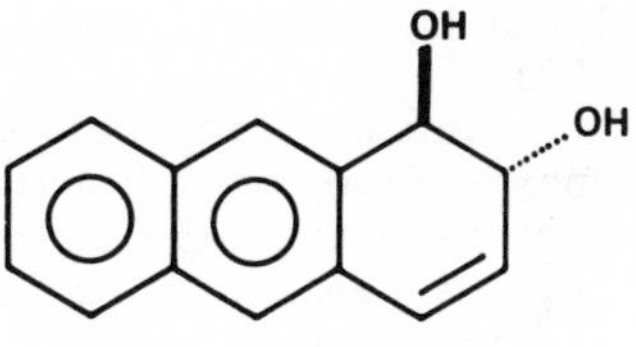

[97]

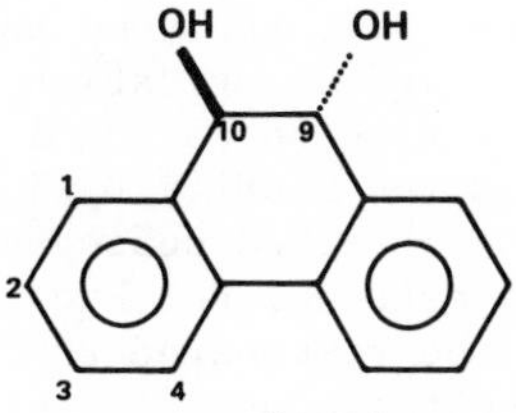

[101]

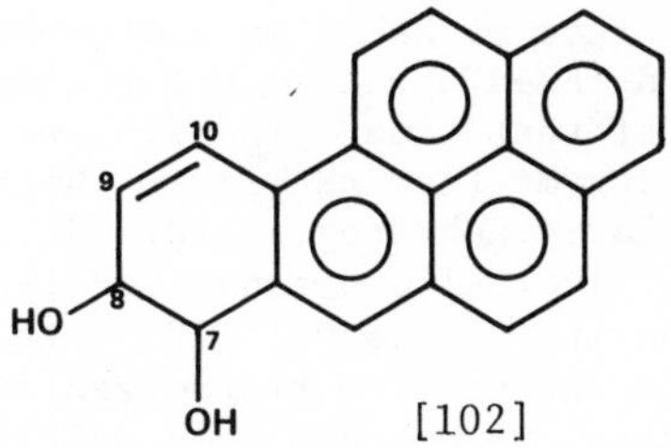

[102]

In all cases where the *relative* stereochemistry of these diols has been ascertained, they were found to be the *trans*-isomers.   The enzymes responsible for their formation (and also those concerned with their further metabolic transformation) seem to operate with curiously low stereospecificity. In intact animals, the diols excreted in the urine usually consist of the racemate together with one or the other of the optical antipodes, whereby marked species-differences seem to exist. Rats, for example, excrete racemic [100] together with some of the (-)-diol, while rabbits produce the racemate and the (+)-antipode (88b); guinea pigs excrete mainly the (-)-diol (88c). Findings of Jerina, Ziffer, and Daly (93) on the enzymatic formation of [100] from naphthalene (or its oxide) by liver microsomes from rat, rabbit, guinea pig, and mouse were more consistent, since the diol isolated from these experiments was levorotatory in every instance; however, the optical purity was relatively low. A sample of (+)-[100] with higher optical rotation had been isolated earlier (94) from action of a homogenate of mouse liver upon naphthalene.

The *absolute* configurations of the (-)-antipodes of [98], [99], [100], and [101] obtained with the use of microsomes have been demonstrated (93) to be the ones shown. The same absolute stereochemistry has been established for the [98] (86) and [99] (87) from rabbit urine; that of (-)- [100] was proved by Miura and co-workers (95) using resolved synthetic material. As formed by liver microsomes, (-)-[100] and (-)-[102] had optical purities of 30-50 and 60-70%, respectively (93).

The dehydrogenation of these diols to catechols by a soluble enzyme from liver is similarly stereoselective rather than stereospecific (93); surprisingly, it is the (+)-antipode, that is, the product formed to a lesser extent, which is preferentially dehydrogenated, at approximately seven times the rate of the (-)-antipode in the case of [98].

This enzymatic dehydrogenation was studied in 1959 by Ayengar and co-workers (96) with a sample of [98] synthesized by the method of Nakajima and co-workers (97), the compound not having been obtained yet through biological oxidation of benzene at that time. The diol was converted to catechol by a soluble enzyme from rat liver which required $NADP^+$. Diol [100] was similarly dehydrogenated. Sato (86) observed subsequently that rat-liver slices converted [98] to catechol but had little action on its *cis*-isomer [78]. It will be remembered that the analogous enzyme from microorganisms shows the opposite stereospecificity and depends upon NAD^+ as cofactor.

The problem of the origin of the oxygen atoms in the diols has been the object of several studies; it is of course fundamental for an understanding of the mechanism by which these compounds are formed. Booth and Boyland (98) found that the

conversion of naphthalene into [100] by liver microsomes requires
$NADP^+$ and molecular oxygen. The enzyme responsible for this
reaction is thus an oxygenase: by definition (99) an enzyme
that catalyzes the incorporation of O_2 into the substrate.
Holtzman and co-workers (94) showed that only one atom of ^{18}O
appears in [100] formed from naphthalene by mouse-liver microsomes in presence of $^{18}O_2$; all the label is located in the
α-hydroxyl. This incorporation of *one* oxygen atom shows that the
enzyme is a "mixed-function oxygenase" in the sense of Mason
(100), that is, one that catalyzes the transfer of one atom
of oxygen from O_2 into the substrate, and reduction of the other
to the level of H_2O, thus functioning simultaneously as both
oxygenase and dehydrogenase. In complementary experiments with
$H_2^{18}O$, Jerina and co-workers (101) demonstrated that the
β-hydroxyl of [100] is furnished by water. These findings about
the origin of the oxygens of [100] give strong support to an
earlier hypothesis, which assumes that the biosynthesis of these
diols from aromatic hydrocarbons proceeds through intermediary
formation of an arene oxide ([88]; cf. also [12]) and subsequent
hydration of the epoxide ring. This possibility was discussed
by Boyland (89a, 102) as early as 1950, (when such cyclohexadiene
epoxides were completely unknown) as an explanation for the
trans-stereochemistry of the hydroxyls in [97], [100], and [101];
opening of the epoxide group of such an intermediate would
necessarily yield a *trans*-diol.

Many aromatic compounds are excreted in the urine as the
conjugates with N-acetylcysteine which are known as mercapturic
acids [103]. (Transformation into derivatives of this kind is
not restricted to aromatic compounds, but the resulting substances are not pertinent to our discussion.) Already at the
time of the discovery (82) of the first mercapturic acids
[103a,b] it was noticed that they can be isolated only after
pretreatment with acid; they are thus presumably artifacts formed
from precursors by the acid. The existence of these precursors
was confirmed (103) by chromatographic observations, and by the
subsequent isolation of crystalline salts of two of them, derived
from naphthalene (104a) and *p*-bromobenzene (104b), respectively.
These compounds were shown to have structures [104a] and [104b];
treatment with acid converts them to the corresponding mercapturic acids [103a,b]. The name "premercapturic acids" was introduced (103b) for compounds of this type.

The steps leading to these compounds in the metabolism of
aromatic substances were elucidated by Boyland and co-workers
(105). Incubation of naphthalene with rat liver slices in
presence of glutathione gave a compound formulated as [105].
This substance lost glutamic acid, apparently by transpeptidation

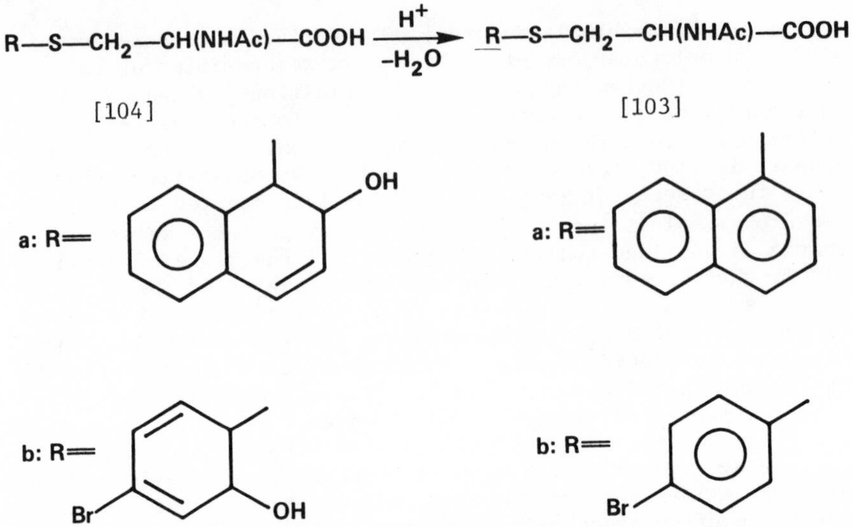

$$R—S—CH_2—CH(NHAc)—COOH \xrightarrow[-H_2O]{H^+} R—S—CH_2—CH(NHAc)—COOH$$

[104]   [103]

a: R=

b: R=

rather than by hydrolytic cleavage. From the resulting product
[106], glycine is removed by peptidase action, and the cysteine
conjugate [107] so formed is acetylated to [104a] (see Scheme
IX). Compounds [104a] - [107] have also been found (106) to
occur in the bile of rats after administration of naphthalene.
Analogous substances derived from 9,10-dihydrophenanthrene were
observed in urine and bile of rats dosed with phenanthrene (107);
indications were also obtained for the presence in the urine of
similar conjugates of 1,2- and 3,4-dihydrophenanthrene.

As in the case of the dihydrodiols, formation of an epoxide
was postulated (104a, 105a) as the first step in the biosynthesis of these sulfur-containing metabolites; this epoxide would
thus be a common first intermediate in the genesis of both types
of compounds.

The hypothesis that epoxides derived from aromatic compounds ("arene oxides") are common obligatory intermediates in
the formation of the two classes of metabolites that have just
been discussed became accessible to experimental scrutiny after
synthetic approaches to such compounds had been developed. The
first two substances of this type, the ones derived from phenanthrene and benz[a] anthracene, were prepared by Newman and Blum
(108). Subsequently, Vogel and Günther (109) developed a generally applicable method for the synthesis of arene oxides,
established their tautomerism with oxepins (cf. [88]⇌[89]),
and observed their rapid isomerization to the corresponding
phenols. Experimental studies made possible by the availability

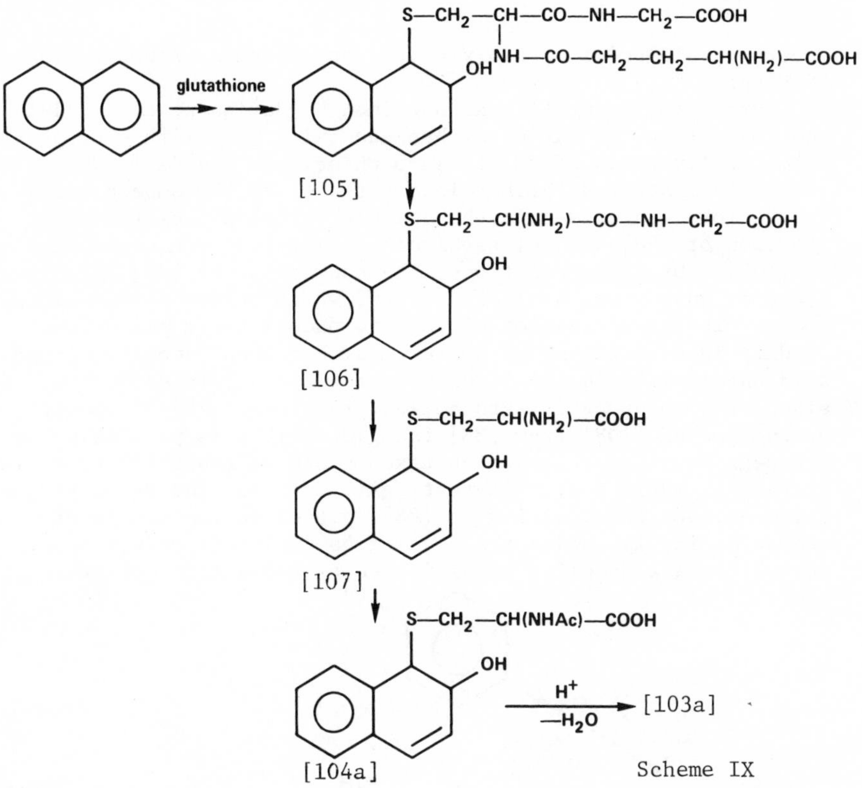

Scheme IX

of those compounds were initiated by Jerina, Daly, Witkop, and their co-workers; the experiments led rapidly to conclusive proof for the key role of arene oxides in the transformations of aromatic compounds, and to a very detailed exploration of the entire field. This research has been the object of numerous reviews, for example (83, 84, 110); they should provide information on many aspects of this fascinating field which can be mentioned only briefly here.

Proof for the function of arene oxides as obligatory intermediates rests on three main types of evidence: (1) demonstration that they are capable of yielding the various compounds that are known to result from the metabolism of aromatic substances in intact animals or tissue preparations; (2) demonstration that arene oxides are actually formed during such metabolism; and (3) proof that the isomerization of arene oxides to phenols proceeds with the remarkable shift of hydrogen (or certain other atoms or groups) to adjacent positions on the aromatic ring which

occurs in biological hydroxylations and which is known as the
"NIH shift."

Proof for point (1) was obtained by Jerina et al. (111),
who found that incubation of benzene oxide [88] with microsomes
from rat liver transforms it into phenol and the diol [98]; the
latter is converted further into catechol by the supernatant.
Furthermore, an enzyme in the soluble fraction promotes the
reaction of [88] with glutathione to give [108], a close analog
of [105]. Of these reactions, the conversion to [98] is catalyzed by an enzyme, epoxide hydrase, which occurs in the microsomes. The isomerization of [88] to phenol takes place very
readily in nonbiochemical systems, either spontaneously or under
acid catalysis. In the liver preparations, it is nonenzymatic,
since it also proceeds with boiled fractions. The enzymatic
production of [108] from [88] and glutathione is paralleled by
a nonenzymatic reaction which takes place at about 1/5 the rate
of the enzymatic one. Interestingly, cysteine and N-acetylcysteine did not react well with [88], either enzymatically or
nonenzymatically, under the conditions employed; this finding
demonstrates a specific significance of the entire structure of
glutathione.

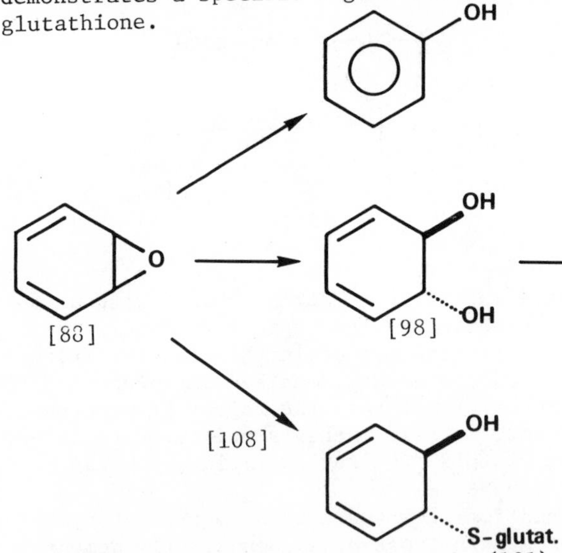

Similar observations were made (101) with the arene oxide
[109] derived from naphthalene. The synthetic (112) oxide was
converted into 1-naphthol [110] (together with a small amount
of 2-naphthol [111]) and (-)- and racemic forms of the *trans*-
diol [100]) by rat-liver microsomes; action of an enzyme in the
soluble fraction gave [105]. Here again, the isomerization of

the oxide was nonenzymatic. The analogous formation of [110]
and [105] from naphthalene itself by liver microsomes had been
observed earlier (98, 113). For conversion of naphthalene into
[105], presence of both microsomes and supernatant was required
(114); this observation points strongly to the existence of an
intermediate that is produced by the microsomes and coupled with
glutathione by a soluble enzyme in liver. 2-Naphthol [111] has
often been observed as product of the metabolism of naphthalene
in vivo (for example, see ref. 88).

For proof of point (2): actual formation of an arene oxide,
the microsomal system that produces [100], [110], and [105] was
used (101). Incubation of naphthalene with this system produced
[109]; while it was not possible to separate the extremely
reactive epoxide completely from unchanged naphthalene, its
presence and identity were demonstrated unequivocally. Similar
evidence was obtained subsequently (115) for the very transient
appearance of epoxides on incubation of carcinogens such as
dibenz[a,h]anthracene with liver microsomes.

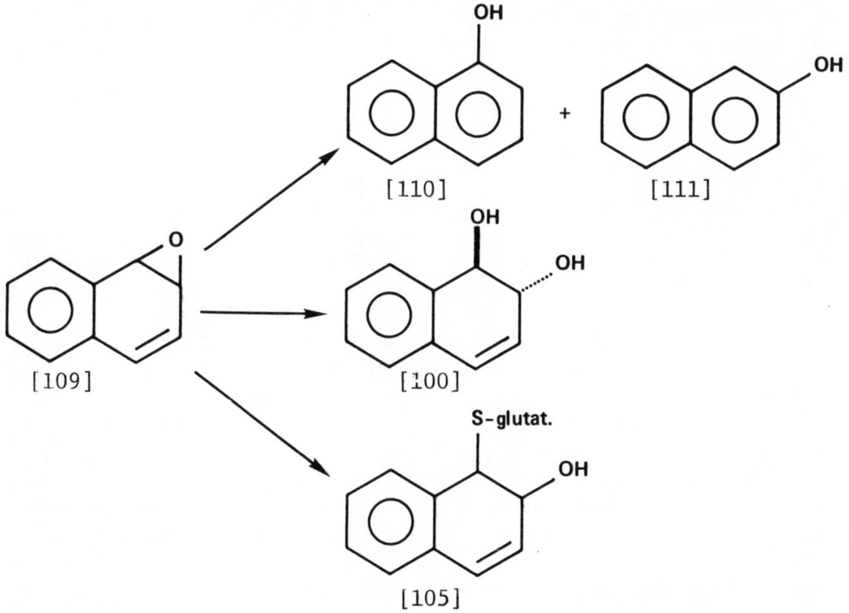

Finally, point (3): occurrence of the "NIH shift" during
the transformation of an epoxide to the isomeric phenol was
established (116). 3,4-Toluene oxide deuterated in position 4
[112] was synthesized and converted into *p*-cresol [113] by
microsomes, by HCl, or spontaneously on standing at room

temperature in carbon tetrachloride. In every instance, substantial retention (37-75%) of deuterium was observed. The label was located in position 3 of [113] by nmr spectroscopy. Comparable retention (56 and 58%, respectively) was found during hydroxylation of toluene-4-^{2}H with rabbit liver microsomes or with peroxytrifluoroacetic acid, a hydroxylation reagent that is known (110) to produce NIH shifts similar to those observed with microsomes.

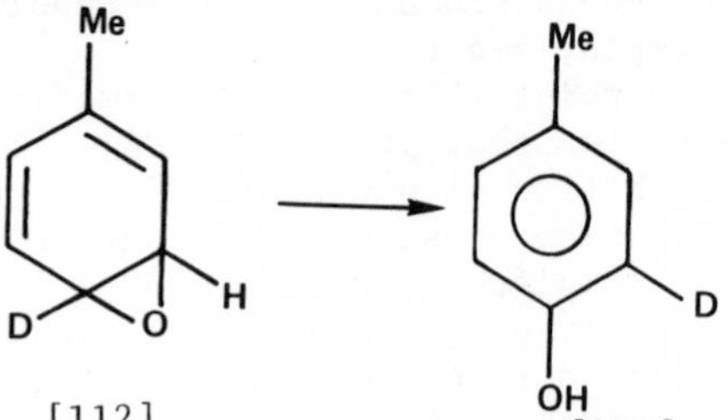

[112]   [113]

The function of arene oxides as intermediates in these metabolic transformations of aromatic rings is thus well established. Because of this important role, their chemical and biological transformations have been studied in great detail (83, 84, 110). Out of the wealth of available observations, a few may mentioned that appear significant to our topic of formation and transformation of aromatic structures.

It has been noted before that arene oxides such as [88] are tautomeric with oxepins [89]. In extreme cases, only one or the other of these tautomers may exist, such as the epoxide [109] (its oxepin tautomer could form only with loss of aromaticity in the other ring) and oxepin [114] (here, it is the corresponding epoxide that would be nonaromatic). Only those arene oxides that can exist at least partly in the epoxide form have so far been found to be involved in metabolic transformations. However, participation of oxepins in biosynthetic processes is documented by compounds [92] - [94].

The hydration of arene oxides to *trans*-1,2-diols (cf. [97]-[102], etc.) proceeds differently in two classes of these oxides which are being distinguished. In those cases where epoxide formation involves the so-called K-region of the parent hydrocarbons (positions 9 and 10 of phenanthrene, and analogous ones in related hydrocarbons), purely chemical hydration takes place normally [84a]; in "non-K-region" arene oxides, it has so far been observed only as an enzymatic reaction, in striking contrast to the isomerization of these compounds to phenols which proceeds nonenzymatically even in biological systems. In the case

of indan-8,9-epoxide [115], the course of the reaction is exceptional, since it yields the diol [116] by 1,6-addition of water,
instead of undergoing the usual 1,2-hydration (83). No biochemical significance for this transformation has been demonstrated
so far, but the occurrence of oxygen- or nitrogen-containing
substituents at position 4 of an indole ring in a very small
number of compounds might be explicable by analogous processes.
Examples of such oxygenated substances are psilocin[117], its
phosphate ester psilocybin, and a few complex alkaloids from
Mitragyna species; nitrogen is attached to this position in the
antibiotic teleocidin B(117).

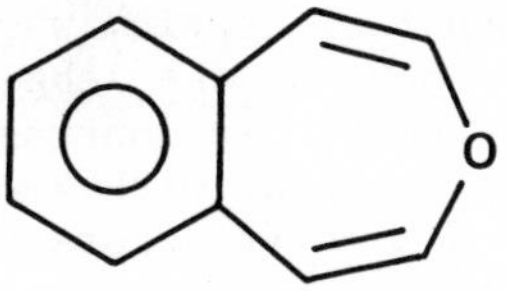

[114]

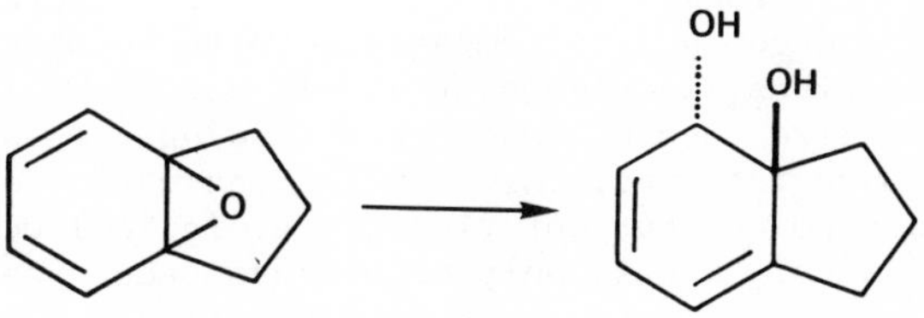

[115]   [116]

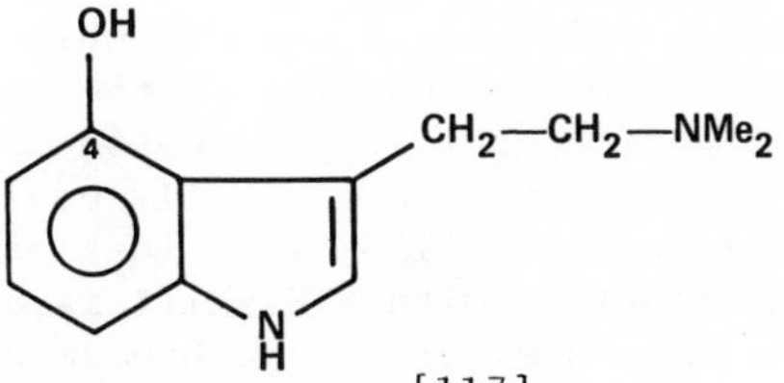

[117]

The isomerization of arene oxides to phenols (cf., e.g., [109]——→ [110] + [111] generally proceeds with very great ease; it takes place under conditions of acid catalysis, or spontaneously in neutral or alkaline solution. A characteristic feature of this isomerization is the occurrence of the NIH shift, which has already been mentioned: the migration of atoms or groups (H, Cl, Br, Me) from the point of attachment of the oxide ring to an adjacent position (cf. [112] ——→ [113]). This shift was discovered by Guroff and co-workers (118) during studies of the enzymatic hydroxylation of 4-^{3}H-phenylalanine to tyrosine. *A priori*, complete displacement of the label and formation of inactive tyrosine would have been expected; instead, much less than the anticipated amount of tritium was liberated, and the tyrosine formed retained most of the label, which was present entirely in positions 3 and 5. In contrast, chemical hydroxyla-tion of the labeled phenylalanine with Fenton's reagent produced inactive tyrosine. The NIH shift occurs generally during hydroxy-lation of aromatic substrates by monooxygenases, including some from microorganisms or higher plants. However, certain chemical reagents likewise hydroxylate aromatic rings with NIH shift. Such shifts take place, for example, during action of photo-excited pyridine-N-oxide and some of its relatives, of chromyl acetate, and of peracids (119). In particular, peroxytrifluoro-acetic acid produces shifts comparable to those observed in the enzymatic reactions; in view of the well-known ability of peroxy acids to epoxidize double bonds, this finding suggests that arene oxides are intermediates. In the enzymatic reactions, chlorine and bromine (but not fluorine or iodine) were found to migrate. Of alkyl groups, only methyl has been observed to undergo the shift, if acted upon by a specific phenylalanine hydroxylase from liver or a *Pseudomonas* sp. (120). This enzyme converts 4-methylphenylalanine into 3-methyltyrosine, together with some 4-methyl-3-hydroxyphenylalanine and much 4-hydroxy-methylphenylalanine. A 3,4-epoxide of the starting material is assumed (84) to be an intermediate in the formation of the first two products. The nonspecific microsomal enzyme system does not induce this migration of methyl groups but leads to hydroxymethyl derivatives exclusively.

Detailed studies (83, 84) of the NIH shift suggest the mechanism shown in Scheme X: opening of the epoxide ring to give an intermediate such as [118]; formation of the keto tautomer of the phenol with shift of R; and finally enolization to the phenol. Re-aromatization of [118] with loss of R would account for the fact that retention with NIH shift is usually not com-plete. Convincing evidence for the mechanism was obtained in the case of the epoxide [119], where NIH shift produced the stable ketone [120] (121).

Scheme X

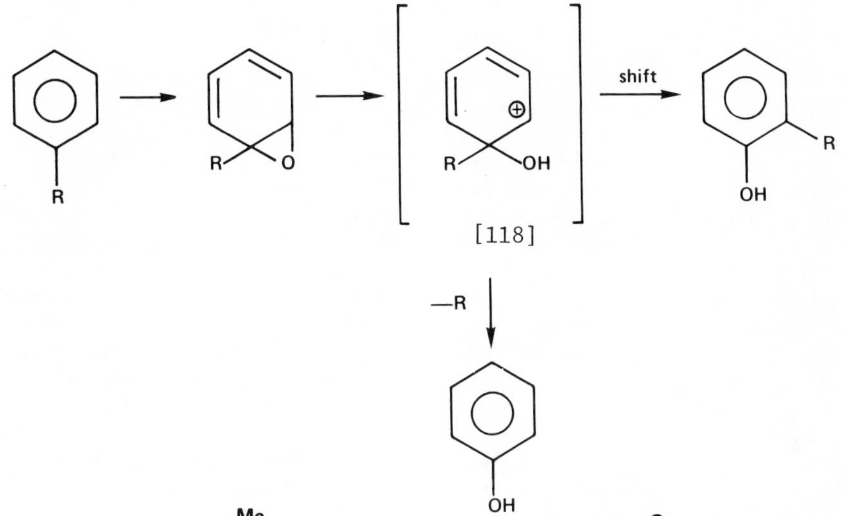

[118]

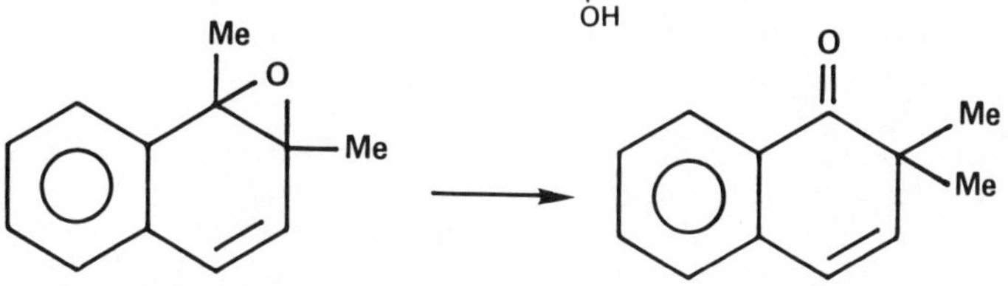

[119]   [120]

Although no actual case seems to have been observed so far, NIH shift of methyl groups or other carbon substituents during hydroxylation of aromatic rings could well play a role in biosynthesis. The long-known migration of the side-chain during the transformation of *p*-hydroxyphenylpyruvic to homogentisic (hydroquinone acetic) acid might seem to provide an example of such a process; actually, however, this rearrangement proceeds through an unrelated mechanism (84).

Arene oxides formed from polycyclic carcinogenic hydrocarbons seem to be the actual active carcinogens (83, 84, and references quoted there).

The formation of arene oxides seems to be catalyzed by two different groups of enzymes (84b): those directed specifically to the hydroxylation of particular endogenous metabolites, and others adapted to transformation of a wide variety of extraneous aromatic substrates. Enzymes of both groups are mono-oxygenases. From the fact that hydroxylation by both types proceeds with NIH shift, it appears that arene oxides are involved in either case, although their biosynthesis by enzymes of the first group does not seem to have been established by other evidence, such as proof of transformation of production of arene oxides (cf. points (1) and (2), above).

The specific enzymes of the first group convert, for example, phenylalanine into tyrosine, tyrosine into 3,4-dihydroxyphenyl-alanine (dopa), tryptophan into 5-hydroxytryptophan (serotonin) (84b, 110, and literature quoted there). The phenylalanine hydroxylases of rat liver have already been mentioned in Section 7-1. As mono-oxygenases, these enzymes incorporate one atom of oxygen into the substrate, while the other one is reduced to water. The specific hydrogen donor is often a tetrahydropteridine. Fe^{2+} and some other metal ions have a stimulating effect. Enzymes of this type are widely distributed: they have been found in microorganisms, animals, and higher plants. The extent of the NIH shift, as measured, for example, by retention of tritium label, depends upon the substrate, but not upon the source of the enzyme: for instance, 95% retention was found for the conversion of phenylalanine to tyrosine by phenylalanine hydroxylase from a *Pseudomonas*, a *Penicillium*, or mammalian liver. This result is consistent with the formation of the same intermediate oxide and occurrence of the shift during its further transformation.

In contrast to these specific enzymes, the enzyme system from liver microsomes introduces oxygen into a wide variety of substances; in most cases where aromatic compounds have been studied, the substrates were exogenous. The enzyme system is inducible in rodents by pretreatment with phenobarbital or 3-methylcholanthrene; the resulting enzymes, however, differ in chemical nature and, to a certain extent, in substrate specificity. Such microsomal mono-oxygenase systems from rodent liver have been solubilized by Lu and Coon (123) and shown to consist of a hemoprotein, cytochrome P-450, an NADPH-dependent cytochrome reductase, and a heat-stable lipid factor subsequently (124) identified as being mainly phosphatidyl-choline. Cytochrome P-450 is present if phenobarbital was the inducer (122); methylcholanthrene induces a different hemoprotein, P-488. The differences in specificity are mostly connected with the presence of one or the other cytochrome. The active enzyme system can be reconstituted from these three fractions (122).

The microsomal epoxide hydrase responsible for conversion of arene oxides (and other epoxides) into *trans*--dihydrodiols has been studied by Oesch, Jerina, and Daly. Like the mono-oxygenases which produce these oxides, the hydrase is inducible by phenobarbital or 3-methylcholanthrene (125); it has been purified about 10 times over the microsomes and shown not to require any cofactors (126a). In liver microsomes it seems to occur as a fairly stable complex with the mono-oxygenases (126b); preparations of cytochromes P-450 and 448 contain much of the hydrase, and reconstituted system is able to convert naphthalene into the oxide, [109], and hence into [100] and [110] (127).

To explain the formation of the arene oxides from aromatic compounds, an initial attack by an enzymatically generated oxygen atom with six valence electrons is postulated (101, 110, 119). The name "oxene" is proposed for this species, which would be isoelectronic with the well-established carbene and nitrenes, and similarly capable of adding to double bonds to give three-membered rings; formation of an arene oxide through reaction with oxene would be analogous to the long-known reaction of carbenes with benzene to give bicyclo[4.1.0]heptadienes (norcaradienes).

REFERENCES

1. H.-J. Hansen, B. Sutter, and H. Schmid, Helv. Chim Acta 51, 828 (1968), and subsequent papers by H. Schmid et. al.; H. Hart, P.M. Collins, and A.J. Waring, J. Amer. Chem. Soc. 88, 1005 (1966).

2. S.L. Stenmark, D.L. Pierson, R.A. Jensen, and G.I. Glover, Nature 247, 290 (1974); R.A. Jensen and D.L. Pierson, ibid. 254, 667 (1975).

3. F. Bohlmann and M. Grenz, Tetrahedron Lett. 1455 (1970); F. Bohlmann and C. Zdero, Chem. Ber. 102, 2211 (1969); 104, 1957, (2354) (1971).

4. D.V. Banthorpe, B.V. Charlwood, and M.J.O. Francis, Chem. Rev. 72, 115 (1972); W.W. Epstein and C.D. Poulter, Phytochemistry 12, 737 (1973); cf. also R.B. Bates and S.K. Paknikar, Tetrahedron Lett. 1453 (1965).

5. H. Flentje, W. Düpke, and P.W. Jeffs, Naturwissenschaften 52, 259 (1965); D.H.R. Barton, R. James, G.W. Kirby, W. Düpke, and H. Flentje, Chem. Ber. 100, 2457 (1967); W. Düpke, H. Flentje and P.W. Jeffs, Tetrahedron 24, 4459 (1968); G. Snatzke and G. Wollenberg, J. Chem. Soc. (C) 1681 (1966).

6. K.H. Baggaley, B. Blessington, C.P. Falshaw, W.D. Ollis, L. Chaiet, and F.J. Wolf, Chem. Commun. 101 (1969).

7. T. Yamashita, N. Miyairi, K. Kunugita, K. Shimizu, and H. Sakai, J. Antibiotics 23, 537 (1970); J.P. Scannell, D.L. Pruess, T.C. Demny, T. Williams, and A. Stempel, ibid. 618 (1970).

8. (a) G.M. Sharma and P.R. Burkholder, Tetrahedron Lett. 4147 (1967);

(b) G.M. Sharma, B. Vig, and P.R. Burkholder, J. Org. Chem. **35**, 2823 (1972).

9. Sir Robert Robinson, The Structural Relations of Natural Products, Clarendon, Oxford, 1955, p. 54.

10. (a) K. Takeda, T. Okanishi, K. Igarachi, and A. Shimaoka, Tetrahedron **15**, 183 (1961);
    (b) K. Igarachi, Chem. Pharm. Bull. **9**, 722 (1961).

11. M.-M. Janot, Ph. Devissaguet, Q. Khuong-Huu, and R. Goutarel, Ann. Pharm. Fr. **25**, 733 (1967).

12. (a) D.H.R. Barton, Chem. Brit. 330 (1967);
    (b) A.R. Battersby, in Oxidative Coupling of Phenols, W.I. Taylor and A.R. Battersby, Eds., Dekker, New York, 1978, Chapter 3;
    (c) T. Kametani and K. Fukumoto, Synthesis 657 (1972).

13. D.H.R. Barton and T. Cohen, Festschrift Prof. Dr. Arthur Stoll, Birkhäuser, Basel, 1957, p. 117.

14. (a) J. Gadamer, Arch. Pharm. **249**, 498 (1911);
    (b) C. Schöpf and K. Thierfelder, Ann. **497**, 22 (1932).

15. M.D. Glick, R.E. Cook, M.P. Cava, M. Srinivasan, J. Kunitomo, and A. I. da Rocha, Chem. Commun. 1217 (1969).

16. (a) L.J. Haynes, K.L. Stuart, D.H.R. Barton, and G.W. Kirby, Proc. Chem. Soc. 208 (1963); 261 (1964); J. Chem. Soc. (C) 1676 (1966);
    (b) K. Bernauer, Helv. Chim. Acta **46**, 1783 (1963); **47**, 2122 (1964).

17. M.P. Cava, K. Nomura, R.H. Schlessinger, K.T. Buck, B. Douglas, R.F. Raffauf, and J.A. Weissbach, Chem. Ind. 282 (1964); M. Shamma and W.A. Slusarchyk, Chem. Rev. **64**, 59 (1964).

18. K. Bernauer and W. Hofheinz, Fortsch. Chem. Org. Naturstoffe **26**, 245 (1968).

19. L.J. Haynes, K.L. Stuart, D.H.R. Barton, D.S. Bhakuni, and G.W. Kirby, Chem. Commun. 141 (1965); D.H.R. Barton, D.S. Bhakuni, G.M. Chapman, G.W. Kirby, L.J. Haynes, and K.L. Stuart, J. Chem. Soc. (C) 1295 (1967).

20. D.H.R. Barton, D.S. Bhakuni, G.M. Chapman, and G.W. Kirby, J. Chem. Soc. (C) 2134 (1967).

21. (a) A.R. Battersby, T.H. Brown, and J.H. Clements, J. Chem. Soc. 4550 (1965);
    (b) A.R. Battersby, T.J. Brocksom, and R. Ramage, Chem. Commun. 464 (1969);
    (c) A.R. Battersby, R.T. Brown, J.H. Clements, and G.G. Iverach, Chem. Commun. 230 (1965).

22. (a) M. Pailer, Fortschr. Chem. Org. Naturstoffe **18**, 55 (1960);
    (b) R. Hegnauer, Chemotaxonomie der Pflanzen, Vol. III Birkhäuser Verlag, Basel and Stuttgart, 1964, p. 191.

23. S.M. Kupchan and J.J. Merianos, J. Org. Chem. <u>33</u>, 3735 (1968).
24. (a) L.A. Maldonado, J. Herran, and J. Romo, Ciencia (Mex.) <u>24</u>, 237 (1966); through Chem. Abstr. <u>65</u>, 15438e (1966); (b) M. Akasu, H. Itokawa, and M. Fujita, Tetrahedron Lett. 3609 (1974).
25. H.R. Schütte, U. Orban, and K. Mothes, Eur. J. Biochem. <u>1</u>, 70 (1967).
26. I.D. Spenser, Lloydia <u>29</u>, 71 (1966); F. Comer, H.P. Tiwari, and I.D. Spenser, Can. J. Chem. <u>47</u>, 481 (1969).
27. L. Kühn and S. Pfeifer, Pharmazie <u>20</u>, 659 (1965).
28. K.L. Stuart, Chem. Rev. <u>71</u>, 47 (1971).
29. A.R. Battersby, Proc. Chem. Soc. 189 (1963); cf. also D. Ginsburg, The Opium Alkaloids, Interscience, New York, 1962.
30. J.W. Fairbairn and S. El-Masry, Phytochemistry <u>6</u>, 499 (1967).
31. D.H.R. Barton, G.W. Kirby, W. Steglich, G.M. Thomas, A.R. Battersby, T.A. Dobson, and H. Ramuz, J. Chem. Soc. 2423 (1965); A.R. Battersby and T.H. Brown, Chem. Commun. 170 (1966).
32. D.H.R. Barton, G.W. Kirby, W. Steglich, and G.M. Thomas, Proc. Chem. Soc. 203 (1963).
33. D.H.R. Barton, D.S. Bhakuni, R. James, and G.W. Kirby, J. Chem. Soc. (C) 128 (1967); K.W. Bentley, Experientia <u>12</u>, 251 (1956).
34. T. Kametani, M. Ihara, and T. Honda, Chem. Commun. 1301 (1969).
35. B. Franck, H.J. Lubs, and G. Dunkelmann, Angew. Chem. Int. Ed. <u>10</u>, 969, 1075 (1967).
36. K.L. Stuart, V. Teetz, and B. Franck, Chem. Commun. 333 (1969); K.L. Stuart and L. Graham, Phytochemistry <u>12</u>, 1967 (1973).
37. K. Takeda, Bull. Agr. Chem. Soc. Japan <u>20</u>, 165 (1956).
38. K.W. Bentley, The Chemistry of the Morphine Alkaloids, Clarendon, Oxford, 1954.
39. G. Stork, in The Alkaloids, R.H.F. Manske and H.L. Holmes, Eds., Vol. II, Academic, New York, 1952, p. 189ff.
40. (a) K.W. Bentley and R. Robinson, Experientia <u>6</u>, 353 (1956); (b) K.W. Bentley, J. Amer. Chem. Soc. <u>89</u>, 2464 (1967).
41. D.M. Hall and W.W.T. Manser, Chem. Commun. 112 (1967).
42. D.H.R. Barton, Pure Appl. Chem. <u>9</u>, 35 (1964).
43. A.R. Battersby, A.K. Bhatnagar, P. Hackett, C.W. Thornber, and J. Staunton, Chem. Commun. 1214 (1968).
44. E. Fattorusso, L. Minale, and G. Sodano, Chem. Commun. 751 (1970); J. Chem. Soc., Perkin Trans. I, 16 (1972).
45. W. Fulmor, G.E. Van Lear, G.O. Morton, and R.D. Mills, Tetrahedron Lett. 4551 (1970).

46. D.B. Cosulich and F.M. Lovell, Chem. Commun. 397 (1971);
    L. Mazzarella and R. Puliti, Gazz. Chim. Ital. 102, 391
    (1972).
47. L. Minale, G. Sodano, W.R. Chan, and A.M. Chen, Chem.
    Commun. 674, 968 (1972).
48. E. Fattorusso, L. Minale, G. Sodano, K. Moody, and
    R.H. Thomson, Chem. Commun. 752 (1970).
49. K. Moody, R.H. Thomson, E. Fattorusso, L. Minale, and
    G. Sodano, J. Chem. Soc., Perkin Trans. I, 18 (1972).
50. E. Fattorusso, L. Minale, K. Moody, G. Sodano, and
    R.H. Thomson, Gazz. Chim. Ital. 101, 61 (1971).
51. J.R.D. McCormick, J. Reichenthal, U. Hirsch, and
    N.O. Sjolander, J. Amer. Chem. Soc. 84, 3711 (1962).
52. D.T. Gibson and D.M. Jerina, private communication.
53. D.T. Gibson, Crit. Rev. Microbiol. 1, 199 (1971).
54. (a) D.T. Gibson, G.E. Cardini, R.C. Maseles, and
    R.E. Kallio, Biochemistry 9, 1631 (1970);
    (b) D.T. Gibson, J.R. Koch, and R.E. Kallio, ibid.
    7, 2653 (1968).
55. D.T. Gibson, M. Hensley, H. Yoshioka, and T.J. Mabry,
    Biochemistry 9, 1626 (1970).
56. D.T. Gibson, J.R. Koch, C.L. Schuld, and R.E. Kallio
    Biochemistry 7, 3795 (1968).
57. N. Walker and G.H. Wiltshire, J. Gen. Microbiol. 8, 273
    (1953).
58. D.M. Jerina, J.W. Daly, A.M. Jeffrey, and D.T. Gibson,
    Arch. Biochem. Biophys. 142, 394 (1971).
59. N. Walker and G.H. Wiltshire, J. Gen. Microbiol. 12, 478
    (1955).
60. L. Canonica, A. Fiecchi, and V. Treccani, Ist. Lombardo,
    Rend. Sci. 91, 119 (1957); through ref. 53.
61. C. Colla, A. Fiecchi, and V. Treccani, Ann. Microbiol.
    Enzimol. 10, 77 (1960); through ref. 53.
62. A.M. Reiner and G.D. Hegeman, Biochemistry 10, 2530 (1971).
63. H. Taniuchi and O. Hayaishi, J. Biol. Chem. 238, 283 (1963).
64. H. Ziffer, D.M. Jerina, D.T. Gibson, and V.M. Kobal,
    J. Amer. Chem. Soc. 95, 4048 (1973).
65. F.A. Catterall, K. Murray, and P.A. Williams, Biochim.
    Biophys. Acta 237, 361 (1971).
66. M. Mori, H. Taniuchi, Y. Kojima, and O. Hayaishi, Biochim.
    Biophys. Acta 128, 535 (1966).
67. R.Y. Stanier and O. Hayaishi, Science 114, 326 (1951);
    O. Hayaishi, H. Taniuchi, M. Tashiro, and S. Kuno,
    J. Biol. Chem. 236, 2492 (1961).
68. F.A. Catterall and P.A. Williams, J. Gen. Microbiol. 67,
    117 (1971).

69. D.M. Jerina, H. Ziffer, and J.W. Daly, private communication; cf. also ref. 93.

70. M. Reiner, J. Biol. Chem. <u>247</u>, 4960 (1972).

71. M.R. Bell, J.R. Johnson, B.S. Wildi, and R.B. Woodward, J. Amer. Chem. Soc. <u>80</u>, 1001 (1958).

72. J. Fridrichsons and A. McL. Mathieson, Acta Cryst. <u>23</u>, 439 (1967).

73. N. Neuss, R. Nagarajan, B.B. Molloy, and L.L. Huckstep, Tetrahedron Lett. 4467 (1968).

74. J.D. Bu'Lock and A.P. Ryles, Chem. Commun. 1404 (1970).

75. N. Johns and G.W. Kirby, Chem. Commun. 163 (1971).

76. G. Lowe, A. Taylor, and L.C. Vining, J. Chem. Soc. (C) 1799 (1966).

77. M.S. Ali, J.S. Shannon, and A. Taylor, J. Chem. Soc. (C) 2044 (1968).

78. R. Nagarajan, L.L. Huckstep, D.H. Lively, D.C. De Long, M.M. Marsh, and N. Neuss, J. Amer. Chem. Soc. <u>90</u>, 2980 (1968).

79. R. Nagarajan, N. Neuss, and M.M. Marsh, J. Amer. Chem. Soc. <u>90</u>, 6518 (1968).

80. D. Hauser, H.P. Weber, and H.P. Sigg, Helv. Chim. Acta <u>53</u>, 1061 (1970).

81. R. Hodges, J.W. Ronaldson, A. Taylor, and E.P. White, Chem. Ind. (London) 42 (1963); J. Fridrichsons and A. McL. Mathieson, Acta Cryst. <u>18</u>, 1043 (1965).

82. E. Baumann and C. Preusse, Ber. <u>12</u>, 806 (1879); M. Jaffé, ibid. 1092.

83. J.W. Daly, D.M. Jerina, and B. Witkop, Experientia <u>28</u>, 1129 (1972).

84. (a) D.M. Jerina, H. Yagi, and J.W. Daly, Heterocycles <u>1</u>, 267 (1973);
    (b) D.M. Jerina and J.W. Daly, Science <u>185</u>, 573 (1974).

85. E. Boyland and A.A. Levi, Biochem. J. <u>29</u>, 2679 (1935).

86. T. Sato, T. Fukuyama, T. Suzuki, and H. Yoshikawa, J. Biochem. (Tokyo) <u>53</u>, 23 (1963).

87. (a) J.N. Smith, B. Spencer, and R.T. Williams, Biochem. J. <u>47</u>, 284 (1950);
    (b) D.M. Jerina, J.W. Daly, and B. Witkop, J. Amer, Chem. Soc. <u>89</u>, 5488 (1967).

88. (a) L. Young, Biochem. J. <u>41</u>, 417 (1947);
    (b) J. Booth and E. Boyland, Biochem. J. <u>44</u>, 361 (1949);
    (c) E.D.S. Corner and L. Young, Biochem. J. <u>58</u>, 647 (1954).

89. (a) E. Boyland and G. Wolf, Biochem. J. <u>47</u>, 64 (1950);
    (b) E. Boyland and P. Sims, Biochem. J. <u>84</u>, 571 (1962).

90. F. Waterfall and P. Sims, Biochem, J. <u>128</u>, 265 (1972).

91. E. Boyland and P. Sims, Biochem, J. <u>91</u>, 493 (1964).

92. P. Sims, Biochem. Pharmacol. _19_, 795 (1970).
93. D.M. Jerina, H. Ziffer, and J.W. Daly, J. Amer. Chem. Soc. _92_, 1056 (1970).
94. J. Holtzman, J.R. Gillette, and G.W.A. Milne, J. Amer. Chem. Soc. _89_, 6341 (1967).
95. R. Miura, S. Honmaru, and M. Nakazaki, Tetrahedron Lett. 5271 (1968).
96. P.K. Ayengar, O. Hayaishi, M. Nakajima, and I. Tomida, Biochim. Biophys. Acta _33_, 111 (1959).
97. M. Nakajima, I. Tomida, A. Hashizume, and S. Takei, Ber. _89_, 2224 (1956).
98. J. Booth and E. Boyland, Biochem. J. _70_, 681 (1958); cf. alos H.S. Posner, C. Mitoma, S. Rothberg, and S. Udenfriend, Arch. Biochem. Biophys. _94_, 280 (1961).
99. O. Hayaishi, M. Katagiri, and S. Rothberg, J. Amer. Chem. Soc. _77_, 5450 (1955); H.S. Mason, W.L. Fowlkes, and E. Peterson, ibid. 2914; H.S. Mason, Ann. Rev. Biochem. _34_, 595 (1965); O. Hayaishi, ibid. _39_, 21 (1969).
100. H.S. Mason, Adv. Enzymol. _19_, 79 (1957).
101. D.M. Jerina, J.W. Daly, B. Witkop, P. Zaltzman-Nirenberg, and S. Udenfriend, J. Amer. Chem. Soc. _90_, 5625 (1968); Biochemistry _9_, 147 (1970).
102. E. Boyland, Biochem. Soc. Symp. _5_, 40 (1950).
103. (a) E. Boyland, P. Sims, and J.B. Solomon, Biochem. J. _66_, 41P (1957);
     (b) R.H. Knight and L. Young, ibid. _66_, 55P (1957); _70_, 111 (1958).
104. (a) E. Boyland and P. Sims, Biochem. J. _68_, 440 (1958);
     (b) B. Gillham and L. Young, ibid. _103_, 24P (1967).
105. (a) J. Booth, E. Boyland, and P. Sims, Biochem. J. _74_, 117 (1960);
     (b) E. Boyland and K. Williams, ibid. _94_, 190 (1965).
106. E. Boyland and P. Sims, Biochem. J. _77_, 175 (1960); E. Boyland, G.S. Ramsay, and P. Sims, ibid. _78_, 376 (1961).
107. E. Boyland and P. Sims, Biochem. J. _84_, 564 (1962).
108. M.S. Newman and S. Blum, J. Amer. Chem. Soc. _86_, 5559 (1964).
109. E. Vogel and H. Günther, Angew. Chem. Int. Ed. _6_, 385 (1967).
110. D.M. Jerina, J.W. Daly, and B. Witkop, in Biogenic Amines and Physiological Membranes in Drug Therapy, Part B, J. H. Biel and L.G. Abood, Eds., Dekker, New York, 1971, p. 413.
111. D. Jerina, J. Daly, B. Witkop, P. Zaltzman-Nirenberg, and S. Udenfriend, Arch. Biochem. Biophys. _128_, 176 (1968).
112. E. Vogel and R.G. Klärner, Angew. Chem., Int. Ed. _7_, 374 (1968).

113. C. Mitoma, H.S. Posner, H.C. Reitz, and S. Udenfriend
      Arch. Biochem. Biophys. 61, 431 (1956).
114. J. Booth, E. Boyland, and P. Sims, Biochem. J. 79, 516
      (1961).
115. J.K. Selkirk, E. Huberman, and C. Heidelberger, Biochem.
      Biophys. Res. Commun. 43, 1010 (1971); P.L. Grover,
      A. Hewer, and P. Sims, Fed. Eur. Biochem. Soc. Lett. 18,
      76 (1971).
116. D.M. Jerina, J.W. Daly, and B. Witkop, J. Amer. Chem.
      Soc. 90, 6523 (1968).
117. N. Sakabe, H. Harada, Y. Hirata, Y. Tomiie, and I. Nitta
      Tetrahedron Lett. 2523 (1966).
118. G. Guroff, D.M. Jerina, J. Renson, S. Udenfriend, and
      B. Witkop, Science 157, 1524 (1967).
119. D.M. Jerina, Chem. Technol. 4, 120 (1973); and refs. quoted
      there.
120. J. Daly and G. Guroff, Arch. Biochem. Biophys. 125, 136
      (1968).
121. D.M. Jerina, N. Kaubisch, and J.W. Daly, Proc. Natl. Acad.
      Sci. U.S. 68, 2545 (1971); N. Kaubisch, J.W. Daly, and
      D.M. Jerina, Biochemistry 11, 3080 (1972).
122. A.Y.H. Lu, R. Kuntzman, S. West, M. Jacobson, and
      A.H. Conney, J. Biol. Chem. 247, 1727 (1972), and literature
      there.
123. A.Y.H. Lu and M.J. Coon, J. Biol. Chem. 243, 1331 (1968);
      A.Y.H. Lu, K.W. Junk, and M.J. Coon, ibid. 244, 3714
      (1969).
124. H.W. Strobel, A.Y.H. Lu, J. Heidema, and M.J. Coon, J. Biol.
      Chem. 245, 4851 (1970).
125. F. Oesch, D.M. Jerina, and J. Daly, Biochim. Biophys.
      Acta 227, 685 (1971).
126. F. Oesch and J. Daly, (a) Biochim. Biophys. Acta 227, 692
      (1971);
      (b) Biochem. Biophys. Res. Commun. 46, 1713 (1972).
127. F. Oesch, D.M. Jerina, J.W. Daly, A.Y.H. Lu, R. Kuntzman,
      and A.H. Conney, Arch. Biochem. Biophys. 153, 62 (1972).

CHAPTER 9
Biosynthetic Conversion of Chorismic Acid to Tryptophan

9-1. ANTHRANILIC ACID

Tryptophan arises from chorismic acid through a series of intermediates of which the first is anthranilic acid. Although the conversion of chorismate to anthranilic acid is a complex reaction, studies with *A. aerogenes* (1, 2), *N. crassa* (3), *E. coli* (4-6), and *S. cerevisiae* (3, 7, 8) have indicated that the process is carried out by a single enzyme (anthranilate synthetase) which, as the first enzyme specific to this pathway, is subject to end-product inhibition by L-tryptophan.

The first two steps in the biosynthesis of tryptophan by *A. aerogenes* (1), *E. coli* (9, 10, 16), and *S. typhimurium* (11-13, 17, 37) are carried out by an enzyme aggregate consisting of two components: anthranilate synthetase (component 1) which converts chorismic to anthranilic acid, and anthranilate-5-phosphoribosyl-pyrophosphate: phosphoribosyl transferase (PR transferase, component 2). These two enzymes remain aggregated during purification, and both activities of the enzyme complex are inhibited by tryptophan; the aggregated enzyme uses glutamine as its nitrogen source for the formation of tryptophan (see below).

By using suitable mutants, the two compnents of the enzyme aggregate can be obtained, and on mixing these two components a fully functional complex can be reconstituted; a complete aggre-. gate can also be formed by mixing suitable extracts of appropriate mutants of different organisms. Several studies of the two component enzymes derived from these three organisms have been undertaken. For example, in *A. aerogenes* (1), *E. coli* (16), and *S. typhimurium* (12, 37, 39), component 2 is active when not aggregated with anthranilate synthetase, but it is no longer inhibited by tryptophan, indicating that the site of tryptophan inhibition must be on component 1. On the other hand, component 1 can use ammonia as a nitrogen source for the formation of

238

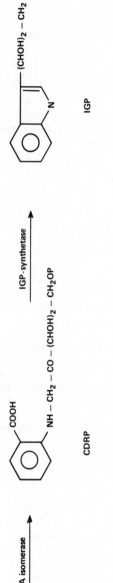

PR = Phosphoribosyl.

5 PRPP = 5-Phosphoribosylpyrophosphate.

PRA = N-(5-phosphoribosyl)-anthranilic acid.

CDRP = 1-(o-Carboxyphenylamino)-1-deoxyribulose phosphate

IGP = Indole-3-glycerolphosphate

tryptophan but cannot utilize glutamine, indicating that the binding site for glutamine is probably on component 2.

From the existence of such an enzyme aggregate, it may be inferred that anthranilic acid may not be an enzyme-free intermediate in the biosynthesis of tryptophan, at least in these organisms. Not all organisms exhibit aggregation between anthranilate synthetase and PR transferase; investigations have failed to show the presence of a similar enzyme complex in *Bacillus subtilis* (40), *Serratia marcescens* (41, 43), and *Pseudomonas putida* (42). *Neurospora crassa* (14, 38) contains an aggregate enzyme analogous to that in the Enterobacteriaceae but containing three rather than two enzymes: anthranilate synthetase, PRA isomerase, and IGP synthetase.

In *E. coli* (18) and *A. aerogenes* (19), the amino group of anthranilic acid has been found to be derived from the amide nitrogen of L-glutamine, which becomes attached at C-6 rather than C-2 of shikimic acid (20). Furthermore, Srinivasan (21) has shown that the carboxyl group of shikimic acid becomes the carboxyl of anthranilic acid and that the aromatization takes place without rearrangement of the carbon skeleton. Finally, ammonium ions in high concentration and at pH>7 can also serve as a source of nitrogen for the production of anthranilic acid in several organisms. Anthranilic acid synthesis with NH_4^+, unlike that with glutamine, is not inhibited by diazooxonorleucine, a known inhibitor of reactions involving glutamine; this fact indicates that the incorporation of ammonia does not proceed through glutamine.

A possible mechanism (shown below) for the transformation of chorismate to anthranilic acid has been proposed by Levin and Sprinson (22); it is of interest that McCormick et al. (23) have isolated *trans*-2,3-dihydro-3-hydroxyanthranilic acid [2] from a strain of *Streptomyces aureofaciens* with a reduced ability to synthesize tetracycline. Compound 2 is closely related to the intermediate assumed for this mechanism; however, it is inactive as a precursor of anthranilic acid in *E. coli* and *A. aerogenes* (24), presumably because its stereochemistry is incorrect for the E_2-*trans*-elimination of step 2.

The formation of anthranilic acid from chorismic acid could also proceed via a cyclic transition state with the side chain now *trans* to the NH_2 group; however, this is less likely since the intermediate [3] would have to be formed via a carbonium ion at C_2 (a normal double S_n2' displacement of the OH by NH_2, as in the first scheme, would give a relative *cis*-configuration), or via some enzyme-bound intermediate such as has been proposed in the conversion of 5-enolpyruvylshikimic acid 3-phosphate to chorismic acid . Zalkin and Kling (12) have reported that a

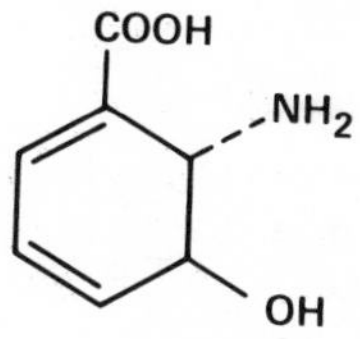

[2]

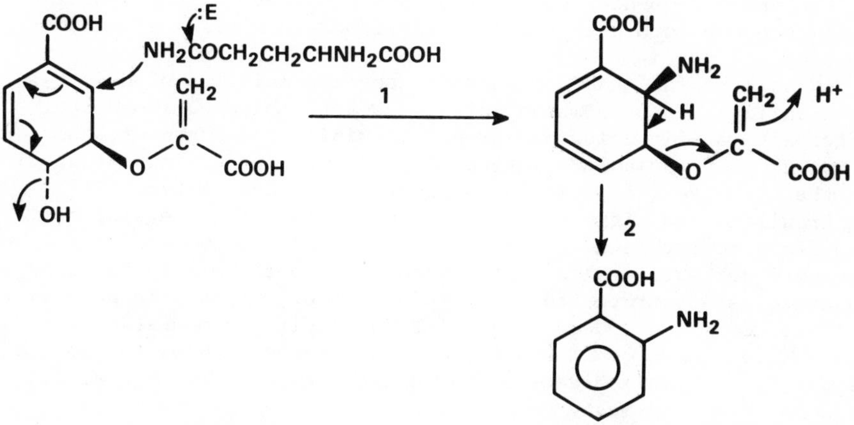

Proton is transferred from chorismic acid to pyruvic acid during the reaction, without interchange with the solvent, and they have postulated the participation of a basic group on the enzyme in the process (i.e., removal of the 6-proton by the enzyme, and subsequent transfer to the methylene group); the enzyme is inhibited by bromopyruvate, and this is interpreted as resulting from alkylation of the basic group that catalyzes the proton transfer. It seems

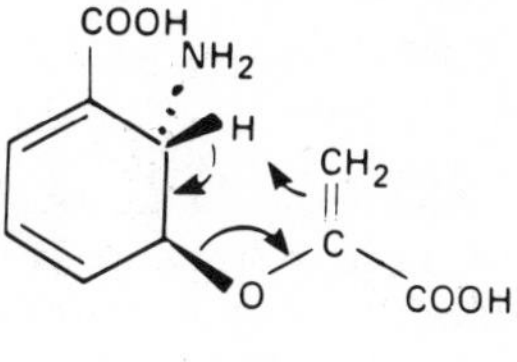

[3]

reasonable to assume that such an enzyme would only be necessary in the case of an E2 *trans*-elimination of pyruvate, in which case the transfer of the proton to the methylene without interchange with the solvent seems very unlikely; indeed, recent observations by Floss (25) and Srinivasan (35) have cast doubt upon the published results already quoted, since they have shown that chorismic acid bearing tritium at C-6 gives rise to unlabeled pyruvic acid. The accumulated evidence is, therefore, compatible with an interconversion similar to the one proposed by Levin and Sprinson, proceeding via an as yet undiscovered intermediate. Lingens et al. (26) have isolated a product of unknown structure from a mutant of yeast, *Saccharomyces cerevisiae*, which could be the intermediate in the above sequence, but they have not succeeded in converting this compound into anthranilic acid either chemically or enzymatically.

Ratledge (27) has demonstrated the accumulation of N-acetyl-anthranilic acid by a mutant of *A. aerogenes* blocked immediately after anthranilic acid, and grown on quinic acid; however, the compound is presumably a metabolite since it is not incorporated by the organism (36). It was suggested (28) that N-pyruvyl anthranilic acid might be an intermediate in the conversion of chorismic to anthranilic acid, and both this compound and the N-glutamyl derivative were synthesized (29); both were found to be inactive as substrates for the production of anthranilic acid by extracts of *S. cerevisiae* (8). Interestingly, fermentations of *Penicillium chrysogenum* and *P. notatum* have been shown to produce N- pyruvylanthranilamide (30), which is converted to anthranilic acid by acid treatment.

Two other chemical observations that might indicate the nature of the proposed (enzyme-bound) intermediate have been made: Somerville and Elford (31) have demonstrated that a hydroxamate is formed during the conversion of chorismic to anthranilic acid by cell-free extracts of *E. coli*, and Zalkin and Kling (12) have shown that anthranilic acid synthesis is inhibited by N-methylhydroxylamine, which acts as an NH_3 analog and gives rise to a hydroxamate in the presence of $FeCl_3$ at acidic pH.

9-2. LATER STAGES

The next step in the biosynthesis of tryptophan is the formation (from anthranilic acid and PRPP) of PRA, an intermediate which has been detected and synthesized only relatively recently (32), but whose existence had long been postulated. An Amadori rearrangement converts PRA to CDRP, which is then cyclized to form the indole nucleus. In the last step of the tryptophan synthesis, indole is not regarded as a true intermediate, since it remains enzyme-bound; however, it can act as a substrate for tryptophan synthesis, and it is excreted by suitable mutants. The pathway from anthranilic acid to tryptophan has been reviewed by Doy (7), and the biochemical and genetic aspects of the pathway from chorismate are covered fully in the reviews by Truffa-Bachi and Cohen (33), and Umbarger (34).

9-3. REGULATION OF THE TRYPTOPHAN PATHWAY

The enzymic progression between chorismate and tryptophan has been the object of detailed study in many organisms, as implied by the many references in the early part of this section; those parts of Chapters 1 and 15 which deal with control mechanisms treat the subject in some detail, but in view of the considerable interest in the biosynthesis of tryptophan we offer an abbreviated discussion here.

In all systems except one that have been investigated so far, the flow of metabolites in the tryptophan pathway is regulated by either one, or both of two mechanisms. The first is feedback inhibition (a rapid process) of the anthranilate synthetase enzyme by tryptophan; in this process the activity of the enzyme is inhibited by the end-product, so that wasteful synthesis of intermediates is prevented. In the second mechanism, control is achieved by repression of the overall enzyme level in the pathway (a much slower control process); this mechanism prevents the wasteful synthesis of enzyme protein. Both these regulatory processes can be initiated by an excess of tryptophan. The exception to these generalizations is the ergot fungus, *Claviceps paspali*, in which tryptophan causes no repression of the enzyme levels, possibly because any excess of tryptophan is rapidly converted into the ergot alkaloids by this organism.

Although feedback inhibition normally affects only the first enzyme specific to the biosynthesis of a metabolite, we have seen that phosphoribosyl transferase is also inhibited by tryptophan in *E. coli, A. aerogenes,* and *Salmonella typhimurium;* since, however, anthranilate synthetase and the transferase enzyme form a complex in these organisms, this difficulty is removed. In *Saccharomyces cerevisiae, Neurospora crassa,* and *Pseudomonas putida,* the synthetase is complexed with other enzymes of the pathway (IGP synthetase and/or PRA isomerase) although the feedback inhibition is unchanged.

The structural genes concerned with tryptophan biosynthesis in the enteric organisms constitute a single contiguous grouping on the chromosome, the so-called *trp*-operon (a group of genes constituting a single functioning unit, together with their controlling sites). The operon is under the control of regulator substances; these are proteins which are coded for by specific regulator genes, and which initiate or prevent transcription of the operon by binding to a special operator region of the chromosome. The operator encompasses both initiating and repressive functions and has both promotor and repressor sites. In other organisms the tryptophan genes are dispersed about the chromosome, sometimes singly, sometimes in groups; however, very similar regulatory mechanisms are thought to operate also in these cases; apparently it is not necessary that the controlling sites and the structural genes be close to one another on the chromosome.

None of the foregoing explains how the enzymic functions of the pathway are controlled by tryptophan; indeed this crucial point is not fully clarified at the moment: In the inhibitory process it is thought that the end-product and the substrate are competitive in their interaction with the enzyme. In view of their structural differences, the substrate and inhibitor probably do not bind to the same enzyme site, but the binding of one compound must interfere in some way with that of the other. One proposed mechanism of repression assumes that the conformation of the repressor protein must be altered by interaction with the end-product (co-repressor) before it can bind to the operator. In repression, all the enzyme levels are depressed to the same extent (coordinate repression) quite in accord with the postulated mechanism.

This account is clearly sketchy at best (e.g., no mention of enzyme induction, or its control) and is certainly not intended for scrutiny by a practicing microbiologist; chemists working with biosynthetic intermediates and biochemical systems may, however, find the simplistic account helpful.

REFERENCES

1. A.F. Egan and F. Gibson, Biochim. Biophys. Acta 130, 276 (1966); Biochem. J. 130, 847 (1972).
2. C.H. Doy, Biochim. Biophys. Acta 118, 173 (1966).
3. J.A. DeMoss, J. Biol. Chem. 240, 1231 (1965); Biochem. Biophys. Res. Commun., 18, 850 (1965).
4. H.S. Moyed, J. Biol. Chem. 235, 1098 (1960).
5. H.S. Moyed, and M. Friedman, Science 129, 968 (1959).
6. J. Ito and I.P. Crawford, Genetics 52, 1303 (1965).
7. C.H. Doy and J.M. Cooper, Biochim. Biophys. Acta 127, 302 (1966).
8. F. Lingens, B. Sprüssler, and W. Goebel, Biochim. Biophys. Acta 121, 164 (1966).
9. J. Ito and C. Yanofsky, J. Biol. Chem. 241, 4112 (1966).
10. T.I. Baker and I.P. Crawford, J. Biol. Chem. 241, 5577 (1966).
11. R.H. Bauerle and P. Margolin, Cold Spring Harbor Symp. Quant. Biol. 31, 203 (1966); G. Smith and R.H. Bauerle, Biochemistry 8, 1451 (1969).
12. H. Zalkin and D. Kling, Biochemistry 7, 3466 (1968).

13. H. Zalkin and E.J. Henderson, Biochem. Biophys. Res. Commun. 35, 52 (1969).
14. F.H. Gaertner and J.A. DeMoss, J. Biol. Chem. 244, 2716 (1969).
15. J. Ito and C. Yanofsky, Abstr. 7th Int. Cong. Biochem. (Tokyo) 4, 669 (1967).
16. J. Ito, E.C. Cox, and C. Yanofsky, J. Bact. 97, 725 (1969); J. Ito and C. Yanofsky, J. Bact. 97, 734 (1969).
17. H. Tamir and P.R. Srinivasan, J. Biol. Chem. 244, 6507 (1969).
18. P.R. Srinivasan and A. Rivera, Biochemistry 2, 1059 (1963).
19. J.M. Edwards, F. Gibson, L.M. Jackman, and J.S. Shannon, Biochim. Biophys. Acta 93, 78 (1964).
20. P.R. Srinivasan, Biochemistry 4, 2860 (1965).
21. P.R. Srinivasan, Fed. Proc. 22, 245 (1963).
22. J.G. Levin and D.B. Sprinson, J. Biol. Chem. 239, 1142 (1964).
23. J.R.D. McCormick, J. Reichenthal, U. Hirsch, and N.O. Sjolander, J. Amer. Chem. Soc. 84, 3711 (1962).
24. A.F. Egan, personal communication.
25. D.K. Onderka and H.G. Floss, Fed. Proc. 28, 668 (1969); Biochim. Biophys. Acta 206, 449 (1970).
26. F. Lingens, W. Luck, W. Goebel, Z. Naturforsch. 18b, 851 (1963).
27. C. Ratledge, Biochim. Biophys. Acta 156, 215 (1968).
28. C. Ratledge, Nature 203, 428 (1964); Biochim. Biophys. Acta 141, 55 (1967).
29. F. Lingens and B. Sprössler, Ann. 702, 169 (1967).
30. P.J. Suter and W.B. Turner, J. Chem. Soc. (C) 2240 (1967).
31. R.L. Somerville and R. Elford, Biochem. Biophys. Res. Commun. 28, 437 (1967).
32. T.E. Creighton, J. Biol. Chem. 243, 5605 (1968).
33. P. Truffa-Bachi and G.N. Cohen, Ann. Rev. Biochem. 37, 79 (1968).
34. H.E. Umbarger, Ann. Rev. Biochem. 38, 323 (1969).
35. H. Tamir and P.R. Srinivasan, Proc. Natl. Acad. Sci. 66, 547 (1970).
36. R.C. Paul and C. Ratledge, Biochim. Biophys. Acta 230, 451 (1971).
37. E.J. Henderson, H. Nagana, H. Zalkin, and L.H. Hwang, J. Biol. Chem. 245, 1416, 1424 (1970).
38. A. Arroyo-Begovich, and J.A. DeMoss, J. Biol. Chem. 248, 1262 (1973).
39. H. Nagano and H. Zalkin, J. Biol. Chem. 245, 3097 (1970); L.H. Hwang and H. Zalkin, J. Biol. Chem. 246, 2338 (1971).
40. J.F. Kane and R.A. Jensen, Biochem. Biophys. Res. Commun. 41, 328 (1970).

41.   H. Zalkin and S.H. Chen, J. Biol. Chem. 247, 5996 (1972).
42.   S.W. Queener, S.F. Queener, J.R. Meeks, and I.C. Gunsalus,
      J. Biol. Chem. 248, 151 (1973) and refs. therein.
43.   F. Robb, M.A. Hutchinson, and W.L. Belser, J. Biol. Chem.
      246, 6908 (1971).

CHAPTER 10

Biosynthetic Conversion
of Chorismic Acid
to *para*-Aminobenzoic Acid

Early work by Davis (1) demonstrated that shikimic acid
could replace PABA as a growth factor for mutants of *E. coli*
unable to synthesize the aromatic amino acids, a finding which
was supported by similar work on mutants of *Neurospora* (2).
Weiss and Srinivasan (3), using a cell-free extract from yeast,
have shown that PABA is produced from shikimic acid 3-phosphate
and L-glutamine, and that the amide nitrogen of L-glutamine is
the most effective source of the amino group. Aspartic acid,
asparagine, and glutamic acid were all poorly utilized, and
NH_4Cl was unable to replace glutamine as the nitrogen source at
either pH 7.4 or 8.2. The reaction was found to be inhibited by
the glutamine analog aza-L-serine. Interestingly, the production
of PABA was stimulated by low concentrations of the inhibitor,
presumably because it blocked the biosynthesis of anthranilic
acid more effectively than that of PABA; this indicates that
there may well be a similar mechanism operating for both reactions. Gibson, Gibson, and Cox (4) have demonstrated the formation of PABA from chorismic acid and L-glutamine using crude
cell extracts of *A. aerogenes* 62-1, the mutant strain that accumulates chorismic acid, grown in the presence of tryptophan to
repress anthranilate synthetase. Under these conditions, *p*-
hydroxybenzoic and prephenic acid are produced, but neither of
these acids was able to replace the substrates used in the formation of PABA.

Huang and Pittard (5) have shown that at least two genes are
operative in the conversion of chorismic acid to PABA in *E. coli*,
and Hendler and Srinivasan (6) have obtained evidence that the
conversion of chorismate to PABA involves at least two enzymes,
by using two mutants of *N. crassa*, both requiring PABA for

247

Chorismic Acid ——→ Unknown Intermediate ——→ PABA

830   H.193

growth, in cross-feeding experiments. When a sterile portion of
the culture filtrate of *N. crassa* H193, grown on minimal medium
supplemented by limiting PABA, was inoculated with *N. crassa* 830,
growth was observed, indicating that *N. crassa* H193 excreted a
substance (the presumed intermediate) which was able to supply
the PABA requirement of strain 830. Furthermore, the same
authors have shown that a partially purified enzyme extract, which
converts chorismic acid to PABA, could be fractionated into two
parts, both of which were necessary for the interconversion.
Very similar findings have been reported by Lingens and his co-
workers in *A. aerogenes* (7); the polyauxotroph 62-1, which
produces chorismic acid, was converted to a PABA-deficient strain
by chemical mutation; this mutant then accumulated an unidentified
compound that could in turn support the growth of a PABA-deficient
mutant and be converted to PABA by heating in solution at pH 3.5.
The intermediate had λ_{max} 271 nm. It had been formulated initially

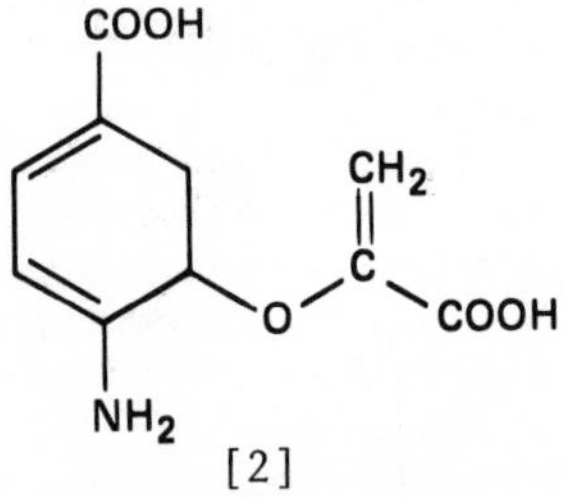

COOH
CH_2
C
O   COOH
NH_2
[2]

as [2] (8), but this interpretation was subsequently revised (10)
to 2,3-dihydro-3-hydroxy-4-aminobenzoic acid on the basis of nmr
evidence. However, the compound is actually 4-glucosylamino
benzoic acid (14), formed nonenzymatically from PABA and glucose.
A prolonged search by Gibson and his co-workers (9) for possible
free or enzyme-bound intermediates in the biosynthesis of PABA
has been without success.

Since, as in the case of the formation of anthranilic acid,
it is the amide nitrogen of glutamine and not the amino N that
is involved in the formation of PABA, it is not feasible to in-
voke the usual pyridoxal-catalyzed transamination of the α-amino
group; a possible mechanism for the conversion could involve

attack of an NH_2^- ion from glutamine on the intermediate carbonium
ion [1] derived from chorismic acid.   The NH_2^- would attack the
intermediate ion from the least hindered side, thus providing
the correct *cis*-stereochemistry for the *cyclic* elimination of
pyruvic acid.   There is no evidence as yet to support the proposed mechanism.

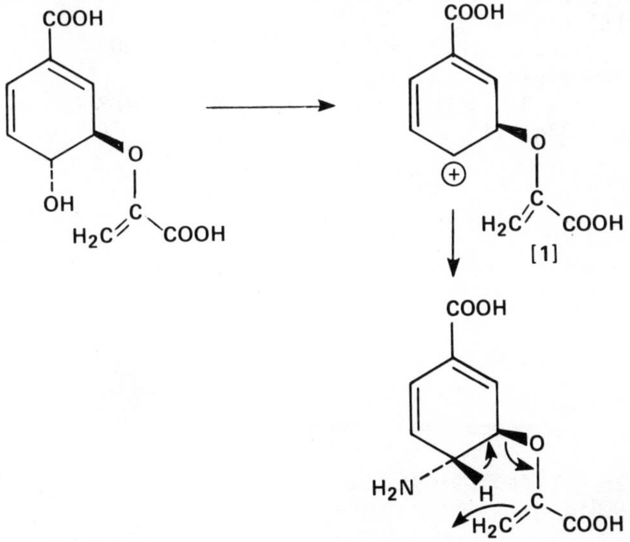

Para-aminobenzoic acid is incorporated intact into the
aromatic groupings of the antifugal antibiotics candicidin (II),
and fungimycin (12), and of the agaritins (glutamylphenylhydra-
zines) in the common commercial mushroom (13).

REFERENCES

1.   B.D. Davis, Experientia 6, 41 (1950).
2.   E.L. Tatum, S.R. Gross, G. Ehrensvärd, and L. Garnjobst,
     Proc. Natl. Acad. Sci. 40, 271 (1954).
3.   B. Weiss and P.R. Srinivasan, Proc. Natl. Acad. Sci. 45,
     1491 (1959); Biochim. Biophys. Acta 51, 597 (1961).
4.   F. Gibson, M. Gibson, and G.B. Cox, Biochim. Biophys. Acta
     82, 637 (1964).
5.   M. Huang and J. Pittard, J. Bact. 93, 1938 (1967).
6.   S. Hendler and P.R. Srinivasan, Biochim. Biophys. Acta 141,
     656 (1967).
7.   K.H. Altendorf, A. Bacher, and F. Lingens, FEBS Lett. 3,
     319 (1969); Z. Naturforsch. 24b, 1602 (1969).
8.   F. Lingens, Angew. Chem. Int. Ed. 9, 384 (1970).

9.   M. Huang and F. Gibson, J. Bact. 102, 767 (1970).
10. K.H. Altendorf, B. Gilch, and F. Lingens, FEBS Lett. 16, 95
     (1971); see also Chem. Abstr. 77, 16447b (1972).
11. C-M. Liu, L.E. McDaniel, and C.P. Schaffner, J. Antibiotics
     25, 116 (1972).
12. C-M. Liu, L.E. McDaniel, and C.P. Schaffner, J. Antibiotics
     25, 187 (1972).
13. H.R. Schütte, H.W. Liebisch, O. Miersch, and L. Senf, Quimica
     68, 889 (1972).
14. A. Bacher, B. Gilch, H. Rappold, and F. Lingens, Z. Natur-
     forsch. 28c, 614 (1973).

CHAPTER 11

Biosynthetic Conversion of Chorismic Acid to *para*-Hydroxybenzoic Acid and Ubiquinone

During some of his mutant studies, Davis (1) showed that certain multiple aromatic auxotrophs of *E. coli* and *A. aerogenes* with complete metabolic blocks grew only very slowly on minimal medium supplemented by the three aromatic amino acids and PABA, while their growth became normal on addition of either shikimic acid or filtrates from wild-type cultures. This observation suggested that an additional factor derived from shikimic acid was required for growth. Bioautography of the filtrates demonstrated the presence of such a factor, and screening tests with 50 available compounds possibly related to shikimic acid showed that one of them, *p*-hydroxybenzoic acid (POB), had the same chromatographic characteristics as the unknown factor, and permitted almost normal growth of the mutants in concentration as low as 0.01 μg/ml. None of the other compounds had any growth-promoting activity. These findings strongly suggested the identity of POB with the growth factor; however, the natural compound has never been isolated. Only one of the several quintuple aromatic auxotrophs had an absolute POB requirement, but the relative requirement of the remaining strains became absolute in the presence of very low concentrations of L-aspartic acid (2-10 μg/ml). D-Aspartic acid had no such effect. The POB requirement is also pH-dependent, being strong at pH 7 but minimal at pH 6. The requirement for POB could be largely replaced by low concentrations of methionine plus lysine (or some of their precursors), or by succinic acid (2). Wyn-Jones and Lascelles (3) have recently reinvestigated the growth-promoting effect of methionine, lysine, and succinate on an aromatic

251

auxotroph of *E. coli* which required POB for aerobic growth. Their results indicate that the replacement of POB by lysine plus methionine is an indirect consequence of coenzyme Q (ubiquinone) deficiency; POB is well known as a precursor of ubiquinone in bacteria (see below), and the *E. coli* mutant produces neither ubiquinone nor vitamin K_2 when grown aerobically in the absence of POB (4). Since ubiquinone is implicated in the synthesis of succinic acid via the tricarboxylic acid cycle, and succinic acid itself is involved in one step in the biosynthesis of both methionine and lysine (5), the organism, growing in the absence of POB, is unable to form these amino acids. Cell extracts also showed diminished activity of α-ketoglutarate dehydrogenase (the enzyme responsible for the formation of succinic acid under conditions of normal aerobic growth)[+]. The addition of POB or succinate is thus seen as a stimulation of lysine and methionine synthesis, and the addition of lysine plus methionine as a sparing of succinate.

The effect of L-aspartic acid and of pH on the relative POB requirement of the mutant is still not understood; neither is the true function of ubiquinone in the Krebs cycle fully explained.

The formation of POB from shikimic acid has been demonstrated in several microorganisms and, although it was suggested that POB might be derived directly from 5-enolpyruvylshikimic acid 3-phosphate (6), it is now certain that POB arises from chorismate by an aromatization step (7).

Chorismic acid can be converted enzymatically to POB (7). The enzyme system has not yet been studied in detail; it seems, however, to be stimulated by NAD. The fact that the enzyme responsible for the aromatization is unaffected by chromatography over DEAE cellulose makes it almost certain that the interconversion involves only one enzyme. Purely chemical conversion of chorismate into POB can be achieved very readily by warming in acidic, basic, or neutral solution (8), or by pyrolysis (9). Two mechanisms appear possible, of which the first, (A), which is a concerted process, seems much more plausible than the alternative (B) involving an improbable *cis*-elimination.

In contrast to the microorganisms mentioned, higher plants and some fungi synthesize POB by either of two related transformations of preexisting aromatic rings: direct *para*-hydroxylation

---

[+]Under anaerobic conditions, succinate is formed from fumarate by succinic dehydrogenase. α-Ketoglutarate dehydrogenase is no longer involved.

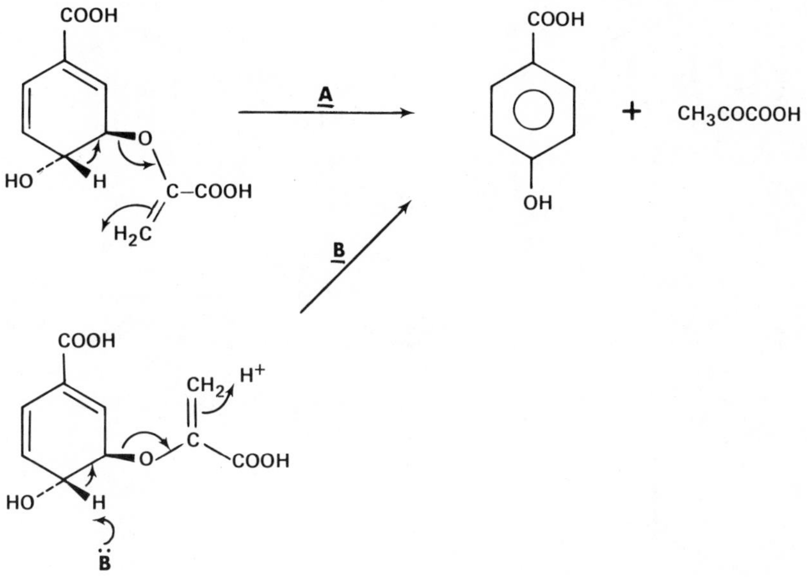

of benzoic acid (10 - 13) (derived from cinnamic acid by oxidative degradation), or β-oxidation of *p*-coumaric acid (10, 14-16, 45) formed either from tyrosine (14) or, more commonly, by hydroxylation of cinnamic acid. In animals POB is formed from tyrosine via *p*-hydroxyphenylpyruvic, *p*-hydroxyphenyllactic, and *p*-coumaric acids. This sequence has been established, *inter alia*, in rat liver slices and mitochondria (46).

   Ubiquinone. The biosynthesis of coenzyme Q has been reviewed in detail by Rudney and Raman (1966) (17) and by Gibson (1973) (65); only a brief outline will be given here. Coenzyme Q is a term used to denote the group of ubiquinones [Q - n] (n = 6 - 10) which are widely distributed in nearly all aerobic organisms, plants, microorganisms, and animals (18), and which are involved in the electron transport and oxidative phosphorylation processes of the cell (19). The ubiquinones have the general structure [1], although in some cases a double bond of the isoprenoid side-chain may be saturated. This is the case for the ubiquinones of *N. crassa* and several *Aspergillus* spp., in which the double bond furtherest from the quinone ring is reduced; these saturated quinones are designated Q-10 (H$_2$), or more correctly Q-10 (X H$_2$) where X denotes the tenth double bond.

   The ubiquinones are derived in part from the acetate-mevalonate pathway, and in part from shikimic acid; thus, although experiments with ^{14}C-acetic acid and ^{14}C-malonate have proved that the quinone ring is not formed from acetic acid (20),

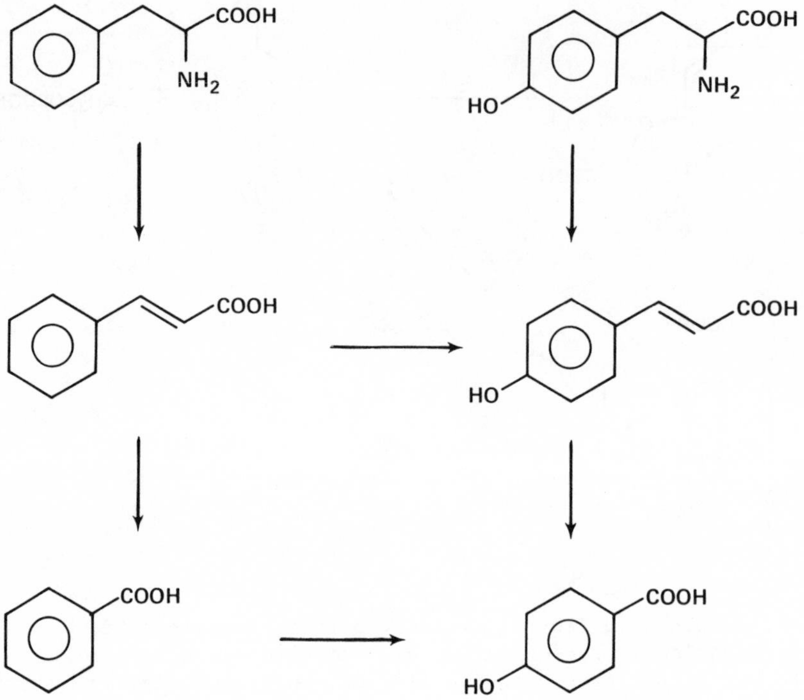

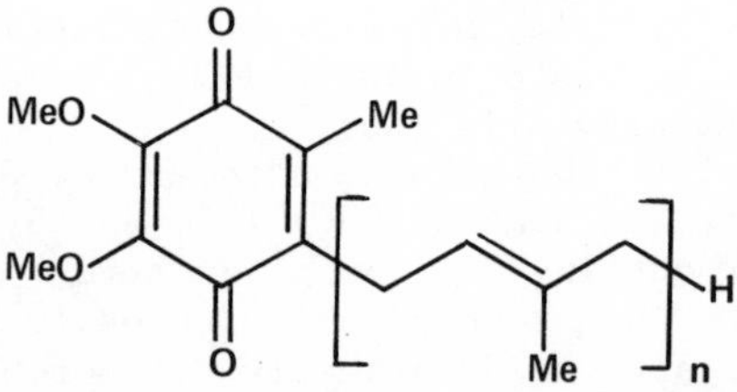

[1]

similar experiments in higher plants (21), microorganisms (22), and rats (23) have shown that the isoprene side-chain is so formed (see the review (1967) by Threlfall and Griffiths (24)). Both the nuclear C-methyl and O-methyl carbons come from S-adenosylmethionine in all three groups of organisms (17, 21-23), and the aromatic ring is formed via the shikimic acid pathway. Preliminary studies in *E. coli* (25), *Azotobacter* (26), yeast (26), and the rat (27) indicate that POB is effectively incorporated into the quinone nucleus with loss of its carboxyl group. These findings were confirmed in higher plants (maize (28)), the green

alga *Euglena gracilis* (29), *Rhodospirillum rubrum* (30), and other
organisms (14, 22, 31).   Shikimic acid (32) *p*-hydroxybenzaldehyde
(17), phenylalanine, and tyrosine (22, 33, 34), have been shown
to be incorporated into the quinone ring as well as POB, and it
now seems likely that ubiquinone is formed in microorganisms from
POB, which may be derived either directly from chorismic acid or
from the metabolism of tyrosine.   In the case of the higher plants,
POB is also formed from *p*-coumaric acid or benzoic acid.   Some of
the foregoing incorporations have been achieved in cell-free systems; thus, an enzyme preparation able to convert POB coenzyme A,
an established precursor (48), into ubiquinone has been isolated
from rat tissue (49).   Two studies (50, 59) have demonstrated
that molecular oxygen is necessary for the conversion of POB to
ubiquinone.   An interesting paper (60) which describes the conversion of cyclohexane carboxylic acid into the aromatic ring of
ubiquinone by rats, more properly belongs to Chapter 25, which
deals with unusual aromatizations.   The observation is presumably
one of nonspecific utilization similar to that encountered in the
conversion of quinic acid to aromatic compounds.

In 1966 Friis et al. (36, 37) published a possible biosynthetic scheme for the formation of ubiquinone, based upon the isolation of a series of phenolic and quinonoid compounds bearing
isoprenoid substituents from *Rhodospirillum rubrum*.   Subsequent
isolation of similar isoprenylated phenols and quinones from a
variety of organisms: *Pseudomonas ovalis* and *Proteus mirabilis*
(50), avian and mammalian species (51), *Aspergillus flavus* (52),
higher plants and *Euglena gracilis* (53), *E. coli* (35, 54, 55, 58),
molds and yeasts (56), and the rat (57, 62), have, on the whole,
supported the proposed scheme, although some of the isolated compounds (50) do not fit in very well.   The biosynthesis of the
compounds has been reviewed by Bentley (1969) (47).   The incorporation of POB and isopentenylpyrophosphate (see Chapter 22-1)
into some of the proposed intermediates has been observed (38);
formation of intermediates has also been reported in cell-free
systems from bacteria (35, 39, 55, 63) the rat (40, 57, 62), and
higher plants (64).

The biosynthesis of the related quinones plastoquinone-9 [2],
tocopherolquinone (vitamin E) [3], and associated compounds
(tocopherols), which occur in higher plants, has been studied in
detail by Goodwin and his co-workers (21, 33, 41-44).   Their
results have shown that POB is not incorporated into these
quinones, but that *p*-hydroxyphenylpyruvic acid, formed from tyrosine (both D and L can be utilized in some cases (61)) or
phenylalanine (except in the case of bean shoots which apparently
lack the hydroxylase enzyme), gives rise to both the aromatic
ring and one of the nuclear methyl groups.

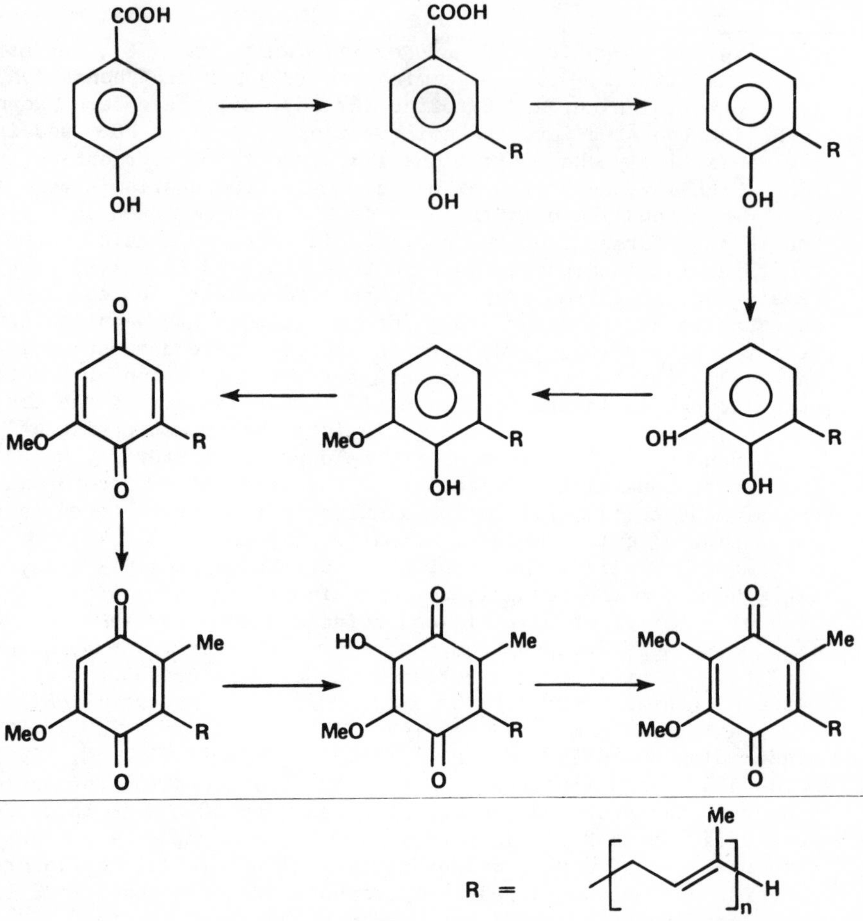

Proposed Biosynthetic Pathway for Ubiquinones in Rhodospirillum rubrum (36,37)

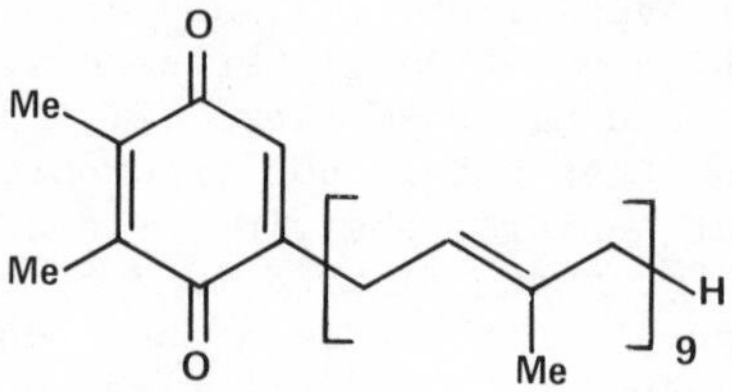

[2]

256

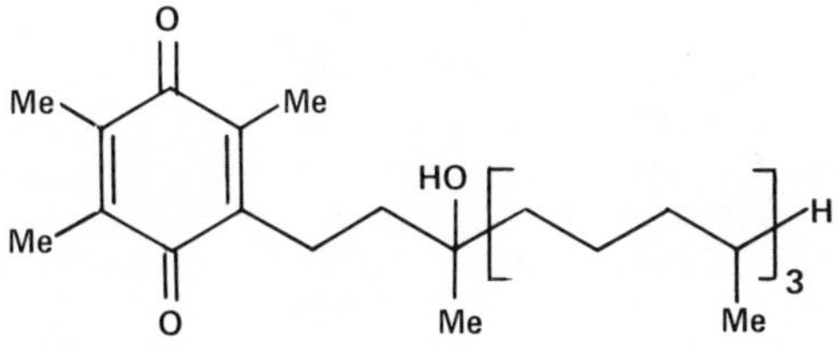

[3]

It has been suggested (44, 61), on the basis of isotopic
competition experiments and the direct incorporation of homogentisic acid, that the incorporation of the β-carbon of tyrosine
is accompanied by an intramolecular rearrangement of the *p*-
hydroxyphenylpyruvic acid to homogentisic acid, similar to that in
animals (Chapter 16), the latter being further transformed to the

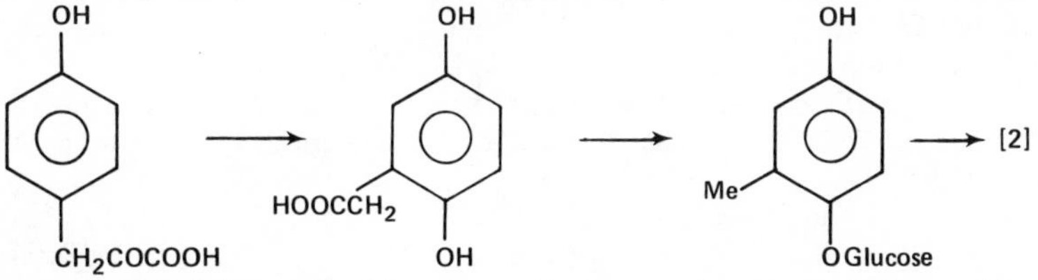

quinones.   The biosynthetic steps between homogentisic acid and
the quinones are by no means clear, although it has been shown
(44, 61) that in maize plants gentisic acid, gentisic aldehyde,
and toluquinol are not involved, and that the α-carbon atom of
homogentisic acid provides the nuclear methyl group *meta* to the
isoprenoid side chain.

REFERENCES

1.   B.D. Davis, Nature 166, 1120 (1950); Harvey Lectures 230
     (1954-1955).
2.   B.D. Davis, J. Bact. 64, 729 (1952).
3.   R.G. Wyn-Jones and J. Lascelles, Biochem. J. 103, 709 (1967).
4.   R.G. Wyn-Jones, Biochem. J. 103, 714 (1967).
5.   J.R. Mattoon, in Biogenesis of Natural Compounds, P. Bernfeld,
     Ed. 2nd ed., Pergamon, Oxford, 1967, p. 1.
6.   W.W. Parson and H. Rudney, Proc. Natl. Acad. Sci. 51, 444
     (1964).
7.   M.I. Gibson and F. Gibson, Biochem. J. 90, 248 (1964).
8.   F. Gibson, Biochem. J. 90, 256 (1964).
9.   J.M. Edwards and L.M. Jackman, Aust. J. Chem. 18, 1227 (1965).
10. S.Z. El-Basyouni, D. Chen, R.K. Ibrahim, A.C. Neish, and
     G.H.N. Towers, Phytochemistry 3, 485 (1964).
11. R.K. Ibrahim, Flora 153, 481 (1964); Chem. Abstr. 62, 8129h
     (1965).

12. H.D. Klambt, Nature, 196, 491 (1962).
13. M.H. Zenk, Z. Naturforsch, 19b, 83 (1964).
14. M.H. Zenk and G. Müller, Z. Naturforsch, 19b, 398 (1964);
     M.H. Zenk, Z. Naturforsch, 19b, 856 (1964).
15. K.O. Vollmer, H.J. Reisener, and H. Grisebach, Biochem.
     Biophys. Res. Commun. 3, 24 (1965).
16. H. Kindl and G. Billek, Monatsh. 95, 1044 (1964).
17. H. Rudney and T.S. Raman, Vitamins Hormones 24, 531 (1966).
18. F.L. Crane, in Biochemistry of Quinones, R.A. Morton, Ed.,
     Academic, New York, 1965, p. 183.
19. E.R. Redfearn, Vitamins Hormones 24, 465 (1966).
20. R. Bentley and W.V. Lavate, J. Biol. Chem. 240, 532 (1965).
21. D.R. Threlfall, G.R. Whistance, and T.W. Goodwin, Biochem.
     J. 106, 107 (1968).
22. G.H. Spiller, D.R. Threlfall, and G.R. Whistance, Arch.
     Biochem. Biophys. 125, 786 (1968).
23. R.E. Olson, Vitamins Hormones 24, 551 (1966).
24. D.R. Threlfall and T.W. Griffiths, in Biochemistry of Chloro-
     plasts, Vol. 2., T.W. Goodwin. Ed., Academic, London, 1967,
     p. 254.
25. G.B. Cox and F. Gibson, Biochem. J. 100, 1, (1966).
26. W.W. Parson and H. Rudney, Proc. Natl. Acad. Sci. 51, 444
     (1964).
27. A.S. Aiyar and R.E. Olson, Fed. Proc. 23, 425 (1964).
28. G.R. Whistance, D.R. Threlfall, and T.W. Goodwin, Biochem.
     J. 105, 145 (1967).
29. R. Powls and F.W. Hemming, Phytochemistry 5, 1249 (1966).
30. W.W. Parson and H. Rudney, J. Biol. Chem. 240, 1855 (1965).
31. F.S. Skelton, K.D. Lunan, K. Folkers, J.V. Schnell,
     W.A. Siddiqui and Q.M. Geiman, Biochemistry 8, 1284 (1969).
32. G.R. Whistance, D.R. Threlfall, and T.W. Goodwin, Biochem.
     Biophys. Res. Commun. 23, 849 (1966).
33. P.H. Gold and R.E. Olson, J. Biol. Chem. 241, 3507 (1966).
34. H. Ozawa and Y. Kaseki, Seikagaku 40, 911 (1968); Chem.
     Abstr. 71, 110533z (1969).
35. G.B. Cox, I.G. Young, L.M. McCann, and F. Gibson, J. Bact.
     99, 450 (1969).
36. P. Friis, G.D. Daves, and K. Folkers, J. Amer. Chem. Soc.
     88, 4754 (1966).
37. G.D. Daves, P. Friis, R.K. Olsen, and K. Folkers, Vitamins
     Hormones 24, 427 (1966).
38. T.S. Raman, H. Rudney, and N.K. Buzzelli, Arch. Biochem.
     Biophys. 130, 164 (1969).
39. H. Rudney, Proc. Biochem. Soc. 113, 21P (1969).
40. M.J. Winrow and H. Rudney, Biochem. Biophys. Res. Commun.
     37, 833 (1969).

41. D.R. Threlfall, W.T. Griffiths, and T.W. Goodwin, Biochem. J. 103, 831 (1967).
42. G.R. Whistance and D.R. Threlfall, Biochem. Biophys. Res. Commun. 28, 295 (1967).
43. W.T. Griffiths, D.R. Threlfall, and T.W. Goodwin, Eur. J. Biochem. 5, 124 (1968).
44. G.R. Whistance and D.R. Threlfall, Biochem. J. 109, 482, 577 (1968).
45. I. More and N. Terashima, Mokuzai Gakkaishi 17, 311 (1971).
46. S. Ranganathan and T. Ramasarma, Biochem. J. 122, 487 (1971).
47. R. Bentley, Lipid Metab. 481 (1970).
48. G.H. Dialameh, G.G. Nowicki, K.G. Yekundi, and R.E. Olson, Biochem. Biophys. Res. Commun. 40, 1063 (1970).
49. B.L. Trumpower and R.E. Olsen, Biochem. Biophys. Res. Commun. 45, 1323 (1971).
50. G.R. Whistance, B.S. Brown, and D.R. Threlfall, Biochem. J. 117, 119 (1970).
51. G.R. Whistance, F.E. Field, and D.R. Threlfall, Eur. J. Biochem. 18, 46 (1971).
52. A. Law, D.R. Threlfall, and G.R. Whistance, Biochem. J. 117, 799 (1970).
53. D.R. Threlfall and G.R. Whistance, Phytochemistry, 9, 355 (1970).
54. I.G. Young, L.M. McCann, P. Stroobant, and F. Gibson, J. Bact. 105, 769 (1971).
55. J.A. Hamilton and G.B. Cox, Biochem. J. 123, 435 (1971).
56. A. Law, D. Threlfall, and G.R. Whistance, Biochem. J. 123, 331 (1971).
57. H.G. Nowicki, G.H. Dialameh, and R.E. Olson, Biochemistry 11, 896 (1972).
58. F. Stroobant, I.G. Young, and F. Gibson, J. Bact. 109, 134 (1972).
59. K. Uchida and K. Aida, Biochem. Biophys. Res. Commun. 46, 130 (1972).
60. A.S. Aiyar, U.V. Gopalaswamy, and A. Sreenivasan, Biochem. Biophys. Res. Commun. 45, 893 (1971).
61. G.R. Whistance and D.R. Threlfall, Biochem. J. 117, 593 (1970).
62. B.L. Trumpower, A.S. Aiyar, C.E. Opliger, and R.E. Olson, J. Biol. Chem. 247, 2499 (1972).
63. I.G. Young, R.A. Leppik, J.A. Hamilton, and F. Gibson, J. Bact. 110, 18 (1972).
64. G. Thomas and D.R. Threlfall, Biochem. J. 134, 811 (1973).
65. F. Gibson, Biochem. Soc. Transactions 1, 317 (1973).

CHAPTER 12

Biosynthetic Conversion of Chorismic Acid to Isochorismic Acid and Related Metabolites

12-1. INTRODUCTION

This section deals with two further growth factor requirements of certain bacterial mutants which, although similar in many aspects, are yet definitely not identical.

The first of these factors was found by Davis (1) to be required for growth even in the presence of the three aromatic amino acids, PABA, and POB. This additional growth factor, called "sixth factor", has never been studied in detail; it is replaceable by shikimic acid and, inefficiently, by a number of 3,4-dihydroxyaromatic compounds, or by inorganic iron.

Knowledge of the second group of compounds, likewise replaceable by shikimate, rests largely upon the observation that certain mutants fail to grow normally even in the presence of all six supplements. This failure has been traced to the lack of certain 2,3-dihydroxy-aromatic compounds, and it has been shown that these substances play a vital role in the utilization and transport of iron. The significance of iron was first noted by Ito and Neilands (2), who isolated 2,3-dihydroxybenzoylglycine from filtrates of *A. aerogenes* and *E. coli* growing on iron-deficient medium. Recent research by Gibson and his associates has shown that these 2,3-diphenols arise from chorismate through a separate branch; chorismic acid is converted to isochorismic acid, which is further modified to 2,3-dihydroxy-2,3-dihydrobenzoic acid and subsequently aromatized. The details of this biosynthesis are by now fairly well understood; the enzymes of this branch of the pathway are apparently regulated in a most unusual way by inorganic iron, rather than through the normal repressive and inhibitory effects of metabolic end-products, (see Chapter 15).

It seems likely that the isochorismate pathway is also responsible for the biosynthesis of the *meta*-carboxyphenyl-amino acids occurring in some higher plants, and for that of salicylic acid in certain organisms.

12-2. "SIXTH FACTOR"

Besides the five aromatic metabolites of shikimic acid that are necessary for the growth of many bacterial auxotrophs, an additional requirement can be demonstrated in some cases (5). Mutants supplied with all five aromatic compounds were found to be unable to grow if the pH of the medium was changed from 7.0 to 7.5; addition of shikimic acid or wild-type filtrates again restored growth. The term "sixth factor" was provisionally introduced for the growth factor thus demonstrated. Further search revealed that the additional requirement was filled to a certain extent by a number of 3,4-dihydroxybenzene derivatives, such as protocatechuic aldehyde, protocatechuic acid, chlorogenic acid, or epinephrine; and, surprisingly, by Fe^{2+} or Fe^{3+} (see below). These early results have been confirmed by Wyn-Jones (3) using a similar *E. coli* mutant. From the ability of these 3,4-dihydroxybenzene derivatives to replace the "sixth factor" requirement, it seems likely that the natural compound should also contain this structural element; however, the growth-promoting substance has never been isolated, and its identity and biological significance remain unknown.

12-3. ISOCHORISMIC ACID

The first indication of a biosynthetic pathway from shikimic acid in which the carboxyl group of the acid was retained, was observed in the early work of Larsen and his associates, who showed that $^{14}COOH$-shikimic acid was metabolized in certain higher plants to the *meta*-carboxylic acids [4] - [7] (see below) with retention of the ^{14}C label. The identification of 2,3-dihydroxybenzoic acid as a bacterial growth factor led Gibson and his co-workers to investigate its biosynthesis, and thus to the structure of isochorismic acid, the intermediate leading to the *m*-carboxyphenyl amino acids.

If chorismic acid is incubated with a protein fraction from *A. aerogenes* 62-1 (the chorismic acid-accumulating mutant) and Mg^{2+}, it is converted into a new compound which has been identified (14) as α-(3-carboxy-2-hydroxy-1,2-dihydrophenoxy)-acrylic acid [1], for which the name isochorismic acid has been proposed.

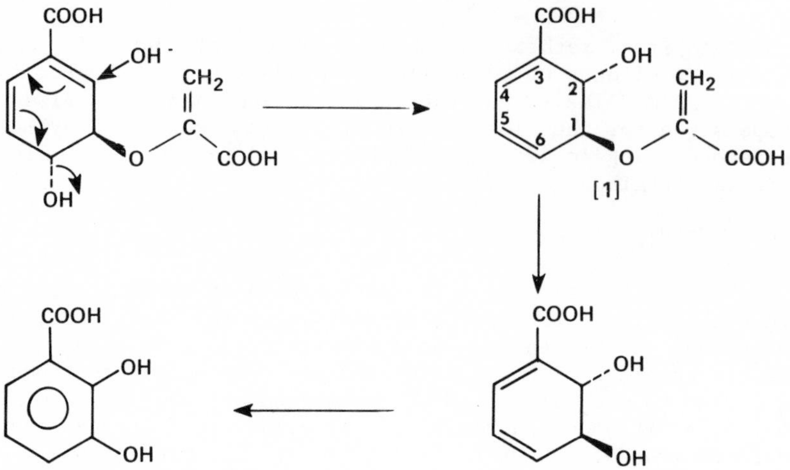

[1]

[2]

Compound [1] was isolated from the acidified reaction mixture by extraction into ethyl acetate, and purified by paper electrophoresis and chromatography over Dowex 1 Cl-ion exchange resin. The acid, which could not be crystallized, has λ_{max} (H_2O) 278 nm ($\varepsilon \sim 13,000$) confirming the presence of a 1,3-cyclohexadiene system with extended conjugation between the diene system and the carboxyl group; the nmr assignments are consistent with the proposed structure and its relationship with chorismic acid (Chapter 6). Ozonolysis of isochorismic acid (5) gave (+)-tartaric acid after oxidative workup, thus demonstrating the relative *trans*-stereochemistry of the two oxygen substituents, and also proving that the absolute stereochemistry of the carbon bearing the enolpyruvyl side-chain is the same in [1] and in chorismic acid. Incubation of [1] with a cell-free extract of *A. aerogenes* 62-1 and NAD converts it to 2,3-dihydroxybenzoic acid; however, if the NAD is omitted, 2,3-dihydro-2,3-dihydroxybenzoic acid [2] is formed.

If isochorismic acid is heated to 100° for 10 minutes at pH 7 it is converted into 3-carboxyphenylpyruvic acid [3]; a possible mechanism for this conversion via isoprephenic acid [2A] has been suggested by Gibson (6). The hypothetical intermediate [2A] has not yet been isolated and described chemically, but its existence, which so neatly explains the formation of the *meta*-carboxylated amino acids (see below), can hardly be doubted. Acid [2] can be isolated (4) by incubation of chorismate with a cell-free extract of *A. aerogenes* 62-1 grown on an iron-deficient medium, in the presence of Mg^{2+}, but without NAD. The absolute stereochemistry of [2] was established by ozonolysis and oxidative workup to

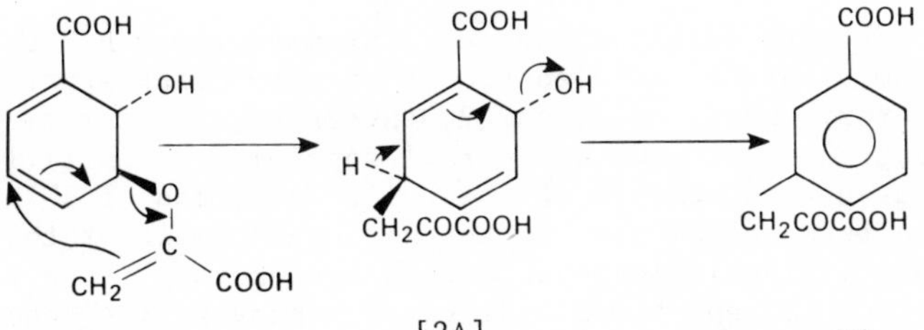

[1]   [2A]   [3]

(+)-tartaric acid (5). Again the results demonstrate the retention of the configuration at the C-5 of shikimic acid. Compound
[2] has been synthesized (7).

When heated in acid or alkali, [2] is converted to *meta*-
hydroxybenzoic acid and a trace of salicylic acid. These interconversions have been reviewed by Gibson and co-workers (8).

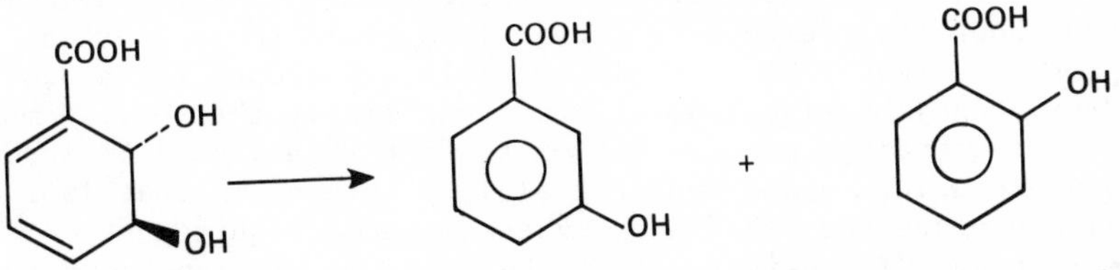

It is likely that the formation of 2,3-dihydroxybenzoic acid from
chorismic acid proceeds by a series of reactions similar to those
considered for anthranilic acid (Section 9-1), in which the initial double Sn$_2$' displacement of the hydroxyl group of chorismic
acid to provide the required *trans*-stereochemistry of [1] would
have to be a two-step process, as discussed for the enzymic formation of chorismate (Chapter 6).

The formation of *m*-hydroxybenzoic acid from [1] (and also
from chorismic acid) may give a hint about the origin of this
acid, which seems to occur widely in nature (e.g., as a fairly
major constituent of the phenolic fraction of human urine); in
view of the scarcity of natural *m*-hydroxyphenyl compounds, its
frequent occurrence is puzzling, although it has been related to
flavonoid degradation (Chapter 21 and Section 7-2).

12-4. 2,3-DIHYDROXYBENZOIC ACID

Still another growth factor was revealed (9) during work
with some strains of multiple aromatic auxotrophs that will not
grow at a normal rate even on a medium supplemented with all six
aromatic compounds; these mutants still require additional small
quantities (0.001%) of yeast extract (10) for normal growth.
Again it was found that shikimic acid would satisfy the growth
requirement, as would supernatants of cultures of an *E. coli*

mutant unable to form ubiquinone. Chromatography of concentrates of this supernatant revealed two growth-promoting substances: 2,3-dihydroxybenzoic acid, and 2,3-hydroxybenzoylserine (10, 11) which, together with the six normal supplements, fulfilled all the aromatic requirements of the cell. Addition of shikimic acid to the culture medium gave no further enhancement of the growth rate, thus making it improbable that yet another trace factor was missing. Except for 2,3-dihydroxybenzoylserine, shown to be formed from 2,3-dihydroxybenzoic acid by enzyme extracts of *A. aerogenes*, and citrate, no organic compounds were found that could replace the requirement for a 2,3-dihydroxybenzoyl derivative shown by these mutants blocked after shikimic acid. The growth stimulation by citrate is thought to be indirect, in that iron is made available to the cell as a citrate-iron complex, which can be taken up by the organism.

During his investigation of the "sixth factor" Davis (1) had shown that ferrous or ferric iron could to some extent replace the requirement of his multiple auxotrophs for a 3,4-dihydroxybenzoyl derivative; it now seems likely that Davis' mutants were responding both to the "sixth factor" and to 2,3-dihydroxybenzoic acid, since Young et al.(10) have shown that iron will also replace the 2,3-dihydroxybenzoic acid requirement of their mutant and, furthermore, that the presence of iron has a repressive effect upon the formation of the enzyme converting chorismic acid to the phenolic acid.

Interestingly, it has been observed (12) that even prolonged, continuous, ether-extraction of wild-type culture filtrates removes only part of their "sixth factor" activity for a mutant growing in the presence of the five normal supplements; it is possible that the solvent removes only the 3,4-dihydroxy compound leaving the more water soluble, but still active 2,3-dihydroxybenzoic acid.

In many cases the formation of *ortho*-dihydric phenols by bacteria is associated with the iron concentration in the medium: When the iron concentration is low the phenols are excreted, and it has been suggested (2) that their function might be to chelate any available iron present in the medium, and make it accessible to the cell. The suggestion has been convincingly vindicated by Peters and Warren (13), who have investigated the connection between the production of phenolic acids and the capacity of the cell for iron transport, using *B. subtilis* grown on an iron-deficient medium; they have shown that growth under conditions of iron deficiency results in a considerable increase in the capacity of the iron transport system, and that the addition of phenolic acids to iron-deficient cells stimulates the level of iron transport by the cell. Mutants unable to produce normal levels of these phenolic acids were shown to have a lower capacity for iron transport. Brot and Goodwin (14) have been able to show that the formation of 2,3-hydroxybenzoylserine synthetase can be controlled by the iron concentration in the medium of a mutant of *E. coli*, requiring methionine and vitamin B12, grown in the absence of iron. Other studies of the relationship between the uptake of iron and the production of 2,3-dihydroxybenzoylserine have been made in *E. coli* (15, 16), and *B. subtilis* (17) (see also Section 12-1 and Chapters 1 and 15).

During their investigations of cultures of *E. coli*, *A. aerogenes*, and *Salmonella adelaide*, grown on iron-deficient medium, Gibson and his co-workers have demonstrated the presence not only of 2,3-dihydroxybenzoate and 2,3-dihydroxybenzoylserine, but also of a linear dimer and trimer, and cyclic trimer (enterochelin) of 2,3-dihydroxybenzoylserine (18, 19). A compound named enterobactin, identical with enterochelin, has been isolated (20) from *Salmonella typhimurium*. Recent evidence suggests that 2,3-dihydroxybenzoate combines with serine to form enterochelin, the true end-product of the pathway and the compound which is biologically active in iron transport; the other conjugates of 2,3-dihydroxybenzoylserine are formed by enzymatic degradation of enterochelin. It is worth noting that other microorganisms produce, and excrete into their culture media, iron-sequestering compounds which do not involve 2,3-dihydroxybenzoic acid; the compounds lusigen from *Fusarium cubense* (21), and schizokinen found in *Bacillus megaterium* (22) are both involved in iron transport in ways similar to enterochelin, but are structurally dissimilar.

2,3-Dihydroxybenzoic acid has been detected in many higher plants (23), as well as in the culture filtrates of *E. coli* (24), *Aspergillus niger* (25), *Claviceps paspali* (26-28), *B. subtilis* (29), *B. megaterium* (29), *Streptomyces rimosus* (30), and *S. griseus* (31).

2,3-Dihydroxybenzoylserine has been isolated from cultures of *A. aerogenes* (9) and *E. coli* (9, 32, 33), and the corresponding glycine derivative was found in the filtrates of *B. subtilis* grown on an iron-deficient medium (2). Another study (34) has shown 2,3-dihydroxybenzoic acid and its glycine derivative to be formed together by *B. subtilis* grown in iron-deficient medium; the lysine derivative has been isolated from *Azotobacter vinelandii* grown again on an iron-deficient medium (35).

The biosynthetic sequences leading to the formation of 2,3-dihydroxybenzoic acid are by no means restricted to the one described from chorismic acid; in the majority of cases, the acid is formed by transformations (hydroxylation, etc.) of salicylic acid (36, 37) anthranilic acid (28, 38-42), 3-hydroxyanthranilic

acid (25, 39) or benzoic acid (43), or possibly by the hydroxylation and degradation of a substituted cinnamic acid (44).

It is of interest to note that the hydroxylation pattern
found in 2,3-dihydroxybenzoic acid and its glycine and serine
derivatives is practically unknown in secondary metabolites
derived from the same biosynthetic pathways. Hydroxylation
patterns are discussed in Chapter 2.

12-5. META-CARBOXY DERIVATIVES

Several *meta*-carboxy-substituted aromatic amino acids have
been found in higher plants (44); for example, L-3-(3-carboxy-
phenyl)-alanine [4], 3-carboxyphenylglycine [5], L-3-(3-carboxy-
4-hydroxyphenyl)alanine [6], and D-3-carboxy-4-hydroxyphenyl-
glycine [7] have all been isolated from seeds of *Reseda lutea*,
and the individual acids have been isolated from other species
(46). These compounds are of interest because of their unusual
meta-substitution pattern. They include two derivatives of
phenylglycine, previously unknown in higher plants, and a D-amino
acid. Incorporation studies have shown that shikimic acid,
glucose, and L-tyrosine are all precursors of [6] in *Reseda*
species (47); significantly, carboxy-labelled tyrosine was incorporated into [6] which became labelled in the aliphatic

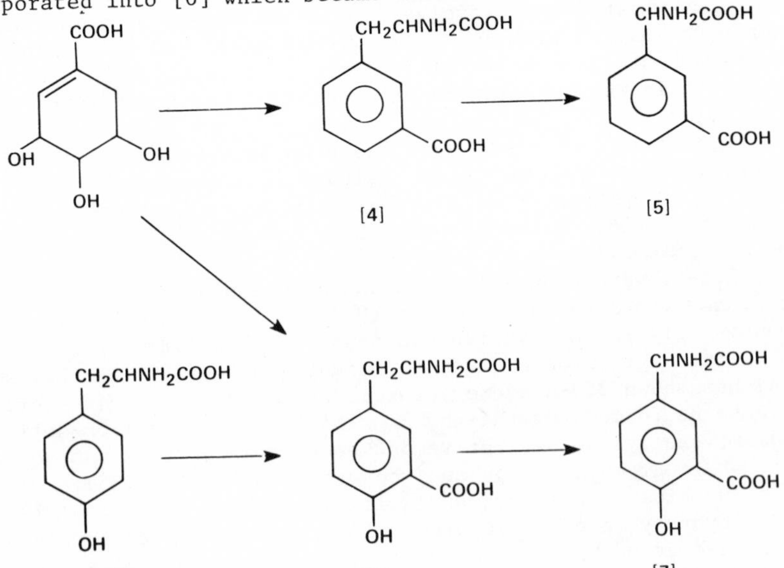

carboxyl group, showing that the C_6-C_3 unit was incorporated intact, wheras 14_C-shikmic acid was transformed into [6] containing 1/7 of the activity in the aromatic carboxyl group. Similar studies (48) have demonstrated pathways from shikimic acid to [4] and [7], but not, surprisingly, from phenylalanine to [5]; however, conversion of [4] to [5] has been shown, and since the incorporation of label is always lower into [7] than [4] it stereospecifically seems likely that the conversion of [4] to [7] is also possible. A subsequent study in which stereospecifically-labeled shikimic acid was converted to the *meta*-carboxy amino acids (49), has shown that these acids are indeed formed by rearrangements that do not involve prephenic acid, thus substantiating to a great extent the scheme outlined on page 262 via isoprephenic acid. The involvement of tyrosine (47) in the biosynthesis is still not

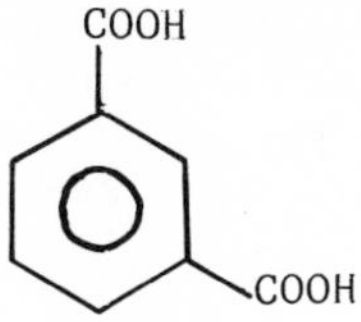

[8]

explained.

It is of interest that isophthalic acid [8] has been isolated from *Iris versicolor* (45); no work on the mode of its biosynthesis has been published.

12-6. SALICYLIC ACID

Although neither *A. aerogenes* nor *E. Coli* accumulates salicylic acid when grown under conditions of iron deficiency, it seems probable that chorismic acid is the precursor of sali-cylic acid in some other organisms. Ratledge (50) has studied the formation of salicylic by species of mycobacteria grown under a variety of conditions and on several carbon sources. Briefly, the amount of salicylic acid formed in the culture media of *Mycobacterium smegmatis* and *Mycobacterium tuberculosis* is greatly increased when the medium is deficient in iron; the acid is thought to be a precursor of certain iron-chelating mycobactins, for example [9], formed by mycobacteria; they are likewise produced in increased amounts under conditions of iron deficiency (51). Salicylic acid is not formed from anthranilic acid, tryptophan, or 3-methylanthranilic acid by these organisms (52); neither is it likely that is should be derived from phenylalanine or phenyl-acetic acid via benzoic acid (53), and Ratledge has suggested that it may be synthesized from chorismate (52.   Formation of salicylic acid is enhanced by incubation of washed cells

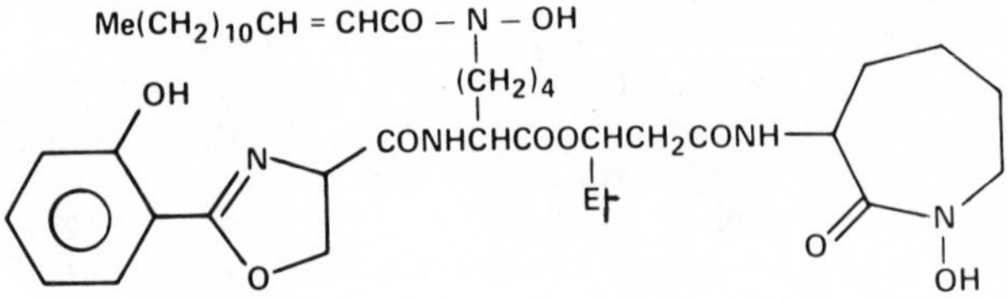

Me(CH$_2$)$_{10}$CH = CHCO − N − OH

[9]

of *M. smegmatis* with shikimic or quinic acids, which indicates that it may well be derived from the shikimic acid pathway,
and preliminary (54) feeding experiments showed that shikimic
acid could be incorporated, although no better than several
other possible precursors. Subsequent work by Ratledge and co-
workers (55), and by Hudson and Bentley (56), has confirmed
that chorismate and isochorismate are the probable precursors of
salicylic acid and of [9] in *M. smegmatis*. The reaction may
well proceed through a simple cyclic transition state.

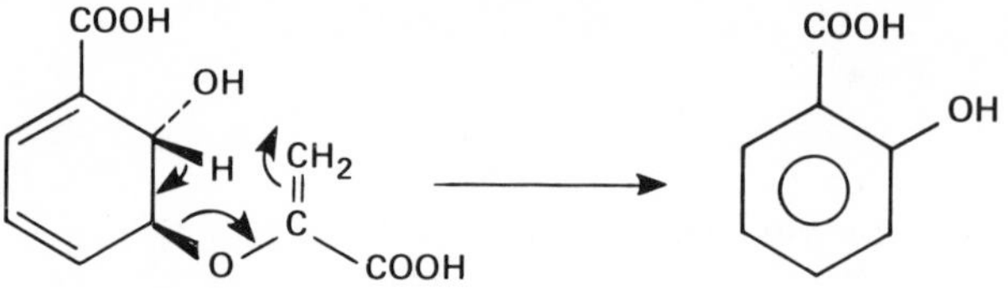

[1]

Again, in higher plants alternative biosynthetic pathways
seem to be favored, and salicylic acid is formed by β-oxidation
of *o*-coumaric acid, or by hydroxylation of benzoic acid (22, 36,
44, 57).

REFERENCES

1. B.D. Davis, J. Bact., 64, 729 (1952); Bull Soc. Chim. Biol.,
    36, 947 (1954); Symposium sur le Metabolisme Microbien, IIe
    Congre's International de Biochimie, Abstract page 32, Paris
    (1952).
2. T. Ito and J.B. Neilands, J. Amer. Chem. Soc. 80, 4645 (1958).
3. R.G. Wyn-Jones and J. Lascelles, Biochem. J. 103, 709 (1967).
4. I.G. Young, L.M. Jackman, and F. Gibson, Biochem. Biophys.
    Acta 177, 381 (1969); 148, 313 (1967).
5. I.G. Young and F. Gibson, Biochim. Biophys. Acta 177, 348
    (1969).
6. I.G. Young, T.J. Batterham, and F. Gibson, Biochim. Biophys.
    Acta 177, 389 (1969); 165 (1968).

7. R.M. DeMarinis, C.N. Filer, S.M. Waraszkiewicz, and
    G.A. Berchtold, J. Amer. Chem. Soc. 96, 1193 (1974).
8. I.G. Young, F. Gibson, and C.G. MacDonald, Biochim. Biophys.
    Acta 192, 62 (1969).
9. A.J. Pittard and B.J. Wallace, J. Bact. 91, 1494 (1966).
10. I.G. Young, G.B. Cox, and F. Gibson, Biochim. Biophys. Acta
    141, 319 (1967).
11. G.B. Cox and F. Gibson, J. Bact. 93, 502 (1967).
12. U. Weiss, unpublished observations.
13. W.J. Peters and R.A.J. Warren, Biochim. Biophys. Acta 165,
    225 (1968).
14. N. Brot and J. Goodwin, J. Biol. Chem. 243, 510 (1968).
15. C.C. Wang and A. Newton. J. Bact. 98, 1142 (1969).
16. G.F. Bryce and N. Brot, Arch. Biochem. Biophys. 142, 399
    (1971).
17. B.R. Byers and C.E. Lankford, Biochim. Biophys. Acta 165,
    563 (1968).
18. I.G. O'Brien and F. Gibson, Biochim. Biophys. Acta 215, 393
    (1970).
19. I.G. O'Brien, G.B. Cox, and F. Gibson, Biochim. Biophys. Acta
    201, 454, (1970).
20. J.R. Pollack and J.B. Neilands, Biochem. Biophys. Res. Commun.
    38, 989 (1970).
21. H. Diekmann and H. Zähner, Eur. J. Biochem. 3, 213 (1967).
22. K.B. Mullis, J. Pollack, and J.B. Neilands, Biochemistry 10,
    4894 (1971).
23. H. Grisebach, and K.O. Vollmer, Z. Naturforsch. 19b, 781
    (1964).
24. A.J. Pittard, F. Gibson, and C.H. Doy, Biochim. Biophys. Acta
    49, 485 (1961).
25. G. Terui, T. Enatsu, and S. Tabata, Hakko Kogaku Zasshi 39,
    724 (1961); Chem. Abstr. 59, 9111d (1962).
26. W.J. Kelleher and R.J. Krueger, Lloydia 32, 527 (1969).
27. F. Arcamone, E.B. Chain, A. Ferretti, and P. Pennella, Nature
    192, 552 (1961).
28. D. Grüger, D. Erge, and H.G. Floss, Z. Naturforsch. 20b, 856
    (1965).
29. B.R. Byers and C.E. Lankford, Bact. Proc. 90 (1966).
30. E.R. Catlin, C.H. Hassall, and B.C. Pratt, Biochim. Biophys.
    Acta 156, 109 (1968).
31. J.R. Dyer, J.R. Heding, and C.P. Schaffner, J. Org. Chem. 29,
    2802 (1964).
32. N. Brot, J. Goodwin, and H. Fales, Biochem. Biophys. Res.
    Commun. 25, 454 (1966).
33. I.G. O'Brien, G.B. Cox, and F. Gibson, Biochim. Biophys. Acta
    177, 321 (1969).

270 Shikimic Acid Pathway

34. W.J. Peters and R.A.J. Warren, J. Bact. <u>95</u>, 360 (1968).
35. J.L. Corbin and W.A. Bulen, Biochemistry <u>8</u>, 757 (1969).
36. S.Z. El-Basyouni, D. Chem, R.K. Ibrahim, A.C. Neish, and G.H.N. Towers, Phytochemistry <u>3</u>, 485 (1964).
37. R.K. Ibrahim and G.H.N. Towers, Nature <u>184</u>, 1803 (1959).
38. H.G. Floss, H. Guenther, D. Grüger, and D. Erge, Arch. Biochem. Biophys. <u>131</u>, 319 (1969).
39. P.V.S. Rao, K. Moore, and G.H.N. Towers, Biochem. Biophys. Res. Commun. <u>28</u>, 1008, (1967).
40. E. Tyler, K. Mothes, D. Grüger, and H.G. Floss, <u>Tetrahedron Lett</u>. 593 (1964).
41. P.V.S. Rao, N.S. Sreeleela, R. Premkumar, and C.S. Nardyanathan, Biochem. Biophys. Res. Commun. <u>31</u>, 193 (1968).
42. N.A. Sreeleela, P.V.S. Rao, R. Premkumar and C.S. Nardyanathan, J. Biol. Chem. <u>244</u>, 2293 (1969).
43. H. Kindl and G. Billek, Monatsh. <u>95</u>, 1044 (1964).
44. G.H.N. Towers, A. Tse, and W.S.G. Maass, Phytochemistry <u>5</u>, 677 (1966).
45. R. Hegnauer, Chemotaxonomie der Pflanzen Vol. 2, Birkhäuser Verlag, Basel and Stuttgart 1963, p. 257.
46. P.O. Larsen, Biochim. Biophys. Acta <u>93</u>, 200 (1964).
47. P.O. Larsen, Biochim. Biophys. Acta <u>115</u>, 529 (1966).
48. P.O. Larsen, Biochim. Biophys. <u>Acta</u> <u>141</u>, 27 (1967).
49. P.O. Larsen, D.K. Onderka, and H.G. Floss, Chem. <u>Commun</u>. 842 (1972).
50. C. Ratledge, in Biosynthesis of Aromatic Compounds, G. Billek, Ed., Pergamon, Oxford, 1966, p. 61.
51. G.A. Snow, Bact. Rev. <u>34</u>, 99 (1970).
52. C. Ratledge and F.G. Winder, Biochem. J. <u>101</u>, 274 (1966).
53. H. Taniuchi, M. Hatanaka, S. Kuno, O. Hayaishi, M. Nakajima, and N. Kurihara, J. Biol. Chem. <u>239</u>, 2204 (1964).
54. C. Ratledge. Biochim. Biophys. Acta <u>192</u>, 148 (1969).
55. C. Ratledge and M.J. Hall, FEBS Lett. <u>10</u>, 309 (1970); B.J. Marshall and C. Ratledge, Biochim. Biophys. Acta <u>230</u>, 643 (1971); <u>264</u>, 106 (1972).
56. A.T. Hudson and R. Bentley, Biochemistry <u>9</u>, 3984 (1970).
57. H.D. Klambt, Nature <u>196</u>, 491 (1962).

CHAPTER 13

Biosynthesis of Vitamin K

The Vitamins K are a series of related compounds which are isoprenylated derivatives of 2-methylnaphthoquinone (MK-0), produced by animals, plants, and bacteria; they play a role in blood clotting, photosynthesis, and electron transport, respectively, in these organisms.

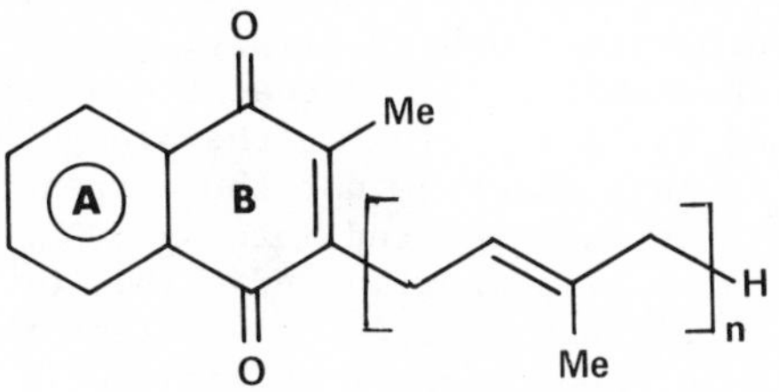

[1] n = 4, 6, 7, 8, and 9

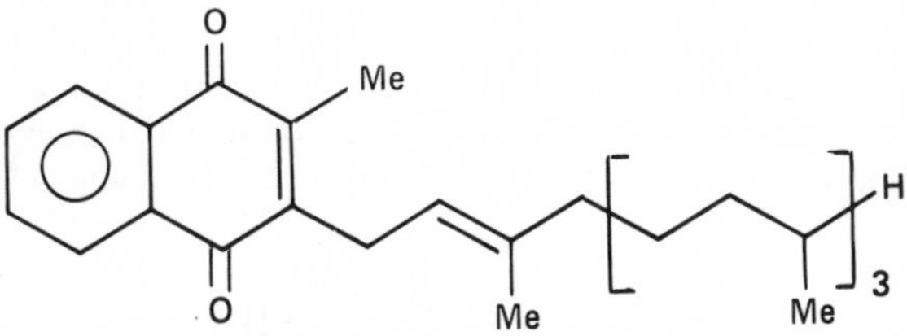

[2]

Vitamin K_1 or phylloquinone (abbreviated K) [2] occurs mainly in the green parts of plants, while K_2 or menadione (abbreviated MK) is the vitamin found in animals and microorganisms. The structure [1] indicates the general formula of the K_2 group of compounds; the side-chain may be hydrogenated in one

271

of its double bonds, as in one of the quinones, MK-9(H$_2$) of
Mycobacterium phlei, and a quinone MK-8(H$_2$) found in *Corynebacterium diphtheriae*. The position of the saturated bonds is not
known.

Most aspects of the biosynthetic pathway to Vitamin K$_2$ have
been clarified, at least in broad outline; the incorporation of
the methyl group of methionine into the ring methyl group has
been reported in *Mycobacterium phlei* with whole cells and cell-
free preparations (1, 5), in *E. coli* and *A. aerogenes* (2, 3),
and *Fusiformis nigrescens* (4). In 1964 Cox and Gibson showed
that U-^{14}C-shikimic acid was incorporated into Vitamin K2 by
E. coli (6), and degradation studies (2) showed that incorporation took place only into the benzene ring (A) of the naphthoquinone. Furthermore, the incorporation of label from shikimate
was lowered in the presence of unlabeled 3,4-dihydroxybenzalde-
hyde, but was not affected by 4-hydroxphenylpyruvic, protocatechuic, 2,3-dihydroxybenzoic, or phenylpyruvic acids, or catechol;
finally, a multiple auxotroph of *A. aerogenes* was found, which
was unable to convert 5-enolpyruvylshikimic acid 3-phosphate to
chorismic acid, and unable to form K$_2$ except in the presence of
added 3,4-dihydroxybenzaldehyde, implying that K$_2$ derives
directly from chorismic acid and that the aldehyde or a closely
related compound has a function in the biosynthetic scheme.
However, neither this aldehyde nor the corresponding acid (protocatechuic acid) are incorporated into K$_2$ by *E. coli* (7-9),
Mycobacterium phlei (7), *B. megaterium* (8), or several other
bacteria (26). Leistner *et al.* (8) verified the effect of the
aldehyde on the incorporation of shikimic acid in *E. coli* and
showed that it is incorporated as a C$_7$ fragment. The carboxyl
group of the uniformly labeled acid is incorporated specifically
into C-4 of the quinone ring by *M. phlei* (7,22, 27).

Although early experiments had shown some incorporation of
acetate into ring B both by *E. coli* (2) and *M. phlei* (7), the
origin of the remaining three carbon atoms remained a mystery
for some time. However, based upon the observation by Campbell
(14) that alanine and aspartic acid contributed specifically

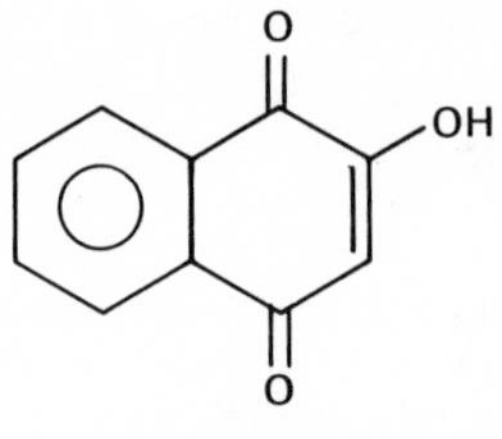

Lawsone

to the quinone ring of the naphthoquinone lawsone (see
Chapter 16), and another additional finding that U-^{14}C-glutamate
labeled carbons 1(4),2, and 3 of the menaquinones of *M. phlei*
and *E. coli* (20, 23), it was proposed that *ortho*-succinylbenzoic
acid [4] (from shikimic, and glutamic acids) might be an intermediate in the formation of menaquinones of *M. phlei*, *E. coli*,
and *A. aerogenes*.   Compound [4] has been shown to be a good
precursor (23), as has 2-oxoglutaric acid (24). That the C_3 unit
is specifically attached at C-2 of shikimate has been shown in
work by Scharf and Zenk (21), who have demonstrated that
2-[3]H-[14]COOH shikimic acid fed to *B. megaterium* is converted into
vitamin K_2 with total loss of the tritium label.   Taken together,
these findings establish a biosynthesis of the vitamins K in a
variety of organisms which must approximate the one shown in
the scheme.

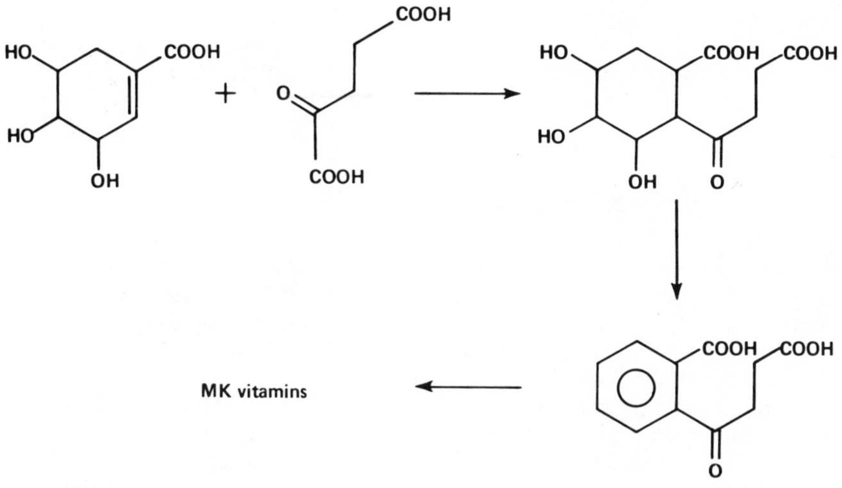

[4]

     Several derivatives of naphthalene have also been impli-
cated in vitamin K biosynthesis by various organisms.   Thus,
α-naphthol has been shown to be a precursor of K_2 in *B. megaterium*
(8) and *S. aureus* (10), carbon 1 being transformed into the keto
groups of the naphthoquinone molecule (but see below).   Taken
together with the fact that the bacterium *Fusiformis nigrescens*
is able to convert 1,4-naphthoquinone to its 2-methyl derivative,
and that 2-methylnaphthoquinone has been shown to be incorporated by *F. nigrescens* (4) and *S. aureus* (10), a tentative scheme

for K_2 biosynthesis via 1,4-naphthoquinone was proposed (11).
The C-methylation of desmethylmenaquinones has been studied by
Samuel and Azerad (12). While this scheme looks attractive,
it must be remembered that Ellis and Glover (9) have reported
that α-naphthol is not incorporated into K_2 by *E. coli*. In

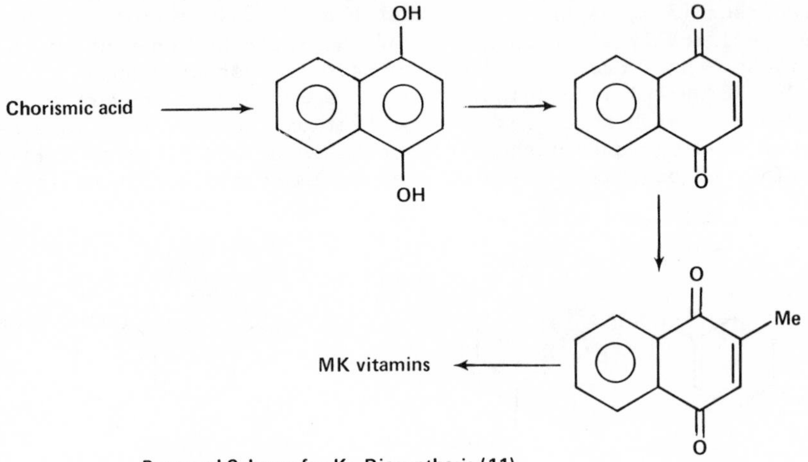

Proposed Scheme for K_2 Biosynthesis (11)

addition, Brown and co-workers (13) have repudiated the earlier
claim of its incorporation by *B. megaterium*. and have demonstrated the nonincorporation of the naphthol by many species of
bacteria. Similar negative findings have been published by
Bentley and co-workers (22).

The growth-promoting effect of 3,4-dihydroxybenzaldehyde
and its influence upon the incorporation of shikimic acid into
K_2 are still unexplained observations, and the involvement of
chorismic acid implied by early experiments (6) seems unlikely
in view of experiments on juglone biosynthesis in which the
acid was readily incorporated into aromatic amino acids, but
not into the naphthoquinone (25).

The fact that K_2 is derived directly from shikimic acid via
a pathway which in all probability does not involve compounds
formed by metabolism of the amino acids, explains why it is a
vitamin and ubiquinone is not; ubiquinone is biosynthesized
from POB, which in animal tissue is formed from tyrosine,
whereas the precursors of K_2 cannot be made by animals.

The biosynthesis of the related plant quinone phylloquinone (Vitamin K$_1$) [2] has been investigated by Goodwin and co-workers (15-18), who observed the incorporation of shikimic acid into ring A, mevalonic acid into the isoprene side-chain, and methionine into the 2-methyl group. However, dilution and labeling experiments (19) with 3,4-dihydroxybenzaldehyde and protocatechuic acid indicated that these compounds are probably not involved in the biosynthesis of K$_1$ in maize shoots (*Zea mays*). α-Naphthol, naphthoquinone, and 2-methyl-1, 4-naphthoquinone (menadione) are inactive as precursors, and *ortho*-succinylbenzoic acid [4] has been shown to be well incorporated (28). A biosynthesis of [2] very similar to that of the bacterial menaquinones and some naphthoquinones (Chapter 16) is clearly indicated.

REFERENCES

1. R. Azerad, R. Bleiler-Hill, and E. Lederer, Biochem. Biophys. Res. Commun. 194, 19 (1965).
2. G.B. Cox and F. Gibson, Biochem. J. 100, 1 (1966).
3. L.M. Jackman, I.G. O'Brien, G.B. Cox, and F. Gibson, Biochim. Biophys. Acta 141, 1 (1967).
4. C. Martius and W. Leuzinger, Biochem. Z. 340, 304 (1964).
5. R. Azerad, R. Bleiler-Hill, and E. Lederer in Biosynthesis of Aromatic Compounds, G. Billek, Ed., Pergamon, Oxford, 1966, p. 75.
6. G.B. Cox and F. Gibson, Biochim. Biophys. Acta 93, 204 (1964).
7. I.M. Campbell, C.J. Coscia, M. Kelsey, and R. Bentley, Biochem. Biophys. Res. Commun. 28, 25 (1967).
8. E. Leistner, J.H. Schmitt, and M.H. Zenk, Biochem. Biophys. Res. Commun. 28, 845 (1967).
9. J.R.S. Ellis and J. Glover, Biochem. J. 110, 22P (1968).
10. R.K. Hammon and D.C. White, J. Bact. 100, 573 (1969).
11. M.H. Zenk and E. Leistner, Lloyida 31, 275 (1968).
12. O. Samuel and R. Azerad, FEBS Lett. 2, 336 (1969).
13. B.S. Brown, G.R. Whistance, and D.R. Threlfall, FEBS Lett. 1, 323 (1968).
14. I.M. Campbell, Tetrahedron Lett. 4777 (1969).
15. G.R. Whistance, D.R. Threlfall, and T.W. Goodwin, Biochem. Biophys. Res. Commun. 23, 849 (1966).
16. D.R. Threlfall, W.T. Griffiths, and T.W. Goodwin, Biochem. J. 103, 831 (1967).
17. D.R. Threlfall, G.R. Whistance, and T.W. Goodwin, Biochem. J. 106, 107 (1968).

18. G.R. Whistance, D.R. Threlfall, and T.W. Goodwin, Biochem. J. 105, 145 (1967).

19. G.R. Whistance and D.R. Threlfall, Biochem. J. 109, 577 (1968).

20. D.J. Robins, I.M. Campbell, and R. Bentley, Biochem. Biophys. Res. Commun. 39, 1081 (1970).

21. K.H. Scharf and M.H. Zenk, unpublished results, quoted by M.H. Zenk in Pharmacognosy and Phytochemistry, H. Wagner and L. Hörhammer, Eds., Springer-Verlag, Berlin, 1971, p. 322.

22. I.M. Campbell, D.J. Robins, M. Kelsey, and R. Bentley, Biochemistry 10, 3069 (1971).

23. P. Dansette and R. Azerad, Biochem. Biophys. Res. Commun. 40, 1090 (1970). See also M.M. Leduc, P.M. Dansette, and R.G. Azerad, Eur. J. Biochem. 15, 428 (1970).

24. D.J. Robins and R. Bentley, Chem. Commun. 232 (1972).

25. E. Leistner and M.H. Zenk, Z. Naturforsch 23b, 259 (1968).

26. M. Guerin, M.M. Leduc, and R.G. Azerad, Eur. J. Biochem. 15, 421 (1970).

27. R.M. Baldwin, C.D. Snyder, and H. Rapoport, J. Amer. Chem. Soc. 95, 276 (1973); Biochemistry 13, 1523 (1974).

28. G. Thomas and D.R. Threlfall, Phytochemistry 13, 807 (1974).

CHAPTER 14

Generality of the Shikimate Pathway

In the preceding chapters of Part 2, the pathway leading from glucose via shikimic acid to aromatic metabolites has been described and a detailed account of the intermediates involved in it has been given. Most of the information contained in these earlier pages is based upon work with a limited number of microorganisms widely used for such research, especially *Escherichia coli*, *Aerobacter aerogenes*, and *Neurospora crassa*. While results obtained with other organisms have been mentioned in a number of instances, it still seems necessary to question whether the knowledge gained from work with these three species can also be applied to other organisms which are able to produce aromatic metabolites from glucose. In particular, three aspects of this problem of general validity have to be explored, and data pertinent to them should be assembled: (1) What is the evidence for the general functioning of the shikimate pathway? (2) What variants occur, if any? (3) Are there any known instances where metabolites usually formed via the shikimate pathway are biosynthesized through other, unrelated routes?

(1) General validity of the shikimate pathway. Suggestive evidence for the general occurrence of the shikimate pathway outside the range of organisms in which it was first established has already been given in a number of places. Many of the intermediates (in addition to the long-known ones such as shikimic and anthranilic acids) have been detected in a wide variety of higher plants and microorganisms; their biosynthetic utilization has been proven repeatedly, and the presence of the required enzymatic apparatus has been demonstrated in many instances. It is not necessary to list those cases here, but some particularly convincing pieces of evidence should be quoted. For much additional information on the functioning of the shikimate pathway in higher plants, the review by Yoshida (1) is valuable.

Proof for the occurrence of the shikimate pathway in higher plants was first obtained by Brown and Neish (2), who found incorporation of generally labeled shikimic acid into lignin in wheat and maple; this observation is particularly significant in view of the enormous amounts of this material being produced by plants.   More detailed evidence for the nature of this biosynthesis was secured from incorporation studies with specifically labeled precursors, and subsequent proof by degradative methods that label was localized more or less exclusively at the positions predicted from the work with *E. coli* and *A. aerogenes*. Thus, Eberhardt and Schubert (3) showed that shikimic acid biosynthesized by *E. coli* and hence labeled in positions 2 and 6 (see Chapter 2) was incorporated by sugar cane (*Saccharum officinarum*) into lignin with the activity in the aromatic rings restricted to the corresponding carbons 2 and 6; this result proved incorporation of the intact ring of the precursor.   The same labeling pattern was found for lignin biosynthesized by spruce (*Picea abies*) from glucose labeled in position 1 or 6 (4), and for methyl *p*-methoxycinnamate produced from glucose $1-^{14}C$ or $6-^{14}C$ by the basidiomycete *Lentinus lepideus* (5).

Among more recent investigation, the work of Rinehart and co-workers (6) on the biosynthesis of chloramphenicol (chloromycetin, $p-O_2N-C_6H_4-CHOH-CH(CH_2OH)-NH-CO-CHCl_2$) by *Streptomyces venezuelae* deserves mention.   The biosynthesis of this antibiotic proceeds in an unusual manner in the stages beyond chorismic acid (see below), but the earlier ones conform to the customary pattern: with $[6-^{14}C]$-glucose as precursor, labeling of the *p*-nitrophenyl ring is almost entirely confined to C-2 and C-6, the expected positions.

Finally, it has been established that the shikimate pathway in a higher plant, *Reseda lutea* (Resedaceae) agrees with the one in *E. coli* also with respect to details of stereochemistry. Olesen Larsen, Onderka, and Floss showed (7) that this plant incorporates shikimic acid, labeled stereospecifically with tritium on C-6, into phenylalanine and tyrosine with retention of label from the *pro-6S*-position, and with loss from the *pro-6R*-position.   The same stereospecificity had previously been found (8) for the conversion of enolpyruvylshikimic acid 3-phosphate into chorismic acid in *E. coli* and *A. aerogenes*. The findings with *Reseda* thus prove complete agreement of this part of the pathway in a higher plant and in the two bacteria.

Interestingly, Leistner and Zenk (9) observed that chorismic acid is efficiently incorporated into phenylalanine and tyrosine by the walnut tree, *Juglans regia* (Juglandaceae). The nutritional unavailability of this acid, which had been observed with bacteria (10), is thus not general.

(2) <u>Variants of the shikimate pathway</u>. Modifications of
the usual sequence have indeed been observed in a number of
instances. The most striking of these was discovered (11) very
recently in the biosynthesis of tyrosine from prephenic acid.
     It has long been known that tyrosine can be formed in two
different ways, both of them parts of the shikimate pathway:
by hydroxylation of phenylalanine (cf. Section 7-1), or from
prephenic acid by the separate branch of the sequence that
involves aromatization of the cyclohexadienol ring to p-hydroxy-
phenyl (presumably via the cyclohexadienone), and conversion of
the pyruvyl side-chain to the α-amino acid structure by transamination (cf. Chapter 7, pages 153 and 170. In the
organisms in which this latter sequence was originally established, the two steps, aromatization and transamination,
follow each other in this order, p-hydroxyphenylpyruvic acid
appearing as an intermediate. It has, however, now been found
(11) that blue-green bacteria, for example *Agmenellum
quadruplicatum*, produce tyrosine from prephenic acid by a
variant of this process where the order of the individual steps
is reversed. Transamination occurs first, yielding the new
intermediate pre-tyrosine, the amino acid analog of prephenic
acid. This compound then gives tyrosine directly through
transformation of its cyclohexadienol moiety into the
p-hydroxyphenyl ring, undoubtedly by a reaction quite analogous
to the formation of p-hydroxyphenylpyruvic acid from prephenic
acid. (Pre-tyrosine has already been discussed in Section 7-2,
page 171.
     Tyrosine can thus be formed in three different ways:
through two variants of the separate pathway branching off from
prephenic acid, or through hydroxylation of the preformed
aromatic ring of phenylalanine. This existence of alternatives
within the prearomatic part of the shikimate pathway seems
unusual; no second instance has come to our attention. In
contrast, the other dichotomy, biosynthesis of a given metabolite either from an alicyclic intermediate or by transformation
of a precursor already containing an aromatic ring, is fairly
common. This is true, for example, of the various hydroxylated
benzoic acids, such as p-hydroxybenzoic, protocatechuic, gallic,
salicylic, and 2,3-dihydroxybenzoic acid. For each of these,
formation through aromatization of a nonaromatic precursor has
been observed: 3-dehydroshikimic acid in the case of protocatechuic and gallic acids (cf. Section 4-4), chorismic acid
in that of p-hydroxybenzoic acid (cf. Chapter 11); isochorismic
acid for salicylic and 2,3-dihydroxybenzoic acids (cf. Sections
12-6 and 12-4, respectively). In every case, however, alternative modes exist that yield the particular acid through

shortening of the side-chain of a C_6-C_3 compound (usually a cinnamic acid or phenylalanine), through introduction of additional hydroxyls into a less highly hydroxylated benzoic acid, through removal of OH from a more strongly substituted one (for formation of C_6-C_1 compounds in general, see Chapter 16, for removal of phenolic hydroxyl Chapter 7, or through some other modification of a preformed aromatic structure.

A few examples for these transformations of aromatic rings may be quoted; no complete coverage is intended. Much information on these questions is contained in ref. 1.

In higher plants, p-hydroxybenzoic acid is formed mainly from p-coumaric acid by degradation of the side-chain; this process was studied in detail by Zenk (12) in *Catalpa* (Bignoniaceae). Direct hydroxylation of benzoic acid was likewise observed. Conversion of caffeic and sinapic acids (3,4-dihydroxy- and 3,5-dimethoxy-4-hydroxycinnamic acids, respectively) into p-hydroxybenzoic acid was found to occur in wheat, *Triticum*, but it constitutes only a minor reaction; it may well proceed by dehydroxylation to p-coumaric acid (13). Dehydroxylation of protocatechuic acid in the rat yields p-hydroxybenzoic acid together with larger amounts of the *meta*-isomer (14); reactions of this type are actually carried out by the intestinal microflora, (see Chapter 16).

Protocatechuic acid is usually formed from 3-dehydroshikimic acid, but biosynthesis through hydroxylation of p-hydroxybenzoic acid has been reported (15). Alternatively, the acid can also be formed from caffeic acid, and its 3-methyl ether vanillic acid is produced from sinapic acid (13), again to a minor extent. Hydroxylation of protocatechuic acid to gallic acid was found to proceed with high yields in *Geranium* (Geraniaceae) (13), but *Phycomyces* was not able to perform this reaction (16). Conversion of 3,4,5-trihydroxycinnamic acid into gallic acid seems likewise to occur (17).

Salicylic and 2,3-dihydroxybenzoic (o-protocatechuic acid can both be formed through the isochorismic acid pathway, either from this acid itself or from 2,3-dihydroxy-2,3-dihydrobenzoic acid (cf. Chapter 12). Both of these acids also yield salicylic acid *in vitro* (18). Its formation from isochorismic acid by cellfree extracts of *Mycobacterium smegmatis* has been studied in detail (19). On the other hand, salicylic acid is formed from o-coumaric (2-hydroxycinnamic) acid by shortening of the side-chain; this has been observed, for example, in wintergreen (*Gaultheria procumbens*, Ericaceae) (20). Alternatively , *ortho* hydroxylation of benzoic acid takes place in certain higher plants (21) such as sunflower (*Helianthus annus,* Compositae), potato (*Solanum tuberosum,* Solanaceae), and pea (*Pisum sativum,* Leguminosae).

Finally, salicylic acid can be formed through oxidation
of salicylaldehyde under the influence of an NAD-dependent
enzyme. This reaction has been observed (21A) in *Pseudomonas*
spp. growing on naphthalene, from which the salicylaldehyde is
formed via *cis*-1,2-dihydroxy-dihydronaphthalene, 1,2-dihydroxy-
naphthalene, and *cis*-*o*-hydroxybenzalpyruvic acid (cf. Chapter 8,
page 211.

In the case of 2,3-dihydroxybenzoic acid, biosynthesis by
hydroxylation of salicylic acid has been observed in leaf discs
of several plants (22), as has its formation from benzoic and
cinnamic acids (13, 23). In certain microorganisms, 2,3-dihy-
droxybenzoic acid is derived from tryptophan. This has been
demonstrated in *Claviceps paspali* (24, 25) and in *Aspergillus
niger* (26); the formation of the C_7 acid seems to proceed by
way of kynurenine and 3-hydroxyanthranilic acid.

Another example of duality of pathways is found in the
formation of phenylpyruvic acid. This intermediate in the
biosynthesis of phenylalanine is normally produced by the
well-known aromatization of prephenic acid (see Section 7-2),
but certain anaerobic microorganisms can synthesize it from
phenylacetic acid and CO_2 under the influence of reduced
ferredoxin (27). This variant has been discussed in Section 7-2,
page 172.

The transformations of chorismic acid into anthranilic
and *p*-aminobenzoic acids (Sections 9-1 and Chapter 10,
respectively) constitute the normal mode of attachment of
nitrogen to positions 2 and 4 of an aromatic ring. However,
alternatives to those reactions seem to occur in the biosynthesis of certain microbial metabolites.

The variant in the introduction of nitrogen into position
4 occurs in the biosynthesis of chloramphenicol; the evidence
obtained prior to 1968 on the process has been summarized by
Vining and co-workers (28); see also ref. 6, above. Formation
of the entire carbon skeleton of the antibiotic through the
shikimate pathway is solidly established (l.c.); the substance
is undoubtedly a typical C_6-C_3 compound, and can hence not be
biosynthesized by way of *p*-aminobenzoic acid. Neither this
acid, nor phenylalanine or tyrosine are incorporated intact,
although the latter two compounds label the amino acids in the
proteins of the chloramphenicol-producing streptomycete. The
antibiotic is thus apparently formed through a separate branch
of the shikimate pathway which starts at a C_6-C_3 compound of
the common sequence. It has been made probable quite recently
that this intermediate is chorismic acid (29). Evidence for
the nature of this branch comes from the high and specific

incorporation of p-aminophenylalanine, *threo-p*-aminophenylserine, and N-dichloroacetyl-*D-threo-p*-aminophenylserinol (28). That these compounds are actual intermediates is made probable by the fact that the first and third one occur in small amounts in chloramphenicol-producing cultures of the mold; cf. also Chapter 16.

The pathway yielding the antibiotic thus seems to convert chorismic acid to p-aminophenylalanine. This process is still unexplored. It would presumably involve introduction of the amino group prior to aromatization in such a way that it would not be removed during formation of the benzene ring; the possibility of reactions of this type is suggested by the occurrence of the cyclohexadienylamine ring of the mold metabolite stravidin (29) (see Chapter 8, page 187). The next steps in the sequence would then be introduction of the hydroxyl in position C-2' of the side-chain, reduction of the carboxyl to $-CH_2OH$, acylation of the amino group of the side-chain, and finally oxidation of the amino group on the ring to the nitro group. These reactions need not be further discussed here.

It is interesting that p-aminophenylalanine has been found in the seeds of *Vigna vexillata* (Leguminosae) (30), and that the N-α-methyl derivative of 4-dimethylaminophenylalanine occurs in the peptide antibiotic ostreogrycin B (31).

Introduction of nitrogen into position 2 without intermediacy of anthranilic acid constitutes a variant that seems to occur in the biosynthesis of the microbial phenazine pigments. More than 30 such compounds are known, of which pyocyanine (the betaine of 1-hydroxy-N(5)-methylphenazine), its precursor (32) phenazine 1-carboxylic acid, and iodinine (the di-N-oxide of 1,6-dihydroxyphenazine) are the most carefully investigated ones. Their biosynthesis is discussed elsewhere (see Chapter 16); it presents a difficult problem which has not yet been completely solved. It is clear, though, that the compounds (pyocyanine, iodinine, etc.) are derived from shikimic acid (33), which was shown (34) to supply all the carbon atoms of the phenazine skeleton. This ring-system must thus be built from two shikimate-derived units joined together by the two nitrogens. Recent work with [6-^{14}C]-shikimic acid has established (35) that the nitrogen atoms become attached to positions 6 of the shikimic acid structures in the biosynthesis of phenazine 1-carboxylic acid, and consequently also in that of pyocyanine (the situation is more involved in the case of iodinine). Several of the phenazines carry carboxyl or hydroxyl groups in position 1, or in 1 and 6 (cf. phenazine 1-carboxylate, pyocyanine, and iodinine); the hydroxyl of pyocyanine is known (32) to replace a carboxyl.

On the basis of these structural arguments, and of the experiments with specifically labeled shikimic acid (35a,b), anthranilic acid would qualify nicely as a precursor of the phenazine pigments. It seems, however, that it is not involved. It is poorly incorporated (see ref. 35b and references therein); it does not dilute the activity of pyocyanine biosynthesized on labeled 2-ketogluconic acid as carbon source (34); and a mutant of *Pseudomonas aeruginosa* blocked at the level of anthranilate synthetase is still capable of normal production of pyocyanine (36). It is therefore generally concluded that the phenazine pigments are formed through a separate pathway branching off from one of the alicyclic intermediates and not involving anthranilic acid (cf., e.g.,ref. 35b). The branch-point metabolite is probably chorismic acid (36, 37).

These examples strongly suggest the existence of processes, other than the ones discussed in Section 4-4 and Chapters 7 through 13, for derivation of aromatic metabolites from the alicyclic intermediates. The existence of such additional pathways was to be expected; the reactions described in the earlier parts of this volume are far from exhausting the potential for chemical transformations of compounds as reactive and versatile as these intermediates. Many other reactions can be anticipated, and some have been studied in the laboratory; see, as one example, the work of Plieninger and Schneider (38) on the conjugate addition of ammonia to the double bond of shikimic acid. Many other additions of this kind could occur with shikimic and especially with 3-dehydroshikimic acid. Other possibilities include allylic shifts of double bonds (actually occurring in the transformation of chorismic into isochorismic acid, but not necessarily restricted to such dienic systems), Diels-Alder reactions of chorismic or isochorismic acids, and many others. It would be surprising if none of these feasible and smooth reactions were utilized in biosynthesis. In many cases, the resulting pathways would lead away from aromaticity; the findings of de Rosa, Bu'Lock, and their co-workers (39) on the derivation of the cyclohexyl ring of the ω-cyclohexyl fatty acids from shikimic acid in a thermophilic bacterium (see Section 5-1, page 113) shows that such transformations do actually exist. The search for additional examples should be rewarding.

(3) Formation of usually shikimate-derived metabolites by alternative pathways. The possibility of entirely different alternatives to the shikimate pathway has to be considered seriously, since well-established examples of such duality exist; for instance, both lysine (40a) and nicotinic acid (40b) can be formed through two radically different sequences.

Existence of such alternative modes of the biosynthesis of phenylalanine and tyrosine has indeed been postulated a few times. For instance, findings on the biosynthesis of these two amino acids by *Trichophyton rubrum* have led Zussman, Vicher, and Lyon to doubt the operation of the shikimate pathway in this fungus. These authors observed (41a) that phenylalanine and tyrosine are formed from glucose in *T. rubrum*. However, the activities of phenylalanine biosynthesized from either [1-^{14}C] or [6-^{14}C] glucose were almost equal, while it would be expected from the work of Sprinson and co-workers (see Chapter 2, page 55) that on biosynthesis via the normal shikimate pathway, glucose labeled at C-6 should have yielded a more strongly active amino acid on account of the preferential labeling of C-2 of shikimic acid from C-6 of glucose. Also, neither shikimic acid, nor shikimic acid 3-phosphate, quinic, or protocatechuic acid could be detected in the fungus by sensitive methods (41b). Furthermore (41c), shikimic acid was poorly utilized for growth when present as sole carbon source, although it was shown to penetrate into the cells; in presence of glucose, neither it nor quinic acid was taken up from the medium; and isotope dilution studies proved that no shikimic acid was formed by growing *T. rubrum*. These results seem to indicate that the biosynthesis of phenylalanine in this fungus does not proceed via the shikimate pathway, a conclusion which seems further supported by the finding (41d) that label from [1-^{14}C] acetate was efficiently incorporated into phenylalanine. The nature of this alternative pathway remains obscure and puzzling.

Similar conclusions have been reached from studies with quinic acid in intact higher plants. Here it was found, for example, that phenylalanine and tyrosine incorporate more label from quinic than from shikimic acid (42); in the oak, *Quercus pedunculata*, quinic acid yields both phenylalanine and shikimic acid, but throughout the duration of the experiment, the former is consistently more strongly labeled than its supposed precursor (43); the ratio of activities of phenylalanine and tyrosine is much higher after the administration of generally labeled quinic acid than after that of similarly labeled shikimic acid (44). All these results are taken to indicate a pathway via quinic acid to phenylalanine which does not proceed by way of shikimic acid. However, many complications may be encountered in such work with intact plants, and the problem remains sub judice.

REFERENCES

1. S. Yoshida, Ann. Rev. Plant Physiol. 20, 41 (1969).
2. S.A. Brown and A.C. Neish, Nature 175, 688 (1955).

3. G. Eberhardt and W.J. Schubert, J. Amer. Chem. Soc. 78,
    2835 (1956).
4. S.N. Acerbo, W.J. Schubert, and F.F. Nord, J. Amer. Chem.
    Soc. 82, 735 (1960); K. Kratzl and H. Faigle, Z. Naturforsch.
    B 15, 4 (1960).
5. H. Shimazono, W.J. Schubert, and F.F. Nord, J. Amer. Chem.
    Soc. 80, 1992 (1958).
6. W.P. O'Neill, R.F. Nystrom, K.L. Rinehart, Jr., and
    D. Gottlieb, Biochemistry 12, 4775 (1973).
7. P. Olesen Larsen, D.K. Onderka, and H.G. Floss, J. Chem.
    Soc., Chem. Commun. 842 (1972).
8. R.K. Hill and G.R. Newkome, J. Amer. Chem. Soc. 91, 5893
    (1969); D.K. Onderka and H.G. Floss, ibid. 5894; H.G. Floss,
    D.K. Onderka, and M. Caroll, J. Biol. Chem. 247, 736 (1972).
9. E. Leistner and M.H. Zenk, Z. Naturforsch. B 15, 259 (1968).
10. F. Gibson and J. Pittard, Bact. Rev. 32, 465 (1968).
11. S.L. Stenmark, D.L. Pierson, R.A. Jensen, and G.L. Glover,
    Nature 247, 290 (1974).
12. M.H. Zenk and G. Müller, Z. Naturforsch. B 19, 398 (1964);
    M.H. Zenk, in Biosynthesis ot Aromatic Compounds, G. Billek,
    Ed., Pergamon, 1966, p. 45.
13. S.Z. El-Basyouni, D. Chen, R.K. Ibrahim, A.C. Neish, and
    G.H.N. Towers, Phytochemistry 3, 485 (1964).
14. J.C. Dacre and R.T. Williams, Biochem. J. 84, 81P (1962);
    J. Pharm. Pharmacol. 20, 610 (1968).
15. W.C. Evans, Biochem. J. 41, 373 (1947).
16. E. Haslam, R.D. Haworth, and P.F. Knowles, J. Chem. Soc.
    1854 (1961).
17. M.H. Zenk, Z. Naturforsch. B 19, 83 (1964).
18. I.G. Young, L.M. Jackman, and F. Gibson, Biochim. Biophys.
    Acta 177, 381 (1969); I.G. Young, T.J. Batterham, and
    F. Gibson, ibid. 389.
19. B.J. Marshall and C. Ratledge, Biochim. Biophys. Acta 230,
    643 (1971).
20. H. Grisebach and K.-O. Vollmer, Z. Naturforsch. B 18, 753
    (1963); 19, 781 (1964).
21. H.D. Klaemt, Nature 196, 491 (1962).
21A. J.I. Davis and W.C. Evans, Biochem. J. 91, 251 (1964).
22. R.K. Ibrahim and G.H.N. Towers, Nature 184, 1803 (1959).
23. H. Kindl and G. Billek, Monatsh. Chem. 95, 1044 (1964).
24. F. Arcamone, E.B. Chain, A. Ferretti, and P. Pennella,
    Nature 192, 553 (1961); F. Arcamone, E.B. Chain,
    A. Ferretti, A. Minghetti, P. Pennella, and A. Tonolo,
    Biochim. Biophys. Acta 57, 174 (1962).

25. V.E. Tyler, Jr., K. Mothes, D. Grüger, and H.-G. Floss,
     Tetrahedron Lett. 593 (1964); D. Grüger, D. Erge, and
     H.-G. Floss, Z. Naturforsch. B 20, 856 (1965).

26. P.V. Subba Rao, K. Moore, and G.H.N. Towers, Biochem.
     Biophys. Res. Commun. 28, 1008 (1967).

27. M.J. Allison, Biochem. Biophys. Res. Commun. 18, 30 (1965);
     M.J. Allison and I.M. Robinson, J. Bact. 93, 1269 (1967);
     U. Gehring and D.I. Arnon, J. Biol. Chem. 246, 4518 (1971).

28. L.C. Vining, V.S. Malik, and D.W.S. Westlake, Lloydia 31,
     355 (1968).

29. A. Emes, H.-G. Floss, D.A. Lowe, D.W.S. Westlake, and
     L.C. Vining, Can. J. Microbiol. 20, 347 (1974);
     K.H. Baggaley, B. Blessington, C.P. Falshaw, W.D. Ollis,
     L. Chaiet, and F.J. Wolf, Chem. Commun. 101 (1969).

30. G.A. Dardenne, M. Marlier, and J. Casimir, Phytochemistry
     11, 2567 (1972).

31. F.W. Eastwood, B.K. Snell, and Sir A. Todd, J. Chem. Soc.
     2286 (1960).

32. M.E. Flood, R.B. Herbert, and F.G. Holliman, J. Chem. Soc.
     Perkin I, 622 (1972).

33. Cf., inter alia, (a) R.C. Millican, Biochim. Biophys. Acta
     57, 407 (1962); (b) M. Podojil and N.N. Gerber, Biochemistry
     6, 2701 (1967); 9, 4616 (1970).

34. W.M. Ingledew and J.J.R. Campbell, Can. J. Microbiol. 15,
     535 (1969).

35. a. U. Hollstein and D.A. McCamey, J. Org. Chem. 38, 3415
         (1973);
     b. U. Hollstein and L.G. Marshall, ibid. 37, 3510 (1972).

36. D.H. Calhoun, M. Carson, and R.A. Jensen, J. Gen. Microbiol.
     72, 581 (1972).

37. R.P. Langley, J.E. Halliwell, J.J.R. Campbell, and
     W.M. Ingledew, Can. J. Microbiol. 18, 1357 (1972).

38. H. Plieninger and K. Schneider, Chem. Ber. 92, 1587 (1959).

39. M. de Rosa, A. Gambacorta, L. Minale, and J.D. Bu'Lock,
     Biochem. J. 128, 751 (1972).

40. a. J.R. Mattoon, in Biogenesis of Natural Compounds, 2nd
         ed., P. Bernfeld, Ed., Pergamon, 1967, p. 27;
     b. E. Leete, ibid. p. 692.

41. R.A. Zussman, E.E. Vicher, and I. Lyon, (a) Mycopathol.
     Mycologia Appl. 32, 194 (1967); (b) ibid. 37, 104 (1969);
     (c) ibid. 42, 1 (1970); (d) ibid. 37, 86 (1969).

42. L.H. Weinstein, C.A. Porter, and H.J. Laurencot, Contrib.
     Boyce Thompson Inst. 20, 121 (1959).

43. A. Boudet, C. R. Acad. Sci., Paris, Ser. D. 269, 1966
     (1969).

44. R. Rohringer, A. Fuchs, J. Lunderstädt, and D.J. Samborski,
     Can. J. Bot. 45, 863 (1967).

CHAPTER 15
Regulation of the Shikimate Pathway

Some general considerations of the need for regulation of branched biosynthetic pathways, and of the various means by which this control is achieved, have been given in the Introduction.

The necessity of such regulation is nowhere more evident than in the shikimate pathway. Here, a continuous chain of seven common intermediates exists, branching out to approximately the same number of primary aromatic metabolites (which range from major building blocks of cell constituents to trace factors) and involving a further important subdivision in one of the branches. Such a sequence is in obvious need of delicately balanced regulatory mechanisms, if the cell is to be supplied at all times with precisely the required amounts of aromatic metabolites, and yet wasteful overproduction of any of them is to be avoided. The existence and perfection of these controls are indicated, for example, by the observation (1) that wild-type *Escherichia coli* excretes only negligible amounts of amino acids and most other metabolites into the growth medium.

It has been pointed out in Chapter I that these controls usually operate at the level of the first specific reaction of a given pathway, or branch of a multiple pathway, and that two main, by no means mutually exclusive, regulatory mechanisms operate: influence by the end-product of the pathway or branch upon the activity of the enzyme that catalyzes the first step, or influence upon its formation; that is, feedback *inhibition* or *repression* by the end-product. (Other possibilities exist, and will be discussed later on: action not by an end-product but by a branch-point intermediate in the chain; activation by the end-product of an alternative chain, and control by an inorganic ion, rather than the organic end-product of a biosynthesis.)

It may not be too surprising that in a process as complex as the biosynthesis of aromatic compounds via shikimic acid, all these possiblities should be found to occur. What does at first

287

glance seem surprising, however, is the observation that the
regulation of the enzymes promoting the same chemical step should
be achieved in such strikingly diverse ways by different organisms. However, in view of the need of these organisms to adapt
to widely differing environments, and of the resulting genetic
and metabolic diversity, the exploitation of a substantial
range of possible steering mechanisms may become understandable
(cf. ref. 2).

This diversity of regulatory modes makes it quite impossible
to present a comprehensive discussion within the limits of the
present volume, especially since they, and the underlying genetic
problems, have been the object of intense study in recent years;
only an abbreviated account of some of the salient features, and
of some unusual findings, can be given. For obvious methodolog-
ical reasons, these studies have been mostly performed with
microorganisms. Fortunately, many excellent reviews exist, of
which a number may be quoted. Regulations of branched pathways
in bacteria have been reviewed by Datta (2), those observed
in the metabolism (mostly the biosynthesis) of amino acids by
Umbarger (3) and by Truffa-Bachi and Cohen (4). Multienzyme
complexes have been treated by Ginsburg and Stadtman (5), pertinent questions of genetics by Martin (6), and by Epstein and
Beckwith (7). The review by Calvo and Fink (7A) on regulations
of biosynthetic pathways in bacteria and fungi became available
after completion of this chapter. The mechanisms controlling
the shikimate pathway are discussed specifically in reviews
and articles by Gibson and Pittard (8), and by Doy (9). Gollub,
Zalkin, and Sprinson (10) treat the regulations in *Salmonella*,
Pittard and Wallace (11) and Brown (12) those in *E. coli*, and
Metzenberg (12A) the ones in *Neurospora*. Of those, refs. 8 and
9 seem particularly pertinent to the question of the multiplicity
of steering mechanisms which have arisen in different microorganisms for the regulation of identical biosynthetic steps
in the shikimate pathway.

This multiplicity can perhaps be exemplified best in the
first specific reaction of the pathway, the formation of
3-deoxy-D-*arabino*-heptulosonate 7-phosphate (DAHP) from enol-
pyruvate phosphate and erythrose 4-phosphate under the influence
of the enzyme DAHP synthetase. In *E. coli* (and also (10) in
Salmonella typhimurium) this reaction is catalyzed by three
distinct DAHP synthetase isoenzymes. Those isoenzymes are inhibited by phenylalanine, tyrosine, and tryptophan, respectively;
inhibition by first two of these is almost complete (95%), while
that by tryptophan amounts maximally to about 60% and had
initially escaped detection (see ref. 8b and literature quoted
there). In addition, the isoenzymes inhibited by tyrosine and

tryptophan are also repressed by these amino acids; repression of
the phenylalanine-sensitive isoenzyme seems less pronounced and
presents complications (8b).   In wild-type cells growing on
minimal medium, the last-named isoenzyme accounts for a much
larger portion of the total activity than the two others.   The
three enzymes are formed individually under the control of
three distinct genes widely spaced on the chromosome.   The
situation in the ascomycete, *Neurospora crassa*, seems to be
similar; in yeast (*S. cerevisiae*), another ascomycetous fungus,
two isoenzymes inhibited, but *not* repressed by phenylalanine and
tyrosine, respectively, have been described (13a,b, 14).   While
thus the situation in the two ascomycetous fungi resembles that
found in the two enterobacteria, the regulation is achieved in
an entirely different way in another bacterium, *Bacillus subtilis*,
which has been studied in great detail by Nester, Jensen, and
their associates.   Jensen and Nester (15) found that *B. subtilis*
achieves the control with only one DAHP synthetase, subject to
feedback inhibition, not so much be the aromatic amino acids
as by the branch-point intermediates, chorismate and prephenate.
The enzyme is, however, repressed by the aromatic amino acids,
particularly tyrosine (16).   DAHP synthetase in *B. subtilis*
forms a complex with chorismate mutase, the enzyme responsible
for the conversion of chorismic to prephenic acid, and,
surprisingly, with shikimate kinase (see below).   Existence of
such aggregates of several enzymes has been observed repeatedly;
they are the subject of several reviews (5, 22).   It is usually
considered that the significance of these complexes lies in a
role they may play in regulatory processes; however, no actual
proof for such a function seems to have been obtained.   Conversion of prephenic acid to phenylalanine and tyrosine, in its
turn, is under feedback control by these amino acids.

   In *B. subtilis*, control of the first specific enzymatic
step is thus achieved not through isoenzymes differing in
sensitivity to the three major end-products, but through inhibition by branch-point intermediates, and it is the further
utilization of these which is controlled by the end-products.
This mode of control has been termed "sequential feedback inhibition" by Nester and Jensen (17).

   Whatever its detailed mechanism, feedback control of the
first reaction directly by the end-products (as in *E. coli*) or
indirectly via branch-point intermediates (as in *B. subtilis*)
will result in a diminished synthesis of the intermediates of
the common pathway.   Evidently, however, additional controls
further along the pathway must, and do, exist, to channel the
remaining biosynthetic material into the required directions.
Inhibition of, for example, the phenylalanine-sensitive DAHP

synthetase isoenzyme of *E. coli* or *Neurospora* by feedback control, once the need for phenylalanine has been filled, would be to no avail if the remaining amounts of common intermediates could distribute themselves freely in all the later branches, including the one via prephenate to phenylalanine. Hence the elaborate further controls after branch-point intermediates. The complex interplay of these various interactions also involves influences of end-products on branches other than their own, for example, of tryptophan upon the enzymes controlling the formation of phenylalanine and tyrosine. These involved questions have been studied in much detail, for instance, in *B. subtilis* (18), but cannot be followed here any further.

Theoretically, however, control at the level of DAHP synthetase (e.g., by the three isoenzymes in *Neurospora*) would be sufficient in itself if the flow of intermediates of the common pathway from each of the isoenzymes were to remain separate by some compartmentalization. This, however, has been shown not to be the case in *Neurospora* (19).

The foregoing discussion of the controls at the level of DAHP synthetase has considered only the effects of the three aromatic amino acids, which are the only *quantitatively* significant primary products of the shikimate pathway. Even if they are in adequate supply, however, mechanisms must exist which permit the continued synthesis of the vitally important trace factors to proceed. A study of this problem by Wallace and Pittard (20) has shown that *E. coli* strain K-12, even under conditions of maximal repression of all three isoenzymes of DAHP synthetase, still produces enough chorismic acid to satisfy the needs for the trace factors. The ability to synthesize these factors in a controlled fashion thus rests upon the fact that feedback control by the three amino acids is not quite complete, rather than on the existence of some further isoenzymes.

From their position in an unbranched chain of biosynthetic reactions, the next steps of the common pathway (from the transformation of DAHP into dehydroquinic acid to the synthesis of chorismic acid) would not be expected to be under feedback control, and indeed enzymes of this part of the sequence have often been found to be only little inhibited or repressed by later intermediates or end-products (cf., *inter alia*, 9, 10, 16).

There is at least one peculiar exception, though. In *B. subtilis*, shikimate kinase (the enzyme that converts shikimic acid into its phosphate) has been found by Nester and co-workers (21) to be under feedback inhibition by prephenate and chorismate. The situation is actually very involved. The enzyme in this species forms a complex with DAHP synthetase and one isoenzyme of chorismate mutase, of which the former, too, is inhibited by

chorismate and prephenate.   In addition, *B. subtilis* actually
produces two different molecular forms of shikimate kinase, one
of which takes part in the complex.   It has already been mentioned
that formation of such multienzyme complexes is usually considered
to be connected with regulatory processes, although the actual
function of these aggregates is none too clear at present (2);
the possible influence of isolation procedures upon their appearance has been pointed out (22).

Whatever the meaning of the association of shikimate kinase
with the two other enzymes, the finding of its feedback inhibition is unexpected in the absence of any known branching of the
pathway at shikimic acid or its phosphate.   In addition, evidence for repression of the kinase by tyrosine and tryptophan in
E. coli has been obtained (8a,b).

The fact that *B. subtilis* contains two different molecular
forms of shikimate kinase is paralleled by the situation in
S. typhimurium, where Morell and Sprinson (23) likewise discovered two forms of the enzyme, which showed widely differing
stability and chromatographic behavior.   Neither of them was
inhibited by chorismate, prephenate, or the three aromatic amino
acids.   It has been pointed out in Section 5-2 that no single-
step bacterial mutants blocked in the phosphorylation of shikimic
acid have been found, with the exception of one strain of
B. subtilis.   However, mutants of this type were obtained from
Neurospora.

The enzyme aggregate just discussed is composed of three
enzymes that catalyze rather widely separated steps of the biosynthetic pathway.   This situation differs markedly from that in
Neurospora.   Here, Giles and co-workers (24) described in 1967
a complex of five enzymes responsible for five consecutive
reactions of the common pathway from the formation of dehydroquinic acid to that of enolpyruvyl shikimate phosphate. This much-
studied aggregate has recently been obtained in highly purified,
homogeneous form (25).   All attempts to resolve it into its
component enzymes failed.   DAHP synthetase, the earliest, and
chorismic acid synthetase, the last one of the enzymes of the
common pathway, were not associated with the complex.   Mutants
were found that were deficient in one single enzyme activity
(with complications in the case of dehydroquinase, see below) or
in all five of them.   Detailed genetic analysis (24, 26) showed
that this enzyme aggregate is coded by a cluster of five adjacent genes; their order on the chromosome is *not* that of the
biosynthetic steps catalyzed by the enzymes they specify.
Mutation of single genes produces a polypeptide chain which lacks
enzymatic activity but which is built into the aggregate.   The
cluster of genes has some properties in common with an operon
but does not include any operator or regulator genes.

An interesting explanation has been proposed for the existence and purpose of this aggregate (24, 27). It is based on the observation that *Neurospora* contains, in addition to the biosynthetic pathway under discussion, also a catabolic pathway from quinic via dehydroquinic to protocatechuic acid. These two pathways have one step in common: the conversion of dehydroquinic to dehydroshikimic acid, but the enzymes promoting this reaction in the two pathways are different; only the biosynthetic enzyme forms part of the aggregate, and is constitutive. The catabolic enzyme is independent, and is induced by quinic or dehydroquinic acid. The two enzymatic activities have been separated and differ widely in certain properties, such as heat sensitivity, the inducible enzyme being much more stable. The existence of these two enzymes explains the initial failure to isolate any mutants deficient in dehydroquinase only. Such mutants were finally obtained (26b) from a strain unable to make appreciable amounts of the inducible enzyme.

The occurrence of the biosynthetic enzyme as part of the aggregate is then plausibly explained by the necessity of protecting the dehydroquinic acid needed for biosynthesis from attack by the dehydroquinase of the catabolic pathway, which would convert it into protocatechuic acid. Such protection could be achieved very well if the intermediates remained bound to the enzyme aggregate. Absence of induction of the catabolic enzyme by biosynthetically formed dehydroquinic acid in wild-type *Neurospora* shows the efficient functioning of the segregation mechanism.

However, this interpretation, attractive though it is, cannot explain all observations. Ahmed and Giles (28) have found that six other fungi, belonging to the phyco-, asco-, and basidiomycetes, contain enzyme aggregates similar to the one in *Neurospora*, as does yeast (*S. cerevisae*). However, only two of these fungi, the ascomycete *Aspergillus nidulans*, and the basidiomycete *Ustilago maydis*, resemble *Neurospora* in having a second dehydroquinase, which is inducible in *A. nidulans* but constitutive in *U. maydis*. In the four other fungi, no evidence for the catabolic pathway could be obtained, and the reason for the existence of aggregates in these species is not clear at present. Explanations based on evolutionary factors have been advanced.

The catabolic pathway in *Neurospora* has very recently been explored (29) in further detail; since it leads to an aromatic metabolite, protocatechuic acid, the salient results of this work may be briefly mentioned. The three enzymes of this pathway, quinate dehydrogenase, dehydroquinase, and dehydroshikimate dehydrase, are under the control of four closely linked genes,

of which the first one appears to regulate the three others,
which encode the three individual enzymes. Some mutants of the
first gene were able to produce the three enzymes even in the
absence of an inducer. The system seems to have many analogies
with those found in bacteria.

No special regulatory mechanisms seem to have been encountered in studies on the last step of the common pathway, the
biosynthesis of chorismic acid. In *Neurospora*, the chorismate
synthetase is outside the multienzyme aggregate.

As was to be expected, the early steps in the individual
pathways branching out towards the various primary aromatic
products are important regulatory sites. The regulations in the
three branches leading to the aromatic amino acids have been
studied in very great detail, and have been found, again, to
vary greatly from one species to the next.

In the regulation of the pathway from chorismic acid via
prephenic acid to phenylalanine and tyrosine, isoenzymes and complexes of the first enzyme chorismate mutase with the ones
catalyzing the next step play a very important role. These
complexes were first observed by Gibson (30) in 1965, and have
been studied in great detail by this author and his co-workers.
Our necessarily much abbreviated description follows essentially
the review by Gibson and Pittard (8). In *A. aerogenes* and
E. coli, the conversion of chorismic to prephenic acid is catalyzed by two different chorismate mutases, called P and T, which
are concerned with the biosynthesis of phenylalanine and tyrosine, respectively. They are separable by chromatographic
techniques from each other, but not from enzymatic activities
responsible for the further conversion of the prephenic acid:
chorismate mutase P is associated with prephenate dehydratase,
the enzyme that converts prephenic acid to phenylpyruvic acid;
chorismate mutase T is similarly inseparable from prephenate
dehydrogenase, which converts prephenic acid to p-hydroxyphenyl-
pyruvic acid. Both activities in either complex are affected
together by mutations and their reversions; both enzymes in the
aggregates P and T are also repressed together by phenylalanine
and tyrosine, respectively. Feedback inhibition by phenylalanine
and tyrosine in *E. coli* affects only the second of the two
activities involved in their respective biosynthesis: phenylalanine that of the prephenate dehydratase of complex P, tyrosine
that of prephenate dehydrogenase of complex T. In *A. aerogenes*,
however, phenylalanine also inhibits chorismate mutase P, and
in one of two strains studied, the situation is further complicated by the existence of a separate prephenate dehydratase A,
not part of any complex, not inhibited by phenylalanine, and
of unknown biological significance. The transaminases which

catalyze the next steps, conversion of the respective α-keto
acids to phenylalanine or tyrosine, are not associated with
the complexes. In *E. coli*, however, a gene has been found
(31) which regulates the synthesis of one of these transaminases
(transaminase A, responsible for conversion of *p*-hydroxyphenyl-
pyruvic acid into tyrosine) together with that of the tyrosine-
sensitive isoenzyme of DAHP synthetase and the two enzymes of
complex T.

Further study of complex T (32, 33) from *Aerobacter* has
resulted in a nearly homogeneous protein; the ratio of its two
enzymatic activities remained constant throughout the various
purification steps. Treatment with acid, urea, or detergents
splits the enzyme reversibly into two identical subunits that have
lost both enzymatic activities.

Among other microorganisms, *S. typhimurium* seems to achieve
the regulation of the biosynthesis of phenylalanine and tyrosine
in a way similar to the one in *E. coli*. The complex of chorismate mutase and prephenate dehydratase has been studied in this
species by Zalkin and co-workers (34). Again, far-reaching
purification did not alter the ratio of the two activities, both
of which were inhibited by phenylalanine. However, for these
reactions, too, other organisms have developed alternative regulations. *Neurospora* (35) and yeast (13a), for example, have only
one chorismate mutase, the former inhibited by both phenylalanine
and tyrosine, the latter only by tyrosine. Regulation in the
ergot, *Claviceps paspali*, is similar to that of *Neurospora* (36).
The *activation* by tryptophan of the enzymes from the latter two
organisms is analogous to cases of activation by end-products of
alternative branches that have been encountered in the stages of
biosynthesis following prephenic acid. These questions will
be discussed below.

In contrast, one strain of *B. subtilis* produces three distinct species of chorismate mutase (21, 37), which can be separated by chromatography. The formation of an aggregate of one of
these with DAHP synthetase and, probably, with shikimate kinase
has been mentioned before.

Since the pathways leading to phenylalanine and tyrosine
finally separate after prephenic acid, occurrence of further
regulations at this branch-point was to be expected. Usually,
either branch is controlled through feedback inhibition by its
own end-product; this type is found, for example, in *B. subtilis*
(17). However, influences more complicated than this simple
pattern have often been observed, and some of those involving
enzymes responsible for earlier steps in the biosynthesis have
already been discussed. These include the existence, in one of
three different strains of *B. subtilis* (mutants resistant to the

inhibitor β-thienylalanine) of two different, separable isoenzymes
of prephenate dehydratase (38), and the unusual cases of feedback
activation already briefly mentioned in Chapter 1 and in the
discussion of chorismate mutase (p. 294). In a *Pseudomonas* sp.,
for example, Cerutti and Guroff (39) observed that prephenate dehydratase is inhibited by phenylalanine, the end-product of its
own branch, but strongly activated by that of the other branch,
viz., tyrosine. Both amino acids increase the heat stability of
the enzyme. The situation observed by Catcheside (40) in *Neurospora* is, as it were, the mirror-image of the preceding one; here,
it is prephenate dehydrogenase that is inhibited by its own end-
product tyrosine, and activated by phenylalanine.

In possible contrast to these cases of *activation* by the
end-product of a related branch stands the *inhibition* by such an
end-product: Rebello and Jensen (18) have observed that, in *B.
subtilis*, prephenate dehydratase is strongly inhibited by tryptophan, and have discussed the significance of this finding. The
action of tryptophan is reversed not only by phenylalanine, but
also by tyrosine.

More surprising than this activating or inhibiting action of
end-products of related pathways is activation (instead of inhibition) of an enzyme by the end-product of its own branch. An instance of such a "regulation-reversal mutation" has been found by
Coats and Nester (38) in two thienylalanine-resistant mutants of
B. subtilis, where prephenate dehydratase is activated by phenylalanine; one of these two mutants is the strain already mentioned
as producing two different prephenate dehydratases. The complexity of such interactions is illustrated further by a number of
cases where influences by end-products of quite unrelated pathways
have been demonstrated. This situation and its biological impli-
cations have been discussed by Jensen (18, 41), who uses the term
"metabolic interlock"; it may be illustrated by the activation of
prephenate dehydratase in *B. subtilis* by leucine or methionine
(41).

The biosynthesis of tryptophan constitutes the other major
pathway branching off from chorismic acid. It consists of five
steps: (1) reactions of chorismic acid with glutamine (or ammonia) to give anthranilic acid; (2) reaction of this with phosphoribosyl pyrophosphate to give N-(5'-phosphoribosyl)-anthranilic
acid; (3) transformation of this compound to 1-(o-carboxyphenylamino)-1-deoxyribulose 5-phosphate by an Amadori rearrangement;
(4) closure of the pyrrole ring to give indoleglycerol phosphate;
and (5) reaction of this with serine to give tryptophan.

The aromatic ring of tryptophan is formed already in the
first step, which is catalyzed by the enzyme anthranilate synthetase. Strictly speaking, only this reaction falls within the

defined scope of this book; consequently, only its regulations will be discussed in any detail, especially since the entire sequence itself has been described elsewhere (Chapter 9).

This branch of the shikimate pathway has been studied more intensely than any other, and a huge amount of information on its functioning in a large number of organisms has accumulated. Neither a complete account of the many different control mechanisms observed during this research, nor an extensive bibliography can be given here; only the more important findings will be discussed. Excellent summaries of these regulations are contained in some of the reviews (3, 4, 8) quoted earlier; especially ref. 4 gives a very detailed account. References to much of the original literature will be found in these reviews.

In *E. coli* and *S. typhimurium* the five genes coding for the enzymes of tryptophan biosynthesis form a much-investigated operon. Genetic analysis showed a linear array of these genes, having the same sequence as the biosynthetic steps; all the enzymes are repressed by the tryptophan in nearly coordinated fashion, that is, to almost the same extent.

Regulation of the tryptophan pathway in these bacteria is achieved through feedback inhibition of the early stages by tryptophan. Surprisingly, not only the first enzyme, anthranilate synthetase, is inhibited, but also the second one, phosphoribosyl transferase ("PR transferase"). These two enzymes form an aggregate, both in *E. Coli* (42) and in *A. aerogenes* (43). In mutants of *E. coli*, PR transferase can act independently of the anthranilate synthetase, but the latter is active only in the presence of the transferase, or of NH_4^+ ions, which can, to a certain extent, replace this enzyme. Both enzymatic activities move together during centrifugation and gel filtration. The aggregate from *E. coli* has been purified (44). An analogous aggregate of the two enzymes from *S. typhimurium* has been isolated in highly purified form (45), as have the unaggregated enzymes from this species (46a). It seems that the unaggregated anthranilate synthetase catalyzes the reaction of chorismate + NH_3 + $Mg^{2+} \longrightarrow$ anthranilate + pyruvate, while aggregation with the transferase alters the specificity in such a way that either ammonia or glutamine can be used. The properties of the two unaggregated enzymes, and of the aggregate, are reported in detail; it was found that also the unaggregated form of the second enzyme, PR transferase, is inhibited by tryptophan, but apparently less so than the aggregated form (46b).

The situation is very different in *Pseudomonas* spp. (47). Here, PR transferase plays no role in the regulation of the biosynthesis of anthranilic acid, but anthranilate synthetase consists of two separable proteins; one of these, AS I, can by itself carry out the synthesis of anthranilic acid from chorismic acid and ammonia, while the presence of AS II (glutamine amide transferase) is needed for the utilization of glutamine. AS I is inhibited by tryptophan.

In *S. cerevisiae* one single genetic locus controls the enzymatic activities of the first and the fourth steps of the pathway, that is, anthranilate synthetase and indoleglycerol phosphate synthetase (48); both enzymatic functions should thus be performed by one single polypeptide chain. However, for anthranilate synthetase activity, a second polypeptide, coded by a different gene, is needed. Tryptophan inhibits the anthranilate synthetase.

Finally, De Moss and co-workers (49) found in *N. crassa* an aggregate of the first, second, and fourth enzyme of the pathway (anthranilate synthetase, PR transferase, and indoleglycerol phosphate synthetase) which retains the ratio of these three activities during 90-fold purification. Only two genes code for these, one of them for anthranilate synthetase, the other for the two remaining enzymes. Here again, anthranilate synthesis seems to require aggregation, while the two other enzymatic activities do not.

The anthranilate synthetase of *Neurospora* is both inhibited (13a, 35a) and repressed (13b) by tryptophan, that of *Claviceps paspali* is uninfluenced (36). This latter fact is ascribed to the flow of tryptophan into ergot alkaloids in the fungus. In a somewhat teleological vein, one might be tempted to argue that this absence of controls at the beginning of the tryptophan pathway militates against the assumption that the clavine alkaloids of this species are otherwise useless products of an elimination process, or of a "luxury metabolism."

The results sketched in the preceding paragraphs--of necessity omitting many findings, some of them very intriguing--show an amazing tendency of anthranilate synthetase to aggregate with some of the enzymes responsible for reactions further along the path towards tryptophan. Almost invariably, these aggregates are inhibited by tryptophan, and usually, synthesis of anthranilic acid from chorismic acid and glutamine is restricted to such complexes, while the unaggregated enzyme is capable of carrying out the same synthesis with ammonia. The goal of extracting some general meaning out of those observations, and of understanding them in terms of chemical mechanisms, of the transformation of the substrates, of mode of enzymatic action, and of regulation of the biosynthesis of tryptophan, has not yet been reached, though; the biological purpose of the manifold aggregations of these enzyme activities remains obscure.

No regulatory mechanism for the conversion of chorismic to p-aminobenzoic and p-hydroxybenzoic acids seem to have been observed. The situation is different, however, for the pathway leading through isochorismic and 2,3-dihydroxy-2,3-dihydrobenzoic acids to 2,3-dihydroxybenzoic acid and its conjugates; this pathway has been described in Chapter 12. Briefly, it involves formation of isochorismic acid through an allylic shift of the secondary hydroxyl of chorismic acid, hydrolysis of isochorismic to 2,3-dihydroxy-2,3-dihydrobenzoic acid, and dehydrogenation of this compound to 2,3-dihydroxybenzoic acid. Further transformations of the last-named acid yield conjugates such as the 2,3-dihydroxybenzoylglycine of B. *subtilis* (50) (the first such compound to be discovered), the corresponding serine derivative in E. *coli*, and others. Formation of the entire series of substances is observed only when the organisms are grown on iron-deficient media (50, 51).

It has been observed (52) that the enzyme systems involved, including the one catalyzing the formation of the serine conjugate in E. *coli*, are repressed by iron (or cobalt). A detailed study of the regulation of this pathway by Young and Gibson (51) has now demonstrated that every one of the individual enzymes of this pathway up to dihydroxybenzoic acid: isochorismate synthetase, 2,3-dihydroxy-2,3-dihydrobenzoate synthetase, and 2,3-dihydroxybenzoate synthetase, is strongly repressed by inorganic iron ($FeSO_4$). Proof that iron itself and not an Fe-chelate of 2,3-dihydroxybenzoic acid, or some conjugate of it, is the actual repressor was furnished with the help of a mutant of E. *coli* having a complete block in the common pathway. In this strain, which is unable to produce any endogenous 2,3-dihydroxybenzoic acid (it had, under certain conditions, an absolute requirement for it), 2,3-dihydroxy-2,3-dihydrobenzoate dehydrogenase was strongly repressed by iron, which was thus effective in the complete absence of 2,3-dihydroxybenzoate. Compared to the effect of iron, 2,3-dihydroxybenzoic acid and 2,3-dihydroxybenzoylserine had only a weak repressing action in A. *aerogenes*. Inhibition by these compounds was likewise slight. The pathway to 2,3-dihydroxybenzoic acid and its conjugates is thus, at every step, controlled through repression by inorganic iron, a very unusual situation apparently connected with the probably role of 2,3-dihydroxybenzoic acid and its relatives; available evidence indicates that these substances function as iron-chelating agents. The enzyme of E. *coli* that conjugates 2,3-dihydroxybenzoic acid with serine is likewise strongly repressed by iron (53).

All regulations described in this section up to now have been found in microorganisms. However, information on these processes in higher plants is not entirely lacking, and a brief discussion of some of these findings is needed.

Some of the enzymes from higher plants were not inhibited by biosynthetic end-products. For instance, Gamborg (54) observed that the activity of quinate dehydrogenase extracted from suspension cultures of mung bean cells (*Phaseolus aureus*) was not influenced by the addition of phenylalanine, tyrosine, or shikimic acid; similarly, the partially purified prephenate dehydrogenase from the cotyledons of the same species was not inhibited by tyrosine or phenylalanine (55). However, the chorismate mutase, extracted and pre-purified by Cotton and Gibson (56) from pea seedlings (*Pisum sativum*) was inhibited by phenylalanine and tyrosine, and activated by tryptophan. The activity of anthranilate synthetase from cell cultures of tobacco (*Nicotiana tabacum*) was found (57) to be under feedback inhibition by L-tryptophan both in extracts and in the intact cells.

Besides such cases of feedback control, also an instance of aggregate formation has been observed; surprisingly, it involves two consecutive enzymes of the common pathway, dehydroquinase and dehydroshikimate reductase, which were found by Boudet (58) to form a complex in extracts from the roots of the oak, *Quercus pedunculata*. All attempts to resolve this complex were unsuccessful.

It may thus be concluded that regulations in higher plants seem to be based upon the same principles as those in microorganisms.

REFERENCES

1. B.D. Davis, Experientia $\underline{6}$, 41 (1950).
2. P. Datta, Science $\underline{165}$, 556 (1969).
3. H.E. Umbarger, Ann. Rev. Biochem. $\underline{38}$, 323 (1969).
4. P. Truffa-Bachi and G.N. Cohen, Ann. Rev. Biochem. $\underline{37}$, 79 (1968).
5. A. Ginsburg and E.R. Stadtman, Ann. Rev. Biochem. $\underline{39}$, 429 (1970).
6. R.G. Martin, Ann. Rev. Genetics $\underline{3}$, 181 (1969).
7. W. Epstein and J.R. Beckwith, Ann. Rev. Biochem. $\underline{37}$, 411 (1968).
7A. J. Calvo and G. Fink, Ann. Rev. Biochem. $\underline{40}$, 943 (1971).
8. F. Gibson and J. Pittard, (a) Bact. Rev. $\underline{32}$, 465 (1968); (b) J. Pittard and F. Gibson, Curr. Top. Cell. Reg. $\underline{2}$, 29 (1970).
9. C.H. Doy, Rev. Pure Appl. Chem. $\underline{18}$, 41 (1968).
10. E. Gollub, H. Zalkin, and D.B. Sprinson, J. Biol. Chem. $\underline{242}$, 5323 (1967).
11. J. Pittard and B.J. Wallace, J. Bact. $\underline{91}$, 1494 (1966).
12. K.D. Brown, Genetics $\underline{60}$, 31 (1968).

12A. R.L. Metzenberg, Ann. Rev. Genetics 6, 111 (1972).

13. F. Lingens, W. Goebel, and H. Uesseler, (a) Biochem. Zeitschr. 346, 357 (1966); (b) Eur. J. Biochem. 1, 363 (1967).

14. P. Meuris, Bull. Soc. Chim. Biol. 49, 1573 (1967).

15. R.A. Jensen and E.W. Nester, (a) J. Mol. Biol. 12, 468 (1965); (b) J. Biol. Chem. 241, 3365 (1966).

16. E.W. Nester, R.A. Jensen, and D.S. Nasser, J. Bact. 97, 83 (1969).

17. E.W. Nester and R.A. Jensen, J. Bact. 91, 1594 (1966).

18. J.L. Rebello and R.A. Jensen, J. Biol. Chem. 245, 3738 (1970).

19. D.M. Halsall and C.H. Doy, Biochim. Biophys. Acta 185, 432 (1969).

20. B.J. Wallace and J. Pittard, J. Bact. 99, 707 (1969).

21. E.W. Nester, J.H. Lorence, and D.S. Nasser, Biochem. 6, 1553 (1967).

22. L.J. Reed and D.J. Cox, Ann. Rev. Biochem. 35, 57 (1966).

23. H. Morell and D.B. Sprinson, J. Biol. Chem. 243. 676, (1968).

24. N.H. Giles, M.E. Case, C.W.H. Partridge, and S.I. Ahmed, Proc. Natl. Acad. Sci. 58, 1453 (1967).

25. L. Burgoyne, M.E. Case, and N.H. Giles, Biochim. Biophys. Acta 191, 452 (1969).

26. a. M.E. Case and N.H. Giles, Genetics 60, 49 (1968);
    b. H.W. Rines, M.E. Case, and N.H. Giles, ibid, 61, 789 (1969).

27. N.H. Giles, C.W.H. Partridge, S.I. Ahmed, and M.E. Case, Proc. Natl. Acad. Sci. 58, 1930 (1967).

28. S.I. Ahmed and N.H. Giles, J. Bact. 99, 231 (1969).

29. J.A. Valone, M.E. Case, and N.H. Giles, Proc. Natl. Acad. Sci. 68, 1555 (1971).

30. R.G.H. Cotton and F. Gibson, Biochim. Biophys. Acta 100, 76 (1965).

31. B.J. Wallace and J. Pittard, J. Bact. 97, 1234 (1969).

32. R.G.H. Cotton and F. Gibson, (A) Biochim. Biophys. Acta 147 222 (1967);
    (B) ibid. 160, 188 (1968).

33. G.L.E. Koch, D.C. Shaw, and F. Gibson, Biochim. Biophys. Acta 212, 375 (1970).

34. J.C. Schmit and H. Zalkin, Biochemistry 8, 174 (1969); J.C. Schmit, S.W. Artz and H. Zalkin, J. Biol. Chem. 245, 4019 (1970).

35. T.I. Baker, (A) Biochemistry 5, 2654 (1966); (B) Genetics 58, 351 (1968).

36. F. Lingens, W. Goebel, and H. Uesseler, Eur. J. Biochem. 2, 442 (1967).

37. J.H. Lorence and E.W. Nester, Biochemistry 6, 1541 (1967).
38. J.H. Coats and E.W. Nester, J. Biol. Chem. 242, 4948 (1967).
39. P. Cerutti and G. Guroff, J. Biol. Chem. 240, 3034 (1965).
40. D.E.A. Catcheside, Biochem. Biophys. Res. Comm. 36, 651 (1969).
41. R.A. Jensen, J. Biol. Chem. 244, 2816 (1969).
42. J. Ito and C. Yanofsky, J. Biol. Chem. 241, 412 (1966).
43. A.F. Egan and F. Gibson, Biochim. Biophys. Acta 130, 276 (1966).
44. J. Ito, E.C. Cox, and C. Yanofsky, J. Bact. 97, 725 (1969).
45. E.J. Henderson, H. Nagano, H. Zalkin, and L.H. Hwang, J. Biol. Chem. 245, 1416 (1970).
46. a. H. Nagano and H. Zalkin, J. Biol. Chem. 245, 3097 (1970); b. E.J. Henderson, H. Zalkin, and L.H. Hwang, ibid. 245, 1424 (1970).
47. S.F. Queener and I.C. Gunsalus, Proc. Natl. Acad. Sci. 67, 1225 (1970).
48. a. J.A. De Moss, Biochem. Biophys. Res. Comm. 18, 850 (1965); b. C.H. Doy and J.M. Cooper, Biochim. Biophys. Acta 127, 302 (1966).
49. a. J.A. De Moss and J. Wegman, Proc. Natl. Acad. Sci. 54, 241 (1965); b. J.A. De Moss, R.W. Jackson, and J.H. Chalmers, Jr., Genetics 56, 413 (1967).
50. T. Ito and J.B. Neilands, J. Amer. Chem. Soc. 80, 4645 (1958).
51. I.G. Young and F. Gibson, Biochim. Biophys. Acta 177, 401 (1969).
52. I.G. Young, G.B. Cox, and F. Gibson, Biochim. Biophys. Acta 141, 319 (1967).
53. J.F. Bryce and N. Brot, Arch. Biochem. Biophys. 142, 399 (1971).
54. O. Gamborg, Biochim. Biophys. Acta 128, 483 (1966).
55. O. Gamborg and F.W. Keeley, Biochim. Biophys. Acta 115, 65 (1966).
56. R.G.H. Cotton and F. Gibson, Biochim. Biophys. Acta 156, 187 (1968).
57. W.L. Belser, J.B. Murphy, D.P. Delmer, and S.E. Mills, Biochim. Biophys. Acta 237 1 (1971).
58. A. Boudet, FEBS Lett. 14, 257 (1971).

CHAPTER 16

The Shikimate Pathway
as a Source of
Aromatic Metabolites

Those features commonly met with in metabolites derived from
shikimic acid have been discussed in Chapter 2, as have the more
frequent exceptions to the criteria by which a compound may be
judged to be formed from this pathway. The true significance of
the shikimate pathway can only fully be appreciated on the basis
of a detailed account of the biosynthetic events leading to the
metabolites produced through its operation. Space limitations
have made it impossible to include such a discussion, but the
omission is filled by the timely appearance of the monograph
"The Shikimate Pathway" by Haslam (1), which contains much of the
needed information, and by the book "Chemistry of the Alkaloids"
edited by Pelletier (2), which covers the nitrogen-containing
metabolites in detail. Here, only a few fairly general remarks
about the biosynthetic origin of some of the more significant
groups of compounds can be given. Literature will be found in
refs. 1 and 2.

The biosynthetic relationships of the simple, nitrogen-free
C_6, C_6-C_1, C_6-C_2, and C_6-C_3 compounds are shown schematically on
page 2. The C_6-compounds elaborated by the shikimate pathway
are generally formed through oxidative decarboxylation of a
benzoic acid; such a decarboxylation of p-hydroxybenzoic acid
yields hydroquinone in *Bergenia crassifolia*; similarly, sinapic
and ferulic acids are converted into the corresponding methoxyquinones by *Triticum vulgare*; see formulae on page 3. The
C_6-C_1 compounds (acids, aldehydes, and alcohols) are either
derived from phenylpropanoid precursors by β-oxidation of the
corresponding cinnamic acid, or by hydroxylation of benzoic acids,

302

303

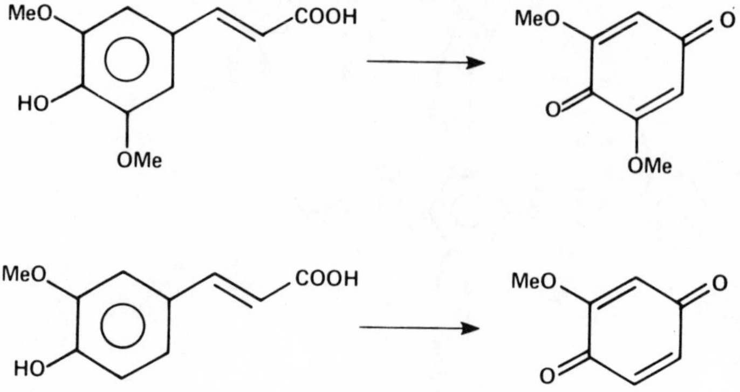

Methoxyquinones of *Triticum vulgare*

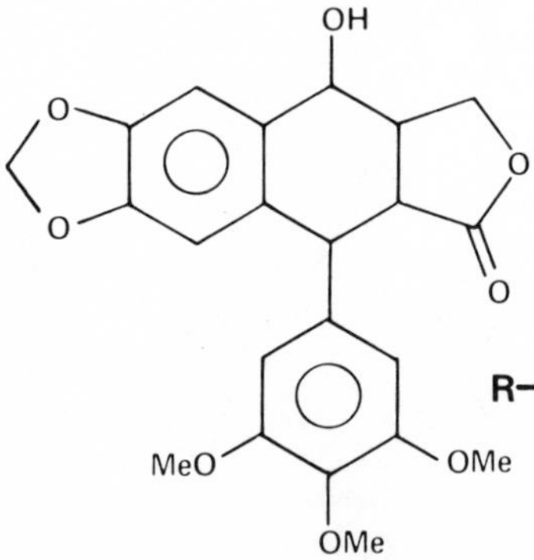

Podophyllotoxin [1]

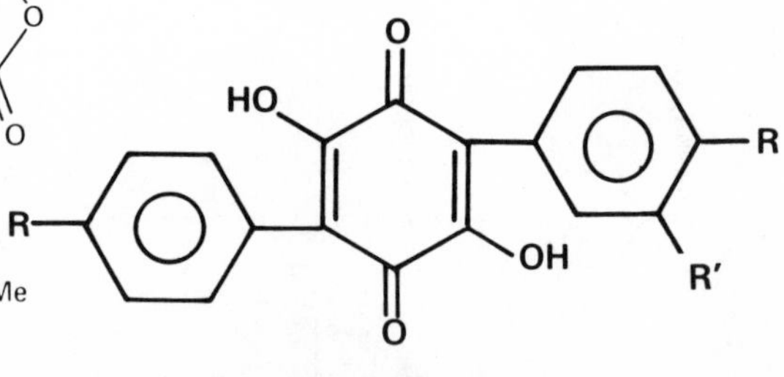

Polyporic acid R=R′=H [2]
Leucomelone R=R′=OH
Atromentin R=OH R′=H

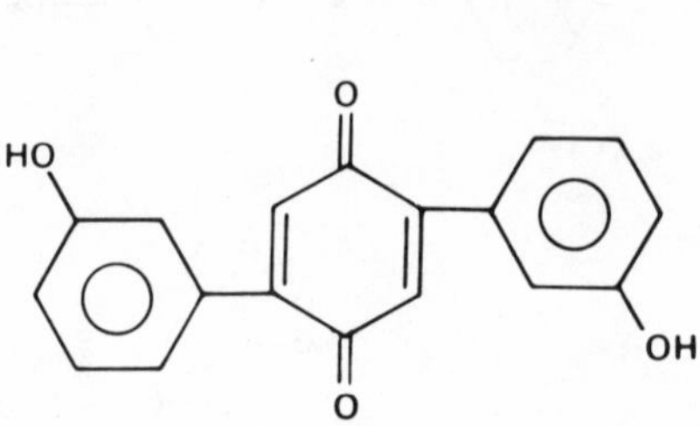

Volucrisporin [3]

304

themselves formed by β-cleavage. The aldehydes and alcohols are derived more or less directly from the C_6-C_3 compounds rather than from the C_6-C_1 acid.

   C_6-C_2 compounds are comparatively rare. Cyanogenic glucosides of this class are formed directly from the aromatic amino acids; thus, tyrosine gives rise to dhurrin, a glucoside of p-hydroxymandelonitrile, the amino nitrogen becoming the cyanide nitrogen and the carboxyl being lost. On the other hand, acetophenone derivatives (e.g. picein, the glucoside of 4-hydroxyacetophenone) and the C_6-C_2 acids (e.g. phenylacetic acid) incorporate both the 2 and 3 carbons of the amino acids.

   As the precursor of phenylalanine and tyrosine, shikimic acid is also the source of the C_6-C_3 compounds found so abundantly in plants and other organisms and of innumerable metabolites formed by their further modification. Here the deamination of phenylalanine to cinnamic acid by the ubiquitous enzyme phenylalanine ammonia lyase is of particular importance; curiously, the corresponding transformation of tyrosine to p-coumaric acid occurs readily only in plants of the Gramineae. The biosynthesis of the variously hydroxylated cinnamic acids is a sequential process operating through a metabolic grid; from these acids, many other secondary metabolites are derived. The lignans (only podophyllotoxin [1] has been studied biosynthetically), the terphenylquinone derivatives (polyporic acid [2], volucrisporin [3], etc.), and their relatives, are all dimeric compounds formed from two molecules of phenylalanine or a C_6-C_3 compound derived from it. The unusual *meta*-orientation of the hydroxyls in [3] is discussed later in this chapter. Lignin, a polymer of high molecular weight, is formed by the coupling of C_6-C_3 molecules at the oxidation level of cinnamyl alcohol.

   Many simple coumarins, such as scopoletin [4], umbelliferone [5], and many others, are formed directly from C_6-C_3 precursors, essentially by the oxidative lactonization of the *cis-p*-coumaric acids; the process is shown for [5] on page 306 . In the 4-phenylcoumarins such as calophyllolide [6] the phenyl ring and the carbons of the lactone are provided by phenylalanine, while in coumaranocoumarins (e.g. coumestrol [7]) the lactone ring is formed by intramolecular transformations of the corresponding isoflavone. Some of the isocoumarins are also of mixed shikimate-acetate biogenesis, with phenylalanine or cinnamic acid providing ring B and carbons 3,4 and 4a, for example in hydrangenol [8].

   The naphthoquinone nucleus can be formed from shikimate metabolites in three ways. In one of these the benzenoid ring is derived from shikimic acid directly (e.g. in lawsone [9] and juglone [10]) via o-succinylbenzoic acid [11] by the pathway already discussed for the Vitamins K (Chapter 13); it is shown

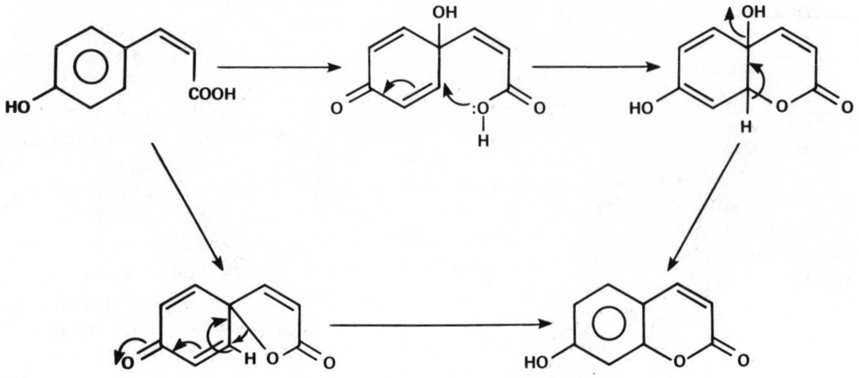

The biosynthesis of umbelliferone [5]

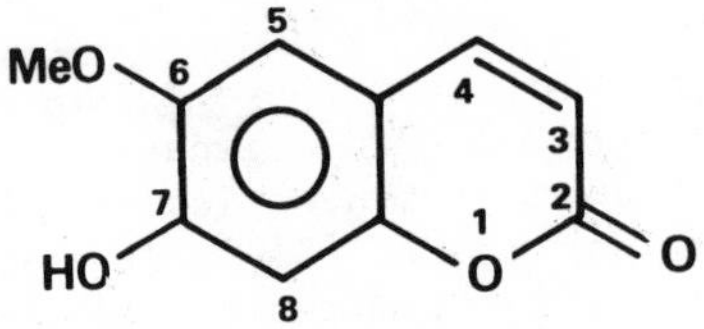

Scopoletin [4]

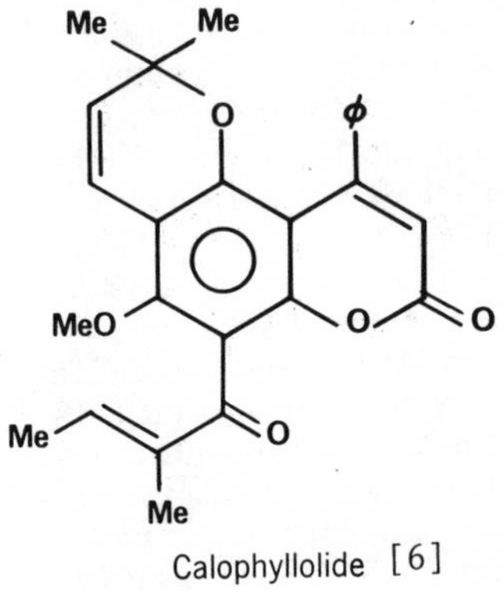

Calophyllolide [6]

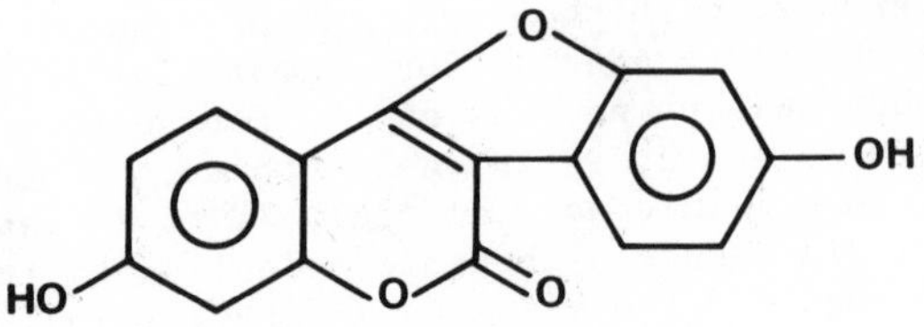

Coumestrol [7]

306

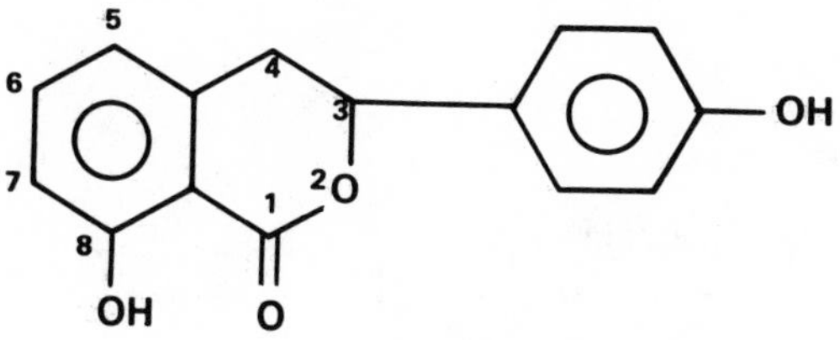

Hydrangenol [8]

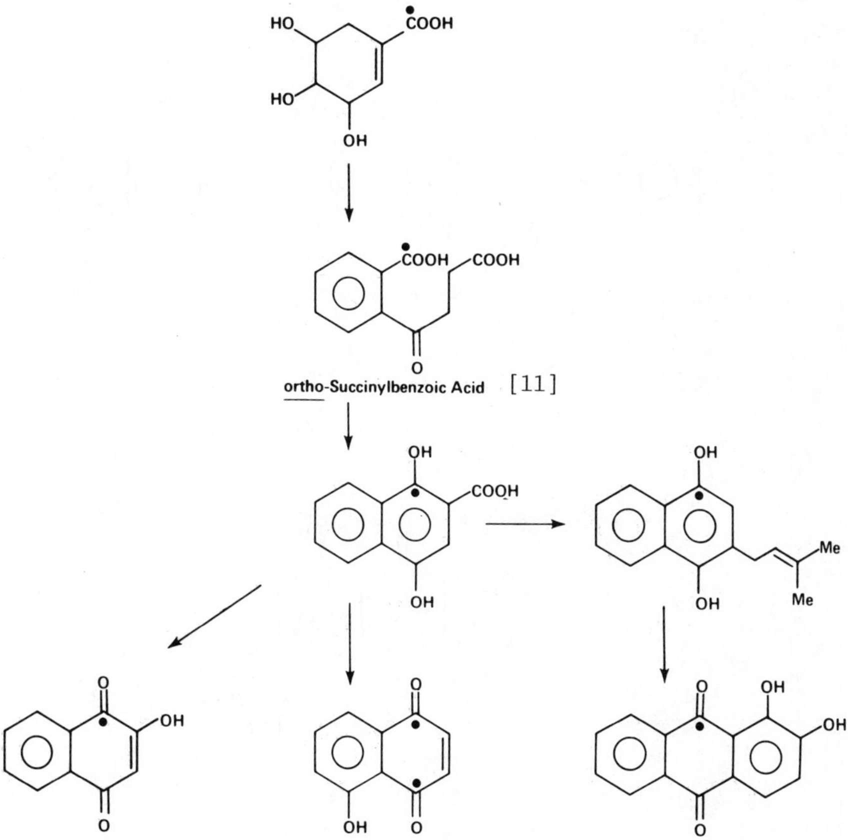

ortho-Succinylbenzoic Acid [11]

Lawsone [9]

Juglone [10]

Alizarin [14]

307

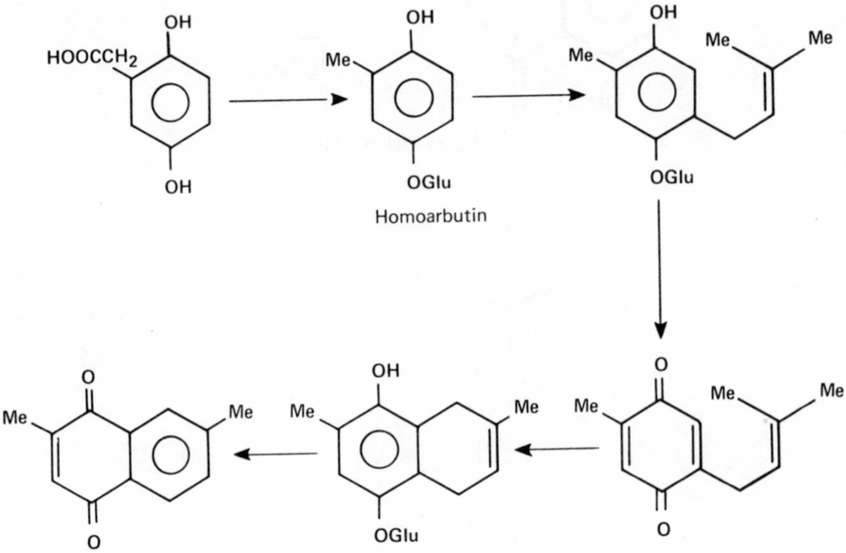

Homoarbutin

Chimaphilin [12]

Renifolin

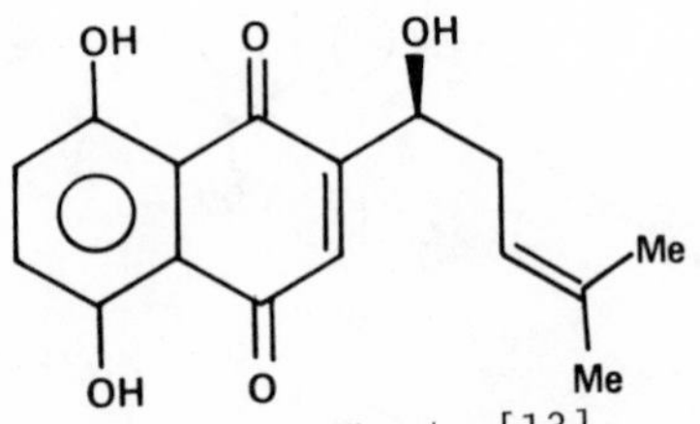

Alkannin [13]

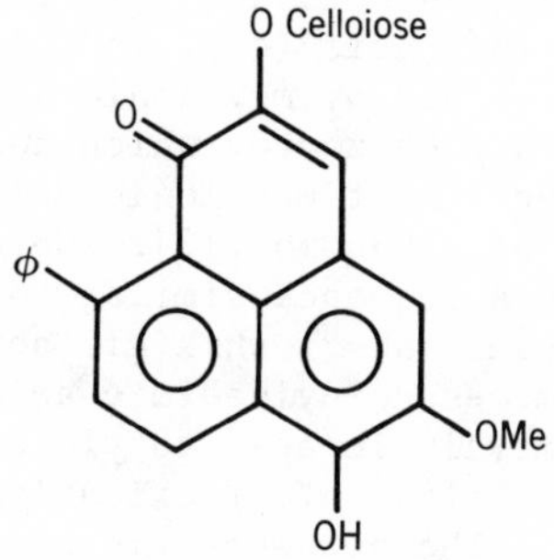

Haemocorin, a plant phenalenone [15]

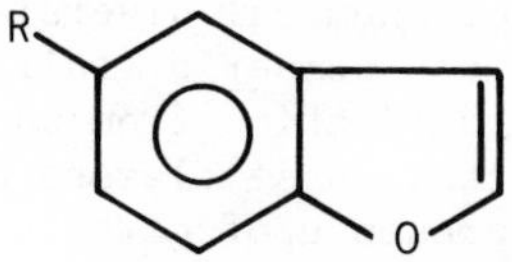

$\overline{R}=OMe$ [16a]
$R=CHOHCHOHCH_3$ [16b]

Benzofurans of *Stereum subpileatum*

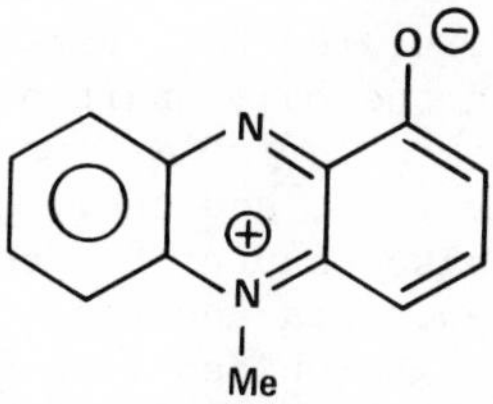

Pyocyanine [17]

on page 307. In the second pathway, this ring is formed from
tyrosine *via* homogentisic acid (e.g. in chimaphilin [12]) with the
additional carbons provided by mevalonic acid; and in the third
it arises from *p*-hydroxy benzoate, which combines with a unit
derived from two molecules of mevalonic acid to form the second
ring, as in the case of alkannin [13]. Certain anthraquinones
are biosynthesized by a sequence similar to the first one
described for naphthoquinones: shikimic acid combines with
glutamic acid to form *o*-succinylbenzoic acid, which reacts further
with a C_5 unit from mevalonic acid to yield the anthraquinone
nucleus (e.g. alizarin [14]) after oxidative ring closure; see
formulae on page 307. The sequences involving mevalonic acid
are discussed in detail in Section 22-6.

Shikimic acid also provides the aromatic ring of a variety
of metabolites whose structures do not fit readily into the
categories described above. Thus the aromatic rings of the
plant phenalenones (e.g. haemocorin [15]), of benzofurans (e.g.
[16a and b]), and of phenazines such as pyocyanine [17], have
all been shown to be biosynthesized from shikimic acid.

In spite of their wide chemical diversity, aromatic alkaloids
can be divided into a limited number of fairly well-defined
classes, based upon their derivation from one or other of the
primary aromatic metabolites; these classes often clearly show
their origin through the presence of certain common structural
features (2). The brief discussion which follows is illustrated
by the diagrams on pages 312-323.

Simple bases such as hordenine [18], mescaline [19], and
psilocybine [20] are formed more or less directly from the
corresponding amino acid; ephedrine [21], on the other hand,
incorporates a benzaldehyde unit and a C-C-NMe group.

An enzymic Mannich reaction between an aldehyde and a
substituted phenethylamine leads to the isoquinoline alkaloids.
Biosynthesis of simple molecules, e.g. anhalonidine [22] and
pellotine [23], involves the condensation of dopamine and
pyruvic acid which provides the C_2 unit; the more complex
1-benzylisoquinoline alkaloids incorporate a molecule of 3,4-
dihydroxy-phenylpyruvic acid. Barton, Battersby, and their
co-workers have elucidated the major part of the biosynthesis
of the alkaloids of *Papaver somniferum*: papaverine [24],
laudanosine [25], morphine [26], and their congeners; for the
biosynthesis of [26] see also Chapter 8, pages 202 ff. Substantial
progress has been made towards an understanding of that of the
other isoquinoline types: aporphines (see Chapter 8, pages 191ff.)
bis-benzyl-isoquinolines, chelidonine [27] berberine [28],
emetine [29], and many others. The same general biosynthesis
leads to alkaloids which do not contain the isoquinoline nucleus:

colchicine [30] and its relatives, alkaloids from plants of the
family Amaryllidaceae (e.g. galanthamine [31]), and alkaloids
related to mesembrine [32] found in the family Aizoaceae.

With few exceptions (e.g. gramine, gliotoxin, cryptolepine,
the ellipticin-euleine group), all the innumerable indole alkaloids
contain the tryptamine skeleton, i.e. the group-C-C-N attached
to position 3 of the indole ring; explanations for the exceptions
have been suggested. Tracer studies have established the
biogenesis of simple indoles, e.g. folicanthine [33], from
tryptamine, and the elegant work of Battersby, Arigoni, Kutney,
and others, has extended this to the complex Corynanthe-Strychnos,
Iboga, and Aspidosperma types exemplified by ajmalicine [34],
catharanthine [35], and vincamine [36], respectively. Similarly,
the total skeleton of the ergot alkaloids, such as agroclavine
[37] and ergometrine [38], can be derived from one tryptamine
and one mevalonate unit. Less obvious is the formation of
quinine [39], cinchonidine [40], and other *Cinchona* bases from
tryptamine and a C9 fragment closely related to the non-tryptamine
part of the Corynanthe-Strychnos family of alkaloids (cf. [34]);
the biosynthesis of [39] involves enlargement of the indole to
the quinoline system through insertion of the α-carbon of the
tryptamine side-chain into the heterocyclic ring.

There are several families of alkaloids whose aromatic rings
are provided by anthranilic acid. The anthranilic acid may be
derived directly from chorismate, in which case tryptophan does
not contribute to the biosynthesis, or it may be formed by the
normal degradation of tryptophan, in which case the incorporation
of tryptophan will be observed. Arborine [41] and viridicatin
[42], for example, are formed from anthranilic acid and phenylalanine, and arborinine [43] from anthranilic acid and acetate.
Dictamnine [44] and skimmianine [45] incorporate a mevalonate
unit into the furan ring, while the anthranilic acid residue
undergoes a double methoxylation. Anthranilic acid formed by
the degradation of tryptophan is found in the phenoxazinone ring
system which is derived from two molecules of an *o*-amino-phenol;
the system occurs in certain mold pigments (e.g. in the actino-
mycins [46]) and in the eye pigments of insects.

Again, there exist several nitrogen-containing metabolites
derived from skikimic acid, which do not readily fit into any
convenient chemical classes. Incorporation studies have shown
that phenylalanine, or tyrosine, is involved in the biosynthesis
of gliotoxin [47], betanin [48], chloramphenicol (chloromycetin)
[49] (see also Chapter 14), pyrrolnitrin [50], tylophorine [51],
and brevianamide [52].

The structure of volucrisporin [3] requires comment because
of its two *meta*-hydroxyphenyl group; this *meta*-orientation of

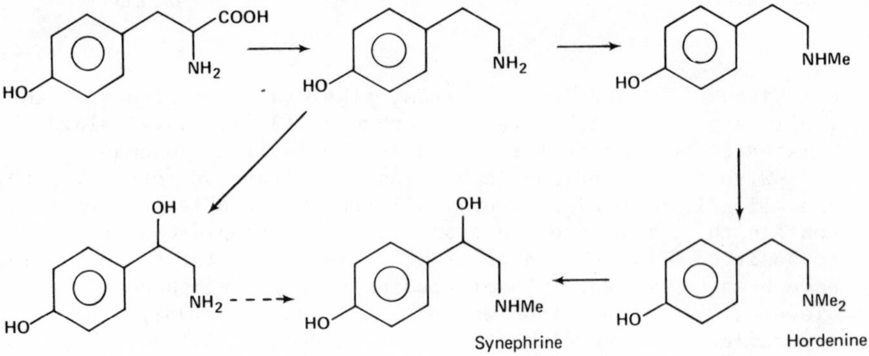
COOH
NH2
HO
NHMe
OH
Synephrine
NMe2
Hordenine

[18]

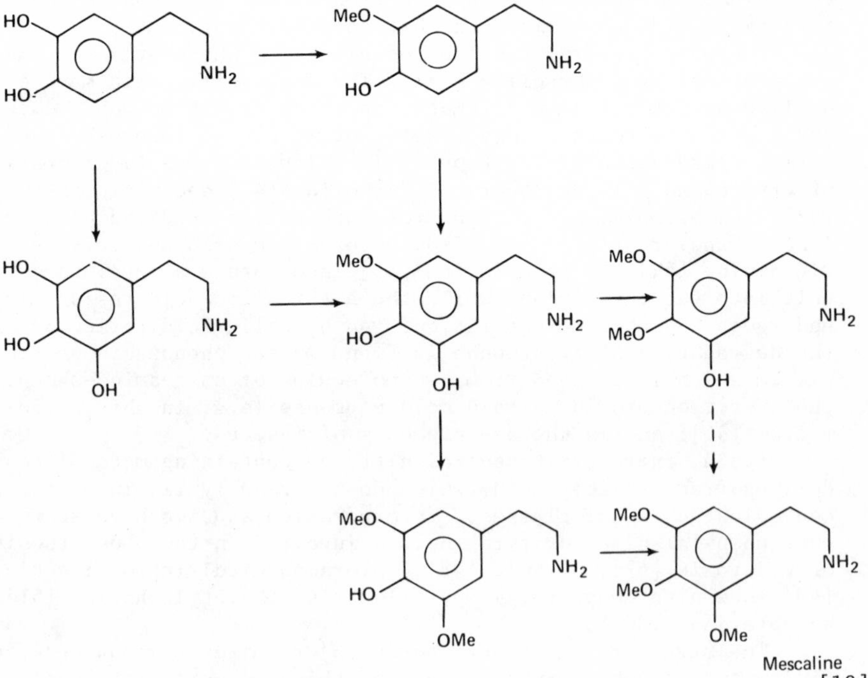
HO
MeO
NH2
OH
OMe
Mescaline

[19]

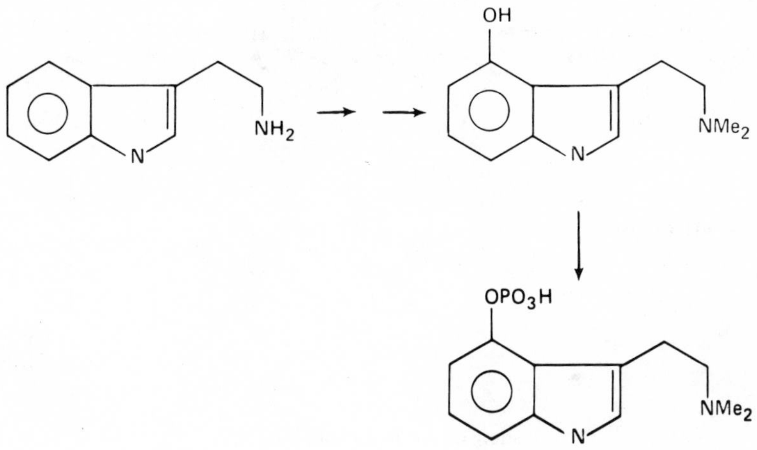

Psilocybine [20]

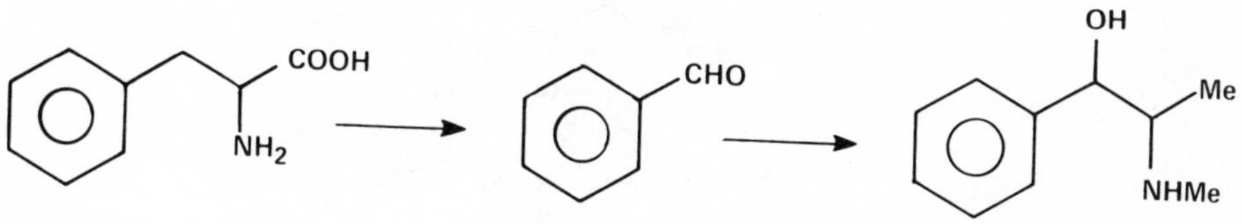

Ephedrine [21]

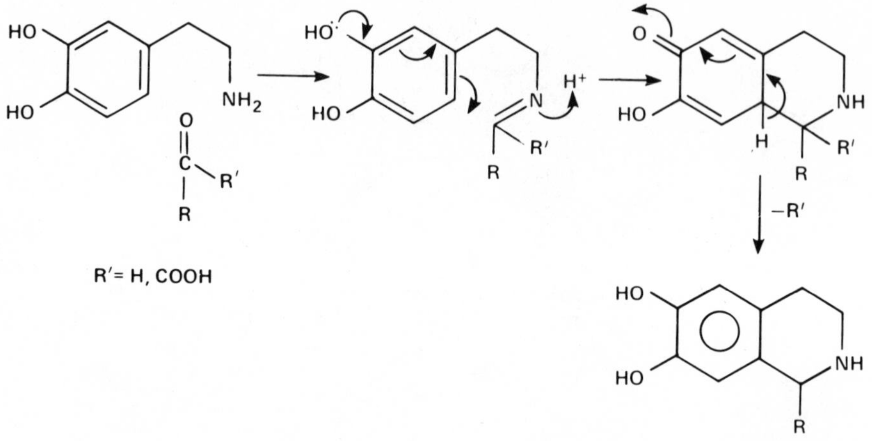

Formation of the isoquinoline nucleus

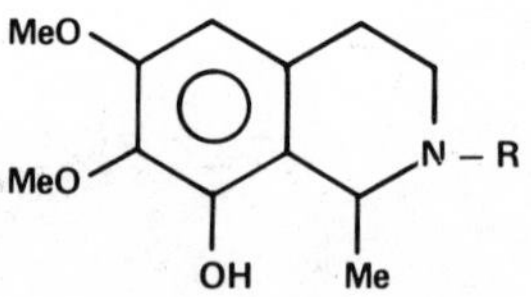

| Anhalonidine | R = Me | [22] |
| Pellotine | R = H | [23] |

314

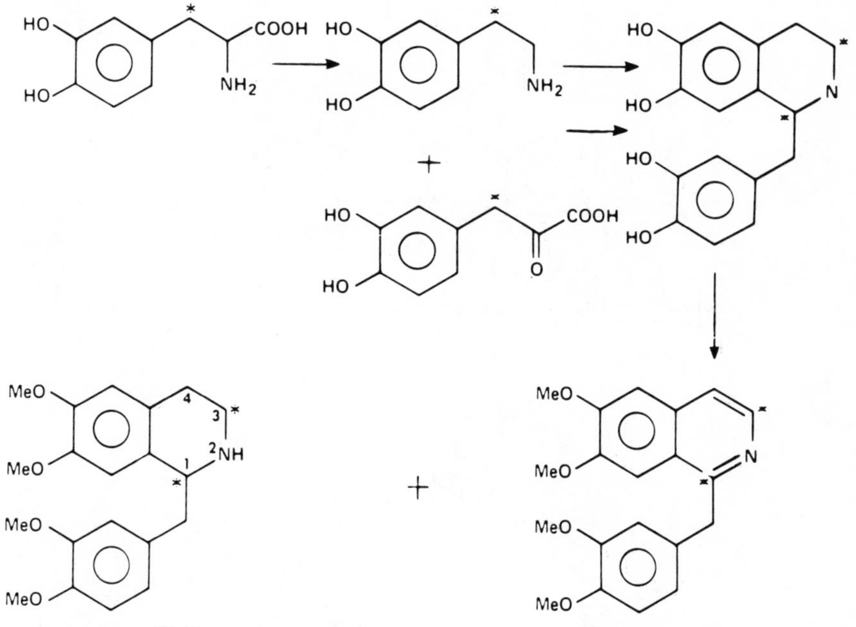

Laudanosine [25]

Papaverine [24]

The biosynthesis of benzylisoquinoline alkaloids

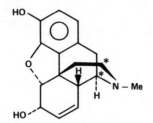

Morphine. Label from 2-¹⁴C-tryosine [26]

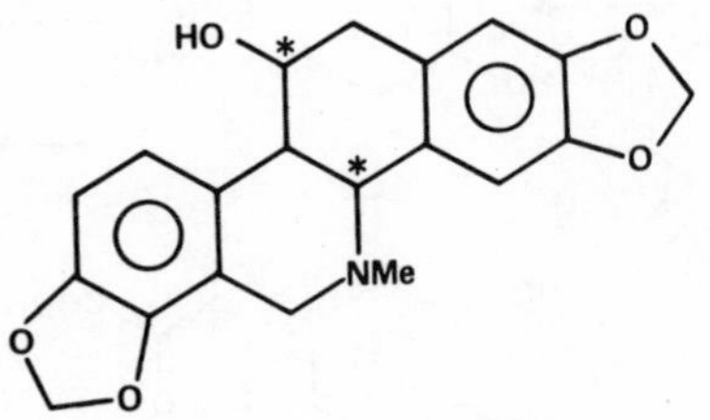

Chelidonine, Label from 2-^{14}C-tyrosine   [27]

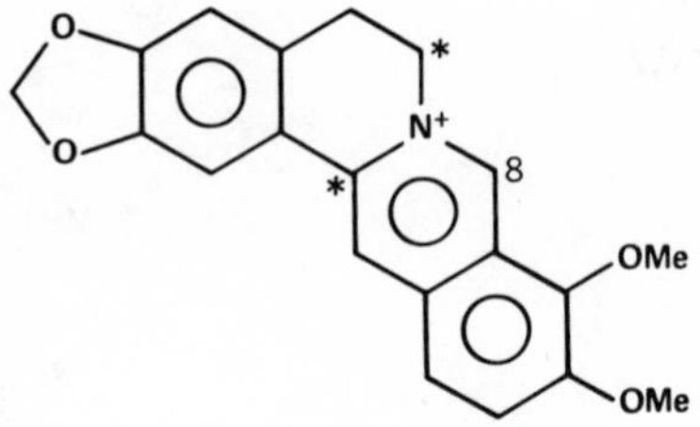

Berberine, Label from C-^{14}C-tyrosine [28]

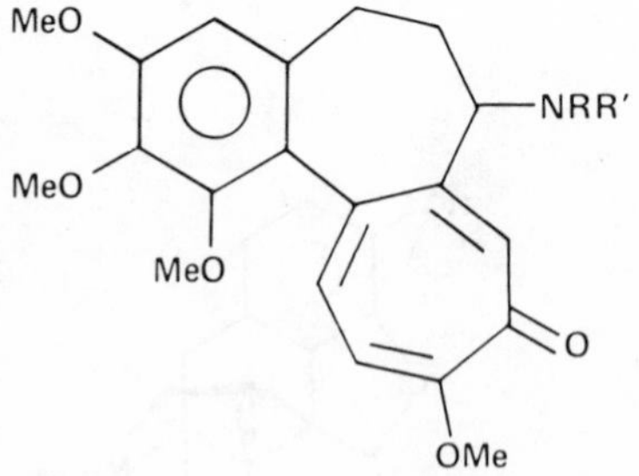

Colchicine   R = COMe, R' = H [30]

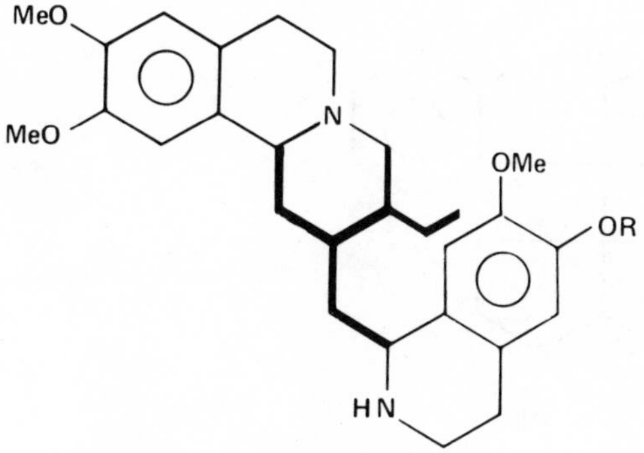

Emetine   R = Me [29]
Cephaeline R = H

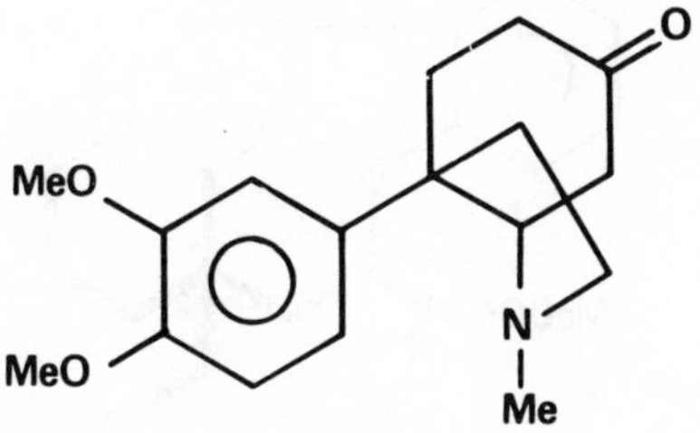

Mesembrine [32]

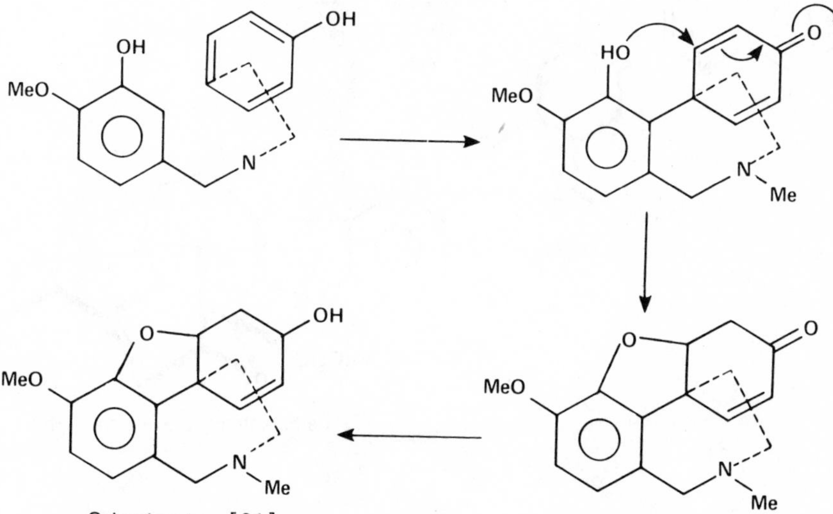

Galanthamine [31]

317

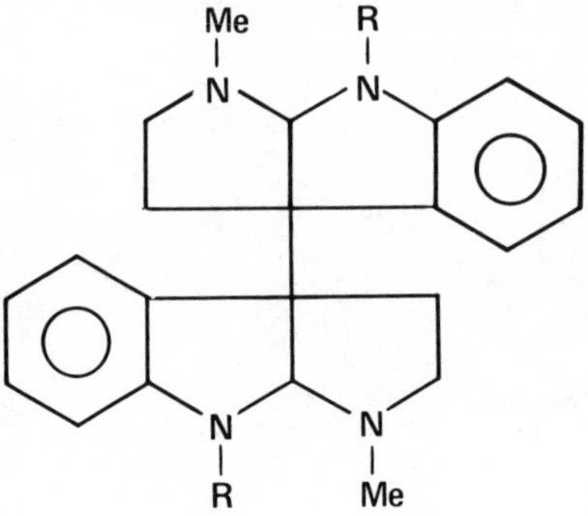

Folicanthine   R = Me [33]

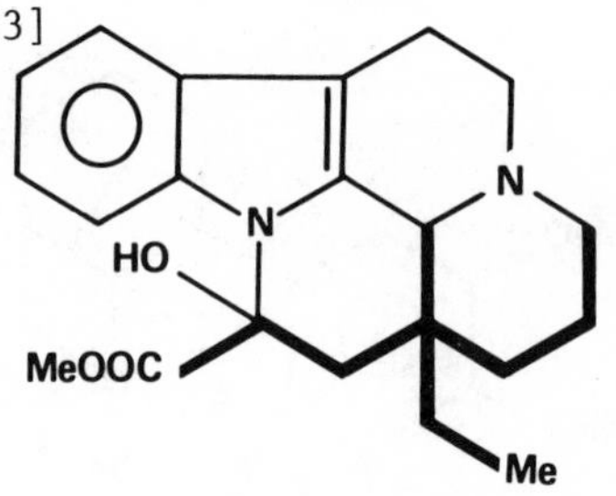

Vincamine-*Aspidosperma* type [36]

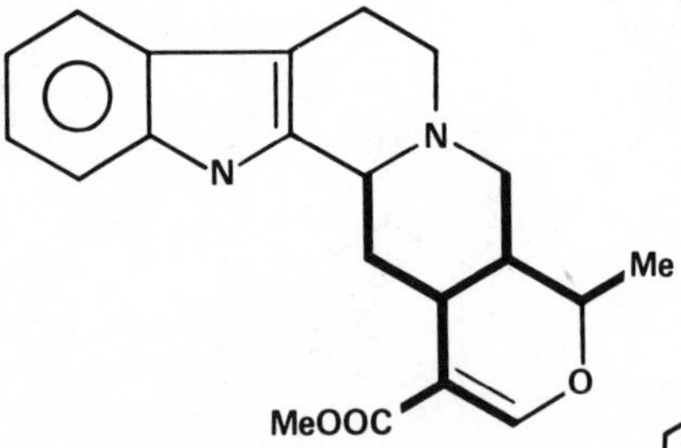

Ajmalicine. *Corynanthe–Strychnos* type
[34]

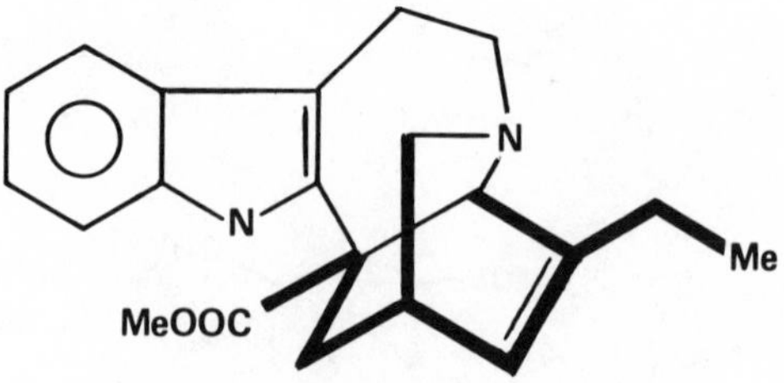

Catharanthine. *Iboga* type [35]

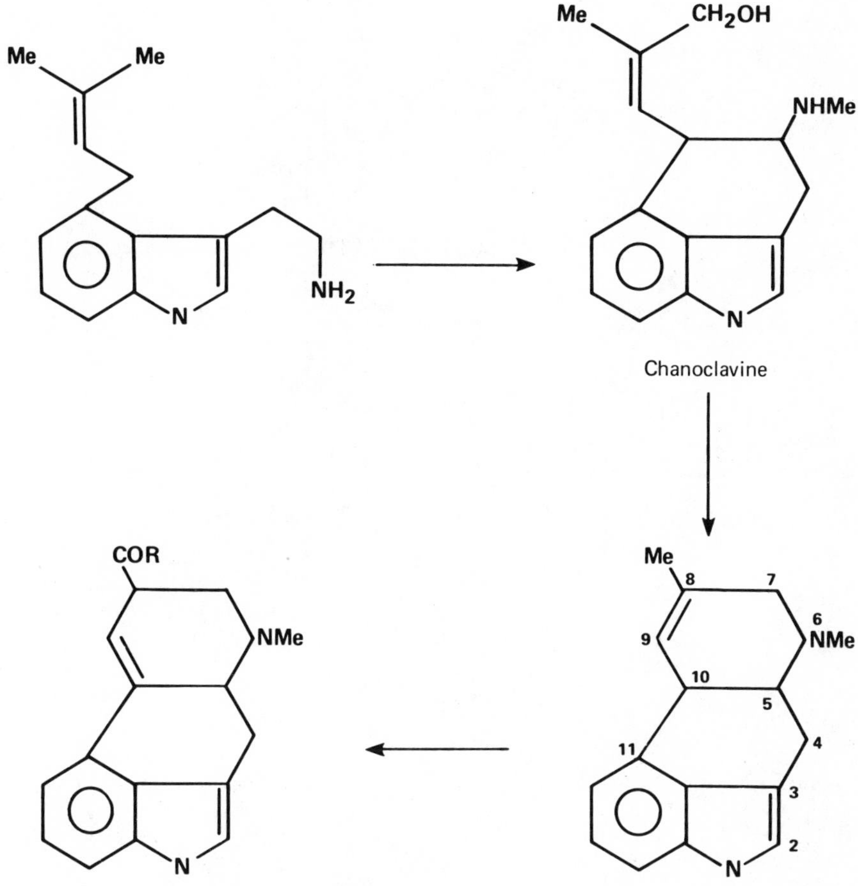

Chanoclavine

Agroclavine [37]

Lysergic acid   **R = OH**
Ergometrine   **R = NH CHCH$_3$CH$_2$OH** [38]

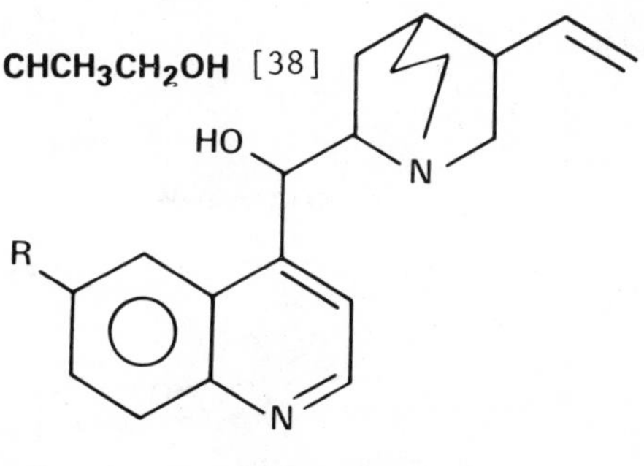

R = OMe   Quinine [39]
R = H   Cinchonidine [40]

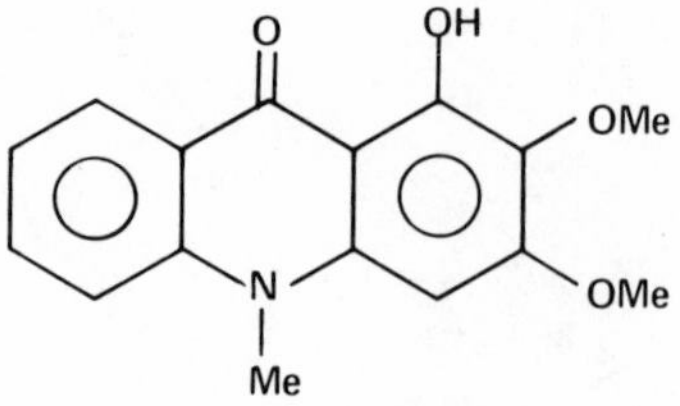

Arborinine [43]

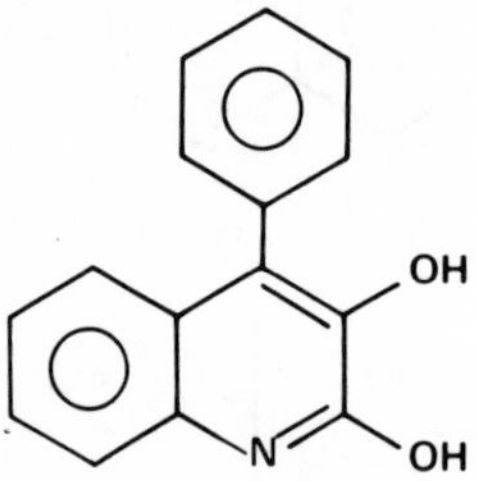

Viridicatin [42]

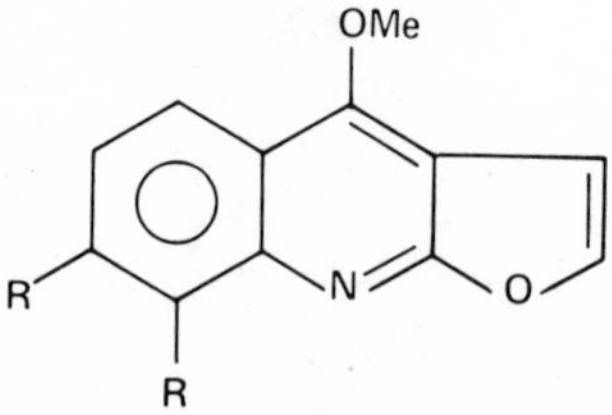

Dictamnine   R = H [44]
Skimmianine R = OMe [45]

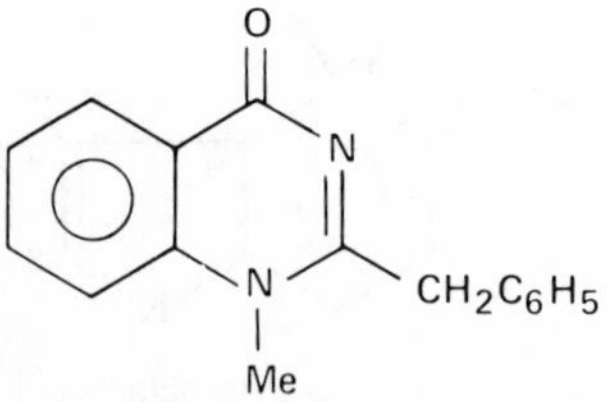

Arborine [41]

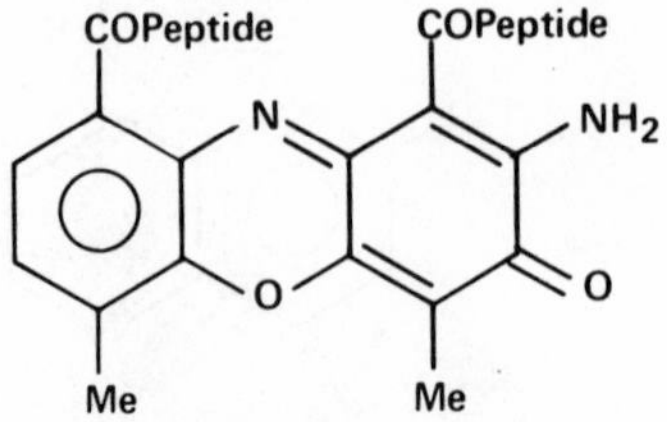

Actinomycin [46]

320

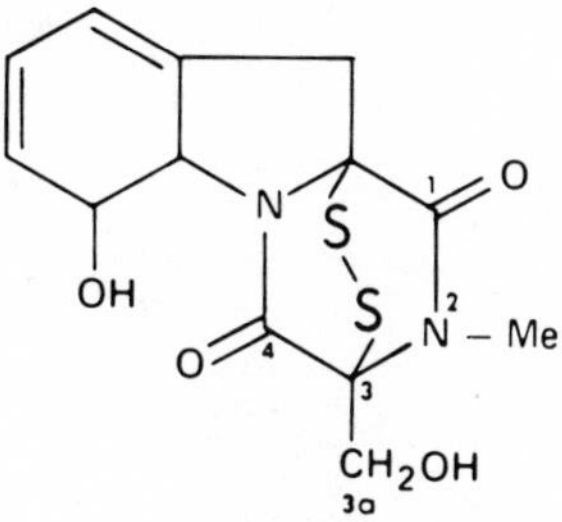

Gliotoxin [47]

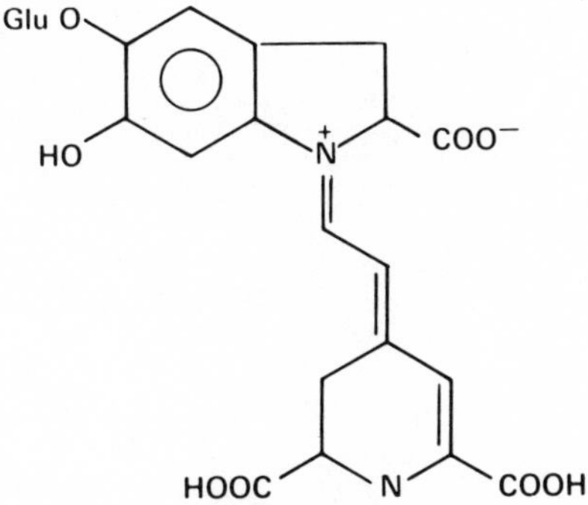

Betanin [48]

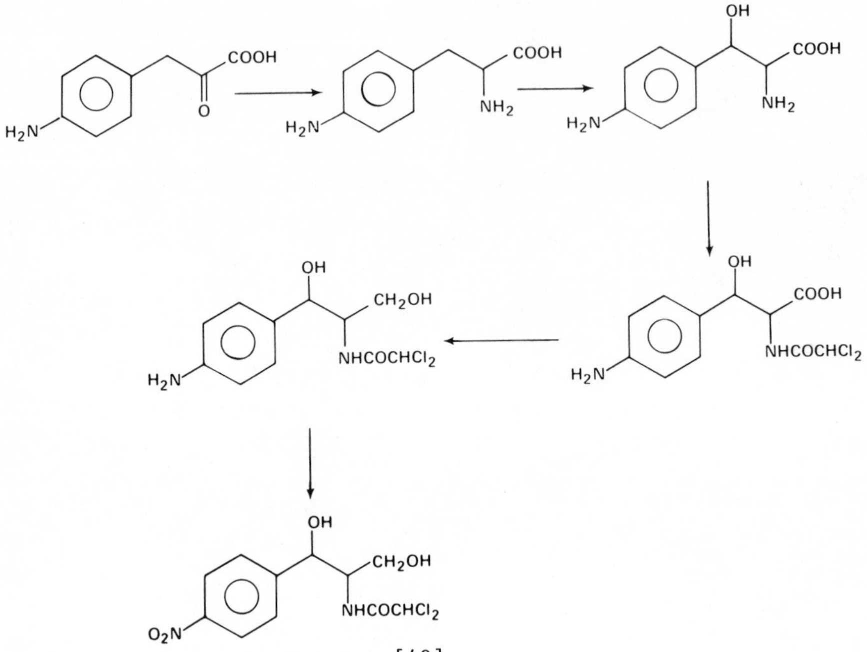

Chloramphenicol [49]

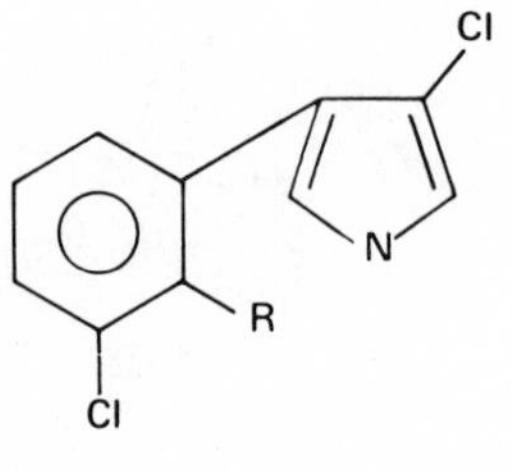

Pyrrolnitrin R = NO$_2$ [50]

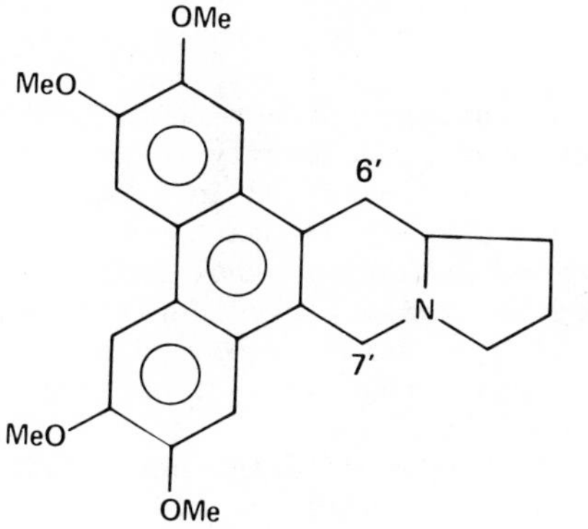

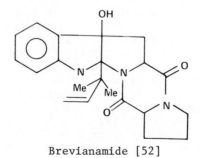

Tylophorine [51]

Brevianamide [52]

the hydroxyls clearly contradicts what has been said in Chapter 2 about the general structural features of shikimate-derived compounds. Origin of [3] from two C_6-C_3 units has been experimentally established.

The case of [3] is rare but not unique; *meta*-orientation of substituents in compounds undoubtedly formed through the shikimate pathway has been observed in a certain number of instances, of which a few may be mentioned. L-*meta*-Tyrosine occurs in the latex of *Euphorbia myrsinites* (3). Zierin, *m*-hydroxyphenylmandelonitrile glucoside, is found in *Zieria laevigata* (Rutaceae). In viridicatol, which occurs together with [42] in Penicillium spp., the C_6H_5- group of [42] is similarly replaced by 3-HO-C_6H_4-.

Such *meta*-orientations may originate in either of two ways: introduction of hydroxyl *meta* to the side-chain of an otherwise unsubstituted phenyl group, or removal of *para*-hydroxyl from the common grouping 3,4-$(HO)_2$-C_6H_3-. Both these modes have been encountered. In addition, the ready conversion of chorismic acid to *m*-hydroxybenzoic acid in vitro (see Chapter 6, p. 139) might provide a starting material for more complex metabolites, but this possibility remains theoretical.

Direct *meta*-hydroxylation of phenylacetic acid by an enzyme specific for this substrate has been observed in the mold *Rhizoctonia solani* (4). The analogous conversion of phenylalanine to *meta*-tyrosine is carried out by homogenates of beef adrenal medulla as a minor variant of the dominant *para*-hydroxylation to tyrosine, and the formation of dopa (5).

The alternative possibility: removal of phenolic hydroxyl in *para*-position to a side-chain, is likewise well documented. While the dopa formed from phenylalanine in beef adrenal medulla is not dehydroxylated to *meta*-tyrosine in this system (5), this reaction does take place in rat brain (6). The most commonly encountered type of this dehydroxylation is represented by the urinary excretion of *meta*-oxygenated aromatic substances after administration of the corresponding 3,4-dioxygenated substances to animals or man; similarly, tyrosine is converted to phenylalanine (7), and 3,4,5-tri-oxygenated compounds such as sinapic acid yield 3,5-dioxygenated metabolites (8). These reactions are actually carried out by the intestinal microflora: they are suppressed by administration of certain antibiotics or other intestinal sterilants (neomycin, sulfathiazole). The analogous removal of the *meta*-hydroxyl from 3,4-dihydroxyphenyl compounds has likewise been observed but occurs to a minor extent only. These *meta*- and *para*-dehydroxylations have been reviewed (8,9).

It is not known with certainty how the dehydroxylations just discussed take place. One mechanism for such processes is well established and has been discussed in Chapter 8: conversion of a *para*-substituted phenol to the dienone, reduction to the dienol, and re-aromatization obviously results in loss of the phenolic group in *para*-position to a side-chain. Many examples for the actual occurrence of this sequence are given in Chapter 8. This process could account for the removal of hydroxyl in *para*-position by the intestinal microorganisms, but not for the (minor) *meta*-dehydroxylation which does take place, and it has been shown not to be valid in the case of [3], where it had been assumed to occur: [3] incorporates label from *meta*-tyrosine.

The removal of phenolic hydroxyl from either position 3 or 4 by the intestinal microflora has led to the suggestion that reduction of the 3,4-dihydroxyphenyl group to the dihydrodiol takes place, which could re-aromatize with loss of either hydroxyl. This process, so far hypothetical, would be analogous to the sequence which leads from kynurenic acid via the dihydrodiol to 8-hydroxykynurenic acid; see Chapter 8, page 212.

Loss of phenolic hydroxyl has been much studied in higher plants, where administration of labeled 3,4-di- or 3,4,5-tri-hydroxyphenyl compounds (cinnamic acids have been mostly studied) leads to labeling of the derived 4-hydroxy- or 3,4-dihydroxy-phenyl compounds; the incorporation is usually much inferior to that into the compounds with unchanged oxygenation pattern. The The mechanism of these *meta*-dehydroxylations does not seem to be known (10).

REFERENCES

1. E. Haslam, The Shikimate Pathway, Wiley, New York 1974.
2. S. Pelletier, Chemistry of the Alkaloids, Van Nostrand Reinhold, New York 1970.
3. K. Mothes, H.R. Schütte, P. Müller, M. v. Ardenne, and R. Tümmler, Z. Naturforsch. 19b, 1161 (1964).
4. K. Kohmoto and S. Nishimura, Phytochemistry 14, 2131 (1975).
5. J.H. Tong, A. D'Iorio, and N.L. Benoiton, Biochem. Biophys. Res. Commun. 44, 229 (1971).
6. A.A. Boulton and L. Quan, Canad. J. Biochem. 48, 1287 (1970).
7. T.W. Scott, P.F.V. Ward, and R.M.C. Dawson, Biochem. J. 87, 3P (1963); 90, 12 (1964).
8. T. Meyer and R.R. Scheline, Xenobiotica 2, 383 (1972).
9. R.R. Scheline, Acta Pharmacol. et Toxicol. 24, 275 (1966).

PART 2 BIOSYNTHESIS OF AROMATIC COMPOUNDS FROM ACETIC ACID

CHAPTER 17

Aromatic Systems Derived Through the Polyketide Pathway

17-1. INTRODUCTION

The second major biosynthetic pathway+ leading to aromatic rings is furnished by the head-to-tail combination of -CO-CH_2- units derived from acetate and malonate. In many respects this pathway stands in marked contrast to the one discussed in Chapter 2, which leads to aromatic rings from glucose via shikimic acid. In the latter pathway, a long series of individual steps involving well-defined chemical compounds yields a limited number of primary aromatic metabolites of relatively simple structure; it is the further elaboration of these primary benzenoid substances that provides a wide variety of final products, of which several play an extremely important biochemical role. In contrast, the acetate-polymalonate pathway appears to lead quite directly from the starting materials to a surprisingly wide array of aromatic structures. Few, if any, intermediates are known, and it is doubtful whether discrete chemical steps between the starting materials and the aromatic products exist. Furthermore, this pathway seems to furnish mostly so-called "secondary" metabolites, that is, compounds whose importance, while often marked, is not concerned with their further metabolic fate.

There are several features that distinguish these secondary metabolites from the compounds more typical of the shikimic acid pathway. The striking taxonomic selectivity of their production,

---

+General reviews of this pathway since 1964 are due to Birch [1966 (1, 2) and 1968 (3)], Shibata [1967 (4)], Bentley and Campbell [1968 (5)], Whalley [1968 (6)], and Turner [1971 (92)].

their impact on regulation of growth processes, ecology, species survival, and other functions are dealt with in the general Introduction (Chapter 1).

17-2.   ASPECTS OF THE POLYKETIDE MODE OF BIOSYNTHESIS

a.   General Considerations.

The biosynthesis of aromatic compounds by the head-to-tail linkage of acetate units to form a "polyketide" was suggested over 60 years ago by Collie (7) on the basis of laboratory observations; he found that diacetylacetone [1] would cyclize under strongly alkaline conditions to orcinol, and under less vigorous conditions to the naphthol [2], and that dehydracetic acid [3], obtained by pyrolysis of ethyl acetoacetate, would yield orsellinic acid under similar conditions.   Interestingly,

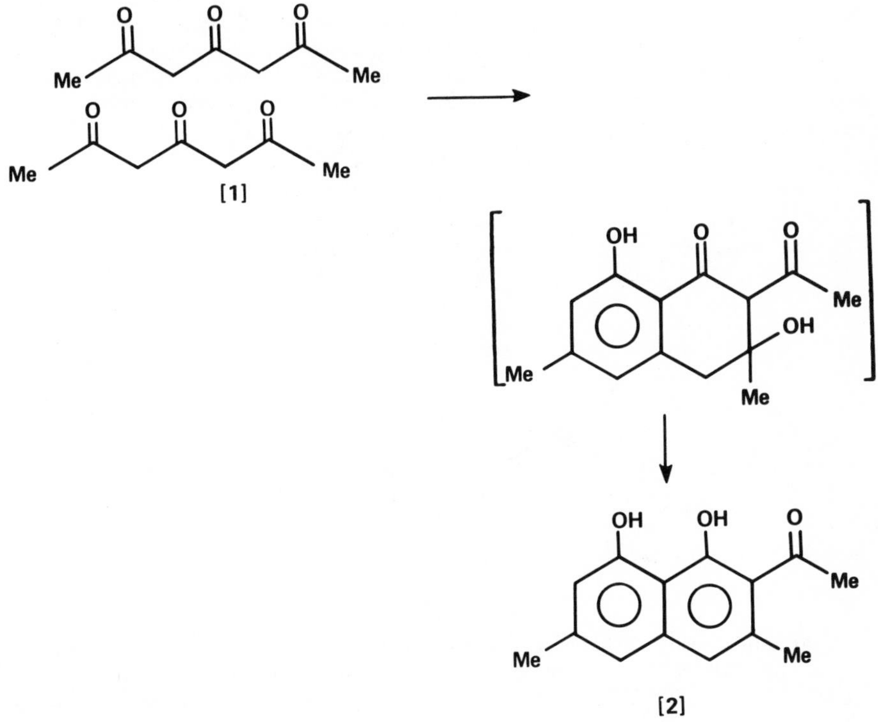

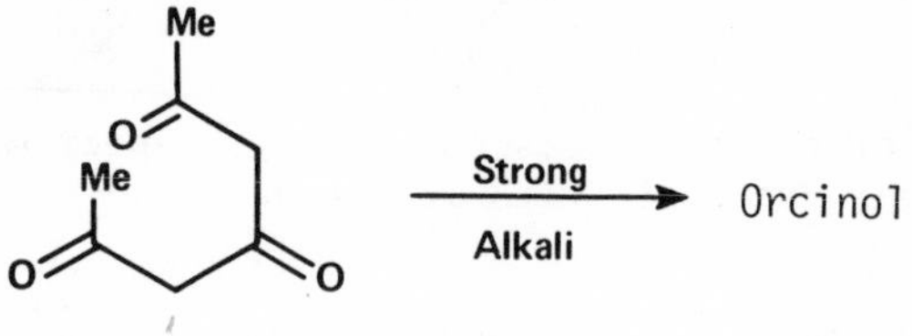

[1]

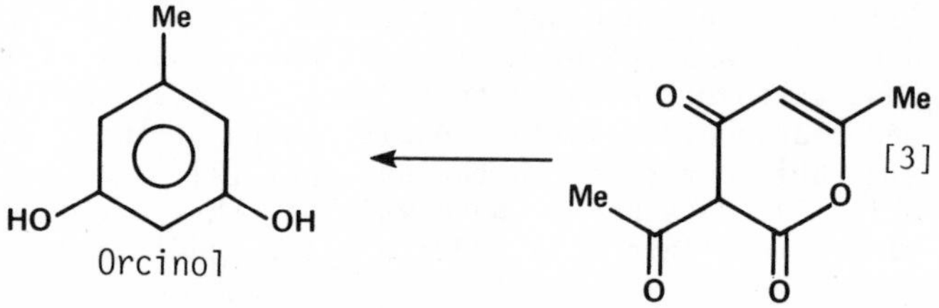

Orcinol

[3]

dianellin [4] (8), 6-hydroxymusizin-8-O-β-D-glucoside [5] (9), and torachrysone [6] (10), which show a striking resemblance to [2], have all been isolated recently from natural sources, while orcinol is, of course, a well-known natural product. Collie also obtained heterocyclic compounds containing oxygen and related to the naturally occuring pyrones.

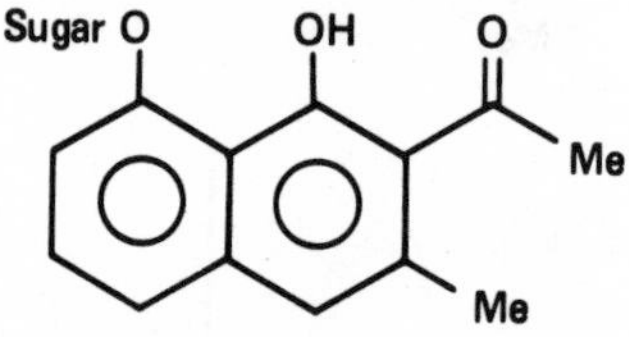

[4]

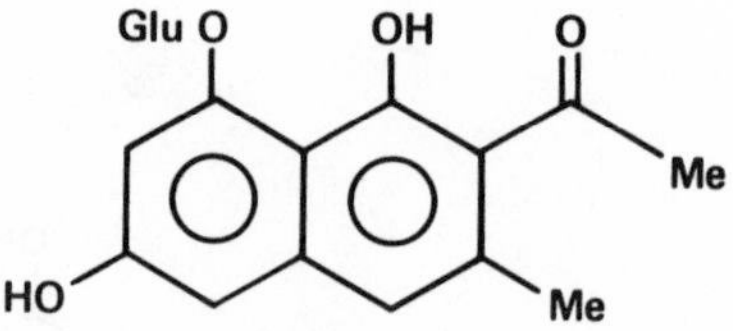

[5]

    At the time, Collie's ingenious ideas failed to provoke the deserved interest, presumably because they could be neither tested experimentally nor correlated with the then known facts of biochemistry. Furthermore, contemporary workers may have felt that practically any organic structure could be built up by a combination of two-carbon units, and that acetic acid, which

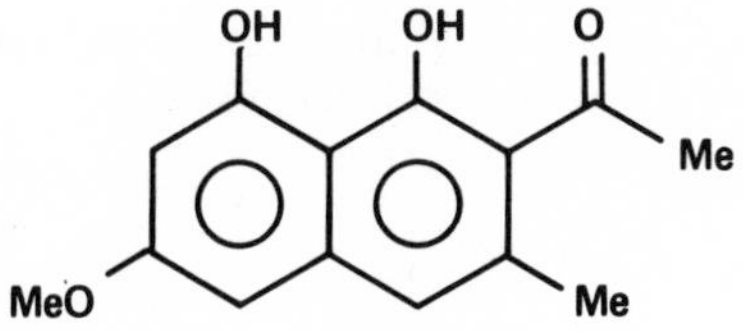

[6]

was known to be a very stable compound, was hardly suitable as a
starting material for a biosynthetic pathway. The subsequent
recognition of the role of acetic acid in the biosynthesis of
fatty acids and terpenoids (11) led A.J. Birch to postulate its
function as a precursor of a wide range of natural aromatic compounds (12-14). The structures, and in particular the location
of the oxygen substituents on alternate carbon atoms in many
such compounds are consistent with a derivation through head-to-
tail junction of -CO-CH$_2$- units. Birch and Donovan (15) suggested
that a preformed polyacetate chain might cyclize in either one of
two general ways to give rise to aromatic compounds, the first,
an aldol-type condensation, yielding β-resorcylic acids [7], and
the second, an internal Claisen acylation reaction, forming

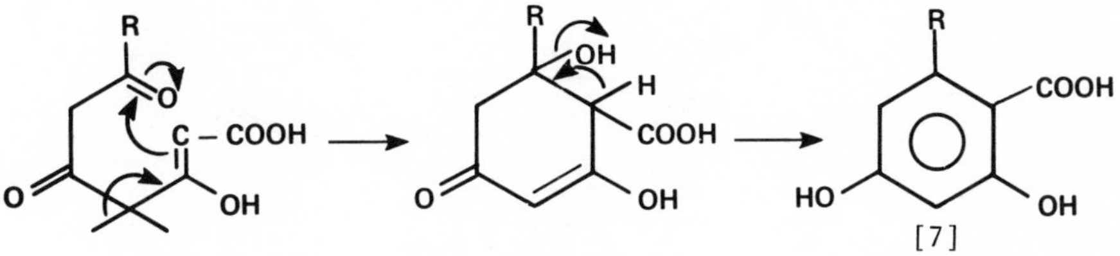

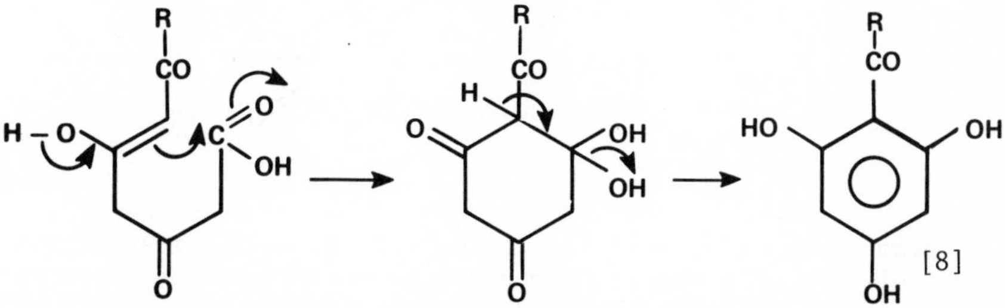

acylphloroglucinols [8]. A nice example of a case in which both
types of reaction occur together in the same organism is mentioned in Chapter 18. Similar, mechanistically acceptable,
cyclizations were assumed to give rise to pyrone ring systems
[9], which also form the basis of a great number of natural
products, and simple lactonization of the polyketide chain would

give rise to the δ-lactone structure found in many metabolites, for example, 4-methoxyparacotoin [10] (16).

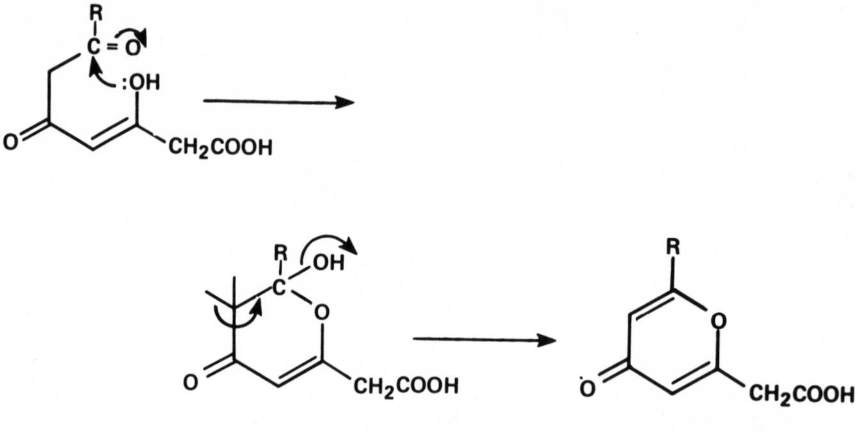

[9]

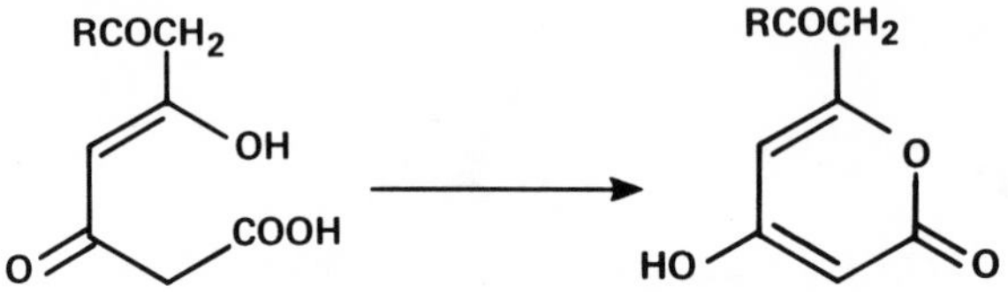

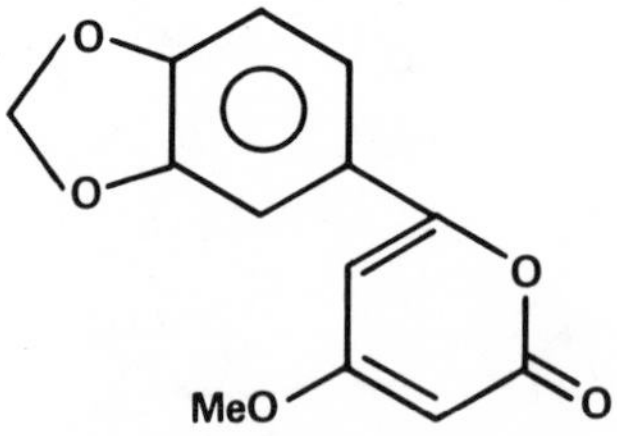

[10]

Before embarking upon a detailed account of the biosynthesis of these classes of aromatic compounds, we present a brief summary of the types of acetate-derived structures that are generally encountered; they are classified here in terms of the inferred modifications that are made to the polyketide chain before it is cyclized. Although several of the examples have been investigated experimentally (Chapter 18), many are included to illustrate specific features of the pathway even though the derivation of

the compounds from acetate is only implicit.   Some of the problems and more esoteric aspects of the pathway are also discussed.
     There are a few examples of aromatic compounds that are
formed by the cyclization of a polyketide without prior modification; in these cases all the oxygen and carbon atoms found in
the metabolite are those originally present in the precursor,
excepting those carbonyl oxygens involved in the process of

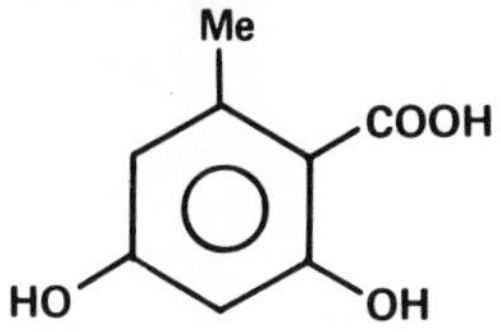

[11]

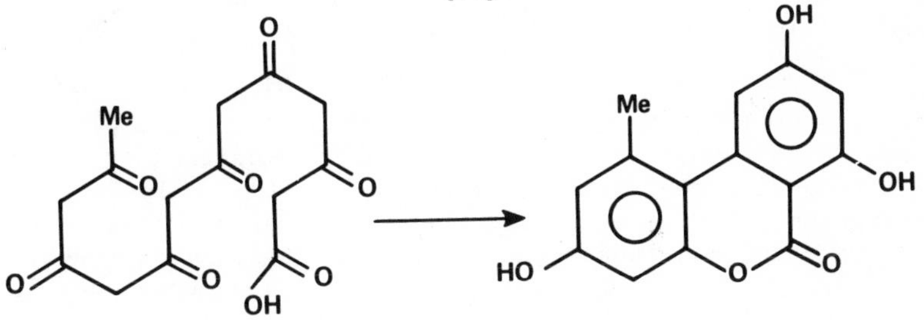

[12]

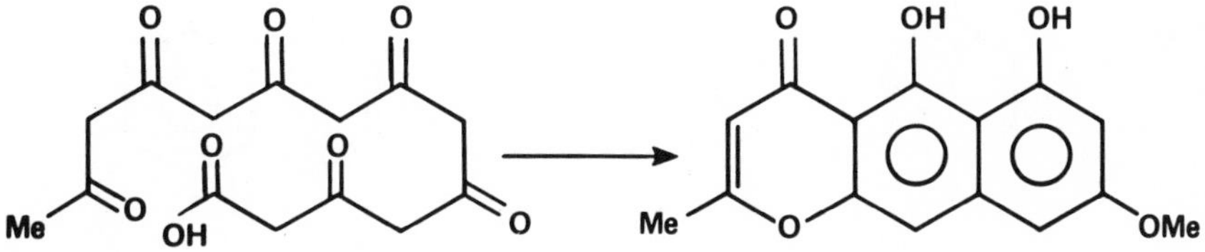

[13]

aromatization.   Orsellinic acid [11], alternariol [12], rubrofusarin [13], and eleutherinol [14], are representative of this
most simple class of compound.
     From a consideration of the structures of many compounds, it
is clear that their formation has taken place with the loss of
the terminal carboxyl group of the polyketide, and in some cases
this is the only modification that is observed; the loss of a
carboxyl group from a β-keto acid is of course quite unexceptional.

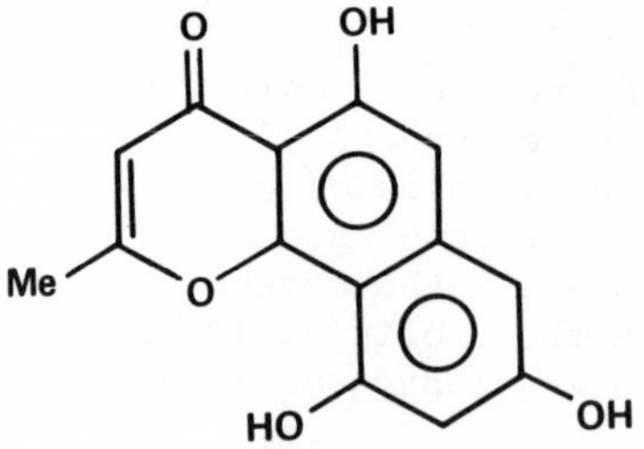

[14]

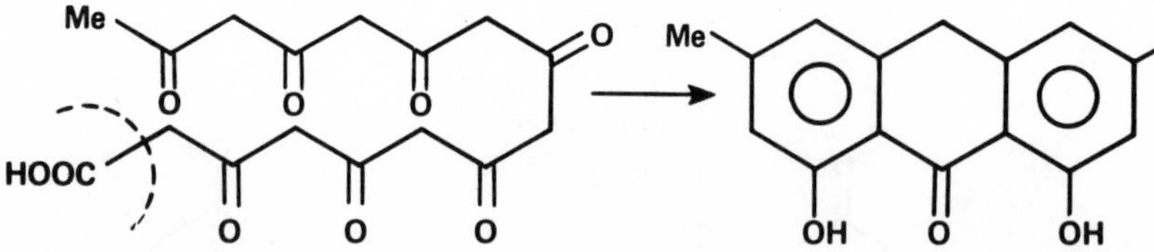

[15]

Torachrysone [6], and emodin-9-anthrone [15] are examples of compounds whose formation involves this minor modification.

A reductive modification of the polyketide chain followed by dehydration prior to or during the process of aromatization leads to compounds lacking one or more of the oxygen functions of the precursor; 6-methylsalicylic acid, the chromanone [16], and paeonol [17], which co-exists with its 5-hydroxy (nonreduced) analog, are simple examples of this loss of an oxygen function.

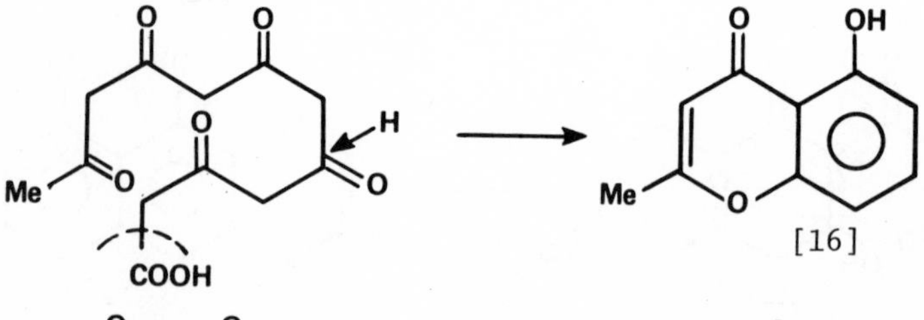

[16]

[17]

A general rule concerning the retention of oxygen has been noted by Turner (92): if a compound is formed as the result of a condensation involving the CH_2 group which is α to the terminal COOH, then the oxygen of the β-keto group is usually retained.

Substituents other than oxygen which are found on compounds derived from this pathway are normally located on those carbons that were originally present as the methylene groups of the polyketide chain; the introduced groups are usually alkyl or oxygen functions, but may also be halogen, or C-glycosyl residues.+ Possibly the most common introduced group is a simple C_1 fragment, as in [18]. The group may also exist in its oxidized forms as $-CH_2OH$ [19], -CHO [20], and -COOH [21]; some of these examples demonstrate more than one modifying influence. There are a few instances (e.g., javanicin [22],) in which a nuclear $-CH_3$ group has been shown to be formed by reduction of a terminal carboxyl. Isoprenoid groups may be isopentenyl [23], or

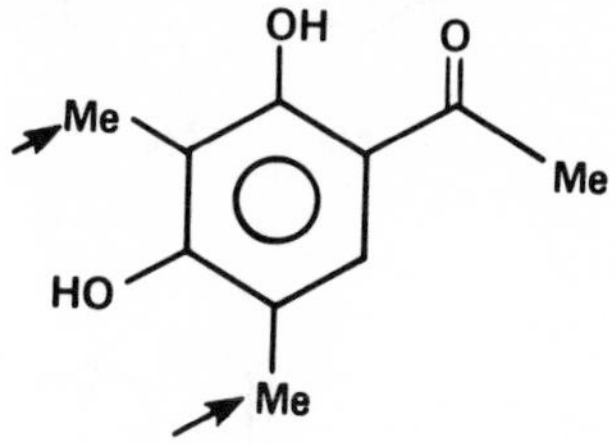

[18]

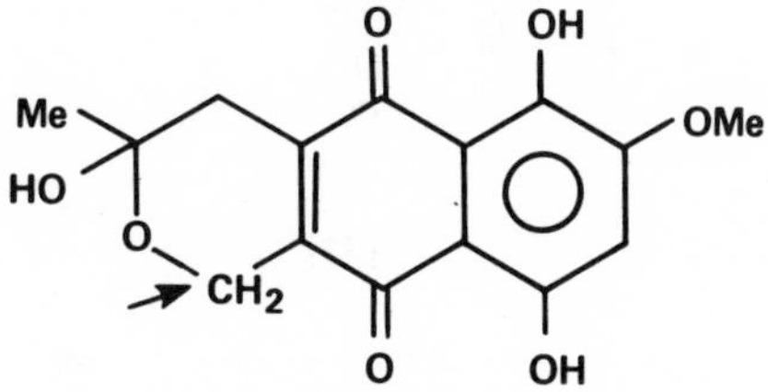

[19]

+Some less usual substituents are also found, for example, N-methylpyrrolidine in ficin and isoficin (90), a cyclic C_9-lactone in obtusifolin (91), and a butyl residue in fusamarin (93).

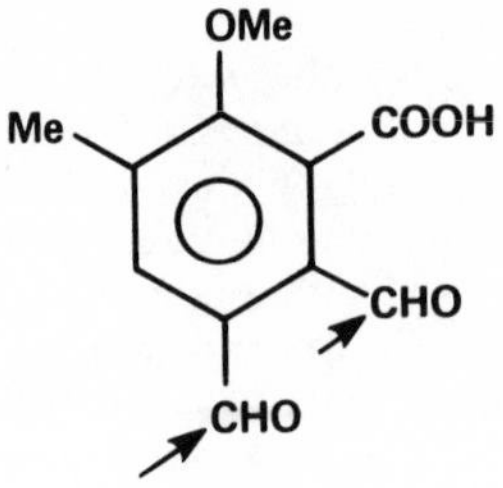
OMe
Me
COOH
CHO
CHO

[20]

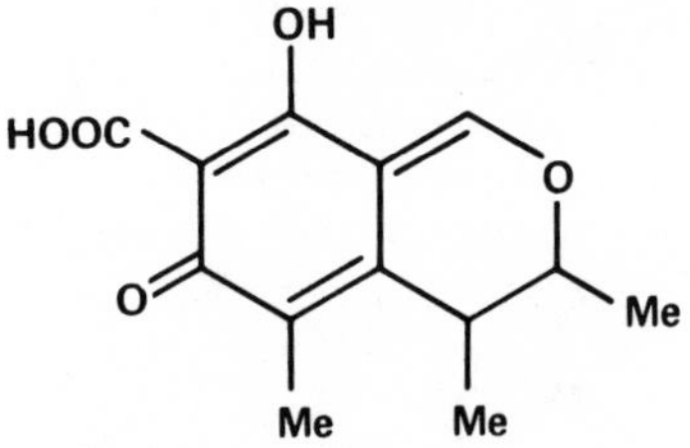
OH
HOOC
O
O
Me
Me
Me

[21]

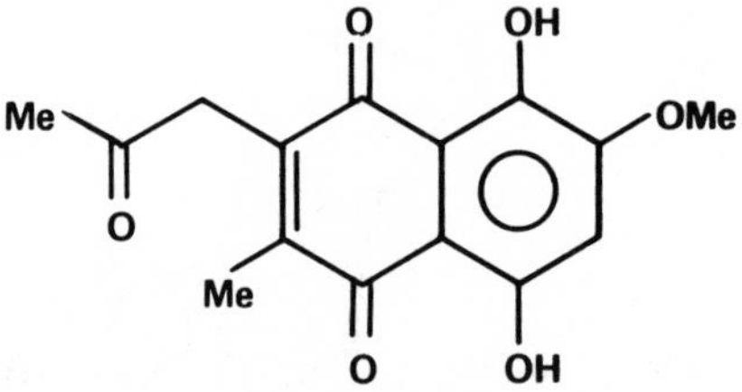
O
OH
Me
OMe
O
Me
O
OH

[22]

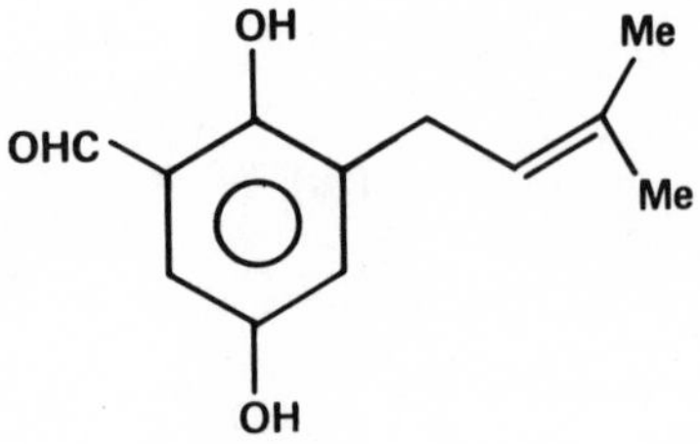
OH
Me
OHC
Me
OH

[23]

[24]

[25]

[26]

poly-isoprenoid [24]; an introduced oxygen function may be phenolic or quinonoid: both are seen in [22]. Nuclear halogen is
found for example in the depside [25], and an introduced C-glyco-
side group in barbaloin [26].

Although in some of Collie's early laboratory experiments
aromatic compounds were generated through combination of two
individual chains, aromatization of a single polyketide chain
seems to be the general rule for naturally occurring compounds.
This is, however, not completely without exceptions, and there
are an increasing number of compounds whose structure can only
be devised from two distinct polyketide precursors, and for
which a two-chain biosynthesis is proposed. As is explained
later in Part 3, evidence for this mode of biosynthesis is
available because of the possibility of a relatively high ^{14}C-
incorporation into a starter unit (i.e., the CH_3CO- group derived
from the acetyl-CoA which initiated the polyketide chain) if
$2-^{14}C$-acetate is fed to an organism in the presence of nonactive
malonate, or alternatively of a low incorporation if $2-^{14}C$-malo-
nate is administered together with nonlabeled acetate.

The biosynthesis of sclerin, citromycetin, mollisin, and
sulochrin, which are fully aromatic compounds of this type, is
discussed in Chapter 18, but a brief mention of the biosynthesis
of a small group of nonaromatic compounds, also derived from two
chains, is included here for two reasons: firstly because of the
novelty of this type of biosynthesis and the techniques involved
in its study, and secondly because the compounds represent an
interesting group of blocked aromatic structures: in each case
the aromaticity of the left-hand ring is prevented by the introduction of the methyl group from the C_1 pool. Incorporation of
^{14}C-acetate and malonate into the fungal metabolites sclerotiorin
[27], (17, 18), rotiorin [28] (19, 18), rubropunctatin [29],
(19, 20), monascin [30] (19), and monascorubrin [31] (20), has
indicated that two polyketide chains are utilized in their
biosynthesis.

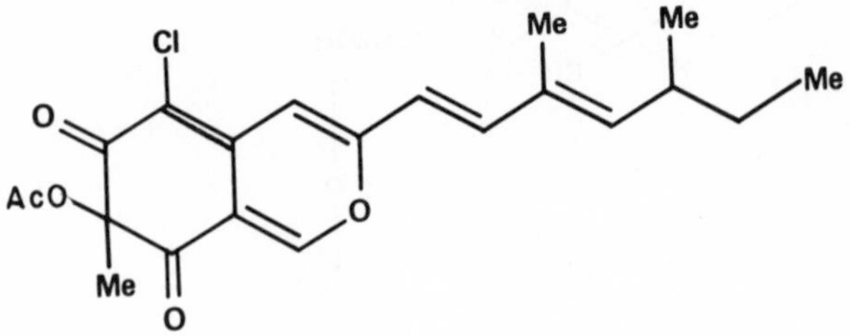

[27]

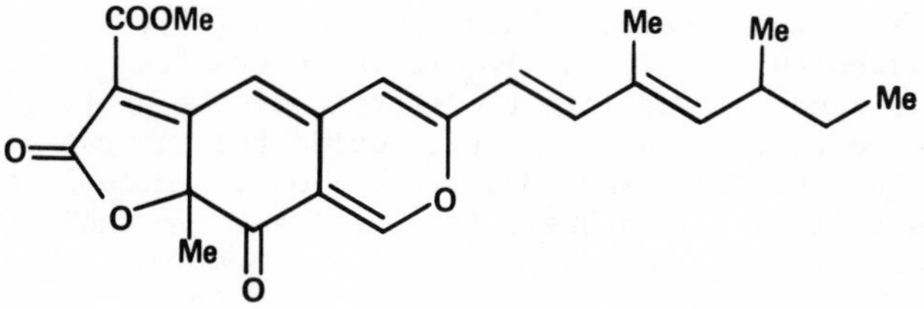

[28]

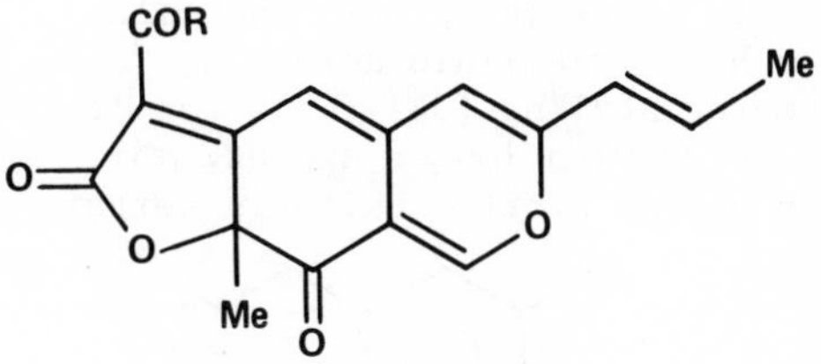

[29] R = C₅H₁₁
[31] R = C₇H₁₅

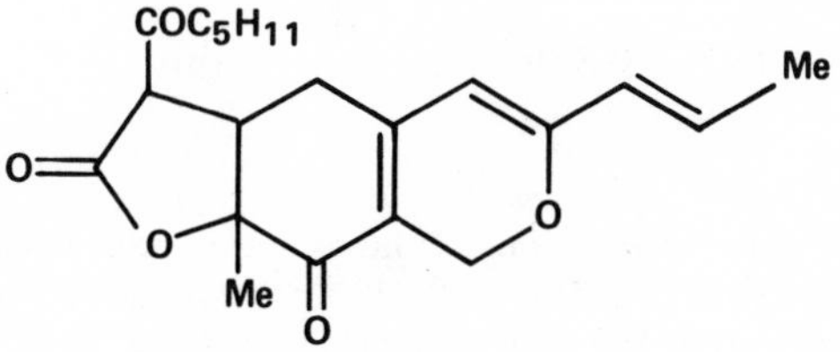

[30]

In each case one chain forms the basic ring system, and the
other the additional carbon chain (CH_3COO in [27], and the β-keto-
lactone system in [28] to [31]). It is interesting that labeled
butyrate is incorporated into the lactone system of [28] and
similarly, hexanoate and octanoate provide the saturated chains
of the metabolites [29] and [31] (20).[+] Other metabolites whose
structures imply a biosynthesis from at least two chains have
been isolated from a *Fusarium* sp. (93) and *Epidermophyton flocc-
osum* (98).

Another interesting feature of the acetate-polymalonate
pathway is the relative nonspecificity of the starter unit of the
polymalonate chain. Acetyl-CoA provides the first unit in a
true polyketide, but it can be replaced by several other acids
(presumably activated as their CoA-esters), as exemplified by the
compounds discussed below; again, experimental verification is
available in some cases and these are referenced; the remaining
cases rest upon inference alone.

Saturated fatty acids (propionic and malonic) provide the
starter units for homo-orsellinic acid [32] (21), rutilantinone
[33] (22), and oxytetracyline [34] (23), and it seems probable
that other fatty acids with longer chains, which may or may not
be saturated, can act as starter units in certain cases. That

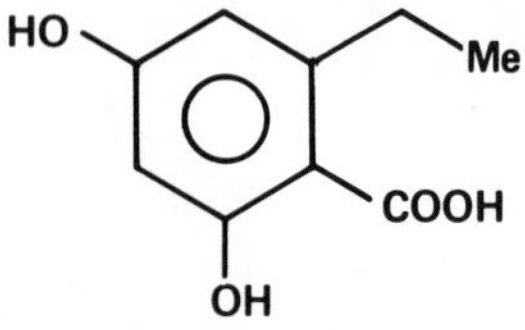

[32]

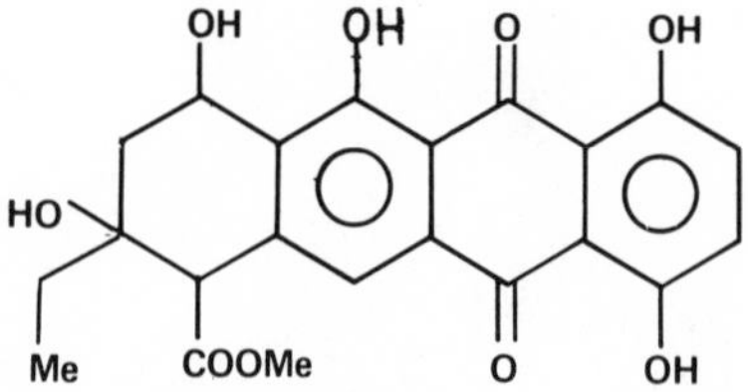

[33]

[+]A similar direct incorporation of hexanoate has been observed
in the biosynthesis of coniine (89). This is an unusual observation, since aliphatic acids are normally metabolized to acetate
and subsequently incorporated with randomization of the ^{14}C-label.

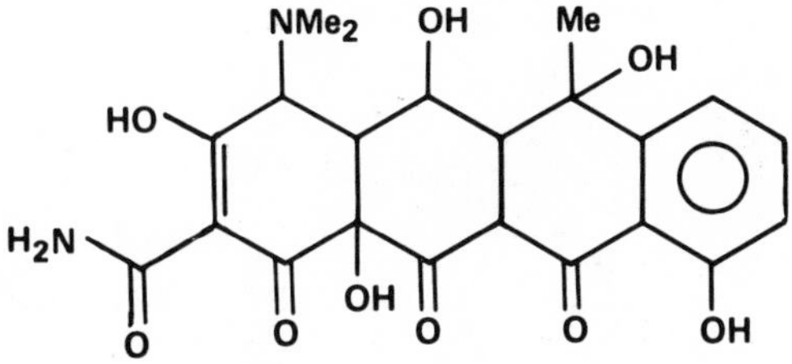

[34]

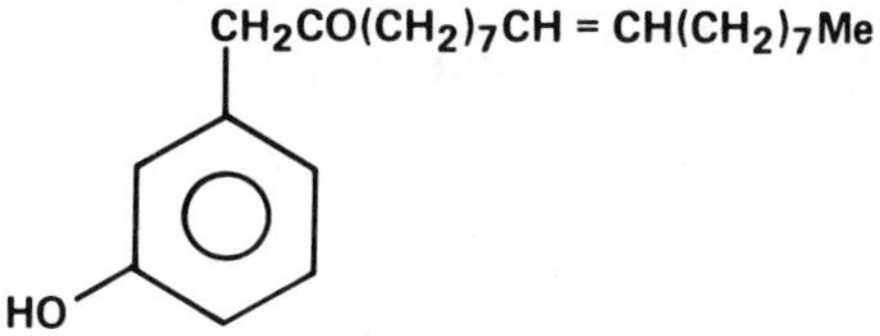

$$CH_2CO(CH_2)_7CH = CH(CH_2)_7Me$$

[35]

$$MeCO(CH_2)_7CH = CH(CH_2)_7Me$$

[36]

oleyl–CoA is involved in the biosynthesis of campnospermonol [35] is made likely by the co-occurrence of [35] and the methyl ketone [36], which would be formed by the condensation and decarboxylation of only one malonate unit with oleyl–CoA instead of the four units required to form [35]; if this is accepted, then the possibility becomes obvious that other fatty acids can act as starter units in the biosynthesis of the large group (24) of simple phenolic acids bearing long–carbon chains (C_5 to C_{18}), which may be saturated or unsaturated. The recently identified (25) turrianes (e.g., [37]) and related compounds (26) isolated from *Grevillea* spp., are interesting examples of the possible condensation of a substituted fatty acid with a short polyketide chain.

   An interesting case is that of baeckeol [38], one of several monocyclic compounds with a side-chain $(CH_3)_2CHCO$, which could well imply isobutyric acid as starter unit; the alternative possibility, double methylation of a terminal CH_3CO group, seems much less likely. These compounds have not been examined by tracer techniques; however, an isobutyric acid starter unit has been implicated in the biosynthesis of piloquinone (Chapter 18).

   Benzoic acid starter units for polyketide chains are relatively rare, and none of the cases which are implied by structure

[37]

[38]

[39]

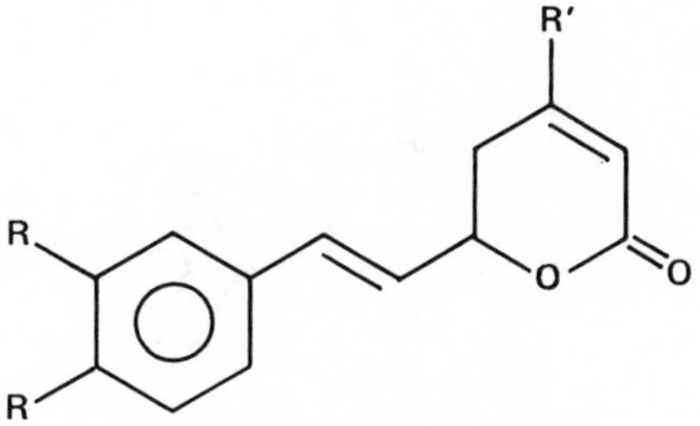

[40] R=H,R'=OMe
[41] R=R'=OH

(for example, piperic acid [39]) has been investigated experimentally.

There are a few compounds that are probably formed by the
cyclization of a short acetate chain attached to a cinnamic acid
starter unit; kawain [40] and hispidin [41] are examples; the
latter has been confirmed by experiment (99).

The biosynthesis of the vast number of compounds formed by
the combination of a cinnamic acid residue and a C_6 polyketide
chain will be reviewed in Chapter 21; here we give only a sample
of the types of structures that can result, with especial reference to the modifications that are seen in the acetate-derived
ring (left-hand, A).   The basic C_6-C_3-C_6 unit can cyclize in
three ways: to give stilbenes [42], arylisocoumarins [43], and
chalcones [44].   The last group of compounds probably represents
the common precursor of the aurones [45], flavones [46],

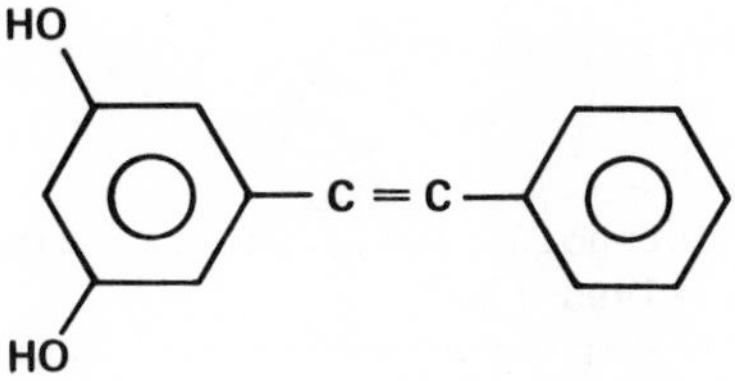

[42]

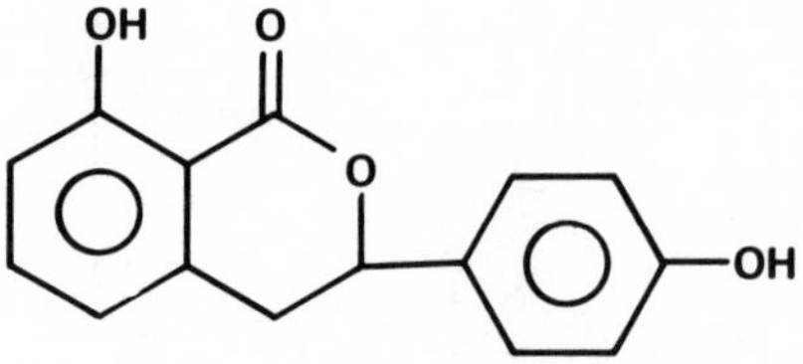

[43] Hydrangenol

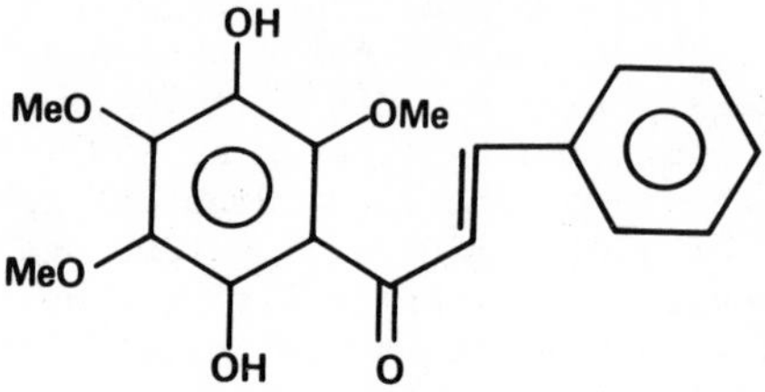

[44] Pedicin

isoflavones [47], rotenoids [48], anthocyanins [49], catechins, and flavanol derivatives [50].

Implication of some more unexpected starter units is also found: in anibine [51] (16), which may be formed from nicotinic acid; the chloro compound [53] (28), perhaps biosynthesized from proline; and the possible involvement of anthranilic acid in the biosynthesis of the acridone alkaloïds (see Chapter 16.)

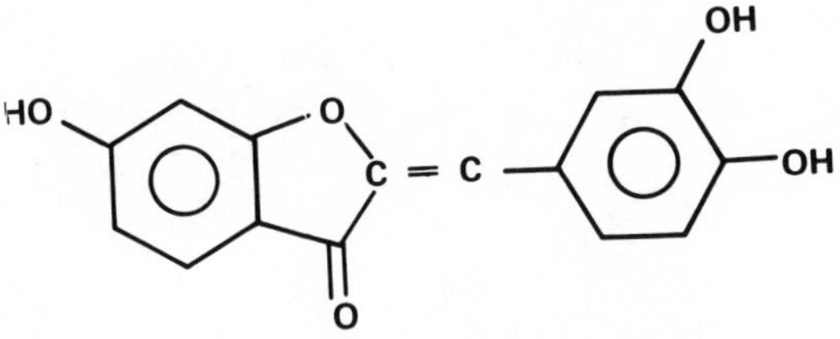

[45] Sulphurein aglycone

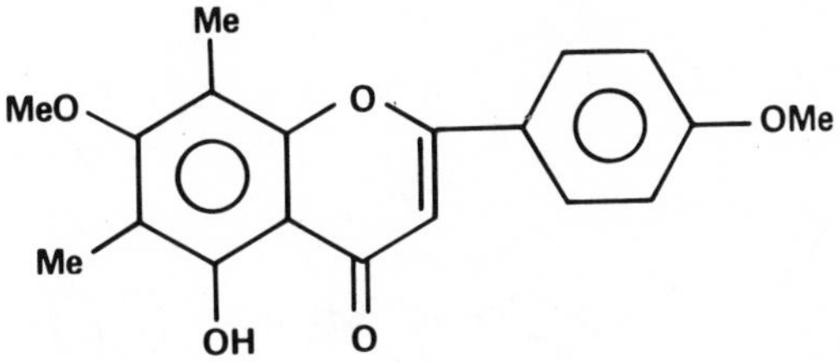

[46] Eucalyptin

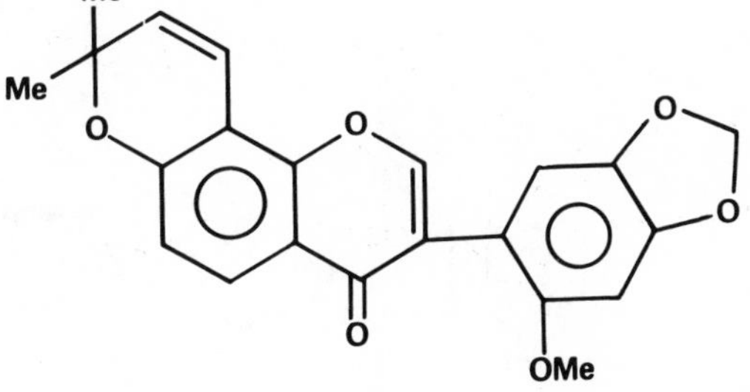

[47] Jamaicin

343

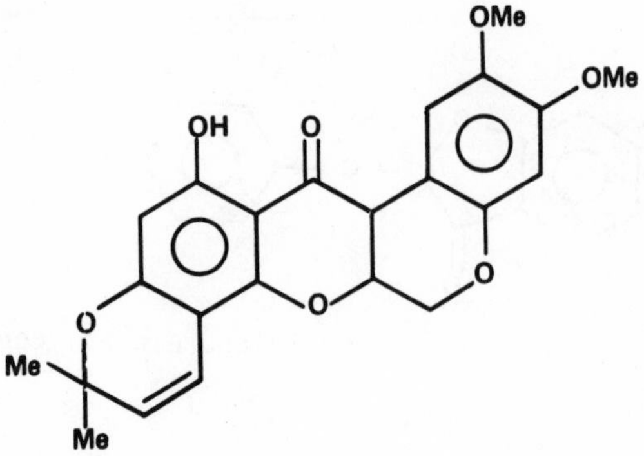

[48] α-Toxicarol

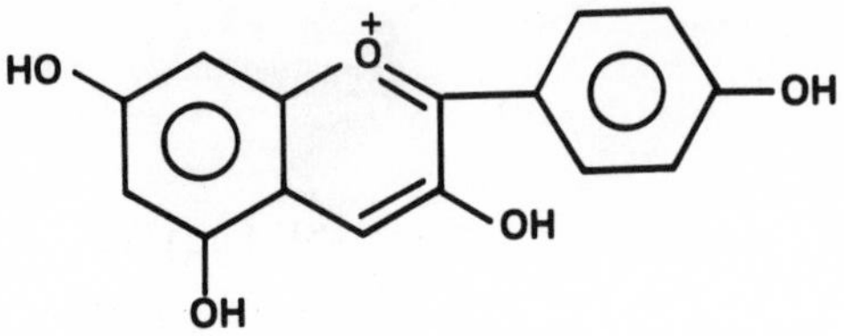

[49] Pelargonidin

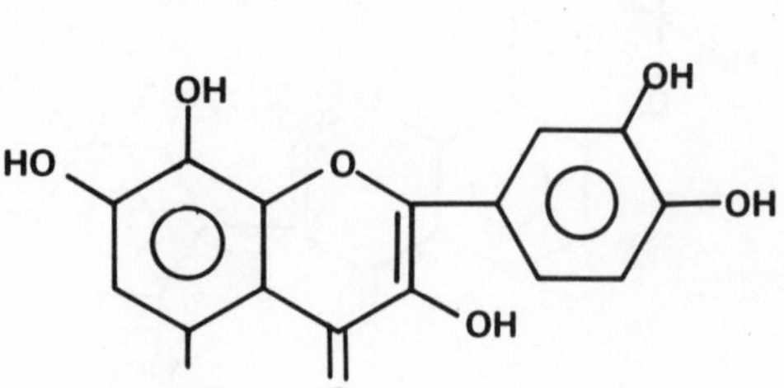

[50] Gossypetin

344

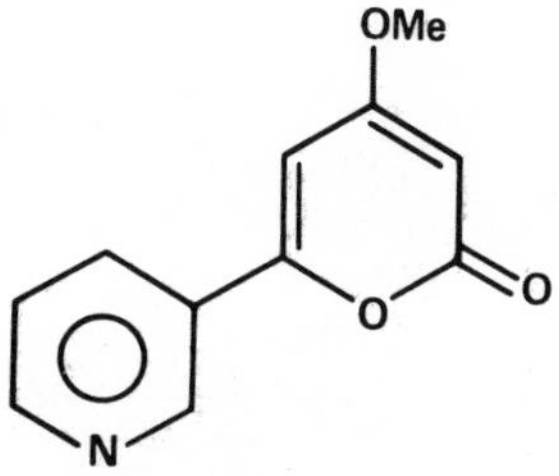

[51]

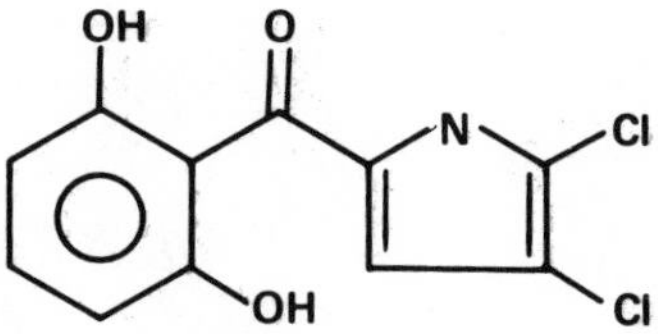

[53]

b.   Formation of the Polyketide Chain.

Acetyl-coenzyme-A was recognized as the "active" form of
acetic acid in 1951 (11), and Birch's original ideas were formulated in terms of active acetate units which condensed to form
CoA-esters of poly-β-keto acid chains; a modification of these
ideas was necessary when the function of malonyl-CoA in fatty
acid biosynthesis was recognized, and it is now clear that the
polyacetate pathway to aromatic compounds is better described
(29) as the acetate-polymalonate pathway. The assembly of the
polyketide chain follows the analogous processes in fatty acid
biosynthesis (30) with the ommission of the reductive steps; it
consists of a succession of combinations of malonyl-CoA units
[54], derived from acetyl-CoA by carboxylation, with an activated
starter unit, which may be acetyl-CoA [55] in a pure polyketide
but which may equally well be the active form of a number of
other acids (see Section 17-2a). Before combining to form the
noncyclized polyketide chain, the basic CoA units are separately
converted to the corresponding thio-esters of a carrier protein
[56] and [57], respectively; the growing chain is presumably
enzyme-bound [58] and is stabilized by a process, as yet poorly
understood, which is discussed in Section 17-2e; in the case of
fatty acid synthesis, the acetoacetyl thio-ester of the carrier

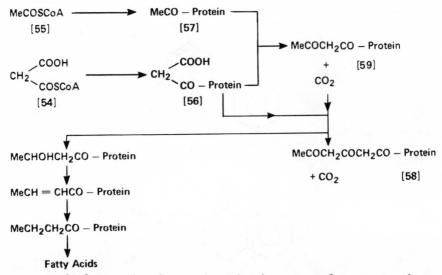

protein [59] is reduced stepwise via the crotonyl ester to the corresponding butyryl ester (CH3CH2CH2CO-protein), which then reacts with another molecule of malonyl-protein [56], and so on. It is one consequence of this synthesis that, in tracer experiments using ^{14}C-acetate, a slightly higher incorporation is observed into those carbons that are formed from the original acetate starter unit than into the carbons added *via* malonate; conversely, a lower incorporation into the starter carbons is observed when ^{14}C-malonate is used, since malonate is incorporated directly as malonyl-CoA and must be decarboxylated if it is to give the acetate starter unit. If suitable degradative reactions are available, it is sometimes possible to make use of this fact in order to establish which carbons in a molecule were originally present in the acetate starter unit. In some cases two starter units can be identified, indicating, as in the case of citromyceitin [60], the probability of a biosynthesis involving the combination of two polyketide chains (see also Chapter 18).

c.   Modification of the Chain.

Although there are a few examples of compounds which are
formed by the cyclization of an unmodified polyketide chain, the
structures of the majority of aromatic metabolities biosynthesized by this process either contain introduced groups, or are
lacking some of the functions of the original polyketide.   In
some cases oxygens are missing, e.g. from [61] when compared with
[62], and it is probable that this is due to the reduction of one
of the carbonyl groups of the polyketide chain, either in its
linear form [63], in a manner analogous to the biosynthesis of
the fatty acids, or in its cyclized form [64].   Both mechanisms
would lead to the "pre-aromatic" structure [65].   The reduction
of a phenolic group after aromatization seems much less likely,
although the *in vivo* dehydroxylation of aromatic rings has been
observed in a number of cases (see Chapter 16).   For example,
sinapic (4-hydroxy-3,5-dimethoxy-cinnamic) acid is converted into
lignin containing coniferyl (4-hydroxy-3-methoxyphenyl) residues
(31), and the same acid can be demethoxylated to ferulic acid in
some plant species (32).
    In many aromatic methobolites, additional groups are found
attached to those carbons that were originally derived from the
methylene groups of malonate; the groups are often methyl or
isoprenoid, but additional hydroxyls, halogen, or C-glycosyl
functions are also found.   It has been demonstrated that the
commonly introduced alkyl groups are derived from methionine or
mevalonate; they are presumably introduced by the interaction of
the alkylating agents S-adenosylmethionine, and the pyrophosphate
esters of dimethylallyl alcohol, geraniol, farnesol, and geranylgeraniol, either with the polyketide (before cyclization or at an
alicyclic stage) or with the preformed aromatic nucleus.
Although it is evident that alkylation of an aromatic ring *can*
take place, as in novobiocin (33) and actinomycin, in which the
methyl group is introduced into an existing aromatic ring, this
process must involve a high-energy Friedel-Crafts-type of reaction; it is likely that in the majority of cases the modification
of the polyketide chain takes place *before* aromatization, and
that the electrophilic reagents preferentially attack the enolic
double bonds of the polyketide either in its linear or alicyclic
form.   For a review of this type of C-Alkylation, see Lederer
(1969) (34).   The formation of C-glycosides [66], and the introduction of additional OH groups [67] (presumably through
electrophilic attack by the biochemical equivalent of OH^+) takes
place by mechanisms similar to the alkylating reactions outlined
above.

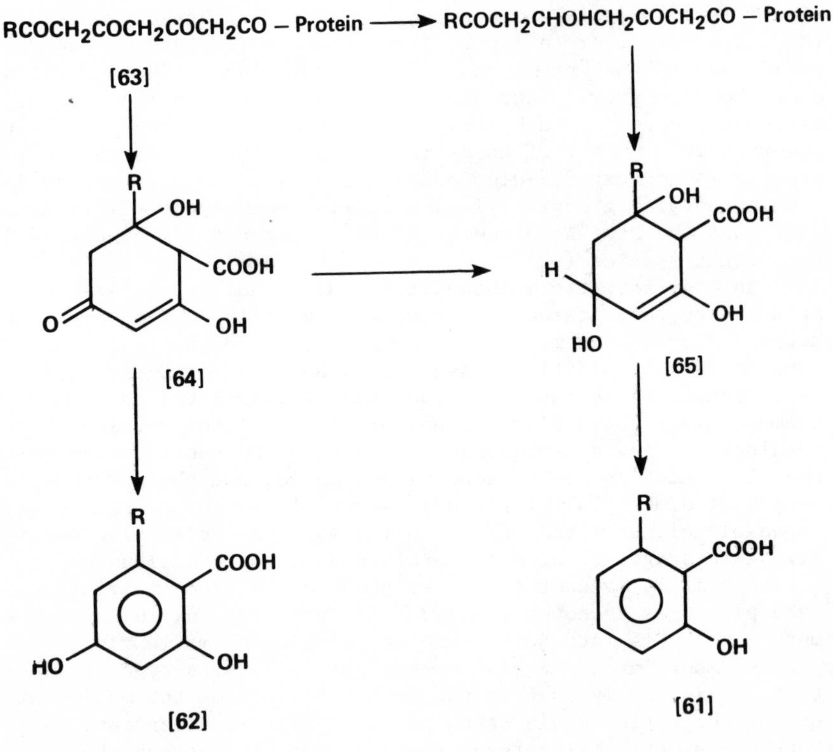
RCOCH2COCH2COCH2CO – Protein
RCOCH2CHOHCH2COCH2CO – Protein
[63]
R
OH
COOH
O
OH
[64]
R
OH
COOH
H
OH
HO
[65]
R
COOH
HO
OH
[62]
R
COOH
OH
[61]

[66]

[67]

The biosynthesis of the flavone O- and C-glycosides in the Lemnaceae (duckweeds) has been studied by Alston and co-workers (35, 36), who obtained evidence that the preformed flavone can, not surprisingly, undergo O-glycosylation but not C-glycosylation. The point at which C-glycosylation occurs is still obscure, although it seems likely that it will be found to be at a prechalcone stage.

Other modifications that take place after aromatization: halogenation [68], esterification [69], O-methylation, and some degradative processes, are normal enzymic reactions that have no bearing on the processes of aromatization.

That the modifications of the polyketide chain take place to some extent before aromatization is supported by the isolation of some incompletely aromatized metabolites from mutant organisms. Certain strains of a tetracycline-producing bacterium, for example, produce [70] instead of the fully aromatic product [71], although both are clearly derived from the same modified

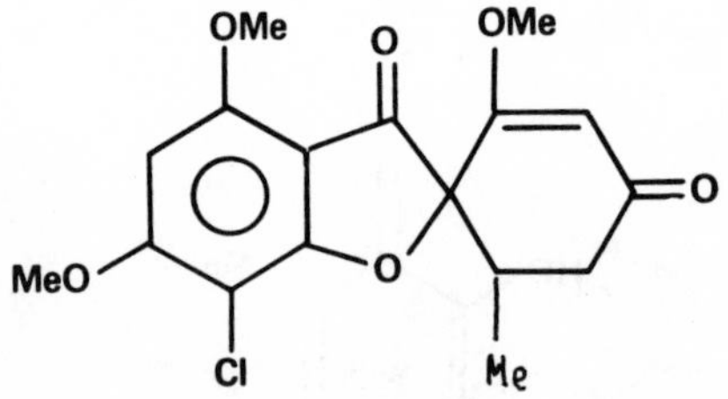

[68]

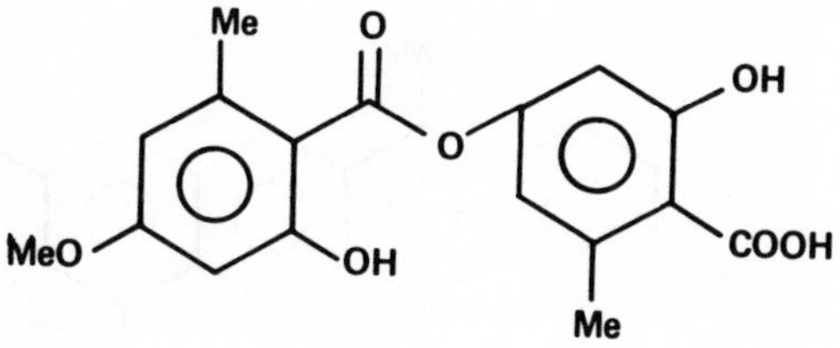

[69]

polyketide (37). Modification of a polyketide chain is also
seen in the blocked aromatic compounds related to sclerotiorin
[72] (see Section 17-2a), which are prevented from aromatization

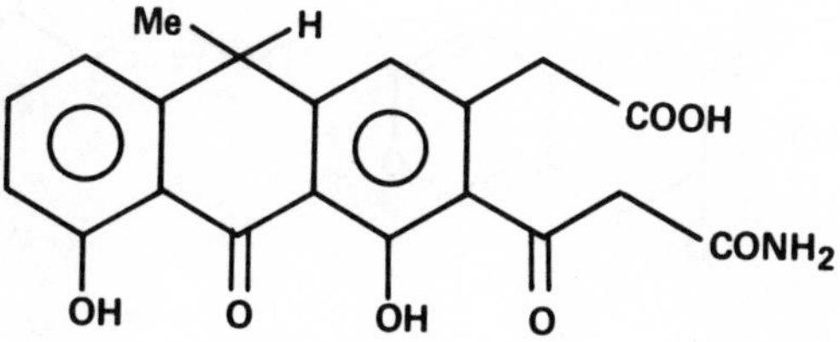

[70]

by the introduction of a methyl group. There are also a large
number of phloroglucinol compounds, for example, aspidinol [73],
and the hop constituents lupulone [74] and humulone [75], which
are certainly acetate-derived and whose aromaticity seems to be
blocked by polyalkylation. However, in this case there is no
evidence to show whether the electrophilic substitutions take
place at the polyketide, hydroaromatic, or fully aromatic stage
in the synthesis; in the case of phloroglucinol derivatives,

[71]

[72]

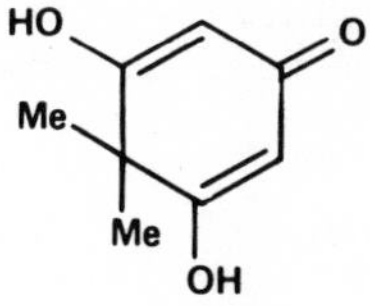

[73]

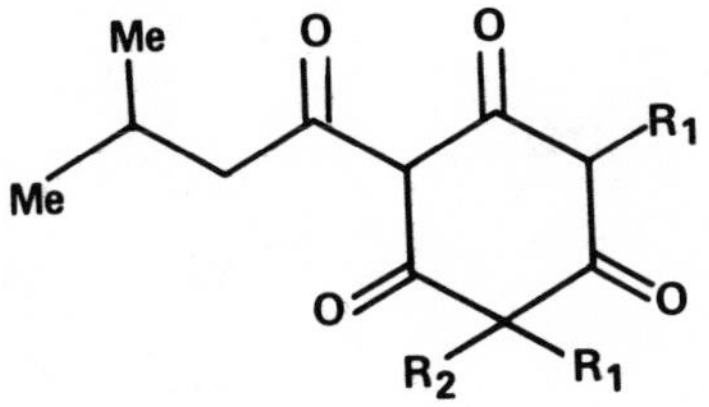

[74] R₁ = R₂ =

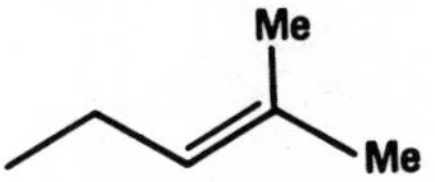

[75] R₁ =

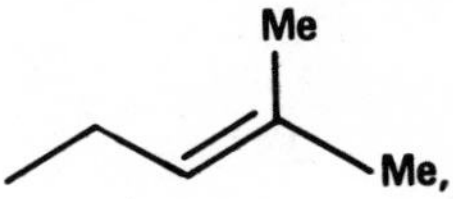

Me, R₂ = OH

353

substitution could certainly take place equally well at the pre-
aromatic or aromatic stages, since these compounds exist to some
extent in the triketo form, and therefore contain active methylene groups. *In vitro* methylation of phloroglucinol and its
derivatives is readily achieved, for example, with methyl iodide
and alkali.

d. Cyclization of the Chain.

When the basic polyketide chain has been formed and modified,
it is cyclized by one of the routes that we have already outlined
(Section 17-2a). There is good reason to believe that the
processes of assembly and modification on the one hand, and
aromatization on the other, are separate aspects of the biosynthetic sequence.

We have seen that several modes of cyclization are available
to a given polyketide chain; thus, for example, the polyketide
[76] can cyclize to give a variety of compounds. A few of these
possibilities are shown in the figure. There is an example

$$Me\ (COCH_2)_6COOH$$

[76]

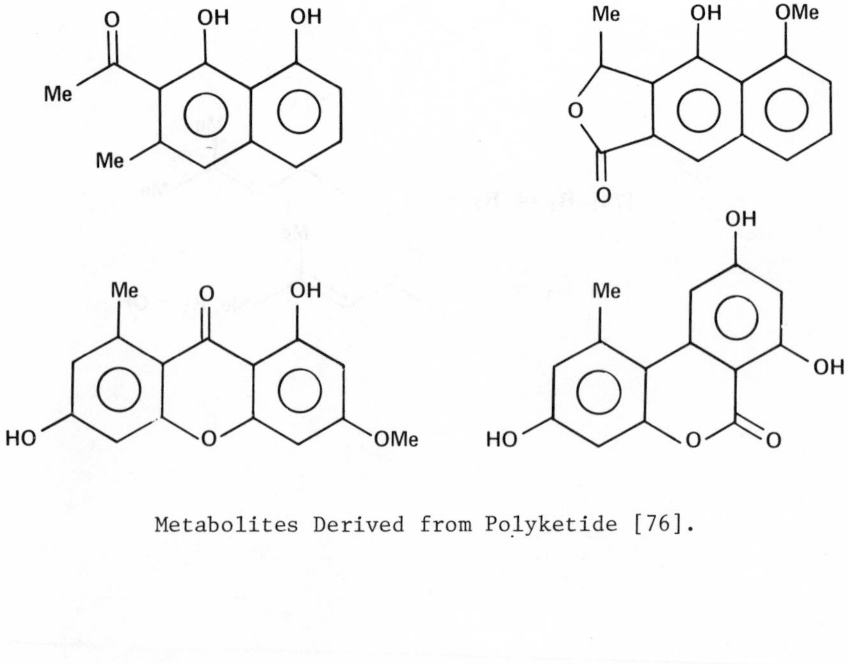

Metabolites Derived from Polyketide [76].

(Section 18-1) of one polyketide chain undergoing both aldol and Claisen cyclization in the same fungus. On the other hand, there are many examples of compounds that have been formed by the same mode of cyclization operating on chains of different length, for example, hydroxypaeonol [77] derived from one acetate plus three malonate residues, and eugenone [78] from one acetate plus four malonates. Other examples are found among the lichen depsides. We have already seen (Section 17-2c) that there are,

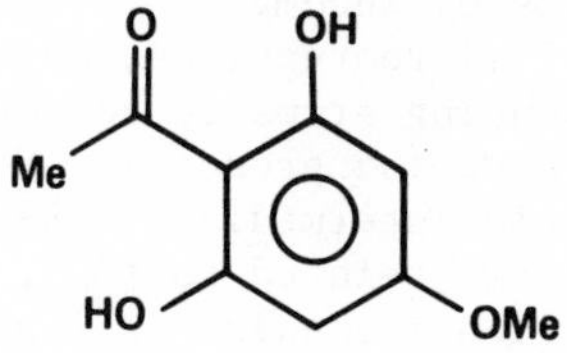

[77]

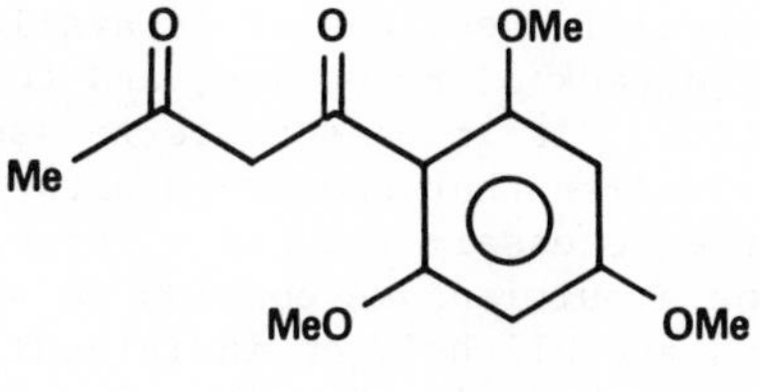

[78]

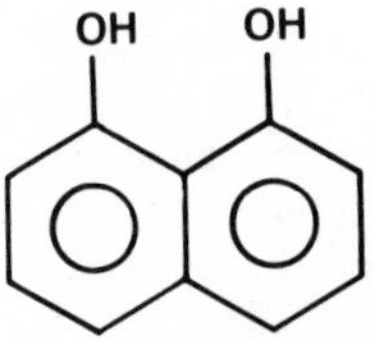

[79] 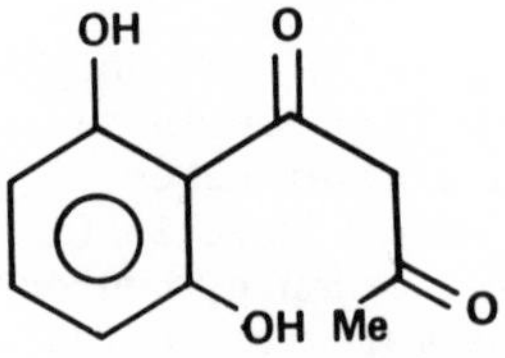   [80]

in addition, examples of incompletely aromatized compounds formed
by mutant strains of microorganisms, which in the wild type
produce the corresponding aromatic compound; for example, the
fungus *Daldinia concentrica* produces 1,8-dihydroxynaphthalene
[79] as a biosynthetic intermediate (38), while a mutant strain
synthesizes metabolites related to [80], which presumably arise
because the cyclization mechanism is impaired which would convert
them to [79].

e.   The Enzyme System.

These rather complex steps of assembly, modification, and
cyclization are enzyme-catalyzed, and several explanations have
been put forward to account for the specificity of the processes
making up this biosynthetic pathway. The enzymic system must be
able to control the length of the growing enzyme-bound polyketide
chain, to make specific sections of it available to modification
by electrophilic attack or reduction, and to predetermine the
mode of cyclization. It is reasonable to assume that the assembly
mechanism is not unlike that concerned with fatty acid biosynthesis without the necessary reductive stages (39, 40). Thus
one could imagine a multienzyme complex in which the acetate and
malonate subunits are all held on individual binding sites at
the enzyme surface; when all the sites are filled, the activated
units are modified and combine with each other to form the
aromatic metabolite in one concerted reaction. The theory would,
however, be hard to reconcile with the evidence presented above
which indicates that assembly and cyclization are two distinct
sequential processes.

An ingenious alternative suggestion has been made by Bu'Lock
(37) who explains the stability and conformational specificity of
the polyketide in terms of specific metal-ion chelation. Given
the fixed coordination geometry of the metal ion, Bu'Lock showed
that there were two ways in which a C_8 polyketide could be held
in threefold coordination with the metal ion; these two conforma-
tions correspond precisely to the two basic modes [81] and [82],
in which such a C_8 chain must be folded in order to facilitate

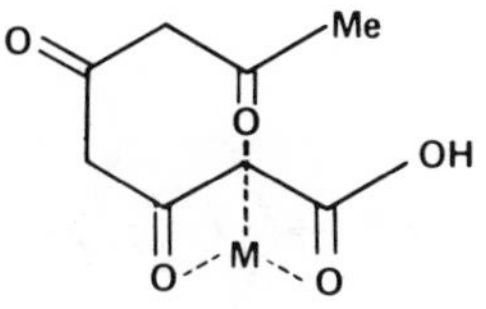

[81] 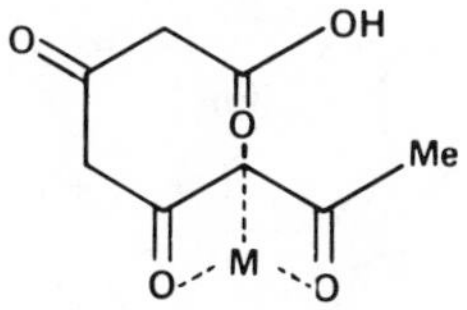   [82]

the formation of orsellinic acid and acetylphloroglucinol,
respectively. Significantly, the keto groups that are not involved in the chelate formation are those that must be reduced
in order to form 6-methylsalicylic acid and 2-acetylresorcinol,
respectively.

An alternative hypothesis has been proposed by Thomas (41)
in which the aldol cyclization (a) leading to, for example,
orsellinic acid is replaced by the cyclization of a poly-β-
enolate (b), where geometry is governed by the *cis* or *trans*
nature of the double bonds; this alternative pathway differs

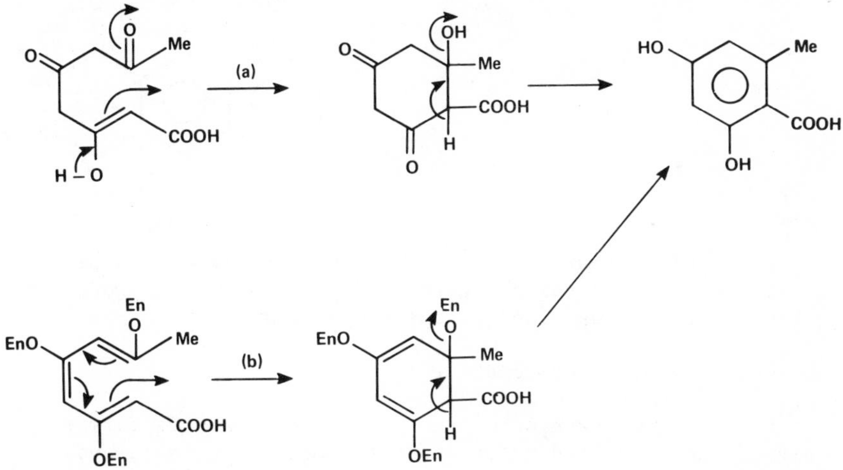

from the normal aldol route in two ways: (1) the primary cyclization is brought about by valency tautomerism, and (2) there are
no keto-groups in the precursor. The poly-β-enolate could be
derived from acetate and malonate directly with either *cis* or
trans geometry depending upon the mode of its formation; the
geometry of the final polyene would then predetermine the mode
of cyclization, as demonstrated for the cases of alternariol and
the naphthol [83], both derived from a C_{14} polyketide.

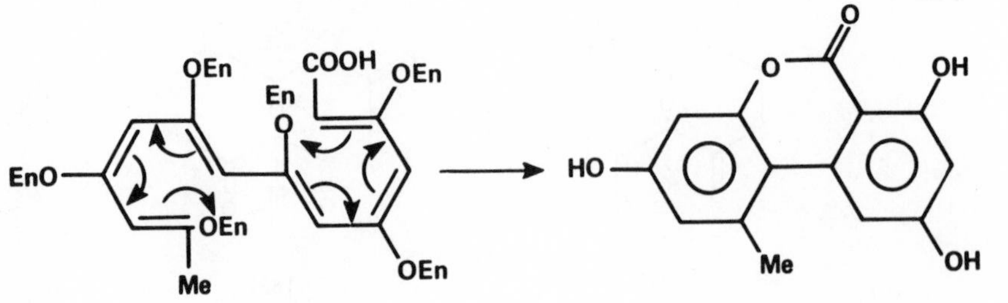

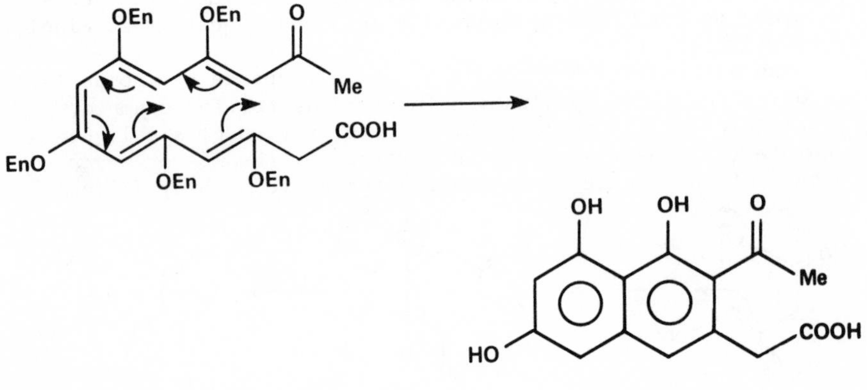

[83]

It is clear that these three hypotheses are not really mutually exclusive; taken together they may give us a good idea of the events taking place on the enzyme surface.

f. Isolation of Intermediates.

Since, as we have seen, the polyketide chain is formed, modified, and cyclized on or about the enzyme surface, and is detached as a more or less final product, it is not too surprising that no real intermediates between the activated precursor units and the ultimate cyclic compounds have been isolated. The nearest compounds to actual polyketides that have been observed so far are a series of lactones, which are simple dehydration products of the polyketides; triacetic lactone [84], tetraacetic lactone [85], and methyltriacetic lactone [86] (42) have been isolated from a tropolone-producing strain of *Penicillium stipitatum* inhibited with ethionine (43); [86] and [84] have also

been isolated from other *Penicillium* spp. (44, 45).   Some evidence also exists for the occurrence of free β-polyketides in
fatty acid biosynthesis: Brodie and co-workers (46) found that
in the absence of TPNH, the purified fatty acid synthetase complex from pigeon liver produced small amounts of triacetic acid
lactone.   Similarly, (47) triacetic acid lactone was produced
by the crude fatty acid synthetase of *E. coli* in the presence
of inhibitory thiols: no compounds containing more than six

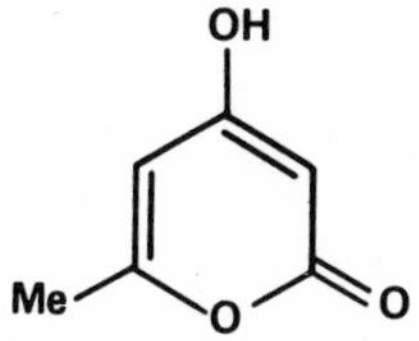

[84]

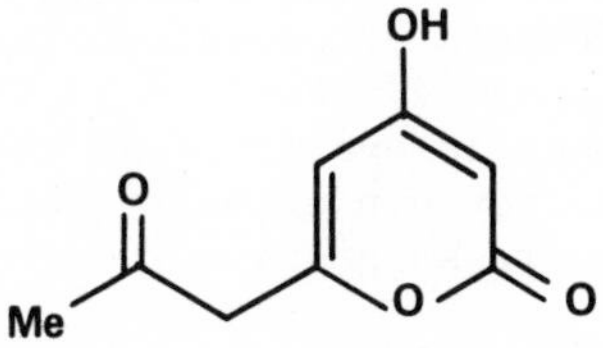

[85]

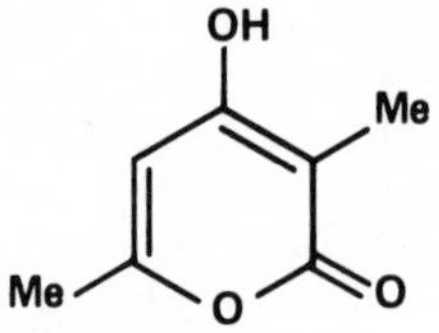

OH

Me

Me   O   O

[86]

carbon atoms were isolated in either case. The formation of [84]
as a product of fatty acid biosynthesis has been discussed by
Lynen (48).   Interestingly, although [84] apparently stimulates
the production of aromatic metabolites in *P. urticae* (49), it is
not well incorporated into 6-methylsalicylic acid by *Penicillium*
spp. (50).   However, no direct evidence for the existence of
polyketides *per se* in biological systems has yet been obtained.

   According to the Birch hypothesis (15), the first cyclic
intermediate formed by the condensation of a polyketide would be
a cyclic aldol (cf. the analogous case in the shikimic acid
pathway, where the aldol 3-dehydroquinic acid is the first cyclic
intermediate).   Several nonaromatic natural structures are known
to fulfill the requirements of precursors of clearly acetate-
derived aromatic compounds synthesized by the same organism.
No experimental evidence is available as yet that the presumed
precursor-product relationship actually obtains in the organism,
but some of the cases are suggestive enough to warrant brief
enumeration.

   The optically active aldol [87], which has been investigated
by Dalton and Lamberton (23), occurs in *Campnosperma* spp.
together with its phenolic analog campnospermonol [88], and the
methyl ketone [89].   The conversion of [87] into [88] occurs
very readily *in vitro*, and the structure of the former is so
precisely the one to be expected for the immediate precursor of
an acetate-derived aromatic compound like campnospermonol that
the natural formation of the latter compound from [87] can hardly be

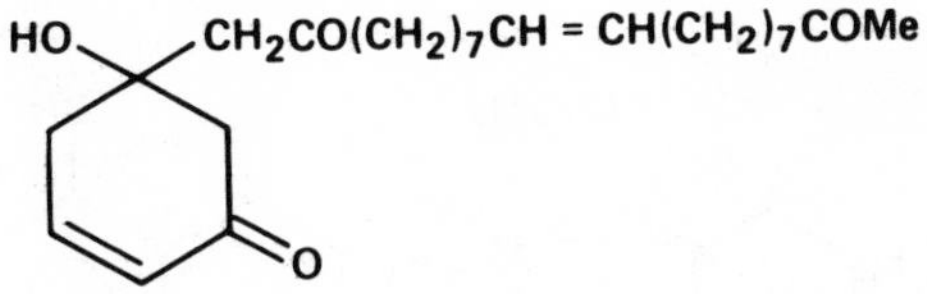

HO— CH$_2$CO(CH$_2$)$_7$CH = CH(CH$_2$)$_7$COMe

[87]

CH$_2$CO(CH$_2$)$_7$CH = CH(CH$_2$)$_7$Me

OH

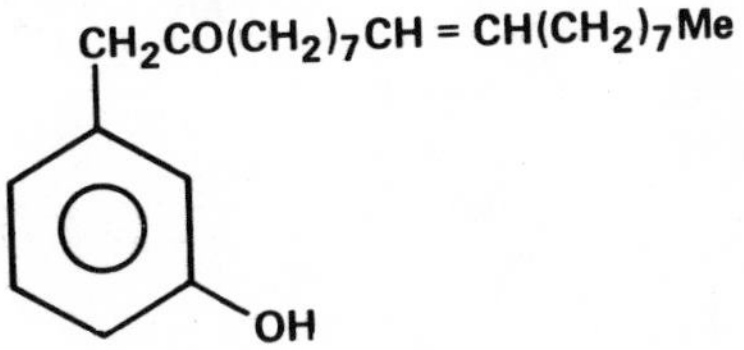

[88]

Me—CO—(CH$_2$)$_7$— CH = CH — (CH$_2$)$_7$ — Me

[89]

doubted.   A similar relationship exists between the ε- and η-
pyrromycinones [90] and [91], which occur together in the mycelium of a *Streptomyces* sp. (51).   The structure of [90] (rutilantinone) is such that it could very well be an intermediate in the
biosynthesis of [91]; the conversion of the hydroaromatic ring
of [90] to the aromatic one of [91] again occurs very easily

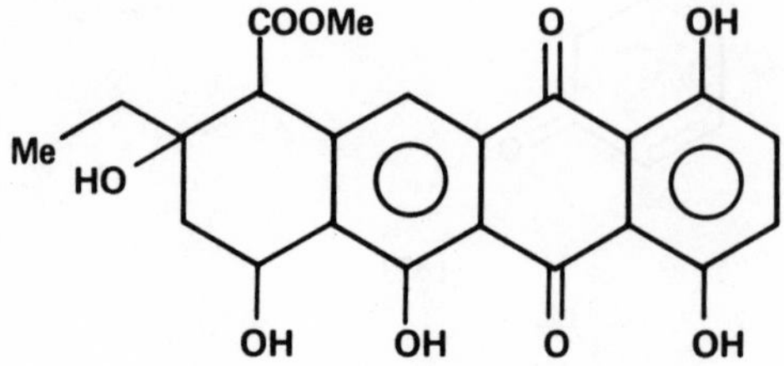

[90]

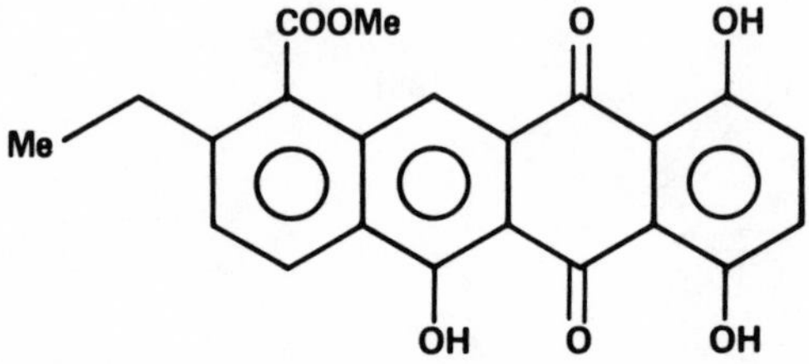

[91]

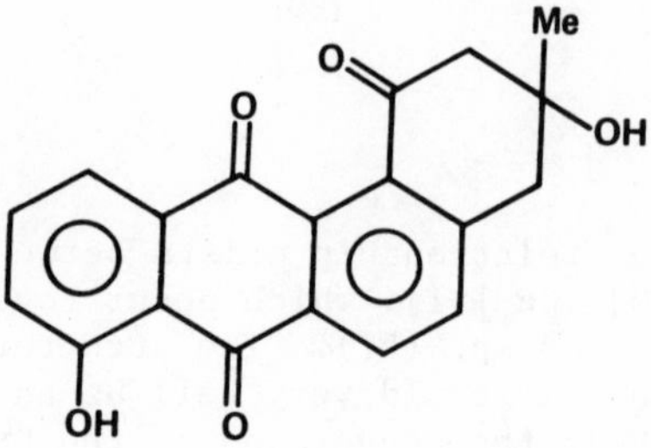

[92]

362

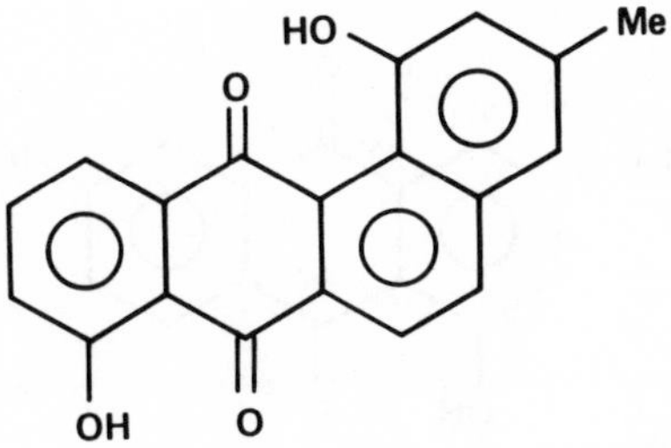

[93]

in vitro.   A similar relationship is apparent in the structures of tetrangulol [93] and tetrangomycin [92], which coexist in fermentations of *Streptomyces rimosus* (52).   Altersolanol-β [94] and the anthraquinone [95] have been isolated from the culture filtrates of *Alternaria solani* (53) along with other related metabolites; again the *in vitro* conversion is readily achieved.   The anthraquinones islandicin [96], and 2-methyl-quinizarin [97], and the quinones [98], and [99], have all been isolated from *Penicillium islandicum* (54); again, the conversion of the hydroaromatic compounds to the corresponding fully aromatic anthraquinone has been demonstrated *in vitro*.

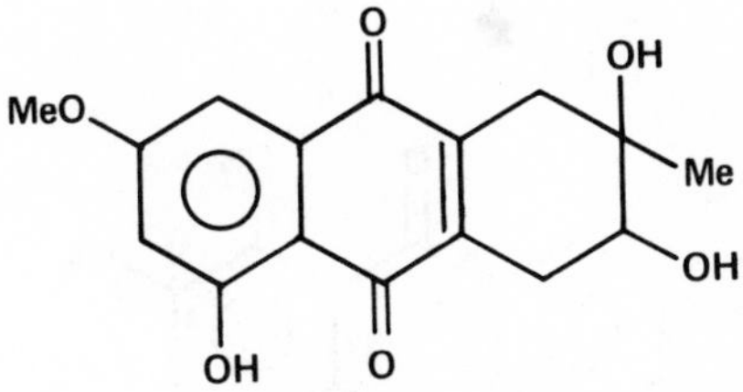

[94]

[95]

[96]

[97]

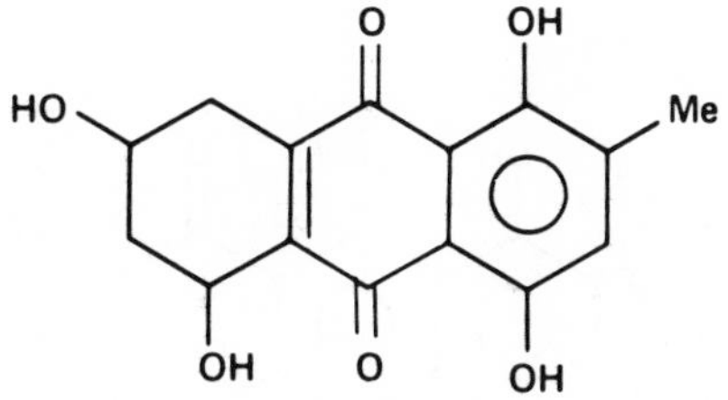

[98]

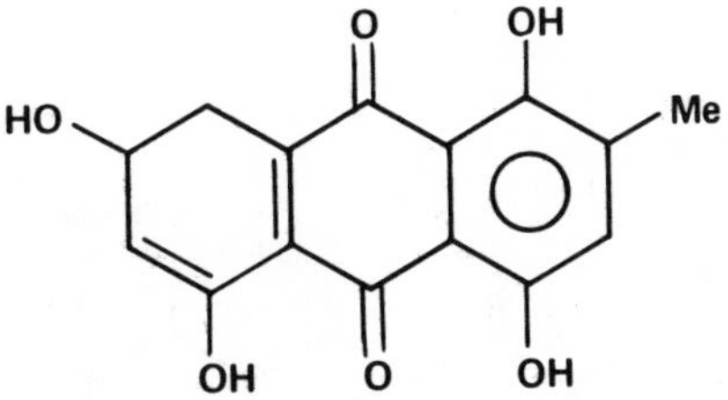

[99]

Flavomannin [100] is a metabolite of *Penicillium wortmanni* which is converted by hot HCl to the aromatic bis-anthraquinone (55); similarly aklavinone [101] can be aromatized with *p*-toluene sulfonic acid (56), as can the hydroaromatic quinone bostrycin [102] isolated from *Bostrychonema alpestre* (57).

Other pairs of metabolites one of which is the aldol-precursor of the other, have been isolated from *Cortinarius odorifer* (96), and *Aspergillus flavus* (97).

Several other incompletely aromatized compounds have been isolated, all of which are suggestive of the aldol-type of intermediate expected in this mode of biosynthesis, although they do not coexist with their aromatic partners. Daunomycin and its aglycone [103] are both found in culture filtrates of *Streptomyces peucetius* (58). Julimycin B-11 [105] isolated from *S. shiodaensis* (59) is a similar example, and Eckardt (60) has listed others elaborated by *S. galilaeus*.

Compounds suggestive of a secondary aromatization involving dehydrogenation are also quite common, although not clearly related to this biosynthetic scheme. Thus, an interesting transient metabolite ramulosin [106] has been isolated (100) together with mellein [107] from a fermentation of *Pestalotia ramulosa*, and

[100]

[101]

[102]

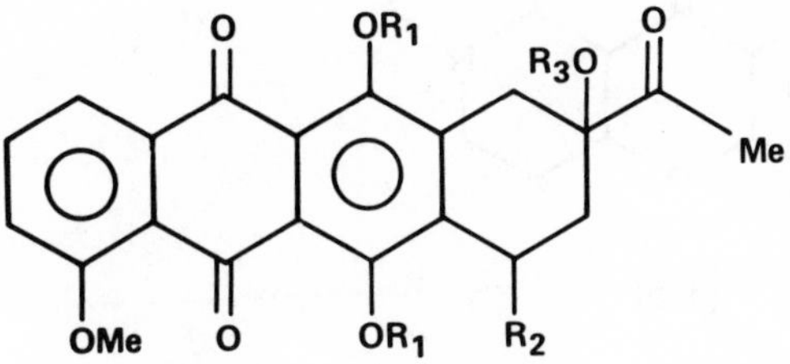

[103] R₁ = R₃ = H, R₂ = OH

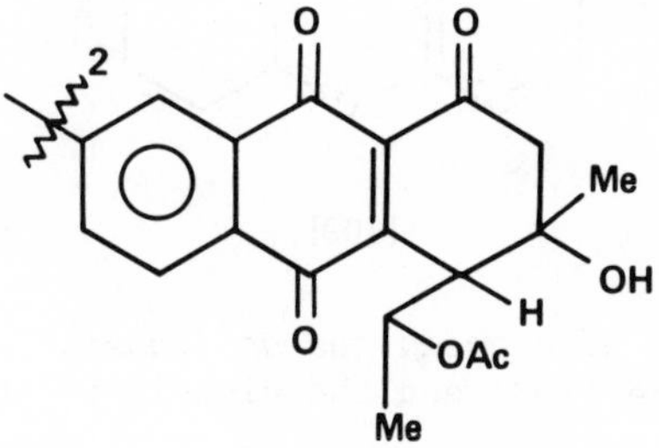

[105]

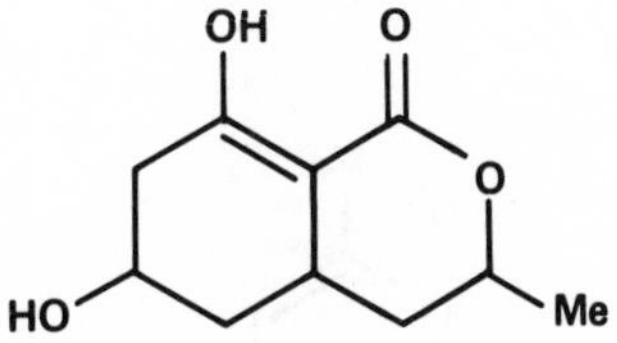

[106]

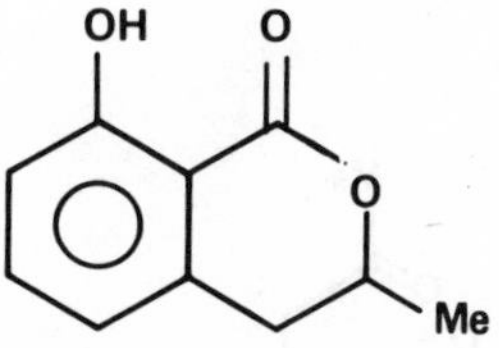

[107]

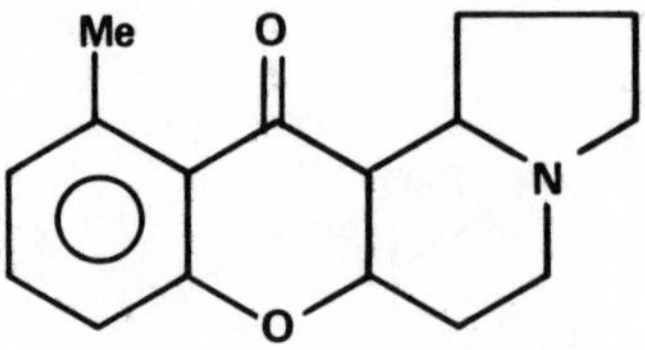

[108]

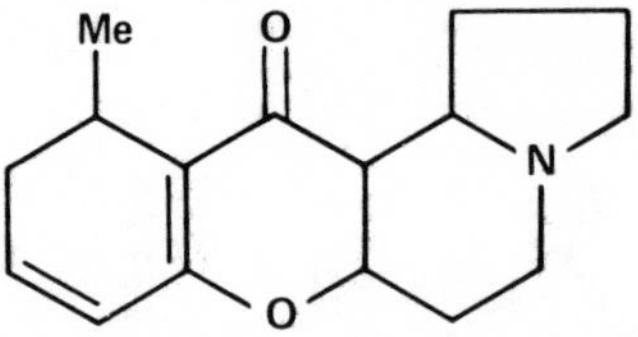

[109]

similarly the leaves of *Elaeocarpus dolichostylis* contain the alkaloid elaeocarpine [108] and the closely related elaeocarpiline [109] (61).

The tricyclic compound barakol [110] has been isolated from *Cassia siamea* (84) and, although its biosynthesis has not been studied, it is undoubtedly of acetate origin; interestingly, the compound has the same skeleton as 6-hydroxymusicin aglycone [111] and torachrysone [112]. On treatment with base [110] is

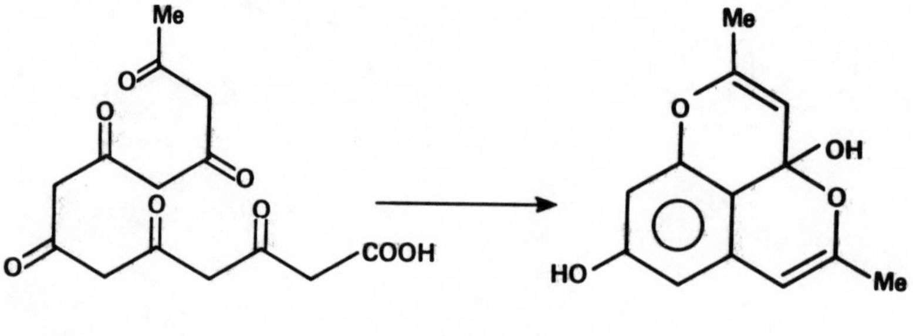

[110]

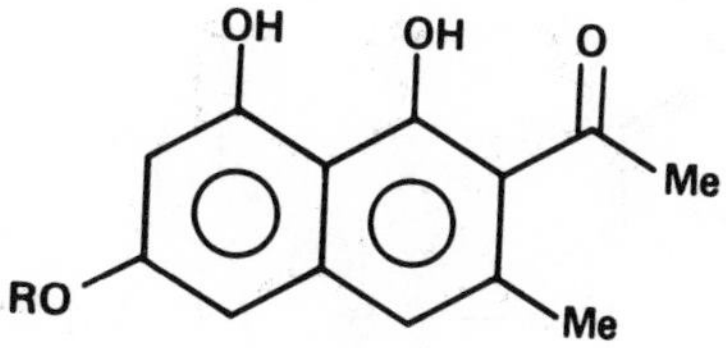

[111] R = H

[112] R = Me

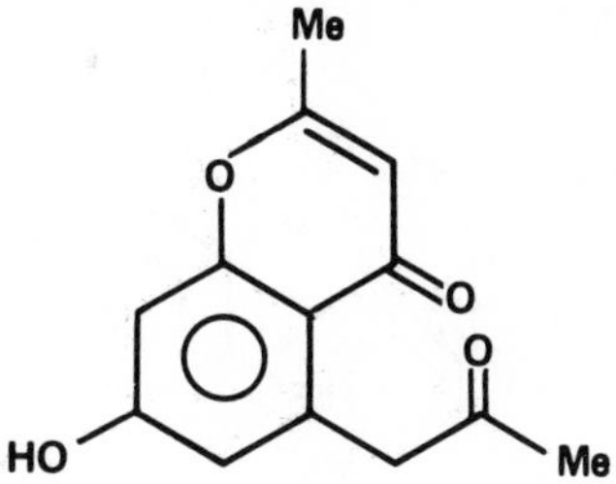

[113]

converted into [111] and the related [113]. It is attractive to
think of barakol as a stabilized polyketide, converted, possibly
enzymically, into aromatic products. If this interconversion
proves to be enzymically possible, the analogous ring-opening
of compounds like eleutherinol [114], which is very similar to
[113], might well lead to the anthraquinones (e.g., emodin [115])
via intermediates such as [116].

g. Model Reactions.

    In order to gain more insight into the early steps of poly-
ketide biosynthesis, there has been in recent years a good deal of
interest in the synthesis of poly-β-keto acids and esters that
might be useful as precursors, and in a study of their cyclization processes from the mechanistic point of view. 1,3-Dicarbonyl
and 1,3,5-tricarbonyl compounds have been accessible for years
(62), but the members of the series capable of cyclizing to yield
aromatic structures, that is, the 1,3,5,7-tetracarbonyl compounds,
have not been available. Harris and his co-workers have succeeded
in preparing seven members of the series [117] by the general

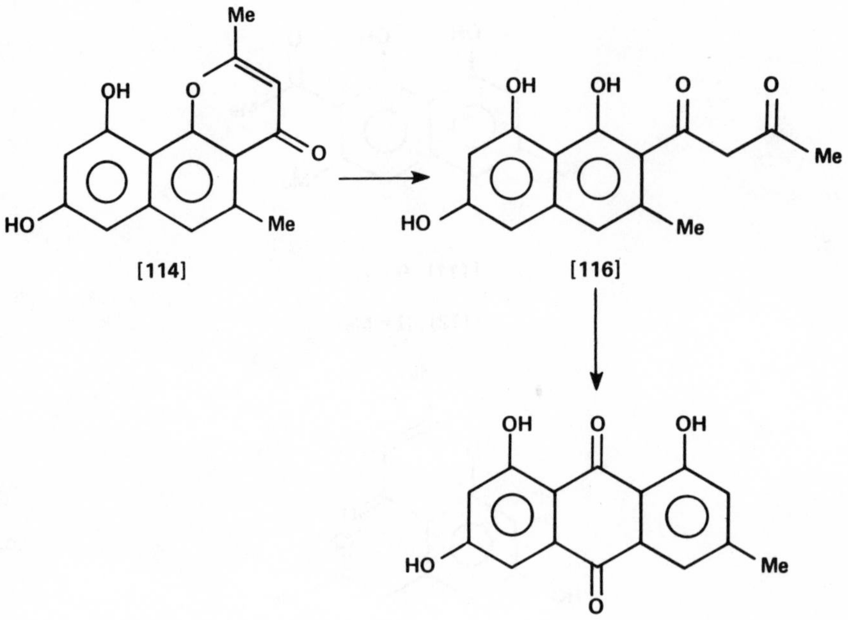

[114] → [116]

[116] → [115]

[115]

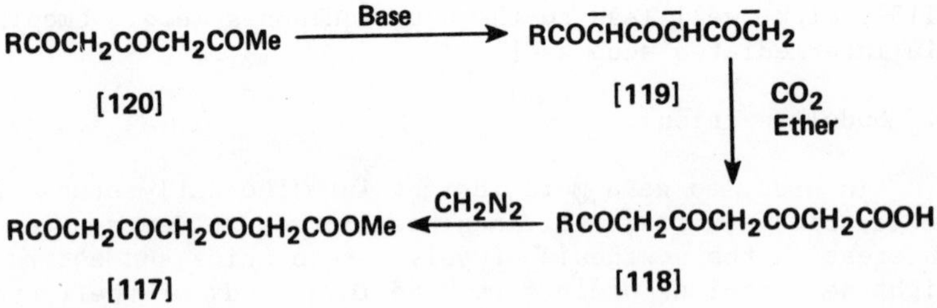

$$RCOCH_2COCH_2COMe \xrightarrow{\text{Base}} RCO\bar{C}HCO\bar{C}HCO\bar{C}H_2$$

[120]   [119]

$$\downarrow \begin{array}{c} CO_2 \\ \text{Ether} \end{array}$$

$$RCOCH_2COCH_2COCH_2COOMe \xleftarrow{CH_2N_2} RCOCH_2COCH_2COCH_2COOH$$

[117]   [118]

R = Me, C_6H_5, $C_6H_5CH = CH$, $C_6H_5CH_2CH_2$, nC_7H_{11}, nC_3H_7

method outlined below: the acids [118] were prepared by carboxylation of the tri-anions [119] of the corresponding triketones
[120].   A variety of methods have been used to generate the ionic
species, usually the action of an excess of sodium or potassium
amide in anhydrous liquid ammonia (62); however, in the case
of [119, R=CH$_3$], the use of lithium diisopropylamide in tetra-
hydrofuran was necessary (63).

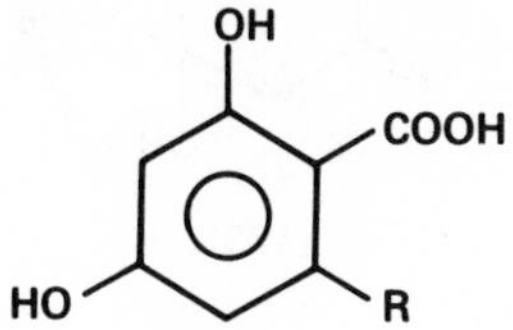

OH

COOH

HO

R

[121]

OH

COR

HO

OH

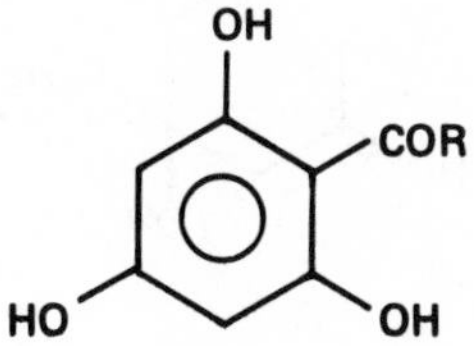

[122]

The keto acids [118] were cyclized to the corresponding
β-resorcylic acids [121] in 0.5 M sodium acetate/MeOH buffer,
pH 5.0; in some cases the esters [117, R = C$_5$H$_6$ and C$_6$H$_5$CH=CH]
could be converted to the acylphloroglucinols [122] with
cold 2M KOH.   The intermediate cyclic aldols (e.g., [123]) have
been isolated in some cases (64); stable intermediates of this
type had been postulated by Birch (68) (for analogous natural
products see Section 17-2f).   It is quite possible that specific
enzymatic reduction of [123] in a manner discussed in Section
17-2c could lead to the substituted 2-hydroxybenzoic acids
(e.g., [125]); reduction could equally well take place at the
polyketide stage.   Methylation of the triketo acid [118, R =
phenyl] gave a mixture of the 3- and 5-methyl ethers of the
corresponding methyl ester.   One ether cyclized spontaneously to
an aldol which readily gave a benzoate ester bearing a methyl
ether (4-methyl-[129]), while the other ether could be cyclized
to 2-methyl-[129] and 4-methyl-[122](R = phenyl) (67).   That
phenolic ethers arise biosynthetically by a similar mechanism
seems unlikely in view of present knowledge, which indicates

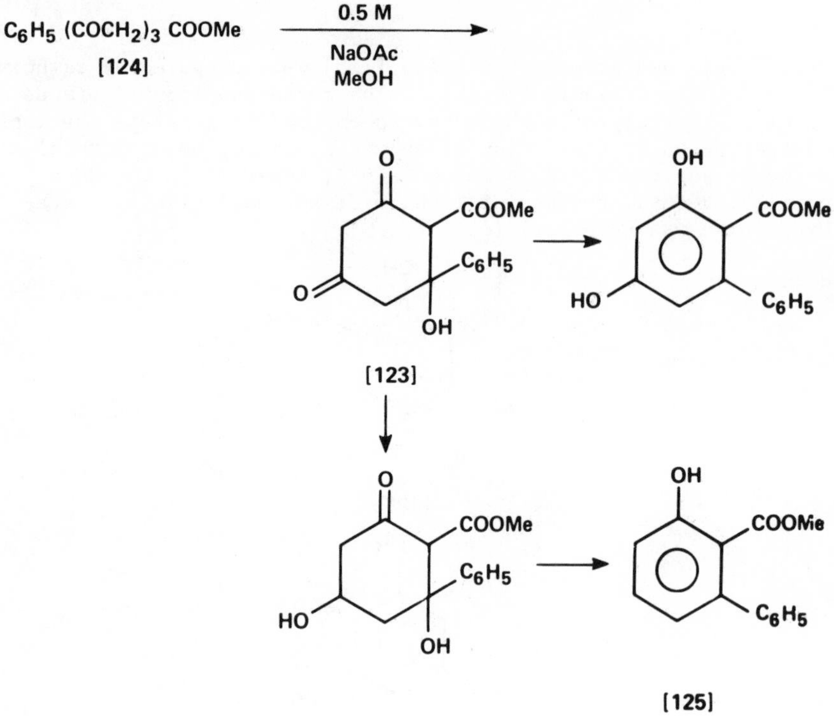

$C_6H_5 (COCH_2)_3 COOMe$ $\xrightarrow[\substack{NaOAc \\ MeOH}]{0.5\ M}$

[124]

[123]

[125]

that methoxyl groups are generated from the parent phenol rather than introduced into the polyketide chain.

In an interesting application of the cyclization outlined above, Hay and Harris (65) have converted 7-(4-orcinyl)-3,5,7-trioxoheptanoate and its dimethyl ether to alternariol and lichexanthone, respectively.

A less direct approach to the problem has been taken by other groups. Thus, Bentley (42) showed that tetraacetic acid lactone, isolated from natural sources, could be converted into orsellinic acid and its decarboxylated product orcinol under very mild alkaline conditions; hydrolysis of the lactone ring presumably gives rise to the poly-β-keto acid, which then undergoes a dehydrative cyclization. Scott (69) has synthesized tetraacetic acid lactone and studied its conversion both into orsellinic acid, and, after reduction (to [126]), into 6-methylsalicylic acid. Both Scott (69) and Harris (70) have synthesized the pyrone [127] and have studied its cleavage and recyclization to aromatic compounds [128] and [129] on treatment with magnesium hydroxide. A magnesium complex of the intermediate triketo ester was isolated in one case (70).

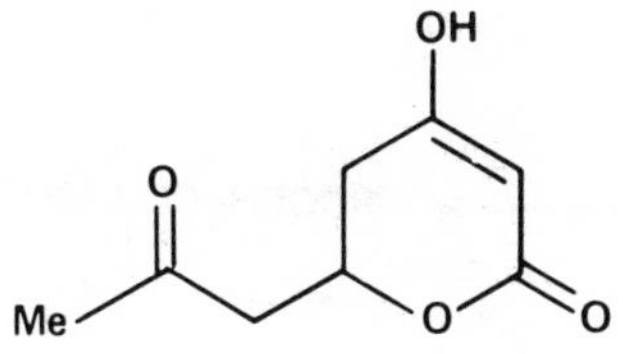

[126]

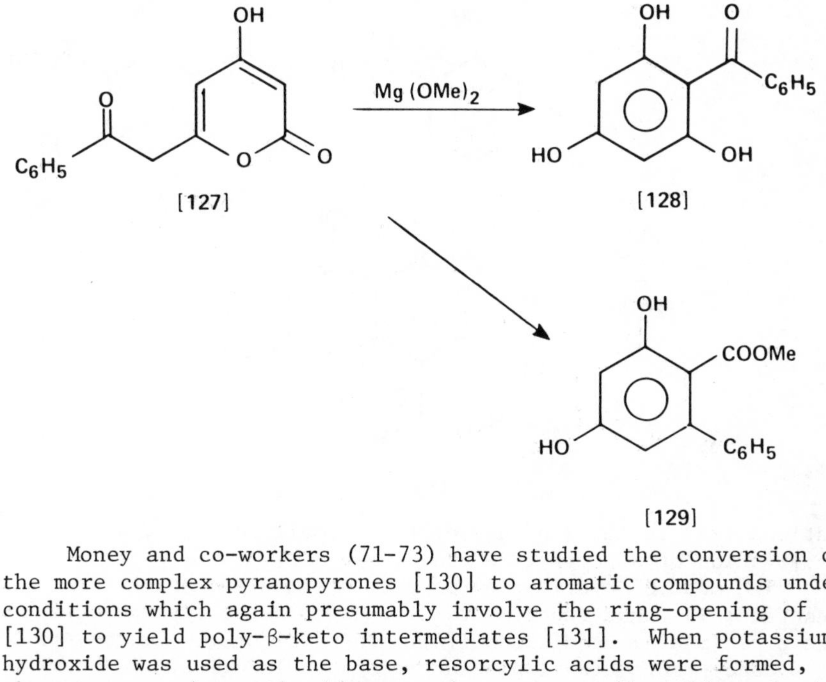

[127]   Mg(OMe)₂ →   [128]

[129]

Money and co-workers (71-73) have studied the conversion of the more complex pyranopyrones [130] to aromatic compounds under conditions which again presumably involve the ring-opening of [130] to yield poly-β-keto intermediates [131]. When potassium hydroxide was used as the base, resorcylic acids were formed, whereas magnesium methoxide gave the corresponding phloroglucinol derivative. Analogous conversions to more complex structural types have been observed when the polypyrone [132] was subjected to similar reaction conditions (74), yielding [132a to d].

A modification of the polypyrone approach has been used by Scott and co-workers (75, 94, 95) in order to study the cycliza-tion of longer polyketide chains. The ketone [133], which contains seven potential carbonyl groups, was treated with KOH/MeOH to yield the xanthone [134] as the only isolated aromatic compound.

The structures of all the observed aromatic products derived from the pyrone and polycarbonyl studies can be

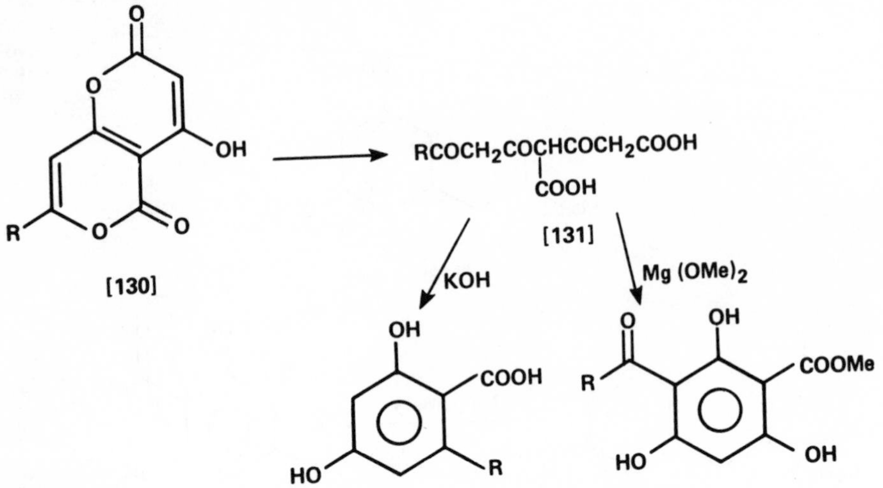

[130]

RCOCH$_2$COCHCOCH$_2$COOH
|
COOH

[131]

KOH

Mg(OMe)$_2$

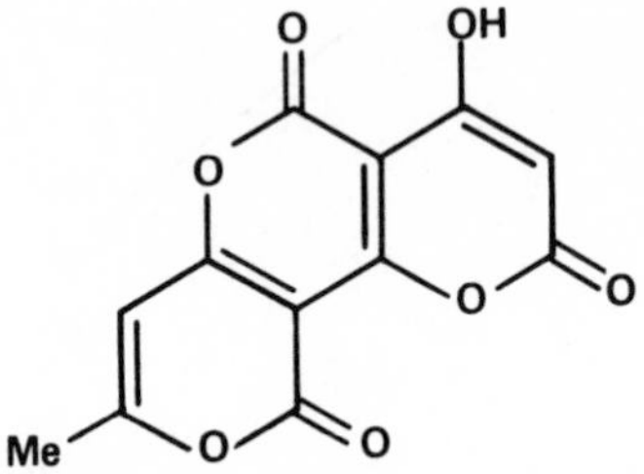

[132]

rationalized in terms of one of the normally accepted modes of cyclization of a polyketide chain.

The mode of cyclization of the intermediate polyketide generated in these experiments depends to a great extent upon the experimental conditions (76). In particular, the conversion of pyranopyrones and simple lactones to phloroglucinols seems to require magnesium ions as catalyst. Crombie and James (77) have proposed a mechanistic explanation for this observation, based upon the formation of metal chelates which give specific geometry to the noncyclized polyketide (see Bu'Lock's ideas, Section 17-2e); however, studies with the more complex [132] have shown that Mg^{2+} exerts less control over the direction of cyclization (74) in this case. No dependence upon metal ions has been observed in Harris' more direct studies (62, 70) using β-ketoacids. A short review by Tanenbaum (78) et al. on the possible role of polyketides in secondary metabolism has some interesting comments upon the pyrone and polyketide studies mentioned above. A review by Money (66) covers the literature through 1969.

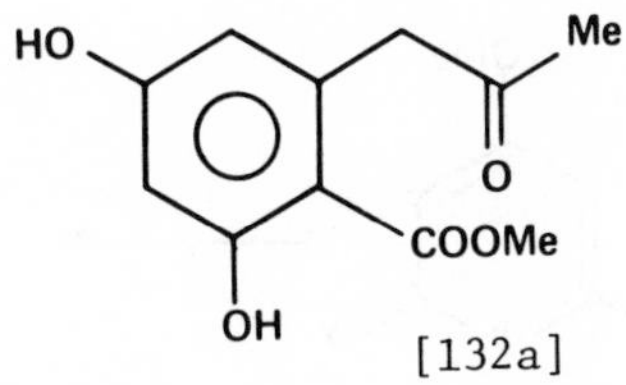

[132a]

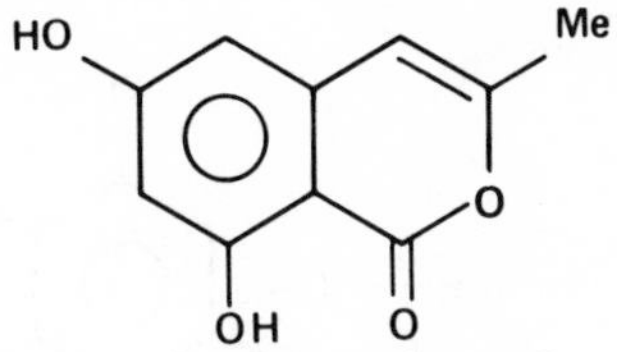

[132b]

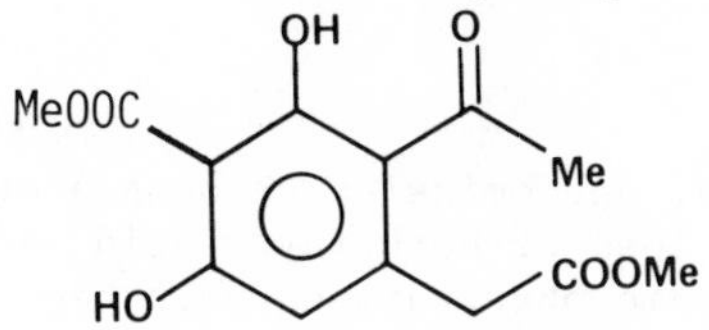

[132c]

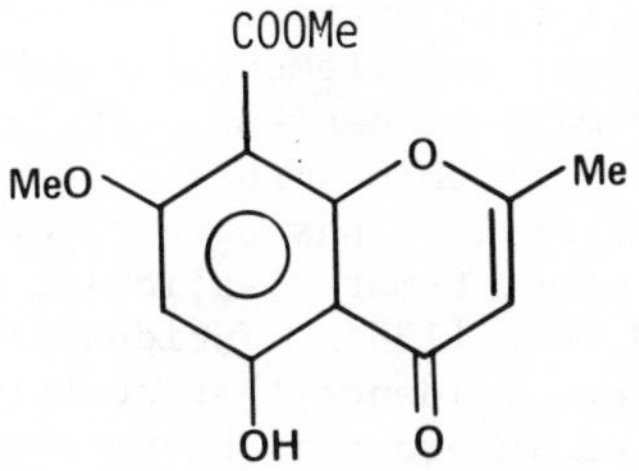

h. Enzyme Studies.   [132d]

While the incorporation of acetate and malonate into aromatic species by intact organisms has been studied in some detail, there is very little published work dealing with these incorporations at an enzymic level; even studies using crude cell extracts are rare. What results are available seem, however, to support the general hypothesis set out in the preceding pages.

A study of those enzymic systems responsible for the production of orsellinic acid by a purified enzyme preparation from the fungus *Penicillium madriti* has been made by Gaucher and Shepherd (80); only actyl-CoA and malonyl-CoA were found to be necessary for the synthesis. When NADPH was added in addition, fatty acid

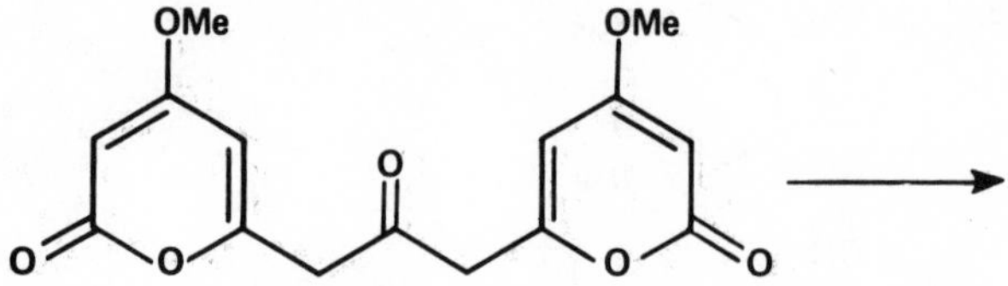

[133]

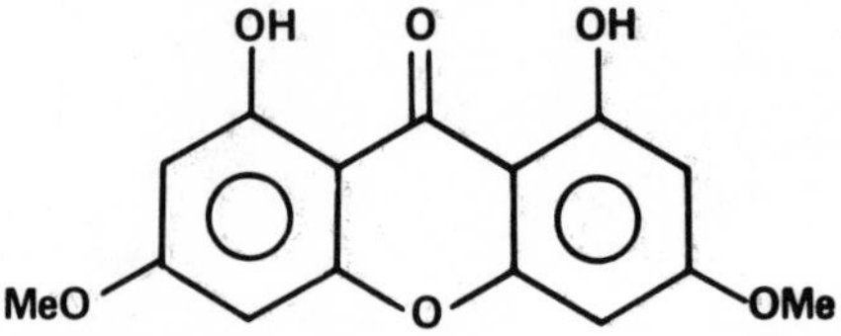

[134]

synthesis was observed, and omission of both acetyl-CoA and NADPH resulted in the loss of both fatty acid and orsellinic acid synthesis. The same observations have been made with regard to the cell-free synthesis of alternariol by *Alternaria tenuis* (81, 82). In the case of the formation of 6-methyl-salicylic acid by a cell-free extract of *Penicillium patulum* (83, 39, 79) an additional requirement for NADPH was observed, which is quite in line with mechanisms postulated in the previous section. The *in vitro* biosynthesis of patulin [135] by a cell-free extract of *P. patulum* has been recorded by Bassett and Tanenbaum (85); acetyl-CoA, 6-methylsalicylic acid, and glucose could all be converted into [135]. Incidentally, this experiment provided the first direct evidence that acetyl-CoA was the active form of acetic acid used in aromatic biosynthesis. The same

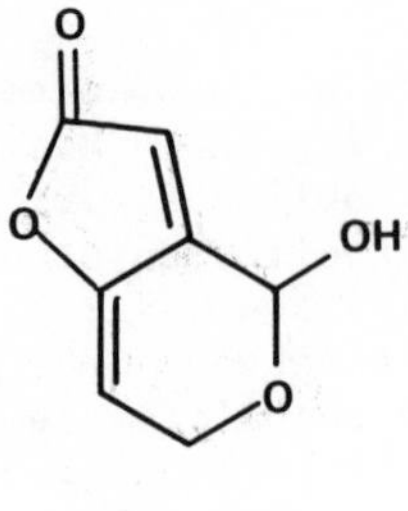

[135]

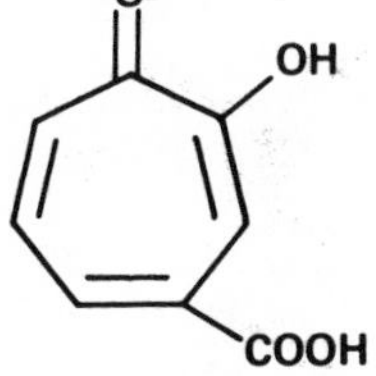

[136]

authors (86) studied the cell-free formation of the nonbenzenoid tropolone stipitatic acid [136] by *P. stipitatum*, and observed the incorporation of acetyl-CoA, malonyl-CoA, methionine, and formate.

Studies on the metabolism of *P. islandicum* (87), directed at the enzymic mechanisms involved in the production of quinonoid pigments (see Section 18-5) formed by this organism, have provided strong support for the reactions already outlined, although no *in vitro* formation of the various pigments has been reported.

These enzymic studies are the subject of a recent review by Light (88).

REFERENCES

1. A.J. Birch, Science 156, 202 (1967).
2. A.J. Birch, in Biosynthesis of Aromatic Compounds, G. Billek, Ed. Pergamon, Oxford, 1966, p. 3.
3. A.J. Birch, Ann. Rev. Plant Physiol. 19, 321 (1968).
4. S. Shibata, Chem. Brit. 3, 110 (1967).
5. R. Bentley and I.M. Campbell, Comprehensive Biochemistry 20, 415 (1968).
6. W.B. Whalley, in Biogensis of Natural Compounds, P. Bernfeld, Ed. Pergamon, Oxford, 2nd ed. 1967, p. 1025.
7. J.N. Collie, J. Chem. Soc. 91, 1806 (1970), and earlier papers quoted.
8. T. Batterham, R.G. Cooke, H. Duewell, and L.G. Sparrow, Aust. J. Chem. 14, 637 (1961).
9. K.S. Brown, D.W. Cameron, and U. Weiss, Tetrahedron Lett. 471 (1969).
10. S. Shibata, E. Morishita, M. Kaneda, Y. Kimura, M. Takida, and S. Takahashi, Chem. Pharm. Bull. 17, 454 (1969).
11. F. Lynen, E. Reichert, and L. Rueff, Ann. 574, 1 (1951).
12. A.J. Birch, in Perspectives in Organic Chemistry, Sir A.R. Todd, Ed. Interscience, New York, 1956, p. 134.
13. A.J. Birch, Prog. Chem. Org. Nat. Prod. 14, 186 (1957).

14. A.J. Birch and H. Smith, Chem. Soc. Spec. Publ. No. 12,
     p. 1 (1958).
15. A.J. Birch and F.W. Donovan, Aust. J. Chem. 6, 360 (1953).
16. W.B. Mors, O.R. Gottlieb, and C. Djerassi, J. Amer. Soc.
     79, 4507 (1957).
17. A.J. Birch, P. Fitton, E. Pride, A.J. Ryan, H. Smith, and
     W.B. Whalley, J. Chem. Soc. 4576 (1958).
18. J.S.E. Holker, J. Staunton, and W.B. Whalley, J. Chem. Soc.
     16 (1964).
19. A.J. Birch, A. Cassera, P. Fitton, J.S.E. Holker, H. Smith,
     G.A. Thompson, and W.B. Whalley, J. Chem. Soc. 3583 (1962).
20. J.R. Hadfield, J.S.E. Holker, and D.N. Stanway, J. Chem.
     Soc. (C), 751 (1967).
21. K. Mosbach, Acta Chem. Scand. 18, 1591 (1964).
22. W.D. Ollis, I.O. Sutherland, R.C. Codner, J.J. Gordon, and
     G.A. Miller, Proc. Chem. Soc. 347 (1960).
23. L.K. Dalton and J.A. Lamberton, Aust. J. Chem. 11, 46, 73
     (1958).
24. J.H. Richards and J.B. Hendrickson, The Biosynthesis of
     Steroids, Terpenes, and Acetogenins, W.A. Benjamin,
     New York, (1964).
25. D.D. Ridley, E. Ritchie, and W.C. Taylor, Aust. J. Chem.
     23, 147 (1970).
26. J.R. Cannon, P.W. Chow, B.W. Metcalf, and A.J. Power,
     Tetrahedron Lett. 325 (1970).
27. F.M. Lovell, J. Amer. Chem. Soc. 88, 4510 (1966).
28. A.J. Birch, P. Hodge, R.W. Rickards, R. Takeda, and
     T.R. Watson, J. Chem. Soc. 2641 (1964).
29. R. Bentley and J.G. Keil, Proc. Chem. Soc. 111 (1961).
30. P.R. Vagelos, Ann. Rev. Biochem. 33, 139 (1964).
31. T. Higuchi and S.A. Brown, Can. J. Biochem. Physiol. 41,
     65 (1963).
32. S.Z. El-Basyouni, D. Chen, R.K. Ibrahim, A.C. Neish, and
     G.H.N. Towers, Phytochemistry 3, 485 (1964).
33. A.J. Birch, R.A. Massy-Westropp, and C.J. Moye, Aust. J.
     Chem. 8, 539 (1955).
34. E. Lederer, Quart. Rev. (London) 23, 453 (1969).
35. J.W. Wallace, T.J. Mabry, and R.E. Alston, Phytochemistry
     8, 93 (1969).
36. R.E. Alston, Rec. Adv. Phytochem. 1, 305 (1968).
37. J.D. Bu'Lock, Biosynthesis and Microbial Development, Wiley,
     New York, 1967.
38. D.C. Allport and J.D. Bu'Lock, J. Chem. Soc. 654 (1960).
39. F. Lynen and M. Tada, Angew. Chem. 73, 513 (1961).
40. H.J. Vogel, in Organizational Biosynthesis, H.J. Vogel, Ed.
     Academic, New York, 1967, p. 443.

41. R. Thomas, personal communication.
42. R. Bentley and P.M. Zwitkowitz, J. Amer. Chem. Soc. 89, 676 (1967).
43. R. Bentley, J.A. Ghaphery, and J.G. Keil, Arch. Biochem. Biophys. 111, 80 (1965).
44. P.E. Brenneisen, T.E. Acker, and S.W. Tanenbaum, J. Amer. Chem. Soc. 86, 1264 (1964).
45. T.M. Harris, C.M. Harris, and R.J. Light, Biochim. Biophys. Acta 121, 420 (1966).
46. J.D. Brodie, G. Wasson, and J.W. Porter, J. Biol. Chem. 239, 1346 (1964).
47. D.J.H. Brock and K. Bloch, Biochem. Biophys. Res. Commun. 23, 775 (1966).
48. M. Yalpani, K. Willecke, and F. Lynen, Eur. J. Biochem. 8, 495 (1969).
49. G. Ehrensvärd, Exptl. Cell. Res., Supp. 3, 102 (1955).
50. R.J. Light, T.M. Harris, and C.M. Harris, Biochemistry 5, 4037 (1966).
51. H. Brockmann, J.J. Gordon, W. Keller-Schierlein, W. Lenk, W.D. Ollis, V. Prelog, and I.O. Sutherland, Tetrahedron Lett. 8, 25 (1960).
52. M.P. Kunstmann and L.A. Mitscher, J. Org. Chem. 31, 2920 (1966).
53. A. Stossel, Can. J. Chem. 47, 767 (1969).
54. J.D. Bu'Lock and J.R. Smith, J. Chem. Soc. (C) 1941 (1968).
55. J. Atherton, B.W. Bycroft, J.C. Roberts, P. Roffey, and M.E. Wilcox, J. Chem. Soc. (C), 2560 (1968).
56. J.J. Gordon, L.M. Jackman, W.D. Ollis, and I.O. Sutherland, Tetrahedron Lett. 8, 28 (1960).
57. T. Noda, T. Take, M. Otani, K. Miyauchi, T. Watanabe, and J. Abe. Tetrahedron Lett. 6087 (1968).
58. R.H. Iwamoto, P. Lim, and N.S. Bhacca, Tetrahedron Lett. 3891 (1968); F. Arcaimone, G. Cassinelli, G. Franceschi, R. Mondell, P. Orezzi, and S. Penco, Gazz. Chim. Ital. 100, 949 (1970).
59. N. Tsuji, Tetrahedron 24, 1765 (1968).
60. K. Eckardt, Chem. Ber. 100, 2561 (1967).
61. S.R. Johns, J.A. Lamberton, and A.A. Sioumis, Chem. Commun. 1324 (1968); Aust. J. Chem. 22, 775, 793 (1969).
62. T.M. Harris and R.L. Carney, J. Amer. Chem. Soc. 88, 2053, 5686 (1966); 89, 6734 (1967).
63. T.T. Howarth, J.P. Murphy, and T.M. Harris, J. Amer. Chem. Soc. 91, 517 (1969).
64. T.T. Howarth and T.M. Harris, J. Amer. Chem. Soc. 93, 2506 (1971).
65. J.V. Hay and T.M. Harris, Chem. Commun. 953 (1972).

66. T. Money, Chem. Rev. 70, 553 (1970).
67. T.M. Harris, T.T. Howarth, and R.L. Carney, J. Amer. Chem. Soc. 93, 2511 (1971); T.M. Harris, C.M. Harris, and K.B. Hindley, Prog. Chem. Nat. Prod. 31, 217 (1974).
68. A.J. Birch, Proc. Chem. Soc. 3 (1962).
69. H. Guilford, A.I. Scott, D. Skingle, and M. Yalpani, Chem. Commun. 1127 (1968).
70. T.M. Harris, M.P. Wachter, and G.A. Wiseman, Chem. Commun. 177 (1969).
71. T. Money, F.W. Comer, G.R.B. Webster, I.G. Wright, and A.I. Scott, Tetrahedron 23, 3435 (1967).
72. T. Money, J.L. Douglas, and A.I. Scott, J. Amer. Chem. Soc. 88, 624 (1966).
73. T. Money, I.H. Qureshi, G.B. Webster, and A.I. Scott, J. Amer. Chem. Soc. 87, 3004 (1965).
74. F.W. Comer. T. Money, and A.I. Scott, Chem. Commun. 231 (1967).
75. G.D. Pike, J.J. Ryan, and A.I. Scott, Chem. Commun. 629 (1968).
76. J.L. Douglas and T. Money, Tetrahedron 23, 3545 (1967).
77. L. Crombie and A.W.G. James, Chem. Commun. 357 (1966).
78. S.W. Tanenbaum, S. Nakajima, and G. Marx, Biotechnol. Bioeng. 11, 1135 (1969).
79. P. Dimroth, H. Walter, and F. Lynen, Eur. J. Biochem. 13, 98 (1970).
80. G.M. Gaucher and M.G. Shepherd, Biochem. Biophys. Res. Commun. 32, 664 (1968).
81. S. Gatenbeck and S. Hermodsson, Acta Chem. Scand. 19, 65 (1965).
82. S. Sjoland and S. Gatenbeck, Acta Chem. Scand. 20, 1053 (1966).
83. R.J. Light, J. Biol. Chem. 242, 1880 (1967).
84. A. Hassanali, T.J. King, and C.S. Wallwork, Chem. Commun. 678 (1969); B.W. Bycroft, A. Hassanali, A.W. Johnson, and T.J. King, J. Chem. Soc. (C), 1686 (1970).
85. E.W. Bassett and S.W. Tanenbaum, Biochim. Biophys. Acta 40, 535 (1960).
86. S.W. Tanenbaum and E.W. Basset, Biochim. Biophys. Acta 59, 524 (1962).
87. Y. Ueno and T. Tatsuno, Chem. Pharm. Bull. 17, 1175 (1969).
88. R.J. Light, Agr. Food Chem. 18, 260 (1970).
89. E. Leete and J.O. Olson, Chem. Commun. 1651 (1970).
90. R. Hegnauer, Chemotaxonomie der Pflanzen, Vol. 5, Birkhäuser Verlag, Basel and Stuttgart, 1969, p. 121.
91. R. Hänsel, D. Ohlendorf, and A. Pelter, Z. Naturforsch. 25b, 989 (1970).

92. W.B. Turner, _Fungal Metabolites_, Academic London (1971).
93. Y. Suzuki, Agr. Biol. Chem. $\underline{34}$, 760 (1970).
94. A.I. Scott, D.G. Pike, J.J. Ryan, and H. Guilford,
     Tetrahedron $\underline{27}$, 3051 (1971).
95. A.I. Scott, H. Guilford, J.J. Ryan, and D. Skingle,
     Tetrahedron $\underline{27}$, 3025, 3039 (1971).
96. W. Steglich and E. Töpfer-Petersen, Z. Naturforsch. $\underline{27b}$,
     1286 (1972); W. Steglich, E. Töpfer-Petersen, and I. Pils,
     ibid. $\underline{28c}$, 354 (1973).
97. J.F. Grove, J. Chem. Soc. Perkin I, 2406 (1972).
98. F. Blank, C. Buxtorf, O. Chin, G. Just, and J.L. Tudor,
     Can. J. Chem. $\underline{47}$, 1561 (1969).
99. G.M. Hatfield and L.R. Brady, Lloydia, $\underline{36}$, 59 (1973).
100. S.W. Tanenbaum, S.G. Agarwal, T. Williams, and
     R.G. Pitcher, _Tetrahedron Lett._, 2377 (1970).

CHAPTER 18

Biosynthesis of the Various Types of Polyketide

A very impressive body of evidence in support of the polyketide mode of aromatic biosynthesis has been assembled over the past 20 years. The isotope labeling techniques discussed in general terms in Chapter I have been the most widely used, although some information has come from mutant studies, FT ^{13}C-nmr techniques, and from experiments with indirect feeding. The compounds that have been investigated are reviewed in groups according to their chemical similarity. This classification is quite arbitrary from the point of view of biosynthesis, since the same chemical structure can very often be generated by quite diverse biosynthetic processes; however, the alternative Hendrickson scheme, employed by Richards and Hendrickson (1), in which metabolites are grouped in terms of the number of acetate residues found to cooperate in their biosynthesis, seemed to offer little advantage for our purpose.

18-1. SIMPLE PHENOLS

6-Methylsalicylic acid [1] and orsellinic acid [2] represent the simplest aromatic compounds derived from one acetate and three malonate units; because of this, their biosynthesis has been extensively studied. In one of the first incorporation studies, Birch (2) and his co-workers fed 1-14C-acetate to the mold *Penicillium griseofulvum* and isolated labeled [1]; degradation showed that the label was located to an equal extent in alternate carbon atoms, in complete accord with theory. A similar study (3) of the biosynthesis of [2] by *P. baarnense* showed an identical incorporation. The more intimate details of this biosynthesis have been explored to some extent. An early

investigation by Birch (4, 5) showed that C-6 was labeled to a
greater extent than any other carbon in the molecule when 1-^{14}C
acetate and unlabeled malonate were fed to *Penicillium urticae*;
this demonstrated the derivation of [1] from a single polyketide
chain with the 6-methyl group provided by the acetate starter
unit.   The converse experiment (6) showed that all the activity
from 2-^{14}C-malonate was found in the ring carbons.   Similar
studies by Bentley and Keil (7) confirmed these results.   In a
further experiment (8) using ^{18}O-acetate, Gatenbeck and Mosbach
showed that the oxygens in the phenolic and carboxyl groups of [1]
were all labeled in exactly the way predicted by theory.   The
generality of the scheme outlined for the biosynthesis of [1]
has been confirmed by studies in *Mycobacterium phlei* (119, 120)
and *M. fortuitum* (121).   Orsellinic acid has been shown to undergo modification to 1-methoxy-2,5-dichloro-4-hydroxy-5-methyl-
benzoic acid by *Micromonospora carbonacea* during the biosynthesis
of the everninomicin antibiotics (122).   Studies of the enzymic
formation of [2] are discussed in Section 19-2c.

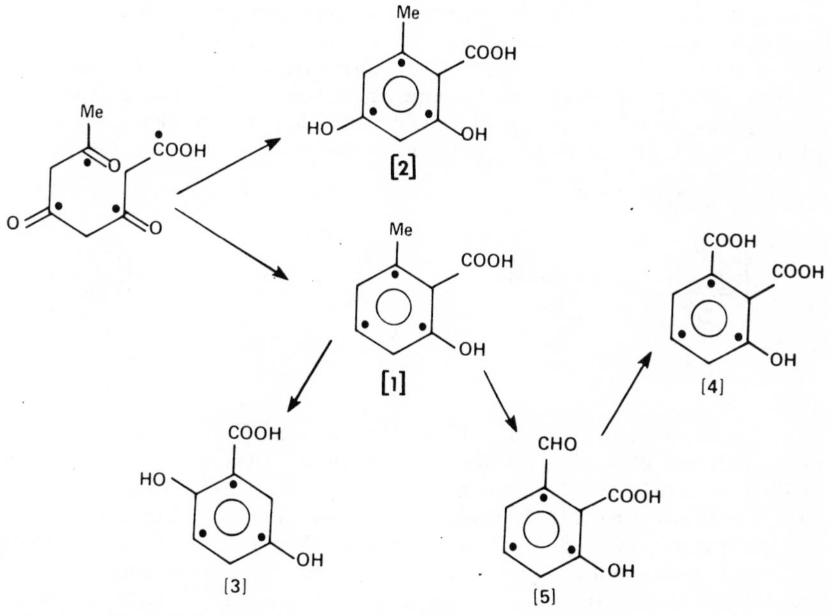

Gentisic acid [3] is produced by *Penicillium* species and
several other organisms. In some cases the acid is biosynthesized
via the shikimic acid pathway, for example, by *Polyporus tumulosus,*
which forms [3] as well as several other shikimate-derived metabolites: 4-hydroxyphenylacetic acid, 3,4-dihydroxyphenylacetic
acid, and 2,5-dihydroxyphenylglyoxylic acid. However, degradation of [3] isolated from *P. urticae* grown on ^{14}C-acetic acid
has shown that the labeling pattern conforms to that expected
for an acetate-derived metabolite. A time-study of the production
of [3] and 6-methylsalicylic acid has provided good evidence that
the former is formed by a modification of the skeleton of
6-methylsalicylic acid (9).

3-Hydroxyphthalic acid [4] is another simple compound derived
from a modified polyketide; degradation of [4] isolated from
Penicillium islandicum which had been grown in the presence of
1-^{14}C-acetate (10) showed the expected distribution of label.
Production by cultures of *P. urticae* is enhanced by the addition of
6-methylsalicylic acid, implying a stepwise oxidation to [4] via
the aldehyde [5], which is also found in the cultures (11).

Clavatol [6] is formed at the same time as 6-methylsalicylic
acid by *Aspergillus clavatus*. Both metabolites are derived from
a C_8 polyketide. The two nuclear methyl groups of [6] are provided by methionine, and there is evidence that the methylation
takes place at the prearomatic stage, since resacetophenone [7]
is not converted into [6] by the organism (12). The co-occurrence
of [7] and 6-methylsalicylic acid provides an instructive example
of two modes of folding of the same polyketide by one organism.

[6]   [7]

Labeled flavipin [8], from *Aspergillus flavipes,* has been
degraded and shown to be acetate-derived (13); it is interesting
that orsellinic acid and the corresponding aldehyde [9] could
also be incorporated, implying the possible methylation and
hydroxylation of a preformed aromatic ring (14). However, there
is still some doubt about the true sequence, since more recent
experiments on cell-free methylation of the aromatic precursor
seem to indicate that orsellinic acid is not methylated, but that
the polyketide is methylated prior to aromatization (15). Incorporation of 1-^{14}C acetate into the closely related cyclopaldic

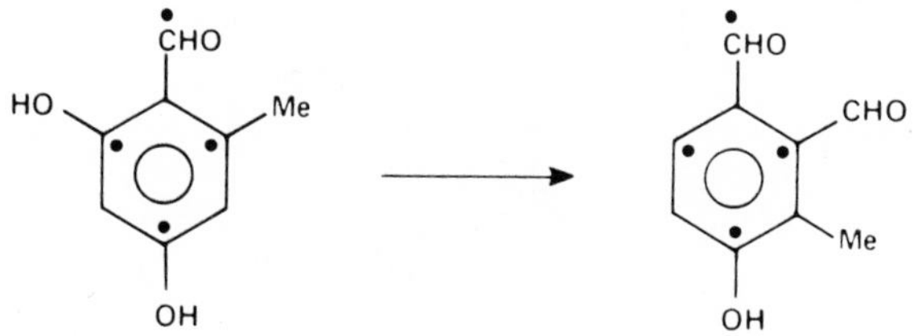

[9]   [8]

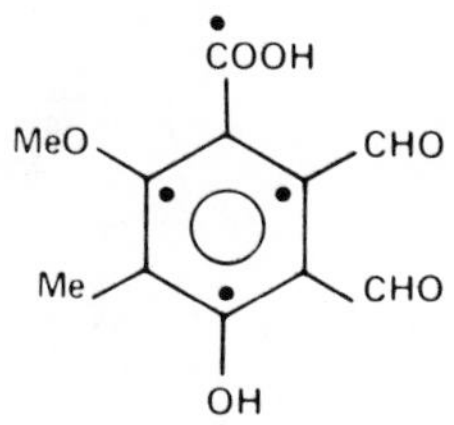

[10]

acid [10] has also been shown to follow the expected pattern
(16).

   The biosynthesis of 7-hydroxy-4,6-dimethylphthalide [11] by
Penicillium gladioli has been investigated by Birch and Pride
(17); the incorporation of acetate was again as anticipated.
Similarly, 1-^{14}C acetate fed to *P. brevi-compactum* gave myco-
phenolic acid [12] having the expected labeling pattern (18).
Tracer studies (112) have indicated that 5,7-dihydroxy-4-methyl-
phthalide [12a] is a precursor of [12]; an isoprenoiol C_{15} unit is
introduced into [12a] by reaction with farnesyl pyrophosphate,

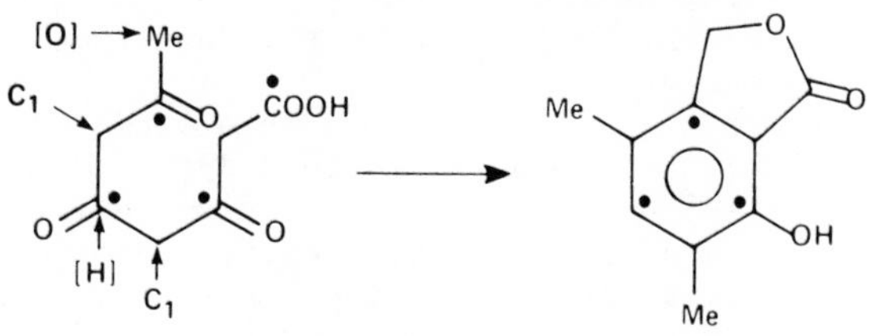

[11]

and the side-chain is then degraded, yielding [12]. Whereas
5-methylorsellinic acid is well incorporated into [12], orsellinic
acid itself is poorly utilized, implying that alkylation probably
takes place at a pre-aromatic stage as expected (137).

   Two groups have investigated the formation of orcylalanine
by the corn cockle *Agrostemma githago*; acetate provides the

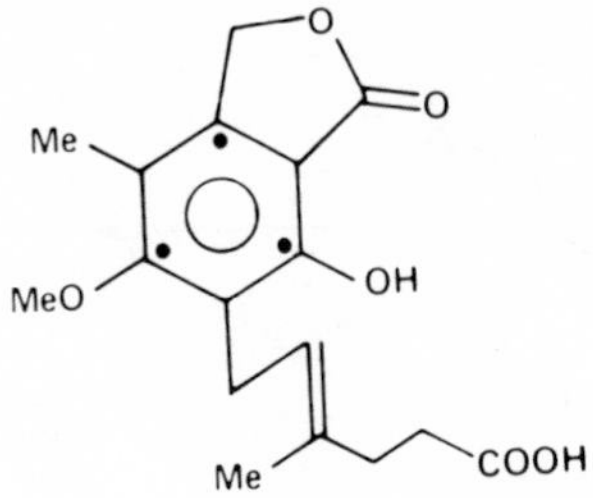

[12]

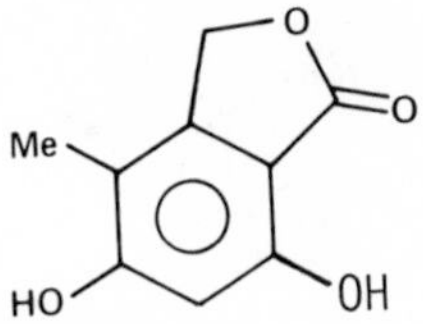

[12a]

[13]

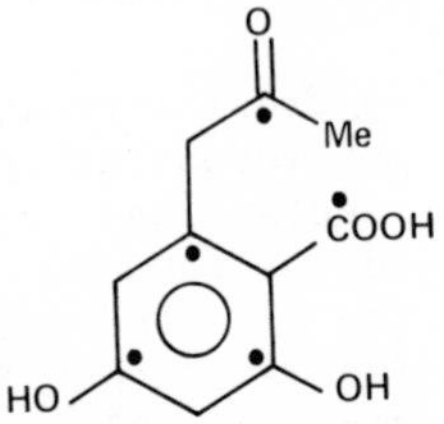

aromatic ring, and serine the side-chain. Details of the biosynthesis are not clear at the present time, but it seems possible
that the mechanism may be related to the formation of dopa from
tyrosine and pyrocatechol (123, 124).

Westley and his co-workers (147) in a study of the biosynthesis of the antibiotic X-537A produced by *Streptomyces* spp.,
demonstrated the formation of the 3-methylsalicylic residue in
the antibiotic from acetate-malonate units. Similarly, a study
of the biosynthesis of curvulic acid (2-aceto-3,5-dihydroxy-4-
methoxyphenylacetic acid) by *Curvularia siddiquii* has shown it
to have a polyketide origin (148).

The phenols mentioned above have all been derived from a
C_8 polyketide; however, several more complex phenols, which are
formed from longer chains, have been investigated.

A simple example is the phenol [13], formed from one acetate
and four malonate units, whose biosynthesis has been studied by
Stickings (19). The formation of [13] provides one more example

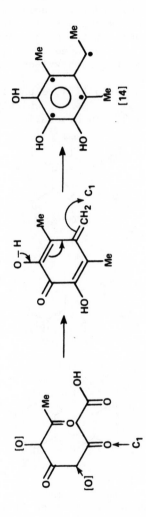

of a compound derived from an unmodified polyketide (see Section
17-2a).   Barnol [14] is another phenol derived from a tetraketide
which has been isolated from an unidentified *Penicillium* species;
its biosynthesis from acetate-malonate has been shown to conform
to the scheme shown (20).   Methionine is not only incorporated
into the aromatic methyl group but also provides the 2' carbon of
the ethyl group; the possibility of a biosynthesis involving
a quinone methide has been pointed out.   The proposed biosynthesis
implies the alkylation of a carboxyl-derived carbon of the polyketide, an unusual event.
     More complex is sclerin [15], a physiologically active meta-
bolite of *Sclerotinia sclerotiorum*, which co-occurs with sclero-
tinins A [16] and B [17].   Incorporation of ^{14}C-acetate and

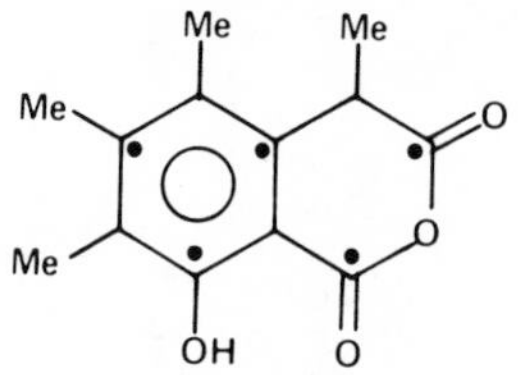

[15]

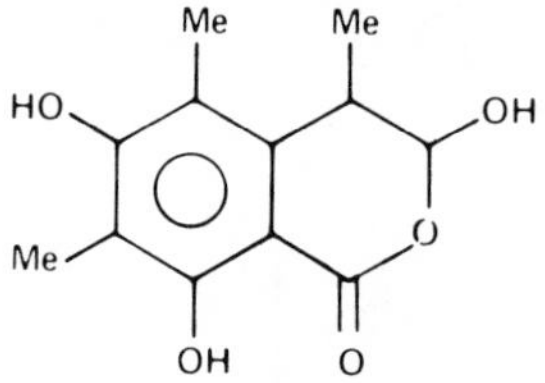

|16|

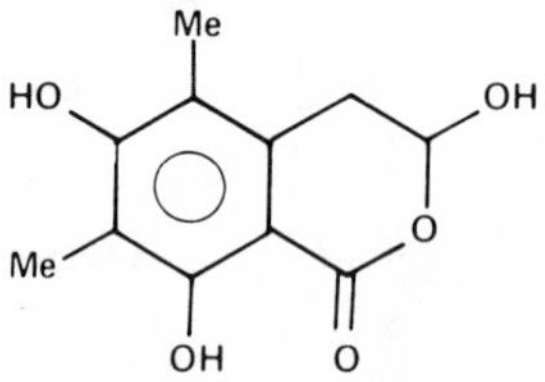

[17]

subsequent degradation of the labeled [15]   indicated the expected
derivation from the acetate-malonate pathway (114).   The number
of labeled nuclear methyl groups found after feeding 1-^{14}C-,
2-^{14}C-acetate and ^{14}C-formate was 0, 1, and 3, respectively; of
the possible modes of formation of [15], only that from two
acetate chains [18] would give this result.   Of the other

possibilities, two, [19] and [20], require the removal of a carbon atom, and the third postulates alkylation at a carbonyl site; these unlikely features make these alternatives much less favorable.

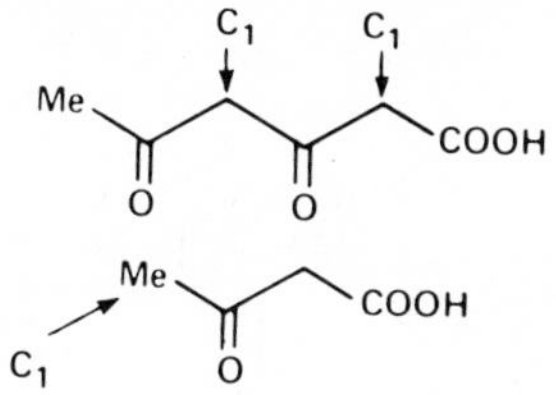

[18]

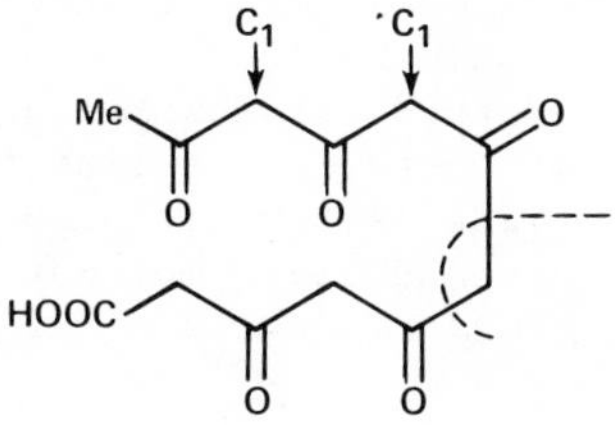

[19]

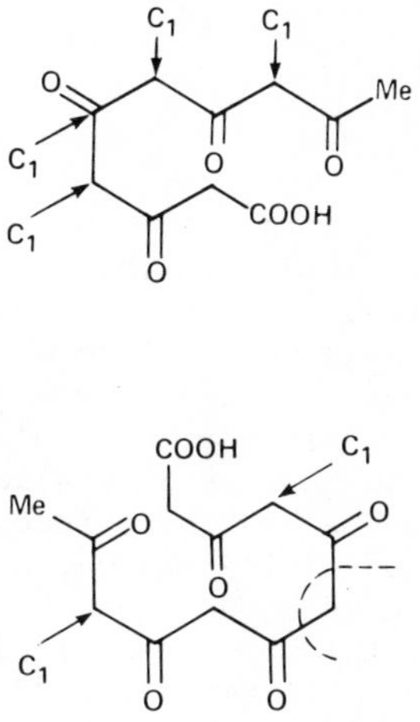

[20]

Auroglaucin [21] is formed, together with flavoglaucin, in which the C_7 side chain is saturated, by *Aspergillus novus* from a C_{14} polyketide and a mevalonoid C_5 residue. Labeling studies have confirmed the biosynthesis (21).

[21]

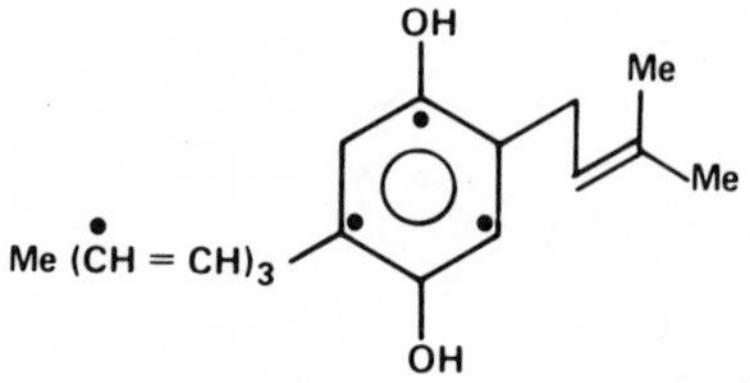

Me $(\overset{\bullet}{C}H = CH)_3$

Feeding of active acetate to a *Curvularia* sp. produces labeled curvularin [22], which on treatment with HBr gives two C_8 fragments, each containing equal radioactivity. This provides strong presumptive evidence for a labeling pattern as shown, although no further attempts to localize the activity have been made (22). The macrolide lactone is probably synthesized by the fungus from one acetate and seven malonate units, as indicated.

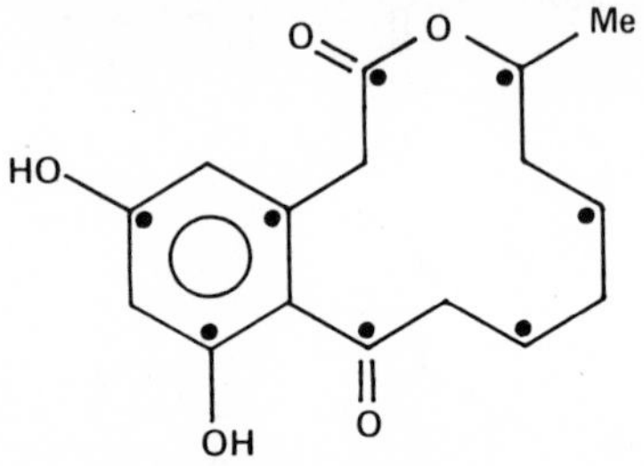

[22]

It is interesting to note that part of the chain has been totally reduced here, except for the one conjugated carbonyl, in a manner similar to that of fatty acid biosynthesis.

The polyketide chain formed from one acetate and six malonate residues is transformed by *Penicillium griseofulvum* into the chlorinated antibiotic griseofulvin [75], in which all the carbon and oxygen atoms have been retained (66). The steps in the biosynthetic sequence have been repeatedly investigated (67), but no really definitive pathway has yet been established; however, there is good evidence that the penultimate step is the oxidative coupling of the benzophenone [76] to the spiran [77] (68). An interesting application of ^{13}C-nmr spectroscopy has shown that the triprenylphenol ascochlorin is formed by the fungus *Nectria coccinea* from a modified farnesyl unit and an acetate-derived phenol (143). A similar mixed biosynthesis has been established for the cochlioquinones produced by *Cochliobolus miyabeanus* (144).

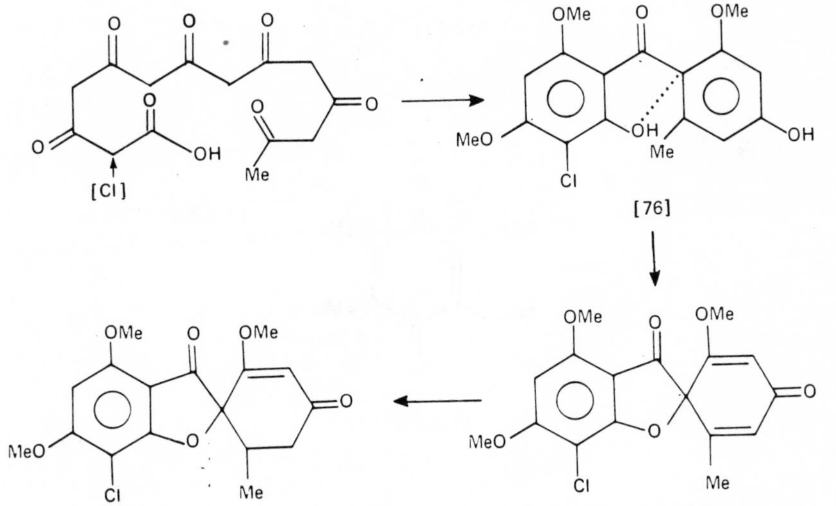

[76]

[77]

[75]

18-2. BENZOQUINONES

The biosynthesis of some of the simple benzoquinones found in fungi has been investigated by isotope labeling techniques. Much of the work has been reviewed by Bentley (1962) (23), Pettersson (1966) (116), and Yamamoto (1970) (125).

Aurantiogliocladin [23] and its quinhydrone rubrogliocladin are two of several related metabolites of *Gliocladium roseum*. Their acetate origin was established in a general way by Birch (24), who also noted that 6-methylsalicylic acid could be incorporated intact into [23]. More recent studies have confirmed that [23] is formed via that acetate-polymalonate pathway (25), and that only one of the methyl groups comes from acetate (26).

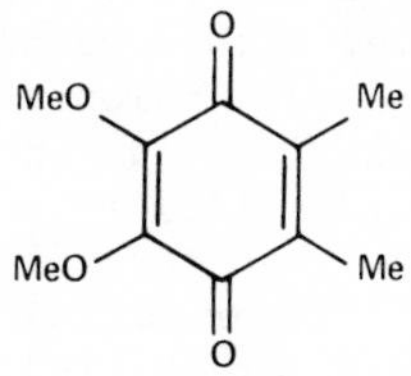

[23]

[25]

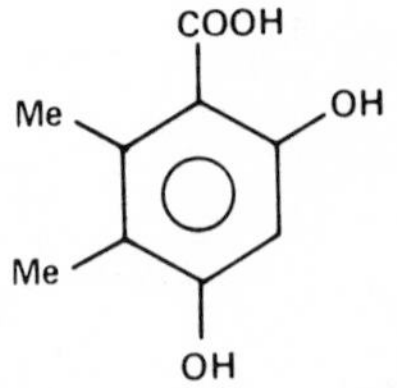

[26]

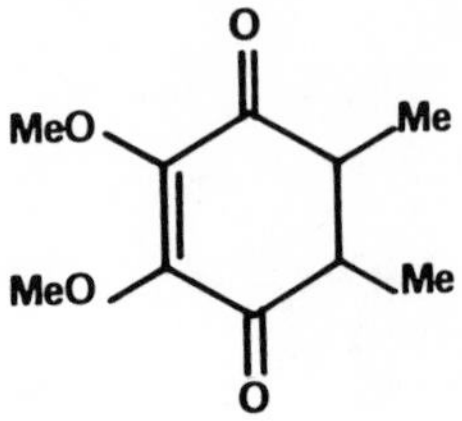

[27]

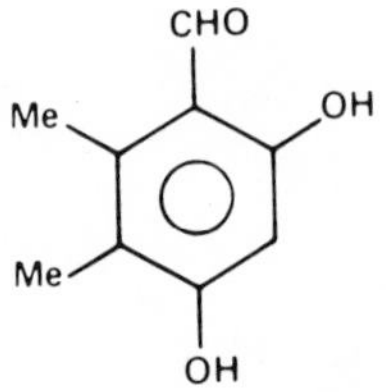

Pettersson (26) has clearly shown that orsellinic acid [2] is
incorporated into [23], as is the acid [25], which has also been
isolated from the culture medium. Some evidence has been presented by Packter to demonstrate that [23], and the analogous
toluquinone, are derived from the dihydroquinone gliorosein [26]
(27), and that 5-methylorcyl aldehyde [27] may be the initial
aromatic precursor of this group of compounds (28). However, an
actual precursor - product relationship between these compounds
has not been established, and in fact these results are totally
at variance with those of Pettersson, who has shown that the
major metabolite of young *G. roseum* cultures is [25] and not
[26] as claimed by Packter.
    The quinones spinulosin [28] and fumigatin [29], produced
by *Aspergillus* spp. together with several other toluquinones,
have been shown to be of acetate-malonate origin (29). Pettersson
has established that orsellinic acid, specifically labeled at
C-2, is incorporated into [29] (30) in such a way that most of the
activity is located at C-5. Packter (31) has found that
A. fumigatus can decarboxylate orsellinic acid to orcinol, which
is itself an excellent precursor of [29]. There has been a
report that 6-methylsalicylic acid can be incorporated into [29]
by the fungus (27), but this is probably only a reflection of the

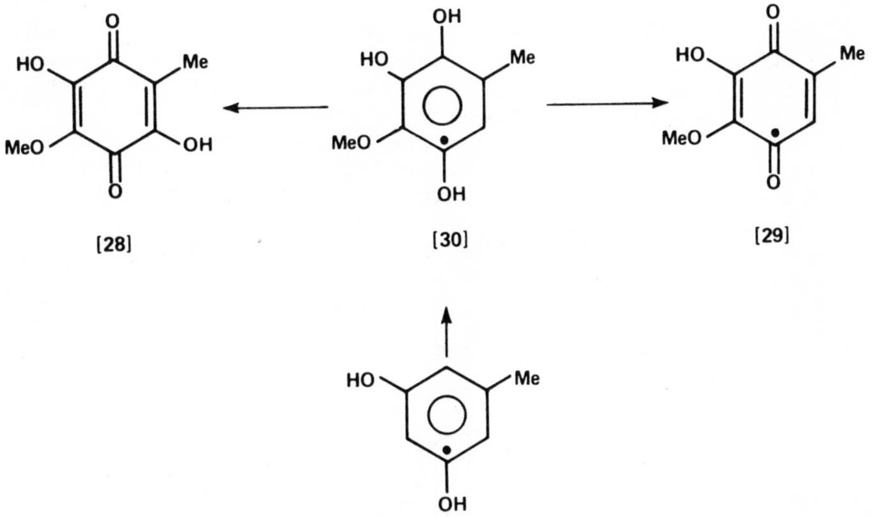

[28]   [30]   [29]

organism's ability to convert 6-methylsalicylic acid to orsellinic
acid (32), and does not imply that 6-methylsalicylic acid is a
natural precursor. Fumigatol [30], the reduced form of [29], is
also formed by *A. fumigatus*, and labeling studies have provided
good evidence that [30] is the immediate precursor of [29] in
the organism (31); the desmethyl homolog, 4,6-dihydroxyorcinol,
has also been incorporated (33). Pettersson has shown that
[30] is also converted to [28] without first being degraded to
acetate. Similarly the incorporation of 6-methylsalicylic acid
into the quinone [31] formed by *Lentinus degener* has been observed

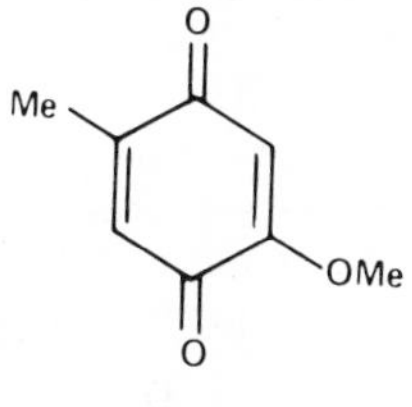

[31]

(32, 24). In contrast to these examples of benzoquinones derived
from acetate units, a recent publication (34) presents apparently
good evidence that 4-methoxytoluquinol, the reduced form of [31],
is derived from the shikimic acid pathway in *L. degener*. Tyrosine
was found to be incorporated, apparently without degradation;
although not entirely unknown, the formation of a ring-methylated

phenol via shikimic acid is rare enough to make this a very singular biosynthesis, if the report is substantiated. The most recent
quinone of this type to be looked at is shanorellin (2,6-dimethyl-
3-hydroxymethyl-5-hydroxy-1,4-benzoquinone) produced by
Shanorella spirotricha; both acetate and methionine contribute
to the biosynthesis, with the hydroxymethyl carbon representing
the C-2 of the acetate starter unit. No likely intermediates
were detected (126, 127).

     6-Methylsalicyclic acid is also involved in the biosynthesis
of the nonaromatic terreic acid [32] by the mold *Aspergillus*
terreus. Labeled acetate is incorporated into both metabolites,

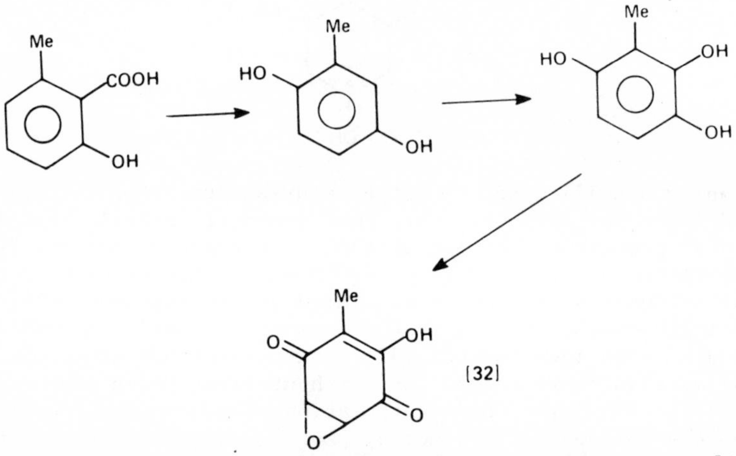

[32]

and active 6-methylsalicylic acid itself is metabolized to [32],
presumably via 2,5-dihydroxytoluene and 3-hydroxytoluquinone.
The specific incorporations clearly demonstrate that [32] is
formed by the acetate-malonate pathway via 6-methylsalicylic
acid. The epoxide oxygen comes from atmospheric oxygen (35).
Epoxidon [32A], an epoxide produced by *Phoma* spp., is likewise
biosynthesized from acetate, probably via gentisyl alcohol (142).

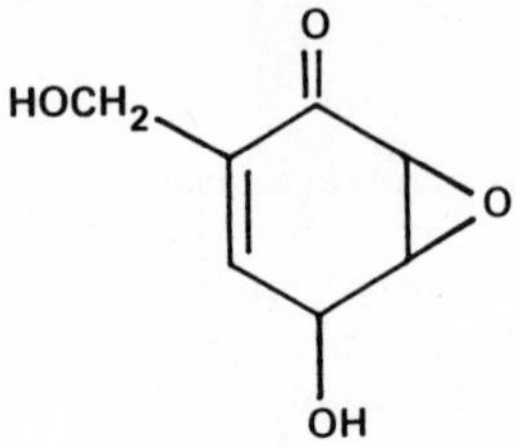

[32A]

An investigation by Meinwald and co-workers (117) into the
formation of some simple benzoquinones in the bombardier beetle
(*Eleodes longicollis*) demonstrated two modes of biosynthesis.
Phenylalanine and tyrosine were metabolized to unsubstituted
benzoquinone, whereas 2-methylbenzoquinone was formed solely from
acetate, and formation of the 2-ethyl derivative involved a propionate starter unit and an acetate chain. That this biosynthesis
was carried out by the beetle rather than by symbionts is debatable (see Chapter 1), but it is interesting that two unrelated
pathways are utilized to achieve the same aromatic system.

18-3. CONSTITUENTS OF LICHENS AND RELATED COMPOUNDS

The lichens are organisms formed through symbiosis of an
alga and a fungus, both of which may contribute the characteristic
metabolites which occur in this class of lower plants. There is,
however, considerable overlap between compounds from lichens and
from fungi; the terphenylquinone thelephoric acid, for example,
is found in both lichens and basidiomycetes. Many of the typical
lichen substances are based upon combination of two or three
units derived from orsellinic acid or related compounds, often
by esterification to form depsides. Five major types of substances result from such combinations. Their chemistry is
discussed in some detail in recent reviews by Mosbach (36) and
Huneck (37). The biosynthesis of representatives of four of
these five groups has been investigated.

The formation of lecanoric acid [33] and atranorin [34] by
the lichen *Parmelia tinctorum* has been studied (38). These two
compounds are depsides derived from orsellinic acid [2] and β-
orcinol carboxylic (3-methylorsellinic) acid, respectively, with
secondary modifications (Me⟶CHO, COOH⟶COOMe) in the case of
[34]. Orsellinic acid was shown to be incorporated into [33] but
not into [34]; conversely, [34] but not [33] was labeled from
β-orcinol carboxylic acid. Orsellinic acid is also converted to
the tridepside gyrophoric acid [35] by the lichen *Umbilicaria
pustulata* (39); the labeling pattern found in [35] indicate that
the orsellinic acid units are all produced in the same manner
and are joined together at a late stage in the biosynthesis.

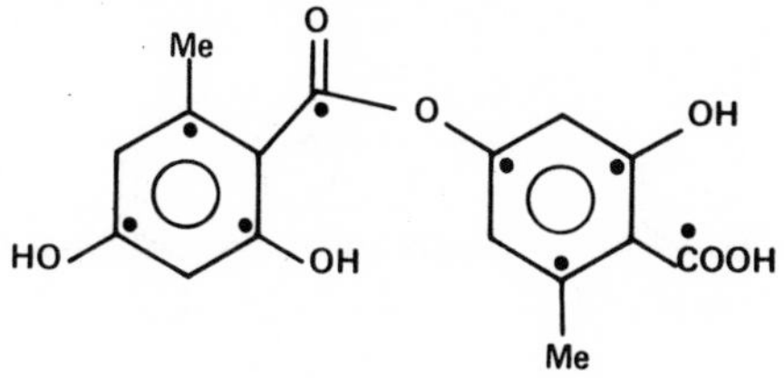

[33]

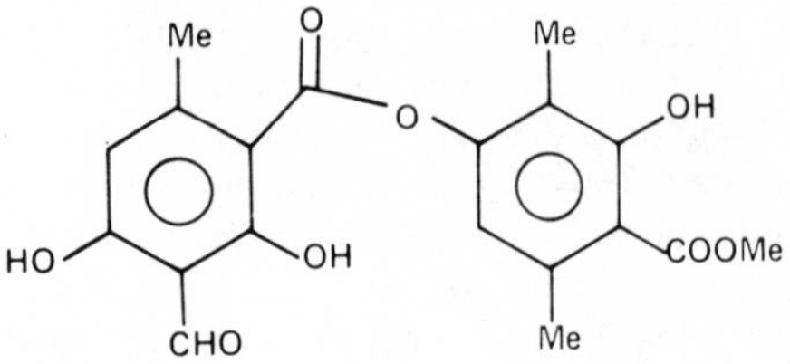

[34]

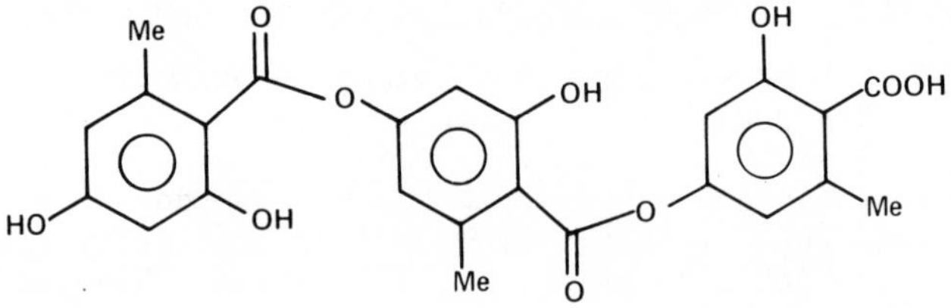

[35]

The bis-benzoquinone oosporein [36] is not a true lichen substance, although it is closely related to pyxiferin [37], which has been isolated from a lichen. However, [36] is a metabolic product of several fungal species; its biosynthesis by a strain of *Beauveria bassiana* has been investigated (40). Orcinol,

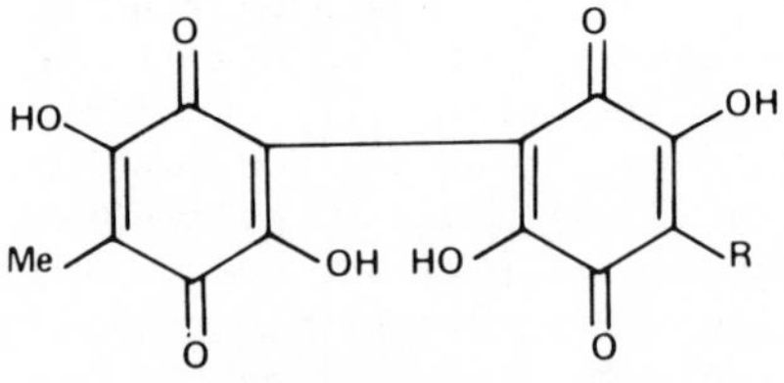

[36] R = Me
[37] R = H

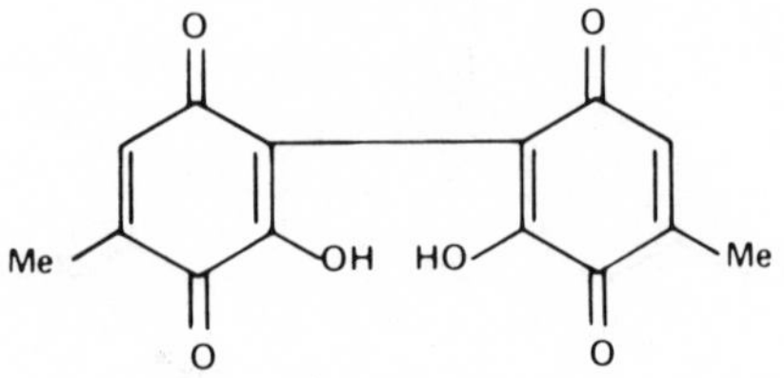

[38]

orsellinic acid and its ethyl ester, and acetic acid were all
incorporated; mevalonic acid, shikimate, and methionine were not.
The results of incorporation and degradation indicate a biosynthesis via orsellinic acid and orcinol, followed by hydroxylation
and oxidative coupling. A more recent study (44) has confirmed
the earlier findings. Similar incorporations have been demonstrated for the biosynthesis of phoenicin [38] by cultures of
Penicillium phoeniceum (42).

Usnic acid [39] is a dibenzofuran biosynthesized in lichens
by the oxidative coupling of two acetylmethylphloroglucinol
groups [40] derived from one acetate and three malonate groups
in the normal way. Isotopically labeled acetate, formate,
malonate, and methylphloracetophenone were all incorporated, but

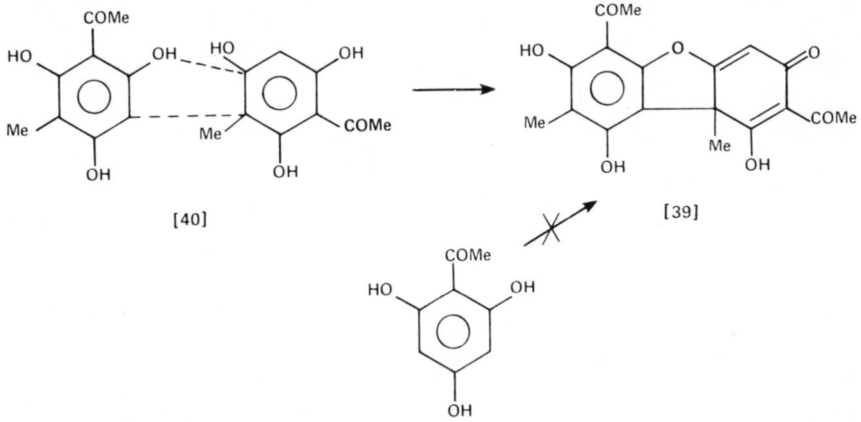

[40]

[39]

[41]

phloracetophenone [41] itself was not, indicating that methylation of the polyketide preceded cyclization (43). The *in vitro*
coupling [40]⟶[39] has been demonstrated by Barton (108) and
the enzymic process, catalyzed by horseradish peroxidase, by
Shibata (43), and by Penttila and Fales (44).

Nidulin [42] is again not a true lichen substance, but its
biosynthesis is clearly related to that of the compounds (e.g.,
[43]) found in the symbiotic organism. The compound, a metabolite
of *Aspergillus nidulans*, was reported to be derived from acetate
and isoleucine (45). Acetate was incorporated about equally into
both rings, with the carboxyl carbon having one fourth of the
activity associated with ring A. Isoleucine was incorporated
mainly into ring B, although some 13% was found in ring A (presumably due to degradation to acetate *in vivo*). Subsequently
(129) it was proposed that ring B and its side-chains are derived
from two isoprene units and one additional C_1 unit; neither of

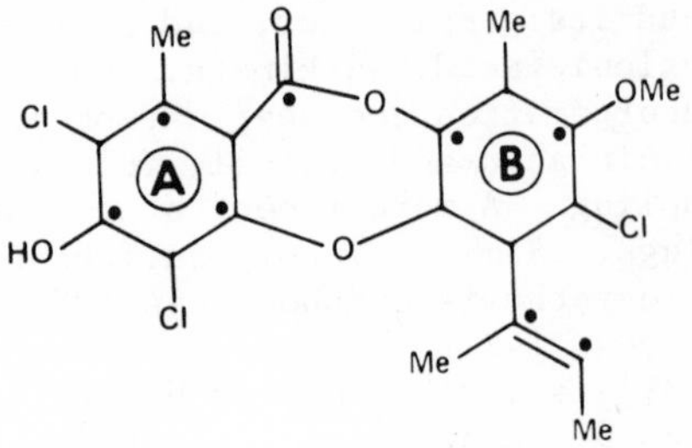

[42]

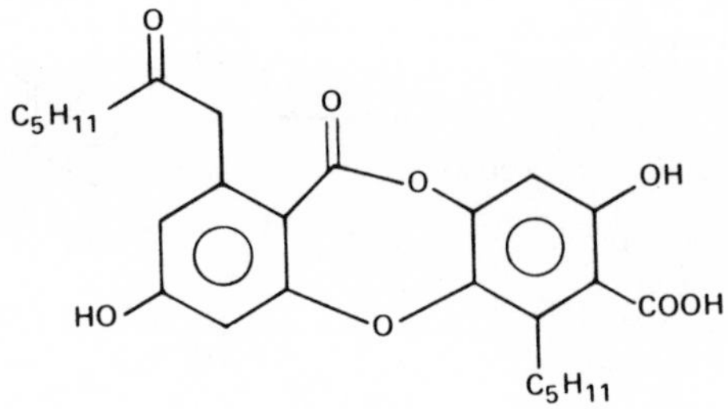

[43]

these two biosynthetic proposals are very convincing, and
Sierankiewicz and Gatenbeck (128) have reinvestigated the situation. The new evidence indicates a biosynthesis from acetate,
malonate, and two one-carbon units.

An interesting tracer study (46) of the biosynthesis of
margaspidin [44] and desaspidin [45] by the fern *Dryopteris*
marginalis has shown that this substituted methylene-bis-phloro-
glucinol is probably formed by an unprecedented process; two
four-carbon units, derived directly from butyrate without prior
breakdown to acetate, are thought to combine to give a C_8 unit
[46], which condenses with one mole of malonate in the normal

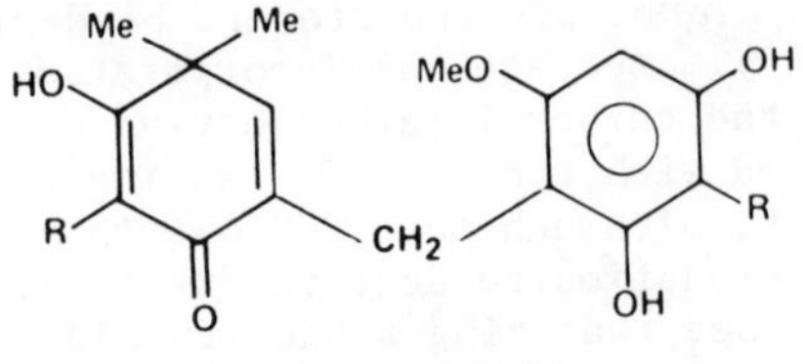

[45] R = CO(CH$_2$)$_2$

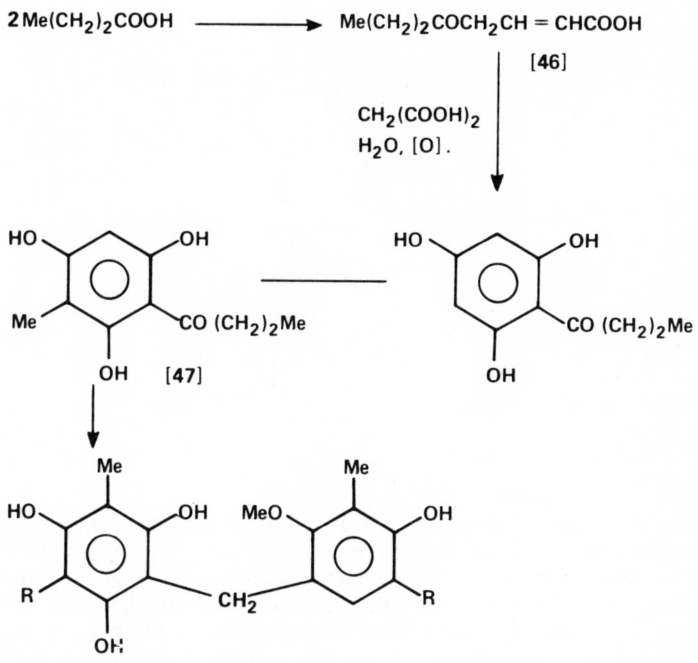

$$2 Me(CH_2)_2COOH \longrightarrow Me(CH_2)_2COCH_2CH=CHCOOH$$

[46]

$CH_2(COOH)_2$
H_2O, [O].

[47]

[44] R = COC$_3$H$_7$

way. Modification, cyclization, and methylation give the monomer [47], which yields [44] and [45] through oxidative dimerization. The scheme is, of course, purely hypothetical, since the true intermediates and sequence of events are unknown, but it has the merit of accounting for the observed labeling pattern in a reasonable way.

18-4. NAPHTHOQUINONES

Javanicin [51] isolated from *Fusarium javanicum* has been shown to be derived from a simple polyketide precursor, with the nuclear methyl group formed by the reduction of the terminal malonate-COOH (90). This relatively uncommon reduction is also seen in the biosynthesis of barnol (Section 18-1). A more recent study has shown that the same polyketide is also cyclized to the naphthoquinone rubrofusarin [52] by *Fusarium graminearum* (109).

The biosynthesis of mollisin [48], the chlorinated naphtho-quinone pigment formed by *Mollisia caesia*, has been investigated

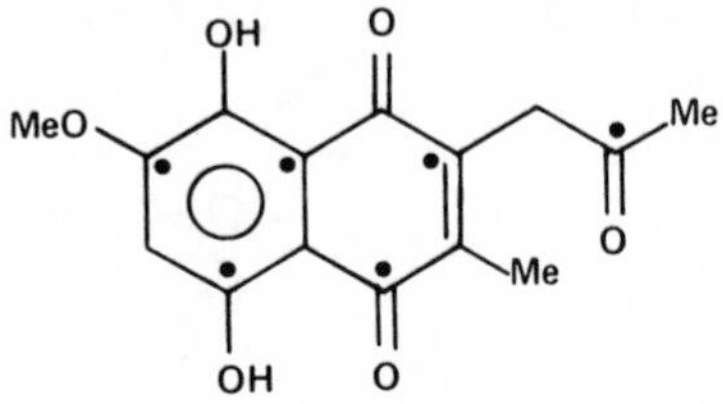

[51]

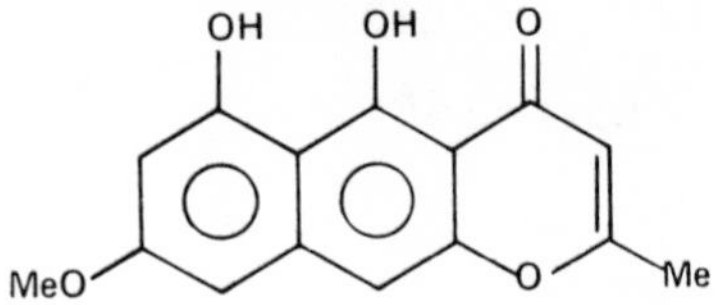

[52]

by Bentley and Gatenbeck (89). Incorporation studies indicated the general acetate-malonate derivation; since neither mevalonic acid (which provides the second methyl group in the case of chimaphilin (see Chapter 16)) nor methionine was incorporated, the second methyl group was thought to be that of a second starter unit and a biosynthesis from two polyketide chains was proposed.

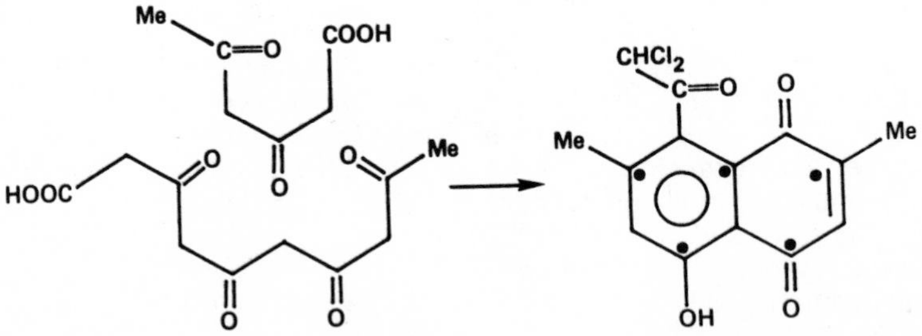

[48]

^{13}C studies have confirmed the radiotracer work (130), and biosynthetic feedings with [1,2-^{13}C]-acetate followed by an analysis of the carbon-carbon couplings in the metabolite have clarified the biosynthesis from two polyketide chains (110).

The biosynthesis of elsinochrome A [53], one of a series of related red pigments elaborated by the fungus *Elsinoe*, has been investigated by Nakanishi (111). The monomer [54] is derived from one acetate and six malonate units, and the perylenequinone is formed by phenolic coupling after decarboxylation, hydroxylation, and O-methylation.

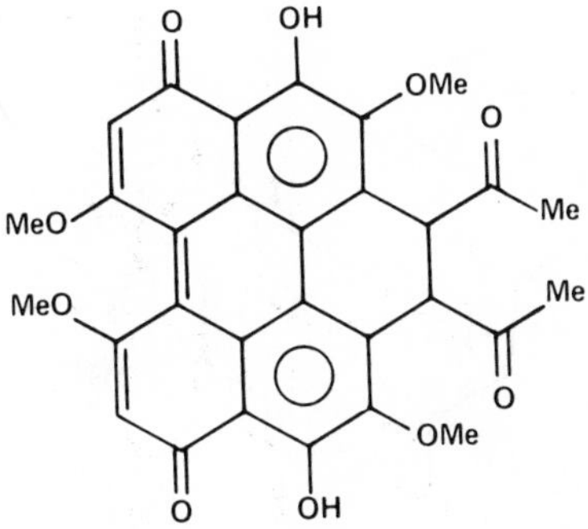

[53]

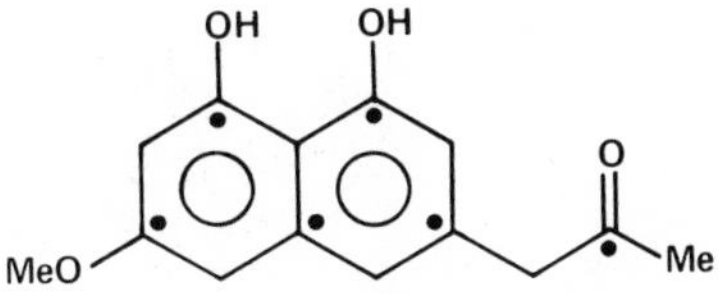

[54]

The spinochromes (e.g., [56]) and echinochrome A [55] form
an interesting group of polyhydroxylated 1,4-naphthoquinones
which have been isolated from sea urchins (131).   Studies on the
biosynthesis of these compounds by Salaque and co-workers (48)
indicate their possible acetate-polymalonate origin and *de novo*

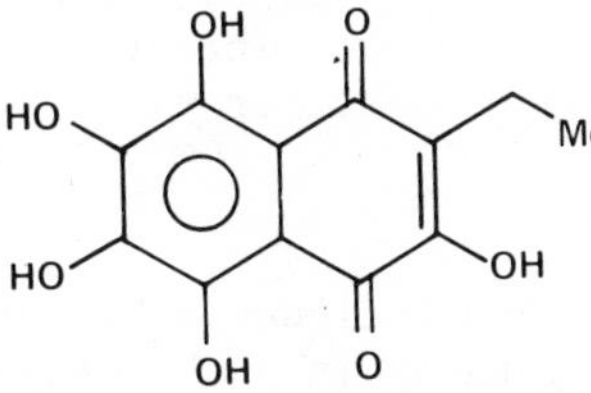

[55]

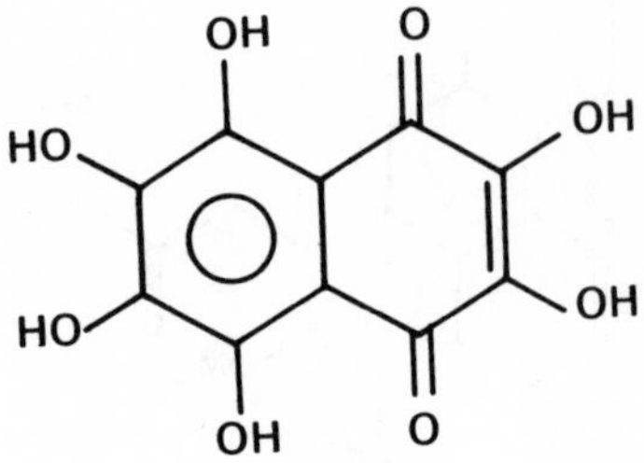

[56]

synthesis by this marine animal. It is of course possible that
the compounds or some aromatic precursors are formed by some
microbial symbiont, or by marine microflora, but if this is not
the case, then this synthesis offers one of the very few examples
of such a biosynthetic process in an animal (cf. Chapter 1).
The authors have proposed a biosynthesis from five or six
acetate units; shikimic acid did not seem to be a precursor,
although it could not be clearly excluded.

Another polyhydroxylated naphthoquinone, mompain (2,5,7,8-
hydroxynaphthoquinone), is formed from acetate in the fungus
Helicobasidium mompa (146). Plumbagin (2-methyl-5-hydroxynaphtho-
quinone), found in *Plumbago europaea*, is acetate-malonate derived
(132); this work is of interest since the acetate pathway has
rarely been established in a higher plant.

18-5. XANTHONES

Although the xanthone nucleus is biosynthesized from a mixed
acetate-shikimic acid pathway in higher plants (Section 21-4),
some fungal xanthones have been shown to be formed from acetate
alone. Thus, degradation (49) of active sterigmatocystin [57],
isolated from a culture of *Aspergillus versicolor* grown on
^{14}C-acetate, has demonstrated that this xanthone is formed via
the acetate-malonate pathway. The labeling pattern implied that

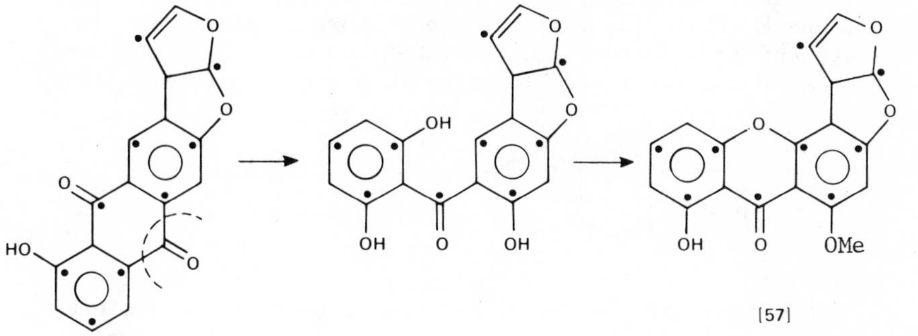

[57a]

[57]

the xanthone nucleus was derived by the normal oxidative ring closure of a substituted benzophenone (50) formed from a polyketide, and that also the furanofuran grouping was formed from two acetate units. A study using ^{13}C-nmr techniques (113) has confirmed these early results, and both investigations demonstrate the truth of a proposal made by Thomas (115) several years ago, that [57] is derived from an anthraquinone [57a] by oxidative ring cleavage, decarboxylation, and recyclization. A similar cleavage gives rise to the hydroxanthone pigments produced by the ergot fungus (see Section 18-10).

18-6. CHROMONES

Serpena neglecta produces the chromones [59] and [60] (steliarin); tracer experiments (52) have shown the anticipated acetate incorporation. Compound [59] is converted into [60] by the organism, implying a novel degradation of the mevalonate-derived side-chain.

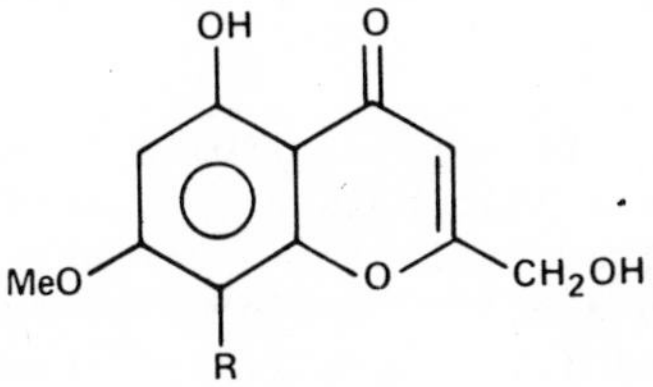

[60] R = C$_4$H$_9$
[59] R = CH$_2$ − CH = CMe$_2$

The fruits of *Ammi visnaga* (Umbelliferae) contain several related furanochromones; of these, khellin [61], visnagin [62] (53), and khellol [63] (54) have been shown to be formed from acetate and mevalonate which furnishes the carbons of the furan ring. A scheme for their biosynthesis has been proposed (133).

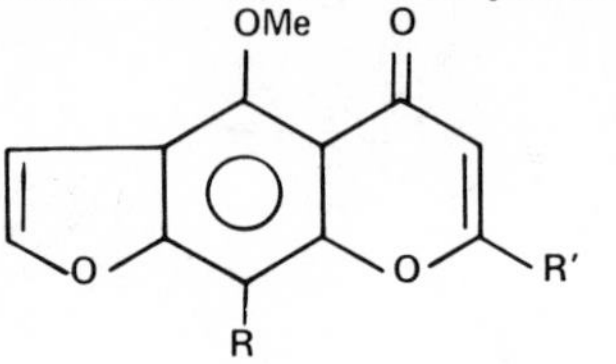

[61] R = OMe, R' = Me
[62] R = H, R' = Me
[63] R = H, R' = CH$_2$OH

The chromanone [64] is formed from a pentaketide chain by *Daldinia concentrica* (55); acetate labeling gave the expected distribution of radioactivity both in [64] and the partially cyclized product, 2,6-hydroxybutyrophenone. The fungus also produces derivatives of 1,8-dihydroxynaphthalene and perylene (see Section 17-2a).

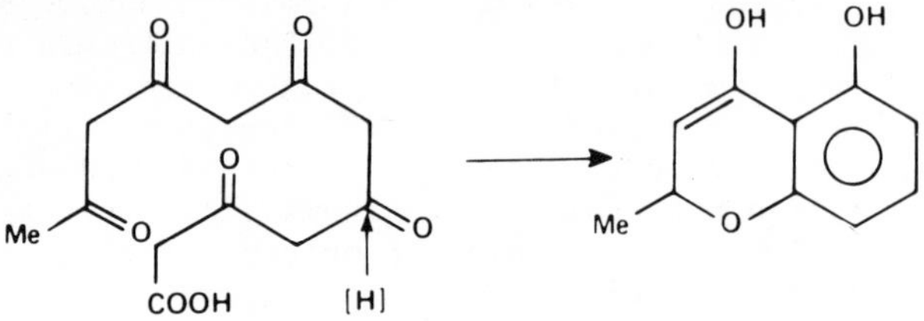

[64]

The acetate origin of citromycetin [65] was established by Birch (56); further investigations (57) have shown that [65] is another instance of a phenolic compound derived from two polyketide chains. Administration of 2-^{14}C-malonate to cultures of *Penicillium frequentans* yielded labeled [65] whose degradation clearly indicated that there were two starter units; the radioactivity of the carboxyl and methyl groups was only about 32% of that of the general malonate level, based upon the total radioactivity of the isolated metabolite.

18-7.   COUMARINS

Many compounds of this class are derived from shikimic acid by a variety of mechanisms (Chapter 16); however, several coumarins are purely acetate-derived. Aflatoxin B$_1$ [66] is one of a series of furano-furanoid metabolites produced by *Aspergillus* spp. Incorporation studies (49, 58, 115, 140) have demonstrated the probability of a biosynthesis from the xanthone sterigmatocystin [57] (see Section 18-5), itself derived from

[65]

[57] → [66]

an anthraquinone, in accord with some earlier biosynthetic proposals by Thomas (115). The steps following the cleavage of
[57], cyclization to a cyclopentenone followed by decarboxylation (115), have not been investigated in detail, but the reported
distribution of label (58) in [66] is certainly in agreement
with the hypothesis. Similarly the distribution of label in the
furano-furan grouping has been variously interpreted (58), but
again Thomas' earlier proposal involving the rearrangement of an
endo-peroxyanthraquinone to [57] (58, 113, 139, 141) will probably prove to be correct.

The isocoumarin system found in canescine [67], a metabolite
of *Penicillium canescens* and *Aspergillus malignus*, is likewise
of acetate origin (59). The derivation of the side-chain is

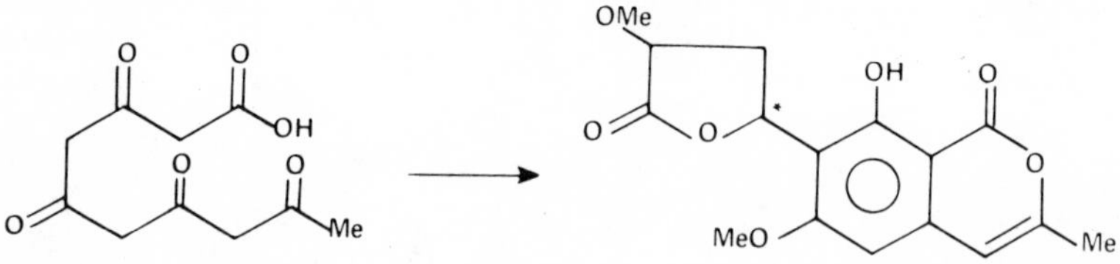

[67]

still not certain, although succinic and malic acids are incorporated; methionine provides the asterisked carbon atom (60),
an incorporation which has not been encountered elsewhere.

Alternariol [68], a metabolite of *Alternaria tenuis*, is one
of the few phenolic compounds that is derived directly from a
polyketide chain without change in the number of carbons or
other modification (Section 17-2a). Thomas (61) has demonstrated

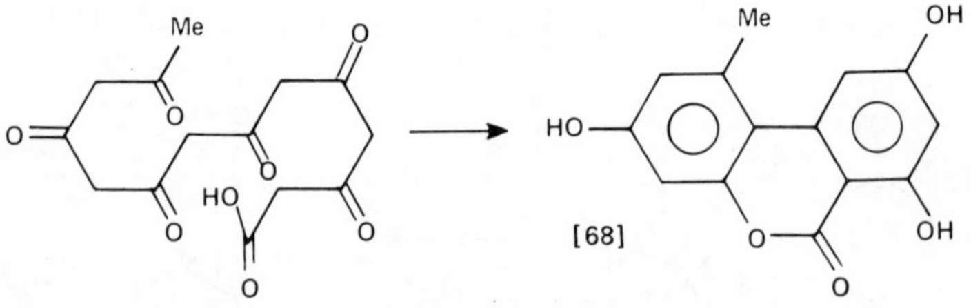

[68]

the specific incorporation of acetate in accord with the biosynthesis from one acetate and six malonate units.

Ochratoxin A [69] contains a phenylalanine system attached
to the isocoumarin ring; feeding studies (62, 118) with
Aspergillus ochraceus have shown that phenylalanine, not surprisingly, is incorporated into the one, and that acetate is well

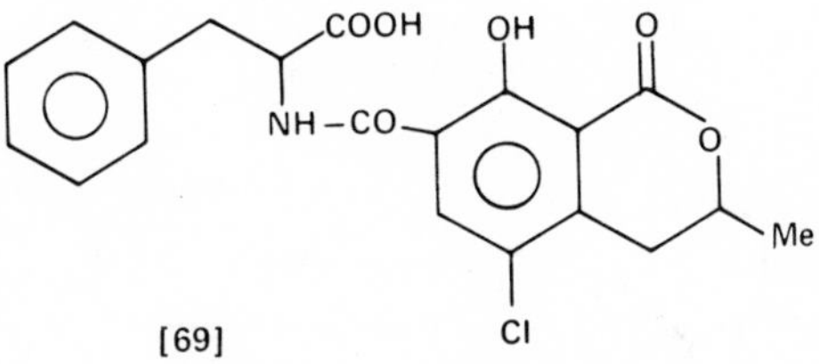

[69]

incorporated into the other. Some doubt had existed concerning
the origin of the methyl group and the amide carbonyl. The
South African workers (118) indicated that the amide carbon
came from methionine, with the methyl group being derived from
acetate, while Searcy et al. (62) reported the reverse.

[70]

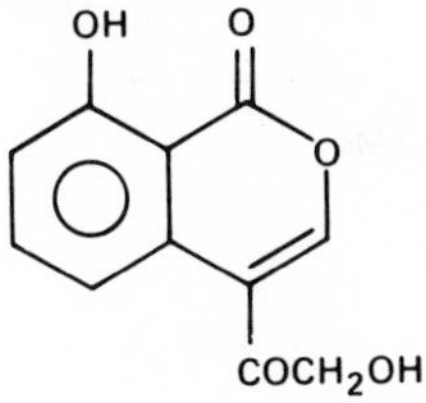

[71]

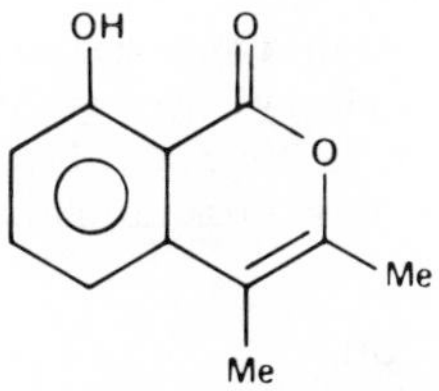

The question has been decided in favor of the former proposal by
^{13}C-nmr studies (134).

A similar origin from acetate has been established (63) for
oosponol [70] and oospolactone [71], two isocoumarins from
Oospora astringens. In these cases the observed incorporations
from the C_1-pool were very singular: C-3 of [70] and the C-4
methyl group of [71] are apparently derived from formate. Both
compounds are clearly formed from the same C_{10} polyketide, which
must be highly oxidized before cyclization to [70].

18-8. QUINONE METHIDES

Citrinin [72] is produced by a variety of organisms; its
biosynthesis has been investigated in *Aspergillus candidus*.
Labeling experiments have shown that the skeleton of [72] is

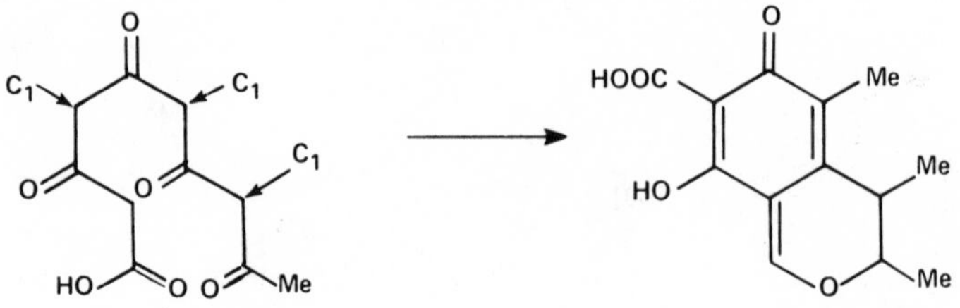

[72]

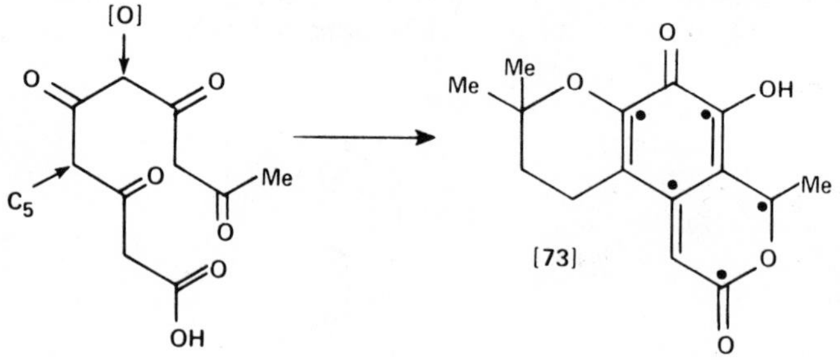

[73]

formed from one acetate and four malonate units (56); the two
methyl groups and the carboxyl group are derived from methionine (64). Similarly, the degradation of fuscin [73], produced
by *Oidiodendron fuscum*, has demonstrated its origin from acetate
(and mevalonate) (21).

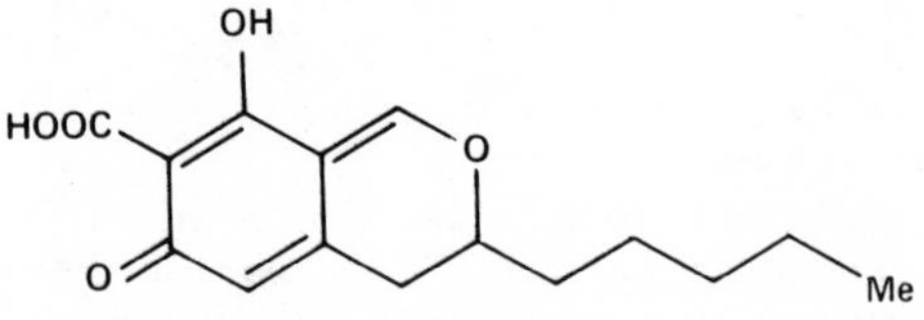

[74]

Another quinone methide, pulvilloric acid [74], a yellow
antibiotic produced by *Penicillium pulvillorum*, is also formed
via the acetate-malonate pathway (65). The carboxyl function
is provided by the C_1-pool; an attempt to implicate a long-chain
fatty acid as starter unit by feeding labeled hexanoate to the
mold was unsuccessful, and resulted only in breakdown of the
acid to acetate.

The biosynthesis of some other quinone methides (e.g.,
atrovenetin) is discussed in the next section.

18-9.   PHENALENONES

The phenalenone ring system is rarely found in natural products. It does occur in atrovenetin [78], herqueinone, and norherqueinone [79], which are produced by some *Penicillium* spp.,

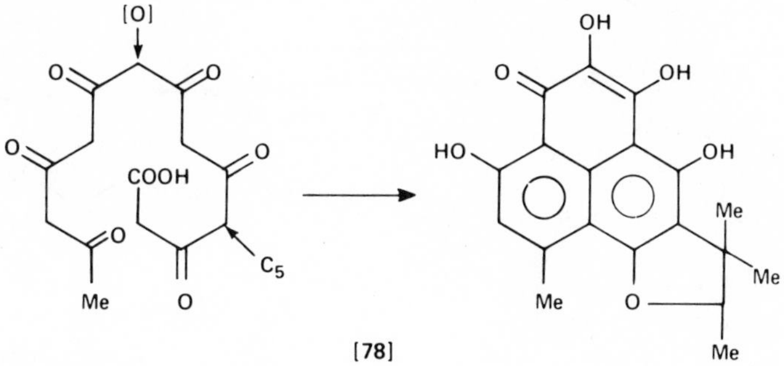

[O]

COOH

C_5

Me   O

[78]

Me   O

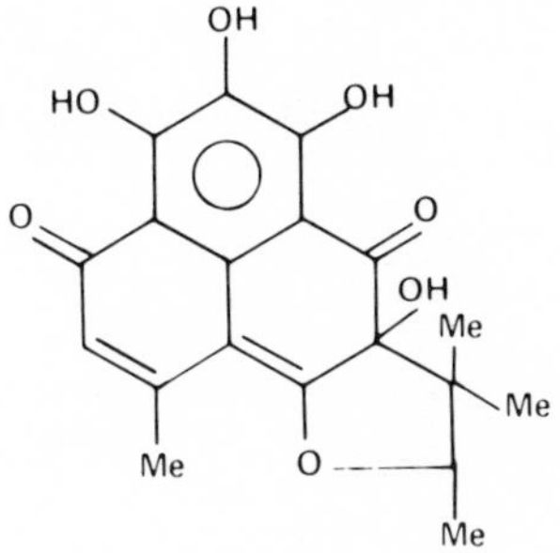

|79|

while a group of phenylphenalenones has been found in higher plants (Chapter 16). Thomas (70) has shown that the perinaphthenone nucleus found in the fungal metabolites is derived from acetate, and that the side-chain comes from mevalonate.

Duclauxin [80] is a bis-phenalenone pigment isolated from *Penicillium duclauxi*. Degradation of [80] isolated from experiments in which the mold has grown on sodium 1^{14}C-or 2^{14}C-acetate has indicated a likely biosynthesis from the modified heptapolyketide [81], followed by cyclization and oxidative dimerization (71).

18-10.   ANTHRAQUINONES

The acetate origin of islandicin [82], and emodin [83], anthraquinone pigments produced by *Penicillium islandicum*, has

[81]

[80]

been demonstrated by Gatenbeck (72). Specific degradation showed
that [82] and [83] were both derived from the same C_{16}-polyketide;
the ^{14}C incorporation was supported by feeding experiments with
^{18}O-labeled acetate, which showed that three oxygen functions in
[82] and four in [83] were formed from carbonyl groups of acetic

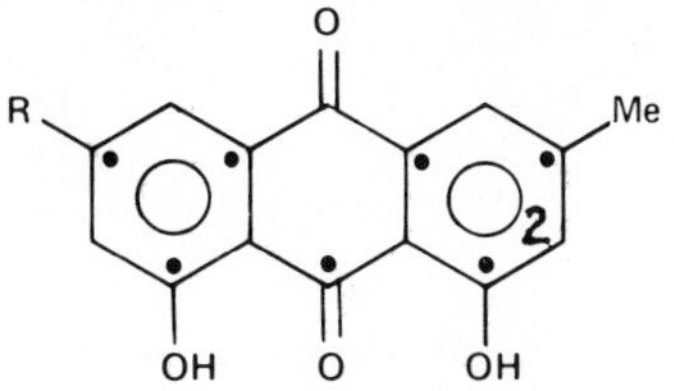

[82] R = H
[83] R = OH

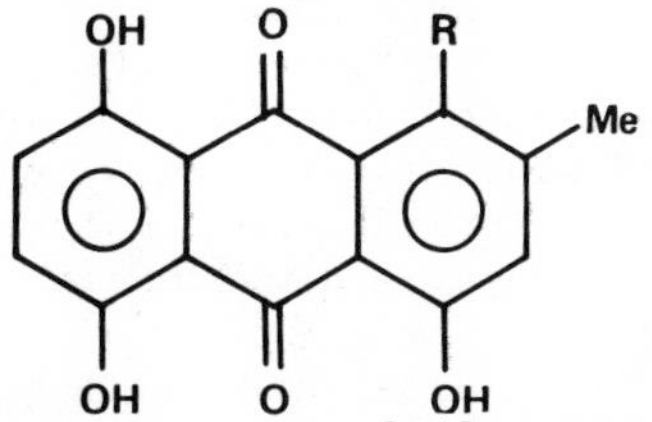

[84] R = H
[85] R = OH

acid. Similarly helminthosporin [84], a pigment from *Helmintho-
sporium gramineum*, is derived from the same C_{16}-precursor suitably modified (73), as is cynodontin [85], a related pigment
from *Phoma terrestris* (74) and other molds. Endocrocin
(2-carboxy-[83]) was tested as a possible intermediate in the
biosynthesis of [83], but found to be inactive (136); it was,
however, readily incorporated into dermolutein and dermorubin,
two anthraquinones having the 2-carboxyl group, though not into
other neutral anthraquinones.

Pachybasin (1-hydroxy-3-methylanthraquinone) has been shown
(137) to be formed from a regular polyketide chain despite its
having an unsubstituted aromatic ring; similarly, rugulosin
[86], a modified bis-anthraquinone formed by *Penicillium brunneum*,
has been shown to be derived from one acetate and seven malonate
units. This is one of the cases in which the acetate starter unit
was identified (75) by feeding experiments with 2-^{14}C-malonate
(see Section 17-2a).

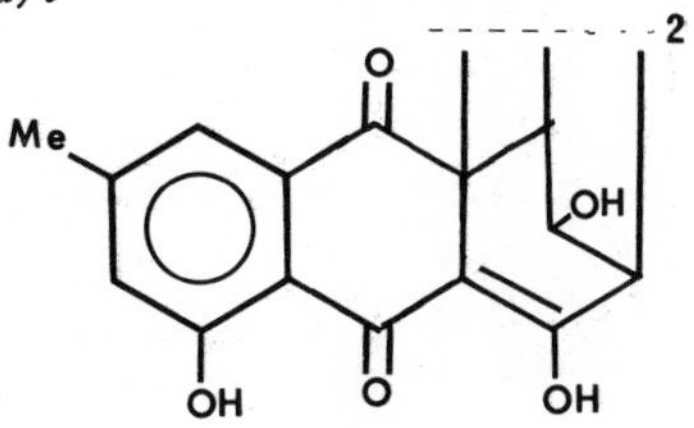

[86]

[87]

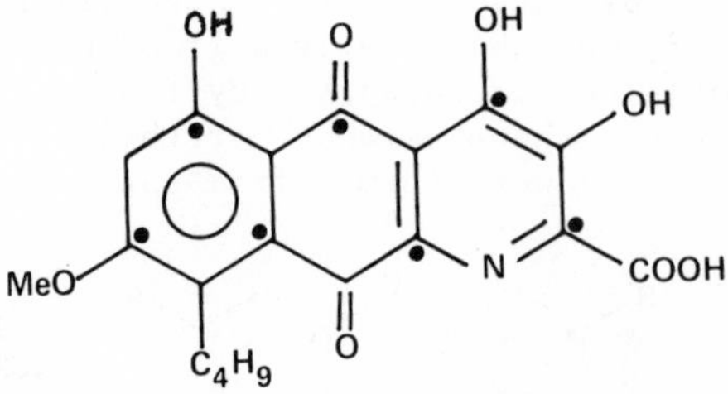

Phomazarin [87], an aza-anthraquinone from the mold *Phoma terrestris*, is an interesting compound from the point of view of biosynthesis. Early incorporation studies (74) showed that $2^{14}C$-acetate was incorporated into [87] in such a way that the carboxyl group carried about one-eighth of the total activity, suggesting that [87] was formed from a polyketide containing eight acetate residues; this idea was modified (145) to accomodate a revised structure.

We have seen (Sections 18-5 and 18-7) that the xanthone and coumarin nucleus can arise by modification of an anthraquinone precursor, and it is becoming recognized that several anthraquinones are modified, usually by an oxidative mechanism, to give additional metabolites. Thus, it has been established that the dimeric hydroxanthone pigments (ergochromes) (76) (e.g., [88]), produced by the ergot fungus, are derived from units formed by an oxidative ring-opening of the simple anthraquinones (e.g., emodin); their basic acetate-origin has also been established (77).

[88]

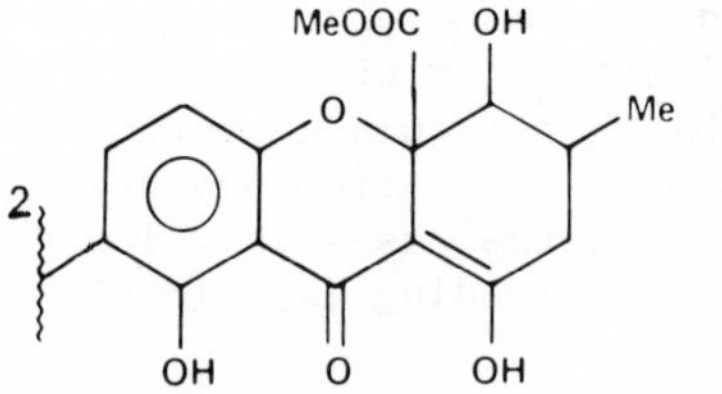

[89]

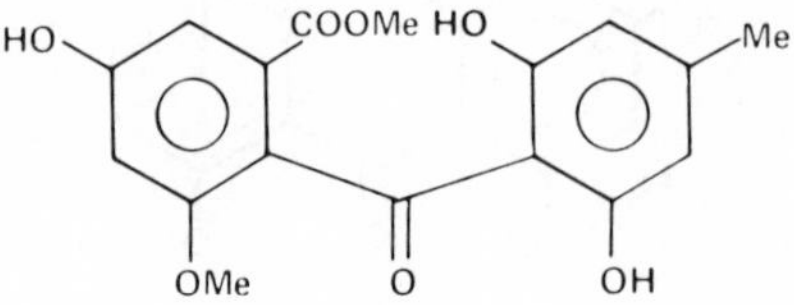

Another interesting case of possible modification of an anthraquinone precursor is that of sulochrin [89]. No linear polyketide precursor for sulochrin can be devised, and degradation of radioactive [89], isolated from cultures of *Aspergillus terreus* grown on active acetate or malonate, indicated that it might be formed from two separate polyketide chains, each derived from an acetate starter unit and three malonates. There was also some mutant evidence for a biosynthesis of [89] by the combination of two units based on *ortho*-orsellinic acid (78); however, incorporation of possible monocyclic precursor units

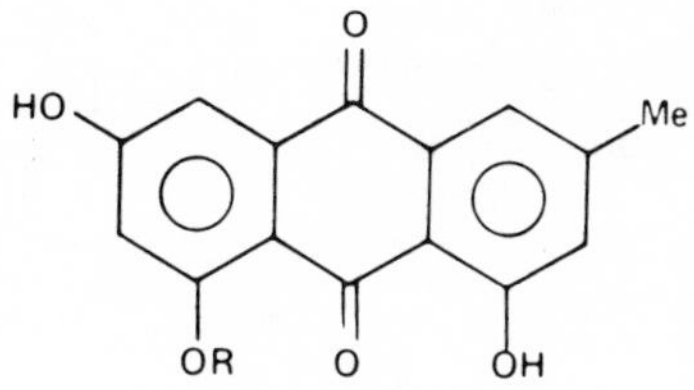

[83] R = H
[90] R = Me

was not successful (79). Gatenbeck (80) suggested that [89] might arise from the oxidative fission of an anthraquinone ([83] or [90]) although attempts to achieve such oxidations *in vitro* had proved unsuccessful (81). This idea has been verified

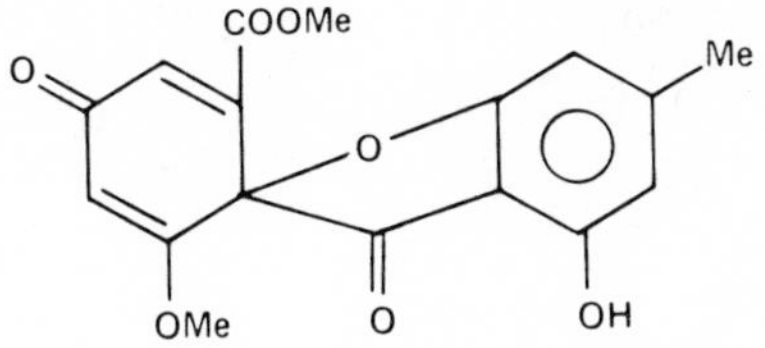

[91]

experimentally both by the incorporation of malonate into ques-
tin [90] and bisdechlorogeodin [91], both co-metabolites of [89]
in *P. frequentans* (82), in such a way as to demonstrate their
formation from a single polyketide chain, and by their further
incorporation into [89] (83, 51).

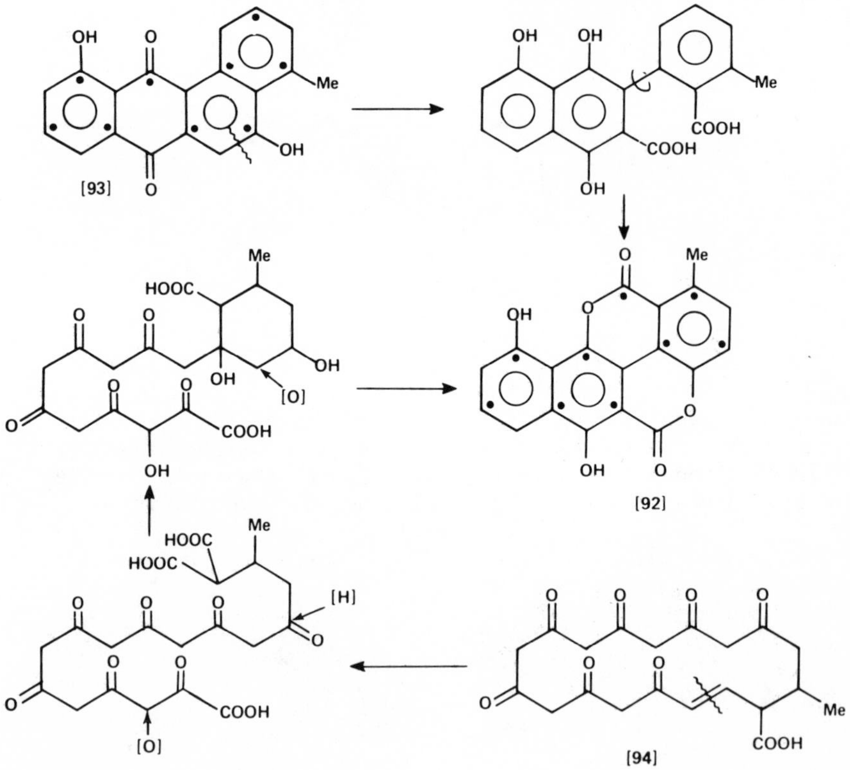

The biosynthesis of chartreusin [92], a metabolite of a *Streptomyces* sp., poses a complex problem. Although the formation of [92] via the acetate-malonate pathway has been established (84), even the localization of the label does not provide a definitive answer to the problem of biosynthesis. There are three possible routes which would account for the observed labeling pattern: (a) the oxidative cleavage of a 1,2-benzanthraquinone precursor [93], (b) biosynthesis from two acetate chains, and (c) the formation of [92] by oxidation and recyclization of a modified polyketide [94]. A second investigation of the biosynthesis has been published (135).

18-11.   PHENANTHRENES

The biosynthesis of some phenanthrenes (e.g., aristolochic acid, q.v.) from shikimic acid has been established, but only one member of this class of compound, piloquinone (see below), has been shown to be formed from acetate units.

Some speculative schemes for the biosynthesis of other phenanthrenes from acetate have been put forward; in particular, it has been suggested that denticulatol [96] and chrysophanol [95], which occur together in *Rheum denticulatum*, could both be formed from the same polyketide derived from eight acetate residues.

Piloquinone [97], a phenanthrenequinone produced by *Streptomyces pilosus*, has been shown to incorporate acetic and

[95]

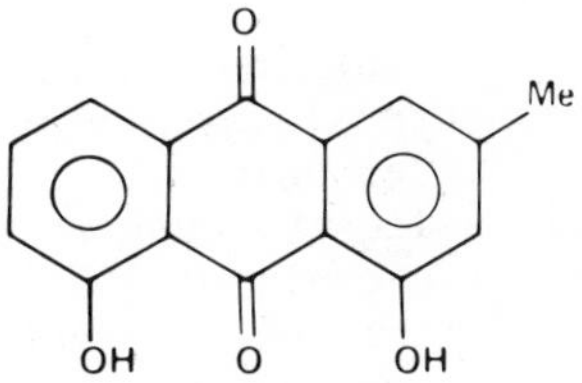

[96]

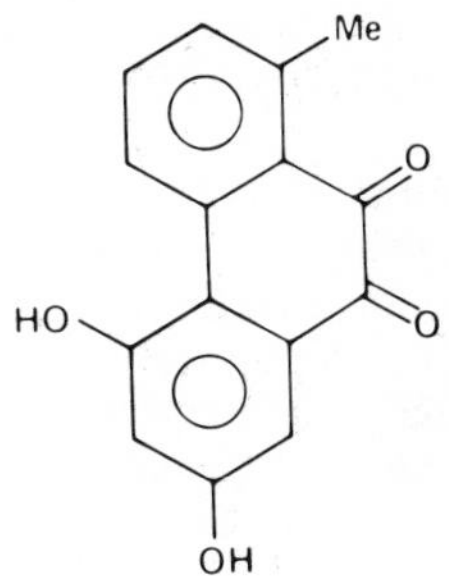

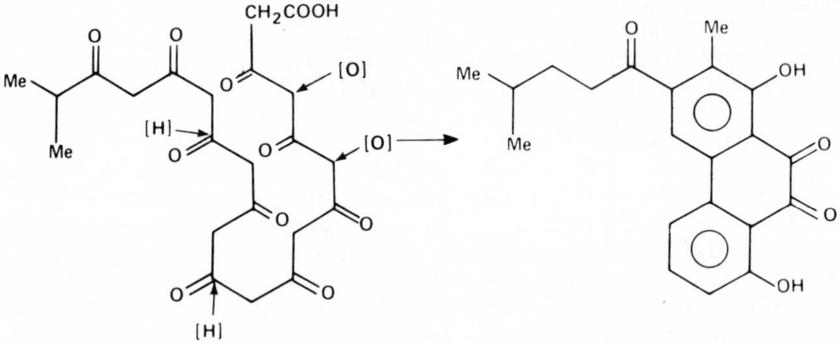

[98]

[97]

isobutyric acids, and valine (85); degradations were shown to be consistent with a biosynthesis from a polyketide [98] having an isobutyryl group as starter unit. These findings are instructive because they emphasize the inherent danger of "paper chemistry"; although the side-chain of [97] can be built up on paper from an isoprene unit, it is clearly not safe to assume that this must be its origin: the chain of valine is a perfect mimic.

18-12. TETRACYCLINE ANTIBIOTICS AND RELATED COMPOUNDS

The tetracyclines are a group of closely related products elaborated by various *Streptomyces* species; the compounds, many of which have considerable therapeutic value as antibacterial agents, are oxygenated hydronaphthacenes, which may contain additional halogen or nitrogen functions. The structures of some of the members of this group of compounds are shown in the table.

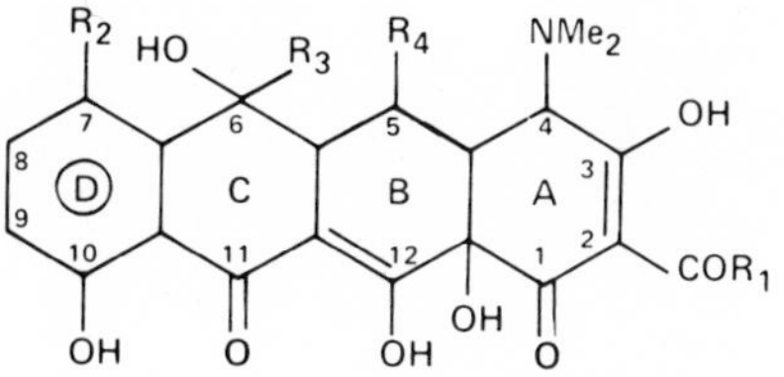

R_1	R_2	R_3	R_4		
NH_2	H	Me	OH	[99]	Oxytetracycline
NH_2	Cl	Me	H	[100]	7-Chlorotetracycline
NH_2	Cl	H	H	[101]	6-Desmethyl-7-chlorotetracycline
NH_2	Cl	Me	OH	[102]	Oxychlorotetracycline
Me	H	Me	OH	[106]	ACETYL-2-DECARBOXAMIDO OXYTETRACYCLINE
NH_2	H	Me	H	[111]	Tetracycline

Early investigations proved that acetate could be incorporated into oxytetracycline [99] (86), and that its incorporation was, in general, in accordance with the head-to-tail union
of the acetate units (87, 88). A detailed study (91) of the
incorporation of acetate into rings C and D of [99] has shown
that the level of incorporation into carbons 6, 7, 9, and 10a is
identical, quite in accord with the accepted biosynthesis from
a single acetate chain. The origin of the substituent methyl
groups from methionine is unexceptional (92); the nitrogen at
C-4 is thought to originate from glutamic acid, via a transamination step involving a C-4 keto group (93); the amide nitrogen
is present in the malonamoyl CoA unit [103] which initiates
the polyketide precursor (see below). The source of the chlorine
atom found in chlorotetracycline [100] and other fermentation
products [101] and [102] is the chloride ion present in the
growth medium (94); if chloride is omitted from the nutrient
medium of a strain that normally produces [100], the corresponding
nonchlorinated compound is obtained. Similarly, addition of bromine to the medium leads to a bromotetracycline (94).

The early stages in the biosynthesis of the tetracyclines
from acetate are still not fully clarified, but it has been
generally accepted that a malonamoyl CoA unit [103] acts as the
starter unit for a C_{20} polyketide, possibly [104], which undergoes modification and cyclization to yield 6-methylpretetramide
[105].

There is no direct experimental evidence for the involvement of [103] in the biosynthesis of the tetracyclines, although
it has been shown by Gatenbeck (88) that an oxytetracycline-

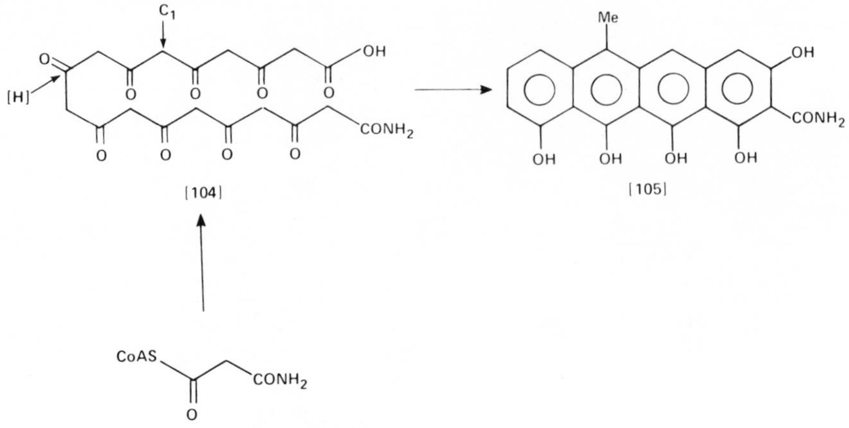

[104]

[105]

[103]

producing strain of *S. rimosus* fed $NaH^{14}CO_3$ produced [99] in
which the radioactivity was specifically localized in the car-
boxamid group of ring A. This result is at least consistent
with the carboxylation of acetyl–CoA and the formation of [103].
The coexistence of 2–acetyl–2–decarboxamidooxytetracycline [106]
and [99] in the same fermentation clearly indicates the position
of the starter unit. That methylation precedes aromatization in
the expected manner is supported by the finding that pretetramid,
which has no 6–methyl group, is converted into tetracyclines
lacking the 6–methyl function by strains which normally produce

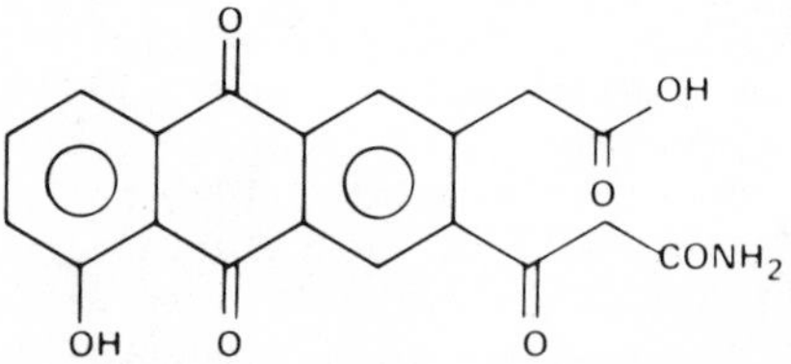

[107]

methylated tetracyclines (95). A mutant of *S. aureofaciens*,
which forms demethyltetracyclines in the wild–type, has been shown
(97) to produce protetrone [107], an incompletely aromatized compound, whose structure indicates that the conformation of the
polyketide precursor is probably [104] as indicated, although
other possibilites have been put forward (96). The structures of
[107] and the related [108], likewise isolated from a *S. aureofaciens* mutant (98), also indicate the early generation of the
carboxamide function, in accordance with Gatenbeck's hypothesis
(88).

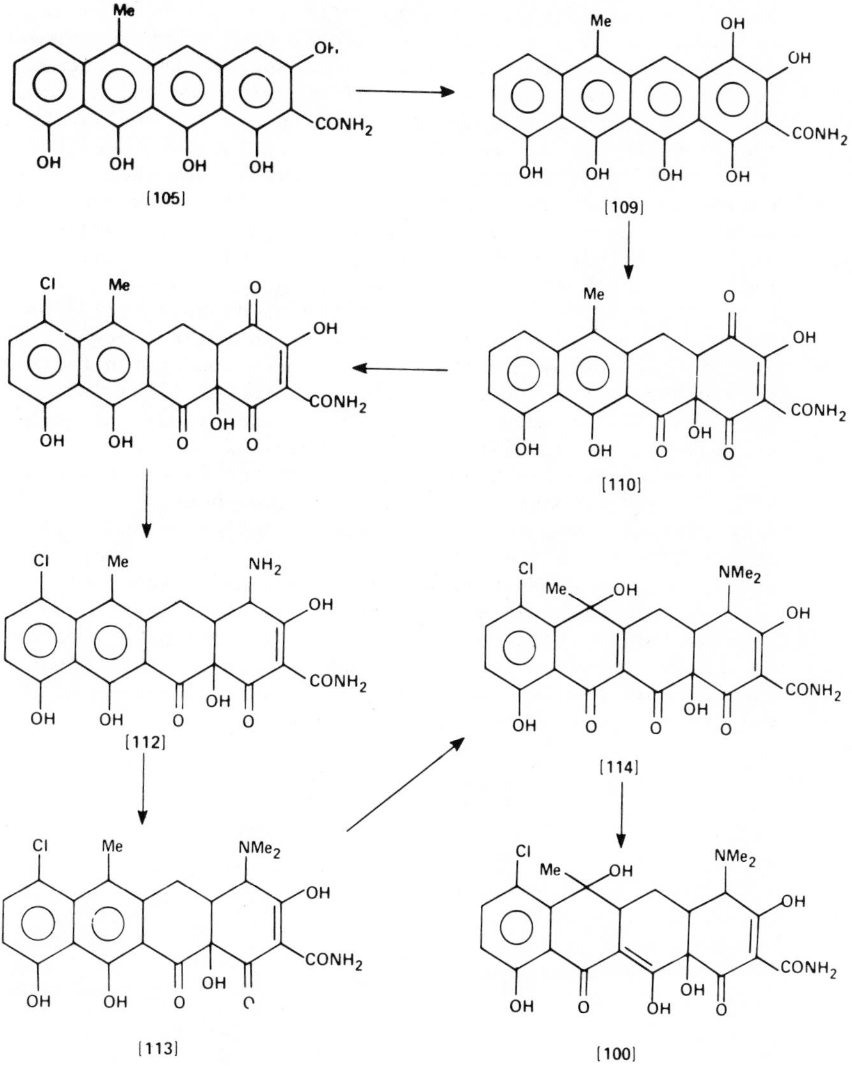

The scheme shown here outlines McCormick's interpretation of the experimental evidence concerning the conversion of the precursor pretetramides into the tetracyclines. Many of these steps have been examined experimentally; 6-methylpretetramide [105] is readily converted to tetracyclines (95), and the conversion of pretetramide to tetracyclines lacking a C-6 methyl group has already been mentioned. 4-Hydroxy-6-methylpretetramide [109] has been identified as an accumulated metabolite in the culture filtrates of a *S. aureofaciens* mutant (99), and the compound has been converted to tetracyclines by other mutant strains. The compound [110] has not been isolated from a natural source, but its intermediacy has been established, since it is converted to tetracycline [111] and 7-chlorotetracycline [100] by *S. aureofaciens* strains (93), which not only implicates [110] as an intermediate, but also indicates the late stage at which halogenation takes place. Growth of *Streptomyces* spp. in the presence of methionine antagonists (e.g. ethionine) causes the accumulation of [112], which can be converted to chlorotetracycline by a cell-free extract of *S. aureofaciens* (100). The evidence that chlorination precedes that transamination step is provided by the observation that the chlorine-free analog of [112] (produced in the absence of chloride ion), is metabolized to tetracycline, where [112] itself yields chlorotetracycline under the same conditions (100). [112] has also been isolated as a metabolite of a mutant of *S. aureofaciens* (103). The methylated analog [113] has been known for some time as a synthetic compound, although it has never been isolated from a natural source; its conversion to tetracyclines has been established (102). Compound [114] is similarly accumulated by a mutant of *S. aureofaciens* (101), and can in turn be converted to [110] by other mutants.

Similar studies have been made of the intermediates in the biosynthesis of oxytetracycline and other pharmaceutically important related compounds; for an account of these and a more detailed review of tetracycline chemistry, the reader is referred to the reviews by McCormick (1967) (93) and Mitcher (1968) (96).

The biosynthesis of ε-pyrromycinone (rutilantinone) [115], a metabolite of various *Actinomyces* spp., which is clearly related to the tetracyclines, has been studied (104). A propionate starter unit provides the ethyl side-chain; the rest of the molecule is formed from nine acetate units.

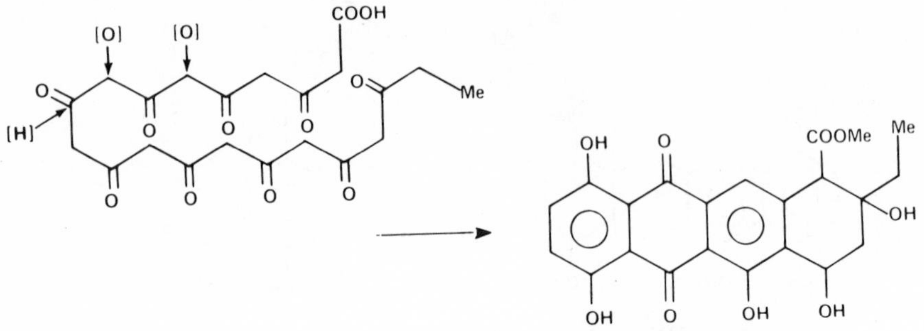

[115]

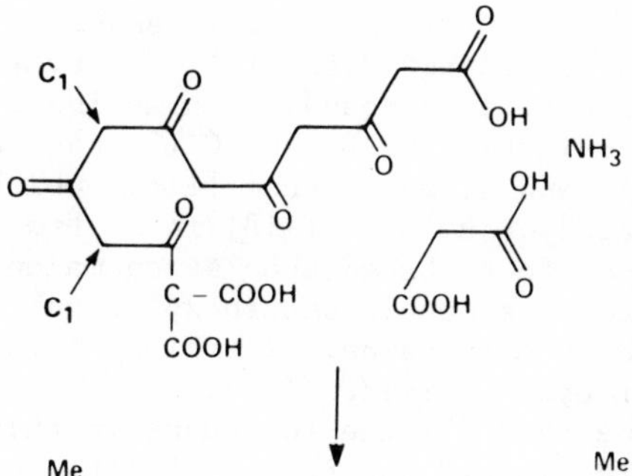

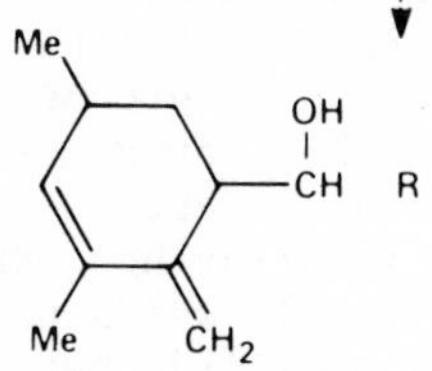

C_1

O

O

OH

O

NH_3

OH

O

C_1

C – COOH

COOH

COOH

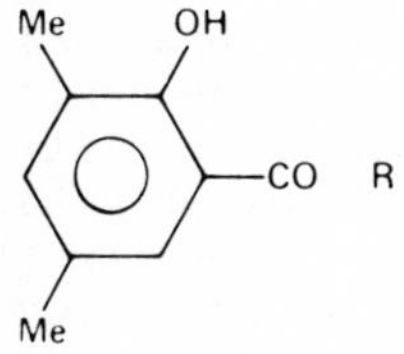

O

Me

OH

CH R   [118]

Me CH_2

Me   OH

CO   R   [116]

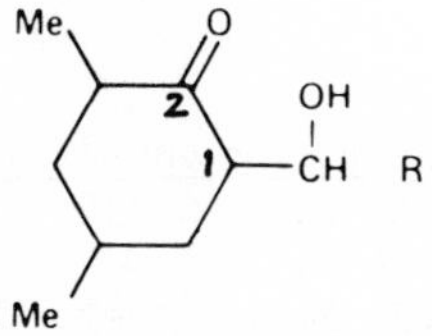

Me

Me

O

2

1   OH

CH R   [117]

Me

OH Me

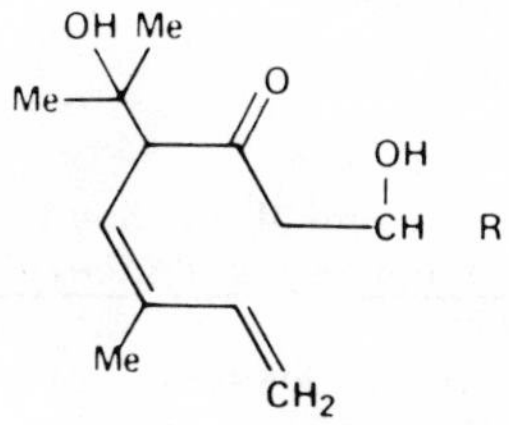

Me

O

OH

CH R   [119]

Me

CH_2

O

R = CH_2

NH

O

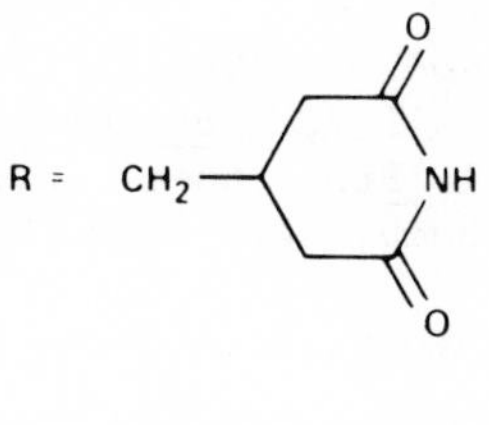

18-13. MISCELLANEOUS ACETATE-DERIVED COMPOUNDS

Actiphenol [116] occurs in *Streptomyces* spp. in association
with the closely related non-aromatic actidione [117]. Several
other compounds of this series have been found, such as inactone
(Δ^1-dehydro-[117]), the streptovitacins A, B, and C (3-, 4-, and
5-hydroxy-[117], respectively), and streptimidone [118], which
lack the aromatic ring. The biosynthesis of [118] in *S. rimosus*
has been studied (105) and found to be based upon seven malonate
units, with one of them undergoing a double decarboxylation; the
two methyl groups are provided by methionine. The biosynthesis
parallels that established for cycloheximide (106).

It is reasonable to assume that all the compounds in this
group have a similar biogenesis; indeed the carbon skeleton of the
various compounds and the positions of the hydroxyl groups in the
streptovitacins are not readily reconciled with a normal origin
from acetic or shikimic acid, although they can be accomodated
by the scheme established for [118]. There is a further compound,
protomycin [119] from *Streptomyces reticuli* (107), which has an
aliphatic system, similar to the ring system of [118], but
containing an interesting isopropyl group; no biosynthetic studies
have yet been reported.

REFERENCES

1. J.H. Richards, and J.B. Hendrickson, The Biosynthesis of
    Steroids, Terpenes, and Acetogenins, W.A. Benjamin,
    New York, 1964.
2. A.J. Birch, R.A. Massy-Westropp, and C.J. Moye, Aust. J.
    Chem. 8, 539 (1955).
3. K. Mosbach, Acta Chem. Scand. 14, 457 (1960).
4. A.J. Birch, A. Cassera, and R.W. Rickards, Chem. Ind.
    (London) 792 (1961).
5. A.J. Birch, Proc. Chem. Soc. 3 (1962).
6. J.D. Bu'Lock, and H.M. Smalley, Proc. Chem. Soc. 209 (1961).
7. R. Bentley, and J.G. Keil, Proc. Chem. Soc. 111 (1961).
8. S. Gatenbeck, and K. Mosbach, Acta Chem. Scand. 13, 1561
    (1959).
9. S. Gatenbeck, and I. Lünnroth, Acta Chem. Scand. 16, 2298
    (1962).
10. S. Gatenbeck, Acta Chem. Scand. 12, 1985 (1958).
11. S. Gatenbeck, Acta Chem. Scand. 14, 296 (1960); 16, 1053
    (1962).
12. S. Gatenbeck, and U. Brunsberg, Acta Chem. Scand. 20, 2334
    (1966).
13. G. Pettersson, Acta Chem. Scand. 19, 35 (1965).

14. G. Pettersson, Acta Chem. Scand. 19, 1724 (1965).
15. S. Gatenbeck P.O. Erikson, and Y. Hansson, Acta Chem. Scand. 23, 699 (1969).
16. A.J. Birch, and M. Kocor, J. Chem. Soc. 866 (1960).
17. A.J. Birch, and E. Pride, J. Chem. Soc. 370 (1962).
18. A.J. Birch, R.J. English, R.A. Massy-Westropp, and H. Smith, J. Chem. Soc. 369 (1958).
19. A.H. Manchanda, and C.E. Stickings, IUPAC Abstr. XIX, 209 (1963).
20. K. Mosbach, and I. Ljungcrantz, Biochim. Biophys. Acta 86, 203 (1964).
21. A.J. Birch, A.J. Ryan, J. Schofield, and H. Smith, J. Chem. Soc., 1231 (1965).
22. A.J. Birch, O.C. Musgrave, R.W. Rickards, and H. Smith, J. Chem. Soc. 3146 (1959).
23. R. Bentley, Ann. Rev. Biochem. 31, 589 (1962).
24. A.J. Birch, in CIBA Foundation Symposium on Quinones in Electron Transport, London, 1961, p. 235; Proc. Chem. Soc. 343 (1958).
25. R. Bentley, and W.V. Lavate, J. Biol. Chem. 240, 532 (1965).
26. G. Pettersson, Acta Chem. Scand. 19, 1827 (1965).
27. N.M. Packter, and M.W. Steward, Biochem. J. 102, 122 (1967).
28. M.W. Steward, and N.M. Packter, Proc. Biochem. Soc., 103, 9P (1967).
29. G. Pettersson, Acta Chem. Scand. 17, 1323 (1963); 19, 1016 (1965).
30. G. Pettersson, Acta Chem. Scand. 17, 1771 (1963).
31. N.M. Packter, Biochem. J. 98, 353 (1966).
32. G. Pettersson, Acta Chem. Scand. 20, 151 (1966).
33. P. Simonart, and H. Verachtert, Bull. Soc. Chim. Biol. 51, 919 (1969).
34. N.M. Packter, Biochem. J. 114, 369 (1969).
35. G. Read, D.W.S. Westlake, and L.C. Vining, Can. J. Biochem. 47, 1071 (1969); Chem. Commun. 935 (1968).
36. K. Mosbach, Angew. Chem. 81, 233 (1969).
37. S. Huneck, in Progress in Phytochemistry 1, 223 (1968).
38. M. Yamazaki, M. Matsuo, and S. Shibata, Chem. Pharm. Bull. 13, 105 (1965); 14, 96 (1966).
39. K. Mosbach, Acta Chem. Scand. 18, 329 (1964).
40. S.H. El Basyouni, and L.C. Vining, Can. J. Biochem. 44, 557 (1966).
41. C.T. Chen, K. Nakanishi, and S. Natori, Chem. Pharm. Bull. 14, 1434 (1969).
42. E.J. Charollais, S. Fliszar, and Th. Posternak. Arch. Sci. Geneva, 16, 474 (1963); Chem. Abstr. 61, 11057f (1964).

43. H. Taguchi, U. Sankawa, and S. Shibata, Tetrahedron Lett. 5211 (1966); Chem. Pharm. Bull. 17, 2054 (1969).
44. A. Penttila, and H.M. Fales, Chem. Commun. 656 (1966).
45. W.F. Beach, and J.H. Richards, J. Org. Chem. 28, 2746 (1963).
46. P.G. Gordon, A. Penttila, and H.M. Fales, J. Amer. Chem. Soc. 90, 1376 (1968).
47. A. Penttila, and H.M. Fales, J. Amer. Chem. Soc. 88, 2327 (1966).
48. A. Salaque, M. Barbier, and E. Lederer; Bull. Soc. Chim. Biol. 49, 841 (1967).
49. J.S.E. Holker, and L.J. Mulheirn, Chem. Commun. 1576 (1968).
50. J.E. Atkinson and J.R. Lewis, Chem. Commun. 803 (1967).
51. R.F. Curtis, C.D. Hassall, and D.R. Parry, Chem. Commun. 1512 (1970).
52. H.C. Forster and M. Puget, personal communication.
53. M. Chen, S.J. Stohs, and E.J. Staba, Planta Medica, 17, 319 (1969); Lloydia 32, 339 (1969).
54. K. Egger, Planta 58, 326 (1962).
55. D.C. Allport and J.D. Bu'Lock, J. Chem. Soc. 654 (1960).
56. A.J. Birch, P. Fitton, E. Pride, A.J. Ryan, H. Smith, and W.B. Whalley, J. Chem. Soc. 4576 (1958).
57. S. Gatenbeck and K. Mosbach, Biochem. Biophys. Res. Commun. 11, 166 (1963).
58. M. Biollaz, G. Büchi and G. Milne, J. Amer. Chem. Soc. 90, 5017, 5019 (1968); 92, 1035 (1970).
59. A.J. Birch, L. Loh, A. Pelter, J.H. Birkinshaw, P. Chaplen, A.H. Manchanda, and M.R. Martin, Tetrahedron Lett. 29 (1965); Aust. J. Chem. 22, 2429 (1969).
60. A.J. Birch, J.J. Wright, F. Gager, L. Mo, and A. Pelter, Tetrahedron Lett. 1519 (1969).
61. R. Thomas, Biochem. J. 78, 748 (1961).
62. J.W. Searcy, N.D. Davis, and U.L. Diener, Appl. Microbiol. 18, 622 (1969).
63. K. Nitta, Y. Yamamoto, T. Inoue, and T. Hyodo, Chem. Pharm. Bull. 14, 363 (1966).
64. E. Schwenk, G.T. Alexander, A.M. Gold, and D.F. Stevens, J. Biol. Chem. 233, 1211 (1958).
65. S.W. Tanenbaum and S. Nakajima, Biochemistry 8, 4626 (1969).
66. A.J. Birch, R.A. Massy-Westropp, R.W. Rickards, and H. Smith, J. Chem. Soc. 360 (1958).
67. W.B. Whalley, in Biogenesis of Natural Compounds, P. Bernfeld, Ed., 2nd ed., Pergamon, Oxford, 1967.
68. T. Kametani, S. Hibino, and S. Takano, Chem. Commun. 131 (1969).
69. A.J. Birch and R.I. Fryer, Aust. J. Chem. 22, 1319 (1969).
70. R. Thomas, Biochem. J. 78, 748, 807 (1961).

71. U. Sankawa, H. Taguchi, Y. Ogihara, and S. Shibata, Tetrahedron Lett. 2883 (1966).
72. S. Gatenbeck, Acta Chem. Scand. 14, 296 (1960); 16, 1053 (1962).
73. A.J. Birch, A.J. Ryan, and H. Smith, J. Chem. Soc. 4773 (1958).
74. A.J. Birch, R.I. Fryer, P.J. Thomson, and H. Smith, Nature 190, 441 (1961); A.J. Birch, Proc. Chem. Soc. 3 (1962).
75. S. Shibata and T. Ikekawa, Chem. Pharm. Bull. 11, 368 (1963).
76. B. Franck, Angew. Chem. 81, 269 (1969).
77. B. Franck, F. Hüper, D. Grüger, and D. Erge, Angew. Chem. 78, 752 (1966).
78. R.F. Curtis, P.C. Harries, C.H. Hassall, J.D. Levi, and D.M. Phillips, J. Chem. Soc. (C), 168 (1966).
79. R.F. Curtis, C.H. Hassall, and R.K. Pike, J. Chem. Soc. (C), 1807 (1968).
80. S. Gatenbeck, Svensk. Kem. Tidskrift. 72, 188 (1960).
81. B. Franck, V. Radtke, and U. Zeidler, Angew. Chem. Int. Ed. 6, 952 (1967).
82. L.Y. Misconi and C.E. Stickings, IUPAC Symposium on Natural Products, London, 1968, p. 184.
83. S. Gatenbeck and L. Malmströ̈m, Acta Chem. Scand, 23, 3493 (1969); IUPAC Symposium on Natural Products, London, 1968, p. 114.
84. R. Ginsig, Ph.D. Thesis, University of Zürich (1963); H. Schmid, Angew. Chem. 75, 347 (1963).
85. J. Zylber, E. Zissmann, J. Polonsky, and E. Lederer, Eur. J. Biol. 10, 278 (1969).
86. P.A. Miller, J.R.D. McCormick, and A.P. Doerschuk, Science 123, 1030 (1956).
87. J.F. Snell, A.J. Birch, and P.L. Thomson, J. Amer. Chem. Soc. 82, 2402 (1960); 84, 425 (1962).
88. S. Gatenbeck, Biochem. Biophys. Res. Commun. 6, 422 (1962).
89. R. Bentley and S. Gatenbeck, Biochemistry 4, 1150 (1965).
90. S. Gatenbeck and R. Bentley, Biochem. J. 94, 478 (1965).
91. E.R. Catlin, C.H. Hassall, and D.R. Parry, J. Chem. Soc. 1363 (1969).
92. A.J. Birch, J.F. Snell, and P.J. Thomson, J. Chem. Soc. 425 (1962).
93. J.R.D. McCormick, in Antibiotics, D. Gottlieb and P.D. Shaw, Eds., Vol. II, Springer-Verlag, New York, 1967, p. 113.
94. A.P. Doerschuk, J.R.D. McCormick, J.J. Goodman, S.A. Szumski, J.A. Growich, P.A. Miller, B.A. Bitler, E.R. Jensen, M. Matrishin, M.A. Petty, and A.S. Phelps, J. Amer. Chem. Soc. 81, 3069 (1959).

95. J.R.D. McCormick, S. Johnson, and N.O. Sjolander, J. Amer. Chem. Soc. 85, 1692 (1963).

96. L.A. Mitscher, J. Pharm. Sci. 57, 1633 (1968).

97. J.R.D. McCormick and E.R. Jensen, J. Amer. Chem. Soc. 90, 7126 (1968).

98. J.R.D. McCormick, E.R. Jensen, N.H. Arnold, H.S. Corey, U.H. Joachim, S. Johnson, P.A. Miller, and N.O. Sjolander, J. Amer. Chem. Soc. 90, 7127 (1968).

99. J.R.D. McCormick and E.R. Jensen, J. Amer. Chem. Soc. 87, 1793, 1794 (1965).

100. P.A. Miller, A. Saturnelli, J.H. Martin, L.A. Mitscher, and N. Bohonos, Biochem. Biophys. Res. Commun. 16, 285 (1964).

101. J.R.D. McCormick, P.A. Miller, J.A. Growich, N.O. Sjolander, and A.P. Doerschuk, J. Amer. Chem. Soc. 80, 5572 (1958).

102. J.R.D. McCormick, P.A. Miller, S. Johnson, N. Arnold, and N.O. Sjolander, J. Amer. Chem. Soc. 84, 3023 (1962).

103. J.R.D. McCormick, E.R. Jensen, S. Johnson, and N.O. Sjolander, J. Amer. Chem. Soc. 90, 2201 (1968).

104. W.D. Ollis, I.O. Sutherland, R.C. Codner, J.J. Gordon, and G.A. Miller, Proc. Chem. Soc. 347 (1960).

105. J. Cudlin, M. Puza, Z. Vaněk, and R.W. Rickards, Folia Microbiol. 14, 406, 499 (1969).

106. J. Cudlin, M. Puza, M. Vondraček, Z. Vaněk, and R.W. Rickards, Folia Microbiol. 12, 376 (1967).

107. H. Umezawa, Recent Advances in Chemistry and Biochemistry of Antibiotics, Microbial Chemistry Research Foundation, Tokyo, 1964, p. 40.

108. D.H.R. Barton, A.M. Deflorin, and O.E. Edwards, J. Chem. Soc. 530 (1956).

109. B.H. Mock and J.E. Robbers, J. Pharm. Sci. 58, 1560 (1969).

110. H. Seto, L.W. Cary, and M. Tanabe, Chem. Commun. 867 (1973).

111. C.T. Chen, K. Nakanishi, and S. Natori, Chem. Pharm. Bull. 14, 1434 (1966).

112. L. Canonica, W. Kroszczynski, B.M. Ranzi, B. Rindone, and C. Scolastico, Chem. Commun. 257 (1971).

113. M. Tanabe, T. Hamasaki, H. Seto, and L. Johnson, Chem. Commun. 1539 (1970).

114. T. Kubota, T. Tokoroyama, S. Oi, and Y. Satomura, Tetrahedron Lett. 631 (1969); J. Chem. Soc. (C), 2703 (1971).

115. R. Thomas, in Biogenesis of Antibiotic Substances, Z. Vaněk and Z. Hoštalek, Eds., Academic, 1965, p. 115.

116. G. Pettersson, Svansk Kemisk Tidskrift 78, 349 (1966).

117. J. Meinwald, K.F. Koch, J.E. Rogers, and T. Eisner, J. Amer. Chem. Soc. 88, 1590 (1966).

118. P.S. Steyn, C.W. Holzapfel, and N.P. Ferreira, Phytochemistry 9, 1977 (1970).

119. A.T. Hudson, I.M. Cambell, and R. Bentley, Biochemistry 9, 3988 (1970).

120. J.G. Dain and R. Bentley, Bioorg. Chem. 1, 374 (1971).

121. A.T. Hudson, Phytochemistry 10, 1555 (1971).

122. A. Sattler and C.P. Schaffner, J. Antibiotics 23, 210 (1970).

123. L.A. Hadwiger, H.G. Floss, J.R. Stoker, and E.E. Conn, Phytochemistry 4, 825 (1965).

124. H.R. Schlütte and P. Mueller, Biochem. Physiol. Pflanz. 163, 518 (1972).

125. Y. Yamamoto, M. Shinya, and Y. Oohata, Chem. Pharm. Bull. 18, 561 (1970).

126. C.K. Wat and G.H.N. Towers, Phytochemistry 10, 103 (1971).

127. C.K. Wat and G.H.N. Towers, Phytochemistry 10, 1355 (1971).

128. J. Sierankiewicz and S. Gatenbeck, Acta Chem. Scand. 27, 2710 (1973).

129. A. Kamal, Y. Haider, R. Akhtar, and A.A. Qureshi, Pakistan J. Sci. Ind. Res. 14, 79 (1971).

130. M. Tanabe and H. Seto, Biochemistry 9, 4851 (1971).

131. J.W. Mathieson and R.H. Thomson, J. Chem. Soc. (C) 153 (1971).

132. R. Durand and M.H. Zenk, Tetrahedron Lett. 3009 (1971).

133. P.G. Harrison, B.K. Bailey, and W. Steck, Can. J. Biochem. 49, 964 (1971).

134. Y. Maebayashi, K. Miyaki, and M. Yamazaki, Chem. Pharm. Bull. 20, 2172 (1972); Tetrahedron Lett. 2301 (1971).

135. J.R. Brown, M.S. Spring, and J.R. Stoker, Phytochemistry 10, 2059 (1971).

136. W. Steglich, R. Arnold, W. Lösel, and W. Reininger, Chem. Commun. 102 (1972).

137. R.F. Curtis, C.H. Hassall, and D.R. Parry, Chem. Commun. 410 (1971).

138. C.T. Bedford, J.C. Fairlie, P. Knittel, T. Money, and G.T. Phillips, Chem. Commun. 323 (1971); Can. J. Chem. 51, 694 (1973).

139. M.T. Lin and D.P.H. Hsieh, J. Amer. Chem. Soc. 95, 1668 (1973).

140. D.P.H. Hsieh, M.T. Lin, and R.C. Yao, Biochem. Biophys. Res. Commun. 52, 992 (1973).

141. M.T. Lin, D.P.H. Hsieh, R.C. Yao, and J.A. Donkersloot, Biochemistry, 12, 5167 (1973).

142. K. Nabeta, A. Ichihara, and S. Sakamura, Chem. Commun. 814 (1973).

143. M. Tanabe and K.T. Suzuki, Chem. Commun. 445 (1974).

144. L. Canonica, B.M. Ranzi, B. Rindone, A. Scala, and C. Scolastico, Chem. Commun. 213 (1973).

145. A.J. Birch, D.N. Butler, and R.W. Rickards, Tetrahedron
      Lett. 1853 (1964).
146. S. Natori, Y. Inouye, adn H. Nishikawa, Chem. Pharm. Bull.
      15, 380 (1967).
147. J.W. Westley, R.H. Evans, D.L. Pruess, and A. Stempel,
      Chem. Commun. 1467 (1970).
148. A. Kamal, T. Begum, and A.A. Qureshi, Pakistan J. Sci. Ind.
      Res. 14, 86 (1971).

CHAPTER 19

Aromatic Polyacetylenes

19-1. NATURAL ACETYLENES IN GENERAL

Substances containing one or several triple bonds form a rapidly increasing group of natural compounds. While some representatives have been found among mevalonate-derived natural products such as carotenoids [for examples, see the review by Weedon (1)] and terpenoids [e.g., freelingyne [1] (2) and chamaecynone [2] (3a)], the vast majority of natural mono- and polyacetylenes are straight-chain aliphatic compounds, or products derivable from these by various cyclization processes,

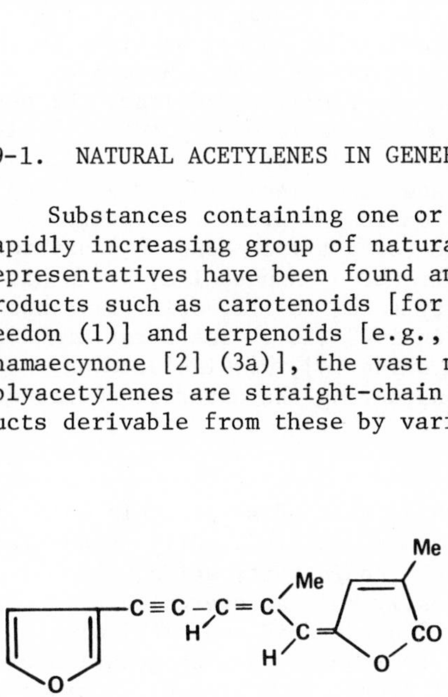

[1] 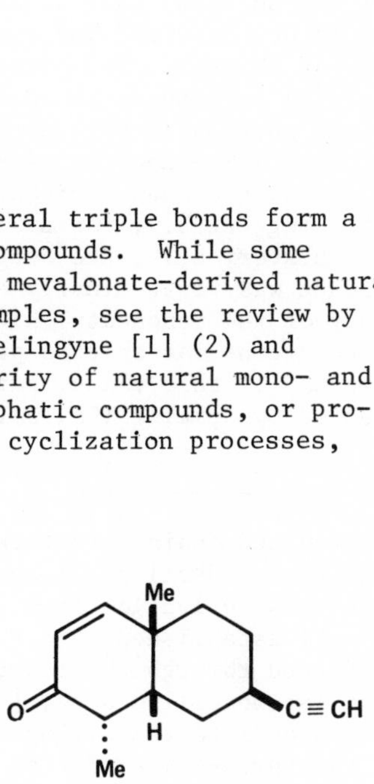   [2]

+The term "polyacetylenes" is here used a little loosely. Actually, several of the substances to be discussed--including carlina oxide [4], the first *aromatic* acetylenic compound to have been the object of detailed chemical study--contain only *one* triple bond. But their biosynthetic relationship with genuine polyacetylenes is manifest.

including formation of benzene rings. The unbranched[+] structure, and the frequent occurrence of carboxy groups, free or esterified, strongly suggest (cf. 5b) a biosynthetic relationship with normal fatty acids and most probably a derivation from them. This interpretation is supported by much experimental evidence, which will be discussed below.

Some valuable reviews of the field of natural polyacetylenes exist (4-7); we wish to acknowledge particularly those by Anchel (4) and Bohlmann (5a) as having been extremely helpful. They should be consulted for many details that cannot be given in the present chapter. For plant polyacetylenes, which have been found in particular abundance in the family Compositae, the pertinent section in Hegnauer's *Chemotaxonomie der Pflanzen* (7) is valuable. The comprehensive volume, *Naturally Occurring Acetylenes*, by Bohlmann, Burkhardt, and Zdero (46) has appeared too late, regrettably, for utilization in the present chapter.

Historically, the occurrence of a triple bond in a natural product was first clearly established by Arnaud (8a, b) in his study of the monoacetylenic fatty acid, tariric acid [3], whose structure he was able to prove in 1902 (8B). The first *aromatic* compound of this class, carlina oxide [4], was isolated and studied by Semmler (9a, b). Considering the natural occurrence

---

[+]A branched-chain structure such as that of sterculynic acid [i] from the seed oil of *Sterculia alata* (3b) is readily reconciled with this derivation from straight-chain fatty acids, since it is well established that the closely related sterculic acid, [ii], is formed through the reaction of oleic acid and methionine to give lactobacillic acid, [iii], followed by dehydrogenation of the cyclopropyl to the cyclopropenyl group. The branching carbon atom of the three-membered ring is thus a secondary addition to a normal straight chain.

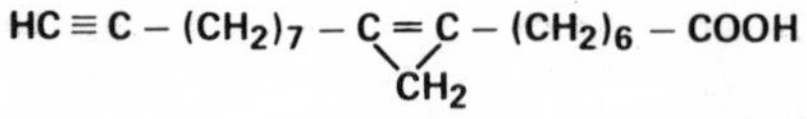

$$HC \equiv C - (CH_2)_7 - \underset{\underset{CH_2}{\diagdown\diagup}}{C = C} - (CH_2)_6 - COOH \qquad [i]$$

$$Me - (CH_2)_7 - \underset{\underset{CH_2}{\diagdown\diagup}}{C = C} - (CH_2)_7 - COOH \qquad [ii]$$

$$Me - (CH_2)_7 - \underset{\underset{CH_2}{\diagdown\diagup}}{CH - CH} - (CH_2)_7 - COOH \qquad [iii]$$

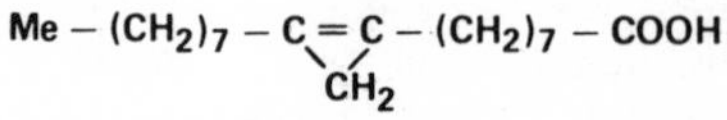

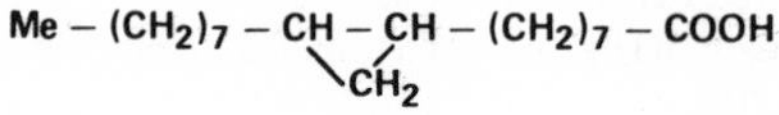

of a triple bond unlikely, Semmler and Ascher (9) proposed the
allenic formula corresponding to [4] for carlina oxide; the correct structure was given by Gilman, Van Ess, and Burtner (10) in
1933. Actually, authentic allenic compounds are much less common
in nature then acetylenic ones; the first such substances to be
discovered (11), the antibiotic mycomycin [5], is both an allene
and a diacetylene. It has been pointed out that this C_{13} compound manages to contain every type of carbon-carbon unsaturation
that is possible in an aliphatic chain: acetylenic, allenic,
and both *cis* and *trans* ethylenic. The organism producing this
compound, initially classified as an actinomycete, seems in
reality to be a basidiomycete (4a).

The structural elucidation of a naturally occurring *poly*-
acetylene was first achieved by Williams, Smirnov, and Gol'mov
(10a) in 1935, who recognized the lachnophyllum ester, isolated
by them from *Lachnophyllum gossypium* (Compositae), as the methyl
ester of dec-2-ene-4,6-diynic acid.

$$Me - (CH_2)_{10} - C \equiv C - (CH_2)_4 - COOH \qquad [3]$$

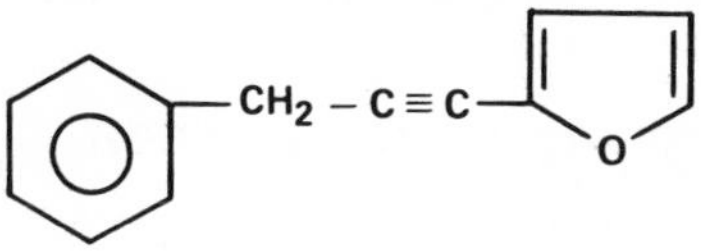

[4]

$$HC \equiv C - C \equiv C - CH = C = CH - CH \overset{c}{=} CH - CH \overset{t}{=} CH - CH_2 - COOH^x$$

[5]

Since 1941, the work of Sörensen (12), Bohlmann (5), and
their co-workers has uncovered an unexpected wealth of polyacetylenes in plants of the family Compositae, and the research of
Anchel (4a), Sir E.R.H. Jones (13), and their associates has
yielded a substantial number of such compounds from basidiomycetes. The total number of natural polyacetylenes known in 1971
is given as approximately 650 (5d). Occurrence of such compounds
is not restricted to these two groups of organisms. About 15
families of higher plants are stated (4a, 5a) to contain poly-

---

[x]*Cis*-and *trans*-stereochemistry of ethylenic bonds is indicated
in the formulae by "c" and "t", respectively.

acetylenes, and at least one more family, the Campanulaceae, has
been added (14) more recently. Besides the Compositae (where
only the plants of the tribe Cichorieae and some genera of the
alkaloid-rich tribe Senecionae seem to be devoid (5a, 12b) of
these compounds), the Umbelliferae are rich in polyacetylenes.
    Natural polyacetylenes with all chain-lengths between C_{18}
and C_6 have been encountered. Those with the longer chains,
down to C_{10}, tend to occur in higher plants, the shorter ones in
microorganisms, which produce mainly compounds in the C_{14} - C_6
range, with C9 and C_{10} preponderant. There is a certain overlap
in the acetylenic compounds biosynthesized by higher plants and
by fungi. The C_{10} polyacetylene *trans*, *trans*-matricarianol [6],
for example, occurs in a number of Compositae (5b), but also in
the basidiomycetes *Polyporus anthracophilus* and *Tricholoma
grammopodium* (15); crepenynic acid [7], a key intermediate in
the conversion of oleic acid to acetylenes (see below), had been
discovered in *Crepis* sp. (Compositae) but is also produced by
some polyacetylene-producing higher fungi (16).

$$Me - CH \overset{t}{=} CH - [C \equiv C]_2 - CH \overset{t}{=} CH - CH_2OH \qquad [6]$$

$$Me - (CH_2)_4 - C \equiv C - CH_2 - CH \overset{c}{=} CH - (CH_2)_7 - COOH \qquad [7]$$

    Within this rather limited range of unbranched chains of
18–6 carbon atoms, an amazing number of structural variants has
been encountered: every conceivable arrangement of double and
triple bonds (cf. [5]), a wide variety of oxygen-containing
groups, MeS-substituents, and numerous cyclic groupings,
including benzene rings and oxygen and sulfur-heterocycles.
(No alicyclic or nitrogen-containing ring systems appear to have
been encountered so far; they may be expected to be found eventually).
    The various chain-lengths are represented with very differ-
ent frequencies, C_{18}, C_{17}, C_{14}, C_{13}, C_{10}, C9 being numerous,
others (C_{16}, C_{15}, C_{11}) quite uncommon (see ref. 6, Fig. 3).
    Among group properties of polyacetylenes, the highly charac-
teristic many-banded uv-spectrum (5b,c, 17) and the extreme
sensitivity of many of them (4) may be mentioned: polymerization
to insoluble, black or colored, poorly characterized products
often takes place on exposure to light, or on attempts to concen-
trate their dilute solutions. It has been proposed (ref. 7, p.
492), very plausibly, that the black, insoluble "phytomelanes,"
characteristic of many of the Compositae, may be related to the
black transformation products of polyacetylenes; it is suggestive
of such a relationship that both types of product are absent from

plants of the tribe Cichorieae of the family Compositae. From
the ready cyclization of certain types of polyacetylenes to
aromatic compounds, it seems entirely possible that these products may be aromatic.

19-2. GENERAL ASPECTS OF THE BIOSYNTHESIS OF NATURAL
       POLYACETYLENES

a. Origin of the Carbon Chain.

The almost exclusive occurrence of straight carbon chains
from C_{18} down suggests at once that the biosynthesis of the
polyacetylenes is a variant of that of the normal fatty acids
from acetic acid. There is abundant experimental evidence for
the correctness of this assumption. It must suffice to give
here only the most significant results of these investigations;
no complete coverage is attempted, since excellent reviews are
available (4a, 5a-c).

Incorporation of carboxyl-labeled acetate into polyacetylenes has been observed repeatedly, both in fungi (18) and in
higher plants (19). The expected pattern of uniform and almost
exclusive labeling of alternate carbon atoms was elegantly demonstrated for the first time in 1959 by Bu'Lock and Gregory (18a):
systematic degradation of nemotinic acid [8a] biosynthesized
from Me-^{14}COOH by a basidiomycete (presumably a *Poria* sp.) gave
the distribution of relative activities shown below:

$$HC \equiv C - C \equiv C - CH = C = CH - CHOH - CH_2 - CH_2 - COOH \qquad [8a]$$

96	0	103	0	102	8	103	1	96	5	100

The same pattern of labeling of alternating carbon atoms
from Me-^{14}COOH was observed by Bu'Lock and Smith (19) in their
investigation of the biosynthesis of fatty acids by seedlings of
Santalum acuminatum (Santalaceae). Besides oleic [11] and
palmitic acids, this plant produces a remarkable series of
unsaturated C_{18} acids, such a ximenynic acid [9a] and its relatives [9b-f]. Oxidative degradation of [9c] and [9e] showed the
labeling to have occurred in the expected way, and the structures
of these acids, which all share the unit $-[CH_2]_7-COOH$ with oleic
acid, are very suggestive of a biosynthesis starting from this
compound and proceeding by stepwise introduction of double and
triple bonds; the experimental findings are compatible with this
interpretation without, however, proving it. Analogous results,
to be discussed below, have been obtained by Bohlmann and Jente
(20) in their studies of the biosynthesis of phenylpolyacetylenes

in Compositae. Incorporation of acetic acid has been observed
in many other instances but has not been studied as systematic-
ally by stepwise degradations as in the cases quoted.

$$Me - (CH_2)_7 - CH \overset{c}{=} CH - (CH_2)_7 - COOH \qquad [11]$$

$$Me - (CH_2)_5 - CH = CH - C \equiv C - (CH_2)_7 - COOH \qquad [9a]$$

$$Me - (CH_2)_3 - [CH = CH]_2 - C \equiv C - (CH_2)_7 - COOH \qquad [9b]$$

$$Me - (CH_2)_3 - CH = CH - [C \equiv C]_2 - (CH_2)_7 - COOH \qquad [9c]$$

$$Me - CH_2 - [CH = CH]_2 - [C \equiv C]_2 - (CH_2)_7 - COOH \qquad [9d]$$

$$Me - CH_2 - CH = CH - [C \equiv C]_3 - (CH_2)_7 - COOH \qquad [9e]$$

$$H_2C = CH - CH = CH - [C \equiv C]_3 - (CH_2)_7 - COOH \qquad [9f]$$

As in the biosynthesis of fatty acids and polyketides (see
Section 17-2), the formation of the carbon chain of the polyacetylenes is initiated by a starter molecule of acetyl-CoA,
and continued by addition of units of malonyl-CoA, followed by
decarboxylation and removal of the carbonyl oxygen. This mechanism was demonstrated by Bu'Lock and Smalley (15b) for the
biosynthesis of dehydromatricarianol [10] in *Tricholoma grammopodium*, the basidiomycete already mentioned as a producer of
the closely related [6]. When the fungus was grown in presence
of diethyl [2^{14}C] malonate, [10] was labeled. Catalytic reduction and subsequent oxidation to *n*-decanoic acid, followed first
by decarboxylation (to give C-1), then by Kuhn-Roth oxidation
(to give C-9 + C-10), showned C-1 to be inactive; C-9 and C-10
together contained only 3% of the total activity of [10]; the
bulk of the label must thus be contained in carbons 2-8. Scheme I
demonstrates that these results agree completely with those to
be expected from the usual acetate-malonate scheme, if the
unknown distribution of label is assumed to be the one shown.
There is thus no doubt that the biosynthesis of [10] follows the
same fundamental mechanism as that of the other acetate-malonate-
derived metabolites. This result can be extended to the other
acetylenic compounds through the multiple biogenetic interrelations that have been observed (4a, 5b).

It has been pointed out by Bu'Lock (16, 19, 21) that many
naturally occurring acetylenic compounds with 18 or 17 carbon
atoms have structures strongly suggestive of biosynthetic relationships with fatty acids, and particularly with oleic acid [11].
Cases in point are the numerous C_{18} carboxylic acids with multiple
bonds (*cis*-ethylenic or acetylenic) between C-9 and C-10 (cf.
compounds [7] and [9a-f]); some C_{18} aldehydes, and many obviously
related structures with shorter carbon chains will be found in

Scheme I

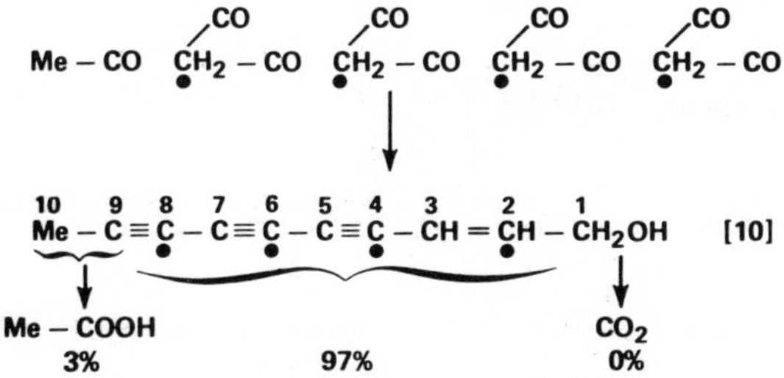

Me – CO $\quad \overset{\displaystyle CO}{\underset{\bullet}{CH_2}} - CO \quad \overset{\displaystyle CO}{\underset{\bullet}{CH_2}} - CO \quad \overset{\displaystyle CO}{\underset{\bullet}{CH_2}} - CO \quad \overset{\displaystyle CO}{\underset{\bullet}{CH_2}} - CO$

$$\underset{10}{Me} - \underset{9}{C} \equiv \underset{8}{\underset{\bullet}{C}} - \underset{7}{C} \equiv \underset{6}{\underset{\bullet}{C}} - \underset{5}{C} \equiv \underset{4}{\underset{\bullet}{C}} - \underset{3}{CH} = \underset{2}{\underset{\bullet}{CH}} - \underset{1}{CH_2OH} \quad [10]$$

Me – COOH CO_2

3%   97%   0%

the reviews already quoted. These considerations led Bu'Lock (21) to postulate a general pathway for the biosynthesis of acetylenic compounds (see Scheme II); his hypothesis has recently found much experimental confirmation. It assumes the formation, from [11], of the ubiquitous linoleic acid [12], which would be transformed into the first acetylenic compound, crepenynic acid [7]. The subsequent biosynthetic events would be the step-wise introduction of double and triple bonds (cf. [13], to be discussed below, and the acids [9a-f]) to give the tri-acetylenic acid [14], from which chain-shortening processes (mostly α- and β-oxidations) and other transformations would yield the various naturally occurring acetylenic compounds.

The first experimental support for this hypothesis came from the work of Bu'Lock and Smith (16) with *T. grammopodium*. Culture filtrates of this fungus were found to contain, besides [11] and [12], also [7] and dehydrocrepenynic acid [13]; the former acetylenic acid had been obtained previously from *Crepis* spp. (Compositae) (22), while the latter was a new substance. Furthermore, the fungus converted $10-^{14}C-[11]$ into [12], [7], and into the C_{10}-compound dehydromatricarianol [10]; in all cases, the label was found by degradation to occupy the expected positions: C-10 in [12] and [7], C-2 in [10].

The fungus is thus able to introduce triple bonds into the chain of [11], and to shorten it, from the carboxyl end as expected, to the C_{10} enetriyne [10].

The general validity of the same scheme also for higher plants has been documented in numerous papers by Bohlmann and co-workers (5a and references there). For instance (23a), *Chrysanthemum flosculosum* (Compositae) incorporates [11], tritiated in the vinylic positions 9 and 10, specifically and with high yields into the spiro-enol acetal [15]. Among the higher fungi, *Clitocybe rhizophora* has been shown in unpublished work from Sir E.R.H. Jones' laboratory (quoted in ref. 23b) to utilize $[9,10-^3H]-[11]$ for the biosynthesis of the C_9 triol [16]. (The incorporation of labeled [11] into *aromatic* acetylenes will

Scheme II

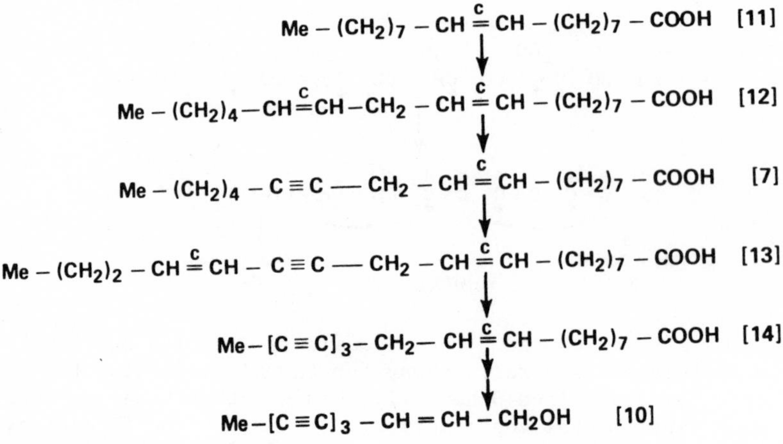

$$Me - (CH_2)_7 - CH \overset{c}{=} CH - (CH_2)_7 - COOH \quad [11]$$

$$Me - (CH_2)_4 - CH \overset{c}{=} CH - CH_2 - CH \overset{c}{=} CH - (CH_2)_7 - COOH \quad [12]$$

$$Me - (CH_2)_4 - C \equiv C - CH_2 - CH \overset{c}{=} CH - (CH_2)_7 - COOH \quad [7]$$

$$Me - (CH_2)_2 - CH \overset{c}{=} CH - C \equiv C - CH_2 - CH \overset{c}{=} CH - (CH_2)_7 - COOH \quad [13]$$

$$Me - [C \equiv C]_3 - CH_2 - CH \overset{c}{=} CH - (CH_2)_7 - COOH \quad [14]$$

$$Me - [C \equiv C]_3 - CH = CH - CH_2OH \quad [10]$$

be discussed below, as will that of some labeled precursors with shorter chains, especially those with 13 and 14 carbon atoms.) Bohlmann and Schulz (24) have observed that the methyl ester of [9-10-^{3}H]-[12] (linoleic acid) is incorporated into [15] by *C. flosculosum* to about the same extent as the ester of [11], and have demonstrated (25) the conversion of the tritiated [11] into [15] in homogenates from leaves of the same species; after centrifugation at 12700g, practically all the activity was found in the sediment.

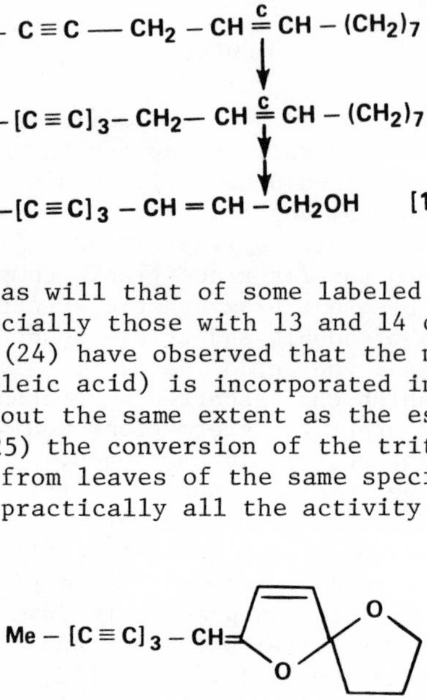

$$Me - [C \equiv C]_3 - CH =$$

[15]

$$Me - CH_2 - CHOH - [C \equiv C]_2 - CHOH - CH_2OH$$

[16]

Evidently, these examples (and others to be discussed later) support Bu'Lock's scheme by showing that the biosynthesis of several polyacetylenes of different chain lengths starts with a C_{18} fatty acid such as [11]. That involvement of [11] itself may not necessarily be obligatory seems suggested by the observation (ref. 5a, pp. 5, 35, 51; ref. 26) that the labeled triyne acids [17, n = 6, 8, and 10] (i.e., the analog of [14] lacking the ethylenic bond, and the corresponding C_{16} and C_{14} acids) are incorporated into *cis*-dehydromatricaria ester [18].

$$Me - [C \equiv C]_3 - (CH_2)_n - COOH \qquad [17]$$

$$Me - [C \equiv C]_3 - CH \overset{c}{=} CH - COOMe \qquad [18]$$

It might be assumed that mechanisms would exist for the biosynthesis of acetylenes from palmitoleic acid, the widely distributed C_{16} analog of [11]. Such a process does not seem to have been observed up to now, though, and can thus hardly be common.

For the steps beyond crepenynic acid [7], Bu'Lock (16, 21) postulated introduction of two additional triple bonds into positions 14 and 16, to give the C_{18} acid [14] (see Scheme II). This compound has apparently never been isolated from any organism; it has, however, been synthesized (27), and the incorporation of variously labeled [14] into a number of naturally occurring substances has been demonstrated several times (see below). Its actual function as an intermediate thus seems probable. In addition, the characteristic terminal grouping $Me-[C \equiv C]_3-$ appears in a substantial number of natural polyacetylenes, making a close relationship plausible; besides [10], [15], and [18], already mentioned, this group of substances includes a variety of other compounds with chain-lengths from C_{17} down to C_{10} (5a). Furthermore, formation of many of the structural units found in natural polyacetylenes can often be explained convincingly through transformations of this particular grouping, and much evidence for the soundness of these explanations has been obtained by incorporation studies. Detailed discussion of these questions can not be our purpose; the interested reader is referred to ref. 5a.

There is much evidence to show that the biosynthesis of the numerous polyacetylenes with 17 to 8 carbon atoms usually occurs through stepwise degradations of [14] by both α- and β-oxidations. Bohlmann and co-workers (27) have shown, for instance, that introduction of triple bonds in the biosynthesis of C_{13} and C_{14} acetylenes ([15] and several aromatic compounds to be discussed later) takes place at the C_{18} level; subsequent shortening of the chain by twice-repeated β-oxidation leads to the C_{14} compounds, while those with C_{13} result from α-oxidation at the C_{18}-level, followed in turn by two β-oxidations. Further transformations of the resulting triacetylenic compounds, [19a,b], into a wide variety of natural polyacetylenes has been established repeatedly by incorporation studies with labeled materials; for details, see ref. 5a and literature quoted there. However, [14] is not necessarily an intermediate in every instance. In

Artemisia vulgaris (Compositae), for example, [114] is a good
precursor of the C_{10} compound [18], while it is only poorly
incorporated into the C_{17} hydrocarbon "Centaur X$_3$" [20] which is,
however, formed from [11] (28). On the other hand, β-hydroxy-
[11], labeled with ^3H in position 2, is an efficient and specific

$$Me - [C \equiv C]_3 - CH_2 - CH \overset{c}{=} CH - (CH_2)_n - COOMe \qquad [19] \quad a: n = 3$$
$$b: n = 2$$

$$Me - [C \equiv C]_3 - [CH = CH]_2 - (CH_2)_4 - CH = CH_2 \qquad [20]$$

precursor of [20]; the introduction of the triple bonds in the
biosynthesis of this C_{17} compound (and several of its relatives)
thus seems to be preceded by β-hydroxylation. Similar utilization of β-hydroxyoleic acid for the biosynthesis of C_{17} polyacetylenes was also observed (28) in *Oenanthe*, a plant of the
family Umbelliferae.

Another mechanism for shortening of polyacetylenic chains
by one carbon atom deserved brief mention, since it seems to
occur quite frequently and appears to play a role also in the
biosynthesis of certain aromatic polyacetylenes. This is the
transformation of the characteristic end group Me-[C≡C]$_3$- into
H-[C≡C]$_3$-, which seems to proceed through stepwise oxidation of
the methyl to HO-CH$_2$-, and HOOC-, with subsequent decarboxylation.
Several suggestive instances of co-occurrence of compounds corresponding to these several steps have been observed. For instance,
nemotinic acid [8a] and its higher homolog odyssic acid [8b] are
both found in a basidiomycete, probably a *Poria* sp. (18a).

$$R - [C \equiv C]_2 - CH = C = CH - CHOH - CH_2 - CH_2 - COOH$$

$$[8] \quad a: R = H$$
$$b: R = Me$$

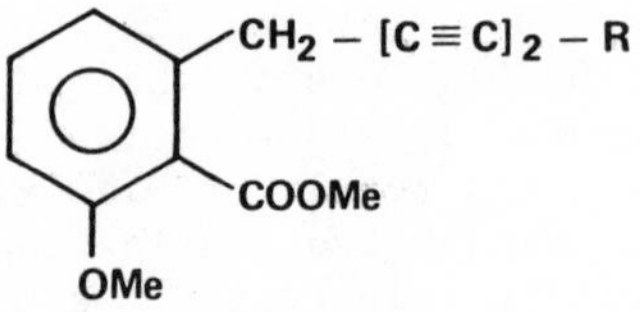

$$[21] \quad a: R = Me$$
$$b: R = H$$

Similarly, the two aromatic compounds, frutescin [21a] and desmethyl-frutescin [21b], occur together in *Chrysanthemum frutescens* (20). After administration of [1-14C] acetate to this plant, the terminal methyl of [21a] is inactive, while the adjacent carbon atom has the expected 1/6 of the total activity; in [21b], it is the terminal carbon which is labeled to about the same extent as C-2 of [21a], to which it evidently corresponds. (The fact that the terminal methyl group of [21a] is *inactive* may appear surprising, since analogy with the ordinary acetate-derived phenols would at first glance suggest that the carboxyl and the methoxylated ring-carbon, and consequently the terminal methyl, ought to be derived from acetate carboxyl. However, the inactivity of the methyl is consistent with the expected formation of the C_{13} skeleton of [21a] from C_{18} fatty acids through shortening by 5 carbon atoms, beginning at the carboxyl end; the carboxyl of [21a] must then originate from C-6 of the C_{18} chain, evidently a carbon derived from acetate methyl. The analogy with the polyketides is thus misleading. The actual mode of the formation of the aromatic ring of [21a] will be discussed later.)

Many of the reactions just discussed have been accessible to direct study in fungi. Culture filtrates of several basidiomycetes can transform the terminal methyl of *trans*-[18] to $HOCH_2$- and HOOC-, and, in the case of *Coprinus quadrifidus* (30a), remove this carbon atom altogether to give the C_9 triynetriol [22], undoubtedly by oxidation of Me- to HOOC-, followed by decarboxylation (and other obvious changes). This decarboxylation of an α,β-acetylenic acid had been shown earlier (30b) to proceed enzymatically in cell-free extracts of *C. quadrifidus*, which convert the C_8 acid [23a] (not a naturally occurring compound) into the C_7 alcohol [23b].

$$Me - [C \equiv C]_3 - CH \overset{t}{=} CH - COOMe \longrightarrow H - [C \equiv C]_3 - CHOH - CHOH - CH_2OH$$

$$[18] \qquad\qquad\qquad\qquad\qquad [22]$$

$$HOOC - [C \equiv C]_2 - CH \overset{t}{=} CH - CH_2OH \longrightarrow H - [C \equiv C]_2 - CH \overset{t}{=} CH - CH_2OH$$

$$[23a] \qquad\qquad\qquad\qquad\qquad [23b]$$

The analogous oxidation of the allylic primary alcohol grouping of *trans*-dehydromatricarianol [10] to the corresponding aldehyde and acid is carried out by the culture filtrate of the fungus *Papulospora polyspora* (29).

b. Origin of the Triple Bond.

A priori, two possibilities exist for the formation of
acetylenic bonds: they could be introduced into carbon chains
already formed by processes which, whatever their detailed mechanism, have the overall stoichiometry of dehydrogenations, or
they could arise during the biosynthesis of the carbon chain
itself. The experimental evidence discussed in the preceding
pages definitely favors the first alternative and establishes
introduction of acetylenic unsaturation at the C_{18} level without,
however, restricting it to this: the possibility of biogenetic
introduction of triple bonds into shorter chains has been demonstrated. Among many existing examples, we may mention the conversion of the C_{14} triyne [24], which occurs naturally in
Centaurea spp. (Compositae) (23a), into the C_{13} tetraynes [25] in
C. ruthenica (31):

Scheme III

[24] $Me - [C \equiv C]_3 - CH_2 - CH \overset{c}{=} CH - (CH_2)_3 - CH_2OH \longrightarrow$

[25] $Me - [C \equiv C]_4 - CH = CH - CHCl - CH_2OR$, $R = H$ or Ac

Nothing further seems to be known with certainty about the
intimate mechanism of the formation of triple bonds, that is,
whether it proceeds through direct removal of hydrogen, or via
the well-known enzymatic introduction of double bonds, but the
essential step: direct dehydrogenation of a double to a triple
bond, does not seem to have any known precedent.
     The other possibility, formation via distinct intermediates,
would evidently involve generating the triple bond through some
kind of elimination process from suitable enolic derivatives.

Such reactions have indeed been proposed (5b, 18a, 32), and have been performed successfully in the laboratory; however, no direct evidence for their actual occurrence in organisms has been obtained so far. The formation of triple bonds from enolic derivatives has been ably reviewed by Cymerman Craig et al. (33). These proposals involve elimination reactions of enol-phosphates of the ketonic intermediates in the biosynthesis of fatty acids; a variant (32) particularly attractive form the mechanistic viewpoint places the formation of acetylenic bonds at the level of the chain-elongation by malonyl-coenzyme A, that is, during the formation of the chain (see Scheme III). As a suitable, biochemically meaningful leaving group X, phosphate or, better, pyrophosphate seems likely.

Scheme III

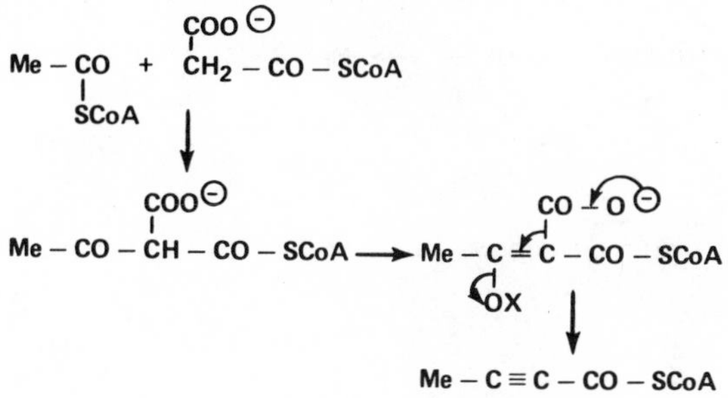

As has been mentioned above, experimentally observed biosynthetic formation of triple bonds has so far always taken place on preformed carbon chains, so the the concerted decarboxylation-elimination mechanism just discussed remains hypothetical. It is not necessary to assume, however, that biosynthetic triple-bond formation always proceeds by the same reactions, and "there may well be more than one answer" (6), here as elsewhere.

The various biosynthetic interrelationships of the natural polyacetylenes - modifications of the chain lengths, introduction and, occasionally, reduction of triple bonds, rearrangements, introduction of oxygen- and sulfur-containing groups, cyclizations, etc., have been studied in great detail in the laboratories of F. Bohlmann and Sir Ewart R.H. Jones, mostly with the use of

specifically labeled precursors obtained by total synthesis.
The synthetic preparation of these often exceedingly unstable
compounds constitutes in itself a very remarkable achievement.
Detailed discussion of the involved interrelations would far
exceed the limitations of this chapter; the work prior to 1967
has been reviewed very completely by Bohlmann (5a). Among more
recent papers not mentioned elsewhere here, the investigation
(34) of the origin of the thio-enol ether groups of [26] and [27]
in *Anthemis tinctoria* (Compositae) deserves brief description.
Feeding of doubly labeled (^{35}S and ^{3}H$_3$C-) methionine showed that
both the S-atom and the methyl group are furnished by the
methionine; surprisingly, they are introduced separately in a
stepwise fashion, since the ratio ^{3}H: ^{35}S is much higher in the
products than in the starting material. Furthermore, [1-^{14}C]-
dehydromatricaria ester [18] was incorporated specifically into
[27], demonstrating the occurrence of both the (formal) addition
of -SMe across a triple bond, and of biological reduction of an
acetylenic to an ethylenic unsaturation.

$$MeS - (CH_2)_2 - CH(NH_2) - COOH \longrightarrow Me - C \equiv C - \underset{\underset{SMe}{|}}{C} = CH - C \equiv C - CH \overset{c}{=} CH - COOMe$$

[26]

$$Me - [C \equiv C]_3 - CH = CH - COOMe \longrightarrow Me - C \equiv C - CH \overset{c}{=} CH - \underset{\underset{SMe}{|}}{C} \overset{c}{=} CH - CH \overset{c}{=} CH - COOMe$$

[18]   [27]

19-3. AROMATIC ACETYLENES

Compared to the widespread occurrence and structural variety
of natural polyacetylenes in general, their aromatic representatives form a restricted and specialized group of compounds. They
have been found so far only in higher plants (mostly, if not
exclusively, in Compositae[+]), and appear to be completely lacking
in fungi; furthermore, only C_{13}, C_{12}, and C_{11} compounds seem to
have been encountered. $C_6H_5 - CH_2 - C \equiv C - C \equiv C - Me$

_____

[28]

[+]Only one instance of alleged occurrence of an aromatic polyacetylene in a plant not belonging to that family has come to our
attention: that of "agropyrene," which has been isolated from
the essential oil of the roots of a grass, *Agropyrum repens*.
However, the root sample was known to be contaminated with roots
(Continued on page 444).

Table I, Selected Aromatic Acetylenes Not Mentioned Elsewhere; from ref. 5a.

A) C_{13} compounds

$$C_6H_5 - [C \equiv C]_2 - CH \overset{t}{=} CH - Me$$

$$C_6H_5 - [C \equiv C]_2 - CH \overset{t}{=} CH - CH_2OH$$

$$C_6H_5 - [C \equiv C]_2 - CH \overset{t}{=} CH - CHO$$

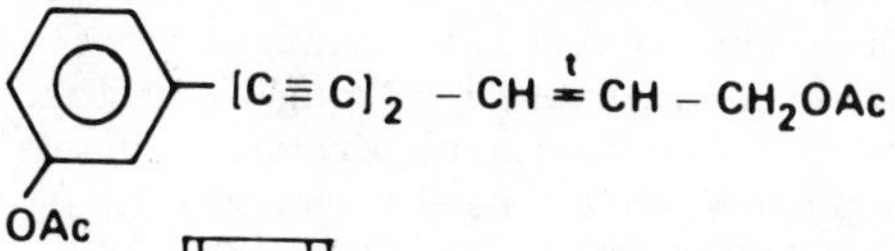

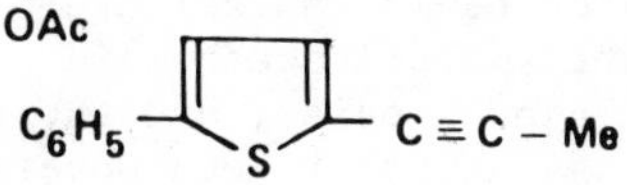

$[C \equiv C]_2 - CH \overset{t}{=} CH - CH_2OAc$ (benzene ring with OAc substituent)

C_6H_5—(thiophene ring, S)—$C \equiv C - Me$

B) C_{12} compounds

$$C_6H_5 - CO - [C \equiv C]_2 - CH_2OH$$

$$C_6H_5 - CO - CH_2 - CH_2 - C \equiv C - Me$$

C) C_{11} compounds

$$C_6H_5 - CH_2 - C \equiv C - C \equiv CH$$

$$C_6H_5 - CH_2 - C \equiv C - CH \overset{c}{=} SMe$$

$$C_6H_5 - CO - C \equiv C - CH \overset{c,t}{=} CH - SMe$$

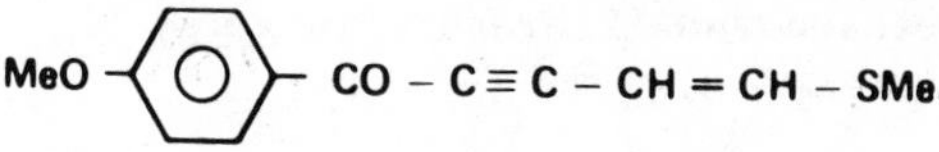

$$MeO - \text{(benzene ring)} - CO - C \equiv C - CH = CH - SMe$$

Even so, the number of known compounds is still too large to make inclusion of a complete listing desirable. Table I gives representative examples. It will be seen from it that the length of the carbon chains is much more limited than are those of related nonaromatic compounds. Features such as enol-lactones and spiro ketals, allenic groups, and polyols are all missing. On the other hand, methyl ether groupings seem to be found almost exclusively here, and the number of modifications that do occur is still substantial.

19-4. BIOSYNTHESIS OF AROMATIC ACETYLENES

The biosynthesis relationship of the aromatic acetylenes to their nonbenzenoid congeners, and hence to the fatty acids, is evident from the exclusive occurrence of structures that are consistent with a genesis through cyclization of straight-chain precursors.[++] Furthermore, aromatic acetylenes and related non-cyclic compounds which could, on structural grounds, be their precursors, frequently occur together in the same plant, or are found in closely related species, and the aromatization of the presumed precursor has actually been observed *in vitro* on several occasions. A striking instance of this has been described

[29] $CH_2=CH-CH=CH-C\equiv C-C\equiv C-C\equiv C-CH\overset{t}{=}CH-Me$

[30] $C_6H_5-C\equiv C-C\equiv C-CH\overset{t}{=}CH-Me$

by Skattebøl and Sørensen (35): a total synthesis designed to give [29] actually yielded a mixture of the expected compound

---

(Continued from p. 442)
of *Artemisia dracunculus* (Compositae). Since capillene [28], a major constituent of *Artemisia* spp., is identical with agro-pyrene, the latter has probably originated in the contaminant. See the discussion and literature in refs. 5b, p. 38, and 12 b, p. 249.

[++]The lone exception: the thio-ether [49], not itself acetylenic but closely related to authentic acetylenes, is demonstrably the product of a rearrangement of a straight-chain acetylenic precursor; see below.

and the aromatization product [30]. Acetylenes [29] and [30] occur together in several *Coreopsis* species.

This result indicates a cyclization of the dienyne end-group of [29] to the phenyl of [30], and the co-occurrence of the compounds suggests a biosynthetic precursor-product relationship as well, although this latter point has not been experimentally established. A very similar connection exists (35) between carlina oxide [4], the acyclic acetylene [31], and the aromatic acetate [32a]; compounds [4] and [31] are both found in *Carlina acaulis*, while [32a] occurs in the closely related *C. vulgaris* (36). The transformation of the free alcohol, [32b], into [4] has been established by Bohlmann and von Kap-herr (37) through the use of material labeled with tritium in the phenyl group; however, direct experimental proof for the biosynthetic conversion of the diene-yne [31] into [32] or [4] has not yet been adduced.

[31] $H_2C = CH - CH \overset{c}{=} CH - C \equiv C - C \equiv C - C \equiv C - CH \overset{t}{=} CH - CH_2OAc$

[32]   a: R = Ac $C_6H_5 - C \equiv C - C \equiv C - CH \overset{t}{=} CH - CH_2OR$
       b: R = H

[4] $C_6H_5 - CH_2 - C \equiv C -$

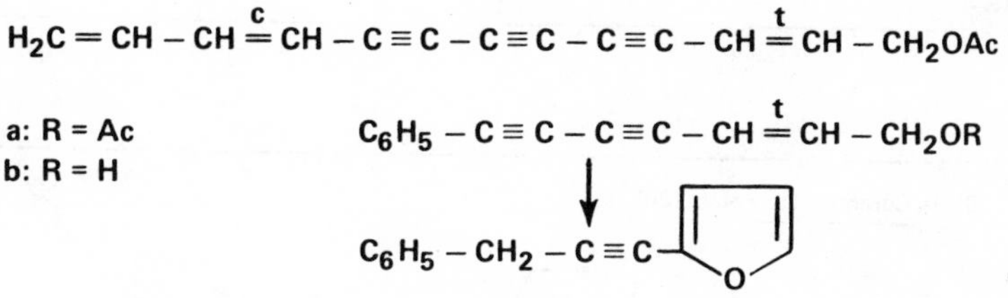

Abundant experimental evidence for the biosynthesis of the aromatic ring in polyacetylenes is available, though. Proof of incorporation of acetate in the expected way, that is, with equal labeling of alternate carbon atoms, was first obtained by Bohlmann and Jente (20): administration of [1-14C]-acetate to *Chrysanthemum frutescens* produced capillol [33a] and frutescin [21a] with labeling patterns summarized in Scheme IV. The distribution of the label was established by systematic degradations, which are indicated in the scheme.

A most interesting result, already briefly mentioned, is the inactivity of the carbomethoxy group of [21a]. This group thus does *not* come from the carboxyl of acetate; it is indeed active if [2-14C] acetate is fed. Extrapolation of the pattern of alternative labeling found in other polyacetylenes permits the conclusion that also the ring carbon bearing the methoxyl is

not carboxyl-derived (the localization of label within the
aromatic ring was not established).   The contrast to the labeling
is structurally analogous polyketides is striking. Similar
results were obtained in feeding experiments with *Coreopsis
lanceolata*, which contains aromatic polyacetylenes different

SCHEME IV

Biosynthesis of Aromatic Polyacetylenes from Acetate.

A) In Chrysanthemum frutescens (20)

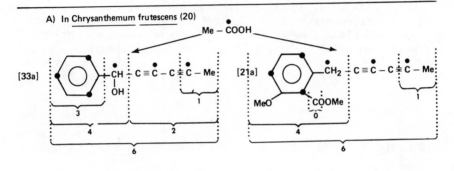

B) In Coreopsis lanceolata (20)

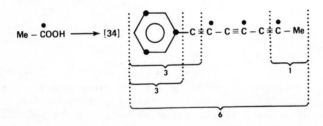

C) In Dahlia (23b)

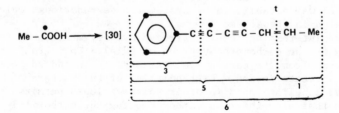

from those of *C. frutescens*. Besides those [21a, 30, 33a, 34] shown in Scheme IV, also some minor aromatic constituents of the latter plant were active, such as capillin [35], capillarin [36] and desmethyl-frutescin [21b]. As has been mentioned before, the terminal carbon atom of the side-chain of [21b] was labeled, suggesting a biosynthesis from frutescin [21a] by loss of the methyl group.

Results very similar to the ones just described were obtained in Sir E.R.H. Jones' laboratory (23b) in an investigation of the biosynthesis of [30] in *Dahlia* roots.

$C_6H_5 - CO - [C \equiv C]_2 - Me$

[35] capillin

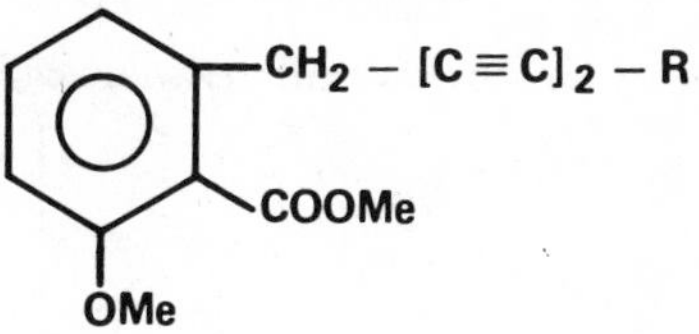

$CH_2 - [C \equiv C]_2 - R$

COOMe

OMe

$CH_2 - C \equiv C - Me$

[21] a: R = Me
b: R = H

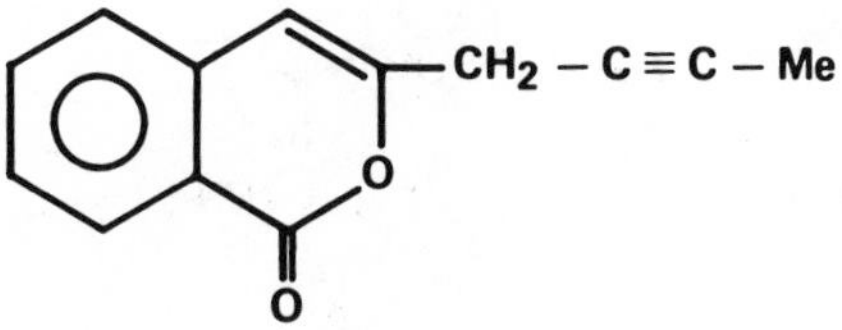

[36] capillarin

Beyond this evidence for the biosynthesis of aromatic acetylenes from acetate, their formation by cyclization of straight-chain acetylenic precursors has been thoroughly established by Bohlmann and his associates; it also follows from their work that here, too, oleic acid [11] is an early intermediate.

In *Chrysanthemum frutescens*, Bohlmann and co-workers (23a) found efficient and specific incorporation of [11] and the C_{13} triynes, [37] and [19a] (labeled with the tritium in the methyl group) into the aromatic acetylenes [21a] (frutescin), [33b] (capillol acetate), and [36] (capillarin) (see Scheme V). Active [33b] was also obtained from [9,10-^{3}H] [11]; interestingly, the C_{16}-triynic acid [38], which lacks the ethylenic bond of [11], [37], and [19a], was incorporated into the straight-chain C_{10} compound dehydromatricaria ester [18], but not into the aromatic substances, [33b] and [21a]. This result points to some specific role of the double bond of [11] and its congeners.

In *Coreopsis lanceolata*, the C_{14}-triynes [39a and b], homologous with [37] and [19a], respectively, and similarly labeled with ^{3}H in the methyl group, were converted to the phenyl heptatriyne [34].

SCHEME V

Biosynthesis of Aromatic Acetylenes from Oleic Acid [11] and from Straight-Chain Acetylenic Precursors (23a).

A) In Chrysanthemum frutescens

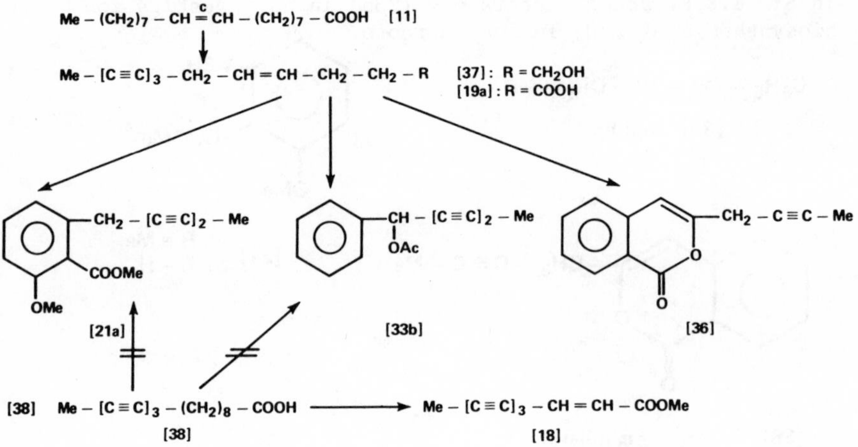

$Me - (CH_2)_7 - CH \overset{c}{=} CH - (CH_2)_7 - COOH$ [11]

$Me - [C \equiv C]_3 - CH_2 - CH = CH - CH_2 - CH_2 - R$ [37]: $R = CH_2OH$
  [19a]: $R = COOH$

$CH_2 - [C \equiv C]_2 - Me$
$COOMe$
OMe
[21a]

$CH - [C \equiv C]_2 - Me$
OAc
[33b]

$CH_2 - C \equiv C - Me$
[36]

[38] $Me - [C \equiv C]_3 - (CH_2)_8 - COOH$ ⟶ $Me - [C \equiv C]_3 - CH = CH - COOMe$
         [38]   [18]

B) In Coreopsis lanceolata

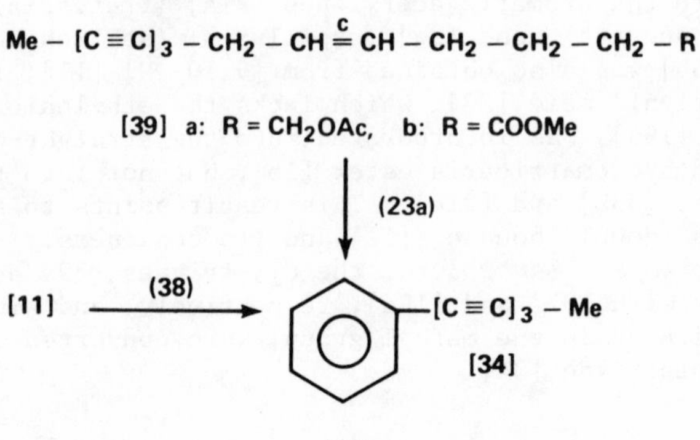

$Me - [C \equiv C]_3 - CH_2 - CH \overset{c}{=} CH - CH_2 - CH_2 - CH_2 - R$

[39]   a: $R = CH_2OAc$,   b: $R = COOMe$

(23a)

[11] $\overset{(38)}{\longrightarrow}$ $[C \equiv C]_3 - Me$
[34]

Details of the formation of [34] from [39] and [11] were studied, again in *C. lanceolata*, by Bohlmann, Bonnet, and Jente (38). Use of [39a], labeled with tritium in positions 2 and 3, or of [9, 10-$^3H_2$]-[11], gave [34] which contained label only in the aromatic ring.

Obviously, in the formation of these compounds [21a, 33b, 34, 36], the aromatic ring originates largely [21a, 33b, 36] or exclusively [34] from the nonacetylenic part of the molecule of the precursor. The situation here is thus quite different from that of the cyclization of dienyne precursor which seems to be involved in the biosynthesis of [4].

For the early steps in the transformation of [11] into the aromatic polyacetylenes, the operation of the general pathway, [11]———→[12]————→[7]————→[14] (see Scheme II), with subsequent shortening of the chain from the carboxyl end, is very probably followed, as was to be expected. Thus, Bohlmann and Schulz (24) found that *Coreopsis lanceolata* and *Chrysanthemum frutescens* incorporate methyl-[9,10-3H]-linoleate [12] into [34] to about the same extent as [11]. (It will be remembered that the same result was also obtained for the biosynthesis of the enol-ketal [15] in the same species). The subsequent events: stepwise introduction of triple bonds to give [14], followed by shortening of the chain from the carboxyl end, have been explored in the work of Bohlmann, Jente, and Reinecke (27), which has been briefly discussed above. These authors studied the incorporation of the methyl esters of oleic acid and its lower homologs [40, n = 2-7], labeled with tritium in the vinylic positions, and of the corresponding triacetylenic acids, [41, n = 2-7], i.e. [14] and its lower homologs, labeled with ^{14}C in the methyl groups, into a variety of polyacetylenes, including the aromatic substances [34] and [35] (capillin). *Coreopsis lanceolata* utilized, among the mono-unsaturated acids, only [11], = [40, n = 7] for the biosynthesis of the C_{13} compound [34]; of the triacetylenic acids [41], only those with an even number of carbon atoms were incorporated, that is, [14] = [41, n = 7], and those with n = 3 and 5, while the odd-numbered ones, [41, n = 4 and 6], where not available for this biosynthesis. In *Chrysanthemum frutescens*, compound [35] ($C_{12}H_8O$) was similarly formed efficiently only from [11], while of the triacetylenes [41], the one with n = 7 was the only compound with an even number of carbon atoms which was well incorporated; in contrast, the odd-numbered ones (n = 2, 4, 6) were now efficient precursors.

These results are consistent with a sequence for the biosynthesis of numerous polyacetylenes, aromatic and otherwise, in which [11] is converted (see Scheme VI) in the usual way via [12] and [7] into [14] = [41, n = 7]. At this stage, the pathways

[40] $Me - (CH_2)_7 - CH \overset{c}{=} C - (CH_2)_n - COOH$ (n = 7: [11]).

[41] $Me - [C \equiv C]_3 - CH_2 - CH \overset{c}{=} CH - (CH_2)_n - COOH$ (n = 7: [14]).

[34] $C_6H_5 - [C \equiv C]_3 - Me$

[35] $C_6H_5 - CO - [C \equiv C]_2 - Me$

branch. Twice-repeated β-oxidation leads to C_{14} products, from which compounds such as [34] would arise by loss of one carbon atom (in this particular case through decarboxylation). Alternatively, α-oxidation would convert [14] into a C_{17} acid or ester, which in its turn would be shortened by twice-repeated β-oxidation to the C_{13} stage; this would yield, for example, a compound such as [35] again with loss of one carbon as CO_2. Apparently, α-oxidation generally occurs only at the C_{18} stage. While evidently many details remain to be worked out (stages of biosynthesis of [14] from [7], etc.), this sequence appears to provide a generalized explanation for many observed facts, such as the apparently specific role of [11], and the incorporation patterns of [40] and [41]. Some features are still unexplained, though; it is not clear, for example, why almost universally, two β-oxidations should follow each other, going from C_{18} to C_{14}, and from C_{17} to C_{13}. The assumption of such a process is consistent with the observed rarity of C_{16} and C_{15} compounds but does not explain it.

It remains to consider the ways in which the actual formation of the benzene ring from nonaromatic precursor takes place.

The cyclization of conjugated dienynes, which is presumably involved in the biosynthesis of [30], [32], [32a], and [4] has precedents in purely chemical transformations (35, 39); compare, *inter alia*, the formation of both [29] and [30] in a synthesis designed to give the former compound (see above), and analogous reactions described in ref. 35.

It should be remembered, however, that no direct experimental proof is available for the actual involvement of these cyclization

Scheme VI

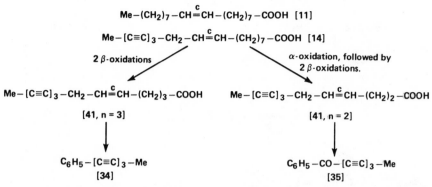

reactions during the biosynthesis of aromatic polyacetylenes. In a number of cases where direct evidence has been obtained by Bohlmann and his co-workers, the benzene ring has been found to originate mostly or exclusively from the *non*acetylenic parts of the precursor molecule.   In these instances, the characteristic terminal group, the Me-[C≡C]$_3$- of the precursor, survives completely [34] or in part [21, 33b, 36] in the structure of the aromatic product.   Insight into the mode of cyclization has been gained by incorporation studies using variously labeled precursors with C_{13} chains, such as [37] and [19a], and their next higher homologs [39a and b].

Compounds [37] and [19a] were found [23a, 27, 38, 40) to be efficient precursors of aromatic acetylenes with 13 and 12 carbons atoms.   The C_{13} compounds were frutescin [21a], [30], and capillarin [36], while desmethylfrutescin [21b], capillene [28], capillol [33a], and capillin [35] represent the C_{12} group. Aromatic acetylenes with 11 carbon atoms seem to be formed from their C_{12} homologs by the loss of the terminal methyl group which has already been discussed; thus labeled [35] is transformed in *Chrysanthemum segetum* into [42] (42):   C$_6$H$_5$-CO-[C≡C]$_2$-Me⟶ C$_6$H$_5$-CO-C≡C-CH=CHSMe

The biosynthesis of some of these C_{12} and C_{13} compounds is formulated by Bohlmann and co-workers (23A, 27, 40) as shown in Scheme VII; they assume β-oxidation of [19a] to the conjugated β-keto acid [43]; this compound could cyclize by a Michael-type reaction involving the first carbon of the triyne system, to give an intermediate formulated as [44].   This in its turn could aromatize, with or without loss of the carboxyl group and/or the carbonyl oxygen, to give [21a], [28], and [36], and their relatives.

Experimental evidence supports this interpretation.   Thus, [37] and [19a], labeled with tritium in the terminal methyl, are incorporated (23a) by *Chrysanthemum frutescens* into [21a], [33b],

Scheme VII.
Biosynthesis of Aromatic Polyacetylenes
in *Chrysanthemum frutescens*

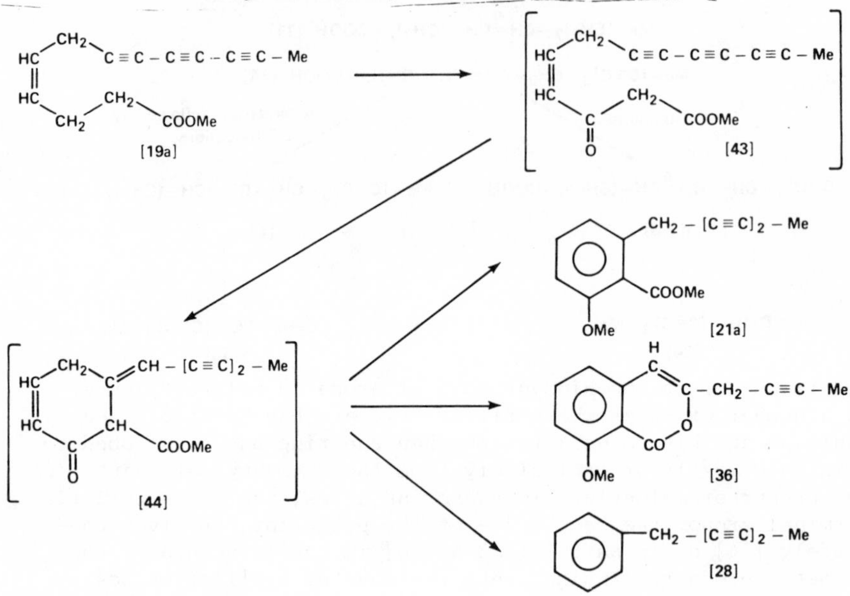

and [36], all of them labeled exclusively in their terminal methyl
groups (cf. Scheme V); compound [39], the higher homolog of [19a],
was only poorly utilized.   In *Lonas annua*, another plant of the
family Compositae, [37], doubly labeled with ^{14}C in the terminal
methyl and with ^{3}H in the vinylic positions, was incorporated into
[35], the ketone corresponding to [28], with essential retention
of the ratio ^{14}C: ^{3}H, and with all the ^{3}H located on the aromatic
ring, as required by the proposed reaction sequence (40).

For the experimentally observed (23A) biosynthesis of the
carboxylfree aromatic C_{13} compounds, such as [34], from C_{14} precursors [39a and b], oxidation of the β-position as well as at the
−CH$_2$− group between the double and the first triple bond is
postulated, (see Scheme VIII).   Initially, oxidation to carbonyl
in both places was assumed (23a), and indeed the resulting
1,5-diketone [45a] would lend itself admirably to ring formation
by aldolization; because of the close analogy with the cyclization step in the shikimate pathway, we would prefer to formulate
the initial cyclic intermediate as [46a] (cf. 3-dehydroquinic
acid!).   Elaboration of the unsubstituted phenyl group of [34]
and its relatives from [46a] would then obviously involve decarboxylation, reduction of the carbonyl, and elimination of two
molecules of water.

Scheme VIII
Biosynthesis of Phenylheptatriyne [34]
in *Coreopsis lanceolata*

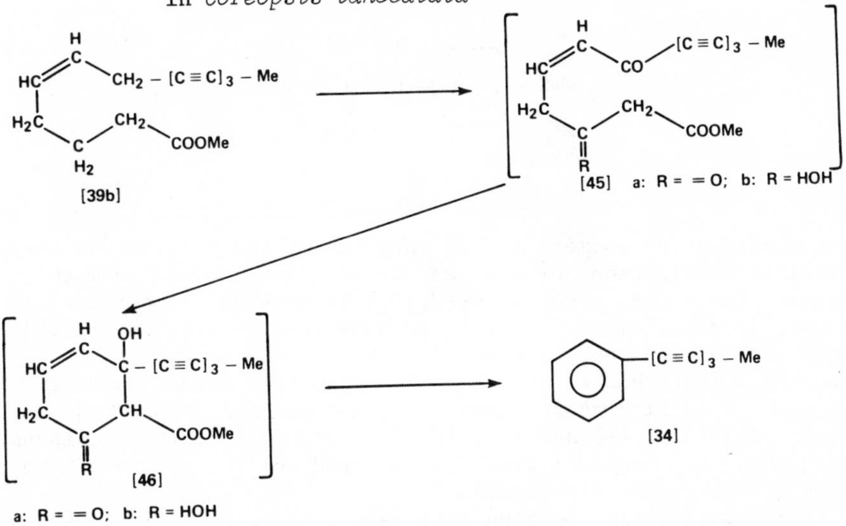

Subsequent studies by Bohlmann, Bonnet, and Jente (38) on the biosynthesis of [34] in *Coreopsis lanceolata* have led to a modification of this scheme. It was found that [39a and b], labeled with ^{14}C in the terminal methyl and with ^{3}H in positions 2 and 3, yield [34] which retains part of the ^{3}H in the aromatic ring. According to the initial formulation of the biosynthesis, all tritium would obviously be lost from position 3 during formation of [45a], and from position 2 during the aromatization step. To accomodate this partial retention of label, the authors assume β-oxidation only to the level of the secondary alcohol, to give [45b]. This modified scheme still furnishes a satisfactory explanation for the formation of the aromatic ring; however, the decarboxylation is now somewhat less readily explained.

There remains one more aromatic compound to be discussed which, while not in itself acetylenic, has been shown to be derived biosynthetically from a nonaromatic acetylene. The substance is unique among this group of compounds in having a *branched* chain.

The product in question is the methyl thioether [49], $C_{12}H_{14}O_2S$, a minor constituent of *Anthemis tinctoria* (41a,b). This plant, and other species of the genus, also contains a fairly large number of obviously related noncyclic acetylenic methyl thioethers of the formula $C_{12}H_{12}O_2S$, of which a few are shown below [50a-d], together with a dihydro compound of this type, [27], which is likewise found in *A. tinctoria*. The other members of the group

$$\text{MeS} - \underset{}{\bigcirc} - \underset{\underset{H}{|}}{C} = \underset{\underset{H}{|}}{C} - \text{COOMe}$$

(with Me above the first C)

[49]

are mostly stereoisomers of the ones shown; the majority of the possible combinations of *cis*- and *trans*- double bonds seem to occur.   The entire group of $C_{12}H_{12}O_2S$ isomers can be derived, on paper, from the widely distributed dehydromatricaria ester [18]; both *cis*- and *trans*-[18] have often been found in *Anthemis* species.   That this relationship is not merely formal follows from the specific incorporation of 1-^{14}C-labeled *trans*-[18][+] into [50a] (41b, 42) and [27] in *A. tinctoria* (34) (this compound, and [50b], had been utilized for the study of the origin of the S-Me group already discussed).

Similarly, [18] labeled with ^{14}C in position 1 (41b) is converted by *A. tinctoria* into the aromatic compound, [49], which was shown by Schmidt degradation to contain almost all the label in the carboxyl group, as expected.   This finding demonstrates that both the aromatic ring and the *branched* side-chain are formed from the *unbranched*, noncyclic [18].   Feeding of [18] labeled with ^{14}C in the terminal methyl group (44) leads to [49] having practically all the label in the C-methyl group, as shown by the degradations outlined below (Scheme IX).   This finding proves that methyl migration does take place.   Together, these results show that [49] is indeed formed from [18], probably via [50b], in a unique reaction sequence involving, besides the formal addition of the SMe to give [50b], aromatization, methyl migration, and reduction ([50b] is $C_{12}H_{12}O_2S$, while [49] is $C_{12}H_{14}O_2S$).   Conversion of a Δ^6-*trans* isomer of [50b] to a cumulene intermediate, [51], $C_{12}H_{14}O_2S$, has been tentatively proposed (44), with further transformation into [49] as shown diagrammatically (Scheme X).   No precedent for such a reaction appears to be known, however, and the details of this very unusual biosynthetic aromatization remain to be established.

[+]The fact that *trans*-[18] is incorporated into [50a] and [27] which have the *cis*-configuration of the corresponding double bond is not significant, since it has been shown (42) that *A. tinctoria* rapidly converts *trans*-[18] into the *cis*-isomer.   Only the latter occurs in this species (43).

[18]   Me − [C ≡ C]$_3$ − CH = CH − COOMe

[27]   Me − C ≡ C − CH $\overset{c}{=}$ CH − C $\overset{c}{=}$ CH − CH $\overset{c}{=}$ CH − COOMe
                                             |
                                            SMe

[50a] a: Me − C $\overset{c}{=}$ CH − C ≡ C − C ≡ C − CH = CH − COOMe
                 |
                SMe

       b: Me − C ≡ C − C = CH − C ≡ C − CH $\overset{c}{=}$ CH − COOMe
                       |
                      SMe

       c: Me − C ≡ C − C ≡ C − C = CH − CH = CH − COOMe
                               |
                              SMe

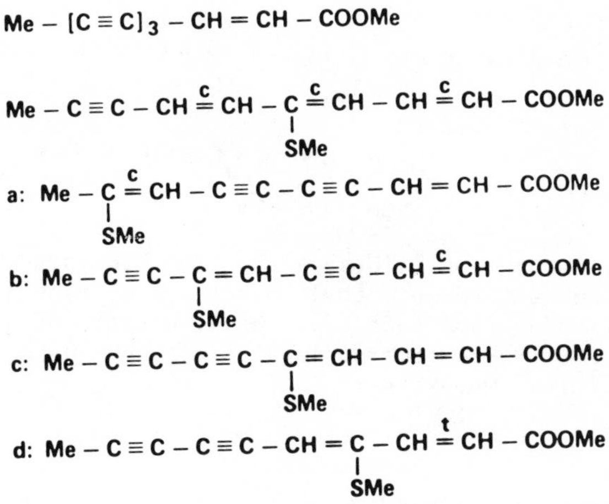

       d: Me − C ≡ C − C ≡ C − CH = C − CH $\overset{t}{=}$ CH − COOMe
                                      |
                                     SMe

Scheme IX

[18] ^{x}Me − C ≡ C − C ≡ C − C ≡ C − CH = CH − COOMe

[49]   MeS— ⬡ —C = CH − COOMe   (with ^{x}Me substituent)

1) perphthalic acid
2) O$_3$

MeSO$_2$— ⬡ —CO (with ^{x}Me) —CrO$_3$→ MeSO$_2$— ⬡ —COOH + ^{x}CO$_2$

(inactive)

455

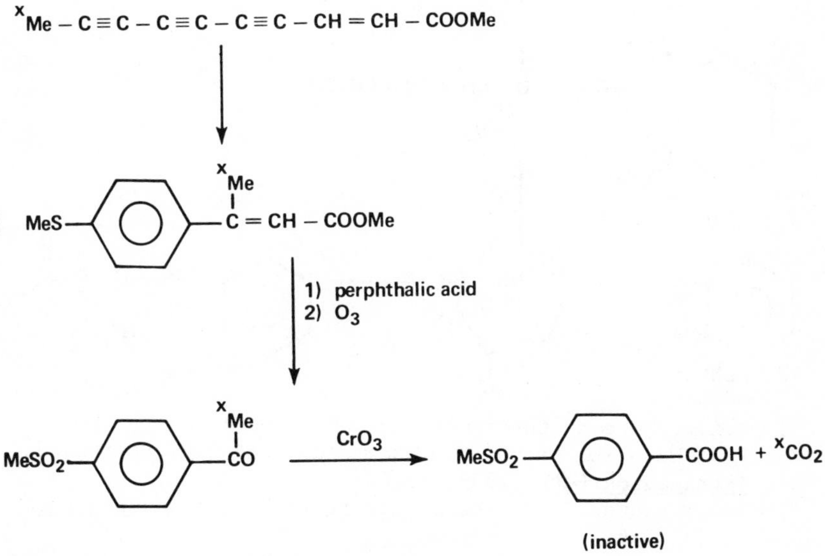

Scheme X

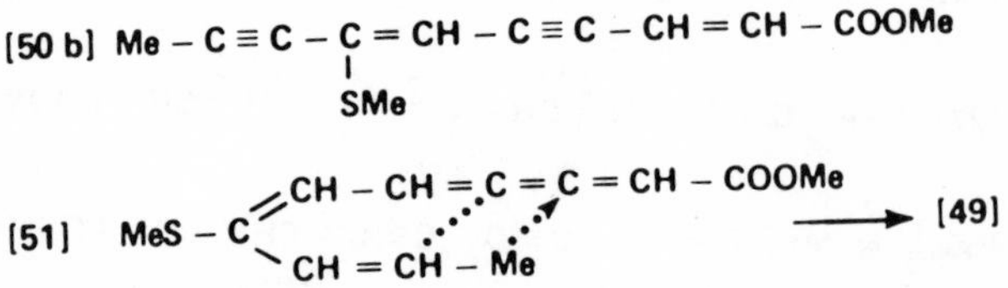

[50 b]   Me – C ≡ C – C = CH – C ≡ C – CH = CH – COOMe
                          |
                          SMe

[51]   MeS – C⟨CH – CH = C = C = CH – COOMe
              ⟨CH = CH – Me   ⟶ [49]

   The assumption (41b, 44) that the biosynthesis of [49] from
[18] proceeds through methyl thioethers of type [50] is supported
by the finding (45) that a synthetic mixture of [50b] with its
three *cis-*, *trans-*isomers, labeled with ^3H in position 2, leads
to active [49] when fed to *A. tinctoria*.

19-5.   ADDENDUM

   Recent unpublished work from Prof. Bohlmann's laboratory
(5d) has shown that the postulated biosynthesis of [4] and [30]
from [31] and [29], respectively, can hardly be correct; no
label from the presumed precursors was incorporated. For the
biosynthesis of [4], the observations suggest the following
process:

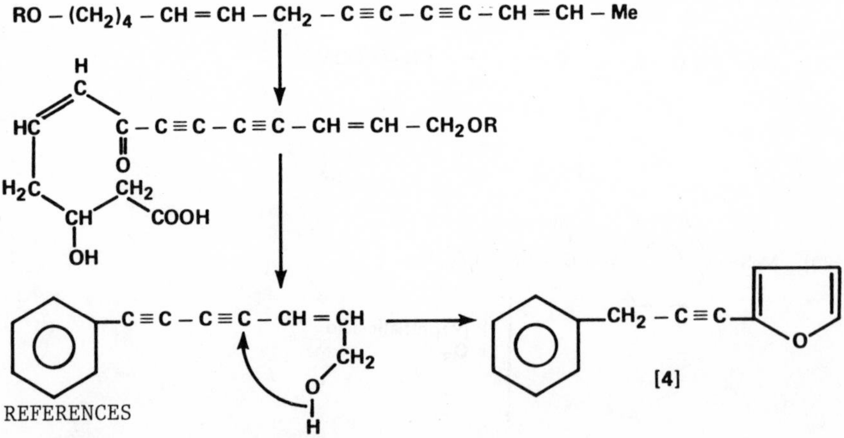

RO – (CH₂)₄ – CH = CH – CH₂ – C ≡ C – C ≡ C – CH = CH – Me

C – C ≡ C – C ≡ C – CH = CH – CH₂OR

C ≡ C – C ≡ C – CH = CH   ⟶   CH₂ – C ≡ C   [4]

REFERENCES

1. B.C.L. Weedon, Chem. Brit. 3, 424 (1967).
2. R.A. Massy-Westropp, G.D. Reynolds, and T.M. Spotswood,
    Tetrahedron Lett. 1939, (1966).
3. a. T. Nozoe, Y.S. Cheng, and T. Toda, Tetrahedron Lett.
        3663, (1966).
    b. A.W. Jevans and C.Y. Hopkins, Tetrahedron Lett. 2167,
        (1968).

4. M. Anchel, in Antiobiotics, Vol. II, Biosynthesis, D. Gottlieb
    and T.D. Shaw, Eds., Springer-Verlag, Berlin-Heidelberg-
    New York, 1967, p. 189.

4A. S.C. Canham, M. Anchel, and G.N. Bistis, Mycologia 62, 599
    (1970).

5. a. F. Bohlmann, Fortschritte Chem. Org. Natur. 25, 1 (1967);
    b. F. Bohlmann and H.J. Mannhardt, ibid. 14, 1 (1957).
    c. F. Bohlmann, H. Bornowski, and C. Arndt, Fortschritte
        Chem. Forschg. 4, 138 (1962); F. Bohlmann, ibid. 6 (1966);
    d. F. Bohlmann, private communication.

6. Sir E. Jones, Chem. Brit. 2,6 (1966).

7. R. Hegnauer, Chemotaxonomie der Pflanzen, Vol. 3, p. 490 ff.
    Birkhäuser Verlag, Basel and Stuttgart, 1964.

8. A. Arnaud, (a) Compt. Rend. 114, 79 (1892); Bull. Soc. Chim.
    Fr. [3] 7, 233 (1892); (b) Compt. Rend. 134, 473 (1902);
    Bull. Soc. Chim. France. [3] 27, 484, 489 (1902).

9. F.W. Semmler, Ber. 39, 726 (1906); F.W. Semmler and E. Ascher,
    Ber. 42, 2355 (1909).

10. H. Gilman, P.R. Van Ess, and R.R. Burtner, J. Amer. Chem.
    Soc. 55, 3461 (1933); cf. also A.S. Pfau, J. Pictet,
    P. Plattner, and B. Susz, Helv. Chim. Acta 18, 935 (1935).

10A. V.V. Williams, V.S. Smirnov, and V.P. Gol'mov, Zhur.
    Obshchei Khim. 5, 1195 (1935); Chem. Abstr. 30, 1176 (1936).

11. W.D. Celmer and I.A. Solomons, J. Amer. Chem. Soc. 74, 1870
    (1952); 75, 1372 (1953).

12. a. N.A. Sörensen and J. Stene, Ann. 549, 30 (1941);
    b. N.A. Sörensen, in Chemical Plant Taxonomy, T. Swain,
        Ed., Academic, New York, 1963, p. 219 ff.

13. Sir E.R.H. Jones, Proc. Chem. Soc. 199 (1960), and numerous
    other papers.

14. R.K. Bentley, J.K. Jenkins, Sir E.R.H. Jones, and V. Thaller,
    J. Chem. Soc. (C) 830 (1969); J. Lam and F. Kaufmann, Chem.
    Ind. 1430 (1969).

15. a. J.D. Bu'Lock and E.R.H. Jones, and W.B. Turner, J. Chem.
        Soc. 1607 (1957).
    b. J.D. Bu'Lock and H.M. Smalley, ibid. 4662 (1962).

16. J.D. Bu'Lock and G.N. Smith, J. Chem. Soc. (C) 332 (1967).

17. K.W. Kausser, R. Kuhn, and G. Seitz, Z. Phys. Chem. B 29,
    391 (1935).

18. a. J.D. Bu'Lock and H. Gregory, Biochem. J. 72, 322 (1959).
    b. J.D. Bu'Lock and D.C. Allport, and W.B. Turner, J. Chem.
        Soc. 1654 (1961).
    c. Sir E.R.H. Jones, G. Lowe, and P.V.R. Shannon, J. Chem.
        Soc. (C) 139, 144 (1966).

19. J.D. Bu'Lock and G.N. Smith, Biochem. J. 85, 35P (1962);
    Phytochemistry 2, 289 (1963).

20. F. Bohlmann and R. Jente, Chem. Ber. 99, 995 (1966).

21. J.D. Bu'Lock, in Comparative Phytochemistry, T. Swain, Ed., Academic, London and New York, 1966, p. 79.

22. K. Mikolajczak, C.R. Smith Jr., M.O. Bagby, and I.A. Wolff, J. Org. Chem. 29, 318 (1964).

23. a. F. Bohlmann, R. Jente, W. Lucas, J. Laser, and H. Schulz, Chem. Ber. 100, 3183 (1967);
     b. J.R.F. Fairbrother, Sir E.R.H. Jones, and V. Thaller, J. Chem. Soc. (C) 1035 (1967).

24. F. Bohlmann and H. Schulz, Tetrahedron Lett. 1801 (1968).

25. F. Bohlmann and H. Schulz, Tetrahedron Lett. 4795 (1968).

26. F. Bohlmann, W. v. Kap-herr, R. Jente, and G. Grau, Chem. Ber. 99, 2091 (1966).

27. F. Bohlmann, R. Jente, and R. Reinecke, Chem. Ber. 102, 3283 (1969).

28. F. Bohlmann and T. Burkhardt. Chem. Ber. 102, 1702 (1969).

29. P. Hodge, J. Chem. Soc. (C) 1617 (1966).

30. a. P. Hodge, Sir E.R.H. Jones, and G. Lowe, J. Chem. Soc. (C) 1216 (1966);
     b. J.N. Gardner, G. Lowe, and G. Read, J. Chem. Soc. 1532 (1961).

31. F. Bohlmann, M. Wotschokowsky, J. Laser, C. Zdero, and K.-D. Bach, Chem. Ber. 101, 2056 (1968).

32. Sir E.R.H. Jones, Chem. Eng. News 39, 46 (1961).

33. J. Cymerman Craig, M.D. Bergenthal, I. Fleming, and J. Harley-Mason, Angew. Chem., Int. Ed. 8, 429 (1969).

34. F. Bohlmann and T. Burkhardt, Chem. Ber. 101, 861 (1968).

35. L. Skattebøl and N.A. Sørensen, Acta Chem. Scand. 13, 2101 (1959).

36. J.S. Sørensen and N.A. Sørensen, Acta Chem. Scand. 8, 1763 (1954).

37. F. Bohlmann and W. von Kap-herr, Chem. Ber. 99, 148 (1966).

38. F. Bohlmann, H. Bonnet, and R. Jente, Chem. Ber. 101, 855 (1968).

39. G. Eglinton, R.A. Raphael, R.G. Willis, and J.A. Zabkiewicz, J. Chem. Soc. 2597 (1964), and literature quoted there.

40. F. Bohlmann, W. Lucas, J. Laser, and P.-H. Bonnet, Chem. Ber. 101, 1176 (1968).

41. a. F. Bohlmann, C. Arndt, H. Bornowski, and K.-M. Kleine, Chem. Ber. 96, 1485 (1963);
     b. F. Bohlmann, D. Bohm and C. Rybak, Chem. Ber. 98, 3087 (1965).

42. F. Bohlmann, W. von Kap-herr, C. Rybak, and J. Repplinger, Chem. Ber. 98, 1736 (1965).

43. F. Bohlmann, K.-M. Kleine, C. Arndt, and S. Kühn, Chem. Ber. 98, 1616 (1965).

44. F. Bohlmann and J. Laser, Chem. Ber. 99, 1834 (1966).
45. F. Bohlmann and W. Skuballa, Chem. Ber. 103, 1886 (1970).
46. F. Bohlmann, T. Burkhardt, and C. Zdero, Naturally Occurring Acetylenes, Academic, New York and London, 1973.

CHAPTER 20

Biosynthesis of Vitamin B₁₂ and Riboflavin

The vitamins B_{12} constitute a family of compounds (cobalamins) which are variants of the same general structure [1]. The biosynthesis of these vitamins has been reviewed by Cheldelin and Baich (1967) (1), Wagner (1966) (2), and Friedmann and Cagan (1970) (3).

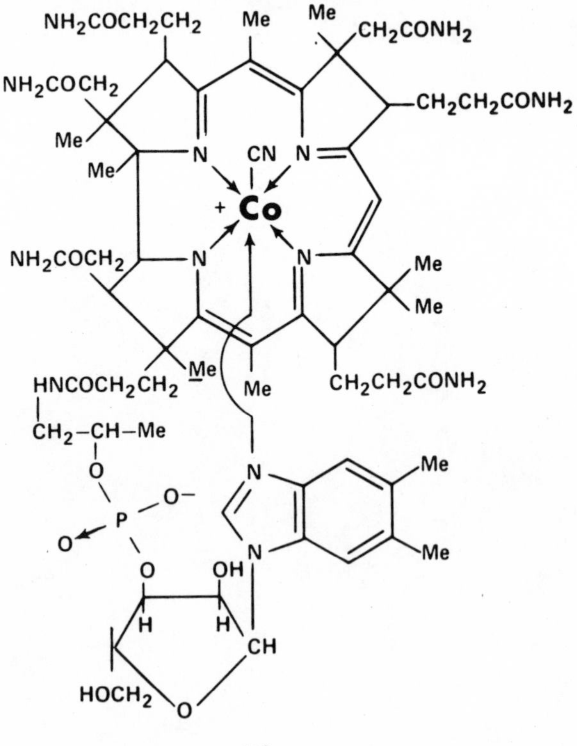

[1]

We are concerned here only with the biosynthesis of the benzimidazole grouping which occurs in several members of the group, since
it is the only aromatic ring system found in the structure.

Some early work by Sahashi (4) and Weygand (5) showed that
5,6-dimethylbenzimidazole [2] could be incorporated as a unit into
the vitamin if it was added to growing cultures of producing
strains of *Streptomyces* species; indeed a whole range of cobala-
mine analogues can be obtained by a similar incorporation of a
variety of bases by B$_{12}$-synthesizing systems (6, 7, 36). In
particular, Perlman and Barrett (6) observed that 1,2-diamino-4,5-
dimethylbenzene [3], and its N-ribityl derivative, could be incor-

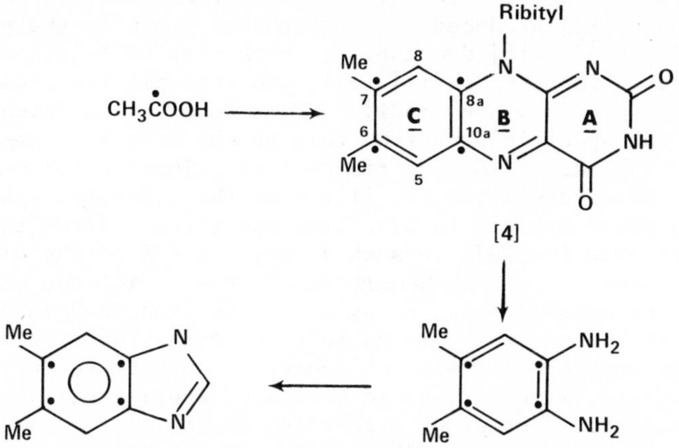

porated into vitamin B$_{12}$ by *Propionibacterium arabinosum*. In
1967, Renz and Reinhold (8), using variously ^{14}C-labeled lactate,
which is converted to acetyl-coenzyme A by propionic acid
bacteria, were able to show that the two methyl groups and the
carbons of the aromatic ring were ultimately acetate-derived;
although no evidence concerning the more intimate steps of the
incorporation was obtained, it was observed that the isotope
distribution pattern was the same as that found for the dimethylbenzene unit of riboflavin (see below), suggesting either that a
similar biosynthetic pathway operates, or that riboflavin might
possibly be a precursor of [2]. Further work by Renz (9) has
shown that uniformly labeled riboflavin can be metabolized to
active [2] by a cell-free preparation of *P. shermanii*. The same
result was obtained when ^{14}C-6-7-dimethyl-8-ribityllumazine [8],
the immediate precursor of riboflavin [4] (see below), was fed
instead of riboflavin itself.

In a comprehensive survey of possible precursors of B$_{12}$ in
Propionibacterium shermanii Alworth, et al. (10, 11) confirmed

the previously observed incorporation of acetate, pyruvate, and lactate, but showed that ribose-1-^{14}C and erythritol-U-^{14}C are even better precursors of the aromatic grouping; the authors postulated a pentose-tetrose intermediate as the most direct precursor of the aromatic ring of B$_{12}$. Forty percent of the label from ribose-1-^{14}C (45% total incorporation into dimethylbenzimidazole) was localized at C-2, indicating a probable role as specific precursor for this carbon atom also. In view of Renz's previous work (9), this result indicated that the ribityl side-chain of riboflavine might well provide this carbon atom.

Subsequent results from Alworth and Renz and their associates have largely clarified the biosynthetic route to [2]. Thus, isolation of [2] produced by *P. shermanii* grown in the presence of 1'-^{14}C, 5-^{15}N-[8] demonstrated both that C-1' is specifically incorporated into C-2 of [2] (34), and that all the atoms of [2] could be provided by [8] (32); similar feeding experiments demonstrated the specific precursor role of the methyl groups of [8] (30, 31, 33), and of C-1 of ribose (29). These results confirm the previous speculation (9, 11), that the 5,6-dimethylbenzimidazole [2] group present in B$_{12}$ forms specifically from 6,7-dimethyl-8-ribityllumanzine [8], in such a way that C-1 of the ribityl group becomes C-2 of the benzimidazole ring. All the atoms of [2] can thus be derived from [8] by a condensation of 2 moles of [8] similar to the one which leads to riboflavine [4]. It is still not clear whether [4] is an obligatory intermediate in the biosynthesis of [2], since [2] could be formed equally well by the condensation of 2 moles of either [4] or [8].

The biosynthesis of riboflavin (vitamin B$_2$) [4] has been studied by several groups; the accumulated evidence has been reviewed by Cheldelin and Baich (1967) (1), and Forrest (1962) (12). We are concerned here again only with the synthesis of the aromatic ring (C), which seems to be formed by an entirely unique process. No parallel for its biosynthesis has been found so far (except for that of vitamin B$_{12}$) and, significantly, the

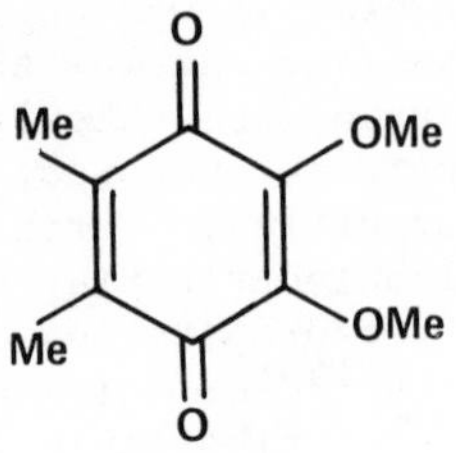

[5]

biosynthesis or aurantiogliocladin [5], which is structurally analogous, bears no resemblance to that of riboflavin.

Ring C of riboflavin can be formed from four acetate units; however, its formation is quite different from the cases discussed in Chapters 17 and 18, since here these units combine in a symmetrical fashion, as indicated in the diagram, rather than in the normal head-to-tail manner.

This fact was shown through the degradations carried out by Plaut (13, 14), which proved an almost exclusive labeling of positions 6, 7, 8a, and 10a from $1\text{-}^{14}C$-acetate.

Perhaps more significantly, the fungus *Eremothecium ashbyii*, which produces so large an amount of [4] that it forms crystals in the growth medium, incorporates $1\text{-}^{14}C\text{-}3$-hydroxy-2-butanone [6] (acetoin; possibly derived from pyruvate) into riboflavin with the expected labeling pattern (15). Masuda has suggested (16) that [6], which is produced by *E. ashbyii* (17), condenses with 4-ribitylamino-5-amino-2,6-dihydroxypyrimidine [7] to give 6,7-dimethyl-8-ribityllumazine [8], a compound which has been isolated from the cultures of several fungi (1) and other organisms (18), and which has been shown to be further metabolized to riboflavine by many biological systems (19). The probable biosynthesis

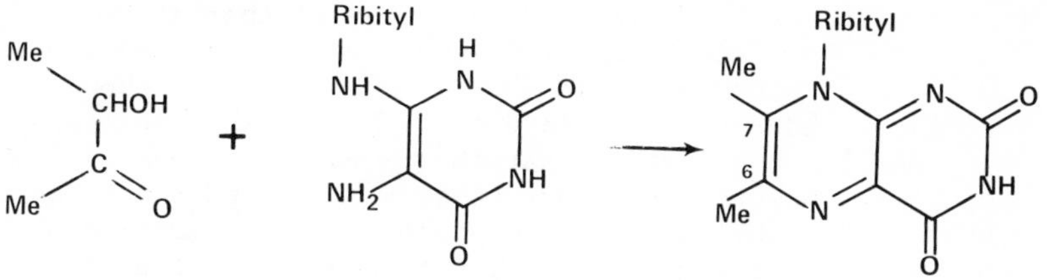

Me—CHOH
|
C=O
Me

[6]

+

Ribityl
|
NH   H
     N
NH ─ ... ─ O
NH₂   NH
  |
  O

[7]

→

Ribityl
|
Me   N
     N ─ O
  7
  6   NH
Me   N
         O

[8]

of [7] from a purine compound has been studied by several groups; the accumulated data are reviewed by Bacher and Lingens (35), but a discussion is considered to be outside the scope of this brief account. Plaut (19, 20) has shown that in the final reaction leading from [8] to the vitamin, two molecules of [8] react to give one molecule each of [4] and [7], proving that a C_4 unit is transferred from one lumazine to another. That 6,7-dimethyllumazine acts both as donor and acceptor of the carbon atoms has been demonstrated by its conversion into [4] by a cell-free system from *E. ashbyii* in the absence of an additional carbon source (16).

The reaction has been studied in a cell-free system from baker's yeast; the name riboflavin synthetase has been proposed for the enzyme involved (21). The second product [7] of the synthetase reaction was identified in the same study.

The details of this final step in the synthesis still remain obscure, although [8] can be converted into the vitamin by refluxing in phosphate buffer at pH 7.3 for 15 hours (22), and it is possible that this conversion may throw some light on the bio-chemical reaction. Wood (22, 23) has suggested that the chemical transformation proceeds via an opening of the pyrazine ring [10] →[11] after the nucleophilic attack of one precursor molecule [8] on the hydroxylated form of the second one [9] to form the adduct species [10]. The subsequent generation (see [11]) of the ketone and its ring closure through an aldol reaction would yield the vitamin and the expected product [7]. There is much sup-porting evidence for the idea that the above mechanism, or one very like it, operates in the biochemical system: besides the fact that it provides the correct stoichiometry and products, deuterium exchange studies (23) have shown that specifically deuteriated [8] is converted into [4] having the distribution of deuterium label which would be predicted from the postulated mechanism. Plaut (24) has also described the synthetase reaction in terms of the addition of 6,7-dimethyl-8-ribityllumazine to two separate binding sites on the enzyme surface, such that one site binds the substrate in such a way that it can act as the donor of the C_4-fragment, and the other so that it can act as an acceptor. In a subsequent investigation of the nonenzymic conver-sion of [8] into riboflavin, Beach and Plaut (25) showed that the reaction could also take place in 0.1M HCl; to accomodate this fact, and the observation that the 7-methyl protons of [8] readily exchange with deuterium under these acidic conditions, as well as at pH 7 (23), the authors have proposed a modification of Wood's scheme not involving a carbonyl intermediate in the final step for the enzymic transformation (26). We find this variant less appealing.

Whatever the ultimate details of the reaction sequence, it is clear that the C ring of riboflavin is built up of carbons 6 and 7 and the methyl substituents from one molecule of [8], together with the same carbons from another molecule. We have already quoted evidence in support of a unique acetate derivation for these carbon atoms (15), but not all the experimental findings are conclusive on this point. On the one hand Plaut (13, 14) showed in the early studies already quoted that acetoin was incorporated better than acetate, and it has been further demon-strated that a cell-free system from E. *ashbyii* can catalyze the condensation of acetoin with 6,7-dimethyl-8-ribotyllumazine (27). On the other hand, Ali and Khalidi (28) have re-investi-gated the incorporation of acetate, pyruvate, and various sugars into riboflavin, and conclude that neither acetate nor acetoin (from pyruvate) is a direct precursor of the xylene ring, and

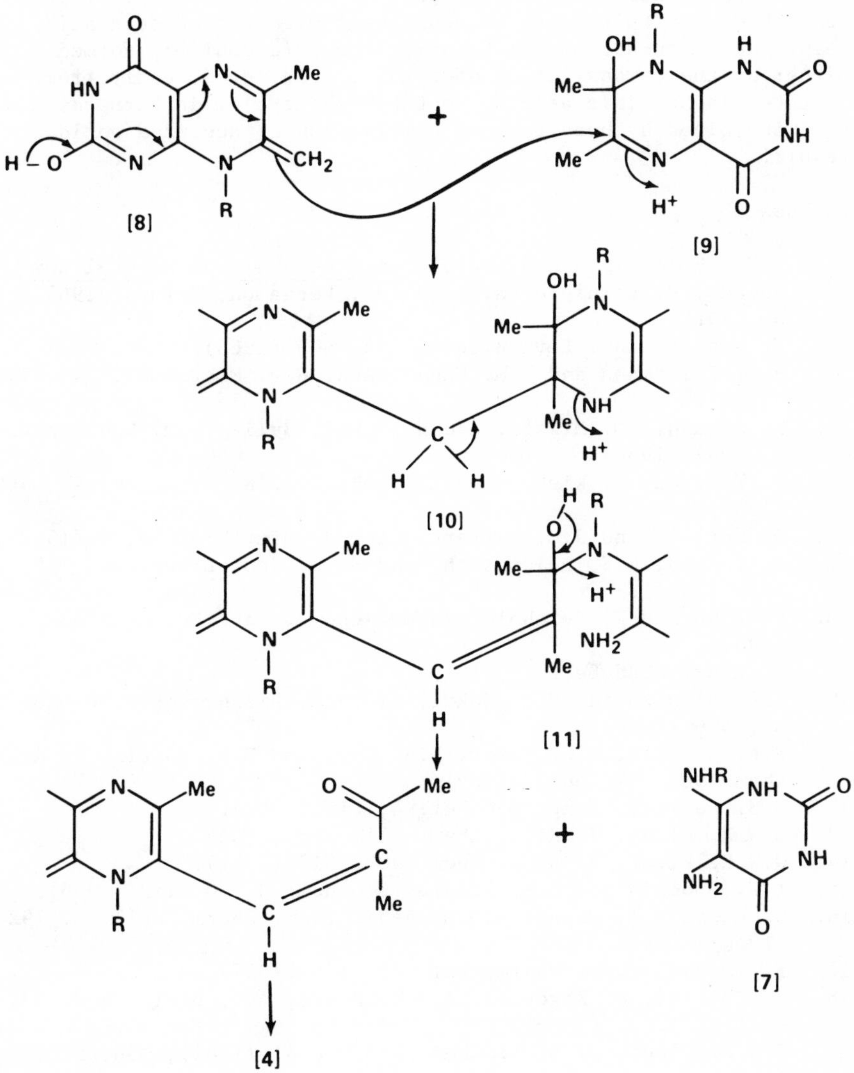
[8]
[9]
[10]
[11]
[4]
[7]

that the immediate precursor will probably prove to be a derivative of compounds found in the pentose phosphate cycle. The
published results seem to be consistent with oxaloacetic acid
being the precursor of the C_4 unit; the acid could be formed
either by the carboxylation of pyruvic acid, or directly from
acetate via succinic acid as in the Krebs cycle; in both cases a
C_4 unit joined by head-to-head combination of acetates would
result.

REFERENCES

1. V.H. Cheldelin and A. Baich, in Biogenesis of Natural Com-
    pounds, P. Bernfeld Ed., 2nd ed., Pergamon, Oxford, 1967,
    p. 679.
2. F. Wagner, Ann. Rev. Biochem. 35, 405 (1966).
3. H.C. Friedmann and L.M. Cagan, Ann. Rev. Microbiol. 24, 159
    (1970).
4. Y. Sahashi, M. Mikata, and H. Sakai, Bull. Chem. Soc. Jap.,
    23, 247 (1950).
5. F. Weygand, H. Klebe, and A. Trebst, Z. Naturforsch. 9b, 449
    (1954).
6. D. Perlman and J.M. Barrett, Can. J. Microbiol. 4, 9 (1958).
7. J.E. Ford, E.S. Holdsworth, and S. K. Kon, Biochem. J. 58,
    XXIV (1954).
8. P. Renz and K. Reinhold, Angew. Chem., Int. Ed. 6, 1083
    (1967).
9. P. Renz, FEBS Lett. 6, 187 (1970).
10. W.L. Alworth and H.N. Baker, Biochem. Biophys. Res. Commun.
    30, 496 (1968).
11. W.L. Alworth, H.N. Baker, D.A. Lee, and B.A. Martin, J. Amer.
    Chem. Soc. 91, 5662 (1969).
12. H.S. Forrest, Comp. Biochem. 4, 615 (1962).
13. G.W.E. Plaut, J. Biol. Chem. 211, 111 (1954).
14. G.W.E. Plaut, J. Biol. Chem. 208, 513 (1954).
15. T.W. Goodwin and D.H. Treble, Biochem. J. 70, 14P (1958).
16. S. Kuwada, T. Masuda and M. Asai, Chem. Pharm. Bull. 8, 792
    (1960).
17. T. Masuda, Chem. Pharm. Bull. 5, 136 (1957).
18. S. Konishi, K. Kageyama, and T. Shiro, Agr. Biol. Chem. 33,
    90 (1969).
19. G.W.E. Plaut, J. Biol. Chem. 235, PC 41 (1960); 238, 2225
    (1963).
20. G.F. Maley and G.W.E. Plaut, J. Amer. Chem. Soc. 81, 2025
    (1959).
21. H. Wacker, R.A. Harvey, C.H. Winestock, and G.W.E. Plaut,
    J. Biol. Chem. 239, 3493 (1964).

22, T. Rowan and H.C.S. Wood, Proc. Chem. Soc. 21 (1963);
J. Chem. Soc. (C), 452 (1968).

23. T. Paterson and H.C.S. Wood, Chem. Commun. 290 (1969);
J. Chem. Soc. Perkin I, 1051 (1972).

24. R.A. Harvey and G.W.E. Plaut, J. Biol. Chem. 241, 2120 (1966).

25. R.L. Beach and G.W.E. Plaut, Tetrahedron Lett. 3489 (1969).

26. R.L. Beach and G.W.E. Plaut, J. Amer. Chem. Soc. 92, 2913
(1970); Biochemistry 9, 760 (1970); G.W.E. Plaut, R.L. Beach,
and T. Aogaichi, ibid. 771 (1970).

27. T. Kishi, M. Asai, T. Masuda, and S. Kuwada, Chem. Pharm.
Bull. 7, 515 (1959).

28. S.N. Ali and U.A.S. al-Khalidi, Biochem. J. 98, 182 (1966).

29. W.L. Alworth, H.N. Baker, M.F. Winkler, and A.M. Keenan,
Biochem. Biophys. Res. Commun. 40, 1026 (1970).

30. S.H. Lu, M.F. Winkler, and W.L. Alworth, Chem. Commun. 191
(1971).

31. W.L. Alworth, S.H. Lu, and M.F. Winkler, Biochemistry 10,
1421 (1971).

32. S.H. Lu, and W.L. Alworth, Biochemistry 11, 608 (1972).

33. H.F. Kuehnle and P. Renz, Z. Naturforsch. 26b, 1017 (1971).

34. P. Renz and R. Weyhenmeyer, FEBS Lett. 22, 124 (1972).

35. A. Bacher and F. Lingens, J. Biol. Chem. 246, 7018 (1971),
and references there.

36. P. Renz and A.J. Bauer-David, Z. Naturforsch. 27b, 539 (1972).

CHAPTER 21

Mixed Acetate-Shikimate Biosynthesis

21-1. INTRODUCTION

The combination of a shikimate-derived C_6-C_3 group with a six-carbon polyketide chain, formed by the union of one acetate and two malonate residues, provides the general precursor for the large class of compounds known collectively as the flavonoids, as well as for several other smaller groups of metabolites. As shown in Scheme I, the polyketide chain may cyclize with loss of CO_2 to form a stilbene [1], or it may undergo ring closure and subsequent lactonization to an arylisocoumarin [2], but the most significant mode of ring formation is that leading to the chalcones [3], which are in turn the precursors of the whole gamut of flavonoid compounds.[+] The scheme gives an outline of the probable relationships between the flavonoid compounds as derived from studies based upon the incorporation of precursors, structural analysis, and some genetic and enzymic evidence; the structures in the scheme may not represent real compounds in all cases, but are presented to show the general structural types and their probable relation to each other. The oxygenation patterns of the two rings usually demonstrate their origin from acetate (ring A) and shikimate (ring B) quite clearly; however, all the rotenoids [4] and many isoflavones have the 2'-oxygen substituent, and ring B of the coumaranocoumarins [5] is apparently of the phloroglucinol type in some instances. The 2'-oxygen in the latter case, however, comes from the carbonyl group

---

[+]There is an interesting example of a bis-anthocyanin (36) which retains a carboxyl group at C-8; presumably the carboxyl of the terminal malonate residue of the polyketide.

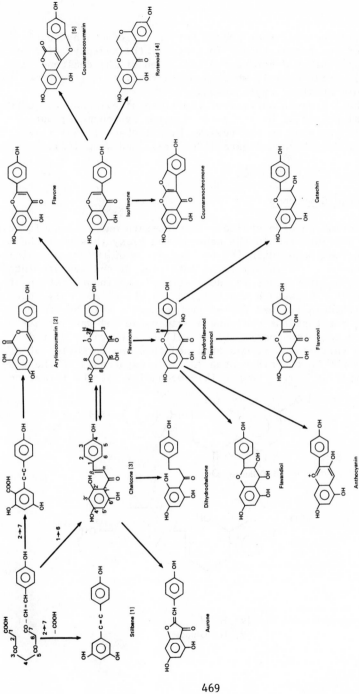

Scheme I. The Biosynthesis of Flavonoid Compounds

of an isoflavone precursor and is not an introduced hydroxyl
function of the normal type. In general, the hydroxylation of
ring A is established before the formation of the chalcone (i.e.,
at the polyketide stage), whereas hydroxylation of ring B may take
place before or after cyclization.

    We shall present sufficient experimental evidence to support
the general scheme of biosynthesis implied by Scheme I; more
detail will be found in the reviews by Grisebach (1965) (1),
(1967) (2), Grisebach and Barz (1969) (3), and Wong (43).

21-2. EXPERIMENTAL EVIDENCE

a. Flavonoid Compounds

    The formation of a chalcone from acetate and a cinnamic acid
was first demonstrated in the case of the biosynthesis of the
dihydrochalcone phloridzin [6] by leaf discs of *Malus*; phenylalanine was shown to be incorporated into ring B, and acetate into
ring A (4); a later study (5) demonstrated the incorporation of
cinnamic acid. Grisebach and Branden (6) showed not only that
isoliquiritigenin [7] was formed from cinnamic acid by *Cicer*
arietinum, but that the chalcone-4'-glucoside [8] was further
metabolized to formononetin [9] by cell-free extracts of red
clover; on this basis, it was proposed that the formation of a
chalcone might be the first step in the biosynthesis of the flavonoid compounds (7). The possible mechanism for the formation

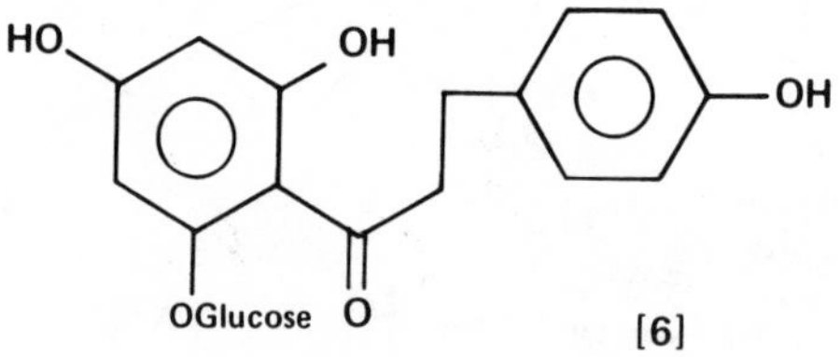

OGlucose O   [6]

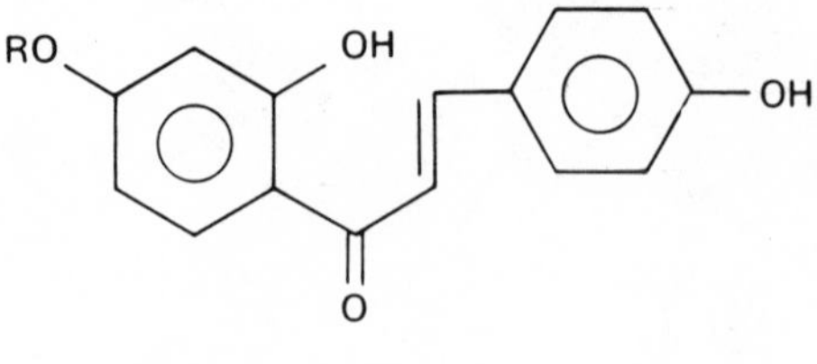

[7] R = H
[8] R = Glucose

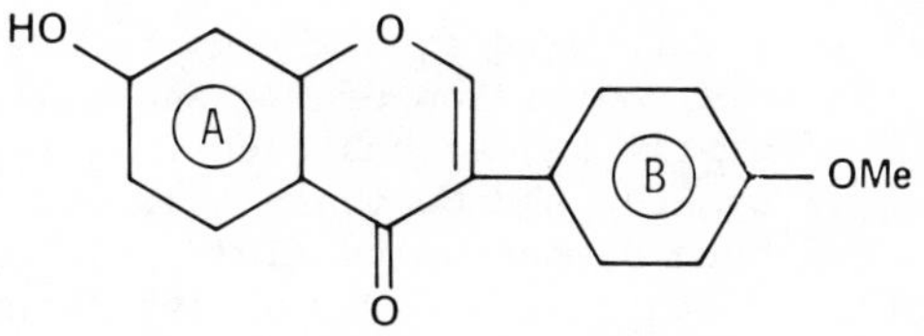

[9]

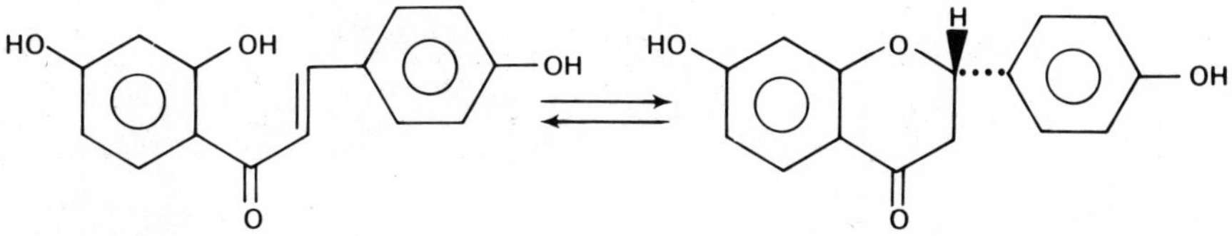

[7]   [10]

471

of chalcones from an activated cinnamic acid and three acetate units has been investigated (8); protein fractions from *Cicer arietinum* have been isolated which catalyze the activation of a variety of aliphatic acids and also of cinnamic and *p*-coumaric acids.

Flavanones and chalcones are isomeric compounds and there exists an enzyme-catalyzed equilibrium between them; an enzyme promoting the interconversion of [7] and (-)-liquiritigenin [10], and of other isomeric pairs, has been isolated from soy bean (*Soja hispida*) by Wong and Moustafa (9), and the reversibility of the system has been demonstrated (10). Similar isomerases have been isolated by other workers, who have generally found them to possess a distinct substrate-specificity, which is by no means always the case with plant enzymes (47, 48). The equilibrium between chalcone and flavanone raises the question of the flava- nones as true intermediates in flavonoid biosynthesis, as implied in the scheme. Grisebach has shown that the naturally occurring (-)-naringenin-5-glucoside [11] is incorporated into quercetin [12] and cyanidin [13] by buckwheat seedlings at a much higher

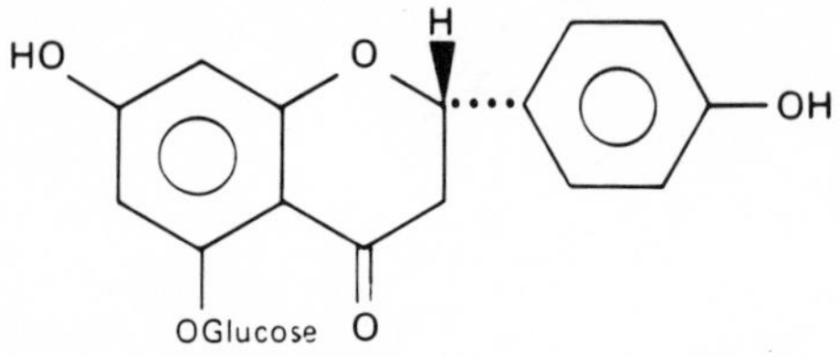

[11]

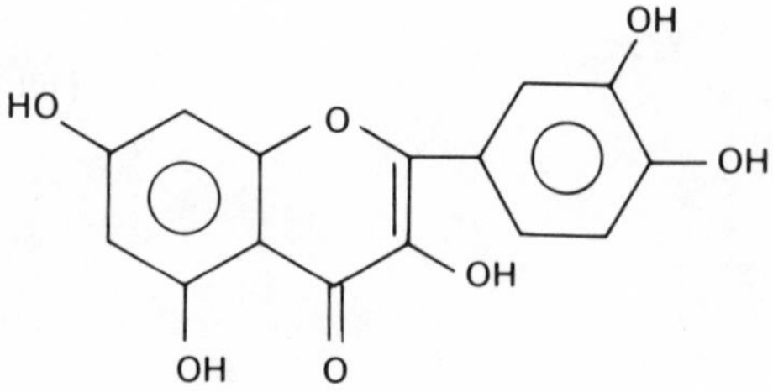

[12]

[13]

rate than the antipodal (+)-flavanone glucoside (11). On the
basis of this result, he placed the flavanone after the corresponding chalcone in the biosynthetic sequence leading to the
flavones, on the grounds that the (+)-flavanone could equally
well yield the chalcone, whereas the chalcone would reasonably
yield the (-)-epimer on ring closure *in vivo*, since all the known
(-)-flavanones have the same *S*-configuration at C-2 (12). That,
however, the flavanone may be enzymically converted to the chalcone (most readily in the case of the (-)-enantiomer) and thence
to the other flavonoids becomes a possibility in view of Wong's
work (10), and it has been postulated (10) that the biosynthetic
sequence might better be presented as in Scheme II. This hypothesis is supported by the relative incorporations of ^{14}C-chalcone

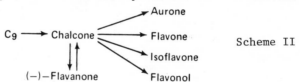

Scheme II

diluted with an equal amount of (-)-flavanone on the one hand, and
^{14}C-(-)-flavanone diluted with an equal amount of chalcone on the
other, into flavonoid compounds by cell-free extracts of *Cicer
arietinum* and red clover. The incorporation was higher in the
former case, indicating the chalcone as the more immediate
precursor (10). In a similar experiment, tritium-labeled
4,2',4'-trihydroxychalcone-^{14}CO and (-)-7,4'-dihydroxyflavanone
were fed together to *Trifolium subterraneum* and *Cicer arietinum*,
to give flavonoid compounds whose T/^{14}C ratio again indicated
that the chalcone was the more direct precursor of the two (13).

In one of the few studies with cell-free systems to have
been reported so far, the formation of naringenin (the aglycone
of [11]) has been found to take place from malonyl-CoA and
p-coumaryl-CoA by an enzyme preparation obtained from parsley
(45).

The aurones represent a small group of compounds related
to the flavones; their occurrence is limited to very few plant

families (Compositae, Leguminosae). The aurones very often
coexist with the corresponding chalcone (e.g., in the Compositae),
and evidence is accumulating for their biosynthesis from the
latter compounds. Chalcones can be transformed into aurones by
oxidation under conditions that could exist in the plant: For
example, okanin [14] is converted smoothly into maritimetin [15]
in an aerated alkaline solution (14), and a similar interconversion has beeen achieved by an extract from the flowers of
Coreopsis lanceolata (15). Wong (16) has isolated hispidol [16]
together with [7] from soybean seedlings, and has shown that
incubation of ^{14}C-[7] with extracts of the seedlings incorporated
radioactivity into the aurone and its 6-glucoside. It is interesting that the conversion of chalcone to aurone can also be

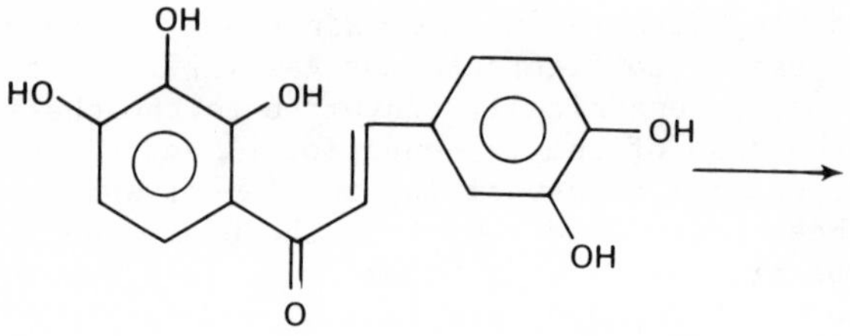

[14]

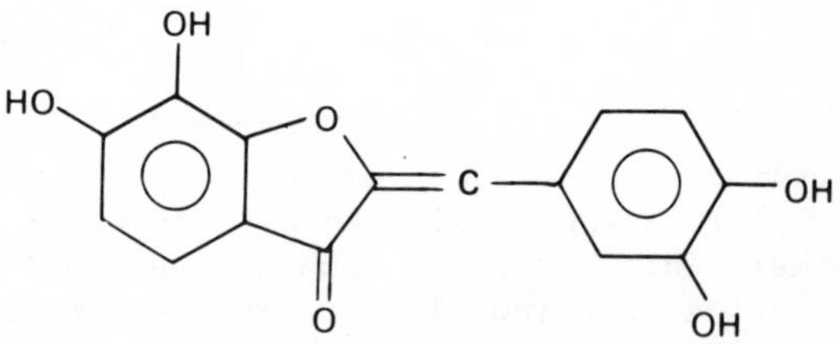

[15]

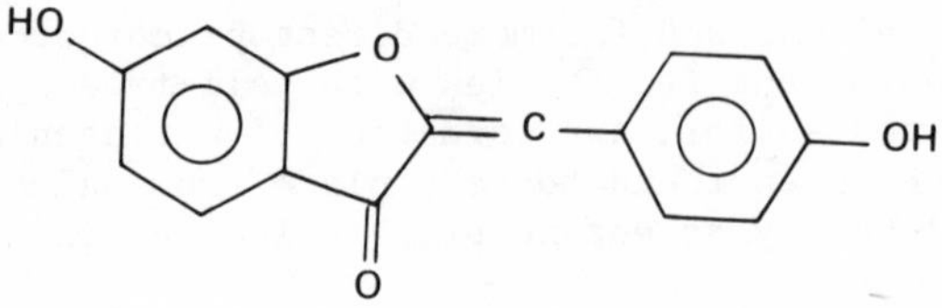

[16]

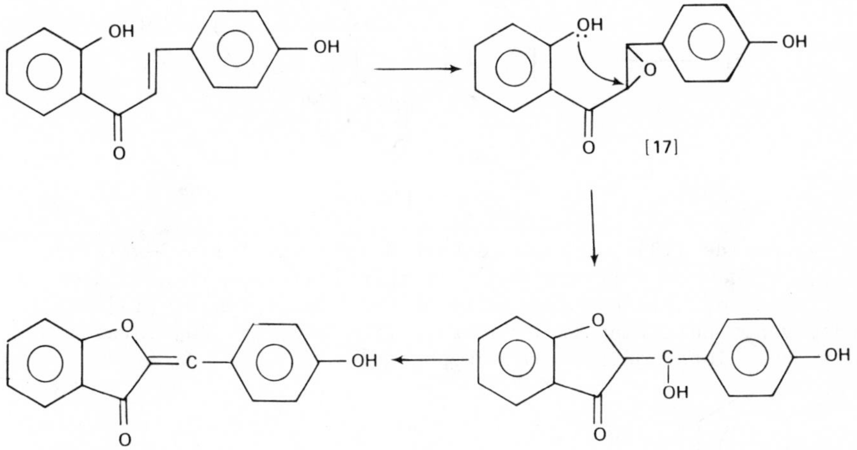

[17]

achieved by oxidation with H_2O_2 (17); if this reaction proceeds via an epoxy intermediate, as it might, it provides a model for one of several instances of an enzymic reaction which could involve a chalcone epoxide [17].

All dihydroflavonols found so far have the stereochemistry shown below; that is, the 2-phenyl ring and the 3-hydroxyl group are *trans* to each other. Laboratory conversions of chalcones to dihydroflavonols via chalcone epoxides [17] (18) lead to the required *trans*-stereochemistry, whereas free-radical oxygenation with, for example, Fenton's reagent, leads to a *cis*-arrangement.

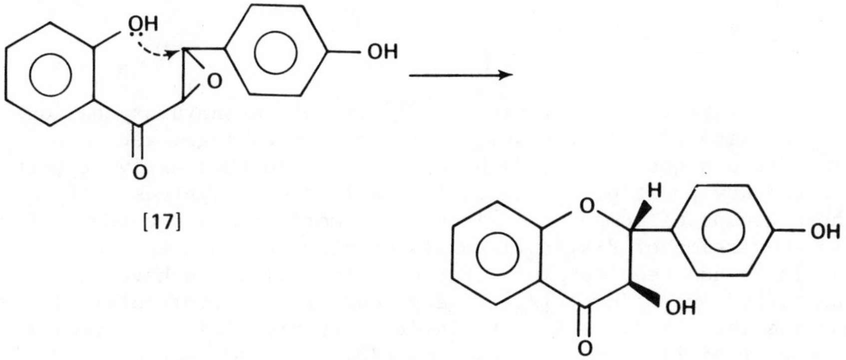

[17]

Hence the possibility of an epoxide type of intermediate exists. Wong (19) has shown that the chalcone isoliquiritigenin [7] is converted into garbanzol [18] by a cell-free extract of *C. arietinum*. In a similar experiment the tetrahydroxychalcone-

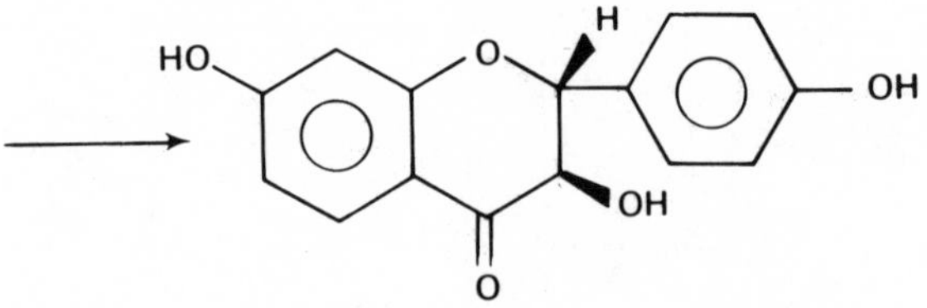

[18]

2'-glucoside [19] has been converted into taxifolin-3-xyloside [20] by *Chamaecyparis obtusa*. In this last experiment, the retention of tritium originally at the β position of [19] was observed, indicating that pathways from [19] to [20] involving a flavone intermediate are not possible.

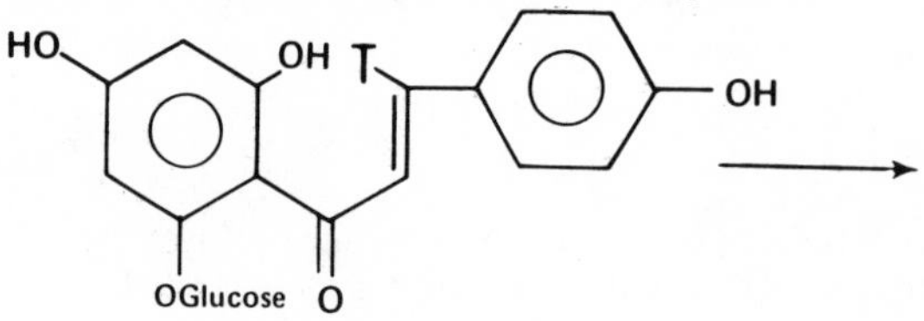

[19]

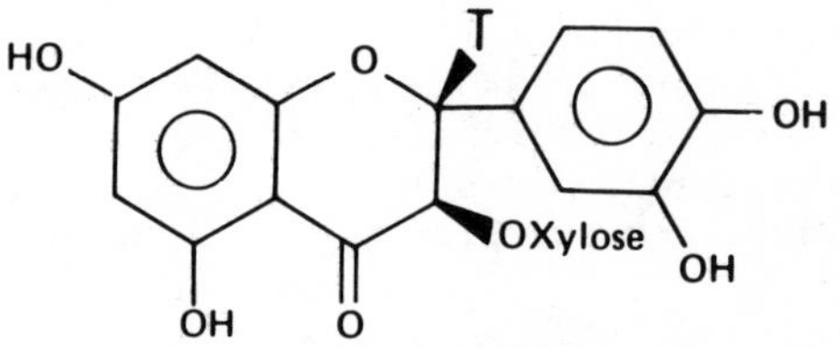

[20]

The intact incorporation of 2',4,4',6'-tetrahydroxychalcone-2'-glucoside [21] into cyanidin [13] by red cabbage, and into cyanidin and quercetin [12] by buckwheat, provided early support for the key position of the chalcones in the biosynthesis of these groups of compounds; since then, more detailed evidence for the conversion of dihydroflavonols to flavonols and anthocyanins in plants has been obtained. Patschke and Grisebach have shown that dihydrokaempferol [22] is a precursor of kaempferol [23] and quercetin [12], and that dihydroquercetin [24] is a specific precursor of [12] in *Pisum sativum* (20); the same authors have also demonstrated that [22] but not [23] is a good precursor of [12] and cyanidin in buckwheat seedlings and in cell suspensions of *Haplopappus gracilis* (52). These experiments are all compatible with the outline shown below. In the laboratory, the conversion

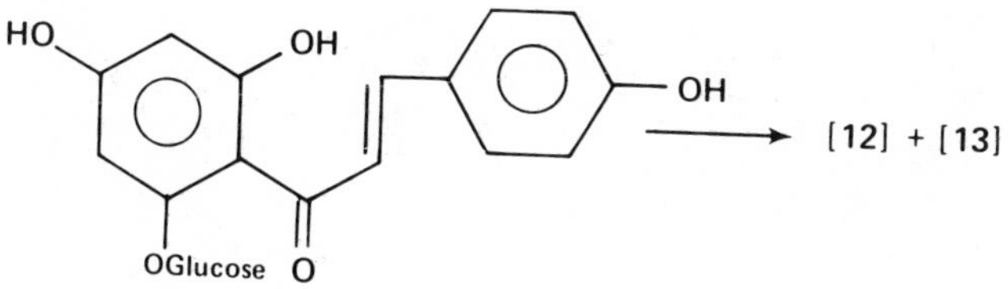

[21]

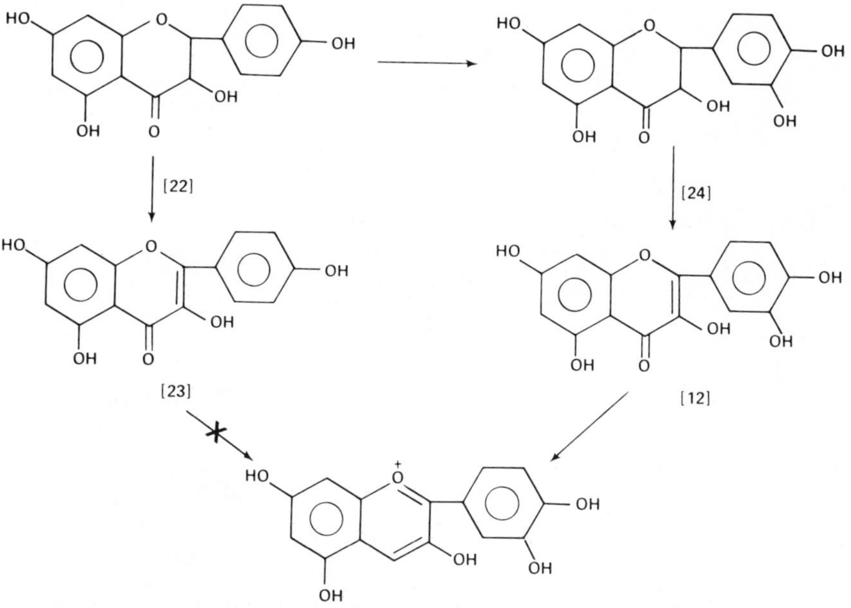

of dihydroflavonols to the corresponding flavanols can be achieved by simple dehydrogenation with iodine and potassium acetate, or alkaline ferricyanide; for example, taxifolin-5,7,3',4'-tetramethyl ether [25] is easily dehydrogenated by these methods (21). Treatment of the dihydroflavanol with acetic anhydride and sodium acetate, followed by alcoholic hydrochloric acid, converts it to the anthocyanin (21). These simple reactions may give an indication of the corresponding enzymic processes (22); however, in the case of the anthocyanins, which in many cases require light for their synthesis, the reactions may well be more complex (23).

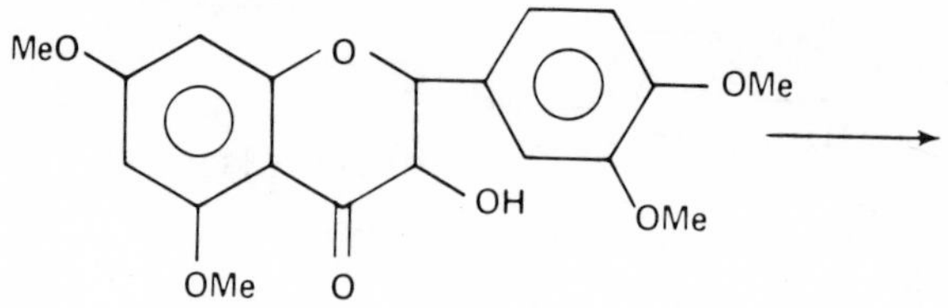

[25]

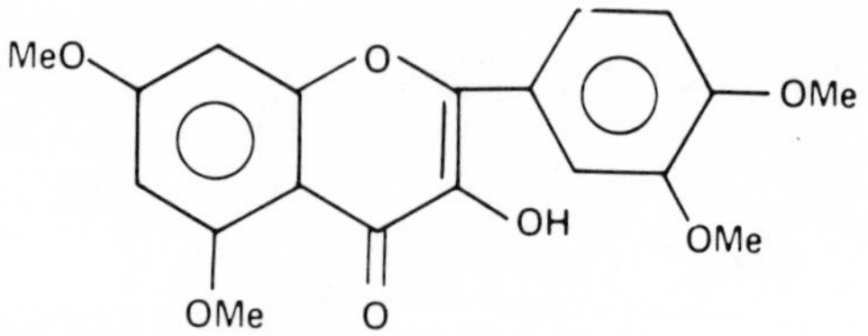

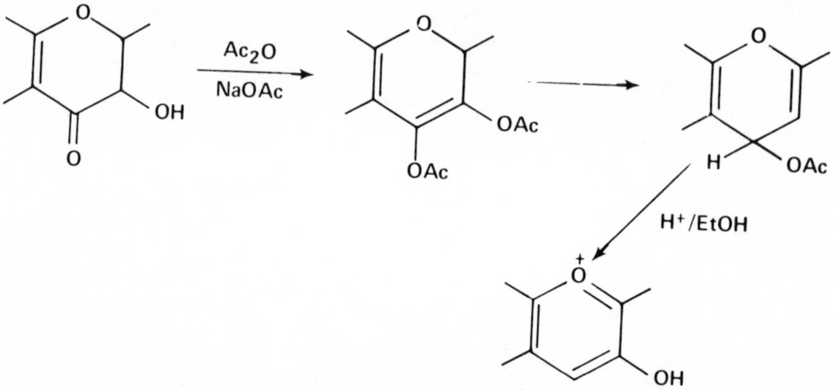

Some early work on the synthesis of catechin [26] in plants established the expected shikimate derivation of ring B, and acetate incorporation into ring A (24). The formation of [26] labeled at C-2 from β-^{14}C-cinnamic acid was also demonstrated (25). More recently Patschke and Grisebach (26) have shown that the chalcone glucoside [27] is incorporated into catechin and epicatechin by tea leaves (*Camellia sinensis*). The intermediacy of the corresponding flavanone and the dihydroflavanol dihydro-kaempferol has been demonstrated (49, 50).

Many plant extracts contain colorless products which yield anthocyanins on treatment with mineral acid. This group of compounds (pro-anthocyanidins) includes 2,3-flavandiols, 3,4-diols, 2,3,4-triols, and other, more complex structures (42). For some time the 3,4-diols were regarded as the probable precursors of

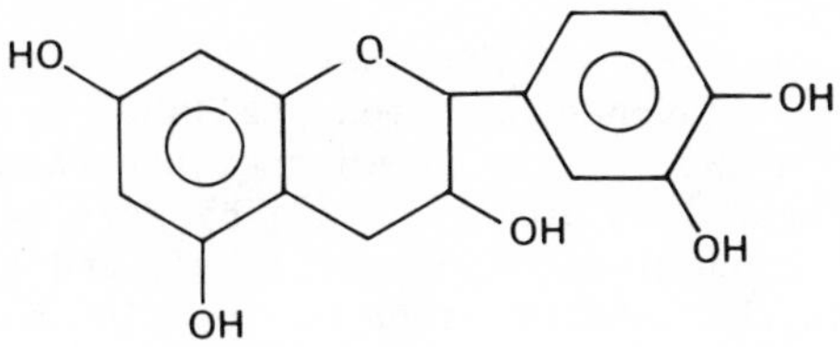

[26]

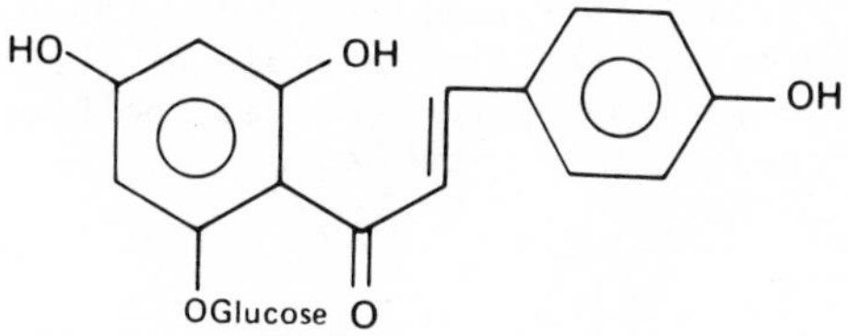

[27]

the anthocyanins, although treatment of synthetic 3,4-diols, or
the naturally occurring colorless leucocyanidin [28], with

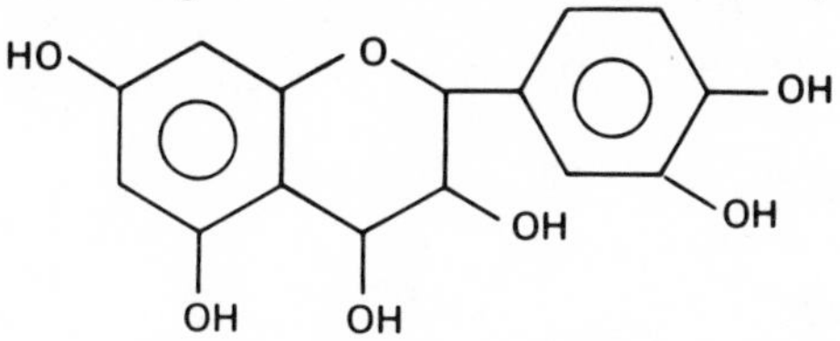

[28]

mineral acids, gives poor conversion to anthocyanin (27). There
is no good evidence to suggest the intermediacy of the flavandiols
in any biosynthetic scheme at present, and it seems likely that
they, like the catechins, are metabolic products that are ultimately incorporated into the condensed tannins by oxidative
polymerization.
     The tannins are a group of complex polyphenolic substances
built up by the condensation of simple phenolic substances (acids
and flavonoids) which are ubiquitous in the plant kingdom. Their
function is not known, although it is unlikely that they are metabolic waste-products; they may play a protective role during the
growth of the plant. The tannins are usually noncrystalline
compounds with astringent properties and the ability to precipitate metal salts and proteins (tanning action). For a review
of those tannins which are derived from flavonoids, see ref.
27A.
FLAVONES: Apart from some earlier work which showed the incorporation of phenylalanine, acetate, and cinnamic acid into fla-

vones (for which see general references), the definitive work on
the biosynthesis of flavones has been published by Grisebach and
his associates. In 1967 it was shown that the chalcone glucoside
[27] was incorporated into apigenin [29] by parsley (*Petroselinum
hortense*) without randomization of the label, and that in competitive experiments the incorporation of the flavanone prunin [30]
into [29] and chrysoeriol [32] is greater than that of the
flavanol [33] (28).
    Subsequently, a rather complete understanding of the enzymic
processes that occur in the biosynthesis of the 7-apiosylgluco-
sides of [29] and [32] has been achieved; both of these flavone
glycosides are present in *P. hortense*.

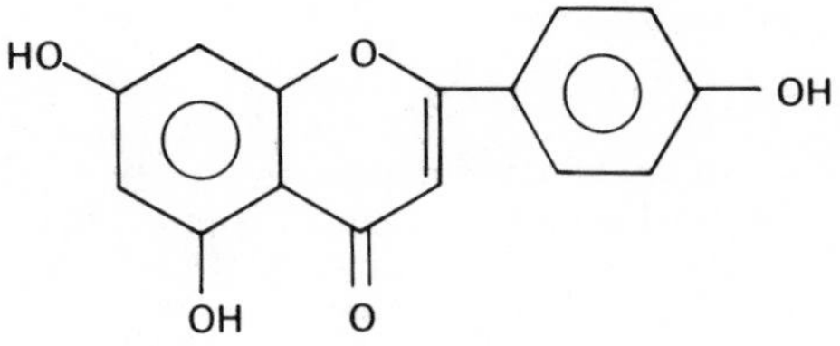

[29]

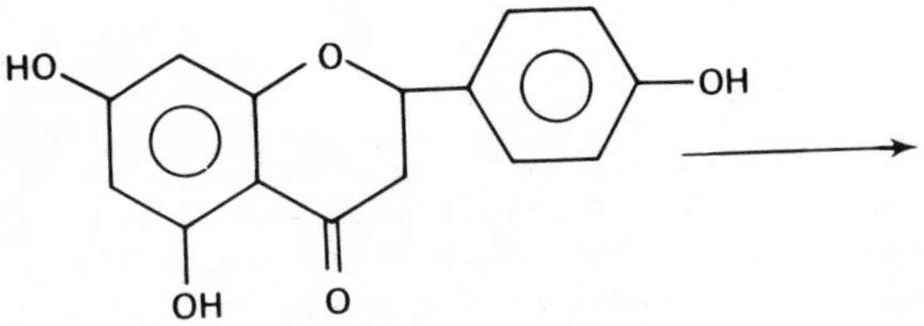

[30]

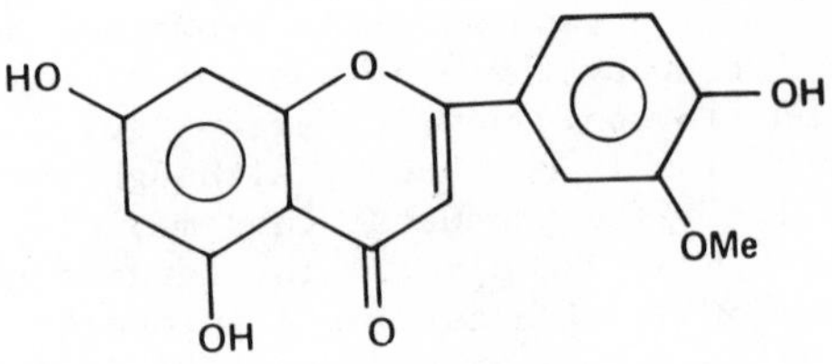

[32]

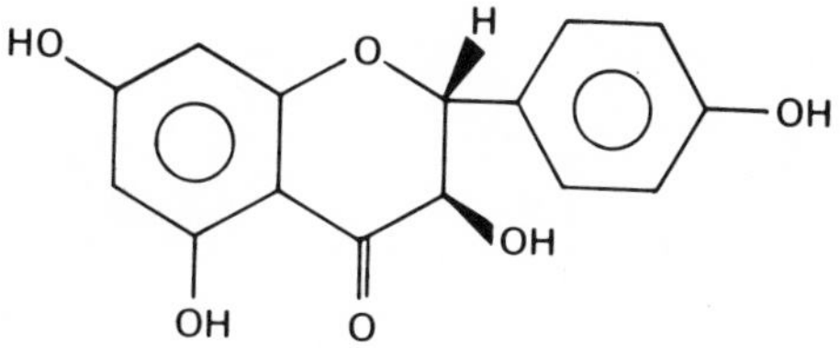

[33]

The ready conversion of prunine [30]-7-glucoside to cosmosiine ([29]-7-glucoside) in flowers of *Cosmos bipinnatus* has also been observed (41). These results indicate that here, as in the case of the isoflavones, the biosynthesis proceeds through dehydrogenation of a flavanone rather than by dehydration of a flavanol. An interesting flavone, chloroflavonin, containing a unique 3'-chloro group, has been isolated from the fungus *Aspergillus candidus*. A biosynthetic study (51) has demonstrated *de novo* synthesis from shikimate precursors, and has also shown that the chlorine atom is introduced into the aromatic ring after the flavonoid skeleton has been established.

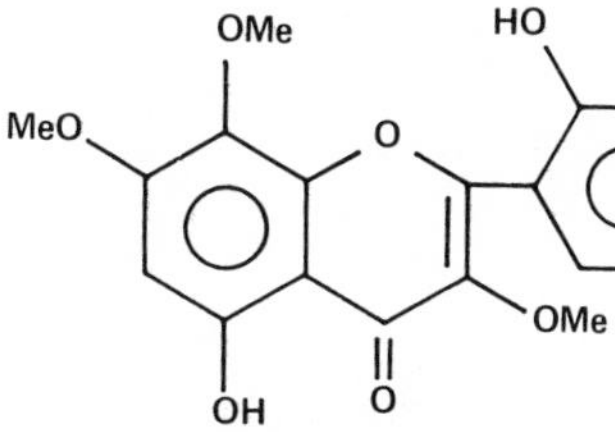

Chloroflavonin

b. Isoflavonoid Compounds

The incorporation of cinnamic acid, phenylalanine, and acetic acid into the isoflavones formononetin [31] and biochanin A [34] by *Cicer arietinum* exactly parallels their incorporation into the flavone series of compounds; in particular, the incorporation of phenylalanine labeled at C-1, C-2, or C-3 shows that the biosynthesis involves a migration of ring B from C-2 to C-3 at some stage (29). More information about these incorporations, and of the stage at which the oxygenation pattern is established were obtained by the administration of isoliquiritigenin-4'-glucoside [35] and 2',4,4',6'-tetrahydroxychalcone-4'-glucoside [36] to *C. arietinum*, when it was found that these compounds were specific precursors of [31] and [34], respectively, but not *vice versa* (30). Finally, although isoliquiritigenin is

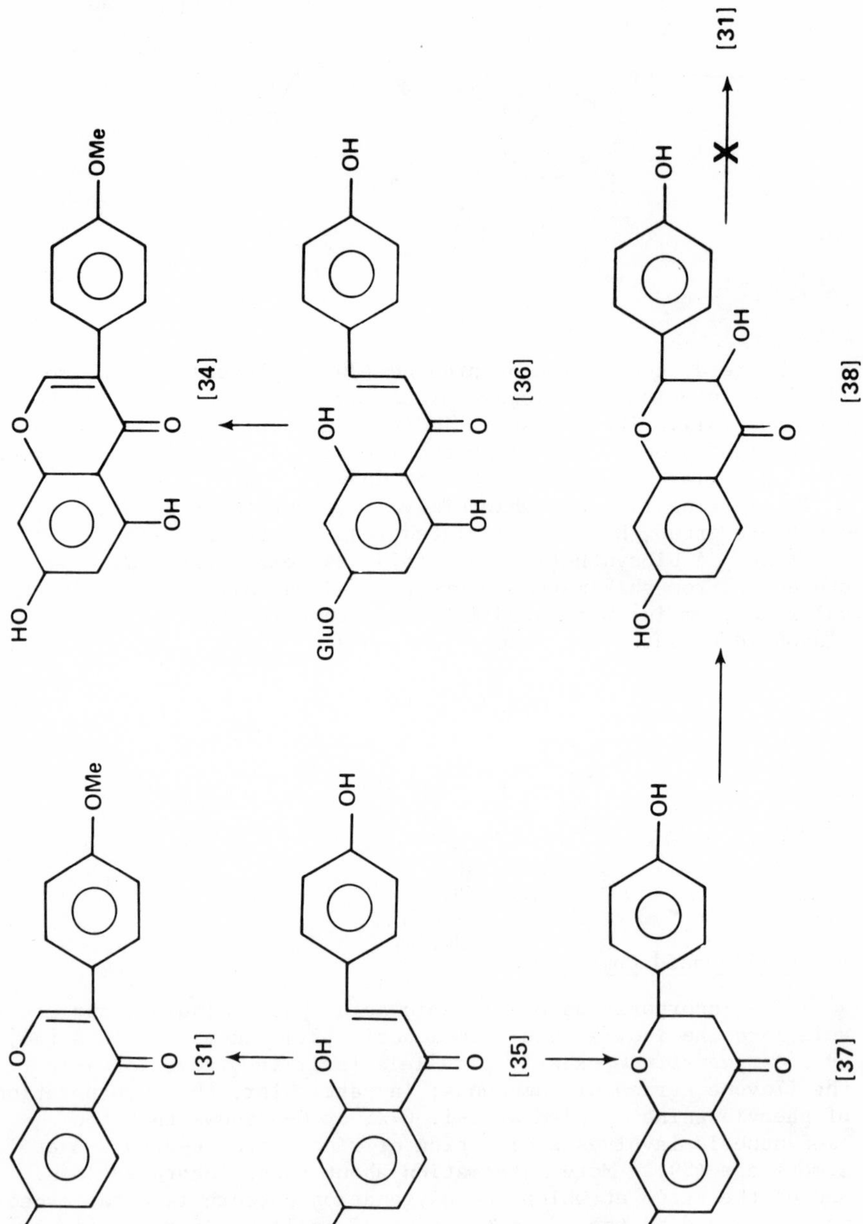
[31]
[34]
[35]
[36]
[37]
[38]
[31]
HO
OH
OMe
GluO
O

incorporated into the flavanone [37], garbanzol [38], and into formononetin by cell-free extracts of the same plant (19), [38] cannot act as a precursor of [31], either *in vivo* or in cell-free extracts. This implies that the biosynthesis of isoflavones proceeds through the dehydrogenation of flavanones, and not by dehydration of dihydroflavonols, and that the rearrangement of the aryl ring from C-2 to C-3 takes place at the flavanone stage. This assumption has been supported by an experiment in which tritium-labeled dihydrokaempferol [39] was fed to seedlings of *Cicer arietinum* both with and without the addition of ^{14}C-phenyl-alanine; the incorporation of [39] into the isoflavones was found to be insignificant (31).

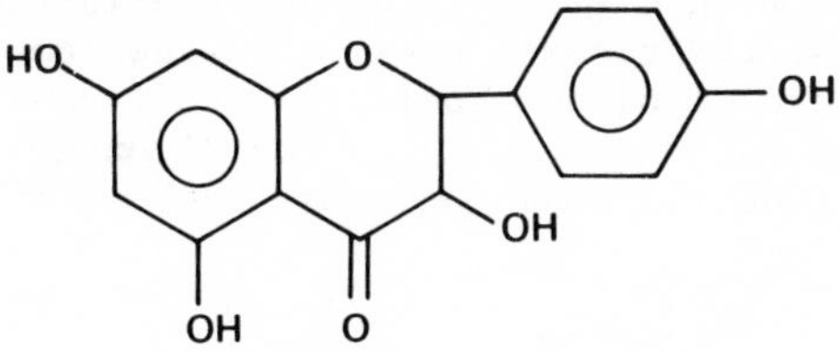

[39]

An hypothetical mechanism for the 2,3-aryl shift can be envisaged, and while it must remain speculative, other proposed mechanisms for the formation of isoflavones via chalcone epoxides, and from 3-hydroxyflavanols, should be discounted in view of the

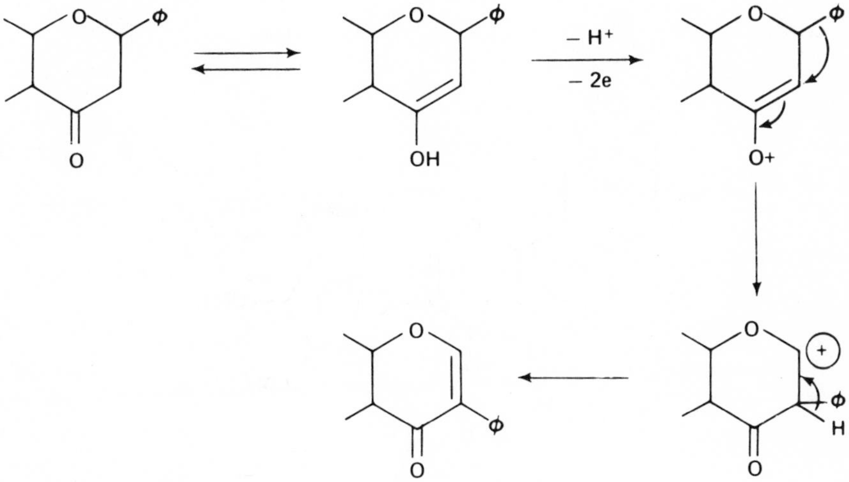

recent experimental evidence. For the possible influence of a
2'-hydroxyl function on this rearrangement, see Section 21-3.

Interestingly, the *in vitro* conversion of 2'-hydroxychalcones
to isoflavones has been achieved by rearrangement of the chalcone
with thallium nitrate and subsequent acid-treatment (44).

The biosynthesis of the coumaranocoumarins (coumestans) has
been considered in conjunction with that of the coumarins
(Chapter 16); it need only be restated here that they are derived
from the isoflavones as immediate precursors by an oxidative
mechanism as yet not fully elucidated.

It has been pointed out by Grisebach and Ollis (32) that
there is good circumstantial evidence for a close biosynthetic
relationship between the rotenoids and the isoflavonoids. In
some cases these two classes coexist in the same plant: for
example, toxicarol [40] and toxicarol isoflavone [41] are both
found in *Derris malaccensis*. Also the 2'-oxygen function of
the rotenoids (and the coumaranocoumarins) is mirrored by its

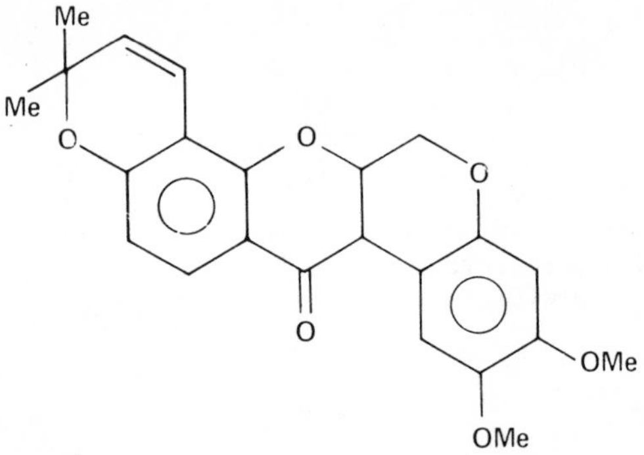

[40]

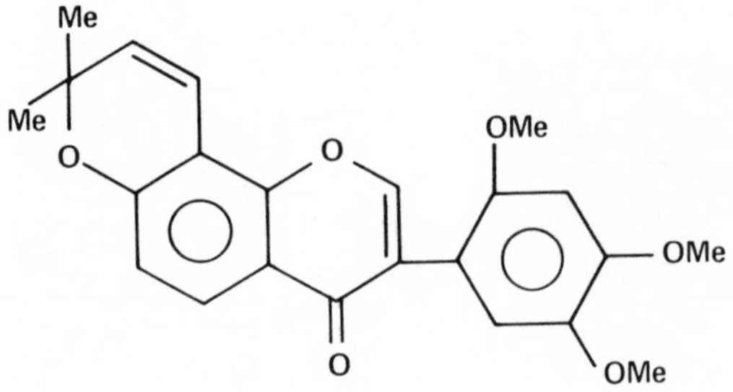

[41]

relatively frequent occurrence in the isoflavonoid but not the
flavonoid compounds.

The first direct evidence for the relationship was provided
by Crombie and Thomas (33), who fed phenylalanine labeled at the
C-1 or C-2 position to *Derris elliptica* plants and showed, by
degradation of the active rotenone [42] so formed, that the C-1
carbon became C-12 of the rotenoid, and that C-2 of phenylalanine
gave rise to C-12a. Further work by the same authors (34) and
by others (35) has shown that 3-^{14}C-phenylalanine labeled [42]
at carbon 6a, and that methyl-^{14}C-methionine provided C-6 and

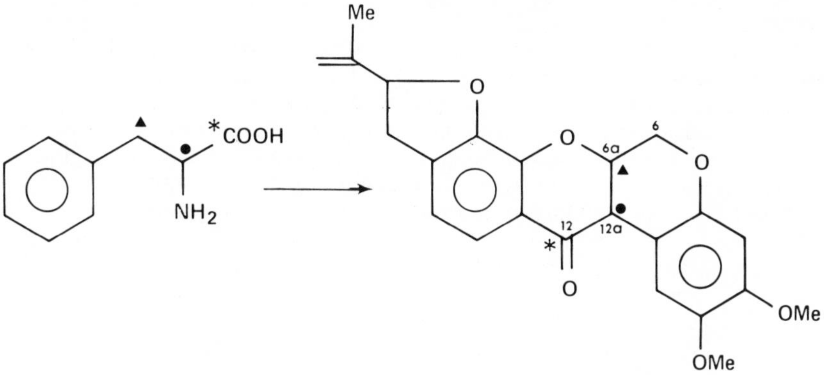

[42]

the two methylether groups. These results indicate that rotenoid
biosynthesis proceeds via a (6a⟶12a)aryl shift, as in isoflavone biosynthesis; it is likely, therefore, that the two pathways
are identical at least as far as the flavanone rearrangement
step, although the coexistence of dolineone [44] and neotenone
[43] (33) suggests the possibility that ring closure and oxidation
might take place at the isoflavanone stage.

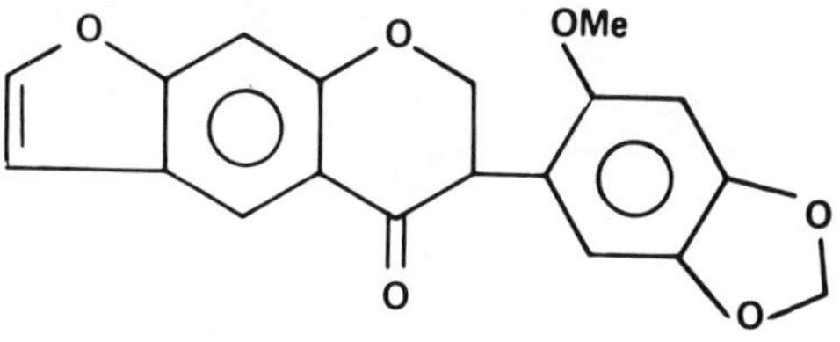

[43]

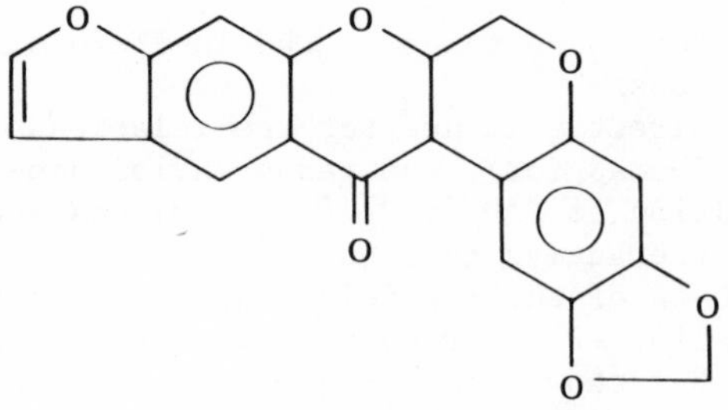

[44]

Subsequent studies have been made by Crombie and his associates
(40), who have studied the biosynthesis of amorphigenin [49] in
germinating seeds of *Amorpha fruticosa*; the isoflavone [47] was
readily incorporated, whereas the corresponding isoflavanone was
not, and the rotenoid normunduserone [48] was established as a
key intermediate.   Subsequent reactions involve isoprenylation,
cyclization, and further modification of the side chain.

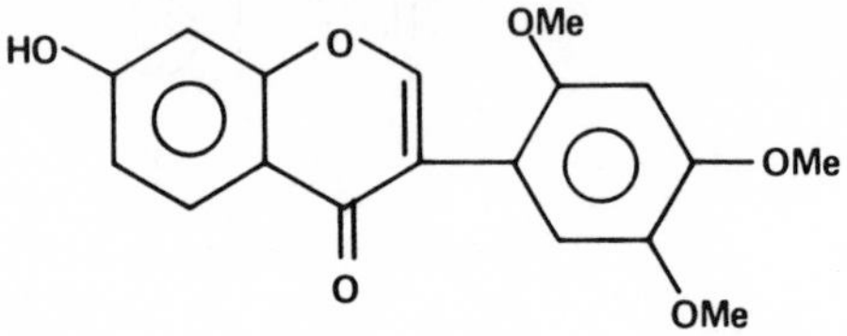

[47]

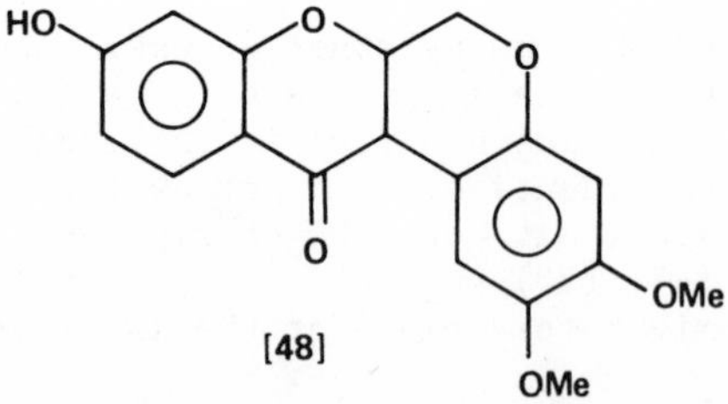

[48]

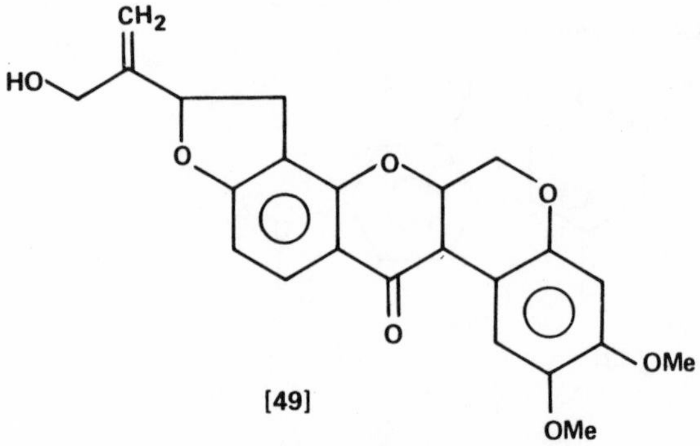

[49]

Pterocarpans, a relatively small group of natural products, are structurally related to the isoflavanoids. A typical representative of the class is phaseolin [45], isolated (37) from *Phaseolus vulgaris*; the compound lacks the C-5 hydroxyl group which is ubiquitous in other flavonoids, thus demonstrating a common feature of the pterocarpans. Another interesting characteristic of these compounds is the occurrence of an oxygen function at C-8, as in [46] from *Swartzia madagascariensis* (38). Biosynthetically, the pterocarpan ring system is possibly derived from that of the corresponding isoflavanone by an oxidative ring closure involving the enol form of the carbonyl group; however, in view of the fact that nearly all of the known isoflavanones have a 2'-hydroxyl group, this may be less likely than a dehydration involving the hydroxyl function. Hess and co-workers (39)

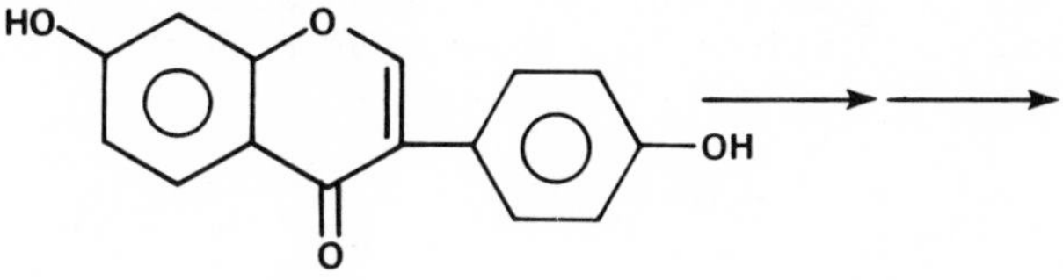

Daidzein

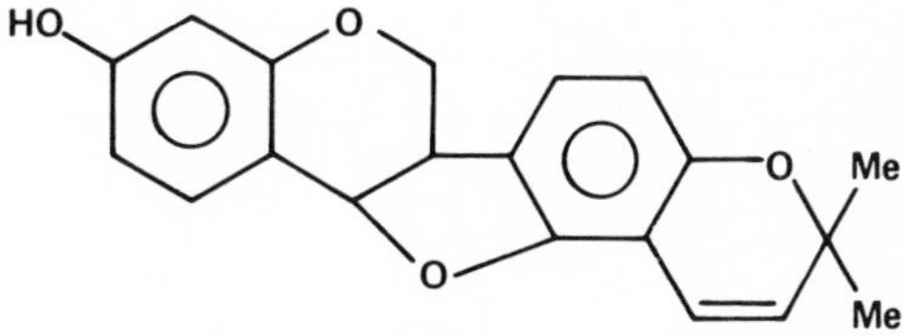

[45]

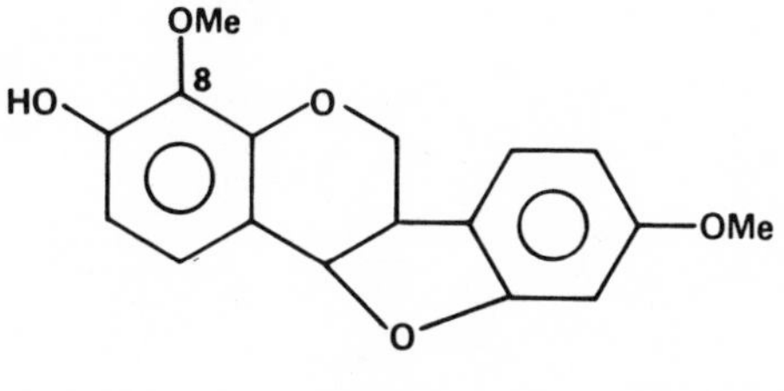

[46]

have investigated the biosynthesis of [45], and have observed
the incorporation of phenylalanine, cinnamic and acetic acids,
and daidzein; interestingly, no incorporation of mevalonic acid
was observed, and the authors suggest that ring D may not be of
isoprenoid origin. At first glance this seems very unlikely.

21-3.   HYDROXYLATION OF THE FLAVONOID COMPOUNDS

If the chalcones (or flavanones) can be accepted as the key
intermediates in flavonoid biosynthesis, then the question arises
whether chalcones with different patterns of oxidation occur as
precursors of the various flavonoids, or whether all flavonoids
are derived from a common intermediate.

We have seen, for instance, that the chalcone [50] is incorporated intact into cyanidin and quercetin as well as kaempferol,
which implies that the 3'-hydroxyl group is introduced *after*
the assembly of the $C_6-C_3-C_6$ intermediate. Vaughan et al. (53)
have shown that enzyme preparations from spinach beet which

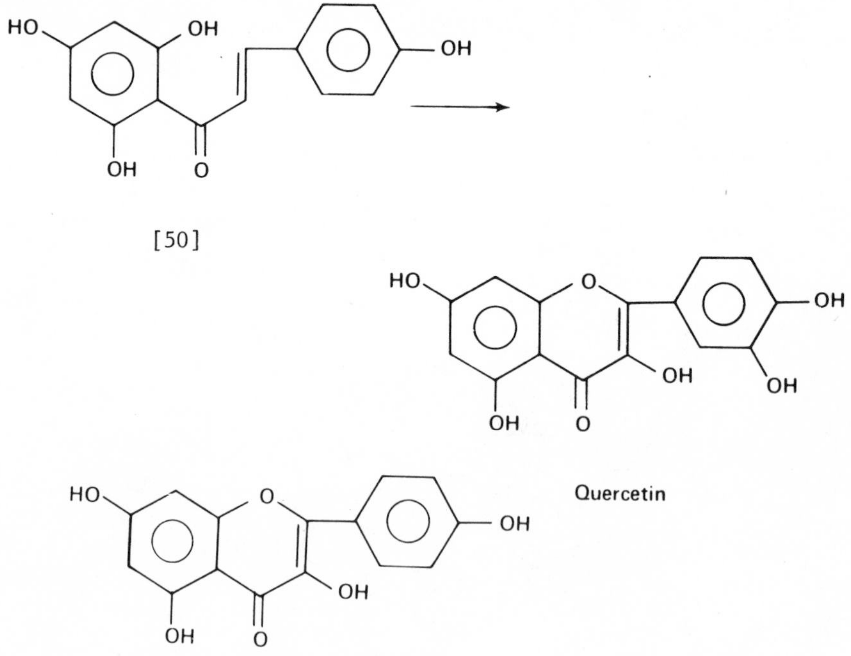

[50]

Quercetin

Kaempferol

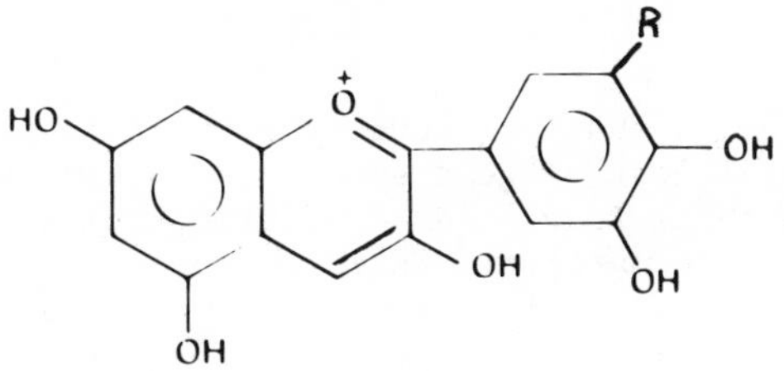

Cyanidin R=H
Delphinidin [53] R=OH

catalyze the conversion of *p*-coumaric to caffeic acid can also
bring about the 3'-hydroxylation of some 4'-hydroxylated flavonoids; naringenin [51], dihydrokaempferol, and kaempferol are

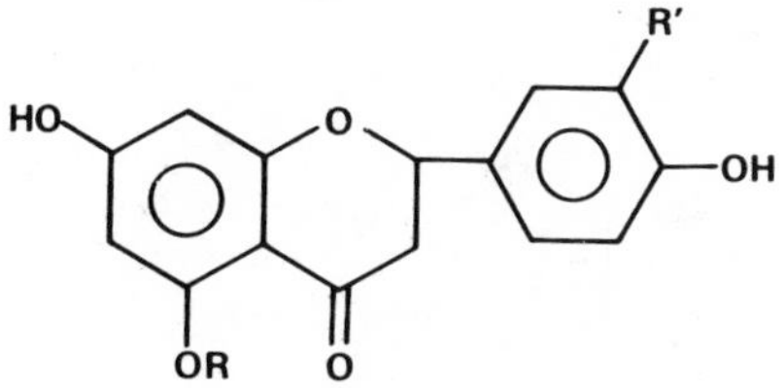

[51] R=H; R'=H
[52] R=H; R'=OH

converted to eriodictyol [52], dihydroquercetin, and quercetin by
the enzyme. Besides showing a high specificity for the hydroxylating system, this result also implies once more the possibility
of a late hydroxylation in flavonoid biosynthesis. Similar evidence for a late hydroxylation comes from work on the flavonoids
of *Datisca cannabina* (61); the plant elaborates 2',3,5,7-tetra-
hydroxyflavone(datiscetin), but does not incorporate *o*-coumaric
acid into the metabolite. It is thought that hydroxylation
takes place at the flavanone or flavanonol stage, because the
corresponding flavone without the 2'-hydroxyl could not be incorporated.

There is, however, abundant evidence that preformed cinnamic
acids can be directly incorporated in other plants. Thus Meir
and Zenk showed (54) that 3,4-5trihydroxycinnamic acid was a
better precursor than either *p*-coumaric or caffeic acids for the
formation of delphinidin [53] by *Campanula media*, and the same
implication, that is, the incorporation of a specific cinnamic
acid into the C_{15} intermediate, has come from work by Hess,
concerned with direct synthesis of anthocyanins from 5-hydroxy-
ferulic acid (55). Steiner (62) has followed the incorporation
of sinapic acid into a variety of anthocyanins by *Petunia hybrida*,

and has established that, although the cinnamic acid is incorporated into several less oxygenated compounds, it is most efficiently
converted into those flavonoids e.g. tricetinidin that show the
3,4,5-hydroxylation pattern. Further support for this latter

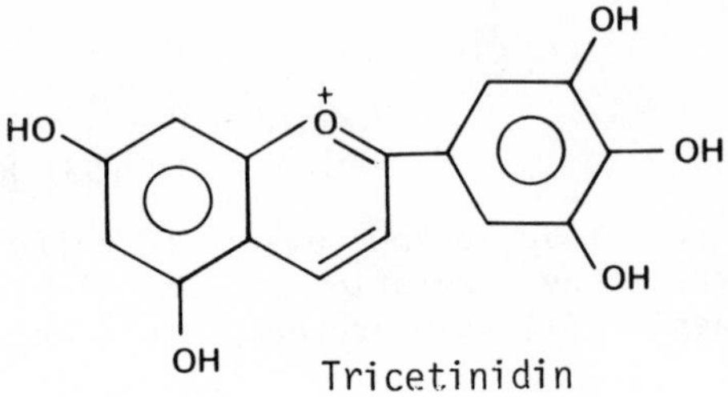

Tricetinidin

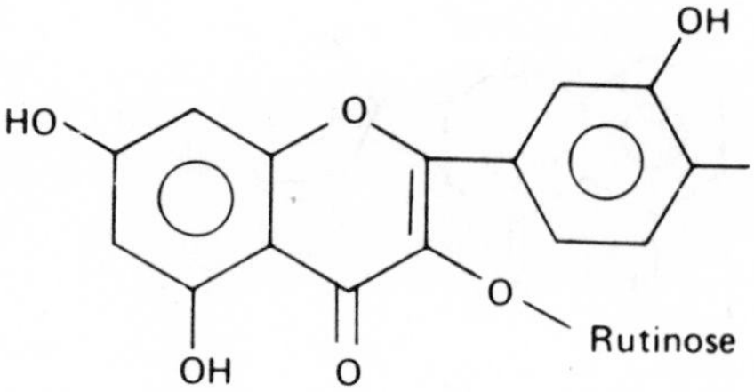

[54]

view has come from the study of Sutter and Grisebach (56), and
Amrhein and Zenk (57), in which the incorporation of 3-^{14}C-4-^{3}H-
cinnamic acid into rutin [54] and cyanidin by buckwheat, into the
flavonols quercetin and kaempferol by *Pisum sativum*, and into the
isoflavones biochanin A and formononetin by *Cicer arietinum* has
been investigated. The retention of tritium in kaempferol,
biochanin A, and formononetin (all 4'-OH) due to the NIH shift
mechanism (Chapter 16) was found to be about that expected for
the corresponding conversion of 4-^{3}H-cinnamic to 3-^{3}H-coumaric
acid (58). If hydroxylation were taking place at the flavonol
stage, [55], the possibility of delocalization of the charge of
the intermediate cation [56] would lead to a reduced retention of
tritium; since the retention was the one anticipated for a simple
aromatic hydroxylation, it is assumed that preformed coumaric
acid is incorporated into the chalcone. Grisebach and associates
have studied the biosynthesis of acacetin (4'-methoxy-5,7-dihydro-
xyflavone) in *Robina pseudacacia;* the direct incorporation of
p-methoxycinnamic acid was readily demonstrated (63).
    It seems likely that the oxygenation pattern of ring B is
established early in the biosynthesis, probably at the cinnamic

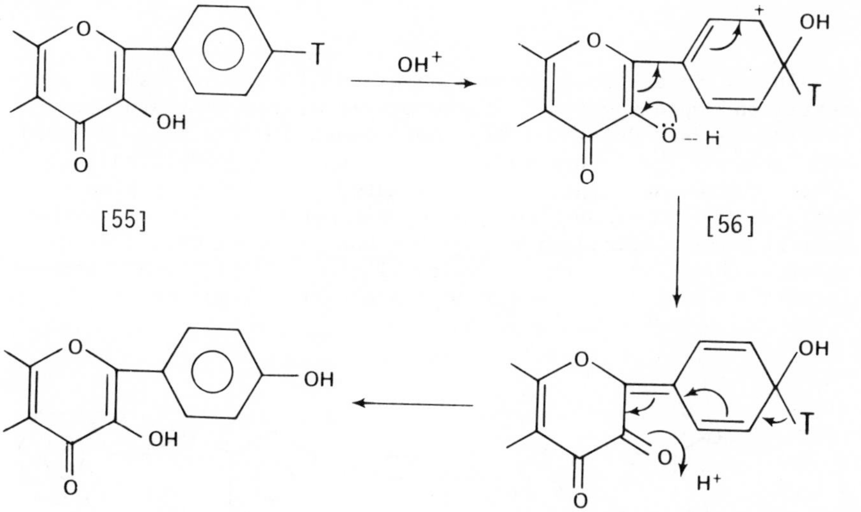

acid stage, unless the ring contains hydroxyl groups at 2' or 6'; the introduction of these groups may well be mediated by special mechanisms, since they occur mainly in either isoflavones or 3-hydroxyflavones.

In all cases so far investigated, ring A of flavonoids comes from acetate; the C-5 and C-7 hydroxylation is thus probably established already at the chalcone stage, as has been indicated, for example, by the specific incorporation of chalcones, suitably hydroxylated in ring A, into biochanin A and formononetin. It is also very unlikely that oxygen functions at C-6 and C-8 are introduced after ring A has been formed, although no direct evidence has been provided so far; it is reasonable to assume oxidation of the methylene groups of the polyketide, as is usually found in acetate-derived polyhydroxylated aromatic compounds.

One of the difficulties in presenting an integrated scheme for the biosynthesis of flavonoid compounds is the remarkable preponderance of the 2'-hydroxyl function in the isoflavonoids. If we accept the biosynthetic scheme outlined in Section 21-2a, then the 2'-hydroxyl group must be introduced either at the chalcone stage by incorporation of an *ortho*-coumaric acid, or later, after rearrangement to the isoflavone, by a hydroxylating mechanism not yet elucidated. In the former case, the presence of the 2'-function in the cinnamic acid could be essential to the subsequent biosynthesis, and assist in some way with those reactions leading to the isoflavonoids (i.e., with the 2,3-aryl shift); in

the latter case, the introduction of the 2'-hydroxyl seems inci-
dental and merely a facet of the enzyme system that produces
isoflavonoids. Unfortunately, not enough studies have yet been
made showing the incorporation of variously hydroxylated chalcones (flavanones) into 2'-hydroxylated flavonoids by plants;
until this is done, any proposals concerning possible mechanisms
must be purely speculative. Pelter has proposed such a mechanism (59), in which the 2'-hydroxyl group is involved in the
formation of an isoflavone from a precursor flavanone; this

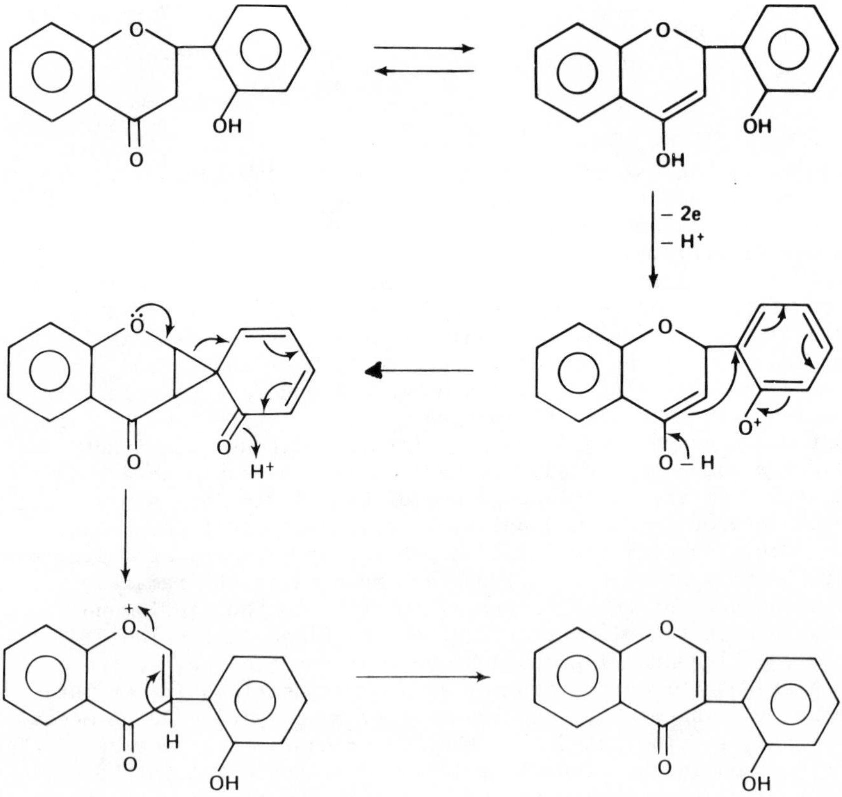

proposal, if proved correct by experiment, would explain very
nicely the appearance of a 2'-function in so many of these
compounds.   The opposite view is taken by Falshaw et al. (60),
who see the isoflavone skeleton favoring the 2'-hydroxylation
process as outlined below.

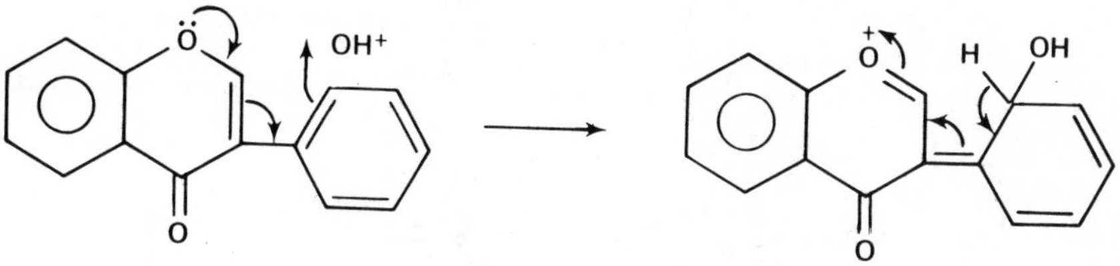

21-4.   METABOLITES RELATED TO THE FLAVONOIDS

The plant stilbenes are a very small group of compounds
found mainly as heartwood constituents in a heterogeneous assembly
of plant species.   In general, the substances have ring A hydroyl-
ation at C3 and C5, implying an acetate derivation, while the
hydroxylation of ring B is compatible with biosynthesis via shikimic acid.   Pinosylvin [57] and piceatannol [58] are typical
examples, although some other compounds are methylated [59] or
bear isoprenoid groupings [60].   Some stilbenes such as oxyresveratrol [61] have the less common 2'-hydroxyl on ring B, and others,
such as 4-hydroxystilbene [62], are so unlike the remainder of the
group that it would be hazardous to draw conclusions on their
biosynthesis.

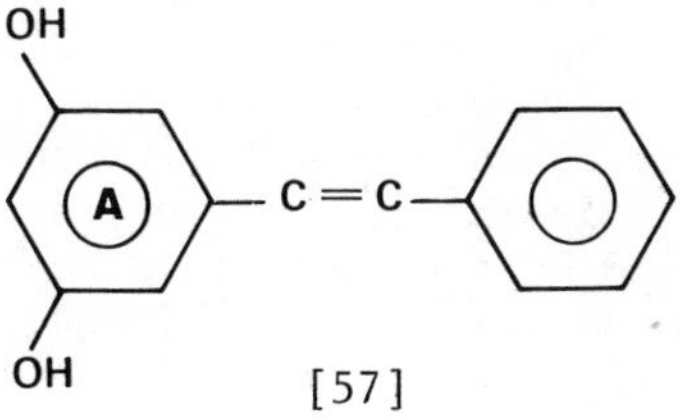

[57]

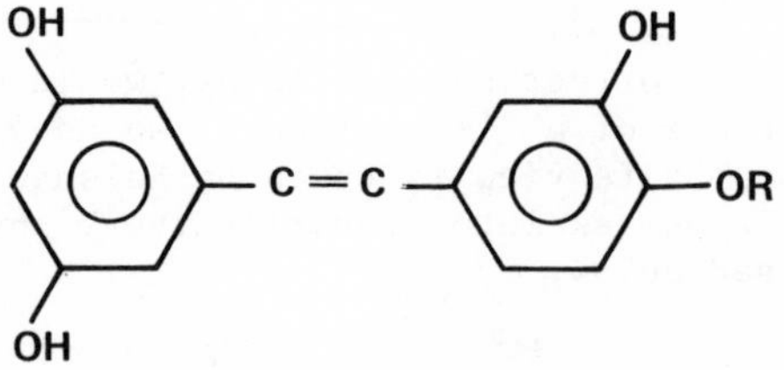

[58] **R = H**
[59] **R = Me**

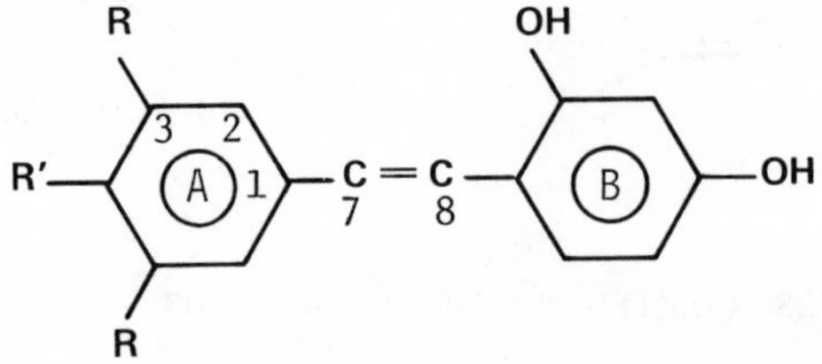

[60] **R = OH, R'** =geranyl(chlorophorin)
[61] R =OH,R'=H

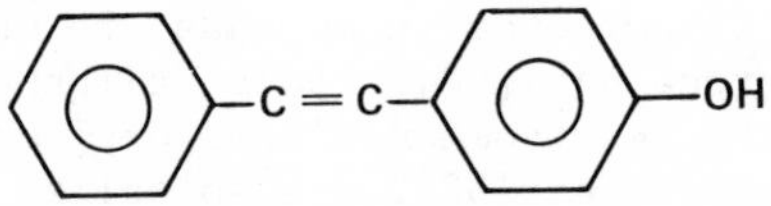

[62]

Some tracer work has been published on the synthesis of stilbenes in plants. It seems likely that for the majority of stilbenes a biosynthesis from a C6–C3 unit plus a C6 polyketide will be established. The first study was made by von Rudloff and Jorgensen (64), who showed that pinosylvin could be formed from phenylalanine and acetate in *Pinus resinosa*, a result confirmed by a similar study in another species of the same genus, *P. sylvestris* (65), and by the further investigation of the formation of two related stilbenes, [59] and resveratrol [63], in *Eucalyptus sideroxylon* (66).

The formation of oxyresveratrol [61] in *Morus alba* has been studied in detail by Billek (67), who found that the biosynthesis followed the expected pattern; acetate-1-^{14}C was incorporated into C-3 and C-5, and *p*-coumaric acid-β-^{14}C into C-8. A subsequent study of the formation of piceatannol [58] in *Laburnum anagyroides*

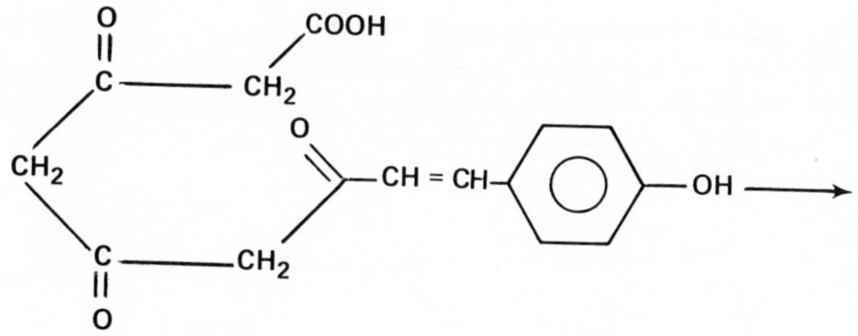

[64]

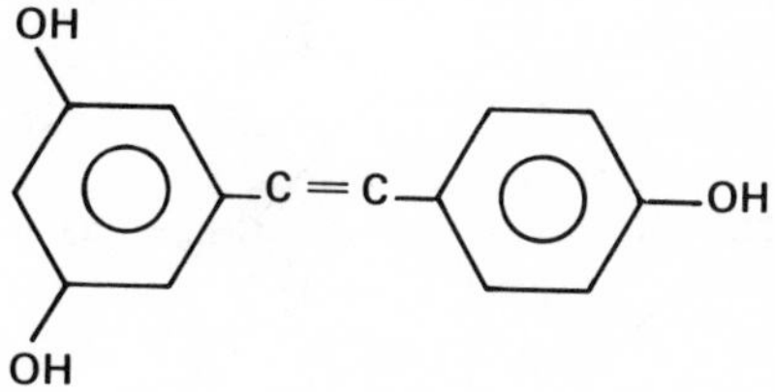

[63]

has further supported the generality of the biosynthetic scheme from acetate and shikimate, and has furnished additional evidence that the fully hydroxylated cinnamic acids are incorporated before ring closure, in preference to a non- or less-hydroxylated precursor (68).

3,4'-Dihydroxydihydrostilbene (lunarin) is formed via the corresponding 2-carboxylic acid (lunularic acid) in the liverwort *Lunularia cruciata* (80). There is a possibility that the biosynthesis proceeds from phenylalanine and acetate first to hydrangenol, the dihydroisocoumarin corresponding to lunularic acid, although the compound could not be detected in the organism (81).

It is interesting that an enzyme in *Eucalyptus sideroxylon* catalyzes the conversion of cinnamoyl triacetic acid [64] into pinosylvin [57]; the crude enzyme system produced only [57] when incubated with [64], whereas the tree itself makes no [57] but rather stilbenes hydroxylated in ring B. The authors speculated that 4'-hydroxylation may be necessary before further hydroxylation can take place (69). Subsequently, acetone powders from the same source have been found to be without pinosylvin synthetase activity (79). The biosynthesis of the 4-hydroxystilbenes remains unknown.

If instead of a cinnamic acid a benzoic acid acts as the starter molecule for a six-carbon polyketide chain [65], subsequent ring closure will yield a benzophenone [66], which on oxidative ring closure would be converted to a xanthone. Although many xanthones are formed from acetic acid alone, (Section 18-5), a

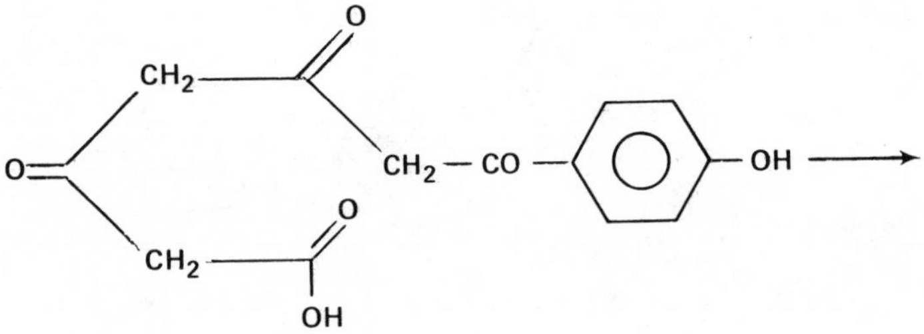

[65]

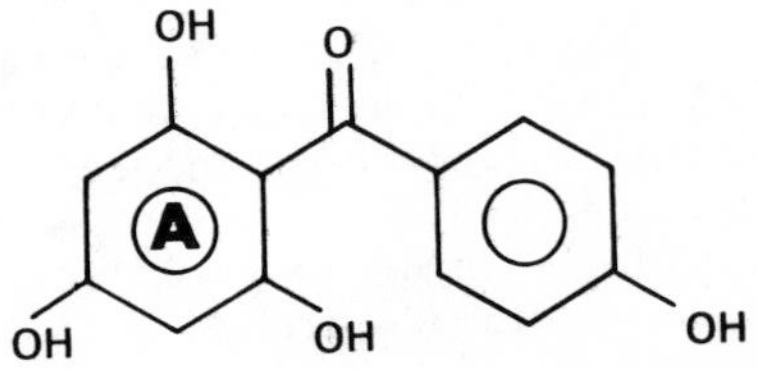

[66]

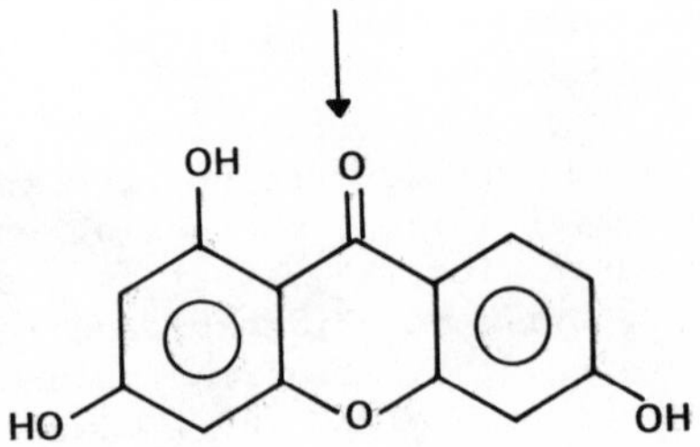

good deal of circumstantial, and some experimental, evidence
exists for the above mechanism. Thus, maclurin [67] and its two
possible cyclization products, 1,3,5,6-tetrahydroxyxanthone [68]
and 1,3,6,7-tetrahydroxyxanthone [69], coexist in the heartwood

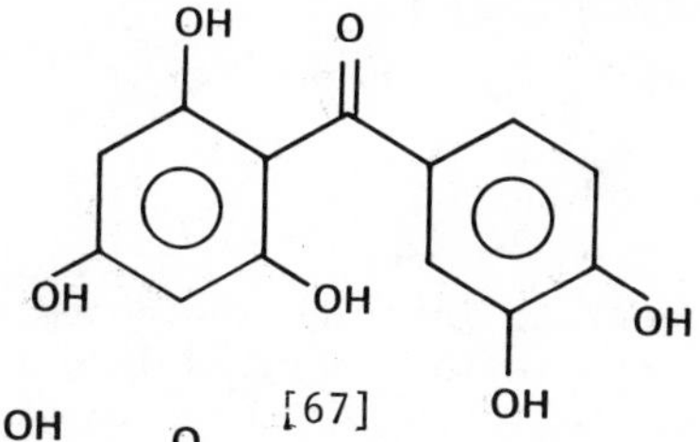

[67]

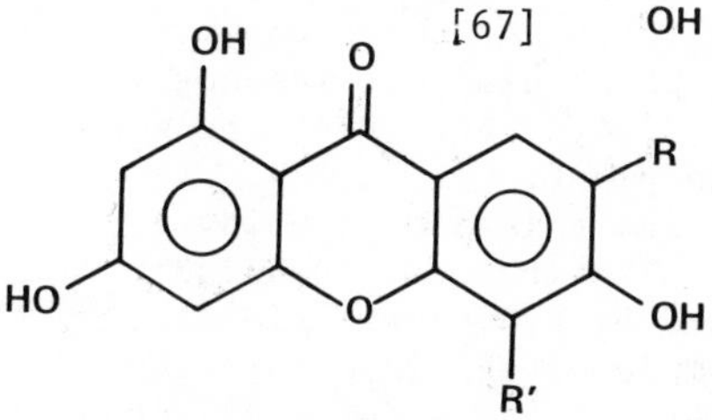

[68] R = H, R' = OH
[69] R = OH, R' = H

of *Symphonia globulifera* (70); similarly, 2,4,6,3'-tetrahydroxy-
benzophenone [70] and 1,3,7-trihydroxyxanthone [71] occur
together in *Gentiana lutea* (71), clearly suggesting the possibility of an *ortho*-ring closure of the benzophenones leading to
the xanthones; indeed such oxidative coupling has been observed
in vitro, in positions *ortho* and *para* to an existing hydroxyl
group (72-74). An investigation (71) of the incorporation of

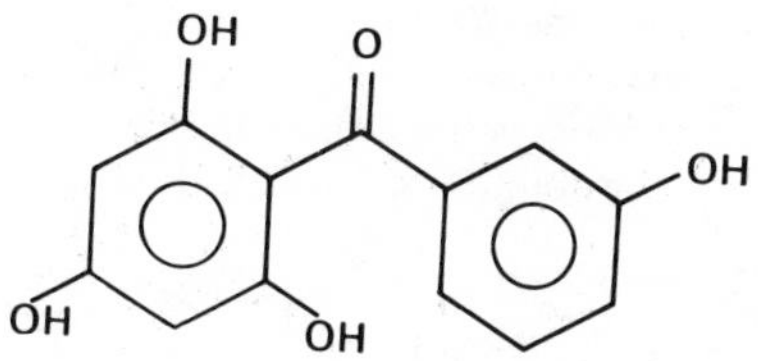

[70]

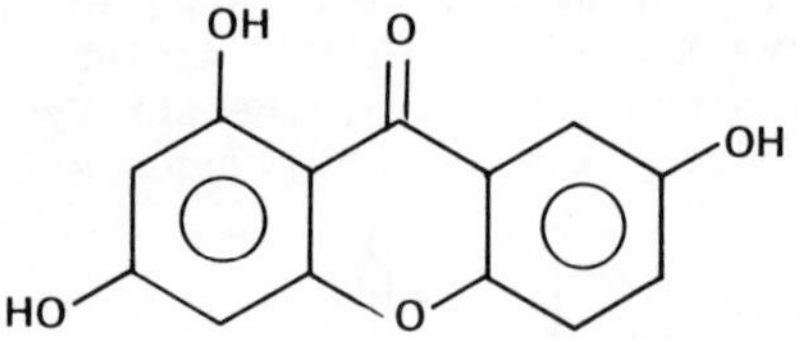

[71]

phenylalanine and acetate into the xanthones of *Gentiana lutea*
has shown that acetate is incorporated into ring A, whereas
phenylalanine provides ring B and the carbonyl carbon; in addition, it was shown that phloroglucinol was not incorporated into
gentisein [71], while 2,3',4,6-tetrahydroxybenzophenone [70] was
a good precursor in this plant.

It is interesting that this sequence involves a *meta*-hydroxyl-
ated ring derived from shikimic acid, presumably via *meta*-tyrosine;
if this assumption should be proved experimentally, the case would
represent the third example of the incorporation of *meta*-tyrosine
(Chapter 16). In statistical survey of the xanthones, published
by Gottlieb (75), a highly speculative biosynthetic scheme based
upon the interaction of dehydroshikimic acid and three acetate
units is proposed. Another review by Carpenter et al. (76) is in
favor of the simpler biosynthesis from a benzoic acid.

Recently, a new class of natural products has been established; several homoisoflavones (e.g., autumnalin [72]) have
been isolated by Tamm and his co-workers (77, 78, 82, 83).
Although these compounds are superficially related to siphulin
[73] (which must be completely derived from acetate), the presence of the 4'-oxygen function in ring B implies a biosynthesis
from shikimic acid. Dewick (84) has investigated the formation
of eucomin, 5,7-dihydroxy-4'-methoxyhomoisoflavone, by *Eucomis
bicolor*, and found it to be derived from shikimate precursors.
The additional carbon is C-3 of phenylalanine, and C-2 of the
flavonoid is provided by the 2-OMe group of a precursor. This
biosynthesis is more analogous to that of the aurones than to that
of isoflavones; it would seem reasonable to accept Wong's suggestion (85) that the compounds be called homoaurones.

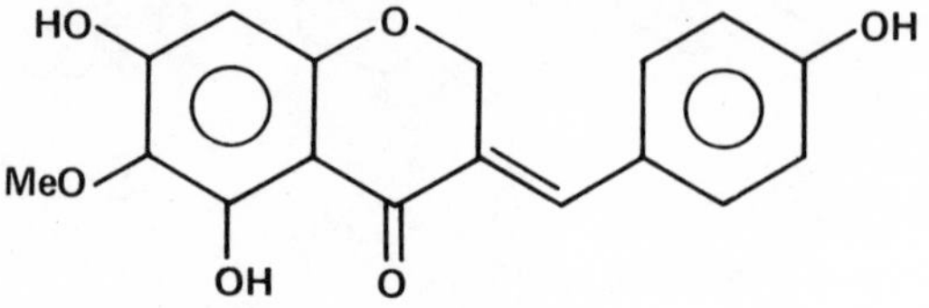

[72]

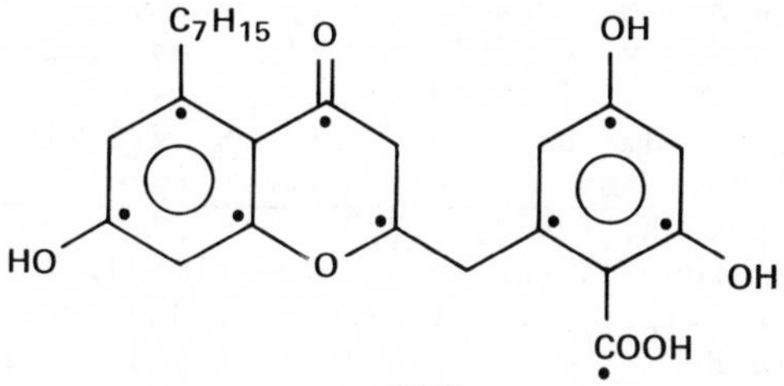

[73]

REFERENCES

1. H. Grisebach, in Chemistry and Biochemistry of Plant Pigments, T.W. Goodwin, Ed., Academic, London and New York, 1965, p. 279.
2. H. Grisebach, Biosynthetic Patterns in Microorganisms and Higher Plants, Wiley, New York, (1967).
3. H. Grisebach and W. Barz, Naturwissenschaften, 56, 538 (1969).
4. A. Hutchinson, C.D. Taper, and G.H.N. Towers, Can. J. Biochem. Physiol. 37, 901 (1959).
5. P.N. Avadhani and G.H.N. Towers, Can. J. Biochem. Physiol. 39, 1605 (1961).
6. H. Grisebach and G. Brandner, Biochem. Biophys. Acta 60, 51 (1962).
7. H. Grisebach, Planta Med. 10, 385 (1962).
8. H. Grisebach, W. Barz, K. Hahlbrock, S. Kellner, and L. Patschke, in Biosynthesis of Aromatic Compounds, G. Billek, Ed., Pergamon, Oxford, 1966, p. 25.
9. E. Wong and E. Moustafa, Tetrahedron Lett. 3021 (1966); Phytochemistry 6, 625 (1967).
10. E. Wong, Phytochemistry, 7, 1751 (1968).
11. L. Patschke, W. Barz, and H. Grisebach, Z. Naturforsch, 21b, 201 (1966).
12. H. Arakawa and M. Nakazaki, Chem. Ind. (London), 73 (1960).
13. E. Wong and H. Grisebach, Phytochemistry, 8, 1419 (1969).
14. M. Shimokoriyama and T.A. Geissman, J. Org. Chem. 25, 1956 (1960).
15. M. Shimokoriyama and S. Hattori, J. Amer. Chem. Soc. 75, 2277 (1953).
16. E. Wong, Phytochemistry 5, 463 (1966).
17. T.A. Geissman and D.K. Fukushima, J. Amer. Chem. Soc. 70. 1686 (1948).

18. F. Fischer and W. Arlt, Chem. Ber. 97, 1910 (1964).
19. E. Wong, Chem. Ind. (London), 1985 (1964); Biochim. Biophys. Acta 111, 358 (1965).
20. L. Patschke and H. Grisebach, Phytochemistry 7, 235 (1968).
21. H.G. Krishnamurty, V. Krishnamurty, and T.R. Seshadri, Phytochemistry 2, 47 (1963).
22. T.R. Seshadri, J. Ind. Chem. Soc. 39, 221 (1962).
23. J.B. Harborne, Comparative Biochemistry of the Flavonoids, Academic, London, 1967, p. 233.
24. M.N. Zaprometov, Biokhimia 27. 366 (1962).
25. P. Comte, A. Ville, G. Zwingelstein, J. Favre-Bonvin, and C. Mentzer, Bull. Soc. Chim. Biol. 42, 1079 (1960).
26. L. Patschke and H. Grisebach, Z. Naturforsch. 20b, 399 (1965).
27. D.G. Roux and M.C. Bill, Nature 183, 42 (1959).
27A. K. Weinges, W. Bähr, W. Ebert, K. Göritz, and H.D. Marx, Progr. Chem. Org. Nat. Prod. 27. 159 (1969).
28. H. Grisebach and W. Bilhuber, Z. Naturforsch. 22b, 746 (1967).
29. H. Grisebach and N. Doerr, Z. Naturforsch. 15b, 284 (1959).
30. H. Grisebach and G. Brander, Experientia 18, 400 (1962).
31. W. Barz and H. Grisebach, Z. Naturforsch 21b, 47 (1966).
32. H. Grisebach and W.D. Ollis, Experientia 11, 1 (1961).
33. L. Crombie and M.B. Thomas, J. Chem. Soc. (C) 1796 (1967).
34. L. Crombie, C.L. Green, and D.A. Whiting, J. Chem. Soc. (C) 3029 (1968).
35. M. Hamada and M. Chubachi, Agr. Biol. Chem. 33, 793 (1969).
36. I.C. du Preez, A.C. Rowan, and D.G. Roux, Chem. Commun. 492 (1970).
37. D.R. Perrin, Tetrahedron Lett. 29 (1964).
38. S.H. Harper, A.D. Kemp, W.G.E. Underwood, and R.V.M. Campbell, J. Chem. Soc. (C) 1109 (1969).
39. S.L. Hess, L.A. Hadwiger and M.E. Schwochau, Plant Pathol. 61, 79 (1971); Chem. Abstr. 74, 72847g (1971).
40. L. Crombie, P.M. Dewick, and D.A. Whiting, Chem. Commun. 1469 (1970); 1182, 1183 (1971); J. Chem. Soc. Perkin I, 1285 (1973).
41. B. Chabannes and H. Pacheco, Bull. Soc. Chim. Fr. Ser. 5, 1486 (1971).
42. R.S. Thompson, D. Jacques, E. Haslam, and R.J.N. Tanner, J. Chem. Soc. Perkin I, 1387 (1972).
43. E. Wong, Prog. Chem. Org. Natl. Prod. 28, 1 (1970).
44. L. Farkas, A. Gottsegen, M. Nogradi, and S. Antus, Chem. Commun. 825 (1972), and refs. there.
45. F. Kreuzaler and K. Hahlbrock, FEBS Lett. 28, 69 (1972).

Flavonoids 501

46. K. Hahlbrock, A. Sutter, E. Wellmann, R. Ortmann, and
     H. Grisebach, Phytochemistry 10, 109 (1971).
47. R. Wiermann, Planta 102, 55 (1972).
48. A.J. Grambow and H. Grisebach, Phytochemistry 10, 789 (1971).
49. M.N. Zaprometov and H. Grisebach, Z. Naturforsch. 28c, 113
     (1973).
50. M.N. Zaprometov and V. Ya. Bukhlaeva, Biokhimiya 36, 270
     (1971).
51. R. Marchelli and L.C. Vining, Chem. Commun. 555 (1973);
     Can. J. Biochem. 51, 1624 (1973).
52. H. Fritsch, K. Hahlbrock, and H. Grisebach, Z. Naturforsch.
     26b, 581 (1971).
53. P.F.T. Vaughan, V.S. Butt, H. Grisebach, and L. Schill,
     Phytochemistry 8, 1373 (1969); 10, 2649 (1971).
54. H. Meir and M.H. Zenk, Z. Pflanzenphysiol. 53, 415 (1965).
55. D. Hess, Planta, 60, 568 (1964).
56. A. Sutter and H. Grisebach, Phytochemistry 8, 101 (1969).
57. N. Amrhein and M.H. Zenk, Phytochemistry 8, 107 (1969).
58. M.H. Zenk, Z. Pflanzenphysiol. 57, 477 (1967); see also
     ref. 5.
59. A. Pelter, Tetrahedron Lett. 897 (1968).
60. C.P. Falshaw, R.A. Harmer, W.D. Ollis, R.E. Wheeler,
     V.R. Lalitha, and N.V.S. Rao, J. Chem. Soc. (C) 365, 374
     (1969).
61. H.J. Grambow and H. Grisebach, Phytochemistry 10, 789
     (1971); 7, 51 (1968).
62. A.M. Steiner, Z. Pflanzenphysiol. 63, 370 (1970).
63. J. Ebel, W. Barz, and H. Grisebach, Phytochemistry 9, 1529
     (1970).
64. E. von Rudloff and E. Jorgensen, Phytochemistry 2, 297
     (1963).
65. G. Billek and W. Ziegler, Monatsh. 93, 1430 (1962).
66. W.E. Hillis and M. Hasegawa, Chem. Ind. (London) 1330
     (1962).
67. G. Billek and A. Schimpl, Monatsh. 93, 1457 (1962).
68. G. Billek and A. Schimpl, in Biosynthesis of Aromatic
     Compounds, G. Billek, Ed., Pergamon, Oxford, 1966, p. 37.
69. W.E. Hillis and N. Ishikura, Phytochemistry 8, 1079
     (1969).
70. H.D. Locksley, I. Moore, and F. Scheinmann, Tetrahedron 23,
     2229 (1967).
71. P. Gupta, and J.R. Lewis, Chem. Commun. 1386 (1969);
     J. Chem. Soc. (C) 629 (1969).
72. J.E. Atkinson and J.R. Lewis, Chem. Commun. 803 (1967).
73. R.C. Ellis, W.B. Whalley, and K. Ball, Chem. Commun. 803
     (1967).

74. B. Jackson, H.D. Locksley and F. Scheinmann, J. Chem. Soc. (C) 2201 (1969).
75. O.R. Gottlieb, Phytochemistry 7, 411 (1968).
76. I. Carpenter, H.D. Locksley and F. Scheinmann, Phytochemistry 8, 2013 (1969).
77. P. Bühler and Ch. Tamm. Tetrahedron Lett. 3479 (1967).
78. W.T.L. Sidwell and Ch. Tamm, Tetrahedron Lett. 475 (1970).
79. W.E. Hillis and Y. Yazaki, Phytochemistry 10, 1051 (1971).
80. R.J. Pryce, Phytochemistry 11, 872 (1972).
81. R.J. Pryce, Phytochemistry 10, 2679 (1971).
82. Ch. Tamm, Arzneimittelforsch. 22, 1776 (1972).
83. R.E. Finckh and Ch. Tamm, Experientia 26, 472 (1970).
84. P.M. Dewick, Chem. Commun. 438 (1973).
85. E. Wong, Prog. Chem. Org. Nat. Prod. 28, 1 (1970).

PART 3 AROMATIC COMPOUNDS DERIVED
FROM MEVALONIC ACID

CHAPTER 22

Aromatic Terpenoids

Among the vast variety of natural compounds of mevalonoid
origin, a fairly large number of aromatic substances has been
observed. Of these, the aromatic steroids (oestrogens, veratramine, viridin, etc.) are dealt with elsewhere in this volume
(Chapter 24), as are the aromatic carotenoids (Chapter 23). The
terpenoids proper, which form the subject of the present discussion, can be divided into those in which the entire aromatic
part of the structure is derived from mevalonic acid, and those
of mixed origin, that is, compounds in which the aromatic ring
system combines parts of terpenoid derivation with others biosynthesized from shikimic acid (e.g., the carbazole alkaloids) or
acetic acid (e.g., cannabinol). The compounds of mixed origin
will be treated separately, after the purely terpenoid aromatic
substances, in Section 22-6.

22-1. GENERAL CONSIDERATIONS

To discuss in any detail the enormous field of the chemistry
and biosynthesis of terpenoid compounds would be out of place in
the present volume, and would be unnecessary, since many excellent reviews exist, of which a few may be quoted (1,2,3,4,5).
Only a brief introductory description of the general principles
underlying the biosynthesis of the various groups of terpenoid
substances will be given as a basis for subsequent detailed
discussion of the aromatic representatives of this class of
natural compounds.

The most general description of terpenoids would define
them as naturally occurring substances with a carbon skeleton
that can, on paper, be built up entirely or at least in part,
from "isoprenoid" units [1][+] (see next page), or which can be derived from such structures by plausible processes such a skeletal
rearrangements, or loss of carbon atoms, singly or in groups. The
initial idea that terpenoids are actually formed by various

503

polymerizations of isoprene [2] itself has long been abandoned,
but use of the term continues to be convenient. The early
observation that the structures of numerous terpenoids can be
dissected into units of [1] is often called the "isoprene rule";
its historical background has been described by Ruzicka, for
instance in his Tilden lecture (6a), which also presents the
further developments of this concept, mostly due to Ruzika and
his associates: the generalization that most terpenoid structures
can be formed by regular head-to-tail arrangements of isoprene
units [I], as in [3], ("biogenetic isoprene rule"), followed by

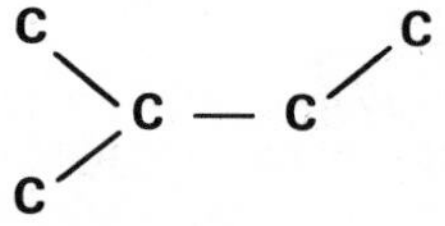

[1] 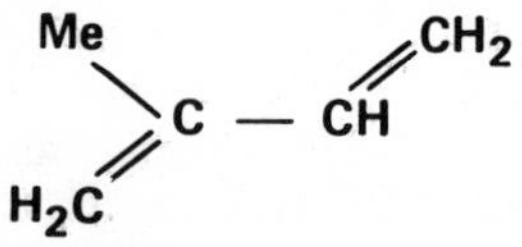   **[2]**

further elaborations, and that exceptions to this rule can usually
be rationalized as resulting from rearrangements belonging to
well-known types, or from the loss of certain atoms or groups.
The correctness of this view has by now been proved experimentally
in an impressive number of cases.

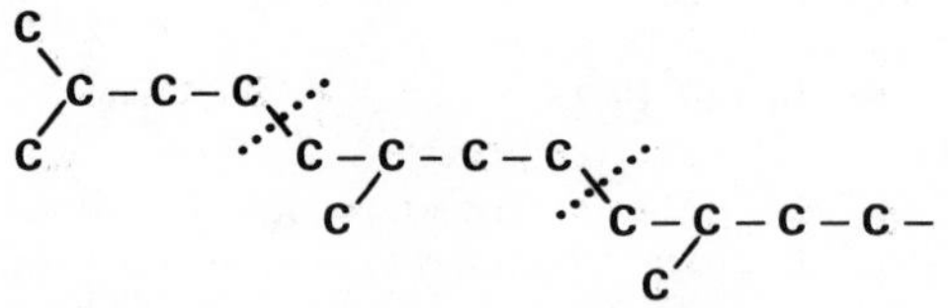

[3]

+In this chapter, references to formulae in the text are indicated
by Arabic numerals in square brackets, such as [2] above, while
references to formulae listed in Tables I - V first cite the
Roman numeral of the table, then the Arabic numeral of the particular compound in that table, again in square brackets, e.g.,
[III-25]. Tables I-V will be found on pages 516, 538, 548, 582,
and 583.

There are, however, a few cases which seem to be explained
more rationally by addition of another isoprene unit (or several
of them) to a pre-existing terpenoid. A structure such as that
of calacone [4] may be a case in point. It can be rationalized
most simply as the product of further isoprenylation of a precursor with the normal carbon skeleton of a monoterpene of the
p-menthane type [20] (see below); admittedly, however, its
carbon skeleton could also be the result of a 1,2 shift of one
of the terminal carbon atoms of the side-chain. A number of

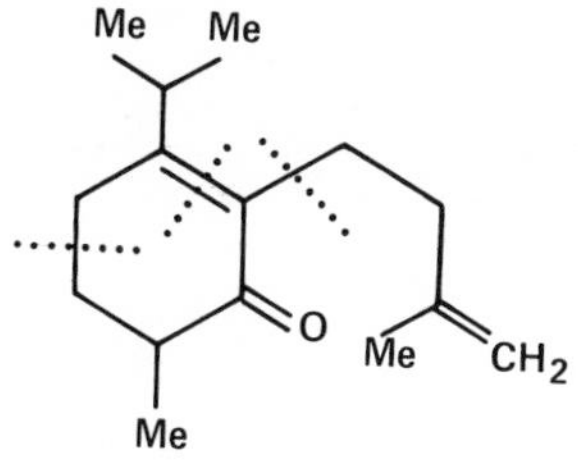

[4]

compounds presumably formed through such secondary isoprenylation will be encountered among the carbazole alkaloids discussed
in Section 22-6.

The C_{10} unit derived from two isoprene residues forms the
basis for nomenclature of the terpenoids, having been recognized
at an early stage as a common structural feature of many constituents of essential oils. Thus, hemiterpene stands for compounds
with one isoprenoid C_5 unit, monoterpenoid for C_{10} compounds,
followed by sesqui-(C_{15}), di-(C_{20}), sester-(C_{25}), tri-(C_{30}), and
polyterpenoids. Again, the possibility of actual biosynthesis
involving the loss of certain carbon atoms has to be considered;
thus, an aromatic diterpenoid, (III-28], to be discussed in
Section 22-4, has only 17 instead of the expected 20 carbon
atoms. The steroids, derived from C_{30} intermediates, form an
enormous class encompassing the entire range of compositions
down to C_{18}. One group of terpenoids, the simaroubalides, not
including any known aromatic representatives and hence not
further discussed here, consists mostly of C_{20} compounds but can
be shown to be derived biosynthetically from triterpenoid C_{30}
precursors (cf. ref. 1, p. 352ff.).

In a general way, the biosynthesis of the isoprenoids
(see Scheme I) can be considered as a special type of acetate-
based process. The first specific intermediate, the C_6 compound
(R)-mevalonic acid [5], is usually formed from three molecules
of acetyl coenzyme A. Two of these can interact to give

Scheme I

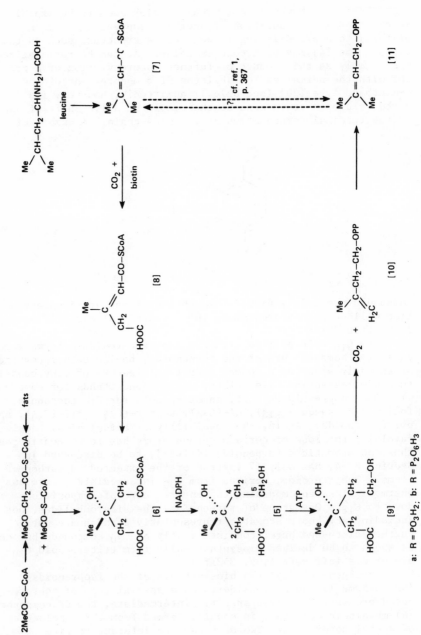

acetoacetyl coenzyme A; reaction of this compound with the third
molecule of coenzyme A gives (S)-3-hydroxy-3-methylglutaryl
coenzyme A [6], which is reduced to [5] in an NADPH- dependent
reaction. The derivation of [5] from acetyl CoA accounts for the
fact that feeding of, for example, Me-14COOH has repeatedly been
shown to yield terpenoids with the labeling pattern of alternate
carbon atoms reminiscent of acetate-based biosynthetic processes.
However, up to [6] the reaction sequence is nonspecific: instead
of being built up from two molecules of acetyl CoA, acetoacetyl
CoA can be formed by degradation of fatty acids. Furthermore,
an alternative pathway for [6] starts with leucine, which is
transformed to β,β-dimethylacryl (senecioyl) coenzyme A [7]; one
of the allylic methyl groups of [7] is next carboxylated in a
biotin-dependent reaction, to give 3-methylglutaconyl coenzyme A
[8], which is hydrated to [6]. This pathway in its early stages,
or some related process, should be operative in the biosynthesis
of a number of substances [mostly C_5 and C_{10} acids] with structures derivable from [1], for whose biosynthesis [5] is not
utilized, while amino acids such as isoleucine or valine are;
for example, isoleucine→tiglic acid (see ref. 1, p. 361 ff).
So far, no aromatic terpenoid has been shown to have this origin,
and thus no detailed discussion is needed. However, participation of the sequence from leucine via [7] and [8] to [5], and
hence to other isoprenoid compounds, can not be excluded at
present, and this path does play a role in the observed stimulation of carotenoid biosynthesis by leucine (see ref. 7,
p. 645a); it may therefore be involved in the biosynthesis of the
aromatic representatives of this class (Chapter 23).

   In contrast, the reduction of [6] to [5] is the only known
mode of formation of the latter, and biosynthesis of terpenoids
its only know function. Furthermore the reaction [6]⟶[5] is
not readily reversible. Mevalonic acid, [5], available in a
variety of labeled forms, is thus particularly suitable for
studies of the biosynthesis of terpenoids. However, complications may occasionally arise from the fact that label from [5]
can enter the pool of general metabolites, apparently through
degradation of terpenoids derived from it (3a). As an example,
see the findings (181) on the incorporation of label from
2-14C-[5] into the acetate ester of the (nonaromatic) monoterpene alcohol sabinol in *Juniperus sabina*: more than 90% of the
activity of the sabinyl acetate biosynthesized by this species
was localized in the esterifying acetate group.

   Because of this position of [5] as the first specific
intermediate, the term "mevalonoid" has come to be used more or
less interchangeably with "isoprenoid." Being more descriptive
of the actual genesis of the compounds in question, it may be

preferable. It is also more specific, and it excludes those substances with isoprenoid structures for whose biosynthesis [5] is not utilized, while amino acids (leucine, isoleucine, valine) are.

In the next step of the biosynthesis, mevalonic acid [5] is phosphorylated by ATP to give stepwise, the mono- and the pyrophosphate [9a and b, respectively]. The latter then yields CO_2 and 3-methyl-3-butenyl pyrophosphate [10], which is isomerized reversibly to the allylic, and hence highly reactive 3-methyl-2-butenyl pyrophosphate (3,3-dimethylallyl pyrophosphate) [11]. Compounds [10] and [11], containing the isoprenoid skeleton [1], are the actual building blocks involved in the further elaboration of terpenoid structures, so that the early "isoprene rule" in fact refers to these then-undiscovered substances. The enzymatic transformation of [10] into [11] proceeds in a stereospecific manner. Consequently, only one of the two structurally identical methyl groups of [11] is labeled from $2-^{14}C-[5]$, which often makes it possible to identify the atoms in mevalonoid compounds that originate in C-2 of [5], provided the atom in question is structurally or stereochemically different from the one derived from the other methyl of [11], that is, from the methyl group of [5].

The first step in the assembly of these various terpenoid structures (see Scheme II) is the reaction of one molecule each of [10] and [11] to give the pyrophosphate [12] of geraniol [12A]. Geranyl pyrophosphate, [12], is the parent of the numerous monoterpenes; likewise derived from it is loganin, the all-important intermediate in the biosynthesis of the nontryptamine moieties of the innumerable indole alkaloids with their many medicinally important representatives. The compound actually involved in the biosynthesis of the *cyclic* monoterpenes must, on structural grounds, be nerol, the *cis*-isomer of geraniol (or some species derived from nerol, e.g. the pyrophosphate); see Section 22-2. A *trans, cis* isomerization of the double bond in position 2 of [12] or [12A] must thus take place at some stage between these aliphatic precursors and the cyclic compounds derived from them. The alternative possibility of a direct biosynthesis of nerol or its pyrophosphate from [10] and [11] is made improbable by the findings of Heinstein and coworkers (187) on the utilization of both *cis, trans*-isomeric pyrophosphates for the biosynthesis of the aromatic bis-sesquiterpenoid gossypol [II-56] (see Section 22-7), and disproven, for the biosynthesis of the glucosides of geraniol and nerol in rose petals, by studies (182) on the incorporation of $2-^{14}C-(4R)-$ and $(4S)-4-^{3}H$ mevalonate.

An exactly analogous reaction between [12] and [11] yields the C_{15} compound farnesyl pyrophosphate [13], which can be furthur elaborated to the sesquiterpenes, a huge class that includes representatives ranging from simple acyclic compounds to amazingly complex structures with systems of three to four interlocking rings.

Scheme II

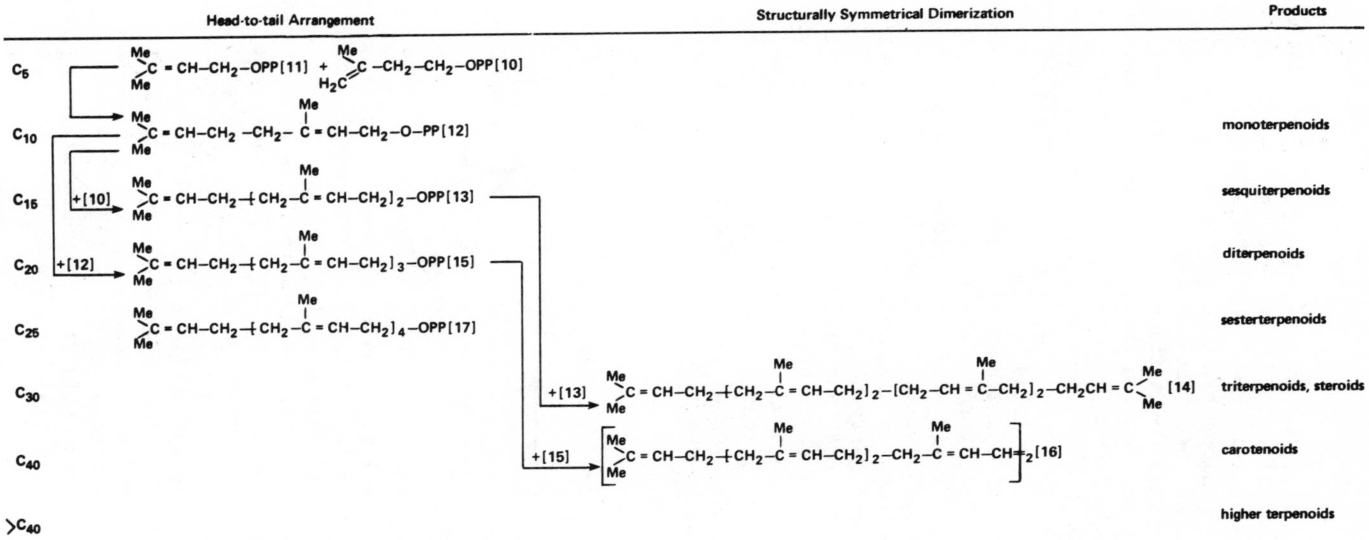

Head-to-tail Arrangement

Structurally Symmetrical Dimerization

Products

C_5

$\text{Me}_2\text{C}=\text{CH}-\text{CH}_2-\text{OPP}[11] + \overset{\text{Me}}{\underset{\text{H}_2\text{C}}{\diagup}}\text{C}-\text{CH}_2-\text{CH}_2-\text{OPP}[10]$

C_{10}

$\text{Me}_2\text{C}=\text{CH}-\text{CH}_2-\text{CH}_2-\overset{\text{Me}}{\underset{}{\text{C}}}=\text{CH}-\text{CH}_2-\text{O}-\text{PP}[12]$

monoterpenoids

$+[10]$

C_{15}

$\text{Me}_2\text{C}=\text{CH}-\text{CH}_2\text{-}[\text{CH}_2\text{-}\overset{\text{Me}}{\underset{}{\text{C}}}=\text{CH}-\text{CH}_2]_2-\text{OPP}[13]$

sesquiterpenoids

$+[12]$

C_{20}

$\text{Me}_2\text{C}=\text{CH}-\text{CH}_2\text{-}[\text{CH}_2\text{-}\overset{\text{Me}}{\underset{}{\text{C}}}=\text{CH}-\text{CH}_2]_3-\text{OPP}[15]$

diterpenoids

C_{25}

$\text{Me}_2\text{C}=\text{CH}-\text{CH}_2\text{-}[\text{CH}_2\text{-}\overset{\text{Me}}{\underset{}{\text{C}}}=\text{CH}-\text{CH}_2]_4-\text{OPP}[17]$

sesterterpenoids

$+[13]$

C_{30}

$\text{Me}_2\text{C}=\text{CH}-\text{CH}_2\text{-}[\text{CH}_2-\overset{\text{Me}}{\underset{}{\text{C}}}=\text{CH}-\text{CH}_2]_2-\text{C}-\text{CH}_2]_2-\text{CH}_2\text{CH}=\overset{\text{Me}}{\underset{\text{Me}}{\text{C}}}[14]$

triterpenoids, steroids

$+[15]$

C_{40}

$[\text{Me}_2\text{C}=\text{CH}-\text{CH}_2\text{-}[\text{CH}_2-\overset{\text{Me}}{\underset{}{\text{C}}}=\text{CH}-\text{CH}_2]_2-\text{CH}_2-\overset{\text{Me}}{\underset{}{\text{C}}}=\text{CH}-\text{CH}_2]_2[16]$

carotenoids

$>C_{40}$

higher terpenoids

Alternatively, two molecules of [13] can react to give the C_{30} compound squalene [14], the parent of triterpenoids and steroids (see Chapter 24). In spite of the symmetrical structure of its product, the biosynthesis of [14] proceeds in a nonsymmetrical manner, as has been shown by the ingenious experiments of Cornforth and co-workers with tritium-labeled precursors (see ref. 3a, b, and literature quoted there). A distinct intermediate, presqalene pyrophosphate, has recently been recognized to be involved in this process (7A). See the Addendum on the structure of this compound.

Assembly of four units of the isoprenoid pyrophosphates, [10] and [11], yields the C_{20} intermediate, geranylgeranyl pyrophosphate [15], which appears to be the precursor of the diterpenoids. This large class of compounds includes, among many others, the resin acids, the gibberellins, the phytol moiety of the chlorophylls, the side-chains of vitamins E and K, and a number of complex alkaloids such as aconitine, delphinine, atisine, and garryine. Furthermore, symmetrical "tail-to-tail" coupling of some C_{20} intermediate, probably [15], leads to the carotenoids (see Chapter 23). The coupling process would seem to resemble that which produces [14] from two molecules of [13], but apparently leads to phytoene [16], a compound differing from a C_{40} analog of [14] by the presence of the central double bond; see, however Section 23-3.

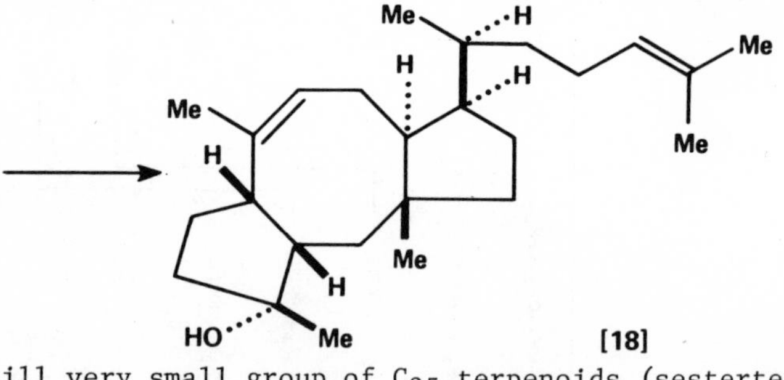

The still very small group of C_{25} terpenoids (sesterterpenes) seems to originate from geranyl-farnesyl pyrophosphate [17], which must be assumed to form from one molecule each of [12] and [13]. Experimental evidence for this biosynthesis has been obtained by Nozoe and Morisaki (8), who found that cell-free extracts of the fungus *Cochliobolus heterostrophus* incorporate both mevalonic acid [5], and synthetic all-*trans*-[17] into the sesterterpene ophiobolin F [18]. Nothing seems to be known so far on the actual biosynthesis of [17], that is, whether it arises through combination of preformed [12] and [13], or by stepwise addition of C_5-units to [12].

The biosynthesis of triterpenoids and steroids from [13] via presqualene and [14] has already been mentioned. The conversion

of the symmetrical structure of [14] to the nonsymmetrical ones
of the steroids and the majority of the triterpenoids proceeds
through squalene-2,3-epoxide in most cases.

Finally, there must exist biosynthetic mechanisms that yield
longer chains of isoprenoid units in regular head-to-tail
arrangement. The C_{45} terpenoid solanesol (from tobacco), the
C_{30}-C_{50} side-chains of some of the ubiquinones and vitamins K,
and high polymers such as rubber and gutta-percha may serve as
examples. No aromatic structures derived from such longer
isoprenoid chains seem to be known.

The occurrence of aromatic substances such as p-cymene
[I-1] and thymol [I-4] among terpenoids was recognized early in

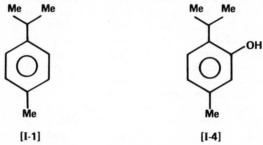

[I-1]   [I-4]

the study of the structures of natural compounds. Compound [I-1]
for example, was isolated from an essential oil in 1841 (9a),
after having been synthesized even earlier, and its correct
constitution was established in 1891 (9b). Not surprisingly, the
rapid development of the study of natural compounds in more recent
times has yielded a substantial number of aromatic mono-, sesqui-,
and diterpenoids. Among the few sesterterpenes that have been
discovered only during the last few years, no aromatic representative has appeared so far; one may expect such compounds to be
found sooner or later. Surprising, however, is the almost total
absence of aromatic structures among the innumerable triterpenoids. Apparently, the only compounds known to belong in this
category are the quinone methides celastrol [IV-1] and pristimerin
[IV-2] (see Section 22-5), which have been found in two closely
related families of higher plants, the Celastraceae and the
Hippocrateaceae. For further representatives see the Addendum.

In the following pages, the structures of the known aromatic
terpenoids are discussed and tabulated. Tables I-IV give the
mono-, sesqui-, di-, and triterpenoids with completely mevalonate-
derived aromatic systems, while Table V lists the structures
of mixed origin. No other compilation seems to exist that treats
the aromatic terpenoids specifically. We have therefore made
every effort to include as many as possible of the natural

compounds isolated so far.   That completeness of such a survey
can at best only be approached asymptotically will be obvious,
but we hope to have given at least an idea of the manifold
types of aromatic terpenoids.   For less well-known or recently
discovered substances, leading references to isolation and
structure proof have been given, but no complete bibliography
on every occurrence or detail of chemistry has been attempted.
No literature references were considered necessary for compounds
that have been known for a long time and studied in great
detail; here, general sources such as the Beilstein handbook
and W. Karrer's reference book on natural compounds (10A) should
suffice.

A few substances have been included whose structures cannot
be completely dissected into isoprene units but show features
which strongly suggest an origin from fully isoprenoid precursors by loss of carbon atoms; australol (p-isopropylphenol),
[I-32], and podocarpic acid, [III-28], may serve as examples.
No experimental evidence for the mevalonoid origin of any of
these defective terpenoids seems to exist, but their structures
are often suggestive enough to justify their inclusion.   In a
number of instances, co-occurrence of such compounds with completely mevalonate-derivable ones provides additional support.

Here, as elsewhere, the status of individual substances as
bona fide naturally occurring compounds may often be somewhat
doubtful.   Aromatic substances could well be formed from suitable
nonbenzenoid precursors during isolation procedures.   Exposure
to higher temperatures during isolation of, for example, sesquiterpenes from essential oils by the usual techniques of fractional
or steam distillation may at times lead to *de novo* formation of
aromatic structures by reactions such as dehydrogenation, disproportionation, or elimination.   For an example of the actual
occurrence of such thermal reactions during the isolation of
certain sesquiterpenes, see the discussion of the chemistry of
the essential oil of *Acorus calamus* in ref. 10, Vol. 2, p. 88.
Many other instances of this complication may actually exist.
Even at room temperature, some hydroaromatic compounds undergo
aromatization on contact with air; see the transformation of
γ-terpinene (p-menthadiene-1,4) into p-cymene [I-1] discussed in
Section 22-7, and the examples of sesquiterpenes in Section 22-3.

Assumption of an isoprenoid (mevalonoid) origin of most of
the tabulated compounds rests entirely upon structural criteria,
which are almost always evident enough.   The actual experimental
evidence, which is available for a few of them, is brought
together in the last section of this chapter (Section 22-7); most
of this evidence consists in proof for incorporation of acetic or
mevalonic acid in the expected manner.   Details of the actual

aromatization step have not yet been accessible to experimental scrutiny in any one terpenoid. Besides the more obvious purely chemical ones, the possibility of photochemical aromatization exists, as has been mentioned in Chapter 1.

The vast majority of the naturally occurring terpenoids are cyclic compounds, often with amazingly complex ring systems. Obviously, the cyclization reaction leading to these systems are of major interest, and have been discussed in great detail in the literature. In formulating the biosynthetic transformations of terpenoids, and in particular those involving ring closures, it is customary, following the lead of Ruzicka (6b), to assume reaction mechanisms which involve intermediate cationic species. These mechanisms indeed lend themselves remarkably well to interpretation of the observed processes, even down to fine details of stereochemistry. They often explain the co-occurrence of terpenoids of different types in the same plant, and have repeatedly suggested novel skeletal structures prior to their actual discovery in natural compounds. In addition, they are in most cases paralleled by *in vitro* transformations of terpenoids under acidic conditions. There is, however, so far no conclusive evidence for the actual participation of such ionic species in biosynthetic events, and alternative radical mechanisms have to be considered, (cf., *inter alia*, ref. 6b, p. 365).

22-2. MONOTERPENOIDS

This relatively small group (see Table I) includes a few hydrocarbons, alcohols, aldehydes, and carboxylic acids. More numerous are phenols, phenol ethers, and quinones; several epoxy compounds have been discovered quite recently. The structures of almost all these substances can be derived from that of *p*-cymene [I-1]. They are thus aromatic representatives of the so-called *p*-menthane [20] type of monoterpenoids. The relationship of [20] with geraniol [12a] and its *cis*-isomer nerol [19] is obvious from the strictly schematic diagram given below. This scheme does not intend to show the actual biosynthetic steps which, incidentally, remain unknown in most cases. Biosynthesis of some representatives (e.g., thymol [1-4]) from acetic and mevalonic acids has been established; these investigations will be discussed in Section 22-7.

As far as our very incomplete knowledge of taxonomic distribution of aromatic monoterpenoids permits us to judge, compounds of this type seem to occur preferentially in a few widely separated plant families.[+] Conifers, and among these especially the

---

[+]See footnote, following page.

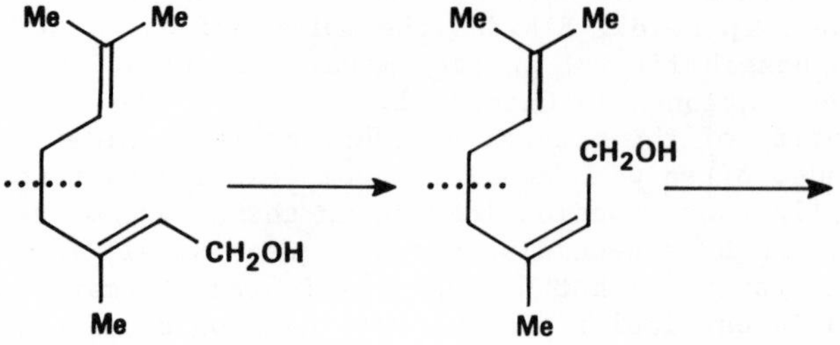

[12a]   [19]

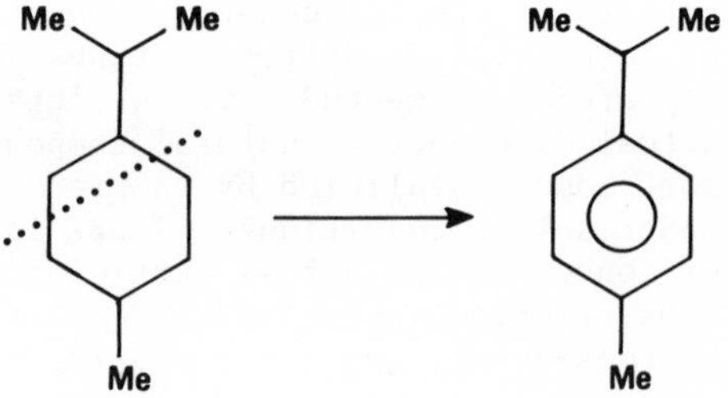

[20]   [I-1]

family Cupressaceae, are rich in such compounds. The chemistry
of constituents of the order Cupressales (Cupressaceae and
Taxodiaceae) has been reviewed by Erdtman and Norin (12). Other
families in which aromatic monoterpenoids seem abundant are the
Labiatae, Myrtaceae, and Compositae. The aromatic monoterpenoids
of the latter have become known only quite recently through the
systematic investigation of its constituents by Bohlmann and his
school (cf. Chapter 19). Besides the lactol [I-18] (13a), *Helenium* species have yielded compounds [I-19 a and b] (13b), while
the analogous pairs [I-20 to I-23, a and b] have been obtained
from the related genus *Gaillardia* (13b). The close relationship
of all these compounds is obvious: with the exception of [I-18],
they are all di-esters, one of the acyl groups being isobutyryl,
the other one the same group in the (a) series, and

+As an oddity, it may be mentioned that the cyanohydrin of
cuminal [I-16] occurs as the glucoside in the millipede *Poly-
desmus vicinus* (12A).

(+)-α-methylbutyryl in the (b) series. This second acyl group
always esterifies a primary hydroxyl in a side-chain. Interestingly, this hydroxyl is borne by the isopropyl group in all the
compounds from *Helenium*, and by the nuclear methyl in those from
Gaillardia. Since mild acid-treatment transforms [I-21a and b]
into [I-22a and b], these latter compounds might be artefacts.

A group of ethers of thymohydroquinone [I-9] occurs in
Libocedrus (Heyderia) decurrens (Cupressaceae) and has been the
subject of a series of papers by Zavarin and co-workers (14).
With the exception of the two thymohydroquinone monomethyl
ethers [I-10] and [I-11], all these compounds are di- [I-24, 25]
and tri-[I-26] nuclear ethers of [I-9]. They must evidently
be formed by phenolic coupling, of which they are classic
examples.

Among the small number of "defective" structures with fewer
than ten carbon atoms, *p*-tolylmethyl carbinol [I-34], isolated
(15a) from the oil of *Curcuma xanthorrhiza* (Zingiberaceae), may
be an artefact formed by degradation of sesquiterpenoids such as
ar-turmerone [II-5] (15b). *o*-Cresol [I-27] and *p*-cresol [I-29]
have structures with too few characteristic features for safe
derivation; they owe their inclusion in Table I to the fact that
they have been found in plants of the family Cupressaceae and
might, therefore, be related to authentic aromatic monoterpenoids
from this family by biosynthetic processes involving loss of the
isopropyl group. The C_{17} diterpenoid podocarpic acid [III-28]
may furnish a precedent (see Section 22-4).

An exceptional structure is that of *o*-cymene [I-3], which
has been isolated by gas chromatography (at 110°) from the tur-
pentine oil from South American *Pinus radiata* (16). If a
genuine natural product, this compound might perhaps be derived
from a precursor with regular head-to-tail arrangement of isoprenoid residues by processes involving the 1,3-shift (or two
consecutive 1,2-shifts) of the methyl or isopropyl groups, or
possibly cleavage of a precursor with the pinene skeleton:

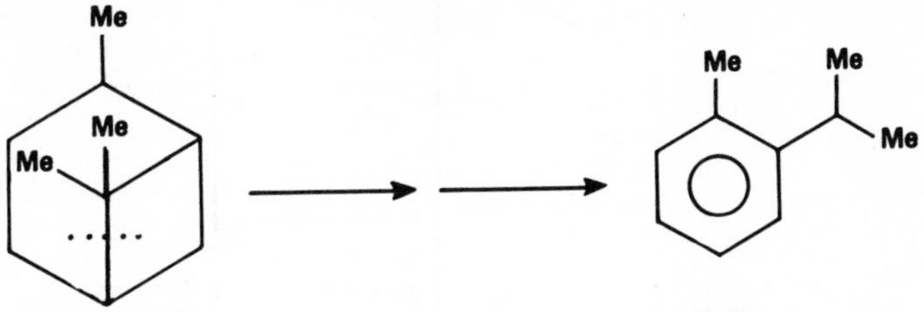

[I-3]

Table I. Aromatic Monoterpenoids

No.	Structure	Name	Occurence	Reference
1		p-cymene	wide-spread	
2			oil of Egyptian hashish (Cannabinaceae)	
3		o-cymene	turpentine oil from Pinus radiata	16
4		thymol	wide-spread	
5		thymol methyl ether	Orthodon spp., Thymus spp., (Labiatae) Crithmum maritimum (Umbelliferae)	
6		carvacrol	wide-spread	
7		carvacrol methyl ether	Cupressus spp., Chamaecyparis spp., (Cupressaceae)	

516

Table I

No.	Structure	Name	Occurence	Reference
8			Chamaecyparis, needle oil	159
9		thymohydroquinone	wide-spread	
10		p-methoxythymol	Heyderia (Libocedrus) decurrens (Cupressaceae)	14a
11		p-methoxycarvacrol	Heyderia (Libocedrus) decurrens (Cupressaceae)	14a
12		thymoquinone	wide-spread	
13		3-hydroxythymo-quinone	Juniperus chinensis (Cupressaceae);	160

Table I

No.	Structure	Name	Occurence	Reference
14	Me Me, HO, O, O, OH, Me	2,5-dihydroxy-thymoquinone	Juniperus chinensis (Cupressaceae); Monarda (Labiatae)	ibid.
15	Me Me, CH₂OH	cuminol	Eucalyptus dives (Myrtaceae); Lavandula sp. (Labiatae)	
16	Me Me, CHO	cuminal	Myrtaceae; Lavandula sp. (Labiatae), etc.	
17	Me Me, OH, CHO	macropone	Eucalyptus spp. (E. cneorifolia); Myrtaceae; Chamaecyparis taiwanensis, Thujopsis dolabrata (Cupressaceae)	161a, c
18	CH₂ CHOH, O, Me		Helenium spp. (Compositae)	13a
19	CH₂ O CH₂, R—O, O – CO – CHMe₂, Me, a, R = CO – CHMe₂, b, R = CO – CHMe – CH₂ – Me		Helenium spp. (Compositae)	13b

Table I

No.	Structure	Name	Occurence	Reference
20	a + b, R as in 19		Gaillardia spp. (Compositae)	ibid.
21	a + b, R as in 19		Gaillardia spp. (Compositae)	ibid.
22	a + b, R as in 19		Gaillardia spp. (Compositae)	ibid.
23	a + b, R as in 19		Gaillardia spp. (Compositae)	ibid.
24		libocedrol	Heyderia (Libocedrus) decurrens (Cupressaceae)	14b

Table I

No.	Structure	Name	Occurence	Reference
25		heyderiol		14d
26		libocedroxythymo-quinone		14c
27		<u>o</u>-cresol	Cupressaceae	12
28		<u>m</u>-cresol	oil of myrrh (Burseraceae); oil of black tea (Theaceae)	10A
29		p-cresol	Thujopsis dolabrata (Cupressaceae)	163
30			ibid.	ibid.

Table I

No.	Structure	Name	Occurence	Reference
31	Me Me / OH		ibid.; *Chamaecyparis taiwanensis* (Cupressaceae)	164
32	Me Me / OH	australol	*Eucalyptus* spp. (Myrtaceae)	161a
33	Me Me / OH / OMe	chamenol	*Chamaecyparis taiwanensis* (Cupressaceae)	164
34	Me OH / Me	p-tolylmethyl carbinol	*Curcuma xanthorrhiza* (Zingiberaceae); perhaps an artefact	15a

22-3.   SESQUITERPENOIDS

Several of the numerous classes of mevalonoid C_{15} compounds have aromatic representatives. The biosynthesis of sesquiterpenes in general, and especially its stereochemical aspects, have been reviewed by Parker, Roberts, and Ramage (17); their classification is followed here as much as possible. Again, as in the case of the monoterpenes, the aromatic sesquiterpenes are frequently associated with related alicyclic analogs, particularly dienic ones, and the ease with which such substances often aromatize (e.g., on mere standing in air) suggests that some of the aromatic compounds that have been isolated may be artefacts.

A small number of compounds of the general type [21] have been observed, in which the aromatic ring is the only cyclic structure. Members of this group have been encountered in a wide diversity of plants belonging to the gymnosperms, mono-, and dicotyledons.

The best-known compound of this class, α- or *ar*-curcumene [II-1], was first obtained from plants of the monocotyledonous family Zingiberaceae (*Curcuma* spp., *Zingiber officinale*), but it occurs in many other families as well (cf. (18) and references there). Both (+) and (−) forms have been found. The related ketone *ar*-turmerone (II-5) occurs in *Curcuma xanthorrhiza*; it has also been found in some Lauraceae (*Nectandra* and *Ocotea* spp.) (19). Its dihydro derivative [II-6] has been isolated from the wood of Himalayan Cedar (*Cedrus deodara*) (20).

Torreya nucifera (Taxaceae) yields the alcohol, nuciferol [II-3] and the corresponding aldehyde, nuciferal (II-4) (21). *Perezia* and *Trixis* spp. of the tribe Mutisiae of the Compositae contain two quinones, perezone (II-9) (22a) and hydroxyperezone (II-10) (22b); the former, known for a very long time and often called pipitzahoic acid in the older literature, had initially been assigned the structure with the alternative position of the phenolic hydroxyl. Its derivative (II-10), much more recently discovered, occurs as the mono-isovalerate (22b).

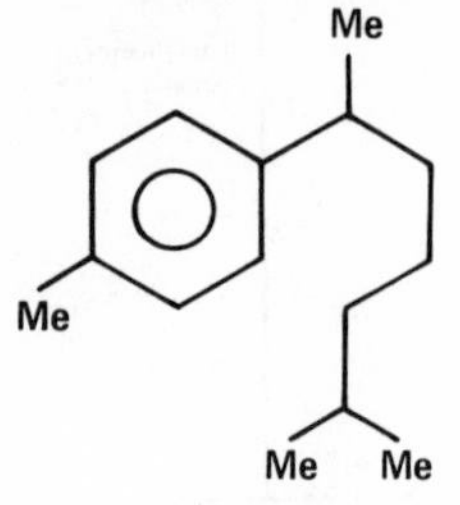

[21]

Finally, *Elvira biflora,* a plant belonging to another tribe of the Compositae, yields (II-11) (23), which is formulated as belonging to the unusual sylvestrene type with *meta*-oriented side-chains. The skeleton of (II-11) can be built from isoprene units, but not in the usual head-to-tail arrangement; its biosynthesis, perhaps involving a 1,2-shift of one of the side-chains, presents an interesting problem. *meta*-Orientation of the side-chains is also present in sesquichamaenol [II-12] isolated recently from *Chamaecyparis formosensis.* Its structure could well be interpreted as resulting from oxidative cleavage of a precursor of the cadinane-type (see below); many nonaromatic compounds with this skeleton have been isolated from the same plant (24).

In two other groups of aromatic sesquiterpenes, the side-chain of [21] has cyclized, forming a cyclopentane ring. In the cuparene group [22] (=[II-13]), this cyclization must have proceeded in a straightforward manner, yielding a skeleton derivable directly from farnesol via compounds of type [21]; the skeleton of the laurene group, [23], can again be constructed from isoprene units, but not in head-to-tail junction, so that shift of a methyl appears very probable.

Cuparene, [22](=[II-13], the parent compound of the first group, was isolated by Enzell and Erdtman (25) from the heartwood of certain conifers of the family Cupressaceae (*Chamaecyparis, Widdringtonia*). Oxidation with nitric acid to terephthalic acid, with ozone to (+)-camphonanic [24] acid, a known degradation product of camphor, proved structure and absolute configuration of [II-13].

At that time, [II-1] was the only known aromatic sesquiterpene, and the co-occurrence of this compound with its cyclohexadiene analogs led to the prediction that similar analogs of [II-13] would be found, a prediction quickly confirmed through the isolation of both cuparene and a dihydroproduct "cuprenene" from the oil of *Thujopsis dolabrata* (Cupressaceae) by Nozoe and Takeshita (26). Actually, their cuprenene was apparently a mixture

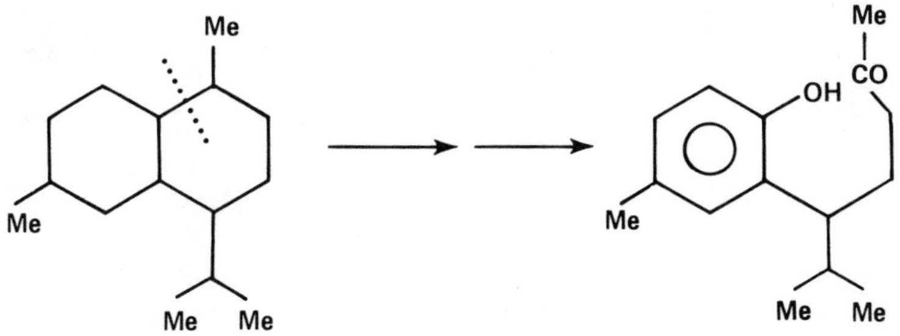

[II-12]

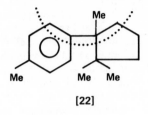

[22]

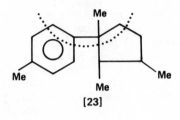

[23]

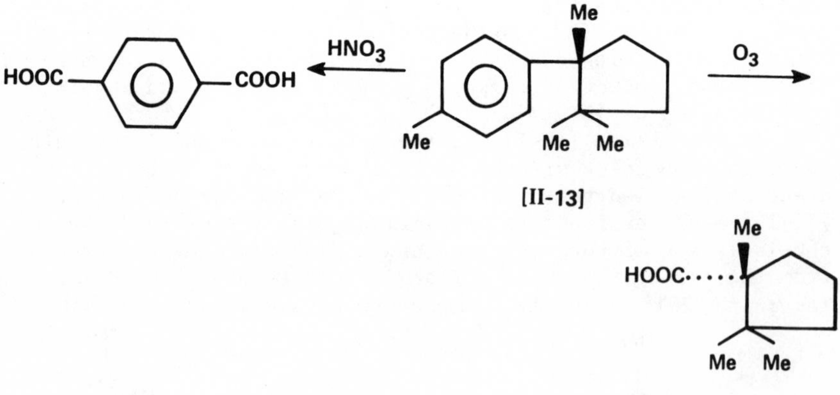

HOOC—⬡—COOH ←(HNO₃)— [II-13 structure] —(O₃)→ [24 structure]

[II-13]

[24]

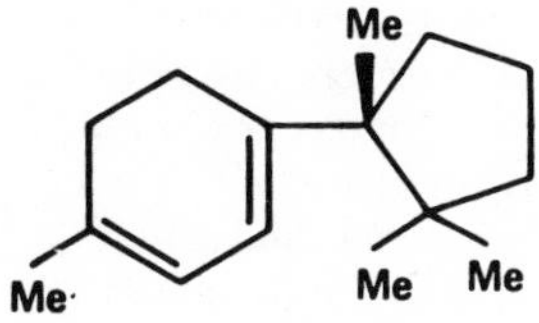

[25a]

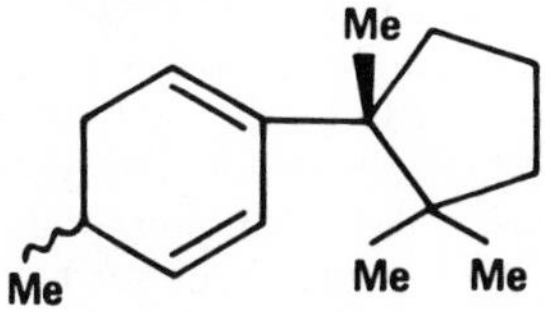

[25b] 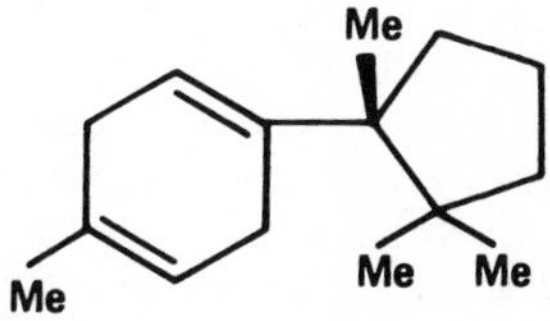   [25c]

of two isomers, α- and β-cuprenene, [25a] and [25b], respectively,
which, however, were not isolated in pure form. Subsequently
[25a] and a new isomer, [25c], were obtained pure from the same
plant source (27), and their structures proved by synthesis from
[II-13], to which both compounds revert on standing in air.
Compound [25b] was not found in this work.

Various oxygenated compounds of this group have been
encountered in plants of the family Cupressaceae. Cuparenic
acid [II-16], the acid corresponding to cuparene [II-13],
occurs together with it (25). The related aldehyde, cuparenal
[II-15], has been isolated (28a) from *Thujopsis dolabrata*,
which also contains the corresponding primary alcohol, γ-cupar-
enol, [II-14] (28b). Two ketones of the series, α-[II-17] and

β-[II-18] cuparenone, occur in the essential oil of a *Thuja* species (29). Both compounds have been reduced to (+)-[II-13] by the Wolff-Kishner method, which proves their configuration (29, 30, 41).

Related secondary alcohols were noticed to occur together with the ketones (29). Three such compounds, α-cuparenol [II-19], α-isocuparenol [II-20], and β-cuparenol [II-21], were subsequently isolated from *Thuja (Biota) orientalis* (30), where they occur together with α-cuparenol [II-14] and the two cuparenones. The epimers [II-19] and [II-20] are both oxidized to [II-17], while [II-21] similarly gives [II-18]. Multiple interrelations have established structures and configurations of these compounds, the constitutions of the group being further proven by the total synthesis of (±) [II-13] (26) and [II-17] (31); the latter had been synthesized already prior to its isolation.

Two quinones with the carbon skeleton of [22], helicobasidin [II-22] (32a-c) and deoxyhelicobasidin [11-23], have been obtained by Nishikawa, Natori, and their co-workers from the phytopathogenic fungus *Helicobasidium mompa* (Tremellales, Basidiomycetes). Interestingly, the stereochemistry of these fungal quinones is antipodal to that of [II-13] and its congeners from higher (cupressaceous) plants: compound [II-22] is oxidized to (-)-camphonanic acid, the antipode of [24]. The experimental evidence for the biosynthesis of [II-22] from acetic and mevalonic acid is given in Section 22-7.

A second, fairly numerous class of aromatic sesquiterpenoids, related to the preceding one but based on [23] rather than [22], is peculiar through its many bromine-containing representatives which, however, frequently occur together with the halogen-free parent compounds.

The first terpenoids of this type to be discovered were aplysin [II-24], debromoaplysin [II-25] and aplysinol [II-26], which were obtained from large sea-slugs of the genus *Aplysia* (Gastropoda, Mollusca), ("sea-hare") by Tamamura and Hirata (34). Irie and his co-workers (33) found subsequently that numerous compounds of this group, including aplysin itself and its two congeners, occur in maritime red algae (Rhodophyta) of the genus *Laurencia* (Rhodomelaceae). Since sea-hares feed on these seaweeds (ref. 35a, footnote 8), it can hardly be doubted that the aromatic terpenoids found in *Aplysia* are formed in the algae.

The genus *Laurencia* turned out to be a treasure-house of unusual bromine-containing compounds, which have been reviewed by Irie (36): two simple dibrominated benzene derivatives, three cyclic ethers with acetylenic side-chains, and a number of brominated aromatic terpenoids together with several of their halogen-free analogs; all the compounds of the last-named type are derived from [23]. The compounds observed so far, in addition to [II-24], [II-25], and [II-26], are the following: laurene [II-27] (35a,b), laurinterol [II-28] (37a,b), debromolaurinterol [II-29] (37a,b), isolaurinterol [II-30] (37b), debromoisolaurinterol [II-31] (37b), and laurenisol [II-32] (38). Structure and stereochemistry of this group rests on chemical and spectroscopic studies and on multiple interrelations, both within the group, and with related compounds of established constitution and configuration. Furthermore, the structures of [II-24] and [II-25] have been confirmed by total synthesis (39), that of [II-27] by total synthesis of its degradation product [23b] (see Scheme III) (35a). The absolute configuration of [II-28] has been established by X-ray crystallography, and interrelation of this key compound with the other oxygen-containing substances of the group in turn defines their absolute stereochemistry. That of the hydrocarbon, [II-27], rests upon the one of cuparene [II-13], with which it has been related, and hence via [24] upon the unequivocally known absolute configuration of camphor.

The more important relationships are presented in Scheme III. Of particular interest are: (a) the acid-catalyzed transformation of [II-28] and [II-30] into [II-24] and of [II-29] into [II-25] (37a,b); (b) the transformation of [II-32] into [II-25] (38); and (c) the correlation of [II-27] via α-cuparenone [II-17] with cuparene [II-13] (41) and hence, via camphonanic acid [24], with camphor [cf. ref. 40b]. All these compounds of the cuparene and laurene series have the same absolute configuration at the quaternary carbon atom, a fact that strongly suggests a biosynthetic relationship and adds interest to the occurrence of members of the antipodal series, [II-22] and [II-23], in *Helicobasidium*. Many of the structural and stereochemical features of this group of compound can be rationalized by assigning a key role in the biosynthesis of this group to [II-28], but the α-configuration of the secondary methyl in [II-27] and [II-32], deduced from their nmr spectra, appears unexpected on this basis; it could be explained in a variety of ways, including a 1,2-shift of one of the *gem*-methyls in a precursor with the skeleton of [II-13], or the assumption that a hypothetical analog of [II-28] with opposite stereochemistry of the cyclopropane ring is an intermediate.

Two major classes of sesquiterpenoids have skeletal structures derived from decalin: the cadinane [26] and eudesmane [27] classes, respectively. Aromatic representatives of both classes have been found; not surprisingly, aromatic sesquiterpenoids of the former are quite numerous, since the skeletal structure of [26] is compatible with straightforward aromatization of either

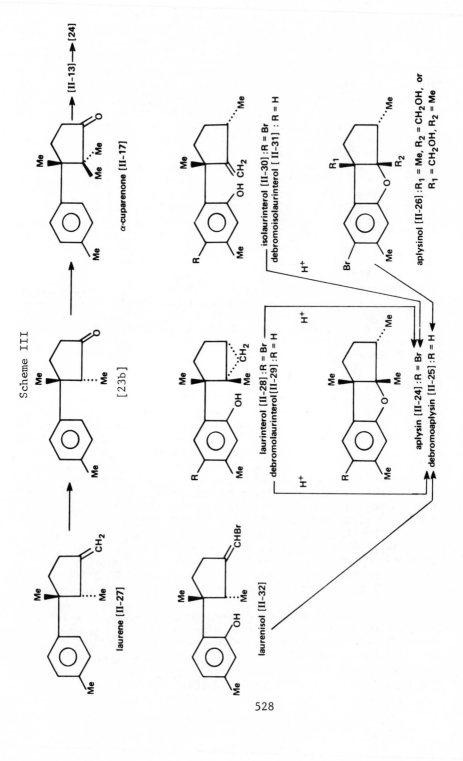

Scheme III

laurene [II-27]

[23b]

α-cuparenone [II-17]

[II-13] ⟶ [24]

laurinterol [II-28] : R = Br
debromolaurinterol [II-29] : R = H

isolaurinterol [II-30] : R = Br
debromoisolaurinterol [II-31] : R = H

laurenisol [II-32]

aplysin [II-24] : R = Br
debromoaplysin [II-25] : R = H

aplysinol [II-26] : R₁ = Me, R₂ = CH₂OH, or
R₁ = CH₂OH, R₂ = Me

one of the rings, or of both. In contrast, aromatization of any precursor with the eudesmane [27] skeleton is possible only with elimination or migration of the angular methyl--obviously a less favorable process, and indeed only two compounds of this type have been found so far. The derivation of [26] and [27] by alternative cyclization of a chain of three isoprenoid units in normal head-to-tail arrangement will be evident from Scheme IV.

Scheme IV

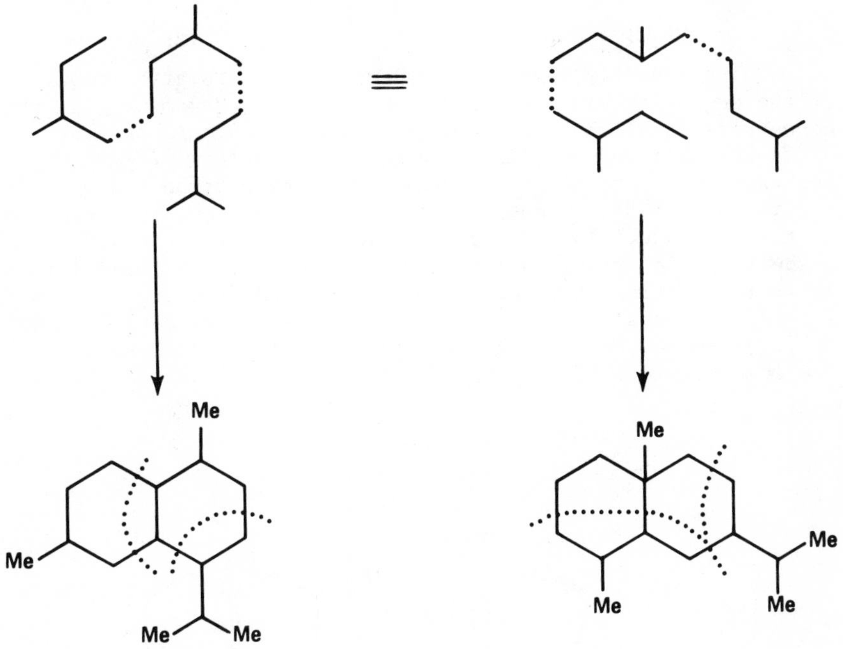

[26]   [27]

Occidol, [II-33], long the lone aromatic representative of the eudesmane type [27], occurs in *Thuja occidentalis*, together with the closely related sesquiterpenoid diene occidentalol [28] (42). The structures of [28] and [II-33] strongly suggest, if not an actual precursor-product relationship, at least a close biosynthetic connection.

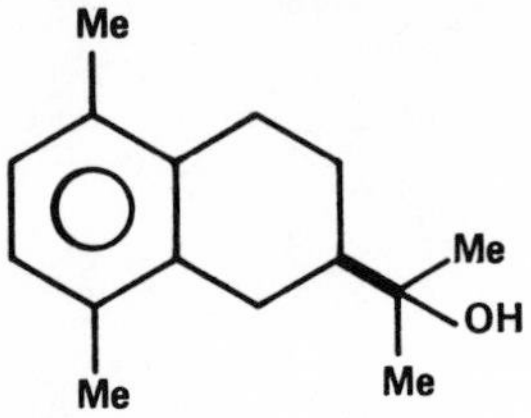

[II-33] 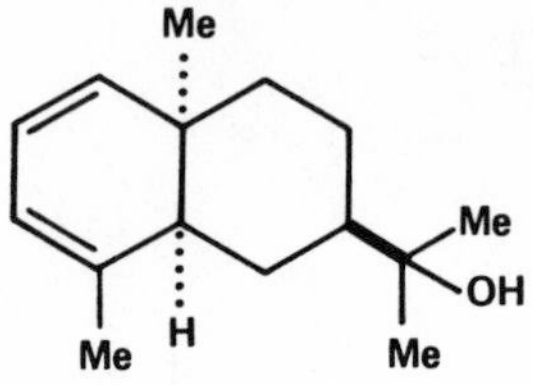   [28]

Quite recently, another closely related aromatic member of the eudesmane class was reported. Katsui, Masamune, and their co-workers (42A) observed that potatoes (*Solanum tuberosum, S. demissum*) infested with the fungus *Phytophthora infestans,* the cause of "late blight," contain the sesquiterpenoid rishitinol [II-34], whose structure was proven by total synthesis of the racemate. The closely related but nonaromatic antifungal sesquiterpenoid rishitin [29] occurs together with [II-34] (42A); it is interesting that these two co-occurring substances exemplify alternative ways in which the blocking quaternary methyl group of [27] can be removed: 1,2-migration in [II-34], loss from the molecule in [29].

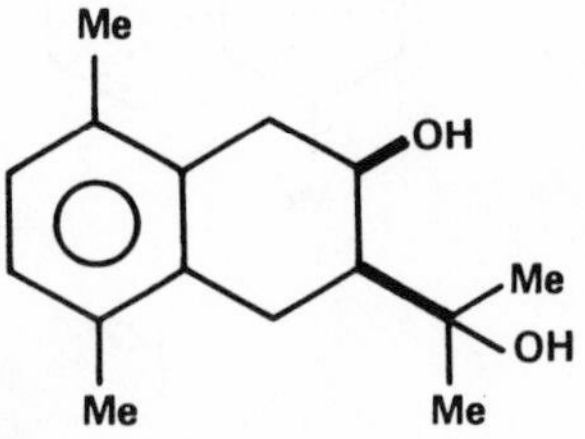

[II-34] 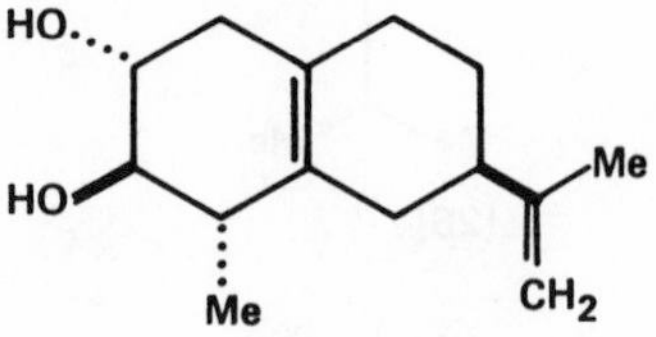   [29]

Aromatic members of the cadinane [26] class are listed in Table II (compounds [II-35] to (II-56]). Calamenene (II-35] (43a) was first isloated from the oil of Sweet Flag, *Acorus calamus* (Araceae), but has subsequently been encountered in plants of many other unrelated families (Piperaceae, Cupressaceae, Melia-ceae, Leguminosae, Compositae, etc.). In many of these plants, [II-35] is accompanied by its dehydro-derivatives, the α-, β-, and γ-calacorenes [II-36, 37, and 38). The first two of these were likewise obtained initially from the oil of Sweet Flag (43b,c), and subsequently from other plant sources; [II-38] has recently been observed in the oil of *Juniperus rigida* and in that of hop (*Humulus lupulus*, Cannabinaceae) (44). The latter oil also contains still another isomer, α-corocalene [II-39] (44).

Oxygen-containing derivatives [II-40 and 41] of calamenene [II-35] occur in the heartwood of many species of elm (*Ulmus*, *Ulmaceae*), which also contains three oxygenated derivatives [II-43, 44, and 45] of cadalene [II-42] (1,6-dimethyl-4-iso-propylnaphthalene) (45). The latter hydrocarbon itself has been observed repeatedly in a variety of plants. Compounds [II-40, 41, 43, 44, and 45] form a very interesting group of closely related substances; the co-occurrence of the analogous methyl- and aldehydo-compounds is noteworthy, as is that of tetralin- and related naphthalene-derivatives, which suggests that the latter might be formed through a secondary aromatization (see Chapter 25) of the former.

The wood of *Mansonia altissima* (Sterculiaceae) contains (46A-D) a similarly close-knit group of quinonoid compounds of the cadinane type. These compounds, which are responsible for the skin-irritant properties of the wood, are called mansonones A-L [II-46 to II-55]. Of these, mansonones A [11-46] and B [II-47] are quinones of the tetralin series; compounds C [II-48] and G [II-49] are 1,2-naphthoquinones, while in the remaining substances, except D, this latter structure is modified by closure of an oxygen-containing six-membered ring, which is saturated in E [II-50], H [II-51], and I [II-52], and unsaturated in F [II-53] and L [II-54]. Finally, mansonone D [II-55] is formulated tentatively (46a) as containing a dihydrofuran ring instead of the dihydropyran of mansonones E, H, and I. Compound [II-53] has a close diterpenoid analog in biflorin [III-2].

Mansonones E [II-50] and F [II-53], which are fungitoxic, have been isolated from the wood of elms (*Ulmus hollandiae*) infested with *Ceratocystis ulmi*, the fungus which causes Dutch elm disease. They were not found in wood from healthy trees or in that from disease-resistant strains (46e). On the other hand, [II-48] seems to be a normal constituent of the wood of some *Ulmus* species (46f, 191).

Another terpenoid obviously derived from [26] is gossypol [II-56] from cotton (*Gossypium* spp.) and some other genera of the family Malvaceae. Its structure, established by R. Adams and his co-workers (47), strongly suggests that this C_{30} compound is not a triterpenoid derived in the usual way from squalene, but that it is formed through symmetrical phenol-coupling of two C_{15} units of the [26] type; the location of the junction between the two moieties is hardly compatible with the former alternative. The actual biosynthesis of [II-56] has been studied in greater detail than that of any other aromatic terpenoid; see Section 22-7.

Volatile oils from fruits and leaves of strawberry (49a), and from peaches (49b), contain the C_{13} compound [II-57], while volatiles from peach leaves yield the same compound together with its dihydro derivative ionene [II-58], long known as an *in vitro* transformation product of ionone. The structures of these compounds, with their *gem*-dimethyl groups, certainly indicate a terpenoid origin. Since the oils are produced by steam distillation of the plant materials, these two compounds might originate as artefacts from ionone-type precursors, and are quoted here with reservations. Alternatively, however, they might be formed from some true sesquiterpenoid by loss of two carbon atoms.

The steam-volatile oil of hops, *Humulus lupulus*, contains 1,6-dimethylnaphthalene and 4,7-dimethyltetralone, together with the three calacorenes [II-36, 37, 38], α-corocalene [II-39], cadalene [II-42], and a number of nonaromatic terpenoids. The structures of the two dimethyl compounds likewise suggest an origin from cadinane-type precursors by loss of the isopropyl group (49c).

In connection with these naphthalenoid compounds it may be significant that mass-spectrometric evidence has been obtained for the occurrence of naphthalene itself and of both its monomethyl derivatives in strawberry volatiles (50).

Several aromatic sesquiterpenoids have been isolated that belong to classes other than those already discussed.

The himachalane [30] class, represented by a small number of sesquiterpenoids from Himalayan cedar (*Cedrus deodara*, Pinaceae), includes the aromatic sesquiterpene *ar*-himachalene [II-59] (51a,b), whose structure has been proved by total synthesis (51b).

The ester [II-60], apparently the first indene encountered in nature, has been isolated from a liver wort, *Calypogeia trichomanis*, and its structure corroborated by total synthesis (52). It is suggested (52) that the compound is formed biosynthetically from a hypothetical precursor [31] with an azulene skeleton by aromatization with rearrangement to form the indene system, accompanied or followed by oxidative shortening of the side-chain. This interpretation is supported by the fact that the plant also contains

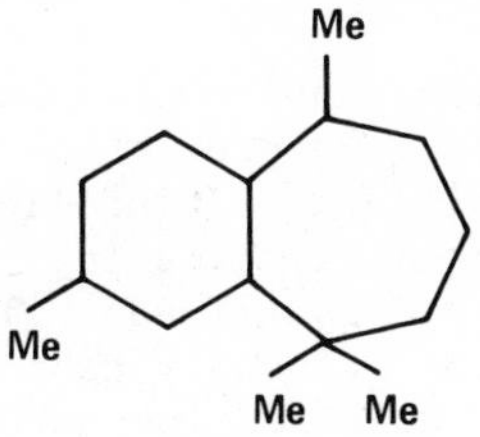

[30]

two authentic azulenes, [32a and b], apparently existing as such
in the plant -- a rare occurrence, since nearly all other azulenes
obtainable from natural sources are isolation artefacts.

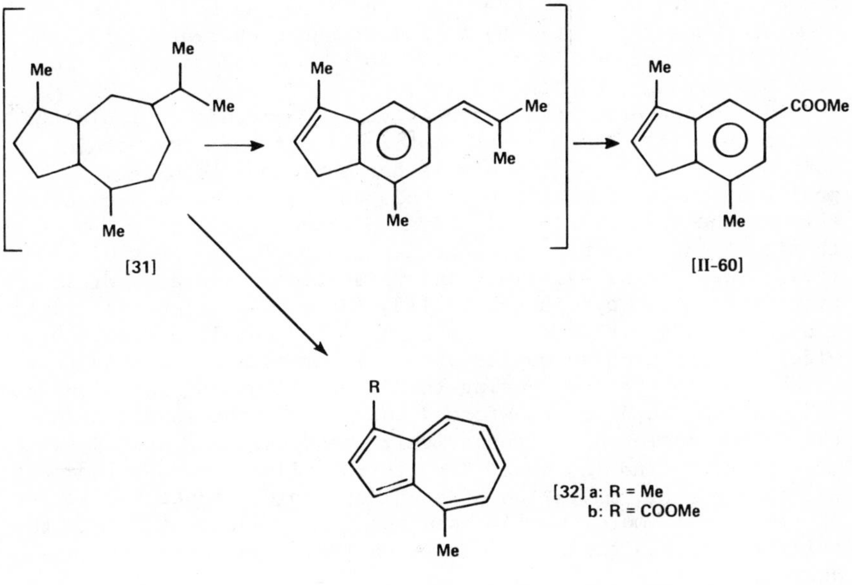

[31]

[II-60]

[32] a: R = Me
b: R = COOMe

The hydrindene skeleton has been shown by Anchel, McMorris, and their co-workers (53) to be present in two metabolites of the Basidiomycete *Clitocybe illudens*, illudalic acid [II-61] and illudinine [II-62], and probably also in a third one, illudoic acid, tentatively formulated as [II-63]. The common carbon skeleton of these compounds evidently cannot be dissected into isoprenoid units. Their probable biosynthesis becomes clear, however, on the basis of that of a number of substances occurring in the same species, and in some other fungi.

Besides [II-61, [II-62], and [II-63], *C. illudens* also contains the illudins M [33a] and S [33b], the corresponding secondary alcohols [34a and b] (54A), and illudol [35] (54). Compounds biosynthetically related to these substances but isolated from other fungi are marasmic acid [36] (57), fomannosin [37], hirsutic acid [38], coriolin [39a], coriolin B [40], and coriolin C [39b] (57A). (For literature references on compounds not specifically documented, see the bibliography of ref. 54.) Compound [33b] has also been isolated from another fungus, *Lampteromyces japonicus*, by Nakanishi and co-workers (55), who called the compound "lampterol." Both *C. illudens* and *L. japonicus* are strongly bioluminescent.

The biogenetic connection of these compounds is indicated by the unit [41] which is present in all of them (except in compounds [37] to [40], where it has undergone secondary modifications). The nature of this connection becomes clear through the scheme which was first proposed (56) for the biosynthesis of [33a and b] and subsequently applied to that of [36] (57). (See Scheme V). This interpretation postulates cyclization of farnesyl pyrophosphate [13] (or an equivalent of it) to a species [42] with the ring system of the humulenes which are widely distributed sesquiterpenes. Intermediate [42] can than be assumed to cyclize, giving the "proto-illudane" skeleton [43] (53), which could be transformed to [35] without modification of the carbon skeleton. Alternatively, rearrangement of [43] with migration of bond (b) would lead to the illudin series [33a and b], [34a and b], migration of bond (a) would produce the skeleton of [36], while more complex changes would generate [37] and the tricyclic carbon skeleton from which [38], [39], and [40] are derived.

[41]

This biosynthetic scheme, proposed before compounds with the carbon skeleton of the hypothetical intermediate [43] were known, gained much in probability by the subsequent discovery of [35] in *C. illudens* (54).

Scheme V

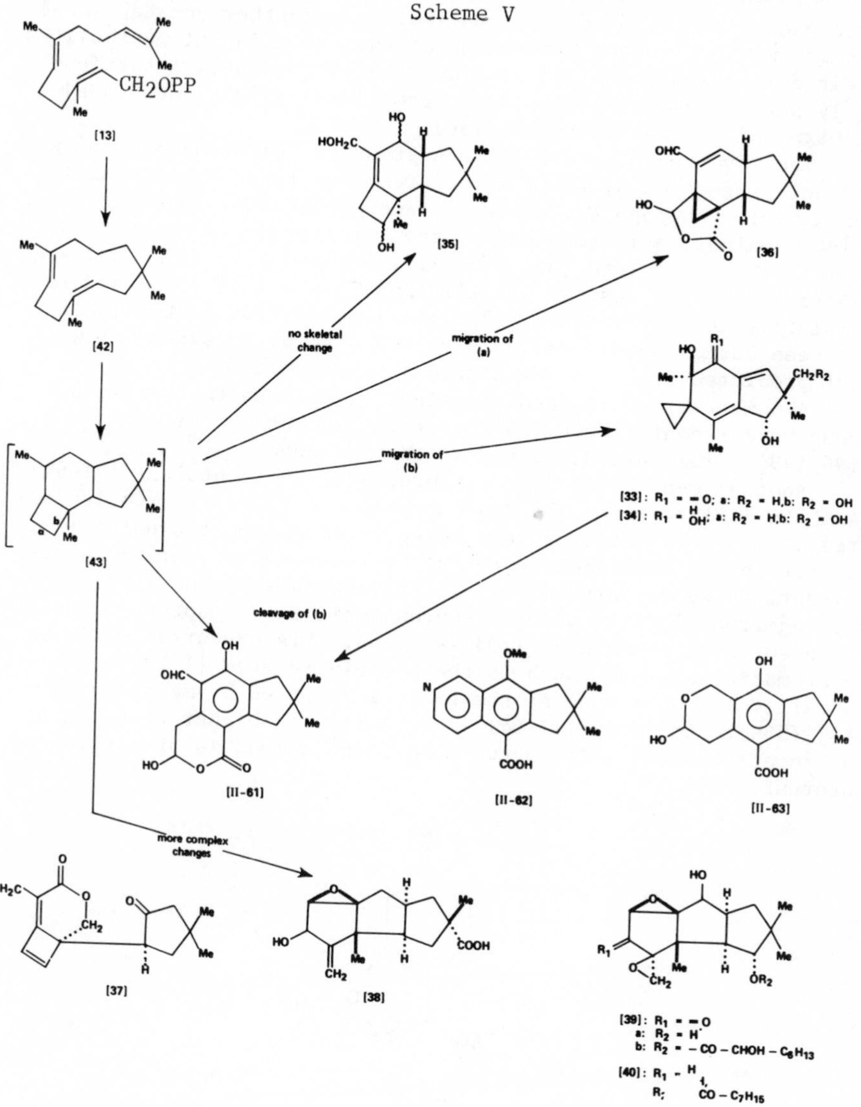

The structures of the three aromatic representatives of the group, [II-61], [II-62], and [II-63], can be rationalized readily by assuming, first, cleavage of bond (b) in [43], or perhaps of the corresponding bond in a precursor with the skeleton of [33], accompanied or followed by aromatization. Further unexceptional changes would give the structures of the three fungal metabolites. Their derivation from [33] or some related compound is particularly attractive, since the transformations of [33a] and [33b] *in vitro* offer much precedent (56).

The mevalonoid nature of this group of compounds (of course strongly suggested by their structures) has found experimental confirmation through the observed incorporation of label from 2-^{14}C mavalonic acid into one of the three aromatic substances, [II-62] (53), and into [33a and b], [35], and [36] (57) [35] and [36](57). In the case of the illudins [33a and b] (56B) and less completely in that of [36] (57), the localization of the label has been established, and has been found to be in agreement with that predicted from the postulated biosynthesis.

A number of hydrindanones evidently related to the illudin group have recently been isolated from two species of fern (196-199). They are discussed in the Addendum.

Another O-heterocyclic C$_{15}$ terpenoid, furoventalene [II-64], has been isolated from the sea-fan *Gorgonia ventalina* (Coelenterata) (58a). Its structure can be built from isoprene units [I], but not in the regular head-to-tail arrangement of normal farnesol-derived sesquiterpenoids. It is best interpreted (58a) as arising from addition of one isoprene unit to a monoterpenoid precursor with p-menthane [20] skeleton; this precursor could be an aromatic compound, such as the closely related [I-18], or and alicyclic substance of the type represented by menthofuran [44] (58b). Compound [II-64) is actually the higher prenolog of an aromatized [44]. Its occurrence in an animal is of particular interest.

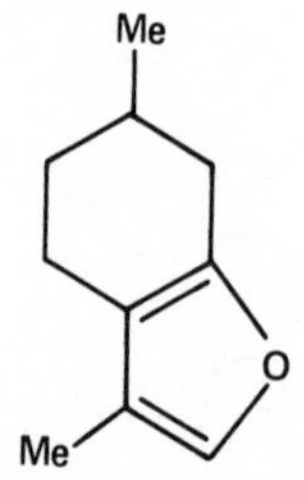

[44]

A further group of aromatic sesquiterpenoids with a furan ring occurs in *Cacalia decomposita* (Compositae). The compounds in question are cacalol [II-65], cacalone [II-66][+], maturinin [II-67], maturin [II-68], maturinone [II-69], and maturone [II-70], apparently all derived from the ring system [45]. In the structures initially assigned to these compounds, their sesquiterpenoid nature was obscured by the incorrect placement of the methyl group at C-5 which had, for seemingly valid reasons, been assumed to occupy position 8. However, simultaneous work in several laboratories (59a) has shown, by degradative and spectroscopic methods and by total synthesis, that maturinone is correctly formulated as [II-69]. Since the other compounds have been related with it, their structures should be similarly modified, although direct evidence for this modification has not been published so far.

Biosynthetically, these benzenoid and naphthalenoid compounds could be derived by migration of the quaternary methyl (followed, in the case of [II-69] and [II-70], by loss of this carbon atom) from a nonaromatic precursor of the normal eudesmane [27] or the rearranged eremophilane [46] class. The second possibility, more

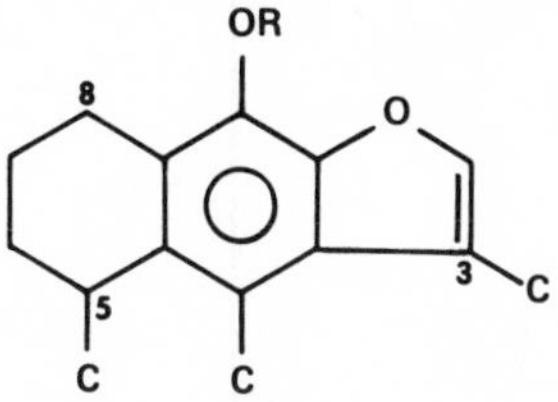

[45]

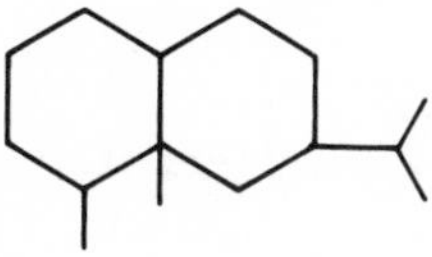

[46]

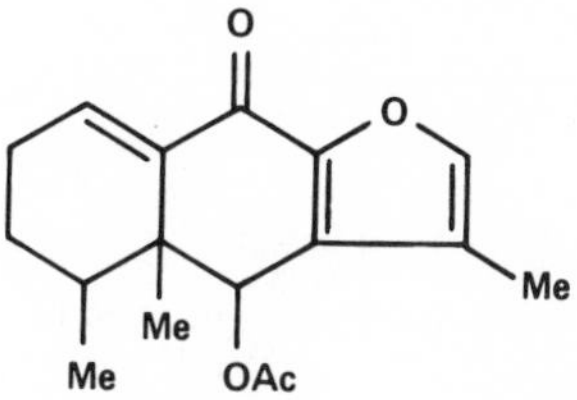

[47]

[+]According to recent evidence, cacalone [II-66] is not an aromatic compound but the 4-hydroxy-cyclohexadienone related to [II-65]; cf. A. Casares and L.A. Maldonado, Tetrahedron Letters 2485 (1976).

Table II. Aromatic Sesquiterpenoids

No.	Structure	Name	Occurence	Reference
a)	Monocyclic compounds			
1	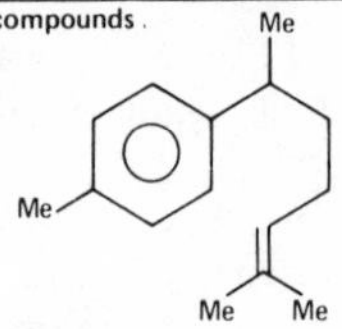	α- curcumene	Curcuma spp., Zingiber officinale (Zingiberaceae) Acorus calamus (Araceae) Solidago canadensis (Compositae), and others	18,43b
2	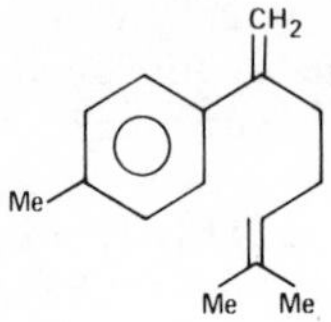	dehydro-ar-curcumene	Thuja orientalis (Pinaceae)	165
3	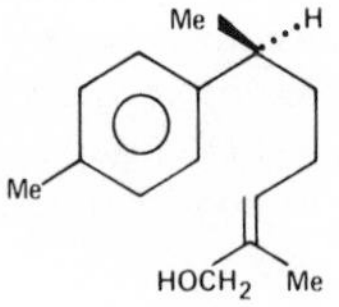	nuciferol	Torreya nucifera (Taxaceae)	21a
4	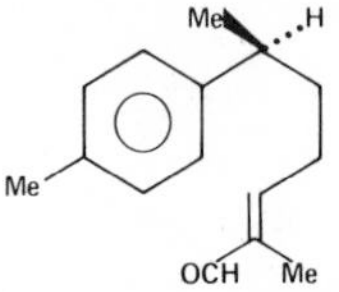	nuciferal	Torreya nucifera (Taxaceae)	21a,b
5	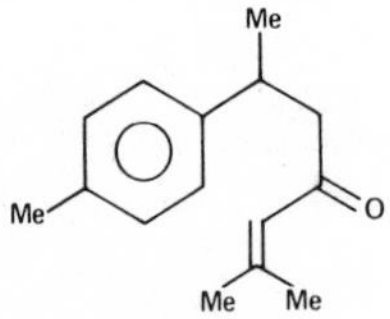	ar- turmerone	Curcuma xanthorrhiza ("Temu Lawak"), Zingiberaceae	166
6	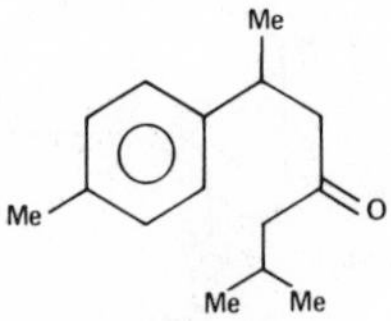	dihydro-ar-turmerone	Cedrus deodara, (Pinaceae)	20
7	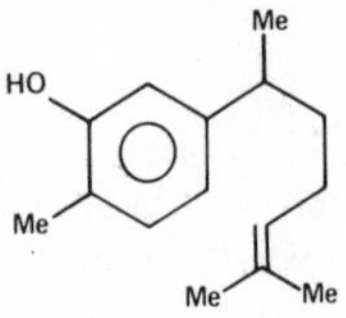	xanthorrhizol	Curcuma xanthorrhiza (Zingiberaceae)	167

Table II

No.	Structure	Name	Occurence	Reference
8	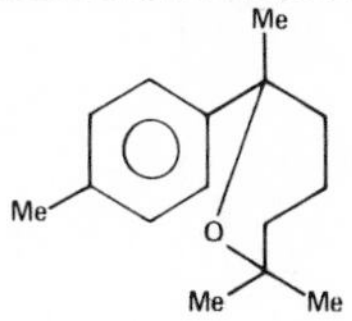		Thuja orientalis	168
9	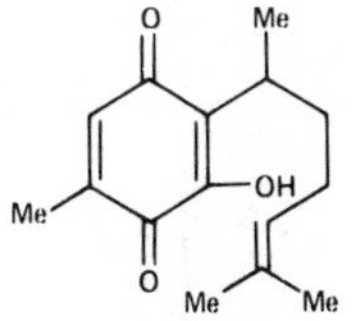	perezone	Perezia spp.; Trixis cacalioides (Compositae)	22a
10	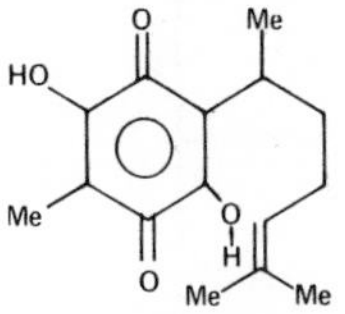	hydroxyperezone	Perezia spp.	22b
11	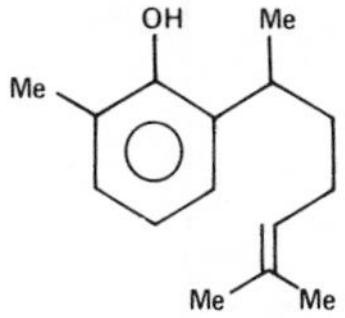		Elvira biflora (Compositae)	23
12	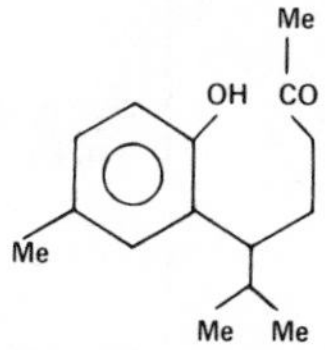	sesquichamaenol	Chamaecyparis formosensis (Cupressaceae)	24

b) Compounds of the Cuparene and Laurene groups.

No.	Structure	Name	Occurence	Reference
13	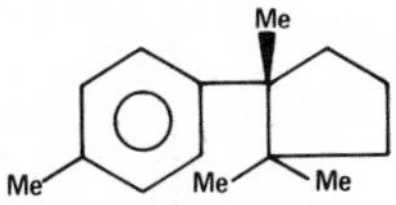	cuparene	Chamaecyparis, Thuja, Thujopsis, Widdringtonia (Cupressaceae)	25, 26
14	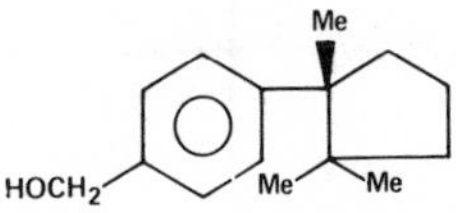	γ-cuparenol	Thujopsis dolabrata	28b

539

Table II

No.	Structure	Name	Occurence	Reference
15		cuparenal	Thujopsis dolabrata	28a
16		cuparenic acid	Chamaecyparis, Thuja, Widdringtonia	25
17		α-cuparenone	Thuja spp.	29
18		β-cuparenone	Thuja spp.	29
19		α-cuparenol	Thuja (Biota) orientalis	30
20		α-isocuparenol	Thuja (Biota) orientalis	30
21		β-cuparenol	Thuja (Biota) orientalis	30
22		helicobasidin	Helicobasidium mompa (Basidiomycetes)	32a,b,c
23		deoxyhelicobasidin	Helicobasidium mompa (Basidiomycetes)	32d

Table II

No.	Structure	Name	Occurence	Reference
23 a			Helicobasidium mompa (Basidiomycetes)	169
24		aplysin	Aplysia kurodai; (Mollusca); Laurencia okamurai (Rhodomelaceae)	33, 39a
25		debromoaplysin	Aplysia kurodai (Mollusca); Laurencia okamurai (Rhodomelaceae)	33, 38
26	$R_1 = Me$, $R_2 = CH_2OH$ or $R_1 = CH_2OH$, $R_2 = Me$	aplysinol	Aplysia kurodai (Mollusca); Laurencia okamurai (Rhodomelaceae)	33, 38
27		laurene	Laurencia glandulifera, L. nipponica	35
28		laurinterol	L. intermedia	37
29		debromolaurinterol	L. intermedia	37
30		isolaurinterol	L. intermedia	37b

Table II

No.	Structure	Name	Occurence	Reference
31		debromoisolaurinterol	*L. intermedia*	37b
32		laurenisol	*L. nipponica*	38
c)	Compounds derived from Tetralin and Naphthalene			
33		occidol	*Thuja occidentalis*	42a,c
34		rishitinol	Potatoes infested with *Phytophthora infestans*	42d
35		(+) or (−) calamenene	widely distributed	45b
36		α-calacorene	widely distributed	43b,c
37		β-calacorene	widely distributed	43c

Table II

No.	Structure	Name	Occurence	Reference
38	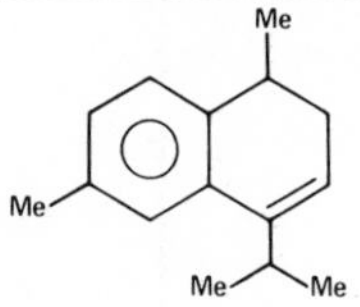	γ-calacorene		44
39	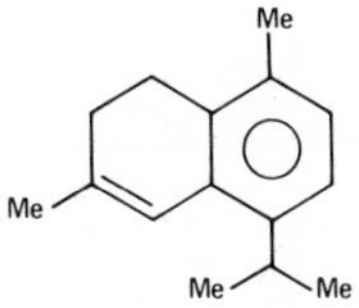	α-corocalene	<u>Humulus lupulus</u> (Cannabinaceae)	44
40	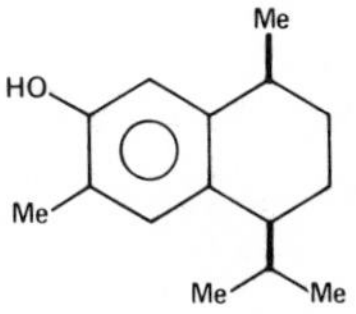	7-hydroxycalamenene	<u>Ulmus</u> spp. (Ulmaceae)	45b
41	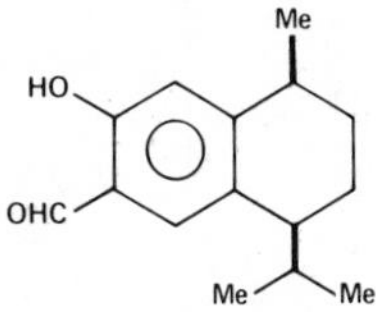		<u>Ulmus</u> spp.	45a,c
42	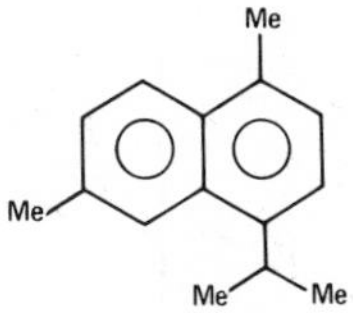	cadalene	widely distributed	
43	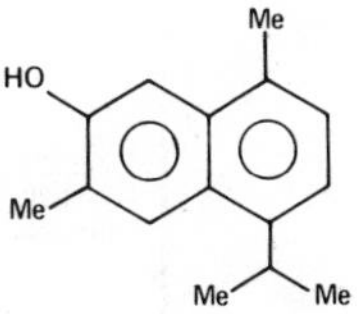		<u>Ulmus</u> spp. (Ulmaceae)	45a,b
44	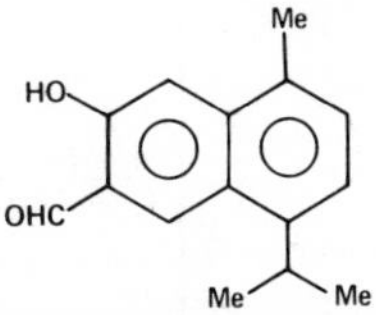		<u>Ulmus</u> spp. (Ulmaceae)	45a,b

543

Table II

No.	Structure	Name	Occurence	Reference
45			Ulmus spp. (Ulmaceae)	45a,b,c
46		mansonone A	Mansonia altissima (Sterculiaceae)	46a,b
47		mansonone B	Mansonia altissima (Sterculiaceae)	46a
48		mansonone C	Mansonia altissima (Sterculiaceae); Ulmus spp. (Ulmaceae)	46a,b 46f,191
49		mansonone G	Mansonia altissima (Sterculiaceae)	46b,c (and ref. inc.)
50		mansonone E	Mansonia altissima (Sterculiaceae); Ulmus spp. (Ulmaceae)	46a,b, 191
51		mansonone H	Mansonia altissima (Sterculiaceae)	46b,c and refs. in 46c

Table II

No.	Structure	Name	Occurence	Reference
52		mansonone I	Mansonia altissima (Sterculiaceae)	46d
53		mansonone F	Mansonia altissima (Sterculiaceae)	46a,f
54		mansonone L	Mansonia altissima (Sterculiaceae)	46c
55	(?)	mansonone D	Mansonia altissima (Sterculiaceae)	46a
56		gossypol	Gossypium sp. (Malvaceae)	47
57			strawberry oil	49a
58		ionene	oil from peach leaves	49b

Table II

No.	Structure	Name	Occurence	Reference
59	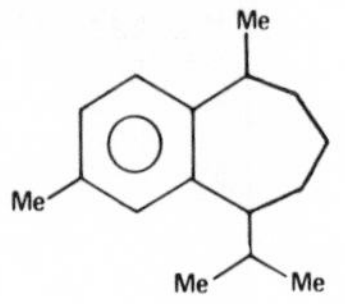	ar-himachalene	<u>Cedrus deodara</u> (Pinaceae)	51
60	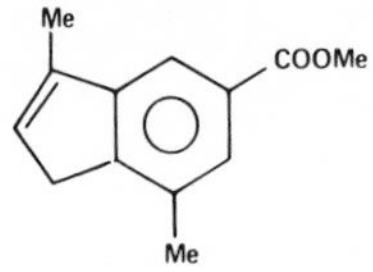		Calopogeia <u>trichomanis</u> (Hepaticae)	52
61	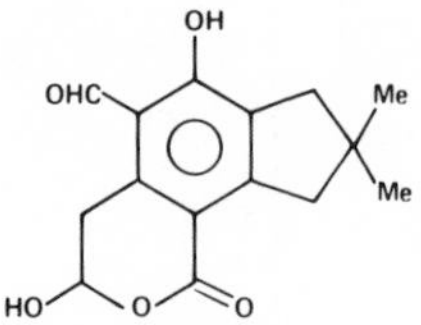	illudalic acid	<u>Clitocybe illudens</u> (Basidiomycetes)	53,54
62	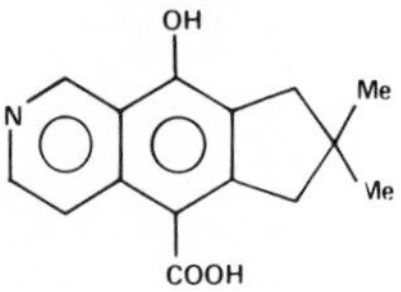	illudinine	<u>Clitocybe illudens</u> (Basidiomycetes)	53,54
63	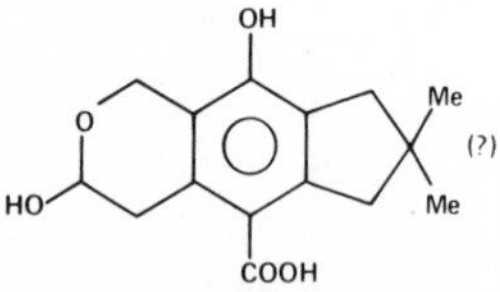	illudoic acid	<u>Clitocybe illudens</u> (Basidiomycetes)	54
64	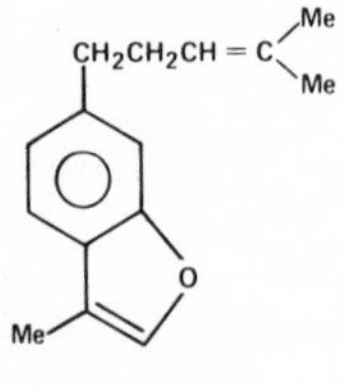	furoventalene	<u>Gorgonia ventalina</u> (Coelenterata)	58a

546

Table II

No.	Structure	Name	Occurrence	Reference
65	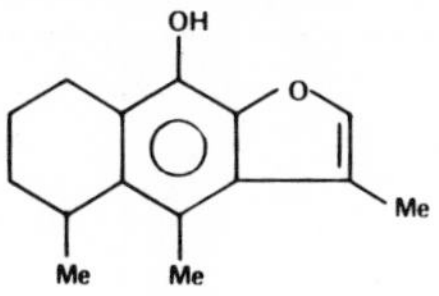	cacalol	*Cacalia decomposita* (Compositae)	59a
66	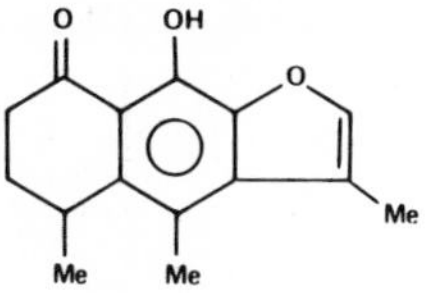	cacalone	*Cacalia decomposita* (Compositae)	59a
67	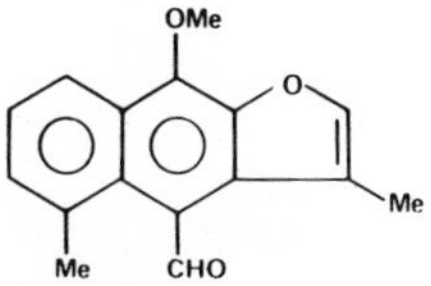	maturinin	*Cacalia decomposita* (Compositae)	59a
68	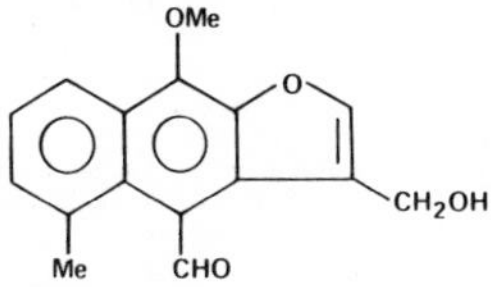	maturin	*Cacalia decomposita* (Compositae)	59a
69	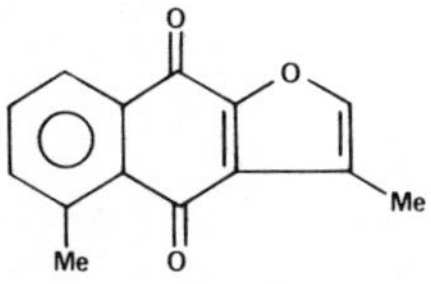	maturinone	*Cacalia decomposita* (Compositae)	59a
70	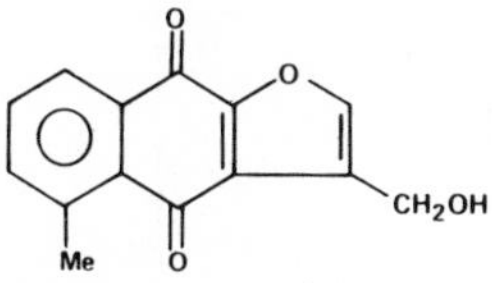	maturone	*Cacalia decomposita* (Compositae)	59a

+According to recent evidence, cacalone [66] is not an aromatic compound but the 4-hydroxy-cyclohexadienone related to [65]; cf. A. Casares and L.A. Maldonado, Tetrahedron Letters 2485 (1976).

Table III. Aromatic Diterpenoids

No.	Structure	Name	Occurrence	Reference
A. Compounds not containing the Phenanthrene Skeleton.				
1		ar - artemisene	Artemisia absynthium (Compositae).	63
2		biflorin	Capraria biflora (Scrophulariaceae)	64
B. Compounds with Phenanthrene Skeleton but not derived from Abietane [48].				
3		cleistanthol	Cleistanthus schlechteri (Euphorbiaceae)	65
4		nimbiol	Melia azadirachta (Meliaceae)	66
C. Compounds derived from Abietane [48] 1) Hydrocarbons				
5		ar · abietatriene	Widely distributed	70, 109c

548

Table III

No.	Structure	Name	Occurrence	Reference
6	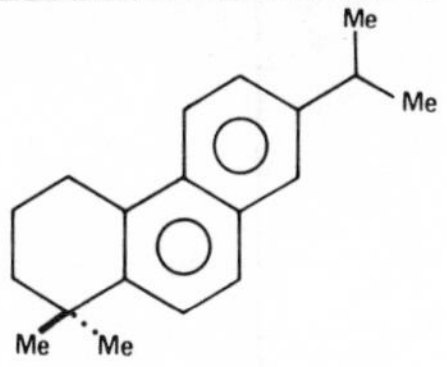	simonellite	In fossilized wood of a taxodiaceous plant	72
7	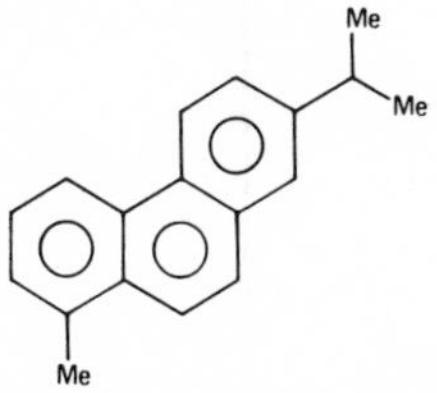	retene	In peat, and in fossil pine trunks in lignite deposits	71

2) Non-pherolic Hydroxyl, Carbonyl, and Carboxy Compounds.

No.	Structure	Name	Occurrence	Reference
8	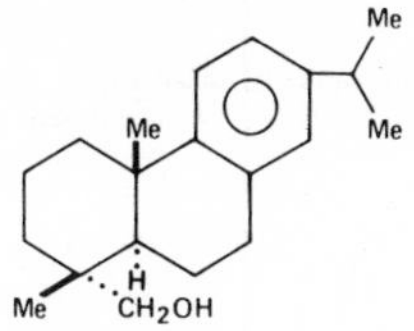	dehydroabietol	Pinus sylvestris, P. banksiana, P. monticola (Pinaceae)	109a,c, 170
9	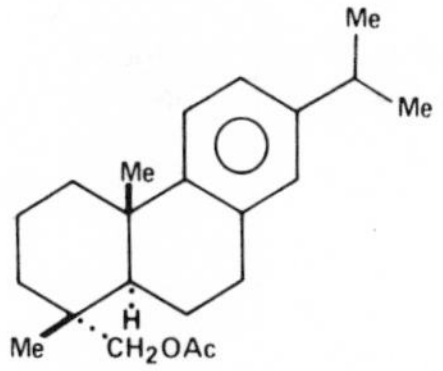	dehydroabietinyl acetate	" " " "	171
10	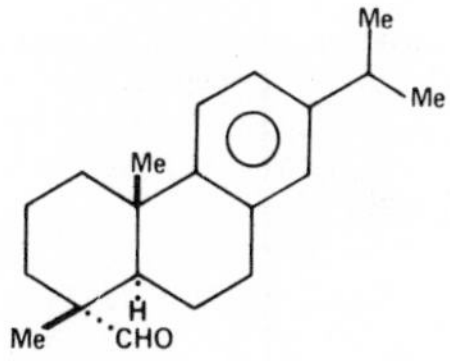	dehydroabietal	" " " "	170, 172
11	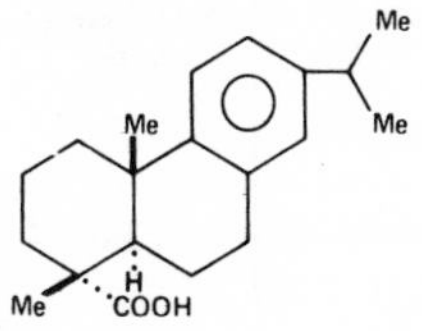	dehydroabietic acid	Pinus spp. (Pinaceae)	173

549

Table III

No.	Structure	Name	Occurrence	Reference
11A			Gum rosin	72A
11B			Distilled tall oil	72Ba
12		15-hydroxydehydro-abietic acid	Agathis spp. (Araucariaceae)	84
13		callitrisic acid, 4-epidehydroabietic acid	Callitris spp. (Cupressaceae)	73
14		7-oxo-4-epidehydro-abietic acid	Callitris columellaris (Cupressaceae)	114
15		teideadiol	Nepeta teydea (Labiatae)	78

Table III

No.	Structure	Name	Occurrence	Reference
3) Phenols				
16		ferruginol	Podocarpus, Dacrydium spp. (Podocarpaceae); Juniperus spp., (Cupressaceae); Inula royleana (Compositae)	75, 81 87
17		dehydroferruginol	Podocarpus, Dacrydium spp. (Podocarpaceae); Juniperus spp. (Cupressaceae)	81
18			Juniperus rigida (Cupressaceae)	86a
19		18-hydroxydehydro-ferruginol, "6-hydroxydehydro-abietinol"	Torreya nucifera (Taxaceae)	175
20		salviol	Salvia miltiorrhiza (Labiatae)	176

Table III

No.	Structure	Name	Occurrence	Reference
21		hinokiol	Chamaecyparis obtusa, and other conifers	177
22		hinokione	" " "	"
23	(?)	shonanol	Calocedrus (Libocedrus) formosana (Cupressaceae)	82, 82A
24			Cupressus sempervirens (Cupressaceae)	79
25		sugiol	Juniperus (Cupressaceae); Podocarpus (Podocarpaceae); Taxodium (Taxodiaceae); Melia azadirachta (Meliaceae)	178 66a
26		dehydrosugiol	Taxodium distichum (Taxodiaceae)	83

Table III

No.	Structure	Name	Occurrence	Reference
27		"xanthoperol precursor" *	Juniperus communis (Cupressaceae)	77
28		podocarpic acid	Podocarpus, Dacrydium spp., (Podocarpaceae)	75
29		methyl podocarpate	Podocarpus dacrydioides (Podocarpaceae)	74a
30		pododacric acid	Podocarpus dacrydioides, P. totara (Podocarpaceae)	74a,b, 85
31		cryptojaponol	Cryptomeria japonica (Taxodiaceae)	80

* Xanthoperol, the 5β-epimer of [III-27], isolated from many conifers, is an artefact.

Table III

No.	Structure	Name	Occurrence	Reference
32		carnosic acid, salvin	Salvia officinalis, Rosmarinus officinalis (Labiatae)	96
33		coleone C	Coleus aquaticus (Labiatae)	95
34		coleone B	Coleus igniarius (Labiatae)	93

4) Quinones and Quinone Methides

No.	Structure	Name	Occurrence	Reference
35		royleanone	Inula royleana (Compositae); Salvia officinalis, Plectranthus sp. (Labiatae); Taxodium distichum (Taxodiaceae)	83, 87, 89, 90 93
36		dehydroroyleanone	Inula royleana (Compsitae); Salvia officinalis, Plectranthus sp. (Labiatae)	87, 89, 90 93

Table III

No.	Structure	Name	Occurrence	Reference
37		7-hydroxy-royleanone, horminone	Horminum pyrenaicum (Labiatae)	83, 88
38		taxoquinone	Taxodium distichum (Taxodiaceae)	83
39		7-acetoxyroyleanone, "9-acetoxy-royleanone"	Inula royleana (Compositae)	87
40		taxodone	Taxodium distichum (Taxodiaceae)	83
41		taxodione	" " "	83
42		fuerstione	Fuerstia africana (Labiatae)	92

Table III

No.	Structure	Name	Occurrence	Reference
43		coleone A	Coleus igniarius (Labiatae)	94
44		miltirone	Salvia miltiorrhiza (Labiatae)	99
45		cryptotanshinone	" " "	97
46		tanshinone II	" " "	97
47		hydroxytanshinone	" " "	98

Table III

No.	Structure	Name	Occurrence	Reference
48		tanshinone II-B	Salvia miltiorrhiza (Labiatae)	96
49		methyl tanshinonate	" " "	98
50		tanshinone I	" " "	97
51		isocryptotanshinone	" " "	98a
52		isotanshinone II	" " "	98a

Table III

No.	Structure	Name	Occurrence	Reference
53	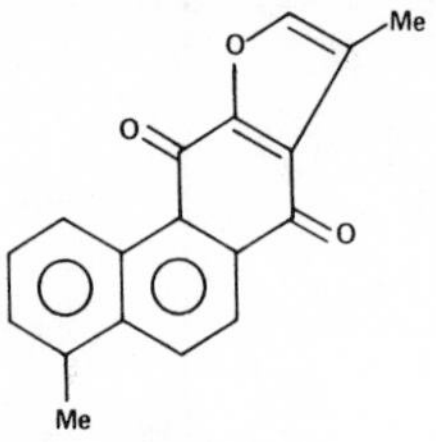	isotanshinone I	Salvia miltiorrhiza (Labiatae)	98a

D) Compounds with rearranged Abietane Skeleton

No.	Structure	Name	Occurrence	Reference
54	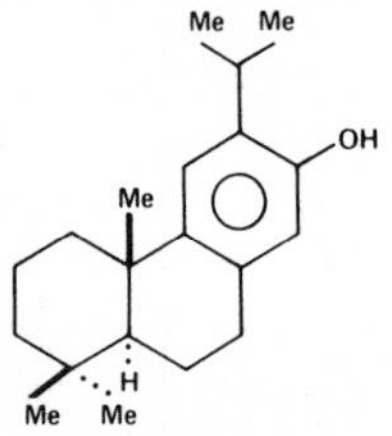	sempervirol	Cupressus sempervirens (Cupressaceae)	101
55	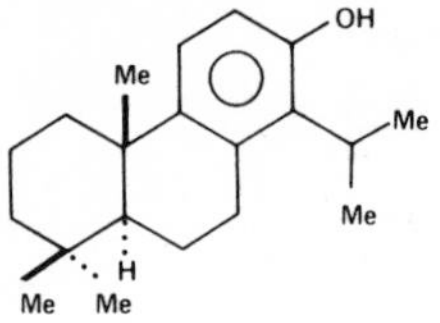	totarol	Podocarpus spp., Dacrydium cupressinum (Podocarpaceae)	100 a,b
56	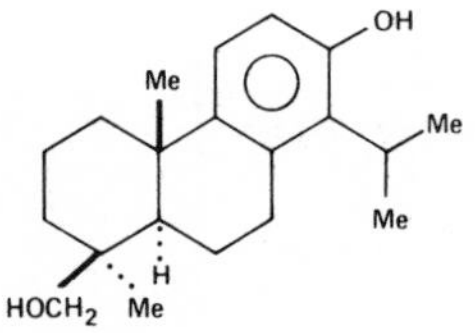	19-hydroxytotarol, "16-hydroxytotarol"	Podocarpus spp., (Podocarpaceae)	104, 105
57	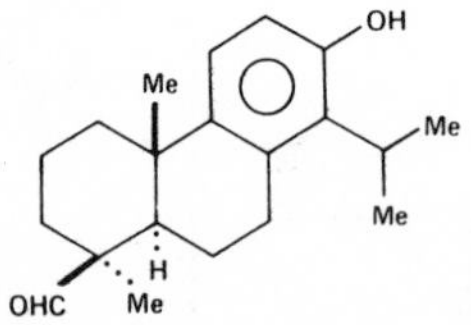	19-oxototarol "16-oxototarol"	Podocarpus spp. (Podocarpaceae)	104, 105
58	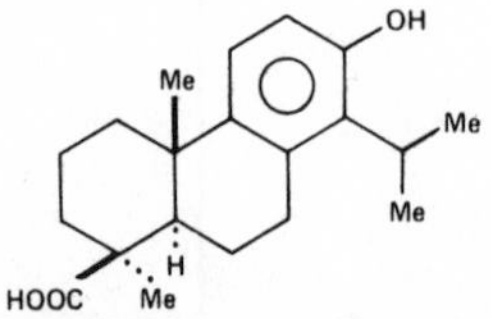	"16-carboxytotarol"	" " "	105

558

Table III

No.	Structure	Name	Occurrence	Reference
59		totarolone	Tetraclinis articulata (Cupressaceae)	103
60		totarolenone	" " "	103
61			Cupressus sempervirens (Cupressaceae)	79
62	 inseparable mixture of both C-7 epimers		Thujopsis dolabrata (Cupressaceae)	106
63		podototarin	Podocarpus totara (Podocarpaceae)	104

Table III

No.	Structure	Name	Occurrence	Reference
64	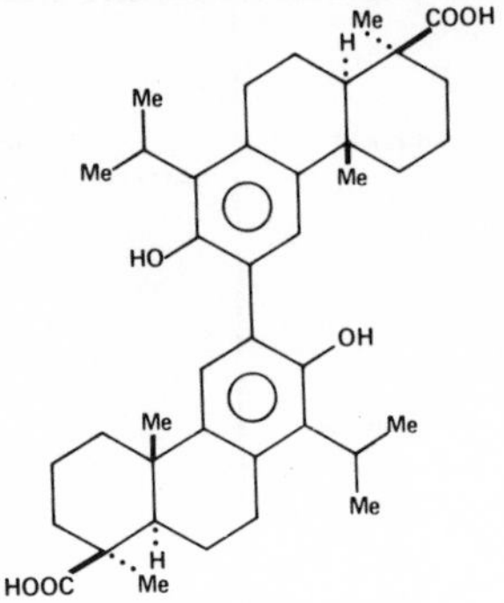	macrophyllic acid	Podocarpus macrophyllus (Podocarpaceae)	107
	E) Appendix*			
65	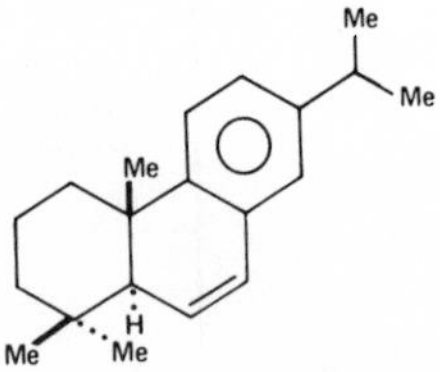	6,8,11,13-abietatetraene	Pinus monticola (Pinaceae)	109a,c
66	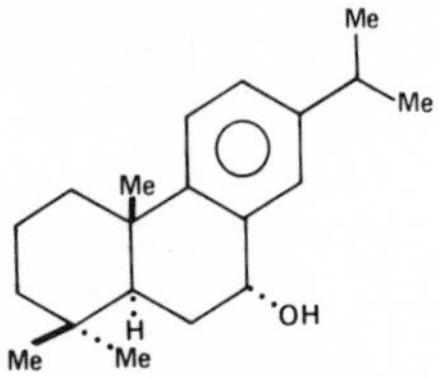	7α-hydroxy-8,11,13-abietatriene	" " "	"
67	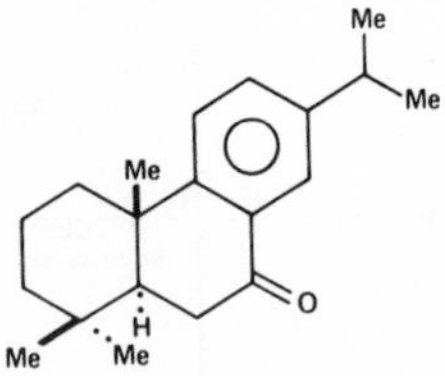	7-oxo-8,11,13-abietatriene	" " "	"
68	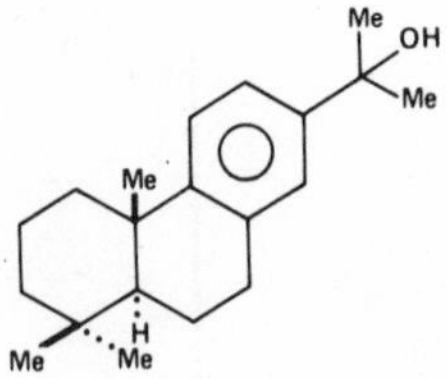		" " "	109a

Table III

No.	Structure	Name	Occurrence	Reference
69		7-oxodehydroabietol	Pinus monticola, P. banksiana (Pinaceae)	109a,b,c
70	a) R = Me b) R = Et	methyl dehydroabietate ethyl dehydroabietate	P. monticola, Pseudotsuga menziesii (Pinaceae); Cistus labdaniferus (Cistaceae) P. monticola (Pinaceae)	109a,b 109A 109a
71		4(18),8,11,13-abieta-tetraene	" "	109a,b,c
72			Pinus monticola (Pinaceae)	109a,b,c
73			" " "	"

Table III

No.	Structure	Name	Occurrence	Reference
74			Pinus monticola (Pinaceae)	109a
75		18-nor-8,11,13-abietatriene-4-ol	Pinus banksiana, P. monticola (Pinaceae)	109b
76		19-nor-8,11,13-abietatrien-4-ol	Pinus banksiana (Pinaceae)	109b

plausible on structural grounds as it requires only a 1,2-migra-
tion, is further supported by the fact that several naturally
occurring furanoeremophilanes are known, one of which, decompos-
tin [47] (59b), actually occurs in *C. decomposita* together with
the aromatic compounds.

22-4. DITERPENOIDS

Like the sesquiterpenoids, the many classes of diterpenoid
compounds (for reviews see, *inter alia,* ref. 60; ref. 1, pp. 291-
311) contain a number of aromatic representatives. However,
in contrast to the diversity of types observed among the aromatic
sesquiterpenoids, their diterpenoid counterparts belong mostly
to one group, the one derived from abietane [48] (see Scheme VI).
It is worthy of note that the carbon skeleton of this group of
compounds, while isoprenoid, does not conform to the normal
head-to-tail arrangement. As will be discussed below, the skeleton of [48] very probably arises through a rearrangement from
precursors of the pimarane [49] type with normal head-to-tail
alignment.

For an understanding of the biosynthetic relationships of
the various classes of aromatic diterpenoids, a general discussion of diterpene biogenesis is needed; of course only the
briefest outline of these much-studied questions can be given.
They have been reviewed recently by Hanson and Achilladelis (61).

A scheme for the biosynthesis of diterpenoids was first proposed by Ruzicka (6b) in 1953; in its main features, it is still
valid and has been confirmed repeatedly by labeling experiments.

Starting from the then-unknown geranylgeraniol [50] (today
replaced by its pyrophosphate [15]), it postulates cyclization to
a bicyclic intermediate with a carbon skeleton found in a number
of non-aromatic diterpenoids; this skeletal type has subsequently
been designated as the "labdane" [51] class. Ring closure between
carbons 13 and 17[+] of a suitable precursor of this type would form
ring C of the tricyclic intermediates, leading at first to the
pimarane [49] skeleton. Rearrangement of this was postulated to
give the abietane system [48].

This scheme needs amplification to include additional
classes, not initially considered, or not yet known in 1953. For
instance, rearrangements of [49], other than the one leading to
[48], could yield types [52] and [53]. No aromatic diterpenoid
derived from [52] seems to have been discovered so far, while

---

[+]See footnote on following page.

[53] is the parent of the phenol, cleistanthol [III-3]. From [48], the kaurane skeleton [54] could be derived by further cyclization, at least on paper; the actual biosynthesis appears to be more complex, and may not involve abietane-type intermediates (cf. ref. 62a,b, and literature quoted there).

Certain aromatic diterpenoids are based on [55] and [56], which seem to arise from [48]. Of [55], designated as totarane (100c), several aromatic derivatives are known; they are exemplified by totarol [III-55]. Type [56] is represented so far only by sempervirol [III-54].

Many further classes of diterpenoids exist (60); not+ having any known aromatic derivatives, they need not be considered here.

Modern experimental research has confirmed the classical scheme of Ruzicka in all its essential features, while at the same time its details have been brought into sharper focus. These developments may be exemplified, *inter alia*, by the studies of West (62a), Hanson (62b), and their co-workers (see Scheme VII). The investigations of West (62a) have shown that an enzyme system from the mycelium of the mold *Gibberella fujikuroi* converts *trans*-geranylgeranyl pyrophosphate [15] into the pyrophosphate [57] of *ent*-8(17), E-13-labdadien-15-ol, or copalyl pyrophosphate; the name "copalol" has been proposed for the parent alcohol. Furthermore, [57] was transformed by the enzyme into (-)-kaurene [58] which, in its turn, was further converted to the gibberellins, (e.g., gibberellic acid [58A]), which are rearranged diterpenoids with 5-membered ring B. The first aromatic representative of this class was discovered only after completion of this chapter; it is discussed in the Addendum. The formation of [58] implies a precursor with a skeleton derived from [49], and indeed [57] was shown to give (+)-sandaracopimaradiene [59], as well as [58] and

---

+ In conformity with prevalent recent usage, and with the recommendations of McCrindle and Overton (60), we have adopted the steroid-type numbering, as indicated in Scheme VI. However, the literature contains many instances where diterpenoids with perhydrophenanthrene skeleton are numbered as phenanthrene derivatives, a circumstance which can lead to confusion; e.g., the phenolic hydroxyl at C-12 (steroid numbering), which is present in many aromatic diterpenoids, would receive number 6 in the numbering system based on that of phenanthrene. In Table III, we have retained the latter numbering in those cases where it appears in the literature, but have placed it in quotation marks *after* the steroid-type designation.

Scheme VI

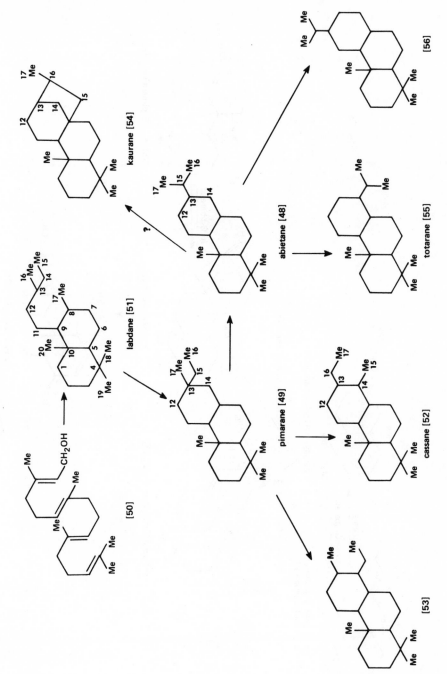

Scheme VII

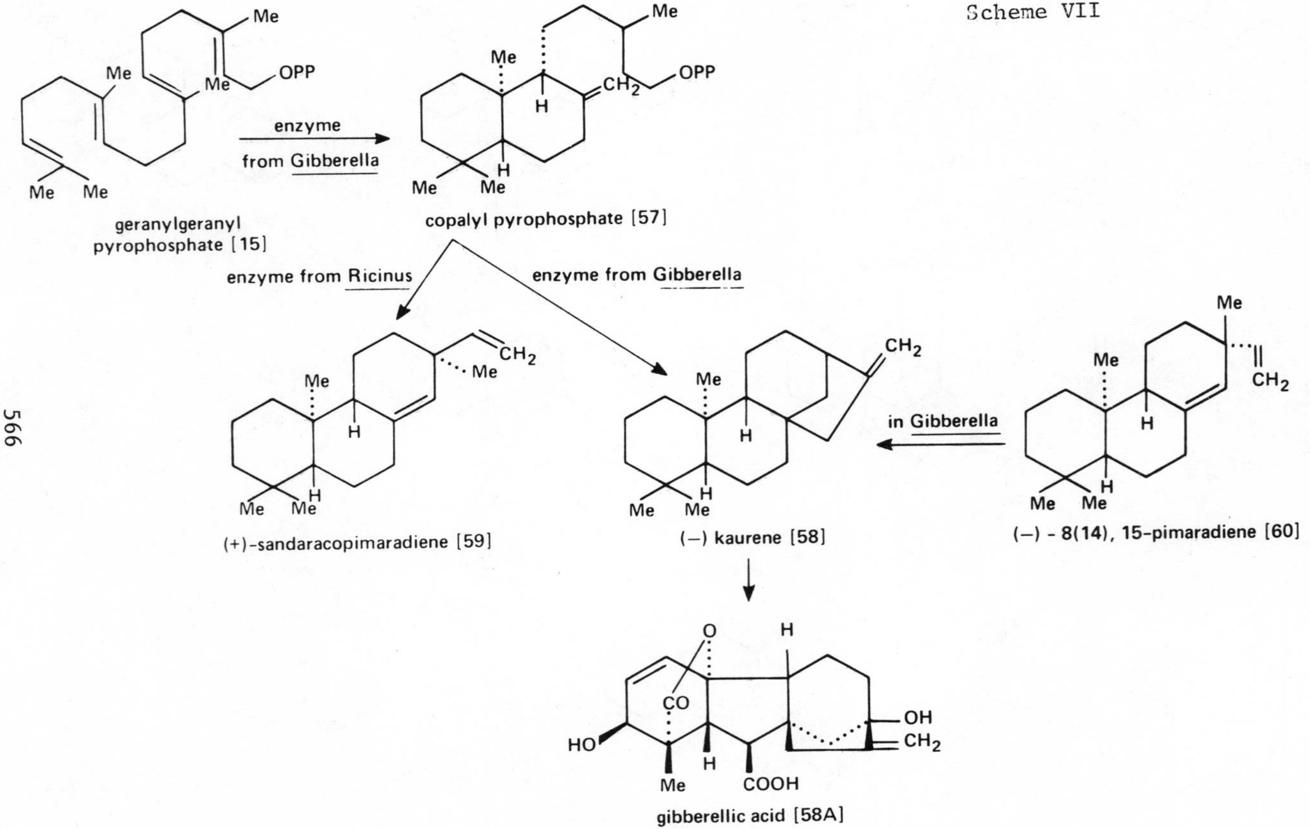

geranylgeranyl pyrophosphate [15]
enzyme from Gibberella
copalyl pyrophosphate [57]
enzyme from Ricinus
enzyme from Gibberella
(+)-sandaracopimaradiene [59]
(−) kaurene [58]
in Gibberella
(−) - 8(14), 15-pimaradiene [60]
gibberellic acid [58A]
Me
OPP
CH2
H
OH
COOH
O
CO
HO

other diterpenoids, under the influence of a soluble enzyme from castor bean, *Ricinus communis*. The related studies of Hanson and White (62b) have demonstrated that [57] (in this case prepared *in vitro* from the corresponding carboxylic acid, a constituent of copal) is converted to [58] (and to gibberellic acid) by *G. fujikuroi* and that labeled (-)-8(14), 15-pimaradiene [60], the C-13 epimer of [59], was specifically incorporated into a kaurane derivative, while its isomers with double bonds at positions 7,8 and 8,9 could be excluded as possible precursors.

Together, these investigations demonstrate a pathway from [15] via [57] to a pimarane-derived intermediate. Further transformations of the latter to the various aromatic diterpenoids remain hypothetical. It will be noticed that the intermediates just discussed: [57], [58], and [59], belong to the series with absolute stereochemistry opposite to the more normal one encountered in all steroids and most diterpenoids, including the aromatic ones (with the exception of cleistanthol, [III-3]). There is no reason to assume the existence of chemical (in contrast to stereochemical) differences in the biosynthesis of the two antipodal classes.

These experimentally established reactions thus confirm the sequence of reactions of Ruzicka's scheme as far as the formation of the pimarane skeleton, [49]. For the further conversions to those skeletal types of which aromatic representatives are known, viz. [48], [53], [55], and [56], and for biosynthesis of certain defective structures, plausible if hypothetical mechanisms are available. They will discussed below.

Obviously outside this scheme are certain diterpenoids that cannot be derived from a labdane-type skeleton--cembrene with a 14-membered ring, the constituents of *Taxus* spp., phorbol--to name a few. We have encountered only two *aromatic* compounds of this kind, the monocyclic *ar*-artemisene [III-1] and biflorin [III-2].

As has been mentioned before, the great majority of diterpenoids is derived from the abietane skeleton [48], or its two modifications [55] and [56]. An exception, among tricyclic compounds, is cleistanthol [III-3], the only known representative of its skeletal type [53].

It is interesting that no aromatic representative of the labdane [51], kaurane [54], or cassane [52] groups seems to have been encountered so far, although the skeleton of [52] would actually permit aromatization of ring C without need for skeletal change. Furthermore, aromatization almost always involves ring C only, although analogous processes involving ring A and/or B (of course with loss or shift of methyl groups) seem entirely possible. The only exceptions (see below) are the tanshinones [III-44 to 53], coleone A [III-43], a deeply modified abietane with rings B and C aromatic, and the two fossil hydrocarbons retene [III-7] and simonellite [III-6], both obviously of doubtful standing as genuine products of biosynthetic aromatization. Even in the compounds just mentioned, however, ring C is always aromatic; no naturally occurring tricyclic diterpenoid seems to be known in which *only* rings A and/or B would be benzenoid. However, in the one known aromatic representative of the *tetra*cyclic diterpenoids ([100]; see Addendum), ring A alone is aromatic.

Aromatic diterpenoids are widely distributed among higher plants, but two taxonomic groups seem markedly favored: the conifers (Cupressaceae, Pinaceae, Podocarpaceae, Taxodiaceae), and the Labiatae; plants of the latter family contain a wide variety of such compounds, which are often strongly oxygenated. No aromatic diterpenoids seem to have been isolated from molds or other microorganisms, but after completion of this chapter, one representative has been found in a *Lycopodium* sp. (see Addendum).

The only known *mono*cyclic aromatic diterpenoid of established structure seems to be the compound $C_{20}H_{30}$ isolated by Šorm and co-workers from wormwood, *Artemisia absynthium* (Compositae) (63a) and subsequently (63b) called *ar*-artemisene [III-1]. *Capraria biflora* (Scrophulariaceae) contains biflorin [III-2], whose structure is unique among diterpenoids (64). It is, however, closely related to the sesquiterpenoids from *Mansonia* (cf. Section 22-3); in fact, [III-2] is the higher prenolog of mansonone F [III-53].

Before discussing the aromatic diterpenoids derived from [48], which constitute the majority of known compounds of this kind, the few benzenoid representatives of other tricyclic classes should be considered.

Cleistanthol [III-3] from *Cleistanthus schlechteri* (Euphorbiaceae) seems to be the only example of the [53] class (65a). The fact that the plant also contains three non-aromatic compounds [61-a,b, and c], which are evidently related to [III-3] but belong to the pimarane class [49], seems suggestive of a biosynthesis of [III-3] from a derivative of [49] by a migration of the two-carbon side-chain. It is interesting that this unique representative of aromatic C_{20} diterpenoids with a skeleton based on phenanthrene, but not of the abietane type, should also be the only one with antipodal stereochemistry.

Another compound that apparently is not related to [48] is the C_{18} terpenoid nimbiol [III-4], which occurs in *Melia azadirachta* (Meliaceae) together with sugiol [III-25], a normal

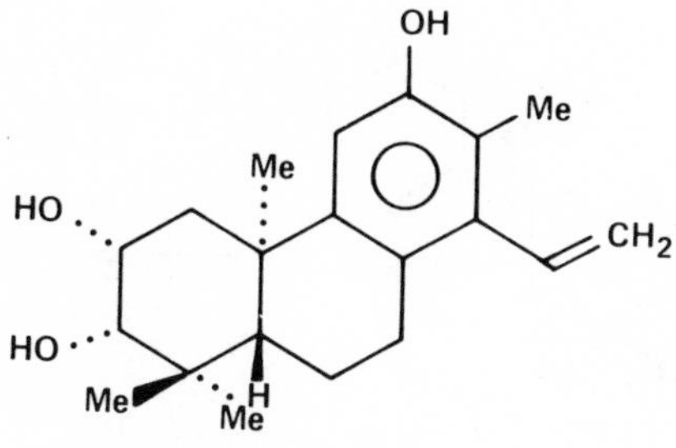

[III-3]

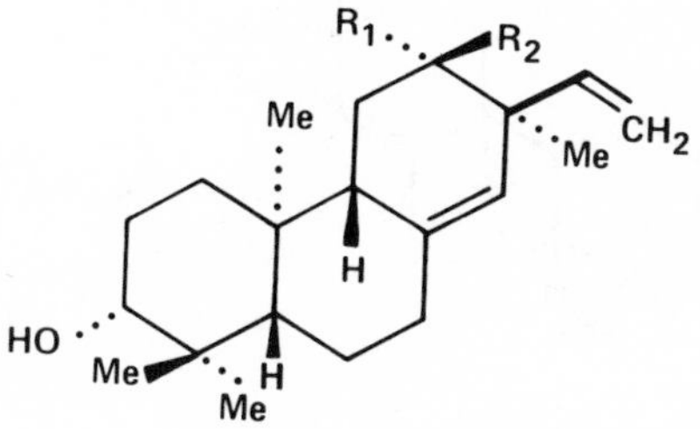

[61]   a: $R_1 = R_2 = H$
        b: $R_1 = OH, R_2 = H$
        c: $R_1 = R_2 = O$

abietane derivative. The structure of [III-4], confirmed by several syntheses, does not permit safe conclusions to be drawn about its biosynthesis. One plausible hypothesis (66a, 67) postulates hydration of the Δ^{15} bond in a 12-ketopimaradiene, followed by loss of C-15 and C-16 through retro-aldolization, as shown on the next page. This interpretation offers the most attractive rationale for the structure of [III-4], although alternative possibilities can hardly be excluded at present; for example, oxidative cleavage of the side-chain double bond in the 12-ketone, followed by loss of the carboxyl from the resulting β-keto acid, or oxidative transformation of the isopropyl in an abietane-type [48] precursor to carboxyl, with subsequent reduction of -COOH to -Me, and aromatization.

Aromatic compounds derived from abietane [48]. This class includes the vast majority of known aromatic diterpenoids. It can be divided into subclasses: the abietane-derivatives proper, which retain the skeleton of [48] without further arrangement, at least of ring C, and two subgroups in which the isopropyl group attached to this ring has shifted to give the structures from which totarol [III-55] and its relatives, and sempervirol [III-54] are derived.

The abietane group also includes a number of compounds in which the fundamental skeleton has undergone other modifications. Cleavage of individual bonds of the ring system, with retention of all the carbon atoms, is encountered in seco-dehydroabietic acid [III-11 B] (C-9 - C-10), the dialdehyde [III-18] (C-6 - C-7), and coleone A [III-43] (C-1 - C-10). Loss of one of the two

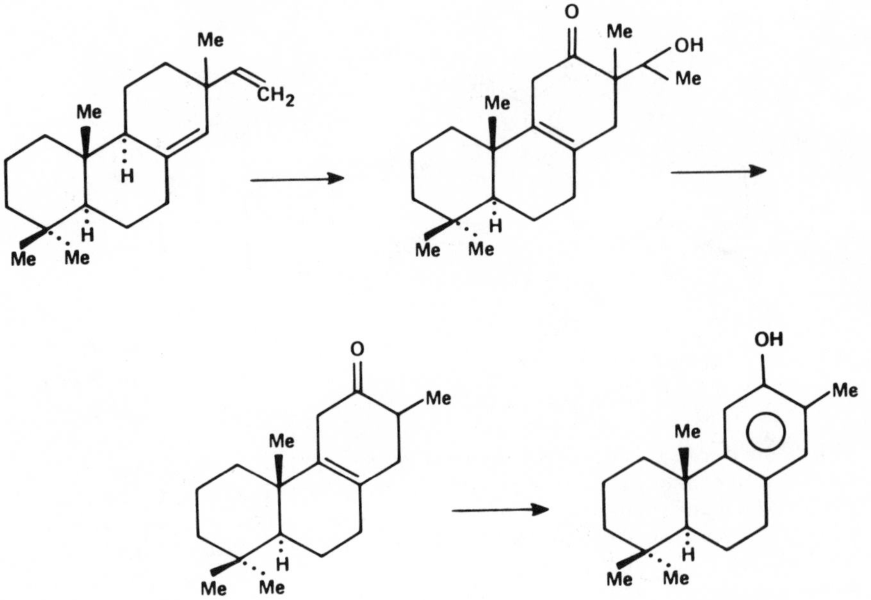

[III-4]

methyls at position 4 (C-18 or C-19) without aromatization of
ring A has taken place in coleone B [III-34] and a number of
recently discovered diterpenoids from *Pinus* spp. [III-71 - 76].
Elimination of C-20, the methyl at position 10, in connection
with aromatization of ring B, has occurred in simonellite [III-6],
miltirone [III-44], and most of the tanshinones [III-45 - 49],
[III-51], and [III-52]. Both these methyls have been lost in
retene [III-7], tanshinone I [III-50], and isotanshinone [III-53].
Finally, podocarpic acid [III-28] has lost the entire isopropyl
group.
      In the present discussion, compounds derived from the unre-
arranged abietane [48] skeleton will be considered first. This
skeleton can be dissected into isoprene units, but their arrangement in ring C departs from the normal head-to-tail sequence;
biosynthesis involving a rearrangement must thus be assumed. A 1,2-
shift of the methyl (C-17) in a precursor with pimarane [49]
skeleton has been proposed, initially by Sandermann (68) in 1938
and by Ruzicka (6) in 1953; their interpretation seems generally
accepted. The rearrangement has laboratory precedent (69) in the

acid-catalyzed conversion of pimaric acid [62] into abietic acid
[63]. However, experimental confirmation for the actual occurrence of this 1,2-shift as a step in the biosynthesis of the
abietanes seems to be lacking.

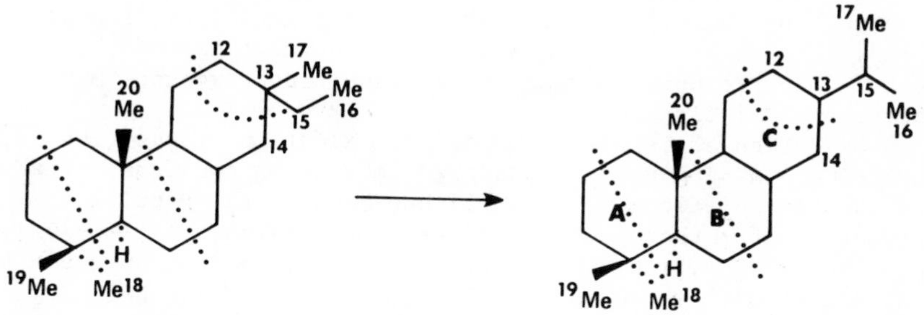

[49]   [48]

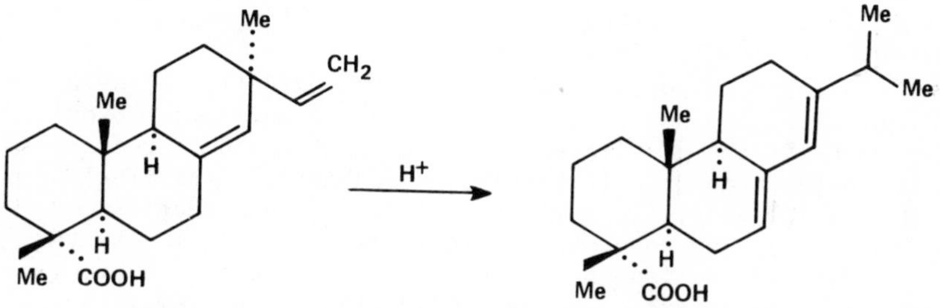

[62]   [63]

The naturally occurring abietane [48] derivatives which we
have been able to find are listed in Table III as compounds
[III-5] to [III-53], and [III-65] to [III-76].

The parent aromatic hydrocarbon of the group, *ar*-abietatri-
ene [III-5], has been encountered (70) only quite recently in
Podocarpus ferrugineus and *Thujopsis dolabrata*, but actually seems
widely distributed (109a). The fully aromatic C^{14} phenanthrene
retene [III-7], often obtained *in vitro* through dehydrogenation

of abietane-type diterpenoids, has been isolated from peat and
lignites (71) and may thus in a way qualify as a natural compound.

In the same category belongs the related naphthalene
simonellite [III-6], which has been encountered in fossil wood
from a plant of the family Taxodiaceae (72).

The other naturally occurring aromatic diterpenoids of the
abietane type can be formally derived from *ar*-abietatriene
[III-5] through introduction of oxygenated functions and, in some
cases, of double bonds. Some of the resulting compounds may
deserve special discussion.

Oxidation at C-18 is common, all oxidation stages (-CH$_2$OH,
-CHO, -COOH) having been encountered, mostly in conifers. The
acid, dehydroabietic acid [III-11] has been observed quite
frequently. For the natural occurrence of its methyl [III-70a]
and, surprisingly, its ethyl ester [III-70b], see the Appendix to
this subsection. The Δ^6-derivative of dehydroabietic acid,
[III-11A] has been isolated from gum rosin pretreated with maleic
anhydride (72A). In compound [III-11B], the abietane skeleton is
modified by breaking of the 9, 10 bond and opposite stereochemistry
at the asymmetric centers corresponding to C-5 and C-10 in [48].
The methyl ester of this compound is identical with one of several
isomeric substances (72Bb) formed on alkali-catalyzed pyrolysis of
methyl laevopimarate, the analog of [III-11] in which ring C contains the $\Delta^8(14),12$-diene grouping.

In contrast to the frequency of the oxidation of C-18 (in
both aromatic and non-aromatic compounds of this class), oxidation of C-19, the β-oriented carbon atom attached to C-4, is rare.
It has been found only in 4-epi-dehydroabietic or callitrisic
acid [III-13] (73), its derivative [III-14], podadacric acid
[III-30] (74), and podocarpic acid [III-28] (75). The latter
compound lacks the isopropyl group; the problem of its biosynthesis will be discussed below. Oxidation of the remaining
quaternary methyl, C-20, has been found only in [III-32] from
Salvia spp. (Labiatae), and very recently in nemorone (see Addendum).

In the alicyclic part of the molecule, oxidation is most
common in the benzylic position 7, as was to be expected; such
compounds as cryptojaponol [III-31], sugiol [III-25], and several
others [III-26, 27, 33, 34, 35], may be mentioned. The 6,7-diketo
compound [III-27] (77) from juniper is the naturally-occuring
precursor of its 5-epimer, xanthoperol, to which it is transformed very readily on treatment with alkali; as a consequence of
this facile conversion, the artifact xanthoperol has been isolated
from the plant much earlier than [III-27]. Oxygenation at C-3
occurs in hinokiol [III-21] and hinokione [III-22]; a hydroxyl
at C-1 is found in teideadiol [III-5] from *Nepeta teidea*

(Labiatae) (78). Compound [III-24] from Cupressus sempervirens
(79) is unusual not only as a 1,3-diketone --an uncommon type--
but also as a phenol methyl ether (however, a methoxyl in the
same position is probably also present in [III-31]), and particularly because of its co-occurrence with the corresponding derivative [III-61] of totarane [55]. The structure of coleone A
[III-43] (see below), which appears to be a 1,10-seco-abietane
(94), makes it probable that a 1-oxygenated compound derived from
[48] is an intermediate in its biosynthesis.

In the aromatic ring, position 12 is the one most frequently
oxygenated; cf. such compounds as sugiol [III-25], ferruginol
[III-16], [III-21, 22, 30], and others. A 12-methoxyl group
occurs in [III-24]; both it, and a free hydroxyl, are probably
present in cryptojaponol, provisionally formulated as [III-31]
(80).

Additional unsaturations occur at several positions:
Δ^6-dehydroferruginol [III-17] regularly accompanies the parent
compound (81); double bonds in conjugation with carbonyl are
present at C-1 in shonanol[+] [III-23] (82), and at C-5 in dehydro-
sugiol [III-26] (83) from Taxodium distichum.

Oxygenation in the isopropyl group has been encountered in
a few cases. 15-Hydroxy-dehydroabietic acid [III-12] (84) occurs
in kauri (Agathis spp.); fuerstione [III-42], which contains the
same grouping, will be discussed later. The unusual pododacric
acid [III-30] has been isolated (74a) from Podocarpus dacrydioides
and P. totara (85); its structure, which has been elucidated only
recently, (74b), is particularly interesting because of the
co-occurrence of this acid with podocarpic acid [III-28]. The
simultaneous occurrence of these compounds lends weight to the
hypothesis that [III-28] may be formed through oxidation of the
isopropyl group to carboxyl in a non-aromatic 12-ketone of the
[48] series, followed by decarboxylation of the resulting β-keto
acid, and aromatization. The alternative possibility of an acid-
catalyzed loss of the isopropyl group has been proposed by Wenkert

---

[+]This compound had at first been formulated tentatively (82) as
a member of the totarane [55] group differing from totarolenone
[III-60] only by the location of the phenolic hydroxyl at C-12
instead of C-13--a biosynthetically unattractive structure (see
below, p. 578). After completion of this section, the two C-5
epimers of this structure were synthesized (82A) and found
different from the natural terpenoid, whose chemistry thus remains
uncertain. For Table III, item 23, we have provisionally adopted
the suggestion (12) that shonanol is a Δ^1-3-keto derivative of
ferruginol [III-16].

and Jackson (86) on the basis of the ready occurrence of this
process on treatment of the nitrile corresponding to [III-11] with
$AlCl_3$. The dialdehyde [III-18] isolated (86A) recently from
Juniperus rigida represents an unusual type, which could arise by
oxidative cleavage of the bond between C-6 and C-7 in a precursor
related to [III-16]; *in vitro,* the compound is obtained by ozon-
ization of dehydroferruginol [III-17].

The group of abietane-derived aromatic diterpenoids also
includes a number of quinones and quinone methides. The plant
family Labiatae is a rich source of such compounds, although by
no means the only one: quinones of this type have also been
encountered in gymnosperms (*Taxodium*) and, at the other end of
the range of vascular plants, in one species (*Inula royleana*) of
the Compositae. The distribution of these quinones, as far as is
known today, thus appears strangely disjointed.

One group of compounds of this type includes a number of
substances which contain a quinone group in ring C, or a quinone
methide grouping there extending into ring B, while the rest of
the molecule is not strongly oxidized. Representatives of this
class: the *p*-quinones royleanone [III-35] and its 7-acetoxy-
[III-39] and Δ^6-dehydro-[III-36] derivatives, together with traces
of ferruginol [III-16], were first observed in *Inula royleana*
(Compositae) by Edwards and co-workers (87). They were soon
followed by the 7α-hydroxy derivative horminone [III-37] from
Horminum pyrenaicum (Labiatae) (88). Quinones [III-35], [III-36],
and [III-39] have also been obtained from the roots of *Salvia
officinalis* (Labiatae) (89), while *Plectranthus* spp., of the same
family, yields [III-35] and [III-36] (90, 93). Finally, *Taxodium
distichum* (Taxodiaceae) has been found by Kupchan and co-workers
(83) to contain [III-35] and taxoquinone [III-38], the 7β-epimer
[III-37], together with sugiol [III-25], 5-dehydrosugiol [III-26],
and the two tumor-inhibiting quinone methides taxodone [III-40]
and taxodione [III-41]. The occurrence of these diterpenoid
quinones in a plant such as *Inula royleana* seems surprising (but
cf. the sesquiterpenoid quinone perezone [II-8] from another,
though not closely related species of the Compositae). The
mechanism of diterpenoid biosynthesis in this plant must, however,
be peculiar also in other respects: along with the quinones, the
species contains (91) several non-aromatic diterpenoid alkaloids
derived from lycoctonine, a type of base otherwise characteristic
of certain genera (*Aconitum, Delphinium,* etc.) of the completely
unrelated family Ranunculaceae. This association of two different
unusual types of diterpenoids is particularly odd since the
juncture of rings A and B in the alkaloids is the mirror image
(5β, 10α) of the normal one (5α, 10β) present in the co-occurring
quinones.

The remaining quinonoid compounds are all obtained from
Labiatae. Carnosic acid or salvin, [III-32], is actually a hydroquinone but may find its place here because of its analogy with
ortho-quinones from other species of the family. It has been
obtained (76) from the leaves of *Salvia officinalis* and *Rosmarinus
officinalis*;its ready autoxidation at the benzylic position 7,
followed by lactonization, leads to carnosol (picrosalvin) (89),
which had previously been isolated from these plants but does not
occur preformed in them. The C-20 carboxyl in [III-32] is so far
unique. Its occurrence may be significant (89) for the biosynthesis of a number of 20-nor-diterpenoid *ortho*-quinones, the
tanshinones (see below), which occur in another *Salvia* species,
S. miltiorrhiza; in these compounds ring B, and in two cases also
ring A, is aromatized with loss of C-20.

More strongly oxidized, dehydrogenated, and otherwise modified compounds from Labiatae have become known through the
research of Eugster and his co-workers. *Fuerstia africana* yields
the quinone methide fuerstione [III-42] (92); the identity of its
chromophore with that of rings A and B in the triterpenoids
celastrol and pristimerin [IV-1], [IV-2] (Section 22-5) has been
stressed.

Another plant of the same family, *Coleus igniarius*, contains
coleone B [III-34] (93), a hydroquinone of the 18- (or 19-)
nor-abietane series, and the even more strongly modified 1,10-
seco-quinone coleone A [III-43] (94). Coleone C [III-33], a
compound closely related to [III-34] but less strongly dehydrogenated, occurs in *C. aquaticus* (95).

Finally, the most widely aromatized diterpenoid quinones,
most of them *ortho*-quinones, form a remarkably coherent group of
constituents of *Salvia miltiorrhiza*. The name "Tan-Shen," under
which the plant has long been used as a valuable drug in China,
forms the basis for naming most of its constituents. Three of
these, tanshinone [III-50], tanshinone II (sometimes (96) referred
to as tanshinone IIA) [III-46], and cryptotanshinone, [III-45]
have been known for many years. A review of the history of their
discovery (Nakao, 1934, Takiura, 1941) and structure elucidation
is given in ref. (97), p. 640 ff., which should be consulted for
detailed references. More recently, other constituents have been
isolated, and their structures elucidated: tanshinone IIB
[III-48] (96), hydroxytanshinone [III-47] (98), methyl tanshinonate [III-49] (98), and miltirone [III-44] (99). Besides the
ortho-quinonoid ring C, all of these compounds have an aromatic
ring B, and in [III-50] even ring A is benzenoid; all of them,
with the exception of [III-44], have the isopropyl group modified
as part of a furan or dihydrofuran ring. Comparison of their
structures (see Table III, compounds 44-50 and p. 577 shows a

remarkable progression from miltirone [III-44], the least modified,
to tanshinone I [III-50], the most strongly dehydrogenated compound
and it is tempting to assume that the sequence of increasingly
hydrogen-poor structures parallels the actual biosynthetic pathway. Oxidative modification of the isopropyl group of [III-44]
would occur first with formation of the dihydrofuran ring [III-45],
followed by conversion to the furan [III-46]; oxidation at ring
A [III-47] or one of the methyls at C-4 [III-48, 49] (or perhaps
of both) follows as a prelude to the loss of C-18 (or C-19) and
aromatization to [III-50]. Of course, loss of C-20 with aromatization of ring B must precede all these transformations; the
hypothesis (89) of an intermediate carboxylic acid of the type of
carmosic acid [III-32] has already been mentioned.

In addition to those *ortho*-quinones, Tan-Shen also contains
three *para*-quinones with the same carbon skeleton. These
compounds, isocryptotanshinone [III-51] isotanshinone [III-52]
and isotanshinone-I [III-53] are closely analogous to [III-45],
[III-46], and [III-50], respectively, differing from then only in
the orientation of the furanoid ring.

Compounds with Rearranged Abietane [48] Skeleton.

The rearrangement that has to be assumed to take place in
the formation of the diterpenoids derived from [48] leads to a
skeleton still compatible with the simple isoprene rule, although
the resulting arrangement departs in ring C from the usual head-
to-tail type. There are, however, two classes of diterpenoids,
based on [55] and [56], respectively, whose skeletal structures,
undoubtedly the result of further rearrangement, are no longer
capable of being dissected into isoprene units. A typical representative of one of these classes, the totarane [55] group, is
totarol [III-55] (100), while the single compound known to belong
the other class is sempervirol [III-54] (101).

The modified compounds tend to occur together with their
normal congeners derived from the unchanged abietane skeleton
[48]; they are found mostly, if not exclusively, in conifers.
For instance, *Cupressus sempervirens*, the species in which
[III-54] was discovered, also contains (101) the abietane [48]
derivative ferruginol [III-16], the totarane [55] derivative
totarol [III-55], and (79) the two 1,3-diketones [III-24] and
[III-61] derived from [48] and [55], respectively. Representatives of all three skeletal types: [48], [55], and [56], are
thus produced in the same plant species. See structures on page 578.

It is significant that all these compounds contain an
oxygenated function (OH or OMe) adjacent to the isopropyl, and
that there is an evident tendency for the co-occurrence of groups
of compounds with structures which are identical except for the
position of these two adjacent substituents on the aromatic rings;

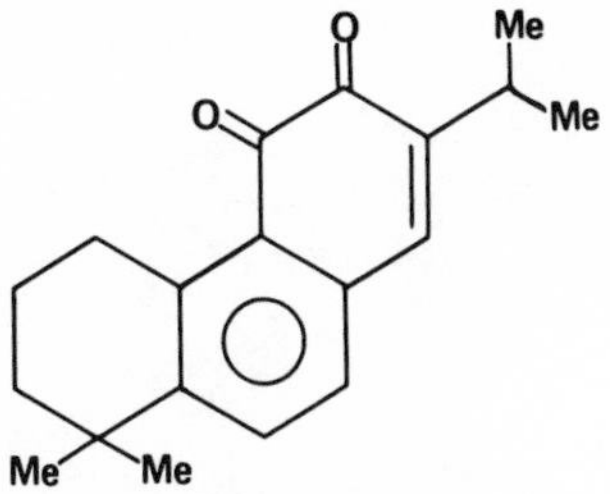

[III-44] miltirone

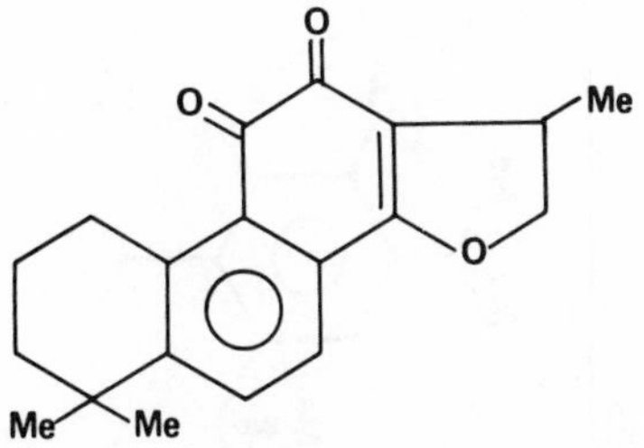

[III-45] cryptotanshinone

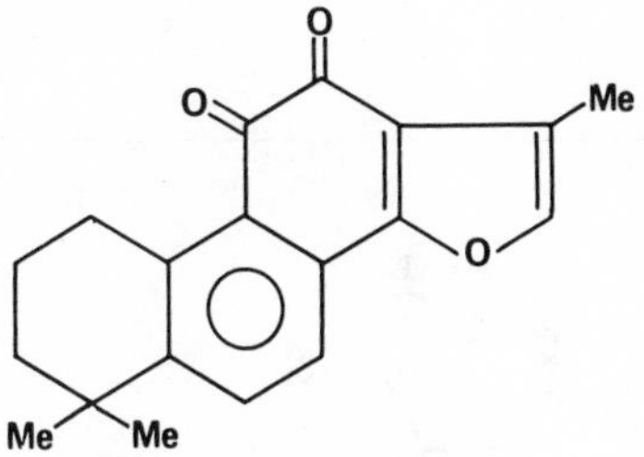

[III-46] tanshinone II

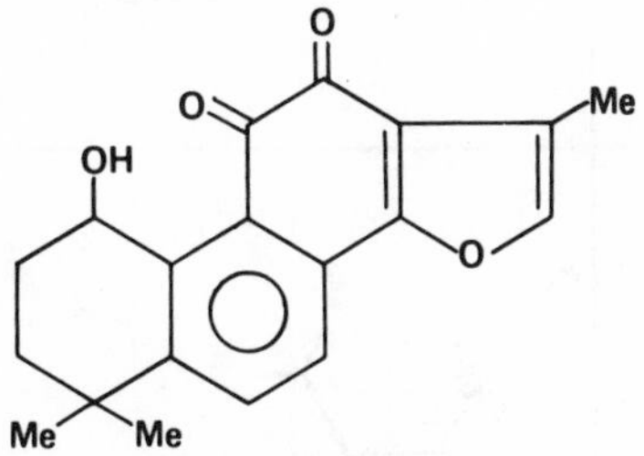

[III-47] hydroxytanshinone

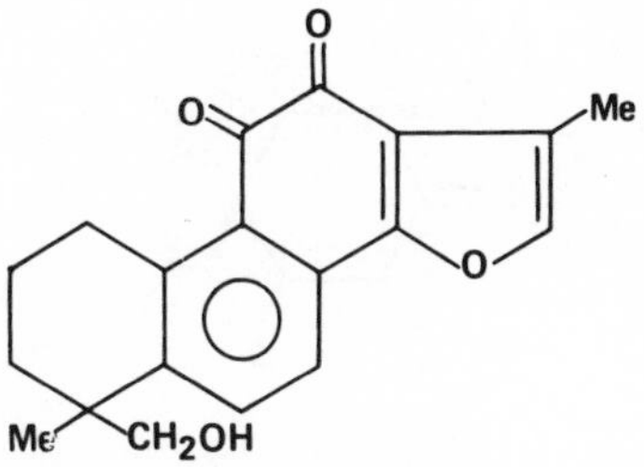

[III-48] tanshinone IIB

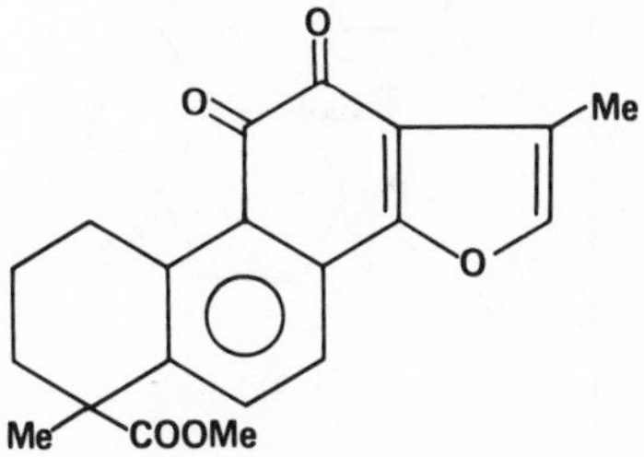

[III-49] methyl tanshinonate

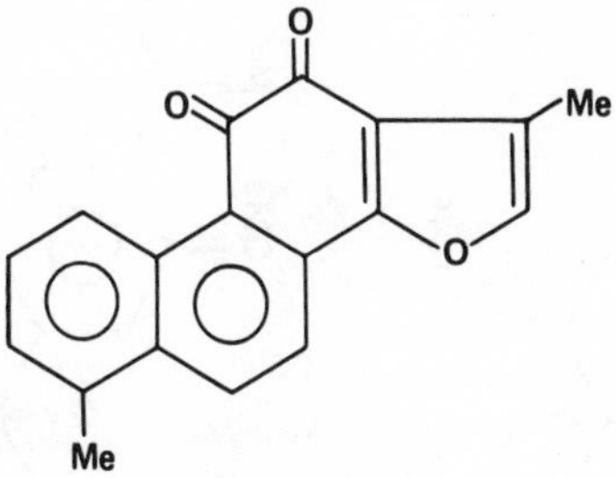

[III-50] tanshinone I

577

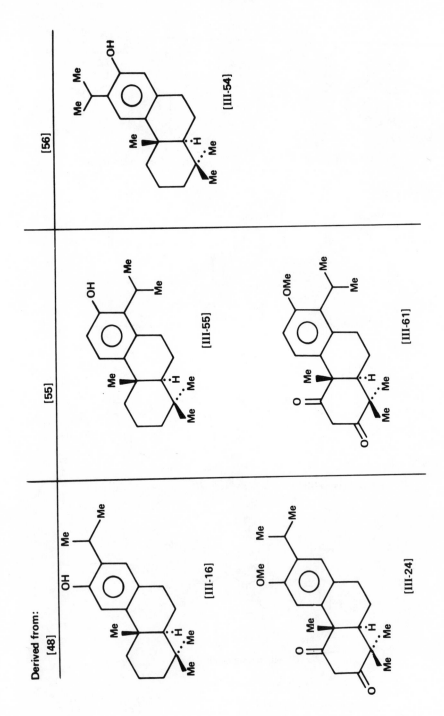

Derived from: [48]

[56]   [55]

[III-54]

[III-55]

[III-61]

[III-16]

[III-24]

578

see the examples just quoted. This fact suggests strongly that
the non-isoprenoid compounds form through rearrangement of normal
abietane-type precursors. While no experimental proof for a
biogenetic relationship of this kind is available, a very attractive hypothetical mechanism for such a transformation of the
skeleton of [48] into that of [55] has been proposed by Wenkert
and Jackson (86); it is shown in somewhat modified form in Scheme
VIII. Rearrangement via path (a) would indeed offer a good
rationale for the structural features encountered in derivatives
of [55], and for the tendency of both classes of compounds to
occur together. Furthermore, the formation of [III-54], not yet
discovered when this mechanism was suggested, can be explained
readily by the alternative migration via path (b). The analogy
of this reaction sequence to the formation of the aporphin alkaloids from the proaporphines will be obvious (see Chapter 8).

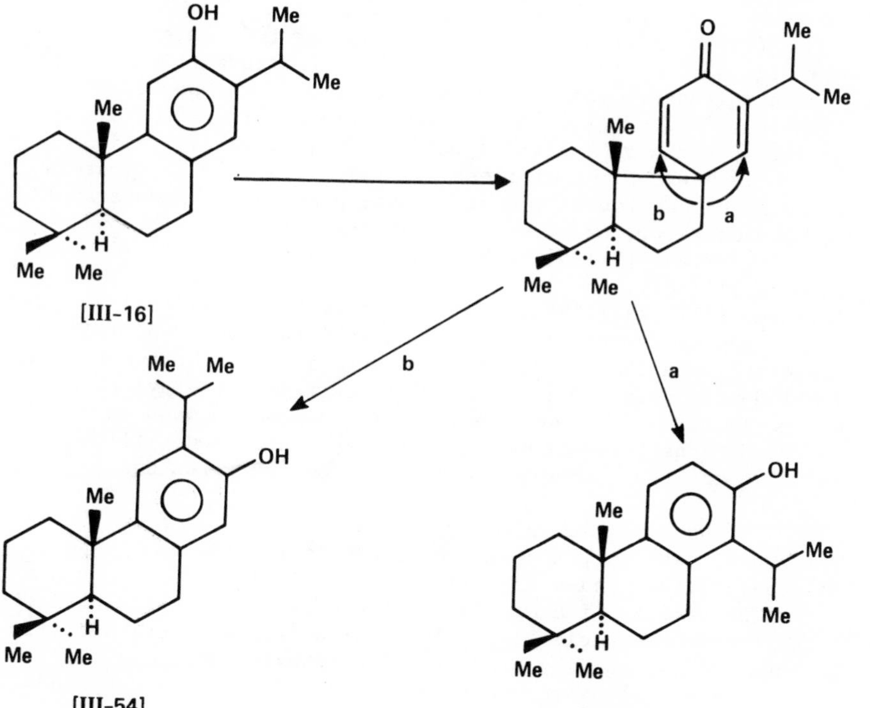

[III-16]

[III-54]

[III-55]

Scheme VIII

Besides compounds [III-55] and [III-61] already mentioned, the totarane group includes the ketone totarolone [III-59] and the corresponding α,β-unsaturated ketone totarolenone [III-60] associated with it in a nonseparable mixture (103), the 19-hydroxy-[III-56] (85, 104, 105), oxo-[III-57] (104, 105) and carboxy- [III-58] (85, 104) derivatives of totarol [III-55] (all from various *Podocarpus* spp.) and the two epimeric 7-hydroxy-totarols, likewise forming an inseparable mixture [III-62], from *Thujopsis dolabrata* (106). Furthermore, two interesting dimeric compounds have been isolated from plants of the genus *Podocarpus*. *P. totara* yields podototarin [III-63] (104), while the corresponding dicarboxylic acid, macrophyllic acid [III-64] occurs (107) in *P. macrophyllus*. These compounds are evidently products of phenol coupling of [III-55] and [III-58], respectively; compound [III-63] has actually been prepared by dimerization of [III-55] under the influence of cell-free oxidase preparations from a basidiomycete, *Polyporus versicolor* (107), and from leaves of a number of podocarpaceous plants (*Podocarpus, Dacrydium, Cunninghamia, Cryptomeria* spp.) (108), yields as high as 58% of pure [III-63] being obtainable.

ADDENDUM

After completion of the manuscript of this subsection, we were informed of the work of Rowe and co-workers (109a-c) on the terpenoid constituents of the bark of two *Pinus* species; this work has yielded a number of new aromatic diterpenoids, which are listed in the Appendix to Table III as numbers [III-65] to [III-76].

The bark of one of these species, *P. monticola* (Western white pine) was found to contain some previously known aromatic terpenoids (e.g., cadalene [II-42], (-)-calamenene [II-35], *ar*-abietatriene [III-5], dehydroabietic acid [III-11], dehydro-abietol [III-8]), and a number of new ones: 6,8,11,13-abieta-tetraene [III-65] (109b), dehydroabietan-7α-ol [III-66] and the corresponding ketone [III-67], 7-oxo-abietatrien-18-ol [III-69], dehydroabietan-15-ol [III-68] and, surprisingly, both methyl and ethyl dehydroabietate [III-70a,b] (occurrence of ethyl esters in nature is rare). In addition, several new nor-compounds were observed: the three isomeric 19-nor-abietatetraenes [III-71, 72, and 73], and the 7α-hydroxy derivative of the first one of these, [III-74].

The bark of jack pine, *P. banksiana* (*P. divaricata*), furnished, besides [III-69], the two epimeric 18- and 19-nor-abieta-8,11,13-trien-4-ols [III-75], [III-76] (109b).

The variety of aromatic diterpenoids found in the bark of these two *Pinus* species is truly remarkable. However, not all of them are necessarily products of synthesis *in vivo*. The various 7- and 15-hydroxy derivatives and Δ⁶-dehydro compounds are logical products of the well-known facile autoxidation at benzylic positions and could form in the bark, where possible precursors are exposed to air (109A). The occurrence of [III-75] and [III-76] might similarly be explained by assuming autoxidation of a parent compound with a C-19 aldehydo group, since Caputo and co-workers (109C) have found that fairly brief (5 hours) exposure to air of a non-aromatic aldehyde of this type (torulosal, a derivative of labdane [51]) leads to replacement of the aldehydo by the hydroperoxy group, both epimeric hydroperoxides being formed; they, in turn, are readily converted to the corresponding hydroxy compounds analogous to [III-75] and [III-76]. The latter two substances may thus well be the products of a similar process.

22-5. TRITERPENOIDS

Although the number of naturally occurring triterpenoids is very large, aromatic structures are conspicuously absent from the normal C_{30} representatives of this class. The only natural compounds belonging biogenetically to this group of terpenoids are the quinone methides celastrol, $C_{29}H_{38}O_4$, [IV-1], and its methyl ester pristimerin [IV-2], mono-nor-diterpenoids with a rearranged, nonisoprenoid carbon skeleton related to that of some C_{30} triterpenoids (particularly cerin). These pigments occur in a number of genera of the closely related families Celastraceae (*Celastrus*, *Tripterygium*, *Denhamia*, *Maytenus*) and Hippocrateaceae (*Pristimeria*).

The research leading to the correct structures of the two compounds is well reviewed by Turner (110). The nature of the extended quinone methide chromophore was recognized by Grant and Johnson (111), that of the carbon skeleton by Cooke and Thomson (112), who tentatively proposed a formula for pristimerin which differed from the definitive one only by placing the carbomethoxy group at C-4, instead of C-20. The correct structures, [IV-1] and [IV-2] were established by Nakanishi and co-workers (113), largely through nmr spectroscopy, and were confirmed by Johnson and co-workers (114). The analogy of the chromophore with that of fuerstione [III-42] has already been mentioned.

Like other quinone methides (e.g., [III-40] and [III-41]), [IV-1]and [IV-2] have significant tumor-inhibiting action but are too toxic for practical use.

For several new representatives of this class, see the Addendum, Section 22-8.

22-6. STRUCTURES OF MIXED BIOSYNTHETIC ORIGIN

In a number of naturally occurring aromatic compounds,

Table IV. Aromatic Triterpenoids

No.	Structure	Name	Occurrence	References
1	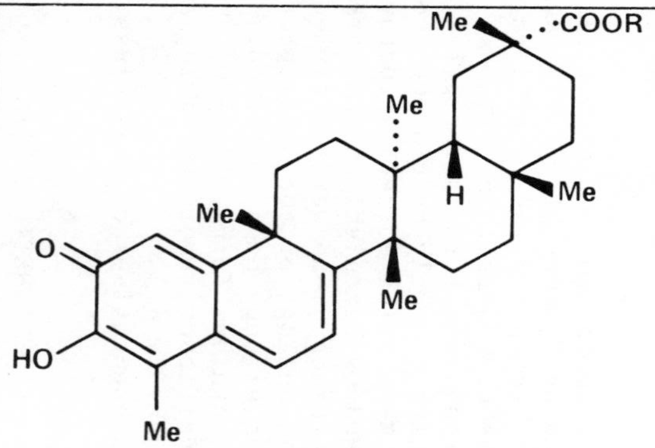 R = H	Celastrol	Celastrus, Tripterygium, Maytenus, etc. (Celastraceae).	110,113
2	R = Me	Pristimerin	Same; Denhamia pitto-sporoides (Celastraceae); Pristimeria spp. (Hippocrat-aceae)	110,113

582

Table V.   Structures of Mixed Biosynthetic Origin

No.	Structure	Name	Occurrence	Reference
I. Naphthoquinones				
1		chimaphilin	Pyrola, Chimaphila, Moneses spp., (Pyrolaceae)	97, p. 199
2		alkannin (free and esterified)	Alkanna tinctoria, Onosma echioides, Arnebia nobilis (Boraginaceae)	97, p. 248
3		shikonin (free and esterified)	Arnebia, Echium, Lithospermum, Macrotomia, Onosma spp. (Boraginaceae); Jatropha glandulifera (Euphorbiaceae)	ibid.
4		alkannan	Alkanna tinctoria (Boraginaceae)	97, p. 252
5		arnebin 2: R = COCH=CHMe$_2$ arnebin 6: R = COMe	Arnebia nobilis (Boraginaceae)	126; 97, p. 251

Table V

No.	Structure	Name	Occurrence	Reference
6	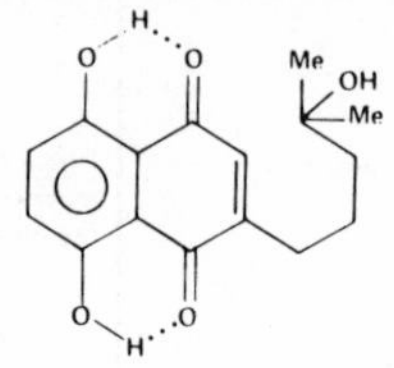	arnebin 5	<u>Arnebia nobilis</u> (Boraginaceae)	126; 97, p. 251
7	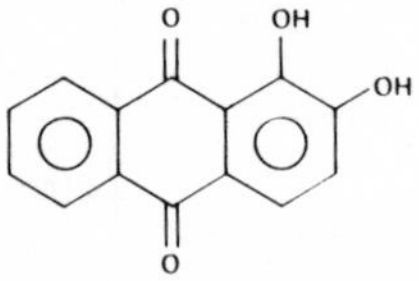	alizarin	In many Rubiaceae, free or as the 2-primveroside ruberythric acid	97, p. 373, 375
8	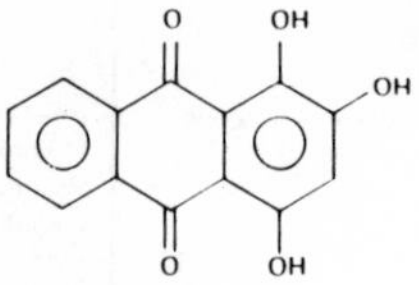	purpurin	In many Rubiaceae	97, p. 407
9	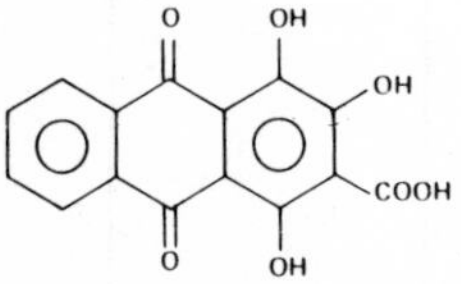	pseudopurpurin	In Rubiaceae, as glycoside, esp. the 1-primveroside, galiosin	97, p. 408
10*	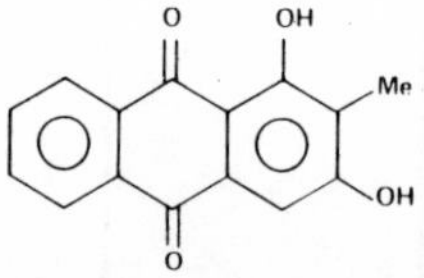	rubiadin	In many Rubiaceae, and in Tectona grandis (Verbenaceae); free and as glycoside.	97, p. 378

*Note: Experimental evidence for a biosynthesis involving mevalonic acid has been obtained so far only for [V-7] – [V-10]. A number of anthraquinones occurring in Rubiaceae have structures so closely related that an analogous biosynthetic origin appears very probable. In the absence of any proof for this, they are here merely listed: both monomethyl ethers of [V-7]; the 1-monomethyl ether of [V-10], and its primveroside, longifloroside; lucidin, the ω-hydroxy-derivative of [V-10], and damnacanthol, its 1-monomethyl ether; nordamnacanthol and damnacanthal, the corresponding aldehyde; and munjistin, the carboxylic acid corresponding to [V-10]. For details on these compounds see ref. 97, p. 378–384.

584

Table V

No.	Structure	Name	Occurrence	Reference
II. Cannabinoids				
11	Me, OH, R, Me, Me, O, $C_5H_{11} - n$, R = H	cannabinol	Cannabis indica (Cannabinaceae)	136
12	same, R = COOH	cannabinolic acid	" " "	"
III. Carbazoles Group 1				
13	CHO, N H, OMe	murrayanine	Murraya koenigii (Rutaceae)	148
14	COOH, N H, OMe	mukoeic acid	" " "	149
15	MeO, Me, N H	glycozoline	Glycosmis pentaphylla (Rutaceae)	146
16	MeO, Me, N H, OMe	glycozolidine	" " "	147, 139B
Group 2				
17	CHO, N H, OH, Me, Me	heptaphylline	Clausena heptaphylla (Rutaceae)	151
18	Me, N H, O, Me, Me	girinimbine	Murraya koenigii (Rutaceae)	140, 152

Table V

No.	Structure	Name	Occurrence	Reference
19		murrayacine	Murraya koenigii (Rutaceae)	152c, 153
20		koenine	" " "	142
21		koenimbine	" " "	141, 154a
22		koenigine	" " "	142
23		koenigicine, koenim-bidine, koenidine	" " "	141, 142, 152c
	Group 3a			
24		mahanimbine	" " "	152c, 154a
25		mahanimbinine	" " "	155

Table V

No.	Structure	Name	Occurrence	Reference
26		mahanine	Murraya koenigii (Rutaceae)	142
27		mahanimbidine, murrayazoline, curryangin	" " "	156, 199
28		cyclomahanimbine	" " "	156
29	or	bicyclomahanimbine	" " "	139, 154c, 156
30		murrayazolidine, curryanine	" " "	157

Table V

No.	Structure	Name	Occurrence	Reference
Group 3b				
31		mahanimbicine, isomahanimbine	Murraya koenigii (Rutaceae)	152c, 158
32	or	bicyclo·mahanimbicine	" " "	139, 158
IV.	Indole Alkaloids			
33		ourouparine	Uncaria gambier (Ourouparia gambir), (Rubiaceae)	158B
34		alstoniline	Alstonia constricta (Apocynaceae)	158A

Table V

No.	Structure	Name	Occurrence	Reference
35		gambirtannine	Gambir from U. gambier	158C
36		dihydrogambir tannine	″ ″ ″	158C
37		oxogambirtannine	U. gambier, and gambir from it	158B,C

mevalonate-derived units participate in the formation of the
aromatic system, together with units biosynthesized from shikimate
or acetate. Most of those compounds belong to either one of two
groups: naphtho- or anthraquinones, and alkaloids derived from
carbazole or indole. Theoretically, benzoquinones of similar
mixed origin could exist; for example, combination of one mevalonate-derived isoprenoid unit with one acetate unit could give
the carbon skeleton of toluquinone. No actual biosynthesis of
this kind seems to have been observed.

The aromatic compounds to be discussed are listed in
Table V.

The knowledge available through mid-1968 on the biosynthesis
of quinones has been summarized by Zenk and Leistner (115), and
Chapter 1 of ref. 97 (which fortunately became available while
this Section was being written) covers the pertinent literature
through October, 1970; these comprehensive reviews should be
consulted for those aspects of the problem which form the background of the present discussion of partly mevalonate-derived
quinones, and for a more detailed bibliography than can be given
here. See also Chapters 11, 13, and 18.

Most naturally occurring naphtho- and anthraquinones are
very typical polyketides; they have accordingly been discussed in
Sections 18-4 and 18-10. In addition, some simple naphthoquinones
from higher plants, such as lawsone and juglone (2- and 5-hydroxy-
1,4-naphthoquinone, respectively) are formed via shikimic acid,
and the same is true of the naphthoquinone system of the vitamins
K (cf. Chapter 13). It is remarkable that in all these cases the
carboxyl of the shikimic acid is retained, forming one or both of
the carbonyls of the quinone ring. The remaining three carbon
atoms seem to be furnished by glutamate (115A). The equal
incorporation of the carboxyl of shikimic acid into both carbonyls
of juglone requires a symmetrical intermediate, probably 1,4-
naphthoquinone (116). For the steps leading to this compound,
transformation of glutamate, via α-ketoglutarate, to succininc
semialdehyde, and reaction of this with shikimic acid, appears
likely (115A b); a further intermediate, o-succinyl-benzoic acid
[64] has been proposed, which is effectively and specifically
incorporated into naphtho- and anthraquinones in certain bacteria
and higher plants (117).

[64]

Recent research has revealed, however, that naphtho- and anthraquinones can also be biosynthesized through another pathway in which mevalonic acid furnishes part of one of the aromatic rings.

The possibility of a biosynthesis of naphtho- and anthraquinones from a benzene or naphthalene precursor and an isoprene unit was first discussed by Robinson (48) for shikonin [V-3] and a few β-methylnaphthoquinones (plumbagin [65a], droserone [65b]), and by Sandermann (118) for β-methylanthraquinones. These proposals were made on the basis of structural features of these compounds, and of the natural occurrence of compounds close to the postulated prenylated intermediates. Both types are often found together (e.g., in teak wood from *Tectona grandis*, Verbenaceae, investigated in great detail by Sandermann and co-workers). Such prenylated benzo- and naphthoquinones could (and *in vitro* sometimes do) cyclize to β-methyl-naphtho- or anthraquinones, respectively.

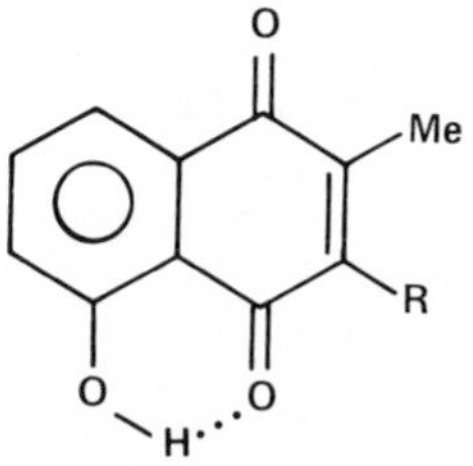

[65] a:R = H
b:R = OH

The biogenetic hypothesis discussed in the preceding paragraph was developed further by Inouye (119) on the basis of the co-occurrence, in *Pyrola incarnata* (Pyrolaceae), of 2,7-dimethyl-1,4-naphthoquinone or chimaphilin [V-1] and of homoarbutin [66a], one of the two possible glucosides of methylhydroquinone. A prenyl derivative [67] of this latter compound could indeed readily give the quinone [V-1]:

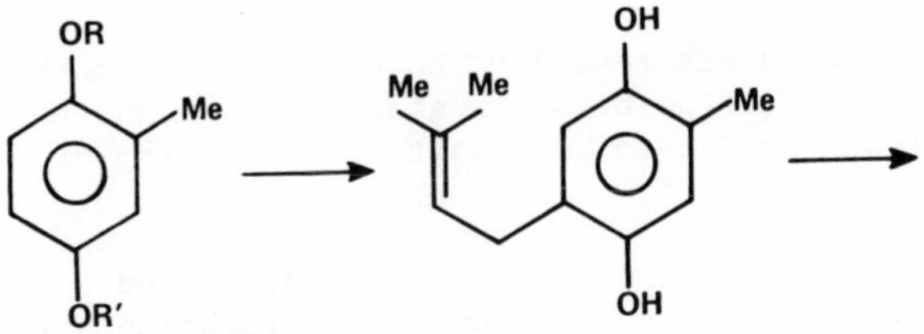

[66] a: R = H, R' = $C_6H_{11}O_5$
     b: R = $C_6H_{11}O_5$, R' = H

[67]

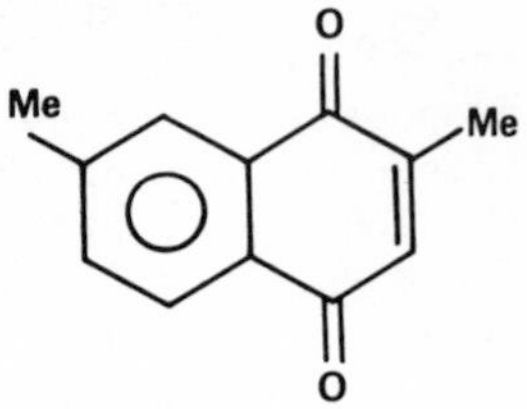

[V-1]

Subsequently, other biosynthetically related compounds were isolated from *Pyrola* spp. by the same author (for literature, see ref. 10, Vol. 4, pp. 73-76, 450-452), such as isohomoarbutin [66b] and renifolin, a glucoside of the 5,8-dihydro derivative of the hydroquinone corresponding to [V-1]. The actual existence of compounds corresponding to the assumed prenylated intermediate [67] was demonstrated by Burnett and Thomson (120), who isolated 2-methyl-5-isopentenyl-1,4-benzoquinone [68] from *P. media*; of possible interest for subsequent discussion of similar pathways to certain naphthoquinones is the occurrence of pyrolatin [69] in the same genus. The presumptive evidence for the actual formation of chimaphilin [V-1] from a precursor related to [68] is thus very strong.

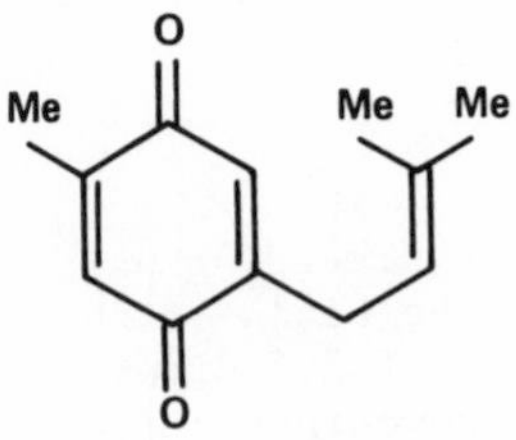

[68]

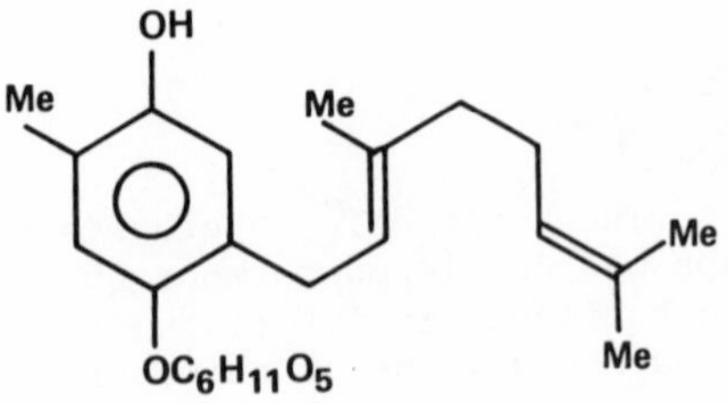

[69]

Experimental proof for this pathway has been adduced by Bolkart, Knobloch, and Zenk (121), who demonstrated the highly specific incorporation of label from $2-^{14}C$ mevalonic acid into the methyl group at C-7 of [V-1]; simultaneously, it was also shown (122) that the quinonoid ring is formed from tyrosine. The structures of [66] and [V-1], and the exclusive labeling (122) of the methyl at C-2 in the latter when $\beta-^{14}C$-tyrosine is the precursor, suggest (115) a reaction sequence involving conversion of tyrosine into homogentisic acid [70]. This pathway is analogous to the one made probable by Whistance and Threlfall (123) for the biosynthesis of the methylated benzoquinone ring of plastoquinone (see Chapter 11).

Convincing evidence for the involvement of [70], and also of methylhydroquinone [71] (both of which occur in *Chimaphila* as the glucosides) has been obtained by Bolkart and Zenk (124). $6-^{14}C$-Shikimic acid ($2-^{14}C$ in the older numbering), fed to *Chimaphila*, yielded [V-1] with 50% of the activity in position 2, the anticipated result if the substrate is converted to 2,6-labeled tyrosine with subsequent shift of the side-chain to give [70]. Since labeled [70] was likewise very specifically incorporated, and since its decarboxylation product [71], the aglucone of the homoarbutin [66a] present in the plant, was active (if weakly) when labeled tyrosine was fed, the biosynthesis of [V-1] can be formulated as shown below:

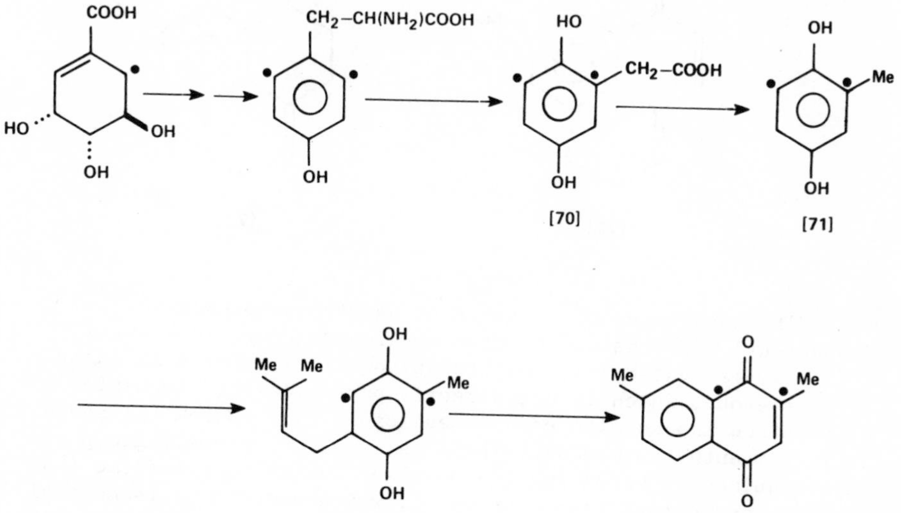

[70]   [71]

[67]   [V-1]

Several other β-methylated naphthoquinones occur in higher plants; a few examples are given below. Their structures can evidently be rationalized (48) on the basis of a biosynthesis analogous to that of [V-1]. In the absence of experimental evidence on the actual biosynthesis of these naphthoquinones, and in view of alternative possibilities (acetate pathway, methylation of naphthalenic precursors), no detailed discussion is warranted. The co-occurrence of [65a], [72], and [73] in *Diospyros* spp. is suggestive enough to deserve mention (125).

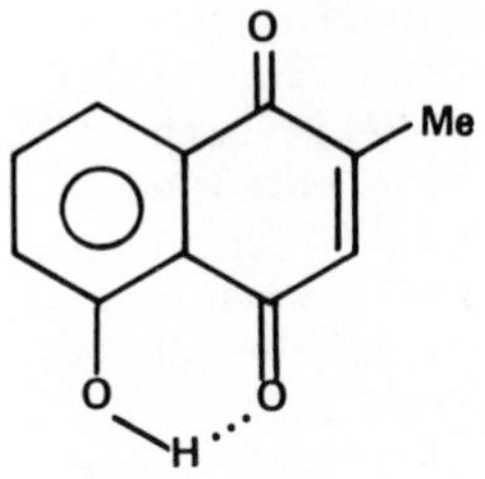

[65a]

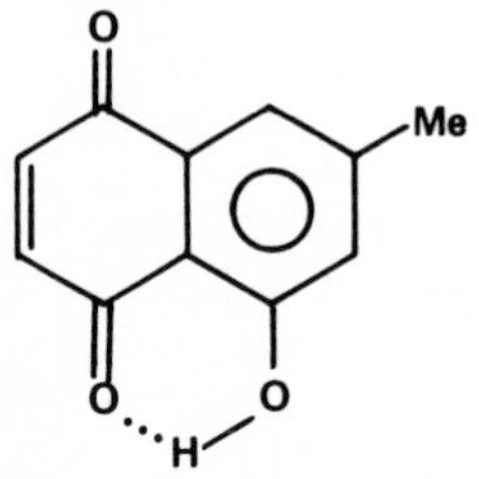

[72]

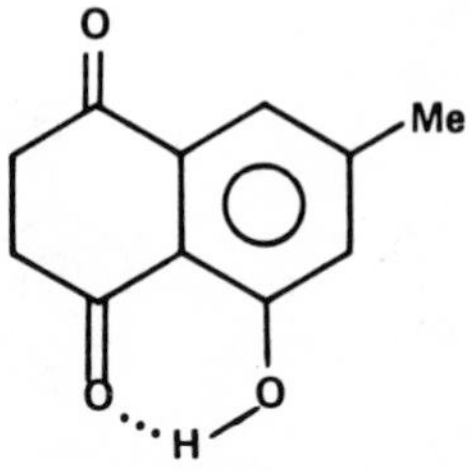

[73]

The occurrence of a number of naphthoquinones with isopentenyl side-chains has been mentioned above; lapachol (2-hydroxy-3-isopentenyl-1,4-naphthoquinone) may suffice as example of this group, which is not directly pertinent for discussion of naphthoquinones with partly mevalonoid ring system (although it is for anthraquinones). There is, however, a small group of naphthoquinones, occurring in the plant family Boraginaceae (and also in one euphorbiaceous species), which contain an iso*hex*enyl or iso*hex*yl side-chain, instead of the usual iso*pent*enyl grouping. These compounds are alkannin [V-2], its optical antipode shikonin [V-3], and alkannan [V-4]; in addition, the related compounds

[V-5], and [V-6], formally arising through hydration of the double bond in the side-chain of [V-2] and [V-4], respectively, have been isolated recently (126) from *Arnebia nobilis* (Boraginaceae). The compounds containing the benzylic hydroxyl seem to occur generally as esters. The unusual presence of a C_6 (rather than C_5) side-chain in these compounds becomes understandable if, following a suggestion by Robinson (48), a benzenoid structure, such as [74], with an isoprenoid C_{10} side-chain is assumed as precursor. This hypothesis finds support in the occurrence of compounds such as [69], whose structure shows precisely the required features (except, of course, for the presence of the additional methyl group in position 5 of the benzene ring). Experimental evidence for the essential correctness of this hypothesis has been obtained very recently (see Addendum, Section 22-8).

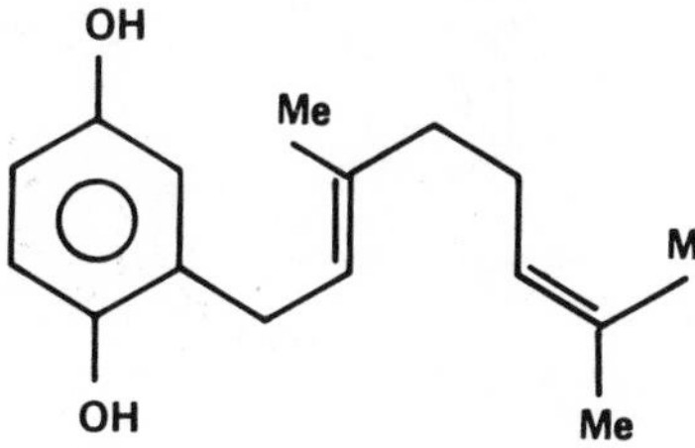

[74] 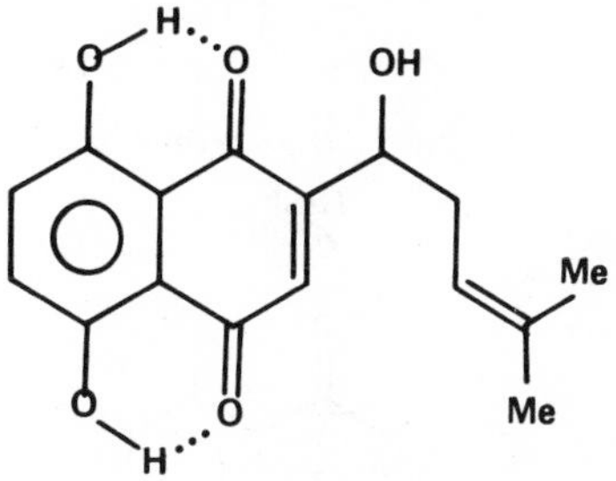   [V-2, 3]

Recent research has shown that representatives of the anthraquinones, a large group of natural products (about 170 known compounds (97)), can be formed through either one of two entirely different pathways. Many natural anthraquinones are typical polyketides; their biosynthesis is treated elsewhere (Section 18-10).

However, certain anthraquinones from higher plants have structures that were early recognized as being inconsistent with the acetate scheme (127). Such compounds occur mainly among the Rubiaceae, Bignoniaceae, Verbenaceae, and Scrophulariaceae; characteristically, their ring A is unsubstituted, and ring C often shows a distribution of oxygenated substituents different from the 1,3-orientation usually found in polyketides. Occasionally, oxygenated functions or carbon side-chains may be entirely absent. It is for some of these compounds that a biosynthesis involving mevalonic acid has recently been established. Typical representatives of this class are β-methylanthraquinone (tecto-quinone) [75] and alizarin [V-7].

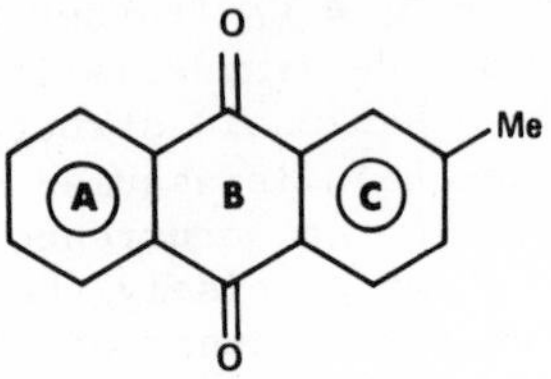

[75]

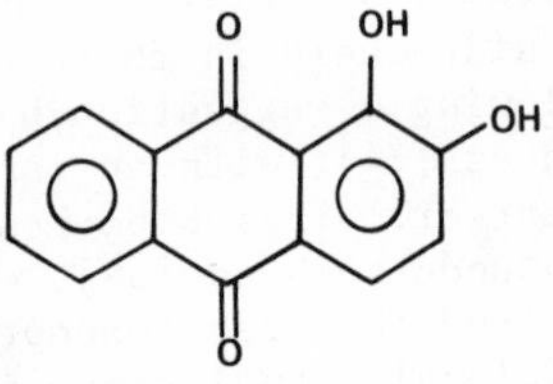

[V-7]

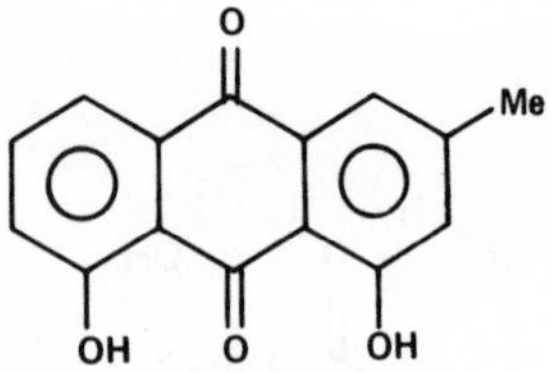

[76a]

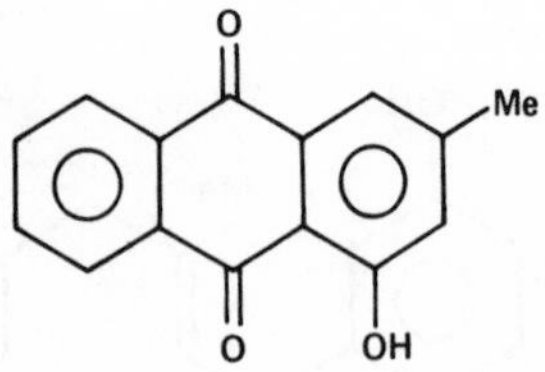

[76b]

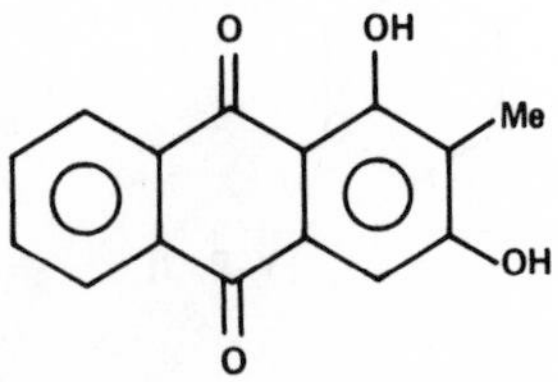

[V-10]

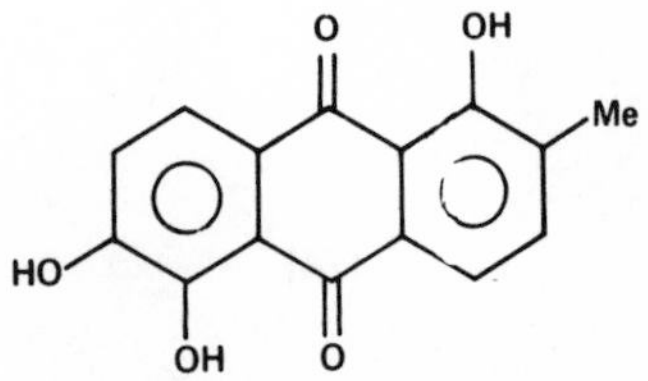

[77]

Interestingly enough, this biosynthetic pathway via meva-lonate seems restricted to higher plants; among fungal anthra-quinones, not only such a typical polyketide as chrysophanol [76a] has been shown to be formed from acetic acid, but also pachybasin [76b] (128), which might, from its structure with unsubstituted ring A, have been assumed to belong to the mevalonate-derived group. Even among the anthraquinones from higher plants, neither of the two structural characteristics just mentioned is a reliable criterion. This is shown by the fact that rubiadin [V-10], in spite of its 1,3-dihydroxylated ring C, has been proven (129) to belong to the mevalonate-derived class, and indicated by such observations as the co-occurrence of morindone [77] with [V-7] and a number of its relatives in *Morinda* spp. (Rubiaceae). For this reason, Table V lists only these anthraquinones for which a bio-synthesis involving mevalonate has been experimentally demon-strated. That, on the other hand, the polyketide pathway operates in higher plants as well as in fungi follows from the proof (130) that [76a] is biosynthesized from acetate and malonate by *Rumex alpinus* (Polygonaceae) and *Rhamnus frangula* (Rhamnaceae).

Anthraquinone pigments have also been isolated from two classes of higher animals: crinoids, and coccid insects. While some of these pigments carry the usual carbon side-chain in β-position [C_2 in case of the insects (ref. 97, pp. 419, 453), C_3 in that of the crinoids (ibid. pp. 443,445,447)], both classes also contain anthraquinones with typical polyketide oxygenation patterns but with carbon side-chains in α-position: C_4 in the case of crinoids (ibid. pp. 492, 507), methyl in position 1, methyl in 1 and carboxyl in 2, or carboxyls in both 1 and 2 in the coccid pigments (ibid. p. 471). The pigments with α-sidechains must arise through a mode of cyclization of a polyketide chain different from that leading to the normal type of plant and fungal anthraquinones with β-sidechains, and the representatives of this type among animal pigments. The origin of the anthraquinones α-sidechains presents an interesting problem, since the metazoan organism does not seem able to produce aromatic rings through the acetate pathway, and compounds of this type appear to be lacking in fungi, and to be very rare in higher plants; it is thus not very probable that the pigments should be furnished by food or symbionts. A few anthraquinones of the coccid type have, however, been encountered very recently in higher plants. For instance, *Aloe saponaria* (Liliaceae) contains (129A) the 3,8-di- and 3,6,8-trihydroxy derivatives of both 1-methylanthraquinone and its 2-carbomethoxy analog. In addition, 3,6-dihydroxy-8-keto-1-methyl-5,6,7,8-tetrahydroanthraquinone and its 2-carbomethoxy derivative are present; those two substances could qualify as the biosynthetic precursors of the fully aromatic ones, which might arise from them through dehydration and dehydrogenation, respectively.

Experimental evidence for participation of mevalonic acid in the biosynthesis of anthraquinones by plants of the families named above was obtained about simultaneously by Burnett and Thomson (129) and by Leistner and Zenk (131). In both cases, Sandermann's postulate (118) of a biosynthesis of β-methylanthra-quinones by cyclization of an isopentenylnapthalin or -naptho-quinone had prompted the experimental study; in the former case

because of new observations on the co-occurrence of anthraquinones
with numerous C- or O-prenylated naphthalenes in Bignoniaceae
(132A) and Rubiaceae (129A, 132B), in the latter because the somewhat earlier work by Leistner and Zenk had shown that shikimic
acid (133) and 1,4-naphthoquinone (134) are specifically incorporated into [V-7] and several of its relatives in *Rubia tinctorum*.
The work of both groups thus suggested a biosynthesis from
naphthalenic precursors, as postulated by Sandermann.

Independent studies by both groups, using madder (*Rubia
tinctorum*), showed good and highly specific incorporation of
$2-^{14}C$-mevalonate into anthraquinones. In the work of Leistner
and Zenk (131), pseudopurpurin [V-9], (which occurs in the plant
as the glucoside, and as the 1β-D-primveroside, galiosin), was
found to contain 85% of the label in the carboxyl group. This
finding shows that the carbon atom in question is derived stereospecifically from the *trans*-methyl group of dimethylallyl pyrophosphate [11], which is known to originate from C-2 of mevalonate. This result was obtained in short-term experiments (30
hours). In the experiments of Burnett and Thomson (129), using
the same plant species but much longer feeding times (15 days),
complete equilibration took place and the label was equally
distributed between the carboxyl and carbon atom 1. Rubiadin
[V-10] was likewise labeled, the activity being equal to that of
[V-9]. As a result of the equilibration, alizarin [V-7] and
purpurin [V-8] incorporated label; the activities of both
compounds were equal and amounted to approximately 1/2 of those
of compounds [V-9] and [V-10], which retain the one-carbon side-
chain. Oxidation of all four compounds gave inactive phthalic
acid.

These results establish the soundness of Sandermann's
initial postulate. The actual naphthalene-type precursor has not
yet been identified. However, it is known from the extensive
studies of Leistner and Zenk that it originates from shikimic
acid. All seven carbons of this acid are incorporated into [V-7]
and [V-8], furnishing ring A and one of the carbonyls (133a,b).
In agreement with this utilization of the entire molecule,
phenylalanine is not utilized; it will be remembered that in its
biosynthesis from prephenic acid, the original carboxyl of shikimic acid is lost (see Chapter 7). When shikimic acid labeled
specifically in positions 1 and 2 (old numbering: 1 and 6) was
used, the entire activity was localized in ring A, showing that
it is the carboxyl of shikimic acid which furnishes the
carbonyl(s) of the anthraquinone.

These findings are entirely analogous to those made for the
biosynthesis of naphthoquinones in higher plants (115), and thus
lend support to the ideas of a naphthalenoid precursor, subsequent

attachment of the isoprene residue, and final cyclization giving
the anthraquinone system. It is also suggestive that certain
simple naphthalenes are present in plants which contain mevalonate-derived anthraquinones: 1-methoxynaphthalene in *Tabebuia
chrysantha* and *Paratecoma peroba* (Bignoniaceae), 4-methoxy-1-
naphthol in *Galium mollugo* and *Asperula odorate* (Rubiaceae)
(132b). In conjunction with the wealth of prenylated naphthalenes
occurring in these plants, and in teak (*Tectona grandis*, Verbenaceae) (118b), simultaneous occurrences of this kind seem significant.

The labeling patterns established by these incorporation
Very recent work by Leistner and Zenk (135) provides detailed
information on the mode of incorporation of both shikimic and
mevalonic acids into [V-7] in *Rubia tinctorum*. Carboxyl-labeled
shikimic acid was administered to the roots. After 30 hours,
[V-7] was isolated and converted to the dimethyl ether. Alkaline
degradation yielded almost inactive veratric (3,4-dimethoxyben-
zoic) acid, together with phthalic acid having the same specific
activity as the ether, and benzoic acid with approximately 50%
of it. These findings prove that the carboxyl of shikimic acid
very specifically furnishes C-9 of [V-7]; a symmetrical intermediate such as 1,4-naphthoquinone is thus definitely excluded,
and the incorporation of this compound (134) must have been an
instance of nonspecific utilization. The biosynthesis of [V-7]
stands in interesting contrast to that of juglone (5-hydroxy-1,4-
naphthoquinone), where a symmetrical intermediate such as 1,4-
naphthoquinone is strongly indicated by the equal distribution of
label from the carboxyl of shikimic acid into both carbonyls
(116).

Additional information on the biosynthesis of ring C of
[V-7] was obtained from studies using 5-^{14}C-mevalonic acid (135).
Conversion of the alizarin to the diacetate of 4-nitroalizarin,
followed by bromopicrin cleavage, showed localization of more
than 80% of the label in position 4.

The labeling patterns established by these incorporation
studies with shikimic and mevalonic acids are shown in Scheme IX.
They are consistent with the assumption that *o*-succinylbenzoic
acid [64] (117) is an intermediate. This compound would cyclize
to 1,4-dihydroxy-2-naphthoic acid [78] (115Ab), which could be
prenylated in position 3, with subsequent decarboxylation and
cyclization. The localization of label from 5-^{14}C-mevalonic acid
shows, somewhat unexpectedly, that the hydroxylated carbons 1 and
2 of alizarin [V-7] originate from carbons 3 and 6 of the precursor. The hydroxyl at position 2 must thus have replaced a carbon
atom derived from one of the methyls group of the isoprene unit.

Scheme IX. Biosynthesis of Alizarin [V-7] in *Rubia tinctorium*

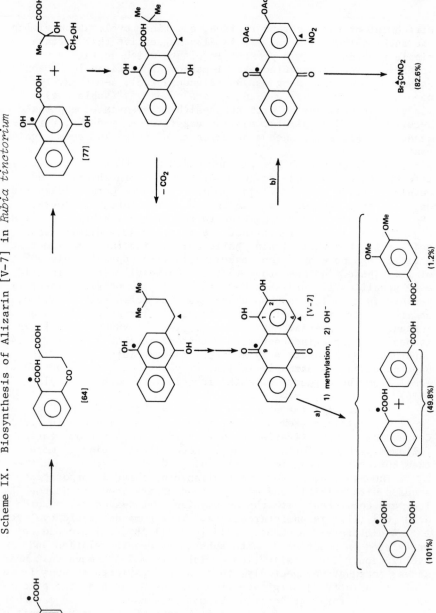

600

Scheme X. Structural and Biosynthetic Relationships of the Main Types of Cannabinoids.

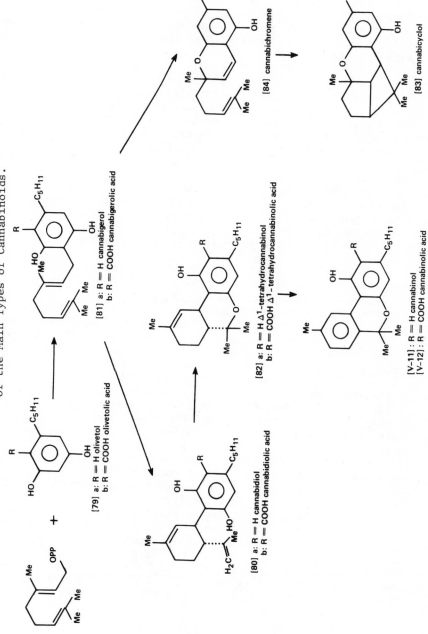

[79] a: R = H olivetol
b: R = COOH olivetolic acid

[81] a: R = H cannabigerol
b: R = COOH cannabigerolic acid

[84] cannabichromene

[83] cannabicyclol

[80] a: R = H cannabidiol
b: R = COOH cannabidiolic acid

[82] a: R = H Δ1-tetrahydrocannabinol
b: R = COOH Δ1-tetrahydrocannabinolic acid

[V-11]: R = H cannabinol
[V-12]: R = COOH cannabinolic acid

Cannabis sativa (Cannabinaceae), (marijuana, hashish)
contains a fairly large number of partly terpenoid compounds, the
cannabinoids, which include two substances in which the terpenoid
part of the molecule is aromatic: cannabinol [V-11] and cannabinolic acid [V-12].

Up-to-date reviews on the chemistry of the cannabinoids have
been provided by Mechoulam (136a) and by Joyce and Curry (136b);
they should be consulted for detailed information on these
compounds, and for complete bibliographic references. Merely as
an illustration of the scope of the field, structures of some of
the major groups of cannabinoids are shown in Scheme X in an
arrangement reflecting their probable biosynthesis, as given in
ref. 136. It will be evident from this presentation that the
skeleton of the cannabinoids appears to result from a combination
of a monoterpene unit derived from *p*-menthane [20] with a phenolic
moiety of typical polyketide structure; this phenolic moiety is
almost always olivetol (5-*n*-pentylresorcinol) [79a] or its carboxy
derivative, olivetolic acid, [79b]; in one instance, the n-pentyl
side chain is replaced by n-propyl to give cannabidivarin, the
C_3H_7 analog of [80a].

No experimental evidence on the actual biosynthesis of the
cannabinoids is available, but the biosynthetic pathway indicated
in Scheme X is plausible, and supported by much *in vitro* analogy.
Reaction of the polyketide moiety with geranyl pyrophosphate [12]
could give cannabigerol [81a], from which the other types (only
a few of which are shown in Scheme X) would arise through uncep-
tional steps. *In vitro*, [81a] is formed from [79a] and geraniol
under the influence of *p*-toluene-sulfonic acid (137).

Besides the two aromatic compounds [V-11] and [V-12], Δ^1-
tetrahydrocannabinol [82a] deserves mention as the main active
constituent, cannabidiolic acid [80b] as the most abundant canna-
binoid, and cannabicyclol [83] because of its structural analogy
with certain of the carbazole alkaloids to be discussed in the
next subsection. Compound [83] can be synthesized from [79a] and
citral (the aldehyde corresponding to geraniol [12a]) in hot
pyridine; it had initially been formulated as [83A], but the
alternative structure [83B] was shown to be the correct one by
NMR (138) and by X-ray crystallography (139).

Several plants of the family Rutaceae, belonging to the
genera *Glycosmis, Clausena,* and especially *Murraya,* produce a
substantial number of alkaloids which are derived from the
fully aromatic carbazole nucleus and have structures suggesting
a biosynthesis through combination of indole with isoprene units,
one of which participates in the formation of the aromatic system.
These alkaloids were discovered only in recent years, mainly as

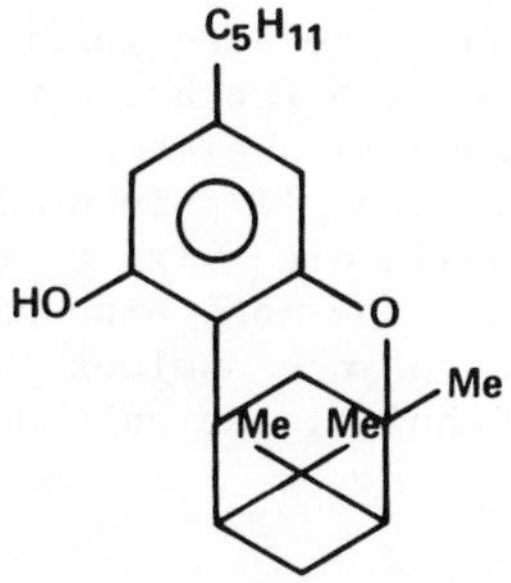

[83A]

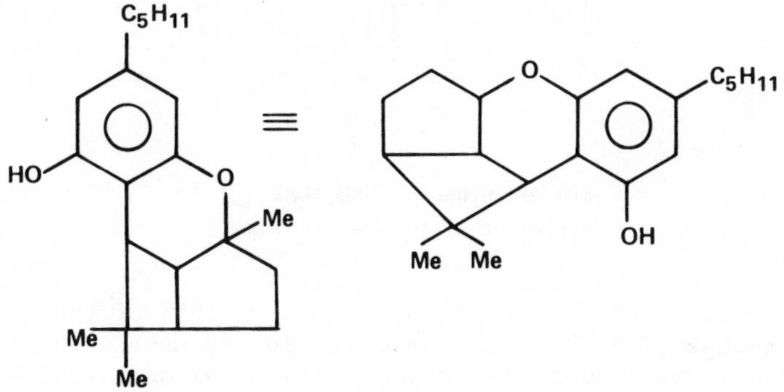

[83B]

the result of research by several groups of workers in India.
They are listed in Table V as [V-13] to [V-32]. A valuable review
by Kapil (139A)--the first one to be published--became available
after completion of the text of this chapter; it has been possible
to incorporate some information from this review.

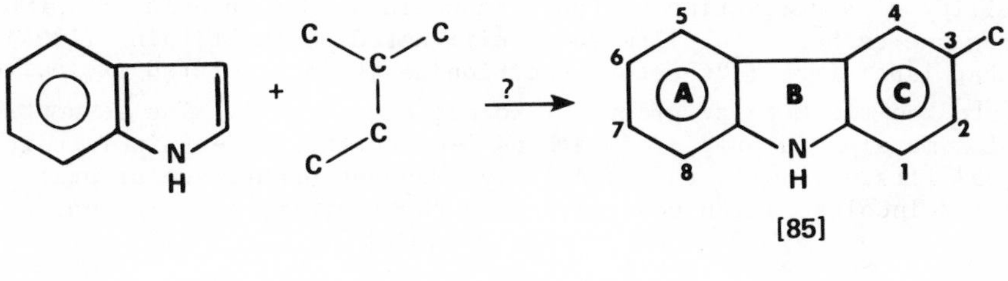

[85]

All alkaloids of this group contain the structural unit [85];
further elaborations add oxygenated functions and, in the
majority of cases, one or two further isoprenoid units which do
not form part of the aromatic skeleton. Carbazoles are otherwise
rare in nature; examples are ellipticine [86] and olivacine [87],
likewise derived from an indole, viz. tryptophan, and an isoprenoid moiety, in this case the monoterpenoid loganin [92] (see
p. 611). A tetrahydrocarbazole nucleus is of course present in
the ring system of strychnine, the calabash curare alkaloids,
aspidospermine, etc.

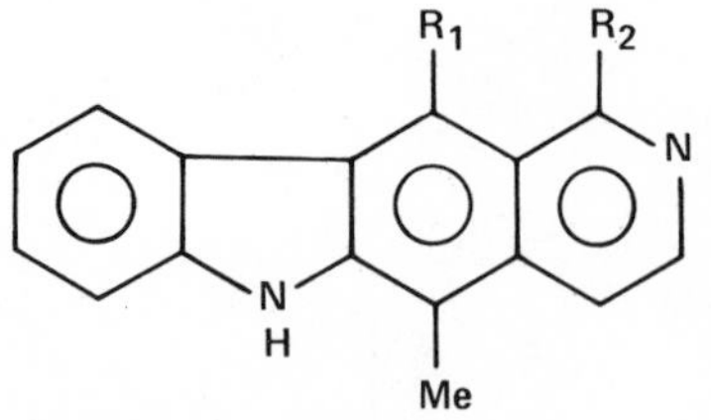

[86] ellipticine: R_1 = Me, R_2 = H
[87] olivacine: R_1 = H, R_2 = Me

The hypothesis that the carbazole unit [85] of the alkaloids
from Rutaceae might be formed from indole and mevalonic acid has
been briefly mentioned by Chakraborty (140) as one of several
alternatives. It was proposed more explicitly by Kureel, Kapil,
and Popli (141), and by Narasimhan and co-workers (142). This
proposal rested upon the dissection of the structures of these
compounds, and upon the occurrence of several other types of alkaloids derived from anthranilic acid or indole in the Rutaceae;
the acridones, furoquinolines, and quinazolones, widely distributed in this family, may be mentioned. In *Glycosmis pentaphylla*,
representatives of all three types occur together with the carbazoles. Derivation of the latter from indole is thus very plausible, and the consistent presence of the unit [85] is well
explained by the postulated biosynthesis. The alternative possibility of a derivation of the side-chain at C-3 in unit [85] from
the one-carbon metabolism seems eliminated by the finding (139A)
that label from [^{14}C-methyl]-methionine is incorporated exclusively
into the two methoxy groups of koenigicine [V-23]. The assumption
of a mevalonoid origin of carbons 1-4 and the side-chain in unit
[85] fits well with the relatively frequent occurrence of prenylated indoles, which contrasts with the scarcity of isoprenoid

units on other shikimate-derived aromatic systems. Long-known
examples are provided by the ergot alkaloids and by echinulin,
recent additions to the list by the isomeric 6- and 7-isopentenyl-
indoles which have been isolated from the liverwort *Riccardia
sinuata* (143). Moreover, isoprenoid units are quite frequently
encountered in the other anthranilate-derived alkaloid types
(acridones, furoquinolines). Particularly striking instances
are provided by atalaphylline and its N-methyl derivative [88a,b],
from *Atalantia monophylla* (Rutaceae) (144), and by the acridones
[89a,b,c] which co-occur with the carbazoles glycozoline [V-15]
and glycozolidine [V-16] in *Glycosmis pentaphylla* (145); the
structural resemblance of [88] and [89] with the carbazoles
heptaphylline [V-17] and girinimbine [V-18], respectively, is
striking.

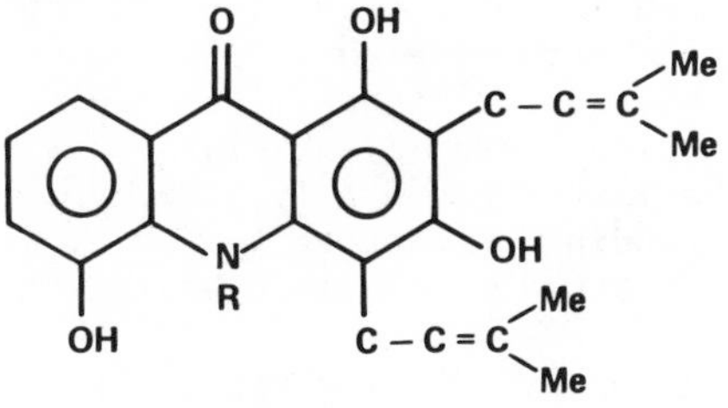

[88] a: R = H
     b: R = Me

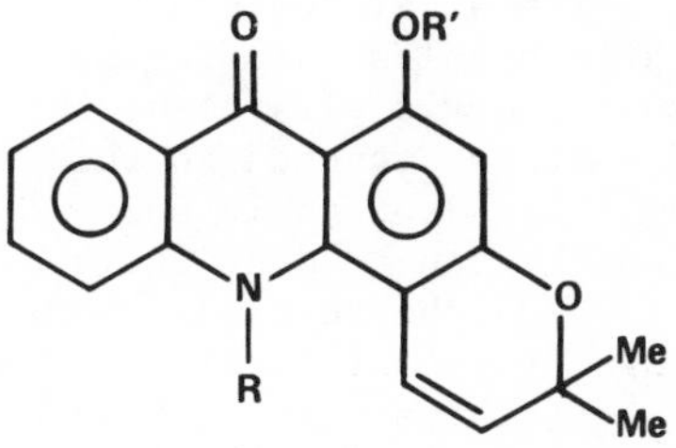

[89] a:R = Me, R' = H
     b:R = R' = H
     c:R = H, R' = Me

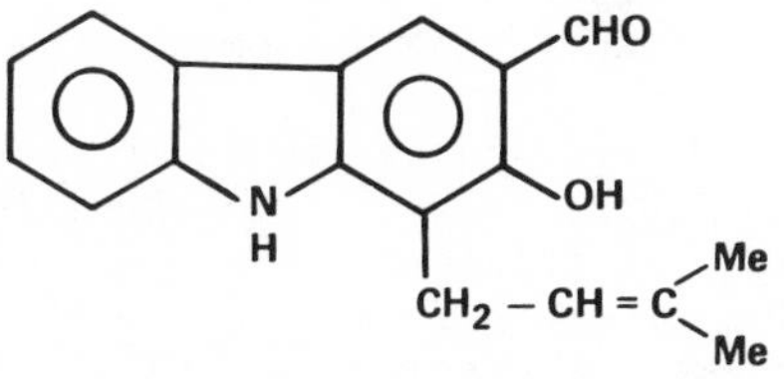

[V-17]

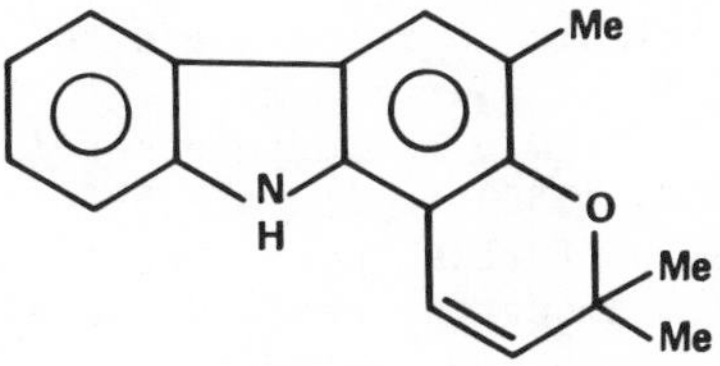

[V-18]

These various lines of argument by analogy are considered
suggestive enough to justify inclusion of the carbazole alkaloids
in this chapter.
    The assumption of a partly mevalonoid origin of ring C of
[85] may receive experimental support from the recent finding

(139A) that labeled mevalonic acid is indeed incorporated into
the carbazole alkaloids [V-21], [V-22], and [V-24] by *Murraya
koenigii*. The fact that the observed incorporations were only
very small (139B) does not necessarily militate against the hypothesis, since poor utilization of mevalonic acid for the biosynthesis of unquestionably mevalonoid compounds has been observed
repeatedly (see Section 22-7). This fact has, however, made
it impossible so far to prove by degradative techniques that the
incorporation of the label was specific.

The structures of the approximately 20 known carbazoles have
been established mostly by spectroscopic methods, and confirmed
by total synthesis in many instances. We have not considered it
necessary to give a complete bibliography on isolation and
synthesis, but are only quoting the papers most pertinent to
structure proof.

The alkaloids can be classified in three groups: (1) simple
C_{13} carbazoles carrying only a one-carbon side-chain and one or
two methoxyls, (2) alkaloids in which an additional isoprenoid
C_5 unit is attached to this C_{13} nucleus, and (3) alkaloids carrying a C_{10} isoprenoid unit. In the latter group only, two types,
a and b, can be distinguished, depending on whether the C_1 side-
chain and the terpenoid unit are attached to the same benzene
ring, or to different ones.

Of the simple carbazoles of group 1, two alkaloids, glycozoline [V-15] (146) and glycozolidine [V-16] (147), occur in
Glycosmis pentaphylla, while the two others, murrayanine [V-13]
(148) and mukoeic acid [V-14] (149) have been isolated from
Murraya koenigii. Glycozolidine [V-16] has been identified very
recently (139b) as 2,6-dimethoxy-3-methylcarbazole by comparison
with synthetic material; it had been formulated initially (140)
as the 5,7-dimethoxy derivative, a location of the methoxyls
difficult to reconcile with a derivation of ring A from anthranilic acid or indole.

It may be worth observing that *M. exotica* has recently been
found (150) to contain 3-formylindole; the biosynthetic significance of this occurrence is not clear at present.

Among the seven known alkaloids of group 2, that is, with
one isoprene unit added to the fundamental C_{13} skeleton of group 1,
heptaphylline, [V-17], from *Clausena heptaphylla*, is the simplest
one (151). The other alkaloids, all constituents of *M. koenigii*,
contain a pyran ring undoubtedly formed by interaction of an
isopentenyl group, as in [V-17], with an ortho-hydroxyl, a

reaction very commonly encountered; it takes place, for example,
in [V-17] under the influence of polyphosphoric acid (151).

Initial difficulties in deciding whether the one-carbon
side-chain of some of these alkaloids is attached to C-3 or C-6
of the carbazole system have now been resolved in favor of the
first alternative. Of the compounds of this group the two simplest ones, girinimbine [V-18] (140,152) and murrayacine [V-19]
(152c,153) deserve special mention; the others differ from [V-18]
only by the presence of oxygenated functions in ring A.

All known alkaloids of group 3 (i.e., C_{13} carbazole plus
two isoprene units) contain the pyranocarbazole ring system of
[V-18], and the C_1 side-chain is always methyl. In most of the
alkaloids of this group, all these substituents are attached to
the same benzene ring; two of them, however, mahanimbicine [V-31]
and bicyclomahanimibicine [V-32], form a subgroup in which the
otherwise unsubstituted benzene ring carries the methyl. The
biosynthetic significance of this divergence will be discussed
later.

In all alkaloids of group 3, the two isoprenoid units
added to the 3-methylcarbazole system occur in the customary
head-to-tail arrangement. However, in one of them, murrayazolidine [V-30], the mode of attachment of the oxygen bridge is
different from that in the other alkaloids. It is, however,
considered possible that [V-30] may actually be identical with
cyclomahanimbine [V-28] (139A).

The structure of mahanimbine [V-24], the most straightforward compound of this group, is simply that of an isopentenyl-
girinimbine (154b). The other alkaloids are derivable from
[V-24] by hydroxylation of ring A, as in mahanine [V-26] (142),
by hydration of the terminal double bond (mahanimbinine [V-25]
(155)), or by various cyclizations, to give cyclomahanimbine
[V-28] (156), bicyclomahanimbine [V-29] (156), mahanimbidine
[V-27] (156), and murrayazolidine [V-30] (157). The assumed
biosynthetic cyclization reactions are unexceptional in having
much precedent *in vitro* and *in vivo*, and the conversion of
mahaminbine [V-24] into [V-29] occurs readily on acid treatment
(156). This fact also explains the simultaneous synthesis of
[V-24] and [V-29] from 2-hydroxy-3-methylcarbazole and citral in
refluxing pyridine (154c); on this basis, [V-29] may actually be
an artefact (156).

Mahanimbicine [V-31] and bicyclomahanimbicine [V-32] (158),
the two alkaloids of the subgroup (b) with the methyl in C-6
rather than C-3, are in every other respect the complete analogs
of [V-24] and [V-29], respectively. Their synthesis from
2-hydroxy-6-methylcarbazole and citral (158) is likewise
completely analogous with that of the two 3-methylated alka-

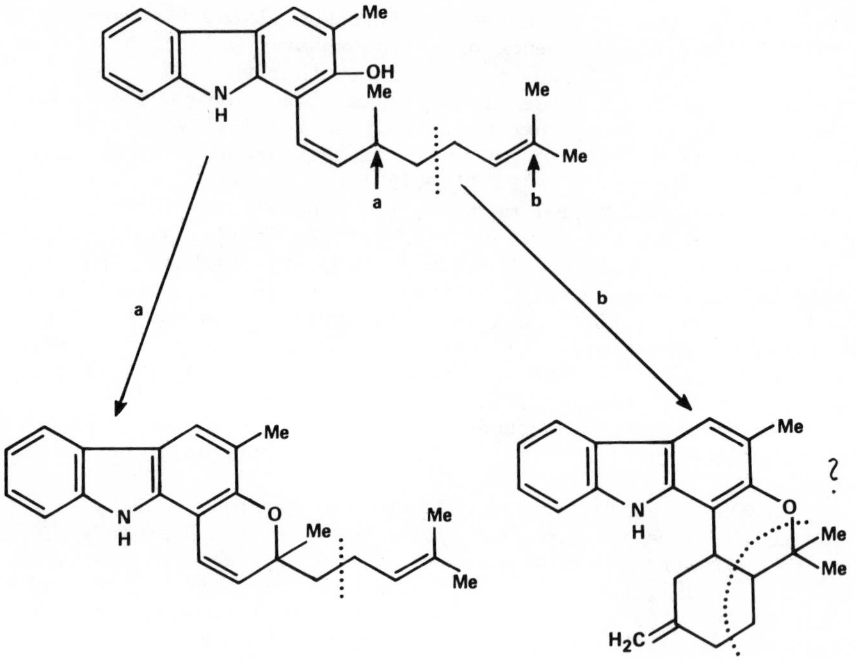

[V-24]   [V-30]

loids from 2-hydroxy-3-methylcarbazole. While the formulations
of all these bases are firmly based on spectrocopic findings,
and in many cases also on total synthesis, that of the bicyclo
bases, [V-29] and [V-32], is open to doubt, in spite of the fact
that these alkaloids have been synthesized. The synthesis used
is in both cases perfectly analogous with that of cannabicyclol
[83] from olivetol [79a] and citral (138), and also the formation
of [V-29] from [V-24] on treatment with acid completely parallels
the cyclization of cannabichromene [84] to [83]. However, the
formulae given for the two bicyclo alkaloids are analogous to
the *older* structure of [83], and are undoubtedly in part based
on it (156). The recent revision of this older structure [83a]
to [83b] on the basis of X-ray evidence (139) suggests that a
similar reformulation of the alkaloids may be needed (139).

The existence of the two subgroups of group 3, with methyl
in C-3 and C-6, respectively, can be neatly rationalized following Kureel, Kapil, and Popli (154c, 158), if the biosynthetic
sequence is assumed to proceed through 3-methylcarbazole [91]
(or a related compound with the same carbon skeleton), which

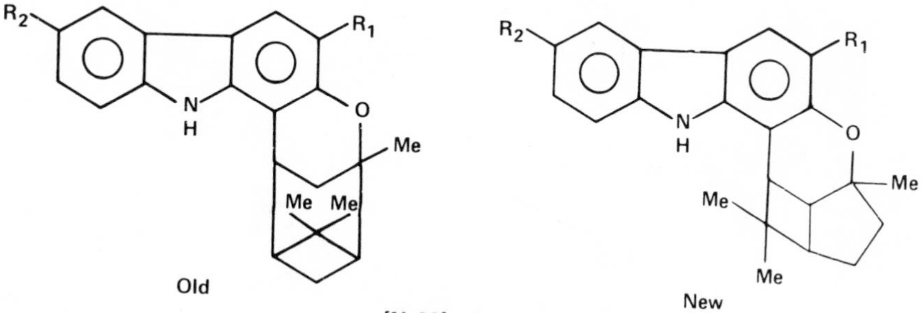

Old   New

[V-29] : R_1 = Me, R_2 = H

[V-32] : R_1 = H, R_2 = Me

could then undergo hydroxylation in either of the two equivalent positions, 2 and 7, respectively. The 2-hydroxy-3-methylcarbazole resulting from the former process would be the progenitor of the alkaloids of groups 2 and 3a, while the latter sequence would yield the two alkaloids of group 3b, [V-31] and [V-32]. On this basis, the absence of representatives of a subgroup 2b, similar to [V-18] etc., but with the methyl at C-6, seems surprising; it may, however, simply be due to our still incomplete knowledge. It is noteworthy that the isoprenoid units in ring C and its C_5 or C_{10} side-chains are *not* arranged in the usual head-to-tail sequence, although the C_{10} unit itself invariably is. This fact evidently supports the hypothetical biosynthesis via [91], with subsequent attachment of one or two isoprene units *ortho* to the hydroxyl in the biosynthesis of the compounds of groups 2 and 3.

The carbon-nitrogen skeleton of the vast majority of the innumerable indole alkaloids is now known to result from the combination of two moieties, of which one originates from tryptophan, the other from the cyclopentanoid monoterpene glucoside loganin [92]. The biosynthesis of these alkaloids is briefly discussed in Chapter 16. The manifold transformations of the loganin-derived moiety result in the several major classes of indole alkaloids.

One of these classes, of which yohimbine and reserpine are typical representatives, includes a small number of alkaloids in which ring E has become aromatic. These are alstoniline [V-34] from *Alstonia constricta* (Apocynaceae) (158A), the closely related ourouparine [V-33] from *Uncaria gambier* (*Ourouparia gambir*) (Rubiaceae) (158B), and three bases which have been isolated (158C) from gambir, a tanning material prepared from the latter plant: gambirtannine [V-35], oxogambirtannine [V-36], and

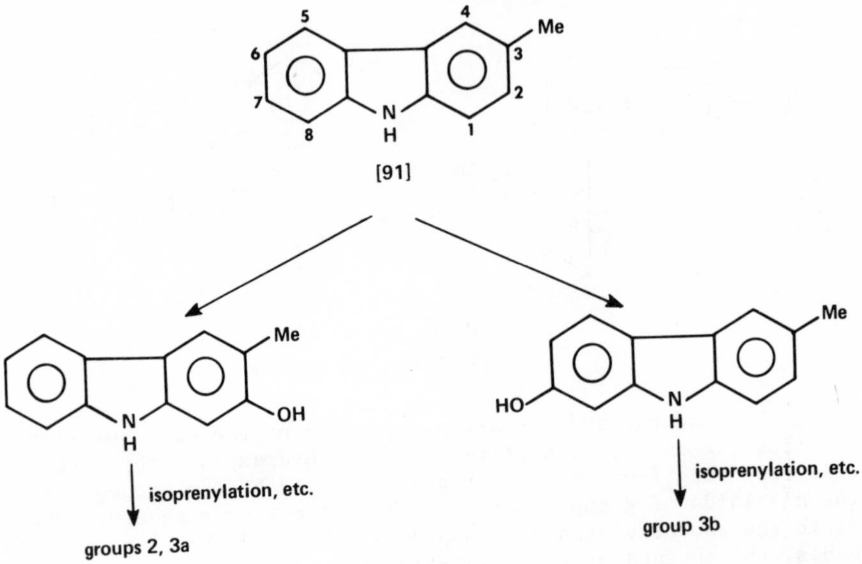

[91]

isoprenylation, etc.

groups 2, 3a

isoprenylation, etc.

group 3b

dihydrogambirtannine [V-37]. It seems uncertain, however, whether the three last-named bases preexist in *U. gambir*, or whether they are artefacts formed during preparation of the gambir, or during isolation from it.

These few alkaloids thus represent one more type of aromatic terpenoid of mixed origin, a type in which the mevalonoid moiety has undergone such deep-seated changes that its origin could not have been deduced from structural criteria, and is recognizable only through the obvious relationship of these bases with other alkaloids of known antecedents. No experimental work on the biosynthesis of the bases with aromatic ring E seems to exist.

22-7. EXPERIMENTAL EVIDENCE FOR THE BIOSYNTHESIS OF
       AROMATIC TERPENOIDS

This section attempts to collect and correlate the actual experimental results pertinent to the biosynthesis of aromatic mono- and sesquiterpenoids that have been obtained so far. Considering the large number of compounds in these two classes, the available evidence is surprisingly scanty; even more surprising is the total absence of any experimental work on the biosynthesis of the aromatic diterpenoids. Evidence concerning the compounds of mixed origin has already been given in

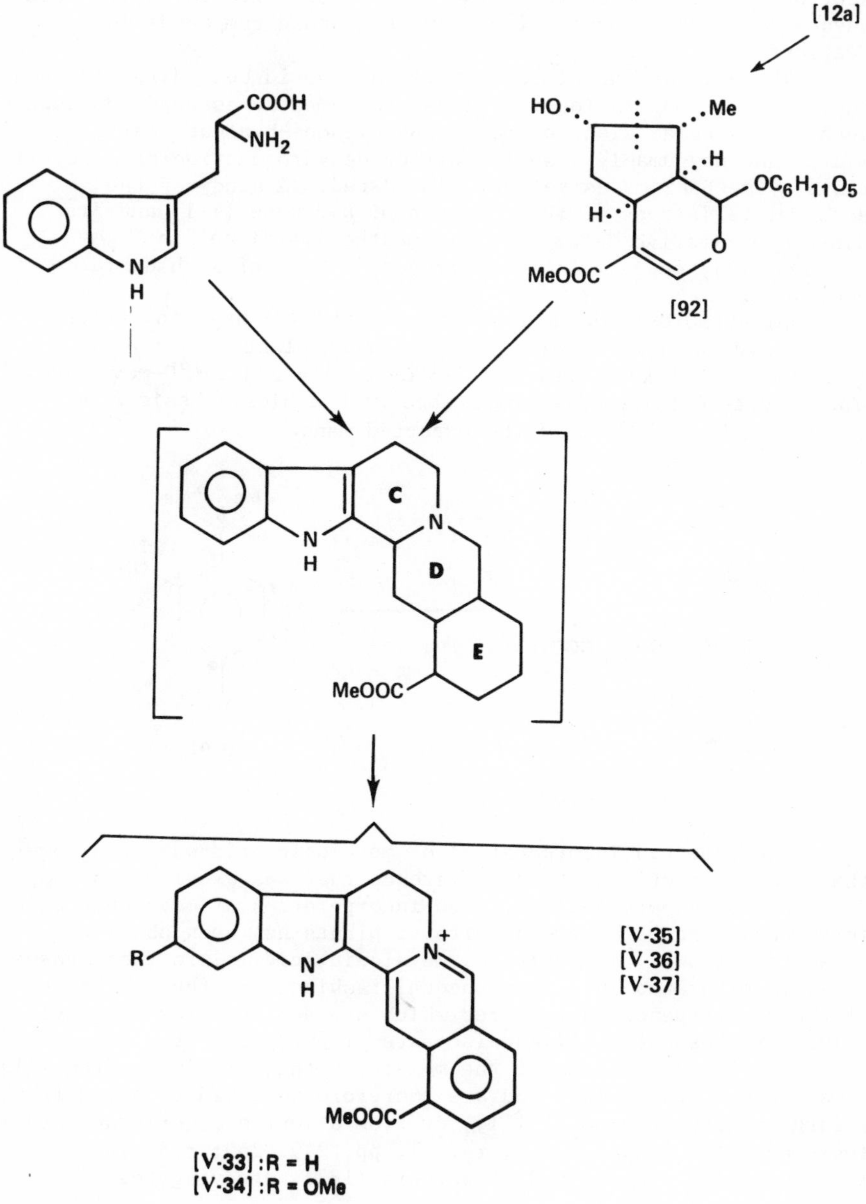
[12a]
COOH
NH2
N
H
HO
Me
H
OC6H11O5
H
O
MeOOC
[92]
C
N
N
H
D
E
MeOOC
R
N
H
N+
MeOOC
[V-35]
[V-36]
[V-37]
[V-33] : R = H
[V-34] : R = OMe

Section 22-6; it would not have been possible to organize that
section on the sole basis of the structural criteria, which were
adequate for the classes of purely mevalonoid compounds in
Sections 22-1 to 22-5.

There is no experimental work that would bear directly upon
the actual aromatization reactions; they may be assumed to involve
such processes as dienone-phenol and dienone-benzene rearrangements, and presumably also the direct dehydrogenation of alicyclic
rings by mechanisms as yet not elucidated. A study of the
seasonal variations of the contents of p-cymene [I-1] and its
dihydro-derivative γ-terpinene (p-mentha-1,4-diene) in *Thymus
vulgaris* (11) suggests the occurrence *in vivo* of such dehydrogenations.

(1) MONOTERPENOIDS: Shibata and co-workers (179) have investigated
the biosynthesis of thymol [1-4] in young plants of *Orthodon
japonicum* (Labiatae), using a 1-^{14}C-acetate and 2-^{14}C-mevalonate [5].
The results obtained are compatible with a biosynthesis from
acetate and mevalonate in the expected manner:

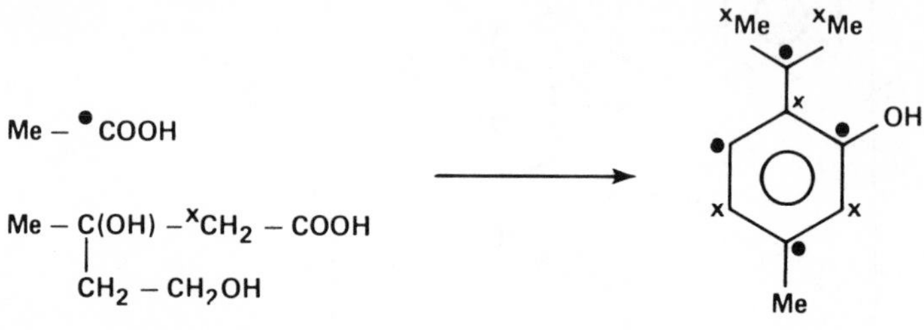

Me – •COOH

Me – C(OH) –^{x}CH$_2$ – COOH
      |
     CH$_2$ – CH$_2$OH

[5]

[I-4]

The observed incorporation of mevalonic acid was much lower
than that of acetic acid, a difference that was ascribed to
differences in permeability. Low incorporation of mevalonic acid
into mono-terpenes by green parts of plants has been observed
repeatedly (180). In addition, mevalonic acid is, in those cases,
often incorporated in a very unequal fashion into the two halves
of the monoterpenoid structures which are derived from dimethylallyl pyrophosphate [11] and isopentenyl pyrophosphate [10],
respectively. The parts of the molecules that originate from [11]
remain almost unlabeled, and are therefore believed to arise from
a large metabolic pool of [11], or from a source other than mevalonic acid (cf., *inter alia*, ref. 1, pp. 249, 250; ref. 181).
Interestingly, Francis and co-workers (182), studying the

biosynthesis of geraniol [12a] and nerol [19] and their β-gluco-
sides in rose petals found rapid, specific, and high incorporation
of mevalonic acid, equally distributed between the [10]- and [11]-
derived moieties of the terpenoids. The low and unequal incorporation thus seems to be characteristic of the green parts of
plants, an interpretation further supported by the work on biosynthesis of gossypol [II-56] in roots and root-extracts of the
cotton plant (see below).

When $2-^{14}C$-malonic acid or $2-^{14}C-$ [5] was administered to
stems or leaves of *Monarda punctata* (Labiatae), labeled p-cymene
[I-1] was obtained (181A).

(2)SESQUITERPENOIDS. The formation of helicobasidin [II-22] by
the basidiomycete *Helicobasidium mompa* has been studied by three
groups of workers (183-185). The earlier papers (183, 184) gave
results consistent with the general mevalonate scheme for
[II-22]: retention of 50% of label from acetate, 67% from
mevalonate on oxidation to (-)-camphonanic acid [24], as expected.
However, Natori and co-workers (183) reported relatively high,
Bentley and Chen (184) very poor labeling from mevalonate.

More recently, Nozoe and co-workers (185) studied the incorporation of $[2-^{14}C, 4-^{3}H]$-mevalonate into [II-22], [II-23], and
[II-23a]. In the former two compounds, the initial 3:3 ratio of
^{3}H to ^{14}C changed to 3:2, while it was retained in [II-23a].
Since conversion of a farnesol precursor [13] to the aromatic
structure of [II-23a] should result in the loss of one H, conversion to the fully substituted quinonoid systems of [II-22] and
[II-23] in the loss of two hydrogen atoms, the findings prove the
occurrence of a hydrogen shift (see next page).

The biosynthesis of the bis-sesquiterpenoid gossypol [II-56]
in the excised roots (and in root extracts) of the cotton plant,
(*Gossypium hirsutum*) (Malvaceae), has been the object of two
studies (186, 187) by Heinstein and co-workers. Initially (186),
efficient incorporation of 1- and $2-^{14}C$ acetate into [II-56] was
observed. Degradation of the resulting gossypol was carried out
by known methods (47) (cf. Scheme XI): methylation of the
hydroxyls, followed by cleavage of the resulting hexamethyl ether
into formic acid (from the aldehyde groups) and apogossypol
hexamethyl ether, and transformation of this compound into acetone
(from the isopropyl groups) and des-apogossypol hexamethyl ether.
The activities found in the C_1 and C_2 fragments, and in the two
apo-compounds, agreed very well with those expected from a biosynthesis of [II-56] from acetic acid via mevalonic acid, [5], as
is shown on the next page.

The observation of excellent biosynthesis of [II-56] in
root homogenates not only from acetate but also from mevalonate
(∿22% incorporation!) led to a subsequent detailed study (187)

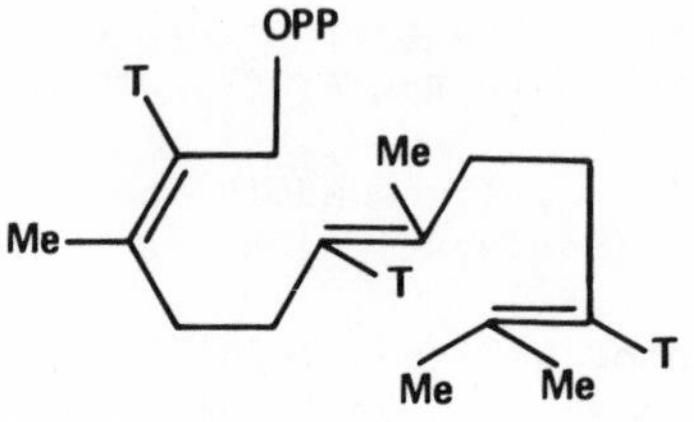

[II-22] : R = OH
[II-23] : R = H

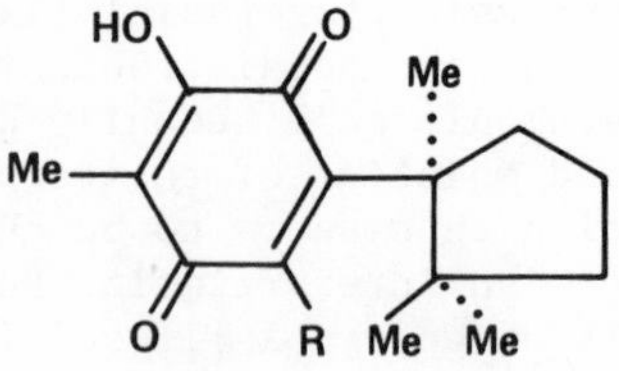

[13]

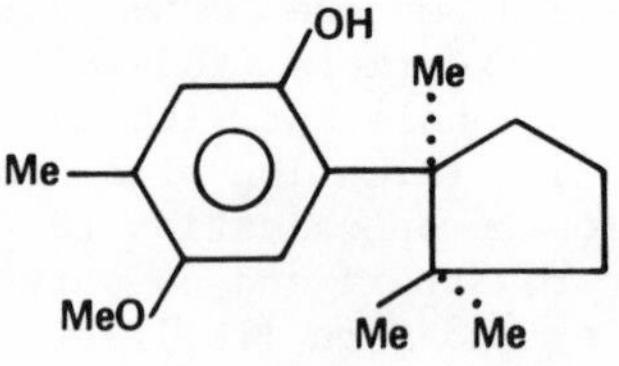

[II-23a]

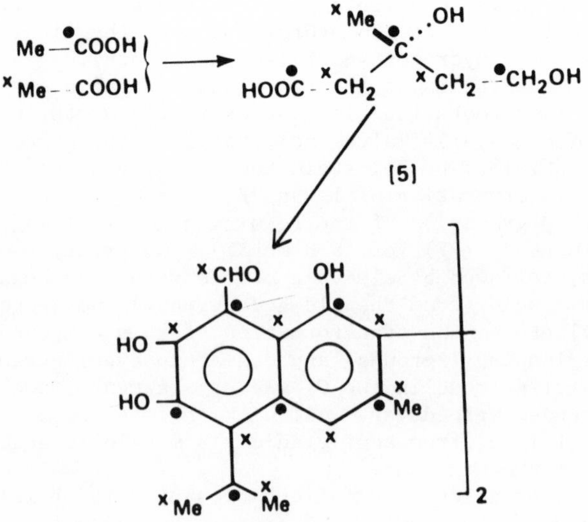

[5]

[II-56]

of the incorporation of various isoprenoid precursors in this system. This investigation is by far the most comprehensive experimental research done on the biosynthesis of an aromatic terpenoid up to now.

The degradation reactions used before were here supplemented by oxidation of the hexamethyl ether of [II-56] to two molecules of gossic acid, that is, with the loss of carbons 2,3, and 11 (see Scheme XI). It was again observed that $2-^{14}C$ mevalonate [5] was remarkably well incorporated by the enzyme system from the roots and that the incorporation occurred in the expected manner; for example, the activities found in des-apogossypol hexamethyl ether and in acetone, respectively, were sufficiently close to the expected 2:1 relationship. The stereospecific nature of this incorporation was shown (a) by the inactivity of the formyl groups, which are thus derived from the methyl, rather than from C-2, of [5], and (b) by the fact that the gossic acid, formed from [II-56] with loss of C-11, retained all of the activity.

The pyrophosphates of geraniol [12A] and its *cis*-isomer nerol [19] were likewise incorporated, whereby the former was utilized to about the same extent as [5], while the latter was used about twice as well. Of farnesol pyrophosphate [13] and its three *cis,trans*-isomers, the *cis,cis*-compound was four times more efficient as a precursor than the *trans,cis*-isomer, while the two remaining stereoisomers with *trans* stereochemistry of Δ^6 (*cis*, *trans* and *trans*, *trans*) were hardly incorporated at all.

Biosynthesis of the cadinane-type naphthalene rings of [II-56] from a C_{15} precursor with the normal head-to-tail arrangement of the isoprene residues could take place in three different ways depending upon the stereochemistry of the double bonds, whereby the label from C-2 of [5] and the precursors with 10 and 15 carbon atoms would be expected to occupy the positions shown on page 617.

The marked preference for C_{10} and C_{15} precursors with *cis*-stereochemistry at Δ^2, and the failure of the two C_{15} precursors with *trans* arrangement at Δ^6 to be utilized, confirms the conclusion from the work with [5] that the biosynthesis proceeds in a stereospecific manner, and suggests that *cis,trans* isomerization is possible at Δ^2 (cf. Section 22-1) but not at Δ^6. Results of degradation experiments provide a clear-cut decision in favor of diagram A (Page 617). Thus, conversion of [II-56] biosynthesized from either of the two C_{10} precursors, to gossic acid (loss of C-2, 3, and 11) resulted in retention of practically all the activity in both cases. On the other hand, the same degradation of [II-56] from the C_{15} precursor with *trans,cis* stereochemistry yielded CO_2 (from carbons 2, 3, and 11) with an acitivity close to the one expected if all the label had been contained in one of these

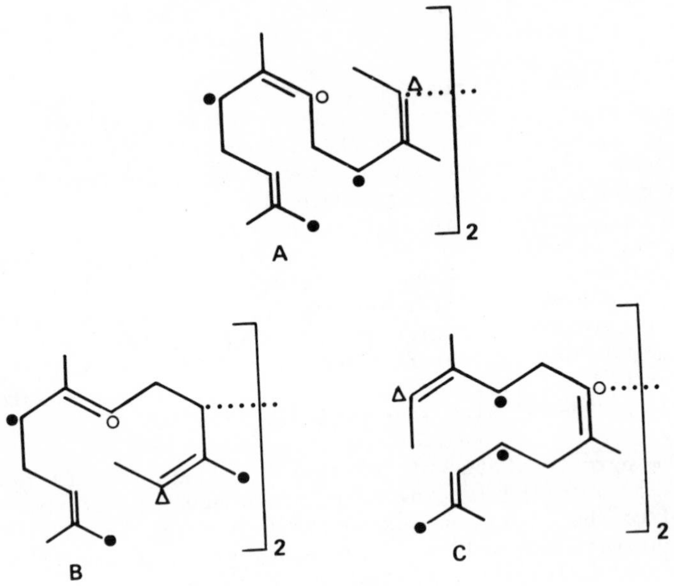

A

B

C

● C – 2 of [5]
○ C – 2 of C_{10} precursor
Δ C – 2 of C_{15} precursor

carbon atoms.   Clearly, those two findings together are compatible only with A and eliminate B and C.

The thermolabile enzyme activity responsible for the biosynthesis of [II-56] by homogenates of *Gossypium* roots was found in the supernatant fraction after prolonged centrifugation at 105,000g; ATP and Mg^{++} (or Mn^{++}) were required, and nicotinamide considerably enhanced the activity of the system.

In this case, no clear indication seems to have been obtained of any unequal incorporation of [5] into parts of the molecule originating from [10] and [11], respectively.   The details of the actual formation of the aromatic rings, and of the other concomitant or subsequent processes (dimerization, introduction of oxygen) remain unexplored.   The results already obtained, however, give highly suggestive information on the stereochemistry of the cyclization, and they prove that such a biosynthesis can proceed with high efficiency in extracts, and is thus accessible to detailed experimental study.

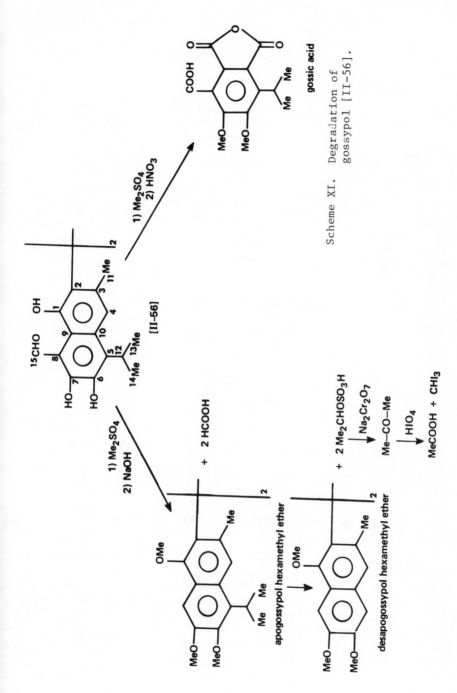

Scheme XI. Degradation of gossypol [II-56].

617

The incorporation of [5] into another aromatic sesquiterpene, illudinine [II-62], has been briefly mentioned (53).

22-8. ADDENDUM

Since the completion of the manuscript of this chapter, several additional publications have come to our attention. A number of these merit brief discussion; no complete coverage is intended.

A compilation by Devon and Scott (188) gives the structures of all naturally occurring terpenoids.

The pyrophosphorylated C_{30} intermediate [93] between farnesyl pyrophosphate [13] and squalene [14], now named "presqualene pyrophosphate," has the structure shown below (189). The mechanism of its biosynthesis, and even more so that of its transformation into [14], which must involve skeletal rearrangement, constitute fascinating problems.

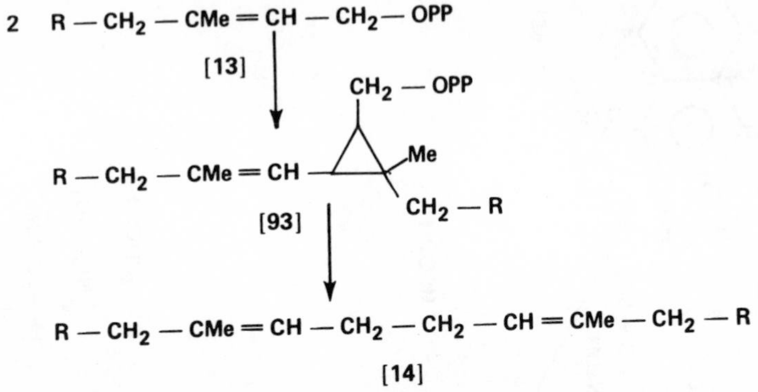

$$2 \quad R - CH_2 - CMe = CH - CH_2 - OPP$$

$$[13]$$

$$R - CH_2 - CMe = CH \overset{CH_2 - OPP}{\underset{CH_2 - R}{\triangle}} Me$$

$$[93]$$

$$R - CH_2 - CMe = CH - CH_2 - CH_2 - CH = CMe - CH_2 - R$$

$$[14]$$

$$R = Me_2C = CH - CH_2 - CH_2 - CMe = CH - CH_2 -$$

The biosynthesis of the *monoterpenoids* has been reviewed by Banthorpe and co-workers (190).

Inula and *Helenium* spp. (Compositae) have yielded several new aromatic monoterpenoids. The isobutyryl esters of 7-hydroxythymol methyl ether [94a] and 7-hydroxythymohydroquinone dimethyl ether [94b] have been isolated from *I. salicina* (191), while *I. viscosa* contains [94a] together with the corresponding isovalerate [94c] (192).

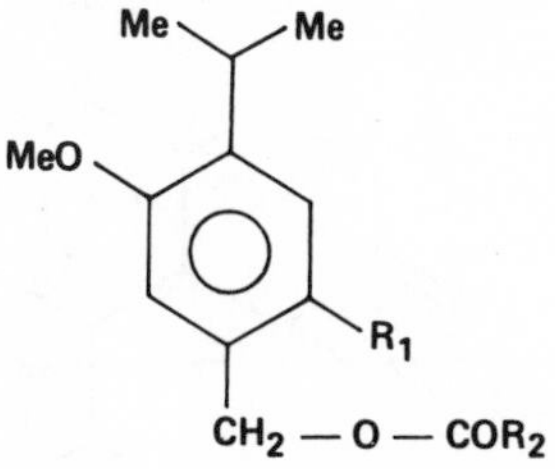

[94]   a: R_1 = H, R_2 = CHMe$_2$
       b: R_1 = OMe, R_2 = CHMe$_2$
       c: R_1 = H, R_2 = CH$_2$CHMe$_2$

The essential oil from the roots of *Arnica montana*
(Compositae) contains thymohydroquinone dimethyl ether [95]
(193A,B) and its Δ^8-dehydro derivative [96], together with [I-4]
and [I-8] (193B). Compound [96] has also been isolated from
tne ayapana oil of *Eupatorium triplinerve* (*E. ayapana*)
(Compositae) (193C).

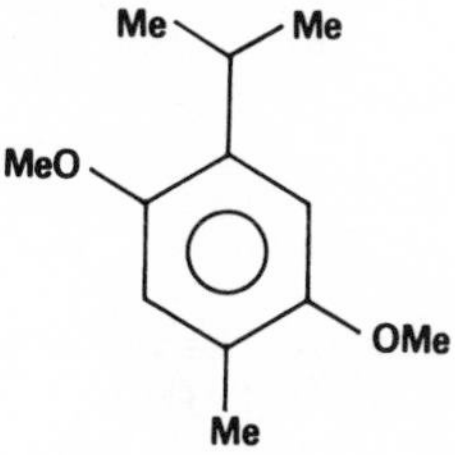

[95] 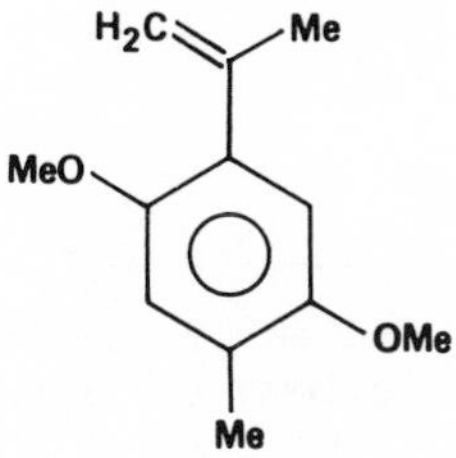   [96]

Three new *sesquiterpenoids* of the cadinane [26] class, the
lacinilenes A, B, and C, have been isolated from the wood of
Ulmus laciniata (194). For compound A, structure [97] of a hydroxy-
nor-cadalene has been proposed, while lacinilene B was tentatively formulated as [98], a structure which does not conform to
the isoprene rule (194a). A similar constitution was advanced
(194b) for lacinilene C. However, this compound and its methyl
ether were subsequently isolated from the bracts of cotton,
Gossypium hirsutum, and were formulated (194A) as 1,7-dihydroxy-
4-isopropyl-1,6-dimethyl-2-(1*H*)-naphthalenone and its 7-methyl
ether, respectively; structures derived from the normal cadalene
skeleton. Lacinilene B, too, may thus turn out to be a normal,
fully isoprenoid derivative of [26].

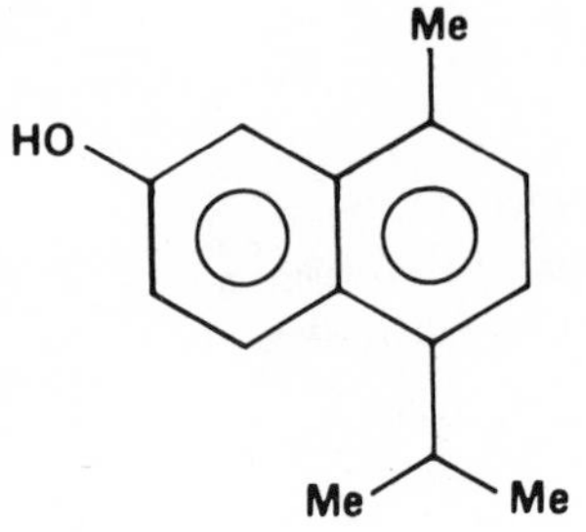

[97]

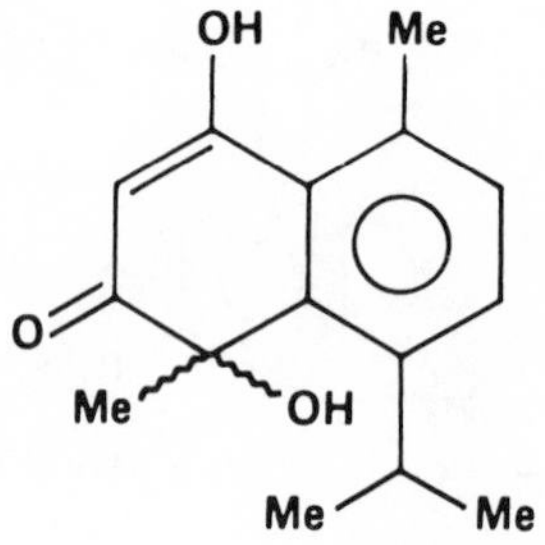

[98]

From studies of the stereochemistry of the calamenenes [II-35], it follows (195) that the widely distributed terpene is the *trans*-compound, while the *cis*-epimer occurs in Alaskan cedar (*Chamaecyparis nootkatensis*).

Recent work on the constituents of bracken, *Pteridium aquilinum* var. *latiusculum* (Pteridaceae) (196a,b) has yielded eight 1-indanones, the pterosins A-F and Z [99a-h], whose structures show clearly that they belong biosynthetically into the illudin group of sesquiterpenoids; the carbon skeleton of several of them [99a-c,i,j] is actually identical with that of illudalic acid [II-61] and illudinine [II-62]. In addition, the plant also contains the β-D-glucosides of [99c,d,e,b, and a], which have been designated as pterosides A,B,C,D, and Z (196B 197). The glucose residue in the pterosides is attached to the hydroxyethyl side-chain. Three additional members of this group, the pterosine J, K, and L, and six acyl derivatives (acetyl, isocrotonyl, palmityl, benzoyl) of pterosins A,B, and C were subsequently isolated from the same plant (198). Another form of the same family, *Hypolepis punctata*, produces, besides Z [99a], two more compounds of this group, pterosins H [99i] and I [99j] (199).

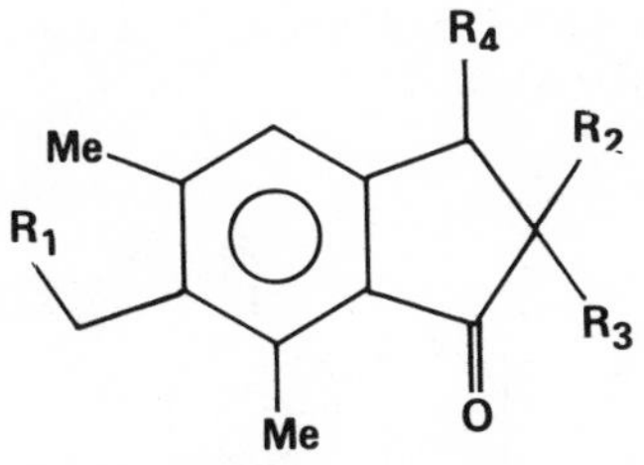

[99]

	R_1	R_2	R_3	R_4	pterosin
a	CH_2OH	Me	Me	H	Z
b	CH_2OH	Me	Me	OH	D
c	CH_2OH	Me	CH_2OH	H	A
d	CH_2OH	Me	H	H	B
e	CH_2OH	Me	H	OH	C
f	OH_2OH	CH_2OH	H	H	G
g	COOH	Me	H	H	E
h	CH_2Cl	Me	H	H	F
i	CH_2Cl	Me	Me	H	H
j	CH_2OMe	Me	Me	H	I

Illudacetalic acid from *Clitocybe illudens* is a mixed acetal of the methyl ether of illudalic acid [II-61] (200); natural occurrence of such an acetal is most unusual.

Pyrocurzerenone [100] has been isolated from the rhizome of *Curcuma zedoaria* (Zingiberaceae) (201).

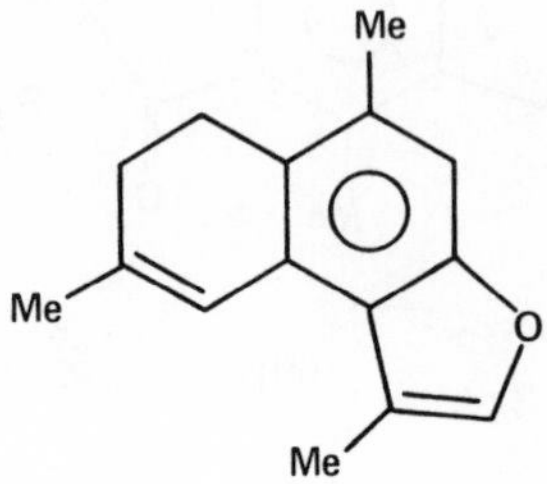

[100]

Lagopodin A [101] and B [102] from culture filtrates of the basidiomycete *Coprinus lagopus* are closely related to helico-basidin [II-22] (202).

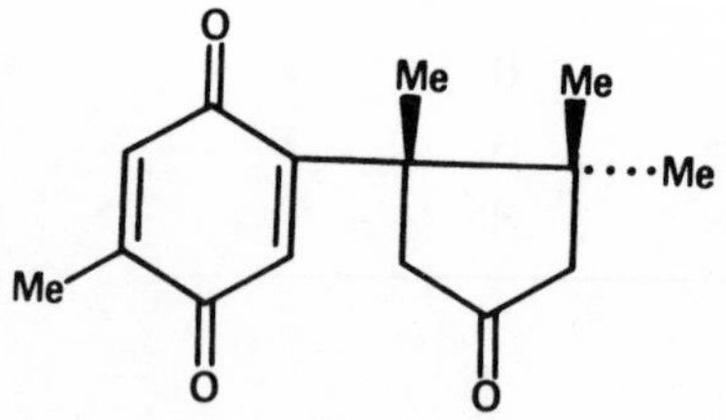

[101] 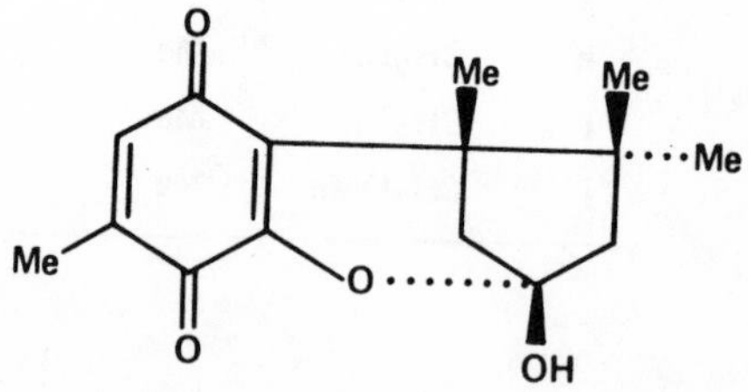   [102]

Compound [103] from *Gibberella fujikuroi* is the first aromatic representative of the gibberellin group (cf. [58A]) of *diterpenoids* (203).

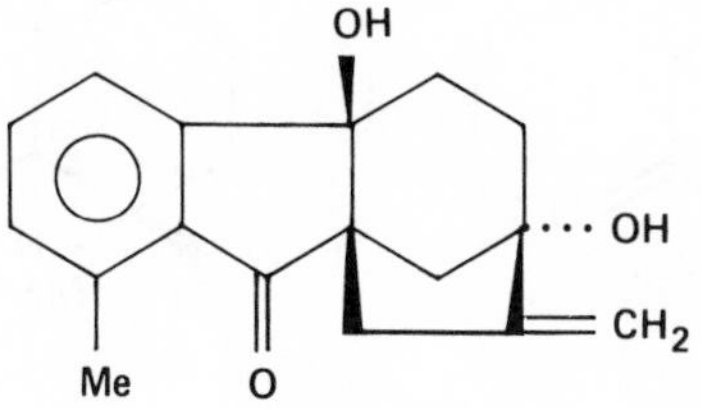

[103]

Several new aromatic diterpenoids of the abietane [48] class have been found. Among them, lycoxanthol [104] (204) from the club moss *Lycopodium lucidulum* (Lycopodiaceae) is particularly interesting both chemically and taxonomically. Its chromophoric system is that of coleone C [III-33], while the dihydrofuran ring connects it with isocryptotanshinone [III-51]. Occurrence of an aromatic diterpenoid in a plant belonging to the Pteridophyta (ferns and their allies) seems unprecedented.

The diterpenoid quinone nemorone [105] has been obtained (205) from *Salvia nemorosa*, where it occurs together with two closely related compounds, horminone [III-37] and the acetate of royleanone [III-35].

The bark of *Podocarpus ferrugineus* contains (206) the methyl ether of sugiol [III-25] and the B-seco-norditerpene lactol [106] together with the previously known aromatic diterpenoids [III-5], [III-16], and [III-31].

Coleone D from *Coleus aquaticus* is the ketonic tautomer of the enol coleone C [III-33] and seems to be its precursor (207a), while coleone E from *C. barbatus* (207b) has the unusual structure [107], undoubtedly formed from some precursor with normal abietane [48] skeleton by rearrangement of both ring A and the isopropyl side-chain.

The rare class of *triterpenoids* has been enriched by several new representatives, all of them closely related to the two previously known ones, and occurring in the same plant families, Celastraceae and Hippocrateaceae. The C_{28} ketone tingenone [108a] and its 20-hydroxy derivative [108b] occur in several plants of these families (208a,b). *Maytenus dispermus* (Celastraceae) contains pristimerin [IV-2] and the new quinone methide

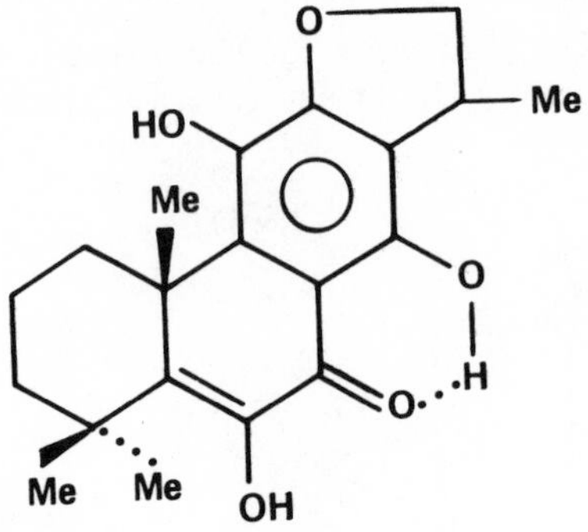

[104]

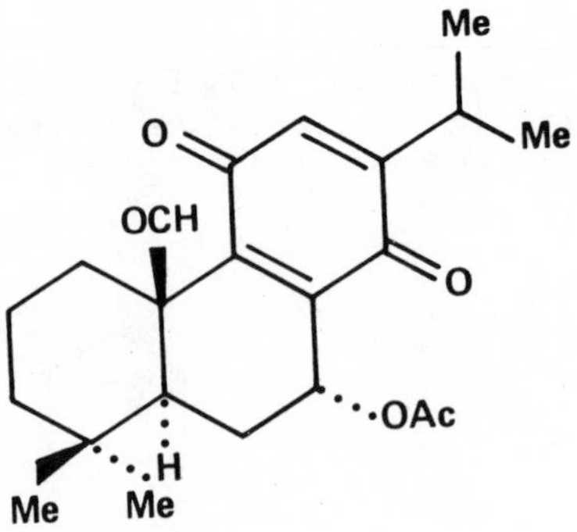

[105]

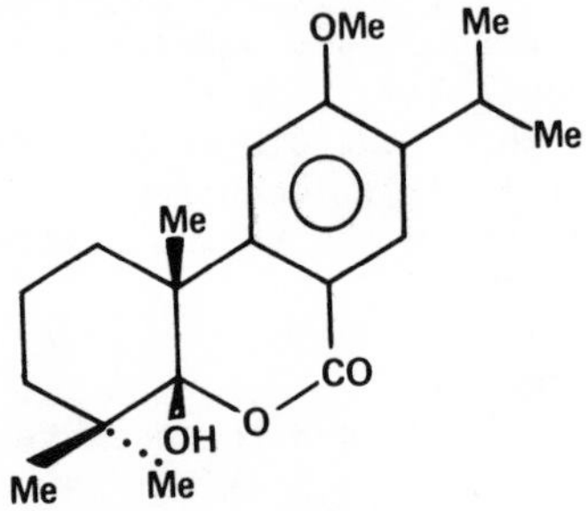

[106]

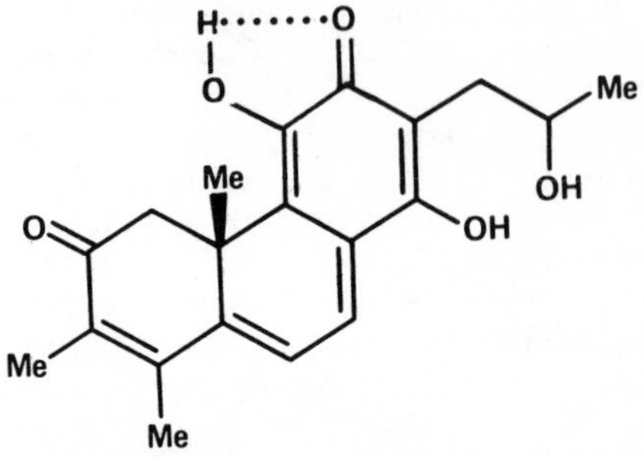

[107]

dispermoquinone, which is the 7-keto derivative of 7,8a-dihydro-
pristimerin. Another quinone methide, maytenoquinone, is apparently the optical antipode of dispermoquinone (208c).

The continued absence of aromatic representatives of the
true C_{30} triterpenoids remains puzzling, as does the restricted
occurrence of nor-triterpenoid quinone methides in two small
families of higher plants only.

Structures of Mixed Origin: One more quinone [109] of the
alkannin [V-2] group has been isolated (209) from *Arnebia nobilis*,
where it occurs together with [V-2] itself, its acetate and
β,β-dimethylacrylate, and compounds [V-5b] and [V-6].

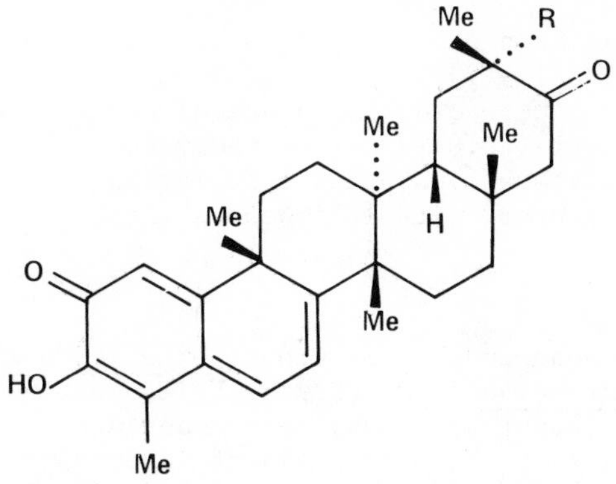

[108] a: R = H
       b: R = OH

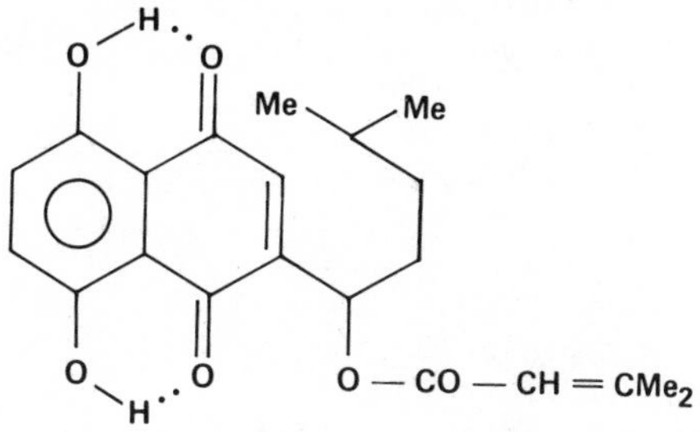

[109]

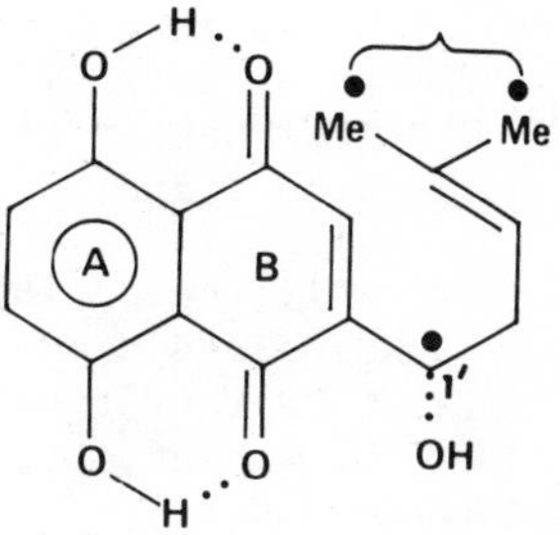

[V-2]

Very recent research by Schmid and Zenk (210) on the *biosynthesis* of alkannin [V-2] in *Plagiobothrys arizonicus* (Boraginaceae) has essentially confirmed the earlier hypothetical scheme. While ^{14}COOH-shikimic acid or U-^{14}C-tyrosine were not incorporated, ring-labeled phenylalanine, cinnamic acid, and particularly 1,2,6-^{14}C-*p*-hydroxybenzoic acid were, activity being confined to ring A of [V-2]. Label from 2-^{14}C-mevalonate was incorporated equally into C-1' (50.08%) and the methyls of the side-chain (50.24%). These results are consistent with a biosynthesis of [V-2] from *p*-hydroxybenzoic acid and two mevalonate units.

The new carbazole alkaloid heptazoline from *Clausena*

heptaphylla has been shown (211) to be the 8-hydroxy-derivative of heptaphylline [V-17].
   Alstonia constricta (Apocynaceae) contains alstonilidine (212), which is 1-(2',6'-dicarbomethoxybenzoyl)-β-carboline; this structure suggests a biosynthetic formation through cleavage of ring D of a precursor with yohimbane skeleton.

REFERENCES

1. T.A. Geissman and D.H.G. Crout, Organic Chemistry of Secondary Plant Metabolism, Chapter VIII, Terpenoid Compounds, Freeman, Cooper & Company, San Francisco, 1969, p. 233ff.
2. H.J. Nicholas, The Biogenesis of Terpenes in Plants, in Biogenesis of Natural Compounds, 2nd ed., P. Bernfeld, Ed., Pergamon, 1967, p. 829ff.
3. J.W. Cornforth, (a) Chem. Brit. 4, 102 (1968);
    (b) Quart. Rev. (London) 23, 125 (1969).
4. J.H. Richards and J.B. Hendrickson, The Biosynthesis of Steroids, Terpenes, and Acetogenins, W.A. Benjamin, Inc., New York and Amsterdam, 1964.
5. R.B. Clayton, Quart. Rev. (London) 19, 201 (1965).
6. (a) L. Ruzicka, Proc. Chem. Soc. 341, (1959).
    (b) Experientia 9, 357 (1953).
7. C.O. Chichester and T.O.M. Nakayama, in Biogenesis of Natural Compounds, 2nd ed., P. Bernfeld, Ed., Pergamon, 1967, p. 641ff.
7A. H.C. Rilling, J. Biol. Chem. 241, 3233 (1966).
8. S. Nozoe and M. Morisaki, Chem. Commun. 1319 (1969).
9. (a) C. Gerhardt and A. Cahours, Ann. 38, 67, 101, 345 (1841);
    (b) O. Widman, Ber. 24, 439 (1891).
10. R. Hegnauer, Chemotaxonomie der Pflanzen, Birkhäuser Verlag, Basel and Stuttgart, 1962.
10A. W. Karrer, Konstitution und Vorkommen der Organischen Pflanzenstoffe, Birkhäuser Verlag, Basel, 1958.
11. P. Granger, J. Passet, and R. Verdier, Compt. Rend. 258, 5539 (1964).
12. H. Erdtman and T. Norin, Fortschritte Chem. Org. Natur. 24, 206 (1966).
12A. E.S. Pallares, Arch. Biochem. 9, 105 (1946).
13. (a) F. Bohlmann, J. Schulz, and U. Bühmann, Tetrahedron Lett. 4703 (1969);
    (b) F. Bohlmann, U. Niedballa, and J. Schulz, Chem. Ber., 102, 864 (1969).
14. E. Zavarin and A.B. Anderson, (a) J. Org. Chem. 20, 443 (1955);
    (b) ibid. 20, 788 (1955);

E. Zavarin, (c) ibid. 23, 1198 (1958);
(d) ibid. 23, 1264 (1958).

15. (a) H. Dieterle and P. Kaiser, Arch. Pharm. 270, 413 (1932);
271, 337 (1933);
(b) P.E. Gumster, thesis, Groningen, 1943.

16. E. Bianchi, Rass. Chim 12, 42 (1960); Chem. Abstr. 55, 14830h
(1961).

17. W. Parker, J.S. Roberts, and R. Ramage, Quart. Rev. (London)
21, 331 (1967).

18. V.K. Honwad and A.S. Rao, Tetrahedron 21, 2593 (1965).

19. Y.R. Naves, Perfum. Essent. Oil Rec. 43, 38, 53 (1952);
Chem. Abstr. 46, 6333f (1952).

20. S.C. Bisarya and Sukh Dev, unpublished work quoted by
J. Alexander and G.S. Krishna Rao, Chem. Ind. (London) 139
(1969).

21. (a) T. Sakai, K. Nishimura, and Y. Hirose, Tetrahedron Lett.
1171 (1963);
(b) G. Büchi and H. Wüest, J. Org. Chem., 34, 1122 (1969).

22. (a) D.A. Archer and R.H. Thomson, J. Chem. Soc. (C) 1710
1967, and refs. there;
(b) T. Garcia, E. Dominguez, and J. Romo, Bol. Inst. Quim.
Univ. Nac. Auton. Mexico 17, 16 (1965).

23. F. Bohlmann and M. Grenz, Tetrahedron Lett. 1005 (1969).

24. M. Ando, S. Ibe, S. Kagabu, T. Nakagawa, T. Asao, and
K. Takase, Chem. Commun. 1538 (1970).

25. C. Enzell and H. Erdtman, Tetrahedron 4, 361 (1958).

26. T. Nozoe and H. Takeshita, Tetrahedron Lett. 14 (1960).

27. W.G. Dauben and P. Oberhänsli, J. Org. Chem. 31, 315 (1966).

28. (a) T. Yanagawa, Y. Hirose, and T. Nakatsuka, Nippon Mokuzai
Gakkaishi 13, 160 (1967); Chem. Abstr. 67, 116966g (1967);
(b) S. Itô, K. Endo, H. Honma, and K. Ota, Tetrahedron Lett.
3777 (1965).

29. G.L. Chetty and Sukh Dev, Tetrahedron Lett. 73 (1964).

30. B. Tomita, Y. Hirose, and T. Nakatsuka, Tetrahedron Lett.
843 (1968).

31. W. Parker, R. Ramage, and R.A. Raphael, J. Chem. Soc. 1558
(1962).

32. (a) H. Nishikawa, Agr. Biol. Chem. 26, 696 (1962);
(b) S. Natori, H. Ogawa, K. Yamaguchi, and H. Nishikawa,
Chem. Pharm. Bull. 11, 1343 (1963);
(c) S. Natori, H. Nishikawa, and H. Ogawa, ibid. 12, 236
(1964);
(d) S. Natori, Y. Inouye, and H. Nishikawa, ibid. 15, 380
(1967).

33. S. Yamamura and Y. Hirata, Tetrahedron 19, 1485 (1963).

34. T. Irie, M. Suzuki, and Y. Hayakawa. Bull. Chem. Soc. Japan 42, 843 (1969).
35. (a) T. Irie, Y. Yasunari, T. Suzuki, N. Imai, E. Kurosawa, and T. Masamune, Tetrahedron Lett. 3619 (1965);
    (b) T. Irie, T. Suzuki, Y. Yasunari, E. Kurosawa, and T. Masamune, Tetrahedron 25, 459 (1969).
36. T. Irie, Nippon Kagaku Zasshi 90, 1179 (1969).
37. T. Irie, M. Suzuki, E. Kurosawa, and T. Masamune,
    (a) Tetrahedron Lett. 1837 (1966);
    (b) Tetrahedron 26, 3271 (1970).
38. T. Irie, A. Fukuzawa, M. Izawa, and E. Kurosawa, Tetrahedron Lett. 1343 (1969).
39. (a) K. Yamada, H. Yazawa, M. Toda, and Y. Hirata, Chem. Commun. 1432 (1968);
    (b) K. Yamada, H. Yazawa, D. Uemura, M. Toda, and Y. Hirata, Tetrahedron 25, 3509 (1969).
40. A.F. Cameron, G. Ferguson, and J.M. Robertson, (a) Chem. Commun. 271 (1967);
    (b) J. Chem. Soc. (B) 692 (1969).
41. T. Irie, T. Suzuki, S. Itô, and E. Kurosawa, Tetrahedron Lett. 3187 (1967).
42. (a) T. Nakatsuka and Y. Hirose, Bull. Agr. Chem. Soc. Japan 20, 215 (1956);
    (b) Y. Hirose and T. Nakatsuka, ibid. 23, 140 (1959);
    (c) M. Nakazaki, Bull. Chem. Soc. Japan 35, 1387 (1962).
42A. N. Katsui, A. Matsunaga, K. Imaizumi, T. Masamune, and K. Tomiyama, Tetrahedron Lett. 83 (1971); N. Katsui, A. Murai, M. Takasugi, K. Imaizumi, T. Masamune, and K. Tomiyama, Chem. Commun. 43 (1968).
43. (a) F. Šorm, K. Vereš, and V. Herout, Collect. Czech. Chem. Commun. 18, 106 (1953);
    (b) F. Šorm, M. Holub, V. Sýkora, J. Mleziva, M. Streibl, J. Plíva, B. Schneider, and V. Herout, ibid. 18, 512 (1953);
    (c) J. Plíva, M. Horák, V. Herout, and F. Šorm, Die Terpene, Sammlung der Spektra and Physikalischen Konstanten, Teil I, Sesquiterpene, Akademie-Verlag, Berlin 1960.
44. Y. Naya and M. Kotake, Bull. Chem. Soc. Japan 42, 2088 (1969).
45. (a) M. Fracheboud, J.W. Rowe, R.W. Scott, S.M. Fanega, A.J. Buhl, and J.K. Toda, Forest Prod. J. 18, 37 (1968);
    (b) J.W. Rowe and J.K. Toda, Chem. Ind. (London) 922 (1969);
    (c) B.O. Lindgren and C.M. Svahn, Phytochemistry 7, 1407 (1968).
46. (a) G.B. Marini Bettòlo, C.G. Casinovi, and C. Galeffi, Tetrahedron Lett. 4857 (1965);
    (b) N. Tanaka, M. Yasue, and M. Imamura, ibid. 2767 (1966);

(c) C. Galeffi, E. Miranda delle Monache, C.G. Casinovi, and G.B. Marini Bettòlo, ibid. 3583 (1969);

(d) K. Shimada, M. Yasue, and M. Imamura, Mokuzai Gakkaishi 13, 126 (1967);

(e) J.C. Overeem and D.M. Elgersma, Phytochemistry 9, 1949 (1970);

(f) V. Krishnamoorthy and R.H. Thomson, ibid. 10, 1695 (1971).

47. R. Adams, T.A. Geissman, and J.D. Edwards, Chem. Rev. 60, 555 (1960), and refs. there.

48. Sir R. Robinson, The Structural Relations of Natural Products, Clarendon, Oxford, 1955.

49. (a) L.P. Stoltz, T.R. Kemp, W.O. Smith Jr., W.T. Smith Jr., and C.E. Chaplin, Phytochemistry 9, 1157 (1970);

(b) T.R. Kemp, L.P. Stoltz, and L.V. Packett, ibid. 10, 478 (1971);

(c) Y. Naya and M. Kotake, Nippon Kagaku Zasshi 91, 275 (1970); Chem. Abstr. 73, 43982s (1970).

50. W.H. McFadden, R. Teranishi, J. Corse, D.R. Black, and T.R. Mon, J. Chromatogr. 18, 10 (1965).

51. (a) T.C. Joseph and Sukh Dev, Tetrahedron 24, 3809 (1968);

(b) R.C. Pandey and Sukh Dev, ibid. 24, 3829 (1968).

52. D. Meuche and S. Huneck, (a) Chem. Ber. 102, 2493, 2502 (1969); (b) ibid. 99, 2669 (1966).

53. M.S.R. Nair, H. Takeshita, T.C. McMorris, and M. Anchel, J. Org. Chem. 34, 240 (1969).

54. T.C. McMorris, M.S.R. Nair, P. Singh, and M. Anchel. Phytochemistry 10, 1611, 3341 (1971).

54A. P. Singh, M.S.R. Nair, T.C. McMorris, and M. Anchel, Phytochemistry 10, 2229 (1971).

55. K. Nakanishi, M. Ohashi, M. Toda, and Y. Yamada, Tetrahedron 21, 1231 (1965).

56. T.C. McMorris and M. Anchel, J. Amer. Chem. Soc. 85, 831 (1963); 87, 1594 (1965); M. Anchel, T.C. McMorris, and P. Singh, Phytochemistry 9, 2339 (1970).

57. (a) J.J. Dugan, P. de Mayo, M. Nisbet, and M. Anchel, J. Amer. Chem. Soc. 87, 2768 (1965);

(b) J.J. Dugan, P. de Mayo, M. Nisbet, J.R. Robinson, and M. Anchel, ibid. 88, 2838 (1966);

57A. S. Takahashi, H. Naganawa, H. Iinuma, T. Takita, K. Maeda, and H. Umezawa, Tetrahedron Lett. 1955 (1971).

58. (a) A.J. Weinheimer and P.H. Washecheck, Tetrahedron Lett. 3315 (1969);

(b) H. Wienhaus, Angew. Chem. 47, 415 (1934); W. Treibs, Ber. 70, 85 (1937).

59. (a) P.M. Brown and R.H. Thomson, J. Chem. Soc. (C) 1184 (1969);

H. Kakisawa, Y. Inouye, and J. Romo, Tetrahedron Lett.
1929 (1969); J. Romo, Pure Appl. Chem. 21, 123 (1970);
(b) L. Rodríguez-Hahn, A. Guzmán, and J. Romo, Tetrahedron
24, 477 (1968); Z. Samek, J. Harmatha, L. Novotný, and
F. Šorm, Collect. Czech. Chem. Commun. 34, 2792 (1969).

60. R. McCrindle and K.H. Overton, Adv. Org. Chem. 5, 47–113(1965).

61. J.R. Hanson and B.A. Achilladelis, Perfum. Essent. Oil Rec.
59, 802 (1968).

62. (a) I. Shechter and C.A. West, J. Biol. Chem. 244, 3200 (1969);
(b) J.R. Hanson and A.F. White, J. Chem. Soc. (C) 981 (1969).

63. (a) F. Šorm, M. Suchý, F. Vonášek, J. Plíva, and V. Herout,
Collect. Czech. Chem. Commun. 16, 268 (1951);
(b) M. Tsutsui and E.A. Tsutsui, Chem. Rev. 59, 1031 (1959),
on p. 1043.

64. J. Comin, O. Gonçalves de Lima, H.N. Grant, L.M. Jackman,
W. Keller-Schierlein, and V. Prelog, Helv. Chim. Acta 46,
409 (1963).

65. (a) E.J. McGarry, K.H. Pegel, L. Phillips, and E.S. Waight,
Chem. Commun. 1074 (1969); J. Chem. Soc. (C) 904 (1971).
(b) H.A. Candy, J.M. Pakshong, and K.H. Pegel, J. Chem. Soc.
(C) 2536 (1970).

66. (a) P. Sengupta, S.N. Choudhuri, and H.N. Khastgir,
Tetrahedron 10, 45 (1960);
(b) P.K. Ramachandran and P.C. Dutta, J. Chem. Soc. 4766
(1960).

67. E. Wenkert, V.I. Stenberg, and P. Beak, J. Amer. Chem. Soc.
83, 2320 (1961).

68. W. Sandermann, Chem. Ber. 71, 2005 (1938).

69. E. Wenkert and J.W. Chamberlin, J. Amer. Chem. Soc. 81, 688
(1959).

70. M. Kitadani, A. Yoshikoshi, Y. Kitahara, J. de Paiva Campello,
J.D. McChesney, D.J. Watts, and E. Wenkert, Chem. Pharm. Bull.
18, 402 (1970).

71. Beilstein, Handbuch der Organischen Chemie V, 683.

72. E. Ghigi and G. Fabbri, Atti Accad. Sci. Ist. Bologna, Classe
Sci. Fis., Rend. Ser. XII, 2 (1965); E. Ghigi, A. Drusiani,
L. Plessi, and V. Cavrini, Gazz. Chim. Ital. 98, 795 (1968).

72A. F.H.M. Nestler and D.F. Zinkel, Anal. Chem. 39, 1118 (1967).

72B. (a) D.F. Zinkel, J.W. Rowe, L.C. Zank, D.W. Gaddie, and
E.F. Ruckel, J. Amer. Oil Chem. Soc. 46, 633 (1969);
(b) H. Takeda, W.H. Schuller, and R.V. Lawrence, J. Org.
Chem. 33, 3718 (1968); 34, 1459 (1969).

73. R.M. Carman and H.C. Deeth, Aust. J. Chem. 20, 2789 (1967);
L.J. Gough, Tetrahedron Lett. 295 (1968).

74. (a) L.H. Briggs, R.C. Cambie, R.N. Seelye, and A.D. Warth,

Tetrahedron, 7, 270 (1959);
(b) R.C. Cambie and K.P. Mathai, Chem. Commun. 154 (1971);
Aust. J. Chem. 24, 1251 (1971).
75. W.P. Campbell and D. Todd, J. Amer. Chem. Soc. 64, 928 (1942).
76. H. Linde, Helv. Chim. Acta 47, 1234 (1964); E. Wenkert, A. Fuchs, and J.D. McChesney, J. Org. Chem. 30, 2931 (1965); C.R. Narayanan and H. Linde, Tetrahedron Lett. 3647 (1965).
77. J. B.-son Bredenberg, (a) Acta Chem. Scand. 11, 927 (1957); (b) ibid. 14, 385 (1960).
78. J.L. Breton Funes, A. Gonzalez Gonzalez, and G. De Leon, An. Quim. 66, 293 (1970); Chem. Abstr. 73, 88048n (1970).
79. L. Mangoni and M. Belardini, Tetrahedron Lett. 2643 (1964).
80. T. Kondo, M. Suda, and M. Teshima, J. Pharm. Soc. Japan 82, 1252 (1962).
81. J.B.-son Bredenberg, Acta Chem. Scand. 11, 932 (1957); cf. also, inter alia, ref. 74A.
82. Y.-T.Lin and K.-T. Liu, J. Chinese Chem. Soc. (Taiwan) 12, 51 (1965); Chem. Abstr. 63, 16772 d(1965); cf. ref. 12, p. 240.
82A. T. Matsumoto, I. Tanaka, T. Ohno, and K. Fukui, Chem. Lett. 321 (1973).
83. S.M. Kupchan, A. Karim, and C. Marcks, J. Amer. Chem. Soc., 90, 5923 (1968); J. Org. Chem. 34, 3912 (1969).
84. R.M. Carman and R.A. Marty, Aust. J. Chem. 23, 1457 (1970).
85. R.C. Cambie and L.N. Mander, Tetrahedron 18, 465 (1962).
86. E. Wenkert and B.G. Jackson, J. Amer. Chem. Soc. 80, 211 (1958).
86A. T. Yanagawa and Y. Hirose, Mokuzai Gakkaishi 15, 344 (1969); Chem. Abstr. 73, 25689b (1970).
87. O.E. Edwards, G. Feniak, and M. Los, Can. J. Chem. 40, 1540 (1962).
88. M.M. Janot and P. Potier, Ann. Pharm. Fr. 22, 387 (1964).
89. C.H. Brieskorn, A. Fuchs, J. B.-son Bredenberg, J.D. McChesney, and E. Wenkert, J. Org. Chem. 29, 2293 (1964).
90. J.H. Gough and M.D. Sutherland, Aust. J. Chem. 19, 329 (1966).
91. O.E. Edwards and M.N. Rodger, Can. J. Chem. 37, 1187 (1959); cf. also A. Khaleque, S.Papadopoulos, I. Wright, and Z. Valenta, Chem. Ind. (London) 513 (1959); S.K. Talapatra and A. Chatterjee, J. Indian Chem. Soc. 30, 437 (1959).
92. D. Karanatsios, J.S. Scarpa, and C.H. Eugster, Helv. Chim. Acta 49, 1151 (1966).
93. M. Ribi, A.C. Sin-Ren, H.P. Küng, and C.H. Eugster, Helv. Chim. Acta 52, 1685 (1969).
94. D. Karanatsios and C.H. Eugster, Helv. Chim. Acta 48, 471 (1965).

95. P. Rüedi and C.H. Eugster, Helv. Chim. Acta 54, 1606 (1971).
96. A.C. Baillie and R.H. Thomson, J. Chem. Soc. (C) 48 (1968), and references quoted therein.
97. R.H. Thomson, Naturally Occurring Quinones, 2nd Ed., Academic New York, 1971.
98. (a) H. Kakisawa, T. Hayashi, I. Okazaki, and M. Ohashi, Tetrahedron Lett. 3231 (1968);
    (b) H. Kakisawa, T. Hayashi, and T. Yamazaki, ibid. 301 (1969).
99. T. Hayashi, H. Kakisawa, H.-Y. Hsü, and Y.P. Chen, Chem. Commun. 299 (1970).
100. (a) W.F. Short and H. Wang, J. Chem. Soc. 2979 (1951);
    (b) J.A. Barltrop and N.A.J. Rogers, Chem. Ind. (London) 397 (1957);
    (c) Ibid., J. Chem. Soc. 2566 (1958).
101. (a) L. Mangoni and R. Caputo, (a) Tetrahedron Lett. 673 (1967);
    (b) Gazz. Chim. Ital. 97, 908 (1967).
102. W. Klyne, J. Chem. Soc. 3072 (1953).
103. Y.-L. Chow and H. Erdtman, Acta Chem. Scand. 14, 1852 (1960); ibid. 16, 1305 (1962).
104. (a) E. Wenkert and P. Beak, Tetrahedron Lett. 358 (1961);
    (b) R.C. Cambie, W.R.J. Simpson, and J.D. Colebrook, Tetrahedron 19, 209 (1963).
105. D.A.H. Taylor, Chem. Ind. (London) 1961, 1712.
106. R. Hodges, J. Chem. Soc. 1961, 4247.
107. S.M. Bocks, R.C. Cambie, and T. Takahashi, Tetrahedron 19, 1109 (1963).
108. S.M. Bocks and R.C. Cambie, Proc. Chem. Soc. 1963, 143.
109. (a) J.W. Rowe, private communication;
    (b) J.W. Rowe, B.A. Nagasampagi, A.W. Burgstahler, and J.W. Fitzsimmons, Phytochemistry, 10, 1647 (1971);
    (c) B.A. Nagasampagi, J.K. Toda, A.H. Conner, and J.W. Rowe, 161st. Amer. Chem. Soc. Meeting, Los Angeles, April 1971, Abstracts Cell-84.
109A. C. Tabacik-Wlotzka, Bull. Soc. Chim. Fr. 618 (1964).
109B. H. Erdtman, B. Kimland, T. Norin, and P.J.L. Daniels, Acta Chem. Scand. 22, 938 (1968).
109C. R. Caputo, L. Mangoni, L. Previtera, and R. Iaccarino, Tetrahedron Lett. 3731 (1971).
110. A.B. Turner, Fortschritte Chem. Org. Natur. 24, 288 (1966).
111. P.K. Grant and A.W. Johnson, J. Chem. Soc. 4079 (1957); cf. also P.K. Grant, A.W. Johnson, P.F. Juby, and T.J. King, ibid. 549 (1960).
112. R.G. Cooke and R.H. Thomson, Rev. Pure. Appl. Chem. 8, 85 (1958).

113. R. Harada, H. Kakisawa, S. Kobayashi, M. Musya,
     K. Nakanishi, and Y. Takahashi, Tetrahedron Lett. 603
     (1962).
114. A.W. Johnson, P.F. Juby, T.J. King, and S.W. Tam, J. Chem.
     Soc. 2884 (1963).
115. M.H. Zenk and E. Leistner, Lloydia 31, 275 (1968).
115A. (a) I.M. Campbell, Tetrahedron Lett. 4777 (1969);
     (b) D.J. Robins, I.M. Campbell, and R. Bentley, Biochem.
     Biophys. Res. Commun. 39, 1081 (1970).
116. E. Leistner and M.H. Zenk, Z. Naturforsch. 23b, 259 (1968).
117. P. Dansette and R. Azerad, Biochem. Biophys. Res. Commun.
     40, 1090 (1970).
118. (a) W. Sandermann and H.-H. Dietrichs, Holzals Roh- und
     Werkstoff 15, 281 (1957); Chem. Abstr. 51, 18953 b (1957);
     (b) W. Sandermann and M.H. Simatupang, ibid. 24, 190 (1966);
     Chem. Abstr. 65, 9342b (1966).
119. H. Inouye, J. Pharm. Soc. Japan 76, 976 (1956).
120. A.R. Burnett and R.H. Thomson, J. Chem. Soc. (C) 857 (1968).
121. K.H. Bolkart, M. Knobloch, and M.H. Zenk, Naturwissenschaften
     55, 445 (1968).
122. K.H. Bolkart and M.H. Zenk, Naturwissenschaften 55, 444
     (1968).
123. G.R. Whistance and D.R. Threlfall, Biochem. Biophys. Res.
     Commun. 28, 295 (1967).
124. K.H. Bolkart and M.H. Zenk, Zeitschr. Pflanzenphysiol. 61,
     356 (1969).
125. R.G. Cooke, H. Dowd, and L.J. Webb, Nature 169, 974 (1952);
     R.G. Cooke and H. Dowd, Aust. J. Sci. Res. 5A, 760 (1952).
126. Y.N. Shukla, J.S. Tandon, D.S. Bhakuni, and M.M. Dhar,
     Experientia 25, 357 (1969); Y.N. Shukla, Thesis, Lucknow
     1970.
127. A.J. Birch and F.W. Donovan, Aust. J. Chem. 6, 360 (1953);
     8, 529 (1955).
128. R.F. Curtis, C.H. Hassall, and D.R. Parry, Chem. Commun.
     410 (1971).
129. A.R. Burnett and R.H. Thomson, (a) Chem. Commun. 1125
     (1967);
     (b) J. Chem. Soc. (C) 2437 (1968).
129A. A. Yagi, K. Makino, and I. Nishioka, Chem. Pharm. Bull. 22,
     1159 (1974).
130. E. Leistner and M.H. Zenk, Chem. Commun. 210 (1969).
131. E. Leistner and M.H. Zenk, Tetrahedron Lett. 1395 (1968).
132. A.R. Burnett and R.H. Thomson, (a) J. Chem. Soc. (C) 850
     (1968);
     (b) ibid. 854.
133. E. Leistner and M.H. Zenk, (a) Z. Naturforsch. 22b, 865
     (1967);

(b) Tetrahedron Lett. 475 (1967).

134. E. Leistner and M.H. Zenk, Tetrahedron Lett. 861 (1968).

135. E. Leistner and M.H. Zenk, Tetrahedron Lett. 1677 (1971).

136. (a) R. Mechoulam, Science 168, 1159 (1970);
      (b) C.R.B. Joyce and S.H. Curry, The Botany and Chemistry
      of Cannabis, Churchill, London, 1970.

137. R. Mechoulam and B. Yagen, Tetrahedron Lett. 5349 (1969).

138. L. Crombie and R. Ponsford, (a) Chem. Commun. 894 (1968);
      (b) J. Chem. Soc. (C), 796 (1971).

139. M.J. Begley, D.G. Clarke, L. Crombie, and D.A. Whiting,
      Chem. Commun. 1547 (1970).

139A. R.S. Kapil, in The Alkaloids, R.H.F. Manske, Ed., Vol. 13,
      (1971), p. 273.

139B. F. Anwer, R.S. Kapil, and S.P. Popli, Indian J. Chem. 10,
      959 (1972); D.P. Chakraborty, A. Islam, and P. Bhattacharyya,
      Abstr. 8th Int. Symp. Chem. of Nat. Prod., New Delhi,
      February 1972, p. 2.

140. D.P. Chakraborty, J. Indian Chem. Soc. 46, 177 (1969).

141. S.P. Kureel, R.S. Kapil, and S.P. Popli, Experientia, 25,
      790 (1969).

142. N.S. Narasimhan, M.V. Paradkar, and S.L. Kelkar, Indian J.
      Chem. 8, 473 (1970).

143. V. Benešová, Z. Samek, V. Herout, and F. Šorm, Collect.
      Czech. Chem. Commun. 34, 1807 (1969).

144. T.R. Govindachari, R.N. Viswanathan, B.R. Pai,
      V.N. Ramachandran, and P.S. Subramaniam, Tetrahedron 26,
      2905 (1970).

145. T.R. Govindachari, B.R. Pai, and P.S. Subramaniam, Tetra-
      hedron 22, 3245 (1966).

146. D.P. Chakraborty, Tetrahedron Lett. 661 (1966); Phyto-
      chemistry 8, 769 (1969).

147. D.P. Chakraborty and B.P. Das, Sci. Culture 32, 181 (1966);
      Chem. Abstr. 65, 13640c (1966).

148. D.P. Chakraborty, B.K. Barman, and P.K. Bose, Tetrahedron
      21, 681 (1965).

149. B.K. Chowdhury and D.P. Chakraborty, Chem. Ind. (London)
      549 (1969); Phytochemistry 10, 1967 (1971).

150. B.K. Chowdhury and D.P. Chakraborty, Phytochemistry 10,
      481 (1971).

151. B.S. Joshi, V.N. Kamat, A.K. Saksena, and T.R. Govindachari,
      Tetrahedron Lett. 4019 (1967).

152. (a) N.L. Dutta and C. Quasim, Indian J. Chem. 7, 307 (1969);
      Chem. Abstr. 71, 3516z (1969);
      (b) N.S. Narasimhan, M.V. Paradkar, and A.M. Gokhale,
      Tetrahedron Lett. 1665 (1970);
      (c) B.S. Joshi, V.N. Kamat, and D.H. Gawad, Tetrahedron 26,
      1475 (1970);

(d) S.P. Kureel, R.S. Kapil, and S.P. Popli, Chem. Ind. (London) 1262 (1970).

153. (a) D.P. Chakraborty, K.C. Das, and B.K. Chowdhury, J. Org. Chem. 36, 725 (1971);
(b) D.P. Chakraborty and K.C. Das, Chem. Commun. 967 (1968).

154. (a) N.S. Narasimhan, M.V. Paradkar, and V.P. Chitguppi, Tetrahedron Lett. 5501 (1968);
(b) D.P. Chakraborty, D. Chatterji, and S.N. Ganguly, Chem. Ind. (London) 1662 (1969);
(c) S.P. Kureel, R.S. Kapil, and S.P. Popli, Chem. Commun. 1120 (1969).

155. S.P. Kureel, R.S. Kapil, and S.P. Popli, Experientia 26, 1055 (1970).

156. S.P. Kureel, R.S. Kapil, and S.P. Popli, Tetrahedron Lett. 3857 (1969).

157. D.P. Chakraborty, A. Islam, S.P. Basak, and R. Das, Chem. Ind. (London) 593 (1970); N.L. Dutta, C. Quasim, and M.S. Wadia, Indian J. Chem. 7, 1168 (1969).

158. S.P. Kureel, R.S. Kapil, and S.P. Popli, Chem. Ind. (London) 958 (1970).

158A. R.C. Elderfield and S.L. Wythe, J. Org. Chem. 19, 683 (1954).

158B. W.I. Taylor and Raymond-Hamet, C.R. Acad. Sci., Paris, Ser. D 262, 1141 (1966).

158C. L. Merlini, R. Mondelli, G. Nasini, and M. Hesse, Tetrahedron 23, 3129 (1967).

159. A.F. Thomas, Helv. Chim. Acta 48, 1057 (1965).

160. C. Pilo and J. Runeberg, Acta Chem. Scand. 14, 353 (1960).

161. (a) A.J. Birch and P. Elliott, Aust. J. Chem. 6, 369 (1953);
(b) ibid. 9, 95 (1956);
(c) Y.-T. Lin, K.-T. Wang and L.-H. Chang, J. Chinese Chem. Soc. (Taiwan) 10, 139 (1963).

162. E. Zavarin and A.B. Anderson, J. Org. Chem. 22, 1122 (1957).

163. T. Nozoe, A. Yasue, and K. Yamane, Proc. Japan Acad. 27, 15 (1951).

164. Y.-T. Lin and K.-T. Wang, J. Chinese Chem. Soc. (Taiwan) 7, 174 (1960).

165. B. Tomita, Y. Hirose, and T. Nakatsuka, Mokuzai Gakkaishi 15, 46 (1969); Chem. Abstr. 71, 50249z (1969).

166. V.K. Honwad and A.S. Rao, Tetrahedron 20, 2921 (1964).

167. H. Rimpler, R. Hänsel, and L. Kochendoerfer, Z. Naturforsch. 25B, 995 (1970).

168. B. Tomita, Y. Hirose, and T. Nakatsuka, Mokuzai Gakkaishi 15, 47 (1969); Chem. Abstr. 71, 50250t (1969).

169. S. Nozoe, M. Morisaki, and H. Matsumoto, Chem. Commun. 926 (1970).

170. H. Erdtman and L. Westfelt, Acta Chem. Scand. 17, 1826 (1963).

171. E.N. Shmidt and W.A. Pantegova, Khim. Prirod. Soedin. 187 (1969).
172. L. Westfelt, Acta Chem. Scand. 20, 2829 (1966).
173. G.C. Harris, J. Amer. Chem. Soc. 70, 3671 (1948).
174. R.M. Carman and H.C. Deeth, Aust. J. Chem. 24, 353 (1971).
175. I. Fukushima, J. Sayama, K. Kyogoku, and H. Murayama, Agr. Biol. Chem. (Tokyo) 32, 1103 (1968).
176. T. Hayashi, T. Handa, M. Ohashi, H. Kakisawa, H-Y. Hsü, and Y.P. Chen, Chem. Commun. 541 (1971).
177. Y-L. Chow and H. Erdtman, Acta Chem. Scand. 16, 1296 (1962).
178. C.W. Brandt and B.R. Thomas, J. Chem. Soc. 2442 (1952).
179. M. Yamazaki, T. Usui, and S. Shibata, Chem. Pharm. Bull. 11, 363 (1963).
180. W.D. Loomis, in Terpenoids in Plants, J.B. Pridham Ed., Academic, New York, 1967, p. 59ff.
181. D.V. Banthorpe, J. Mann, and K.W. Turnbull, J. Chem. Soc. (C) 2689 (1970).
181A. R.W. Scora and J.D. Mann, Lloydia 30, 236 (1967).
182. M.J.O. Francis, D.V. Banthorpe, and G.N.J. Le Patourel, Nature 228, 1005 (1970).
183. S. Natori, Y. Inouye, and H. Nishikawa, Chem. Pharm. Bull. 15, 380 (1967).
184. R. Bentley and D. Chen, Phytochemistry 8, 2171 (1969).
185. S. Nozoe, M. Morisaki, and H. Matsumoto, Chem. Commun. 926 (1970); cf. also P.M. Adams and J.R. Hanson, J. Chem. Soc. Perkin I, 586 (1972).
186. P.F. Heinstein, F.H. Smith, and S.B. Tove, J. Biol. Chem. 237, 2643 (1962).
187. P.F. Heinstein, D.L. Herman, S.B. Tove, and F.H. Smith, J. Biol. Chem. 245, 4658 (1970).
188. T.K. Devon and A.I. Scott, Handbook of Naturally Occurring Compounds, Vol. II, Terpenes, Academic, New York and London, 1972.
189. (a) W.W. Epstein and H.C. Rilling, J. Biol. Chem. 245, 4597 (1970);
(b) J. Edmond, G. Popják, S.-M. Wong, and V.P. Williams, ibid. 246, 6254 (1971).
190. D.V. Banthorpe, B.V. Charlwood, and M.J.O. Francis, Chem. Rev. 72, 115 (1972).
191. Th. Anthonsen and B. Kjøsen, Acta Chem. Scand. 25, 390 (1971).
192. G. Shtacher and Y. Kashman, Tetrahedron 27, 1343 (1971).
193. (a) O. Sigel, Ann. 170, 363 (1873);
(b) K.E. Schulte, J. Reisch, and G. Rücker, Arch. Pharm. 296, 273 (1963);
(c) F.W. Semmler, Ber. 41, 510 (1908).

194. H. Suzuki, S. Yasuda, and M. Hanzawa, (a) Mokuzai Gakkaishi
      17, 221 (1971);
      (b) ibid. 18, 617 (1972).
194A. R.D. Stipanovic, P.J. Wakelyn, and A.A. Dell, Phytochemistry
      14, 1041 (1975).
195. N.D. Andersen, D.D. Syrdal, and C. Graham, Tetrahedron Lett.
      905 (1972).
196. K. Yoshihira, M. Fukuoka, M. Kuroyanagi, and S. Natori,
      (a) Chem. Pharm. Bull. 19, 1491 (1971);
      (b) ibid. 20, 426 (1972).
197. H. Hikino, T. Takahashi, S. Arihara, and T. Takemoto, Chem.
      Pharm. Bull. 18, 1488 (1970); H. Hikino, T. Takahashi, and
      T. Takemoto, ibid. 19, 2424 (1971); 20, 210 (1972).
198. M. Fukuoka, M. Kuroyanagi, M. Toyama, K. Yoshihira, and
      S. Natori, Chem. Pharm. Bull. 20, 2282 (1972).
199. Y. Hayashi, M. Nishizawa, S. Hirata, and T. Sakan, Chem.
      Lett. 375 (1972).
200. M.S.R. Nair and M. Anchel, Tetrahedron Lett. 2753 (1972);
      R.B. Woodward and T.R. Hoye, J. Amer. Chem. Soc. 99, 8007 (1977).
201. H. Hikino, K. Agatsuma, C. Konno, and T. Takemoto,
      Tetrahedron Lett. 4417 (1968).
202. P. Bollinger, Thesis, Eidgenössische Technische Hochschule
      Zürich, 1965; through ref. 97, pp. 131-132.
203. B.E. Cross and R.E. Markwell, J. Chem. Soc. (C) 2980 (1971).
204. R.H. Burnell and M. Moinas, Chem. Commun. 897 (1971).
205. A.S. Romanova, G.F. Prybilova, P.I. Zakharov,
      V.I. Sheichenko, and A.I. Bankovskii, Khim. Prirod. Soedin.
      199 (1971).
206. E. Wenkert, J.D. McChesney, and D.J. Watts, J. Org. Chem.
      35, 2422 (1970).
207. P. Ruedi and C.H. Eugster, (a) Helv. Chim. Acta 55, 1736
      (1972);
      (b) ibid. p. 1994.
208. (a) P.M. Brown, M. Moir, R.H. Thomson, T.J. King,
      V. Krishnamoorthy, and T.R. Seshadri, J. Chem. Soc. Perkin I
      2721 (1973);
      (b) F. Delle Monache, G.B. Marini Bettòlo, O. Gonçalves
      de Lima, I.L. d'Albuquerque, and J.S. de Barros Coêlho,
      ibid. 2725;
      (c) J.D. Martin, Tetrahedron 29, 2997 (1973).
209. Y.N. Shukla, J.S. Tandon, D.S. Bhakuni, and M.M. Dhar,
      Phytochemistry 10, 1909 (1971).
210. H.V. Schmid and M.H. Zenk, Tetrahedron Lett. 4151 (1971).
211. D.P. Chakraborty, K.C. Das, and A. Islam. J. Indian Chem.
      Soc. 47, 1197 (1970); Chem. Abstr. 73, 142126x.
212. W. D. Crow, N. C. Hancox, S. R. Johns, and J. A. Lamberton,
      Austral. J. Chem. 23, 2489 (1970).

CHAPTER 23

Carotenoids in General:
Chemistry and Biosynthesis

This section discusses the small number of natural carotenoids with aromatic rings which have been found so far. As a
background to this discussion, a necessarily brief resumé of the
general features of chemistry and biochemistry of carotenoids is
given; for more detailed information, numerous reviews are available, of which the recent ones by Goodwin (1), Weedon (2),
Chichester (3), and Liaaen-Jensen [4] may be quoted.

After the completion of the text of this section, a comprehensive monograph edited by Isler (5) has appeared, which
provides up-to-date reviews of all aspects of the chemistry and
biochemistry of the carotenoids. Of its chapters, those by
Weedon on Occurrence (5a), by Liaaen-Jensen on Isolation and
Reactions (5b), by Goodwin on Biosynthesis (5c), and by
Straub, List of Natural Carotenoids (5d), are particularly
pertinent.

The carotenoids constitute a fairly large group of isoprenoid compounds; ref. 5d lists about 300 of them. They are
distributed throughout the entire range of organisms, but seem
to be biosynthesized almost exclusively by microorganisms and
higher plants (3a); however, the *de novo* biosynthesis of β-carotene from acetate has been observed in slices and homogenates
from bovine corpus luteum (6).

The typical carotenoids are C_{40} polyenes of a mevalonoid
nature which arise through symmetrical tail-to-tail juncture
of two C_{20} units. Peripheral modifications of this fundamental
structure need not be symmetrical, and often are not. Numerous
modifications occur. They include various cyclizations of the
end-groups to give five- and six-membered alicyclic structures,
and introduction of acetylenic or allenic unsaturations, or of
a wide variety of oxygenated functions (OH, OMe, carbonyl,
carboxyl, epoxy groups, etc.). Other alterations, only recently

observed and recognized (4b,d, 5b), consist in attachment of
additional isoprenoid units near the end of the molecule, to
give C_{45} and C_{50} compounds. Conversely, carotenoids with less
than 40 carbon atoms in the carbon skeleton occur; they are
referred to as apo-carotenoids, if they can be considered
formally as arising through oxidative cleavage of one of the
carbon-carbon double bonds of a normal carotenoid, and as nor-
carotenoids, if other processes are involved (5b).   In the
vitally important cases of vitamin A [26] and the corresponding
aldehyde (retinal), biosynthesis through oxidative cleavage of
the central double bond of β-carotene is well established (3a).

The majority of carotenoids can be represented by the
generalized formula [1].   To exemplify a few actual structures
for later reference, some of the numerous endgroups R, R' that
have been found are indicated in Scheme I.   In a very limited
number of so-called "retro"-compounds, the central C_{18} unit [1]
is replaced by the variant [2].

Most natural carotenoids have been given trivial names at
the time of their discovery, similar or isomeric compounds
often being distinguished by Greek prefixes (e.g., the five
carotenes, $C_{40}H_{56}$, designated as α-ε carotene).   Recently, a
rational system of nomenclature has been proposed (5; 6A);
only those features that are pertinent to our discussion can be
mentioned here.   The rules define carotenoids as a class of
hydrocarbons (carotenes) or oxygenated compounds derived from
them (xanthophylls) consisting of eight isoprene units joined
in such a manner that the arrangement of these units is reversed
in the center, so that the two central methyl groups are in
1,6, the others in 1,5 relationship.   Greek-letter prefixes are
given to the individual end-groups.   The former α-carotene, for
example, becomes β,ε-carotene.   The new prefixes for the end-
groups pertaining to our discussion are given on p. 641.   Since
the new system became available only after the manuscript of this
section was essentially complete, the older names are here used,
but the new ones have been inserted in parenthesis where a
compound is mentioned for the first time.

The presence of the numerous double bonds in the carotenoids
permits the occurrence of *cis-trans* isomerism, the all-*trans*
compound being usually the most stable one among the stereoisomers of one given structure.   Most *cis*-carotenoids are recognizable by a weak absorption band ("*cis*-peak") on the short-
wavelength side of the group of intense bands which is characteristic of the electronic spectrum of most carotenoids.
Natural occurrence of *cis*-carotenoids has been observed repeatedly, but the two aromatic representatives which have been obtained,

Scheme I

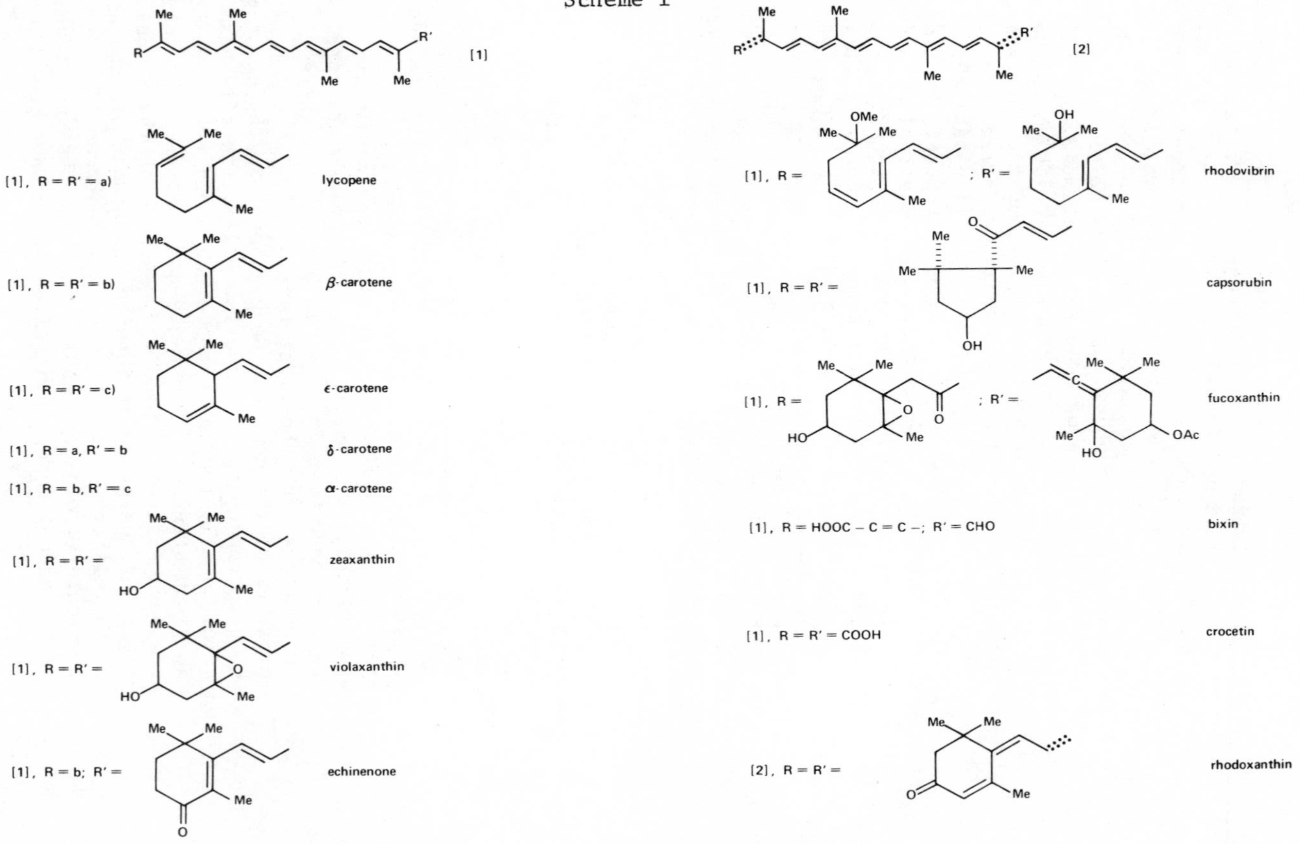

[1]

lycopene   [1], R = R' = a)

β-carotene   [1], R = R' = b)

ε-carotene   [1], R = R' = c)

δ-carotene   [1], R = a, R' = b

α-carotene   [1], R = b, R' = c

zeaxanthin   [1], R = R' =

violaxanthin   [1], R = R' =

echinenone   [1], R = b; R' =

[2]

rhodovibrin   [1], R =   ; R' =

capsorubin   [1], R = R' =

fucoxanthin   [1], R =   ; R' =

bixin   [1], R = HOOC—C = C —; R' = CHO

crocetin   [1], R = R' = COOH

rhodoxanthin   [2], R = R' =

640

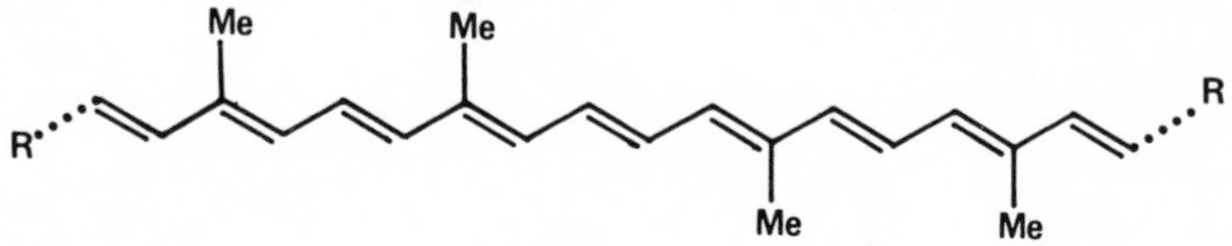

[1]   General Expression for a typical Carotenoid

R =

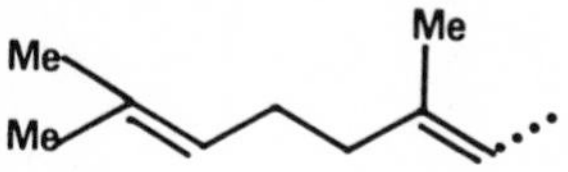

ψ

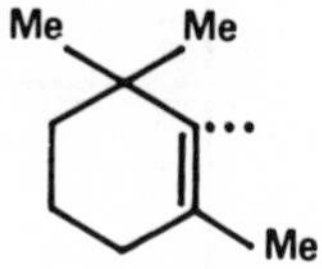

β

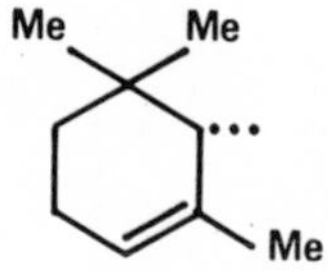

ε

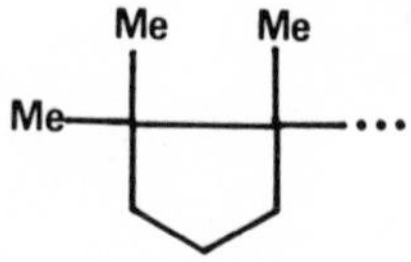

κ

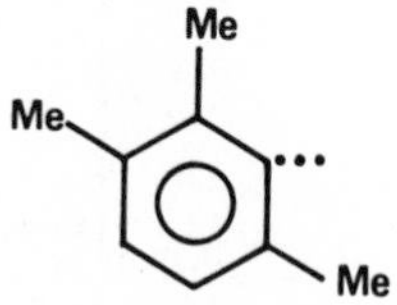

φ

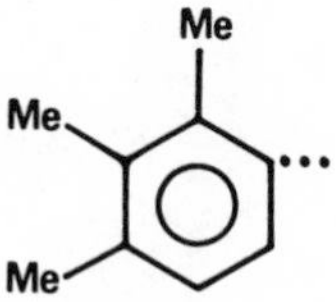

χ

the *cis*-isomers of okenone (see below), seem to be isolation artefacts.

The known naturally-occurring aromatic carotenoids are listed in Table I.

For detailed discussions of the *biosynthesis* of the carotenoids, and for references to much of the extensive original literature, the reviews by Goodwin (1b-d, 5c), Chichester (3a,b) and Porter (7) should be consulted. Only the most general features can be given here.

Table I.   Aromatic Carotenoids.

No.	Structure	Name	Occurence	Reference
	A) Both end-groups aromatic			
7	R = R' 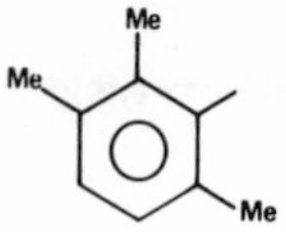	isorenieratene, leprotene	Reniera japonica (10); Streptomyces mediolani (25); Mycobacteria (8,9,24); Phaeobium sp. (22).	8,9,10,11; 17,18,21 23,24.
8	R = R'' 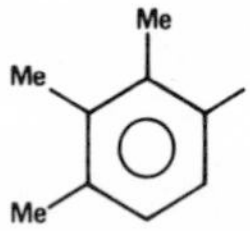	renierapurpurin	Reniera japonica (10)	10,19,21
6	R 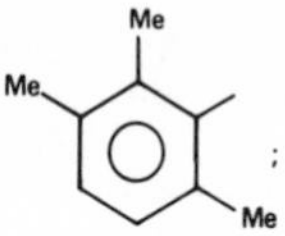 = ; R 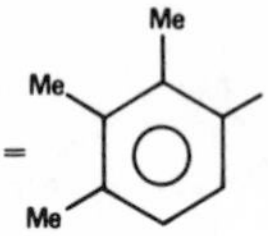 =	renieratene	Reniera japonica (10)	10,13,14, 19,21
19	R 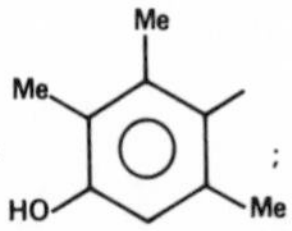 = ; R' 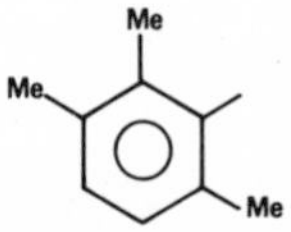 =	3-hydroxy-isore-nieratene	Streptomyces mediolani (25)	25
20	R = R' 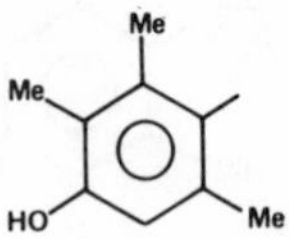 =	3,3'-dihydroxy-isorenieratene	Streptomyces mediolani (25)	25
	B) Only one end-group aromatic			
17	R 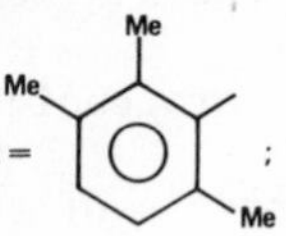 = ; R' 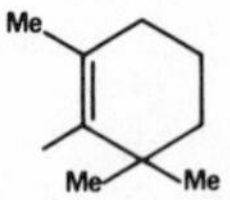 =	β-isorenieratene	Phaeobium sp. (22)	21,22

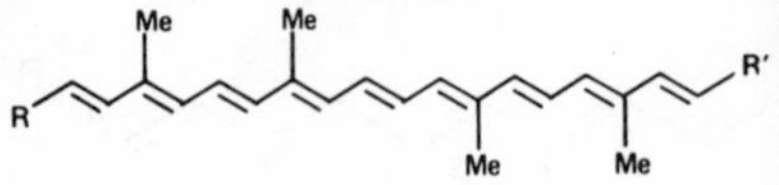

642

Table I.   Aromatic Carotenoids.

No.	Structure	Name	Occurence	Reference
21	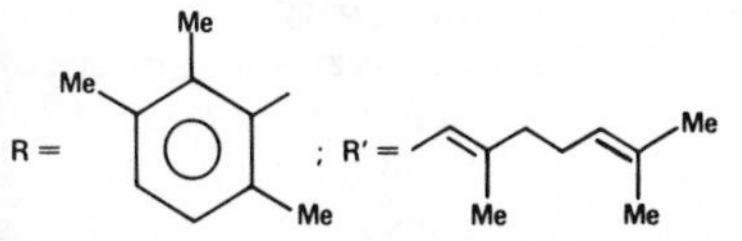 R = ; R' =	chlorobactene	Photosynthetic green bacteria (26); Chloropseudomonas ethylicus; Chlorobium thiosulfatophilum; C. limicola (35), Phaeobium sp. (22)	26,27,35
22	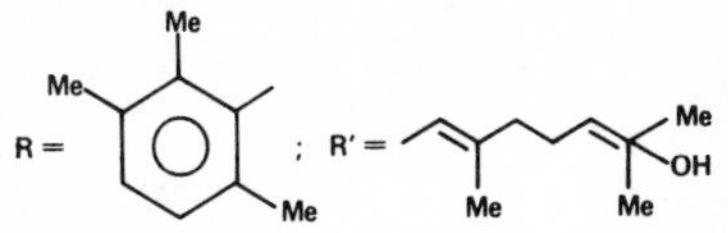 R = ; R' =	OH- chlorobactene	Photosynthetic green bacteria (26)	26,27
23	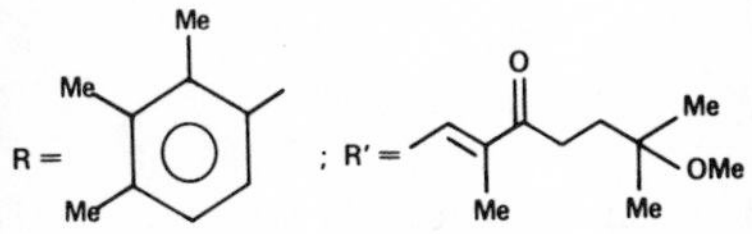 R = ; R' =	okenone	Purple sulfur bacteria (Thiorhodaceae), esp. Chromatium okenii (6, 28).	6,28,29,

C) Compounds presumably derived from vitamin A.

No.	Structure	Name	Occurence	Reference
27	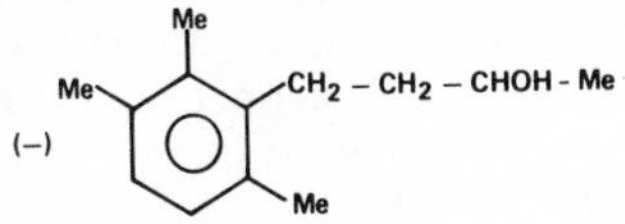 (−) CH$_2$ – CH$_2$ – CHOH - Me		Urine of pregnant mares	30
28	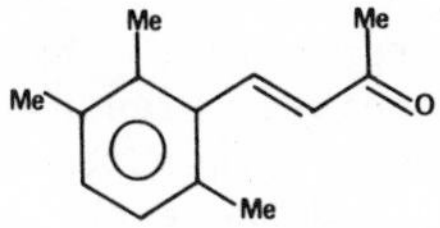		Urine of pregnant mares	31

It seems fairly universally accepted that the symmetrical
C_{40} skeleton of the carotenoids is formed by the coupling of two
molecules of geranylgeranyl pyrophosphate [3] (see Scheme II).
Compound [3] itself is derived from mevalonic acid in a way
quite analogous to the formation of farnesyl pyrophosphate,
the precursor of squalene [4] and hence of the triterpenoids
and steroids (see Chapters 22-5 and 24). Findings on incorporation of label from variously labeled precursors (acetate, mevalonate) into carotenoids are in excellent agreement with a
stereospecific biosynthesis from mevalonate via [3] and symmetrical coupling of two molecules of the latter (1d, 5c). As one
example, the labeling of β-carotene (β,β -carotene) from [2-14C]
mevalonate is shown below:

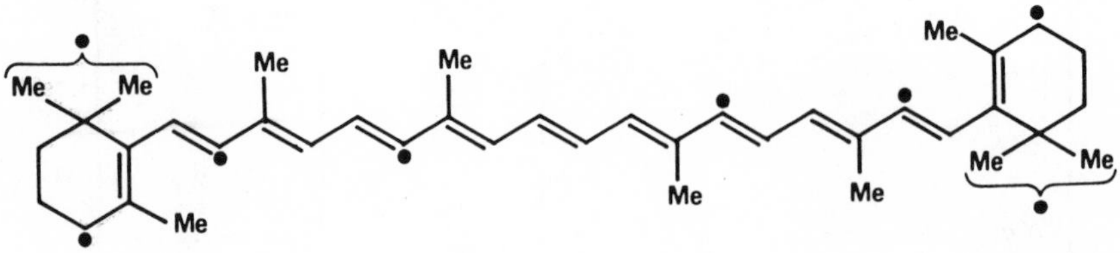

As has been mentioned briefly in Section 22-1, mevalonic
acid can be formed also from leucine via 3-hydroxy-3-methylglu-
taric acid, rather than from acetic acid (1b, 3a, 5c). This
mode of biosynthesis accounts for the marked stimulation of
carotenoid formation by leucine in some (although by no means
all) organisms; furthermore, leucine and β,β-dimethylacrylic
acid (an intermediate in the formation of hydroxymethylglutaric
acid from leucine) are incorporated into carotenoids by certain
organisms (5c).

The fundamental carbon skeleton of the carotenoids is thus,
like squalene [4], a symmetrical structure formed through
tail-to-tail coupling of two identical mevalonoid moieties.
Significant differences emerge, however, in the two dimerization
reactions, 2 C_{15} ⟶ [4] and 2 C_{20} ⟶ carotenoids. The former
proceeds through the nonsymmetrical intermediate, presqualene,
(see Section 22-8) to give [4], an intermediate in which the
two C_{15} units are joined by a single bond, and where the distribution of double bonds in very suitable for the subsequent
cyclization processes which yield the triterpenes and steroids.
In contrast to this, the first[+] C_{40} intermediate in carotenoid

---

[+]However, cf. the Addendum.

Scheme II

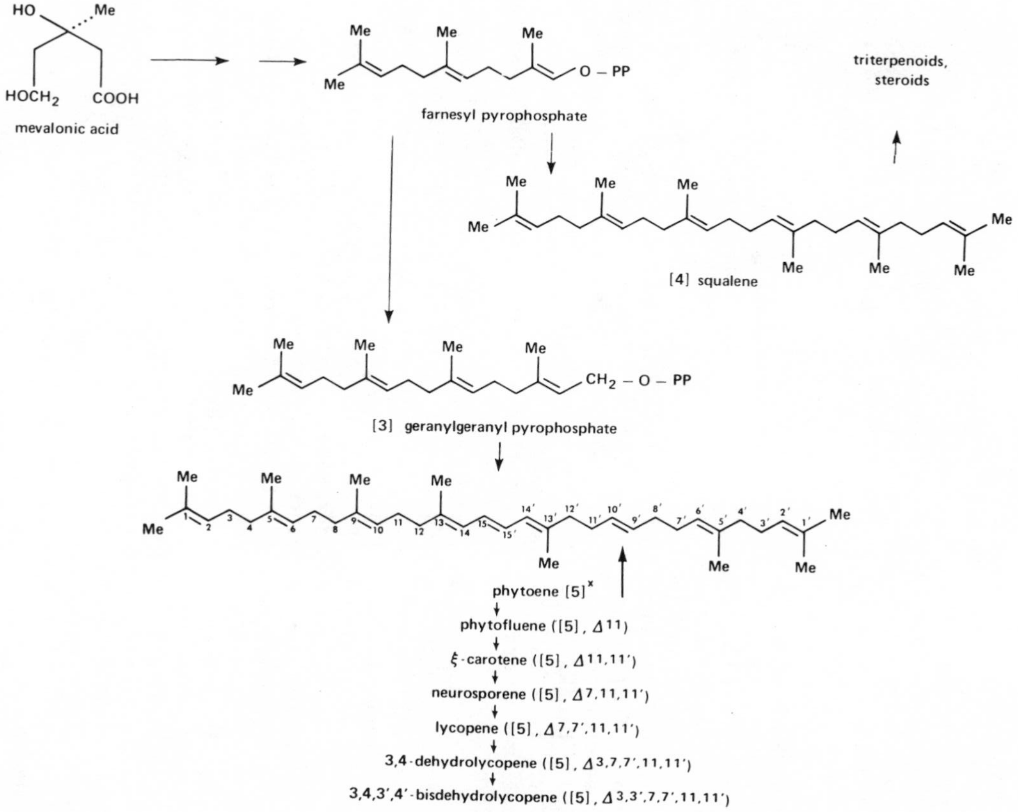

HO
Me
HOCH2
COOH
mevalonic acid
Me
Me
Me
Me
O – PP
farnesyl pyrophosphate
triterpenoids,
steroids
[4] squalene
CH2 – O – PP
[3] geranylgeranyl pyrophosphate
phytoene [5]x
phytofluene ([5], Δ11)
ξ-carotene ([5], Δ11,11′)
neurosporene ([5], Δ7,11,11′)
lycopene ([5], Δ7,7′,11,11′)
3,4-dehydrolycopene ([5], Δ3,7,7′,11,11′)
3,4,3′,4′-bisdehydrolycopene ([5], Δ3,3′,7,7′,11,11′)

biosynthesis is phytoene [5], a compound which already contains
a central group of three conjugated double bonds, and is thus
not capable of giving polycyclic structures by a mechanism
analogous to that of the cyclization of [4]. It may well be
this feature which accounts for the fact that polycyclic C_{40}
compounds comparable to the triterpenoids or steroids have not
yet been found. Lycopersene, the 15,15'-dihydro derivative of
[5], and hence the strict analog of [4], had at one time been
assumed to be the initial C_{40} product in carotenoid biosynthesis; however, this compound has apparently never been observed
to occur naturally[+], and does not seem to be involved in the
biosynthesis of the carotenoids (5c)[+].

Dehydrosqualene, the strict C_{30} analog of [5], has been
isolated from certain microorganisms, such as a colorless
mutant of *Staphylococcus aureus* (or the wild-type grown in the
presence of diphenylamine) (41), and *Halobacterium cutirubrum*
(42). Diphenylamine is a well-known inhibitor of the formation
of the more unsaturated, colored carotenoids from their colorless precursors. The biosynthetic significance of dehydrosqualene is uncertain.

The chain of conjugated double bonds which is so characteristic of the carotenoids appears to be elaborated by stepwise,
symmetrical introduction of double bonds into [5] (see Scheme II).
Dehydrogenation of [5] at positions 11 and 12 gives phytofluene;
further introduction of double bonds occurs symmetrically at 11',
7, and 7', to yield ξ-carotene, neurosporene, and lycopene,
respectively. This sequence of reactions is often called the
"Porter-Lincoln series" (7A). One interesting feature of this
chain of transformations at the C_{40} level is the fact that the
first two links of the chain, phytoene [5] and phytofluene, have
recently been found to have the *cis*-configuration at their central
double bonds, while ξ-carotene and the compounds following it
have the all-*trans* stereochemsitry (cf., e.g., refs. 1d and 7,
and literature quoted there). Elaboration of the fully conjugated
chain of the carotenoids must thus involve a *cis* ⟶ *trans*
isomerization, presumably at the level of phytofluene (see
Addendum).

Additional dehydrogenation of lycopene to 3,4-dehydrolycopene
and the fully conjugated 3,4,3',4'-bis-dehydrolycopene is possible
both in vitro and in vivo, although these compounds, and 3-
dehydrocarotenoids in general, occur rather infrequently.

_____

[+]However, cf. the Addendum.

Cyclization of the end-groups to the cyclohexene structures, β and ε, which are very prevalent in carotenoids (exemplified in Scheme I by β- and ε- (ε,ε)-carotene, respectively), generally occurs at the level of lycopene but can apparently also take place from neurosporene (5c,d). The cyclopentane ring, κ, which occurs in some carotenoids (e.g. capsorubin, 3,3'-dihydroxy-κ,κ-carotene-6,6'-dione; see Scheme I) seems to be formed from six-membered rings by secondary transformations.

23-1. AROMATIC CAROTENOIDS

Historically, the first carotenoid with aromatic rings to be observed was leprotene, isolated in 1937 by Grundmann and Takeda (8) from a bacterium obtained from a leper. The pigment was subsequently isolated from several species of *Mycobacterium* (9). Its structure remained unknown, however, and its benzenoid nature was established only more recently, when its identity with an aromatic carotenoid of known structure was proven (see below).

The occurence of carotenoids with aromatic rings was first recognized by Yamaguchi during his investigation of the pigments of the orange-colored sponge *Reniera japonica*. This species proved an unusually rich source of carotenoids; as many as 15 such pigments have been isolated in crystalline form (10). Two of these were identified as α-carotene (β,ε-carotene) and β-carotene, while the three others, renieratene [6] (the main pigment), a related pigment subsequently (11) called isorenieratene [7], and renierapurpurin [8], appeared unusual**. Their elementary composition, $C_{40}H_{48}$, made them the most highly unsaturated carotenoids known, yet the absorption spectra of [6] and [7] showed great similarity with those of the carotenes, $C_{40}H_{56}$: γ-carotene (β,ψ-carotene) in the case of [6] β-carotene in that of [7]. With a composition indicating 15 double bonds (assuming their structure to embody two rings[+]), [6] and [7] are thus spectro-

---

**Renieraxanthin, $C_{40}H_{56}O_2$, a very minor constituent, is probably related to this group, but was not studied in great detail because of its scarcity; it closely resembles okenone (see below) and may be identical with it (12).

[+]Actually, the maximum number of conjugated double bonds that can be accomodated in a C_{40} carotenoid structure with two rings is 14; 15 is possible only in a completely conjugated noncyclic structure such as 3,4,3',4'-tetradehydrolycopene.

scopically analogous to the carotenes with only 11 conjugated double bonds. Compounds [6] (13, 14) and [7] (11) must therefore contain some strongly unsaturated group which yet has only a small bathochromic influence. Of the limited number of such groups, acetylenic and allenic unsaturations are unlikely because of the absence of corresponding bands in the ir spectrum, leaving aromatic unsaturation as the only reasonable alternative. This conclusion is also supported by the presence of bands at 800–810 cm^{-1} in the ir spectra of [6] and [7]. All subsequent evidence, including nmr spectroscopy (15) and total synthesis, has confirmed the presence of aromatic rings. The structures of [6] and [7] were established by Yamaguchi (11, 13, 14) mostly on the basis of oxidative degradations which are shown diagrammatically in Scheme III.

The structures of [6] and [7] established by the reactions shown are in full agreement with the spectroscopic data; in particular, the positions of the absorption bands at longest wave lengths (in CS$_2$): 532 nm for [6], 520 nm for [7], compared with 541 nm for a synthetic (16) analog with unsubstituted phenyl rings, can be explained through steric interference with planarity by the methyls in both positions *ortho* to the polyene chain in the 2,3,6-trimethylphenyl groups (ϕ) which are present in both [6] and [7]. The fact that [6] has only one such ring, while in [7] both are of this type, explains the difference in position of the absorption bands at longest wavelengths. On this basis, the symmetrical compound in which both rings have methyls in 2,3,4 positions (prepared by synthesis and found identical with [8], see below) should have an absorption spectrum at significantly higher wavelengths then its isomer [6] and [7], in agreement with the experimental results (10) (λ_{max} 544, 504, 475 nm).

Final proof for the correctness of the structures of [6], [7], and [8] was obtained by total synthesis. Yamaguchi (17) synthesized [7] from 4-octene-2,7-dione [12] and the acetylene [13] (see Scheme IV). Khosla and Karrer (18) described an essentially identical synthesis. Using a mixture of [13] and its isomer [14], Yamaguchi (19) obtained [6], [7], and [8], thus incidentally proving the structure of the last-named compound. However, their method, based on the carotenoid synthesis of Karrer and Eugster (20), gave poor yields. See Scheme IV.

Much superior yields were obtained by Cooper, Davis, and Weedon (21), who used the Wittig reaction between crocetindial [11] and the phosphonium bromides [15] prepared from the required aldehydes to synthesize [6], [7], and [8]. Analogous Wittig reactions between [15] (R = 2,3,6-trimethylphenyl and 2,3,4-trimethylphenyl, respectively), and apo-β-carotenal [16], a

Scheme III: Structure Proof of Renieratene [6] (13,14) and Isorenieratene (=Leprotene) [7](11)

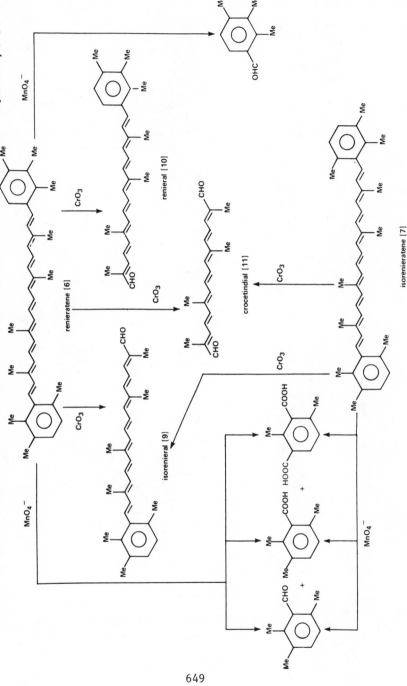

degradation product of β-carotene, gave β-isorenieratene [17]
and β-renierapurpurin [18]. The former was subsequently (22)
found to occur in photosynthetic bacteria of the genus *Phaeobium*
together with β-carotene, [7], and chlorobactene [21] (see
below). See Scheme V.
The identity of leprotene with [7] was demonstrated by
Liaaen-Jensen (23), who compared a remaining trace (0.2 mg) of
the original sample of Grundmann and Takeda (8) with synthetic
(21) [7], and independently by Okukado and Yamaguchi (24), who
used leprotene isolated by them from *M. phlei* and natural [7]
obtained from *Reniera*.

In the new nomenclature system (6A), the naturally occurring
aromatic carotenoids from *Reniera* are all designated as "carotene,"
with the following prefixes: [6],φ,ψ;[7],φ,φ;[8],φ,φ;[17],β,φ.

Recent research has led to the isolation of several other
aromatic carotenoids, and has shown that occurence of compounds
of this type is by no means restricted to *Reniera* or mycobacteria.

Closest to the pigments from *Reniera* are the 3-hydroxy and
3,3'-dihydroxy derivatives of [7], compounds [19] and [20],
respectively, which were shown by Arcamone and co-workers (25)
to occur in *Streptomyces mediolani*, together with [7] itself.
Their structures follow from their oxidation by chromium trioxide
in acetic anhydride to crocetindial [11] (see Scheme III), and
to 2,3,6-trimethyl-4-acetoxybenzaldehyde in the case of [20], a
mixture of this compound and 2,3,6-trimethylbenzaldehyde in
the case of [19]. Total syntheses, shown in Scheme V, completed
the structure proof.

Another related compound, chlorobactene (φ,ψ -carotene),
[21], was shown by Liaaen-Jensen and co-workers (26) to be
the main carotenoid of three species of photosynthetic green
bacteria: *Chloropseudomonas ethylicus, Chlorobium thiosulfatophilum*, and *Chlorobium limicola*. It is accompanied by OH-chlorobactene (1',2'-dihydro-1'-hydroxychlorobactene) [22], and, in
C. ethylicus, by small amounts of the nonaromatic carotenoids
lycopene, γ-carotene, and rhodopin (1,2-dihydro-1-hydroxy-lyco-
pene, 1,2-dihydro-ψ,ψ-caroten-1-ol). Compound [21], which
constitutes more than 90% of the total carotenoids in all three
species, is spectroscopically indistinguishable from γ-carotene
resembles it in many other respects, but can be separated from
it by chromatography on paper with an alumina filler. Structure
[21], which embodies the end-groups of [7] and of lycopene, was
derived for chlorobactene mostly through nmr, ir, and absorption
spectroscopy. OH-Chlorobactene has an absorption spectrum
identical with that of [21]; since it contains one tertiary
hydroxyl and can be dehydrated to [21] with $POCl_3$, structure
[22] follows. Again, total synthesis (27) of [21] and [22]

Synthesis by Yamaguchi (17,19) and Khosla and Karrer (18)

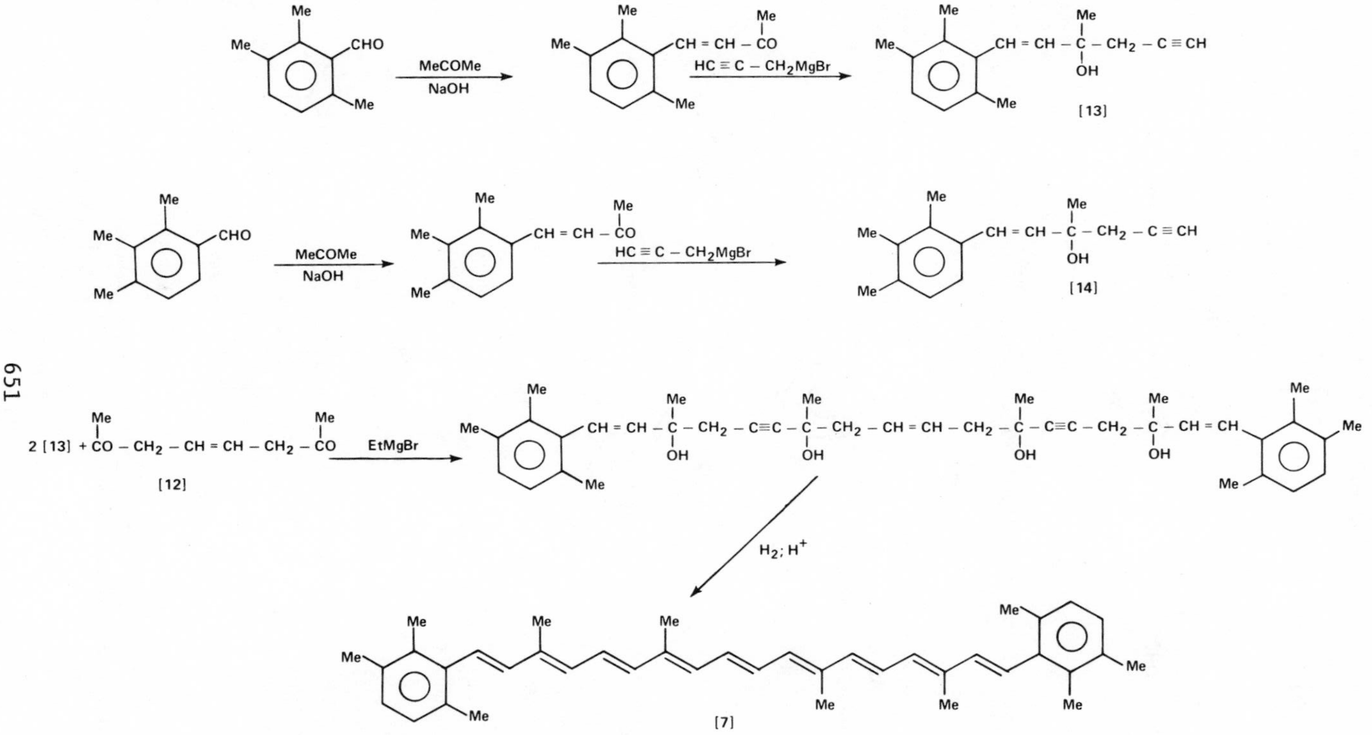

[13] + [14] + [12] $\xrightarrow{EtMgBr}$ $\xrightarrow{H_2;\ H^+}$ [6] + [7] + [8]

Scheme IV: Synthesis of Aromatic Carotenoids from *Reniera*.

Scheme V: Syntheses by Cooper et al. (21), and Arcamone et al. (25).

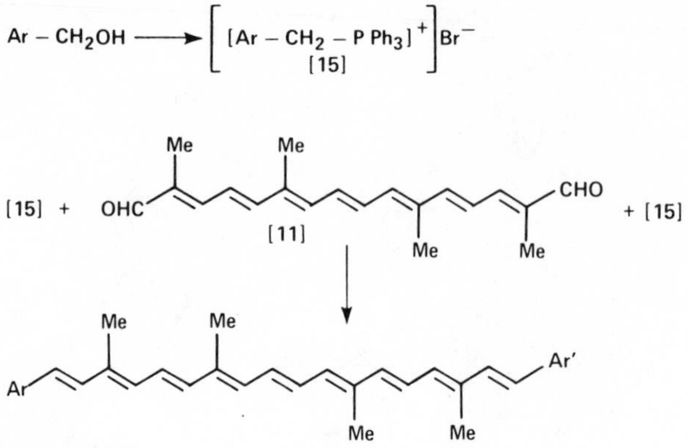

isorenieratene [7], Ar = Ar' =

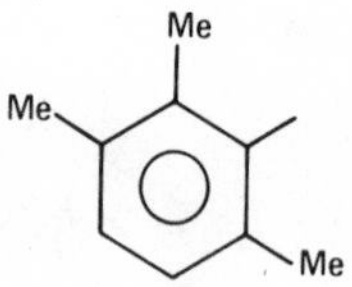

renieratene [6], Ar =

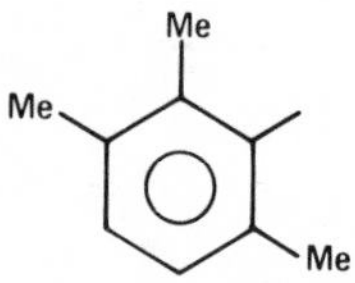

Ar' =

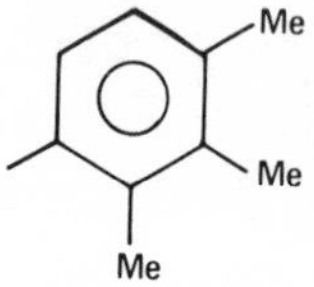

renierapurpurin [8], Ar = Ar' =

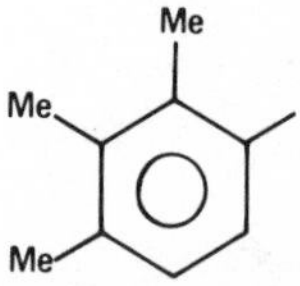

3-hydroxyisorenieratene [19], Ar =

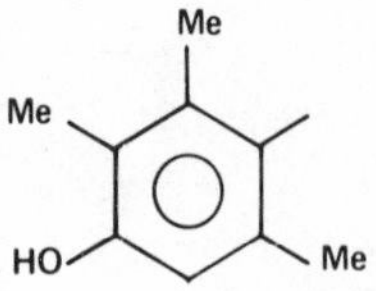

; Ar' =

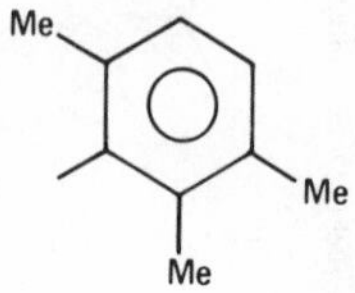

3,3'-dihydroxyisorenieratene [20], Ar = Ar' =

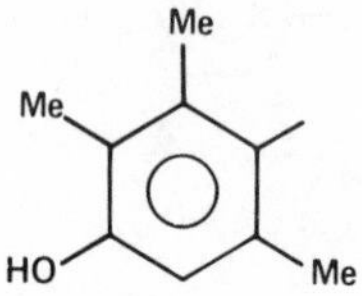

Scheme V: Synthesis by Cooper et al (21) and Arcamone et al (25)

β-isorenieratene [17] : Ar = Me

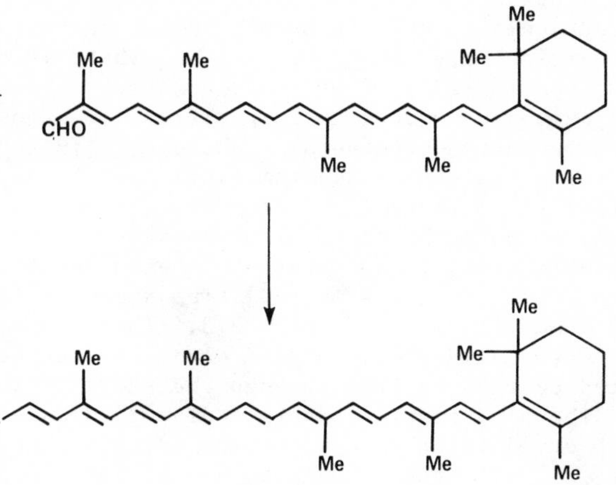

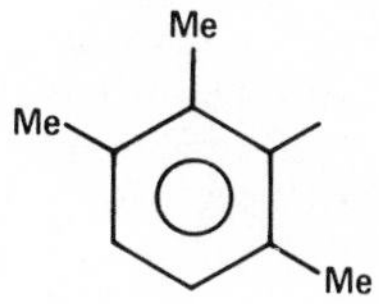

β-renierapurpurin [18] ; Ar = Me

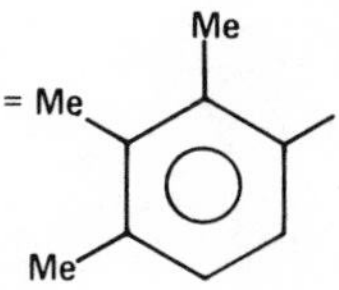

provided conclusive proof. Compound [21] also occurs (22) as a
minor carotenoid in photosynthetic bacteria of the genus
Phaeobium, together with [7] and [17].

A further aromatic carotenoid is okenone, a pigment which
seems to occur only in certain purple sulfur bacteria, Thiorhodaceae, particularly *Chromatium okenii*, where it constitutes
the only carotenoid (28). (The two *cis*-isomers that were
observed on column-chromatographic purification of okenone
appear to be isolation-artefacts). Okenone, $C_{41}H_{54}O_2$, was shown
by Liaaen-Jensen (12) to be an aromatic mono-keto carotenoid of
structure [23] (1'-methoxy-1',2'-dihydro-$\times$,ψ-caroten-4'-one).
This formula, which is based to a large extent upon the results
of spectroscopic studies, was proven by Aasen and Liaaen-Jensen
(29) through total synthesis of [23] from renieral [10] (itself
synthesized from [11] and [15, Ar = $2,3,4-(Me)_3C_6H_2-$], by Wittig
reaction with the ylid [24], to give 4'-desoxookenone [25], which
was converted to okenone [23] through the steps indicated in
Scheme VI. The nonaromatic end-group of [23] does not seem to
have been observed in other carotenoids so far. The possible
identity of [23] with renieraxanthin has already been mentioned
(12).

Because of the well-established (3a) relationship between
vitamin A [26] and the carotenoids containing a β-ionone ring,
some aromatic compounds may be mentioned here which, from their
structures, could well be aromatization products of the vitamin;
however, no experimental proof for this interpretation seems to
exist. Two compounds of this type were found by Prelog and co-
workers in the urine of pregnant mares (used for manufacture of
oestrone). They are (-)-4-(2,3,6-trimethylphenyl)-butanol-3
[27] (30), and 2,3,6-trimethyl-benzalacetone [28] (31). Formation of such compounds from [26] would involve processes quite
analogous to those which must operate in the biosynthesis of the
2,3,6-trimethylphenyl rings in aromatic carotenoids such as [6],
[7], [21], and [22]. An obvious alternative possibility would

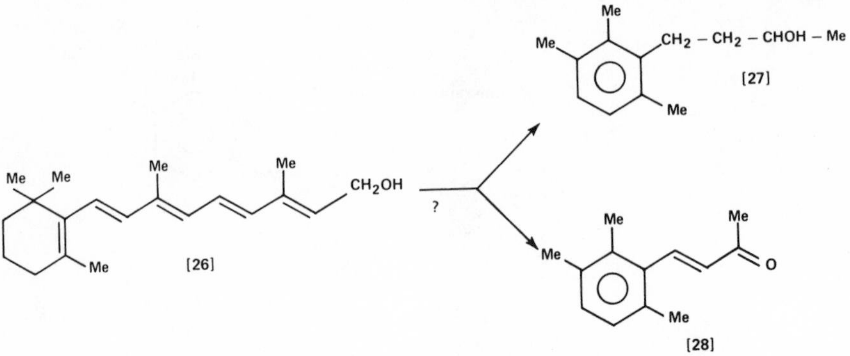

Scheme VI: Total Synthesis of Okenone.

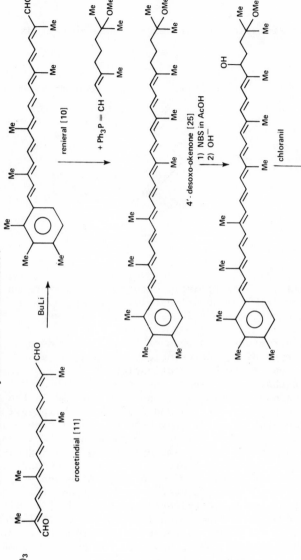

crocetindial [11]

renieral [10]

+ Ph₃P = CH

4′ - desoxo-okenone [25]

1) NBS in AcOH
2) OH⁻

chloranil

okenone [23]

be a derivation from aromatic C_{40} carotenoids through degradative processes.

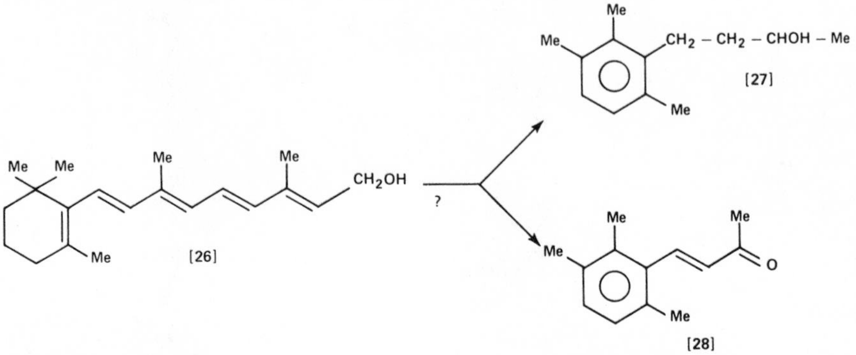

23-2. BIOSYNTHESIS

The biosynthesis of these aromatic carotenoids presents an unusually interesting problem; unfortunately, however, no experimental evidence has been obtained up to now (but see the Addendum).

Obviously, the structural features of these compounds conform to the general type of C_{40} carotenoids, and suggest a formation of the benzene rings by aromatization of normal β-ionone (β) rings. The fact that α- or β-carotene co-occur with many of the aromatic carotenoids is suggestive, and even more so is the joint occurrence of both β-ionone (β) and 2,3,6-trimethylphenyl (φ) rings in β-isorenieratene [17]. Furthermore, the latter endgroup could plausibly arise from a ring through an aromatization involving migration of one of the *gem*-methyl groups by a Wagner-Meerwein rearrangement, for which there is abundant precedent. Some reactions of β-ionone rings *in vitro* proceed in exactly this manner; see, *inter alia,* the conversion of safranol [29] into both safranal [30] and 2,3,6-trimethylbenzaldehyde [31] with MnO_2 (32), and the conversion of β-ionone itself [32] to [33] (the ketone corresponding to [27]) on treatment with N-bromosuccinimide, followed by dehydrobromination (33). However, the occurrence of the 2,3,4-trimethylphenyl rings in renieratene [6], renierapurpurin [8], and okenone [23] cannot be interpreted in such a simple manner, although here, too, the co-occurrence (10) of [6] and [8] with [7], and with α- and β-carotene, seems to suggest related biosynthetic pathways for the trimethylbenzene groups of both types. Yamaguchi (14) has pointed out that the molecule of [6] can be constructed entirely from isoprene residues. However,

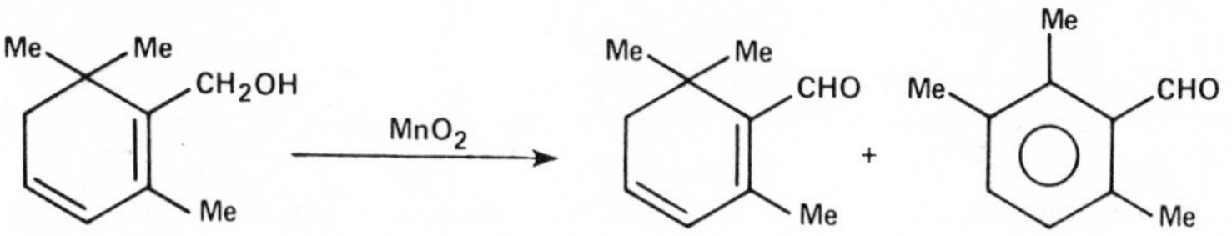

[29]   [30]   [31]

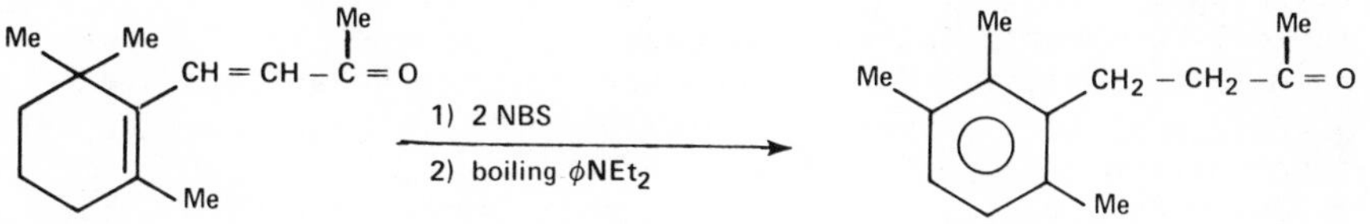

[32]   [33]

β-ionone

the arrangement of these groups in the aromatic rings is not of
the normal head-to-tail type, and consequently a biosynthesis
from unrearranged isoprenoid moieties seems highly unlikely.
Furthermore, the 2,3,4-trimethylphenyl (×) group could not arise
from a normal alicyclic ring (β or ε) by a simple Wagner-Meerwein
rearrangement. It could conceivably be formed through multiple
shifts of methyl groups. Much more plausible than such a sequence
(which seems to have no known precedent in biosynthetic processes)
is the pathway proposed by Liaaen-Jensen(4b) and indicated below:
isomerization of a 2,3,6-trimethylbenzene (φ) group via an
intermediary Ladenburg prism structure. This process is amply
documented in photochemical isomerizations of substituted ben-
zenes (cf., e.g., ref. 34).

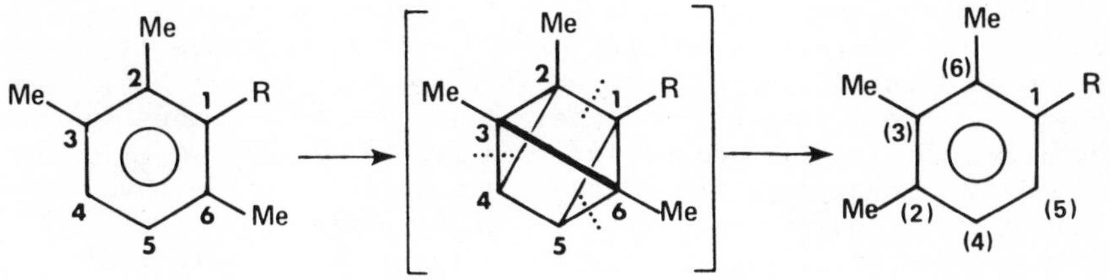

Likewise unsolved is the problem of the locus of the
aromatization which yields the aromatic carotenoids of the sponge,
Reniera. This member of the most primitive phylum of metazoa is
the only nonmicrobial organism in which aromatic carotenoids
have been found so far; the question arises whether the sponge
itself is able to carry out this biosynthesis (or can modify
dietary nonaromatic carotenoids), or whether it gets its pigments
elsewhere. The first alternative seems unlikely but not entirely
impossible, if it is remembered that the biosynthesis of another
mevalonate-derived group of metabolites, the oestrogens, furnishes
the only authentic instance of biosynthetic aromatization in
animals (see Section 24-1). However, the second alternative
appears much more probable in view of the fact that aromatic carotenoids seem widely distributed among microorganisms (*Streptomyces*, mycobacteria, a variety of green, brown, and purple bacteria), but have been encountered only once in the animal kingdom.
    For this reason, it seems likely (35) that the aromatic
carotenoids of *Reniera* are acquired by the sponge from some
microorganisms, which could be part of its food, could inhabit
it, or could be symbionts. This question was investigated by
Eimhjellen (36) in the yellow-brown sponge *Halichondrium panicea*.
Extraction of this animal yielded β-carotene and some other carotenoids, none of them aromatic. However, Eimhjellen isolated
from it the green bacterium *Chlorobium limicola*, a known (26)
producer of chlorobactene [21] and OH-chlorobactene [22]. This
sponge thus harbors a bacterium notoriously capable of synthesizing aromatic carotenoids which, however, are different from
renieratene and its close relatives. This investigation thus
did not furnish direct proof for the microbial origin of the
carotenoids of *Reniera*, and the question of the locus of
aromatization must remain open for the time being.

23-3. ADDENDUM

    Since the completion of the text on the preceding pages,
work in the laboratories of Porter and Rilling has shed light
on the earliest stages of carotenoid biosynthesis, and has
shown that these stages are closely analogous to the ones in
the biosynthesis of squalene [4] (see Section 22-8).
    Geranylgeranyl pyrophosphate [3] was found to be converted
to a new intermediate under the action of extracts from a
Mycobacterium sp. [37], of soluble extracts from tomato fruit
plastids (38a), or of a purified preparation of squalene synthase
from yeast (38a,b). This new compound, termed "prephytoene
pyrophosphate" (37) or "prelycopersene pyrophosphate" (38),
was shown to have structure [34a], which is perfectly analogous

to that of presqualene pyrophosphate [34b]. This formulation is based on a detailed analysis of the mass-spectrometric degradation of [34a] and its relatives (38b), and on the total synthesis of the compound itself (37) and of the primary alcohol formed on removal of the pyrophosphate group (37, 39).

Compound [34a] is transformed into lycopersene (see above, page 647) on incubation with the enzyme systems mentioned (38) in presence of NADPH and Mg^{2+}; both [34a] and lycopersene are further converted enzymatically into phytoene [5] and lycopene (40) by the tomato enzyme supplemented with $NADP^+$, FAD, and Mn^{2+}. This formation and further conversion of lycopersene suggest (38a,b, 40) that this compound is indeed an intermediate in the biosynthesis of carotenoids, whose early stages would then be entirely analogous to those of the biosynthetic formation of [4]. However, this interpretation has been questioned on stereochemical grounds (41), so that the status of lycopersene is still not yet clarified. Mechanisms for the direct conversion of [34a] into *cis-* or *trans-* [5] can be formulated (41).

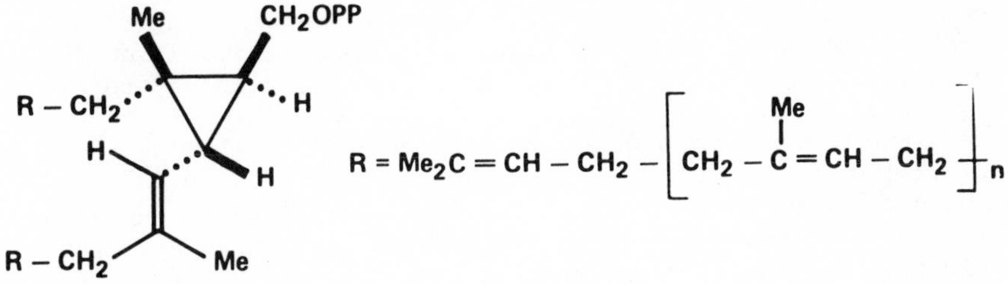

$$R = Me_2C = CH - CH_2 - \left[CH_2 - \underset{\underset{Me}{|}}{C} = CH - CH_2 \right]_n$$

[34]   a: n = 2
        b: n = 1

The soluble enzyme systems from tomato plastids promote the transformation of [5] into *cis-* and *trans*-phytofluene, and the further conversion of the latter compound into *trans-* ξ-carotene, neurosporene, lycopene, and carotenes (42). The first of these steps requires NADP, while the subsequent ones seem to depend upon FAD and Mn^{2+}. In a subsequent series of papers (43), which appeared too late for detailed consideration, all the steps of the entire Porter-Lincoln series to all-*trans* β-carotene and to certain *cis*-carotenoids seem to have been carried out with the use of soluble tomato enzymes.

The question of the stereochemistry of the central double bond in [5] has turned out to be more involved than had been realized. Kushwaha and co-workers (44) observed that the [5] of

Halobacterium cutirubrum contains an appreciable proportion of
the all-*trans* stereoisomer in addition to the well-known *cis*-
[5].   Similar findings have been made with some other microorganisms such as *Mucor hiemalis* (45), *Flavobacterium dehydro-
genane* (46), and *Verticillium agaricinum* (47).   In the case of
F. dehydrogenans, the [5] consists mainly of the all-*trans*
isomer.

Occurrence of several new aromatic carotenoids has been
observed.   *Thiothece gelatinosa* (Thiorhodaceae) produces
okenone [23] as the main carotenoid, together with several minor
ones (48).   Of the latter, one pigment, "*Thiothece*-484," differs
from [23] by replacement of one of the methyls on the aromatic
rings (presumably the one at position 3) by a carbomethoxy group.
The other minor carotenoids all have the same methoxylated non-
cyclic end-group as [23] but contain a β-ionone ring or various
non-cyclic structures instead of the aromatic ring.

Two of the minor pigments of *Reniera japonica* have been
identified (49a) as the acetylenic analogs of renieratene [6]
and isorenieratene [7], respectively.   The triple bond in both
compounds is adjacent to a 2,3,6-trimethylbenzene ring (φ in
the new nomenclature).   The structures assigned mainly on the
basis of spectroscopic evidence have been confirmed by total
synthesis (49b).

Experiments by Moshier and Chapman (50) on the incorporation
of [2-^{14}C]-mevalonic acid into chlorobactene [21] in *Chlorospeudo-
monas ethylica* have furnished the first direct experimental proof
for the mevalonoid origin of the benzene ring in an aromatic
carotenoid.   The findings suggest that a stereospecific migration
of the carbon atom derived from the methyl group of mevalonic
acid furnished the methyl in position 2 of the ring.

REFERENCES

1.   T.W. Goodwin, (a) Distribution of Carotenoids, in Chemistry
     and Biochemistry of Plant Pigments, T.W. Goodwin, Ed.,
     Academic, London and New York, 1965, p. 127; (b) The Biosyn-
     thesis of Carotenoids, ibid. p. 143; (c) Pure Appl. Chem.
     20, 483 (1969); (d) Biochem. J. 123, 293 (1971).
2.   B.C.L. Weedon, Chem. Brit. 3, 424 (1967).
3.   a.   C.O. Chichester and T.O.M. Nakayama, The Biosynthesis of
     Carotenoids and Vitamin A, in Biogenesis of Natural Compounds,
     P. Bernfeld, Ed., 2nd ed., Pergamon, 1967, p. 475;
     b.   C.O. Chichester, Pure Appl. Chem. 14, 215 (1967).
4.   S. Liaaen-Jensen, (a) Pure Appl. Chem. 14, 227 (1967);
     (b) ibid. 20, 421 (1969); (c) Experientia 26, 697 (1970);
     (d) Recent Progress in Carotenoid Chemistry, in Aspects of

Terpenoid Chemistry and Biochemistry, T.W. Goodwin, Ed.,
Academic, New York, 1971, p. 223.
5. O. Isler, Ed., Carotenoids, Birkhäuser Verlag, Basel and
Stuttgart, 1971.
6. B.M. Austern and A.M. Gawienowsky, Lipids 4, 227 (1969).
6A. Biochemistry 10, 4827 (1971); Eur. J. Biochem. 25, 397
(1972).
7. J.W. Porter, Pure Appl. Chem. 20, 449 (1969).
7A. J.W. Porter and R.E. Lincoln, Arch. Biochem. 27, 390 (1950).
8. C. Grundmann and Y. Takeda, Naturwissenschaften 25, 27
(1937).
9. Y. Takeda and T. Ohta, Z. Physiol. Chem. 262, 168 (1939);
265, 233 (1940); J. Pharm. Soc. Japan 64, 67 (1944);
T.W. Goodwin and M. Jamikorn, Biochem. J. 62, 269 (1956).
10. T. Tsumaki, M. Yamaguchi, and T. Tsumaki, J. Chem. Soc.
Japan (Pure Chem. Sect.) 75, 297 (1954); M. Yamaguchi,
Bull. Chem. Soc. Japan 30, 111 (1957).
11. M. Yamaguchi, Bull. Chem. Soc. Japan 31, 51 (1958).
12. S. Liaaen-Jensen, Acta Chem. Scand. 21, 961 (1967).
13. M. Yamaguchi, Bull. Chem. Soc. Japan 30, 979 (1957).
14. M. Yamaguchi, Bull. Chem. Soc. Japan 31, 739 (1958).
15. M.S. Barber, J.B. Davis, L.M. Jackman, and B.C.L. Weedon,
J. Chem. Soc. 2870 (1960).
16. C.H. Eugster, C. Garbers, and P. Karrer, Helv. Chim. Acta
35, 1179 (1952).
17. M. Yamaguchi, Bull. Chem. Soc. Japan 32, 1171 (1959).
18. M.C. Khosla and P. Karrer, Helv. Chim. Acta 43, 453 (1960).
19. M. Yamaguchi, Bull. Chem. Soc. Japan 33, 1560 (1960).
20. P. Karrer and C.H. Eugster, Helv. Chim. Acta 33, 1172
(1950).
21. R.D.G. Cooper, J.B. Davis, and B.C.L. Weedon, J. Chem. Soc.
5637 (1963).
22. S. Liaaen-Jensen, Acta Chem. Scand. 19, 1025 (1965).
23. S. Liaaen-Jensen, Acta Chem. Scand. 18, 1562 (1964).
24. N. Okukado and M. Yamaguchi, Bull. Chem. Soc. Japan 38,
1043 (1965).
25. F. Arcamone, B. Camerino, E. Cotta, G. Franceschi, A. Grein,
S. Penco, and C. Spalla, Experientia 24, 271 (1968);
F. Arcamone, B. Camerino, G. Franceschi, and S. Penco,
Gazz. Chim. Ital. 100, 581 (1970).
26. S. Liaaen-Jensen, E. Hegge, and L.M. Jackman, Acta Chem.
Scand. 18, 1703 (1964).
27. R. Bonnett, A.A. Spark, and B.C.L. Weedon, Acta Chem. Scand.
18, 1739 (1964).
28. a. K. Schmidt, S. Liaaen-Jensen, and H.G. Schlegel, Arch.
Mikrobiol. 46, 117 (1963);

b. K. Schmidt, N. Pfenning, and S. Liaaen-Jensen, ibid. 52, 132 (1965).

29. A.J. Aasen and S. Liaaen-Jensen, Acta Chem. Scand. 21, 970 (1967).

30. V. Prelog and J. Führer, Helv. Chim. Acta 28, 583 (1945); V. Prelog, J. Führer, R. Hagenbach, and H. Frick, ibid. 30, 113 (1947).

31. V. Prelog, J. Führer, R. Hagenbach, and R. Schneider, Helv. Chim. Acta 31, 1799 (1948).

32. E. Büchli and P. Karrer, Helv. Chim. Acta 38, 1853 (1955).

33. P. Karrer and P. Ochsner, Helv. Chim. Acta 31, 2093 (1948); G. Büchi, K. Seitz, and O. Jeger, ibid. 32, 39 (1949).

34. E.E. van Tamelen, S.P. Pappas, and K.L. Kirk, J. Amer. Chem. Soc. 93, 6092 (1971), and refs. quoted there.

35. S. Liaaen-Jensen, in Biochemistry of Chloroplasts, T.W. Goodwin, Ed. Vol. I, Academic, New York, 1966, p. 437.

36. K.E. Eimhjellen, Acta Chem. Scand. 21, 2280 (1967).

37. L.J. Altman, L. Ash, R.C. Kowerski, W.W. Epstein, B.R. Larsen, H.C. Rilling, F. Muscio, and D.E. Gregonis, J. Amer. Chem. Soc. 94, 3257 (1972).

38. a. A.A. Qureshi, F.J. Barnes, and J.W. Porter, J. Biol. Chem. 247, 6730 (1972); b. A.A. Qureshi, F.J. Barnes, E.J. Semmler, and J.W. Porter, ibid. 248, 2755 (1973).

39. L. Crombie, D.A.R. Findley, and D.A. Whiting, Chem. Commun. 1045 (1972).

40. F.J. Barnes, A.A. Qureshi, E.J. Semmler, and J.W. Porter, J. Biol. Chem. 248, 2768 (1973).

41. D.E. Gregonis and H.C. Rilling, Biochemistry 13, 1538 (1974).

42. S.C. Kushwaha, G. Suzue, C. Subbarayan, and J.W. Porter, J. Biol. Chem. 245, 4708 (1970).

43. A.A. Qureshi, A.G. Andrewes, N. Qureshi, and J.W. Porter, Arch. Biochem. Biophys. 162, 93 (1974); A.A. Qureshi, M. Kim, N. Qureshi, and J.W. Porter, ibid. 108; A.A. Qureshi, N. Qureshi, M. Kim, and J.W. Porter, ibid. 117.

44. S.C. Kushwaha, E.L. Pugh, J.K.G. Kramer, and M. Kates, Biochim. Biophys. Acta 260, 492 (1972).

45. R. Herber, B. Maudinas, J. Villoutreix, and P. Granger, Biochim. Biophys. Acta 280, 194 (1972).

46. N. Khatoon, D.E. Loeber, T.P. Toube, and B.C.L. Weedon, Chem. Commun. 996 (1972).

47. L.R.G. Valadon, R. Herber, J. Villoutreix, and B. Maudinas, Phytochemistry 12, 161 (1973).

48. A.G. Andrewes and S. Liaaen-Jensen, Acta Chem. Scand. 26, 2194 (1972).

49. a. T. Hamasaki, N. Okukado, and M. Yamaguchi, Bull. Chem.
     Soc. Japan 46, 1884 (1973);
     b. T. Ike, J. Inanaga, A. Nakano, N. Okukado, and
     M. Yamaguchi, ibid. 47, 350 (1974).
50. S.E. Moshier and D.J. Chapman, Biochem. J. 136, 395 (1973).

CHAPTER 24

Steroids Containing
Aromatic Rings

24-1. OESTRONE AND RELATED COMPOUNDS

The most important group of steroids containing an aromatic ring is a family of related, physiologically active compounds known as the oestrogens, so called because they produce oestrus in females of various species, although their occurrence is by no means limited to females.

The oestrogens are biosynthesized from nonaromatic steroidal precursors, and, apart from the biological significance of the resulting hormones, the process is of great interest as the only authenticated example of the synthesis of aromatic rings in the metazoan organism (see Chapter 1). In the majority of species, the most important compounds so formed are oestrone [1], oestradiol [2], and oestriol [3]. The pregnant mare, in addition to these compounds, produces equilin [4] and equilenin [5], whose biosynthesis seems to present a special problem. There are

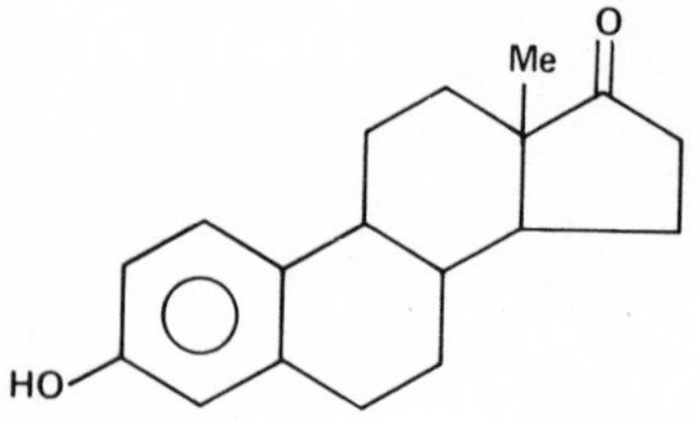

[1]

664

[2]

[3]

[4]

665

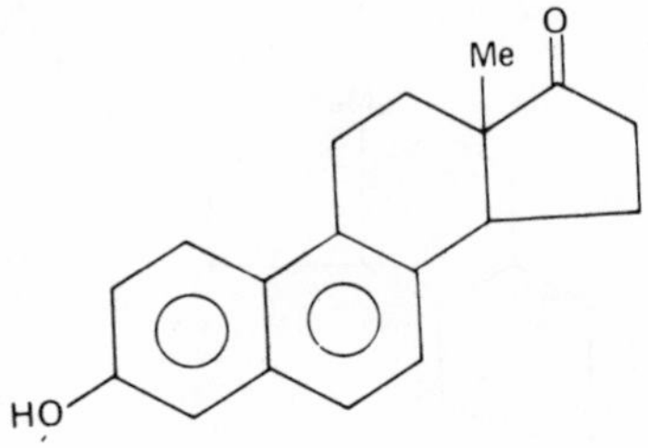

[5]

reports of the isolation of steroidal oestrogens from lower animals and even some plants (1, 2, 116) (however, see a rebuttal
by Jacobsohn and co-workers (3)).

Available information is compatible with the assumption that
in the mammalian organism the oestrogens (except possibly equilin
and equilenin) are derived from cholesterol [6], biosynthesized
in the usual fashion (for a review of steroid biosynthesis,

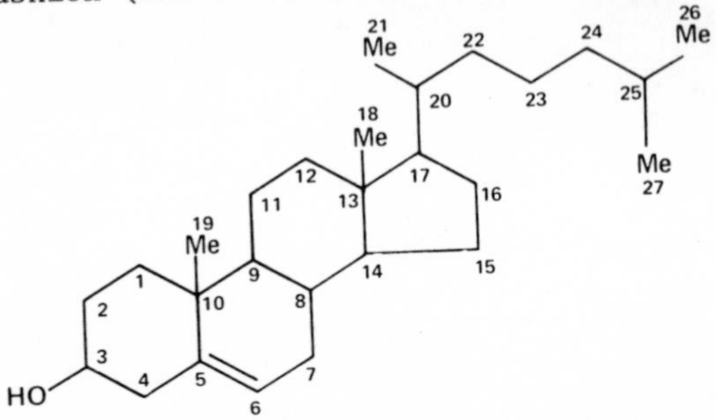

[6]

see Staple (1967) (4), and Morand and Lyall (1968) (5)). This
is suggested by the repeated findings of incorporation of radioactivity into oestrogens from cholesterol or its known precursors,
acetate, mevalonic acid [7], farnesol [8], squalene [9], et seq.
We mention a few such studies without any claim to a complete
coverage of the very extensive literature. Incorporation of
labeled acetate into oestrogens has been observed in perfused
ovaries of the sow (6), bitch (7), and rat (8), human placenta
(9), the pregnant mare (10), rabbit ovarian follicles (11),
and human ovarian preparations (12). Tagged mevalonic acid has
been shown to give rise to radioactive oestrone in human testicular homogenates (13).

In an early experiment, activity from cholesterol was found
to be incorporated into oestrone by pregnant women in a fashion

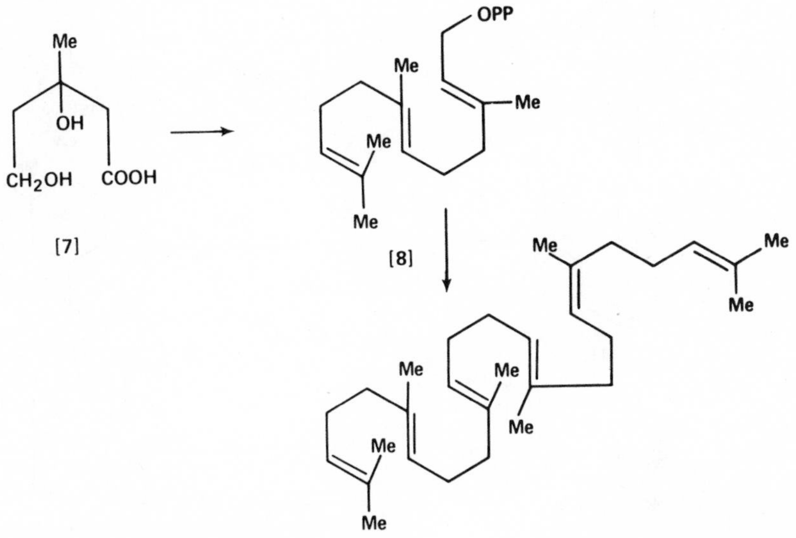

Me
OH
CH$_2$OH   COOH

[7]

OPP
Me
Me
Me
Me
Me

[8]

Me   Me
Me

Me   Me
Me
Me
Me

[9]

considered not to involve fragmentation of the molecule (14);
the same conversion has been established using human placental
slices (15, 16) and the human ovary (17). A study of the incorporation of acetate by human ovarian tissue has shown that
acetate is converted into cholesterol and oestrogens in amounts
consistent with a possible precursor role for cholesterol (18).

The conversion of cholesterol to oestrogens (sometimes
called the Δ^4, or classical pathway) apparently proceeds in
the main by way of the C_{21} compounds, pregnenolone [10], progesterone [11], 17α-hydroxyprogesterone [12], and the C_{19}-androgens,
Δ^4-androstene-3,17-dione [13], and testosterone [14]. Evidence
for the conversion of [11] to [14] via [12] and [13] has been
obtained through work with testicular (19, 20) and ovarian (21)
homogenates. The *in vitro* conversion of pregnenolone to oestradiol by bovine ovarian tissue (22), corpus luteum of pregnancy
(23), and human ovary (24) has been demonstrated; 14C-cholesterol
has been shown to be metabolized to pregnanediol (25), pregnenolone, and progesterone (26). Although the compounds existing
between cholesterol and pregnenolone are still of uncertain
structure, there is a good deal of evidence for the involvement
of 20α-hydroxycholesterol [15A] (27) and 20,22-dihydroxycholes-
terol [16A] (28, 29); other intermediates are mainly hypothetical (28, 29). Despite the obvious chemical appeal of intermediates such as [16A] (cf. the well-known cleavage of vicinal
diols to ketones by periodate or lead tetraacetate), considerable

[15A] R = H
[16A] R = OH

uncertainty over the biosynthetic course of the conversion of cholesterol to pregnenolone still remains (37).

The specific conversion of progesterone into oestrogens has been observed in ovarian tissue (30, 31) and in human and bovine luteal tissue (32), the rat (33), pig (34) and the mouse (35). Interestingly, the *in vitro* conversion of 17α-hydroxyprogesterone [12] to [13] has also been demonstrated in nonmammalian systems: in fishes, birds and frogs (36).

Similar specific incorporation of Δ4-androstene-3,17-dione [13] into oestrogens has been recorded in the rat (38), human ovarian tissue (31, 39, 40), placenta (41-44), and corpus luteum (45). Again, the formation of oestrogens from testosterone [14] has been established in the human ovary (31, 46, 47) and placenta (42, 44, 48, 49), the horse (50), rat (38, 51), and frog (52).

The results so far discussed are clearly compatible with a major pathway of oestrogen biosynthesis proceeding from acetate in normal fashion to cholesterol and thence through C_{21} and C_{19} steroids to the oestrogens.

The pathway outlined above may be considered as the classical route for the biosynthesis of oestrogens from pregnenolone; there is, however, an alternative mode of synthesis for the androgens (the Δ5 pathway), proceeding from pregnenolone [10] to via 17α-hydroxypregnenolone [15] and dehydroepiandrostenone [16] to androstenedione [13]. The first evidence for this alternative pathway was obtained in ovarian tissue (53-57); it has been confirmed in the rat (58, 59), mare (60), bitch (61), and in a corpus luteum of pregnancy (62, 64). *In vivo* studies have been made in the human (65, 66), and the existence of both pathways in various nonhuman primates has been established (67, 68). From the relative amounts of metabolites isolated from incubations, it is postulated that the Δ5 pathway predominates in the human.

There is also evidence for a further variation on this theme, in which pregnenolone [10] is converted directly into 16α-hydroxydehydroepiandrosterone [17], without the intermediacy of dehydroepiandrosterone [16] (69), and thence to oestriol [3] (70).

The experimental results outlined above have provided good evidence for the function of dehydroepiandrosterone [16] and Δ4-androstene-3, 17-dione [13] as important precursors of oestrogens in various systems. However, the ways in which these compounds are converted into the aromatic end-products is still far from clarified; we can conveniently consider the subsequent steps in two parts, viz. the one concerned with modifications made to [13] and [17] before aromatization, and that concerned directly with the aromatization step.

For the first of these, two possible interconnected routes of biosynthesis are shown below; one leads through androstendione

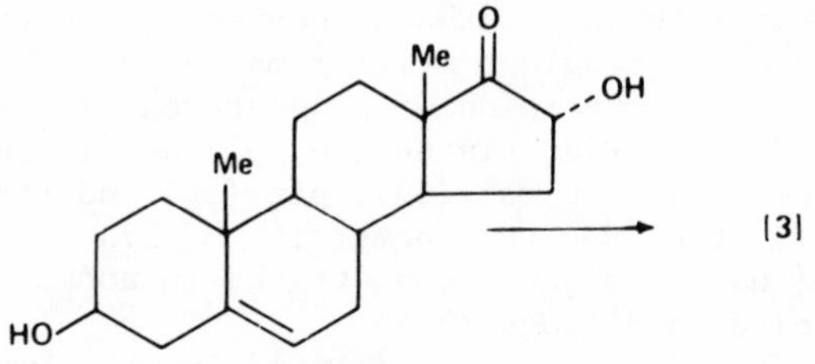

[17]

[3]

[13] and other ketonic intermediates [18], the other through 16α-hydroxydehydroepiandrosterone [17] and other 16α-hydroxy-compounds [19]. The operation of both pathways has been established in human pregnancy (70), but there is considerable debate in the literature over their relative importance in a given situation (71-73, 139).

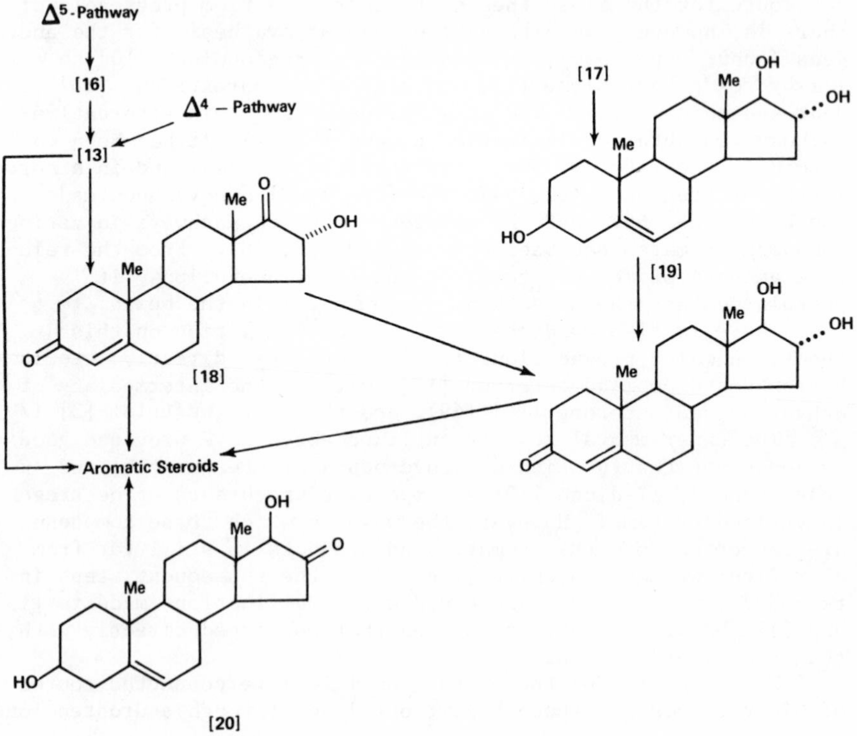

There is also substantial evidence that additional pathways exist (e.g., from 16-oxo-Δ5-androstenediol [20] to oestriol) in some situations (74). It is thus not justified to speak of one pathway, but rather of a family, or network, of alternate paths which operate together in many instances.

The actual aromatization reaction is known to involve hydroxylation at C-19. This was first suggested by A.S. Meyer

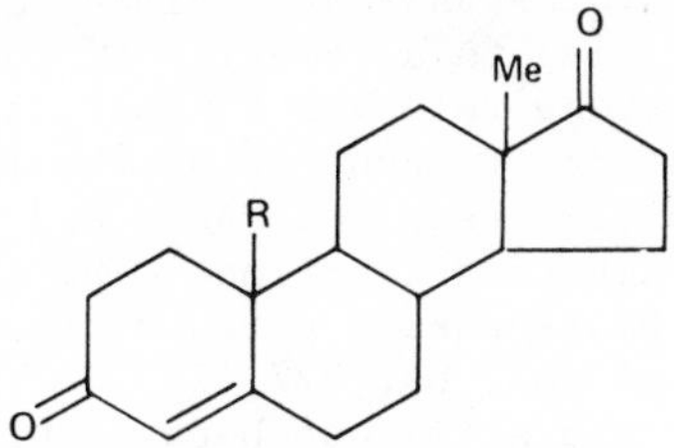

[21] R = CH₂OH
[22] R = CHO

(75, 76), who observed that the conversion of 19-hydroxy-Δ4-androstene-3,17-dione [21] into oestrone by placental tissue proceeds much more readily than that of the parent androstene-dione. Subsequently, Ryan (48) also found that the 19-hydroxy compound was just as readily aromatized as androstenedione or testosterone by the washed microsome fraction from human placenta. Since hydroxylation at C-19 of steroids has been observed in the adrenals (76) and in the fungus *Corticium sasakii* (77), [21] is a biochemically reasonable intermediate. Further evidence for the role of [21] as an intermediate in oestrogen biosynthesis was obtained by Longchampt et al. (78): incubation of andro-stenedione-4-14C with a microsome preparation from human placenta, similar to the one used by Ryan (48), yielded active oestrone together with another compound identified as [21]; furthermore, incubation of this compound with the microsomes gave oestrone. During an *in vivo* study of the metabolism of 14C-testosterone by perfused full-term placentas, Charrcan (79) et al. isolated radioactive [21] as well as other oestrogens, again indicating the probable intermediacy of the compound. A direct conversion of [21] to oestrone and oestradiol by human corpus luteum has been demonstrated by Fahmy and co-workers (80). The hydroxyla-tion reaction has been studied by Wilcox and Engel (43), who presented kinetic evidence in favor of the 19-hydroxy compound as an intermediate in the aromatization brought about by the pla-cental microsomal system. Other work, all of it supporting the intermediacy of [21], is due to Breuer (44), Kelly and Ranucci (81) (an *in vivo* study in the human), Hayano et al. (82), and

Axelrod and Goldzieher (83).   The conflicting results obtained
by Hollander (84), which seem to indicate that [21] is not
an intermediate, appear to require some alternative explanation.
     The next step in the sequence has often been suggested to
be a further oxidation at C-19, and the intermediary role of
the resulting 19-oxo compounds in oestrogen biosynthesis has
been demonstrated in placental microsomes (85); thus, joint
incubation of labeled 19-hydroxyandrostenedione and unlabeled
19-oxoandrostenedione [22] led to a lowering of the incorporation of activity into oestrone relative to that in the absence
of [22], and an incorporation of about 3% of the label into [22].
The conversion of [21] to [22] requires NADPH and oxygen, as would
be expected for such an oxidation reaction.   The aldehyde [22]
has been identified as a biosynthetic product of oestrogen-forming
tissue in a number of cases (83, 85, 127, 132).   The final steps
from the 19-oxo steroids have still not been completely clarified,
although in summary the events leading from the C-19 oxygenated
steroids to the oestrogens are chemically unexceptional; the
C-10 methyl group is eventually lost as formaldehyde (44) or
formic acid (86, 87).   However, the oxydation of the 19-aldehyde
to -COOH seems unlikely under physiological conditions, since the
10β-carboxy compound is reported to be a poor precursor of the
aromatic steroids, and to be decarboxylated to the 19-norsteroid
by the placental system (88).   Aromatization of ring A, which is
presumed to involve oxygenation at either C-1 or C-2 before the
introduction of a double bond, requires NADPH and oxygen, which
are the normal cofactors for an enzymic hydroxylation (88).   It
has been established that the overall reaction from androstenedione
to oestrogen requires 3 moles of oxygen, that is, that three
oxidative steps are involved (140).   It is a fairly simple exer-
cise to develop acceptable chemical mechanisms for these changes
based upon the facts outlined above, or upon the suspected course
of chemical or microbial aromatizations of the steroid nucleus
that are discussed below.
     There are three agencies by which ring A of the steroid
nucleus can be readily aromatized.   Some direct chemical methods
have been investigated, and those which seem to throw light
upon possible biochemical processes are briefly reviewed.
Microbiological systems bring about aromatization in two general
ways depending upon the substrate.   Thus the 19-hydroxy, 19-oxo,
and 19-nor-steroids are converted into oestrogens, whereas the
10-methyl steroids are usually metabolized to the 9,10-seco-3-
hydroxy compounds (see below); finally, the mammalian system is
capable of aromatizing all three classes of steroid.

The chemical aromatization of 19-oxygenated steroids has
been investigated by several authors: for example, treatment of
[23], derived from the cardiac glycoside ouabain, with aqueous
alkali gave rise to formaldehyde and a phenolic product [24]
isolated as its methyl ether (89). This interconversion, and
other similar purely chemical aromatizations (90, 91), have been
rationalized in terms of mechanisms involving a dienone intermediate [25] and loss of the 19-oxo function by a simple reverse
aldolization.

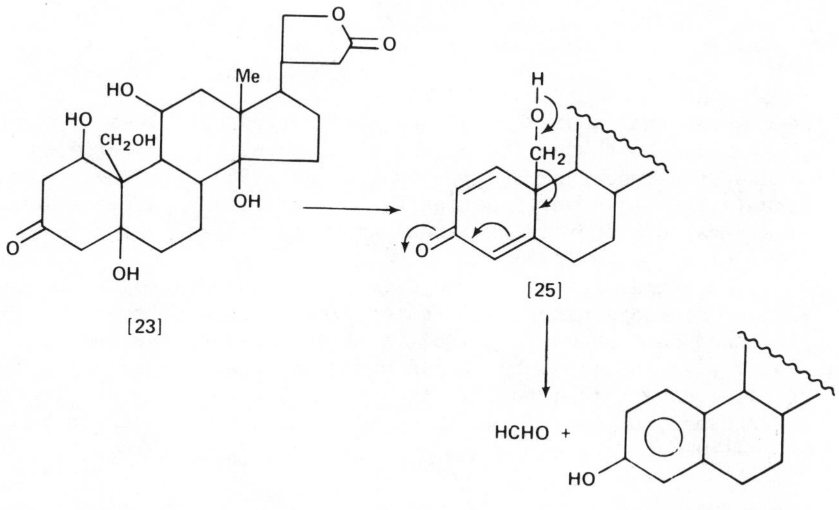

Several microbial systems are capable of bringing about the
aromatization of C-19 oxygenated steroids, and again studies of
these enzymic processes have given some insight into the situation in the mammalian organism. *Nocardia restricta*(92, 101)and
Pseudomonas species (93), both as whole cell and as purified enzyme
systems, will metabolize 19-hydroxyandrostenedione to oestrone,
and an investigation (94) of the stereochemical aspects of the
conversion, using substrate with tritium label at either 1α or 1β,
has shown that the 1α-hydrogen is lost during the conversion.
This loss indicates that the aromatization in microbiological
systems takes place by a mechanisms different from that in the
human placenta (where the 1β-proton is lost, see below). The
mechanism may well involve a C-1,2-dehydrogenation, since it has

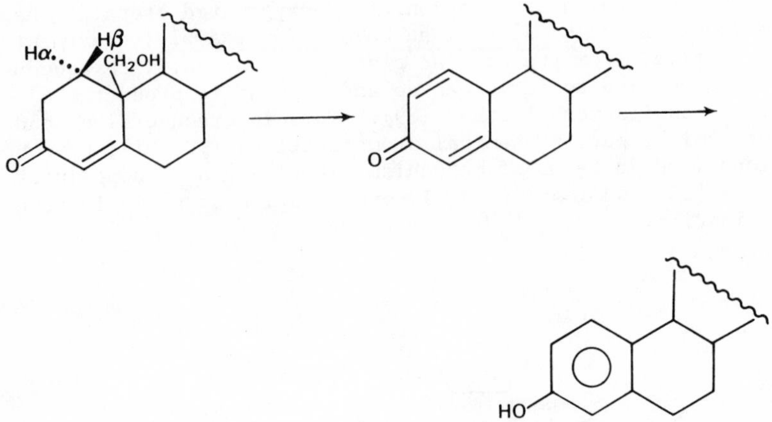

been shown that androstenedione is converted with loss of the
1α-hydrogen to the corresponding Δ^1-compound by the purified
Δ^1-dehydrogenase from *Bacillus sphaericus* (95) and *B. cyclooxi-
dans* (94). It is worth noting that bacterial dehydrogenations
of several C$_{19}$ steroids has been shown to proceed via a 1α-2β-
trans elimination (128).

The aromatization of the C-19 oxo compound strophanthidin
[26] by microorganisms has been studied by Sih and co-workers
(96), who have shown that [26] is metabolized to the two products, [27] and [28], by *N. restricta* and that the organism can
readily convert [27] to [28] in high yield; whether this interconversion proceeds via a 1,2-dehydrogenation has not yet been
established.

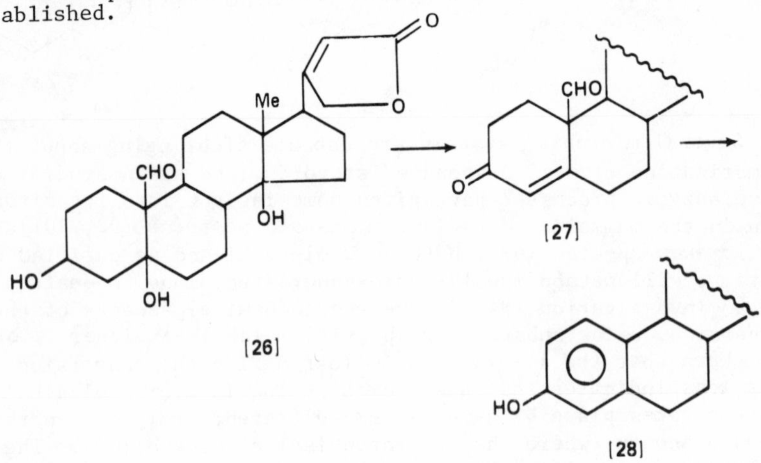

Several 19-nor steroids can be converted into oestrogens by microorganisms (5); for example, the conversion of the 19-nor compounds [29] and [30] to oestrone, and of [31] to equilin, takes place, in *Pseudomonas testosteroni* (97) and *Corynebacterium simplex* (98, 99) presumably again via 1,2-dehydrogenation and

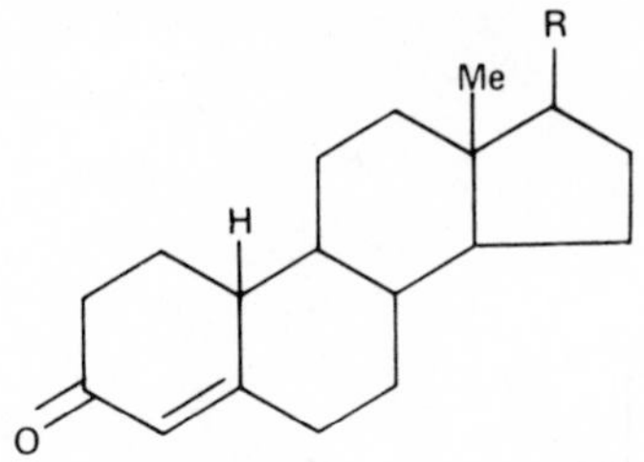

[29] R = O   [30] R = OH

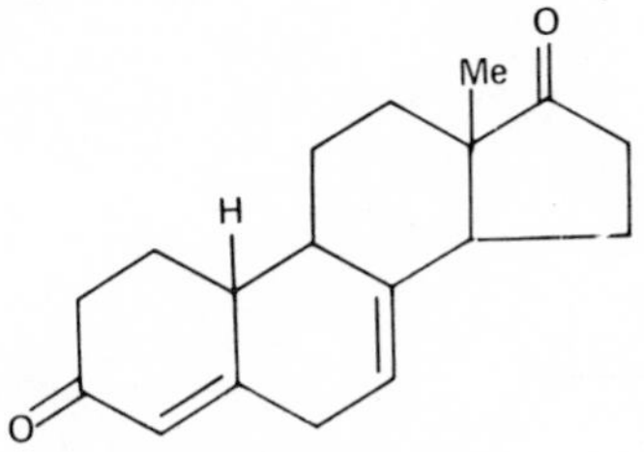

|31|

enolization. Indeed, the stereochemistry of the C-1,2 dehydrogenation during the interconversion of [29] into oestrone has been shown to involve the 1α and 2β hydrogens, that is, a *trans* elimination (129). This stereochemistry is the same as for the C$_{19}$ steroids. A study of the aromatization of the 9-α-hydroxy analog of [29] by *Arthrobacter simplex* demonstrated a conversion both to 9α-hydroxyoestrone and to 9-keto-9,10-seco-oestrone (see below).

The third well-established pathway for the conversion of nonaromatic steroids to aromatic metabolites by microorganisms leads to the aromatization of ring A with concomitant fission of the 9,10 bond. We include a short discussion of the reaction here, even though it results in a drastic change of the steroid skeleton, yielding products which are not closely related to oestrogens; there are, however, obvious similarities in the enzymic system involved, and some insight into the operation of

the mammalian system can be gained from its study. This aromatization proceeds from 3-keto steroids through C-1,2-dehydrogena-
tion and 9α-hydroxylation to a 9,10-seco phenol [34]. Thus,

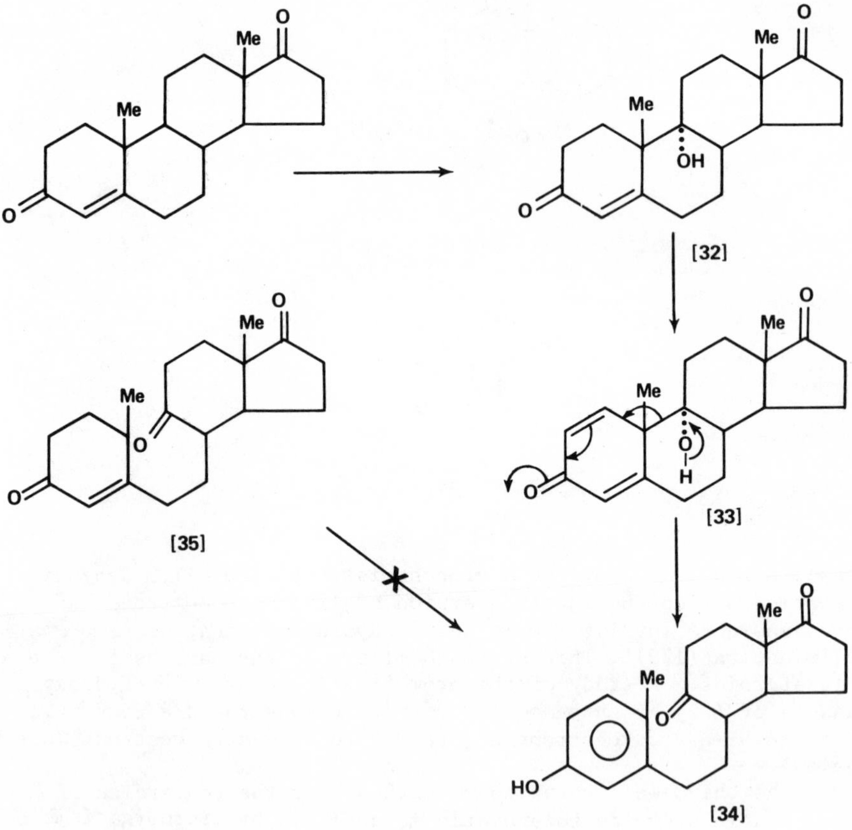

incubation of 9α-hydroxy-4-androstenedione [32] with a purified 1,2-dehydrogenase from *Nocardia restricta* (one of several micro-organisms capable (104) of oxidizing a variety of steroids by this pathway) produced 9,10-seco-3-hydroxy-1,3,5-androstatriene-9, 17-dione [34] in high yield (101), whereas incubation of 9,10-seco-4-androstene-3, 9,17-trione [35] with the same system produced no conversion.

The aromatization thus presumably takes place via 1,2-dehydrogenation to [33] and a vinylogous de-aldolization, quite analogous to the loss of formaldehyde from the 19-oxo steroids. Indeed the enzyme preparation, both from *N. restricta* (92, 101) and *Pseudomonas* spp. (93), which brings about this oxidative degradation, is capable of converting 19-hydroxy-androstenedione to oestrone.

Conversions of this type have been observed in *Pseudomonas* (93) and *Arthrobacter* spp. (93, 100), *N. restricta* (102), and *Mycobacterium smegmatis* (103).

The most significant aromatization is of course that brought about by the mammalian system, which converts the androgens into oestrogens via their 19-oxygenated analogs. The mechanism of this aromatization appears to differ from those of the chemical and microbial conversions that have already been discussed. In 1962 Morato and co-workers (95) showed that when tritiated androstenedione, labeled either at 1α or 1β, was incubated with the microsome system from human placenta together with an NADPH-generating system, conversion to oestrone was accompanied by specific loss of the 1β-H^3 label. The authors proposed two possible mechanisms to account for this observation: In the first (A), a C-2 proton is lost by enolization of the C-3 carbonyl group (models show that the loss of the 2β-axial proton would be favored for steric reasons) and subsequently the C-1β proton is eliminated in a concerted reaction which also removes the oxygenated C-19 as formaldehyde. In this case the C-1 proton would be lost either as a hydride ion or indirectly after its replacement by an activating group. The alternative pathway (B) assumes the formation of a C-1,2 double bond after replacement of the 1β-proton by a leaving group, and subsequent loss of the C-19 function by a reverse aldol reaction similar to the reaction in the microbial system.

In a more recent investigation by Townsby and Brodie (105), the aromatization of 19-nor steroids by the same microsomal system was studied in an attempt to distinguish between the above possibilities. It was established that both the C_{18} and C_{19} compounds are converted into oestrogens in a very similar way. The stereochemistry and cofactor requirements for the aromatization of 19-nor-androstene-3,17-dione [36] and

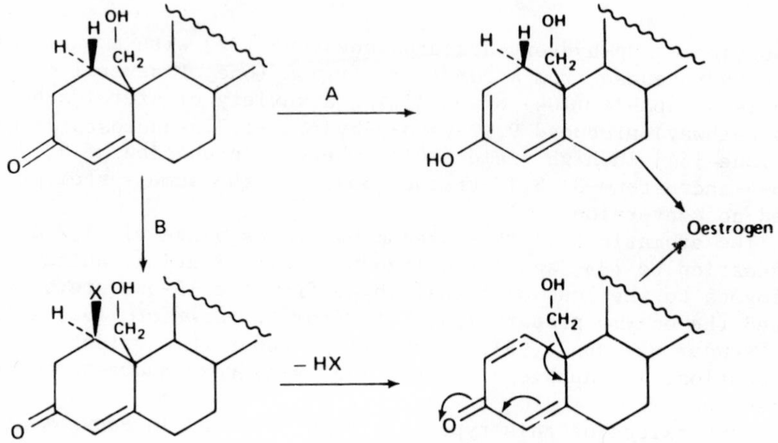

androst-4-ene-3,17-dione [37] were compared, and in each case
tritium-labeling studies showed that the 1β-hydrogen was lost,
and that again NADPH and oxygen were necessary.[+] These facts

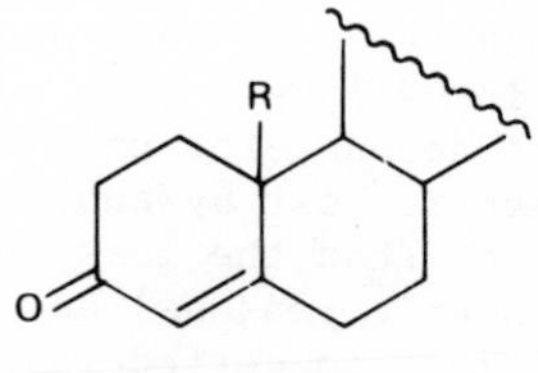

[37] R = Me
[36] R = H

[+]The assumed parallelism between the oxidative aromatization of
the C-19 and C-18 compounds may not be valid, since Meigs and Ryan
(109) have shown that the oxidase enzymes involved are not the
same; aromatization of the C-18 compounds is inhibited by CO,
whereas that of the C-19 homologs is not.

[38]

clearly indicate the probability that ring hydroxylation is a
necessary prelude to aromatization. Since the aromatization of
[36] has the same stereochemical and cofactor requirements as
that of [37] and its C-19 oxygenated analogs, it seems likely that
oxygenation occurs at C-1 (105) or possibly C-2. On the other
hand, although 1β-hydroxy-19-nor-androstenedione [38] has been
isolated from an incubation of [36] with placental microsomes
(105, 106), and has been shown to be formed through replacement
of the C-1β hydrogen by OH, the compound is inactive as an
oestrogen precursor, as are the 2α and 2β-hydroxy compounds
(107). It seems unlikely, therefore, that a *free* hydroxy
compound will be found as an intermediate (107). Similar tritium
(and deuterium (107)) labeling studies in the C-19 series have
established that the loss of hydrogen from C-2 is again highly
stereospecific; here, too, it is the β-hydrogen that is removed
(108). This means that there is an effective *cis* dehydrogenation
between C-1 and C-2. In addition, Fishman et al. (108) showed
that the loss of tritium from C-2β was not reversible, which is
difficult to reconcile with the normal (presumably reversible)
keto-enol tautomerism implied by mechanism (A) above; it seems
likely, therefore, that enolization and aromatization must occur
as one concerted process if mechanism (A) applies. Confirmation
of the general correctness of the proposed mechanisms has been
provided by Milewich and Axelrod (130) working with placental
microsomes from baboons, and Osawa and Spaeth (131) using human
placental microsomes. In both experiments, variously labeled
1,2-^{3}H-[14] was used, and the resulting loss of tritium on
aromatization (70 and 75%) agreed well with expectation (69%).
The distribution of label in the substrate was not uniform, but
the ^{3}H was found predominantly at 1β and 2β. No distinction
between pathways (A) and (B) can be made on the basis of these
experiments. The roles of oxygen and NADPH are still unclear
in these final steps, although they do seem to rule out the
possibility of the direct *cis* dehydrogenation.

The foregoing discussion will have shown that the details
of the aromatization process are far from being completely
understood; multiple possibilities for parallel reactions exist,
and there may not be an exclusive preference for one branch of

this network even in one organsims. The points at issue are:
(1) The oxidation status of C-19 at the time of its elimination.
It could be present as $-CH_2OH$, $-CHO$, or $-COOH$ at that moment;
chemically, the reaction would then be a vinylogous de-aldolization, β-dicarbonyl cleavage, or decarboxylation of a β-keto
acid, all reactions for which ample precedent exists. Attempts
to establish experimentally the oxidation status of the
resulting C-1 fragment are vitiated by the ease with which,
for example, formaldehyde and formic acid are interconverted
by enzymes, and conclusions derived from the ease or extent of
utilization may suffer from possible influence of permeability
differences. (2) Similarly unclear, and perhaps not uniquely
defined, is the time sequence of the loss of C-19 and the desat-
uration of ring A. All three types of reaction that could lead
to the loss of C-19 could take place at the level of the Δ^4-3-
one, or (perhaps more readily?) of a $\Delta^{1,4}$-3-dienone. Again, two
possibilities, loss of C-19 preceding or following establishment
of Δ^1, are possible, and both may occur under given circumstances.
(3) The enzymatic reaction which produces the Δ^1 double bond in
the mammalian system is peculiar and not fully understood. The
requirement for O_2 and NADPH precludes a simple dehydrogenation
analogous to, for example, the conversion of succinic to fumaric
acids, and suggests an oxygenation-elimination sequence, details
of which remain to be established.

24-2. EQUILIN AND EQUILENIN

It is evident that the biosynthesis of equilin [39] and
equilenin [40] in the pregnant mare proceeds by a pathway quite
different from that of oestrone in the same animal. Thus, in
some early experiments, Savard et al. (10) observed a much lower
incorporation of label from acetate into these two steroids than
into oestrogens formed simultaneously, and experiments *in vivo*
have shown that acetate (115) but not cholesterol (114) is incorporated into the steroids by the pregnant mare (133). Similarly,
other established precursors of the more usual oestrogens cannot
be converted into equilin and equilenin; attempts to incorporate
testosterone and 19-hydroxytestosterone (112, 50) and androstenedione (117, 122) have been unsuccessful; furthermore, neither
oestrone nor 17β-oestradiol (110, 111) is incorporated. These
findings indicate that the pathway to equilin and equilenin
probably branches off before cholesterol from that of the
normal oestrogens. There is good evidence that the pathways
diverge prior to the formation of squalene (138).

The biochemical conversion of equilin to equilenin has been established; an enzyme from rat liver has been shown to transform [39] into [40] (113), as have porcine adrenocortical slices (118). There is also a clear indication that the same enzyme exists in equine placental tissue, since this system has been shown to convert 3β-hydroxyandrosta-5,7-diene-17-one [41] into a mixture of both [39] and [40] (119). The interconversion of

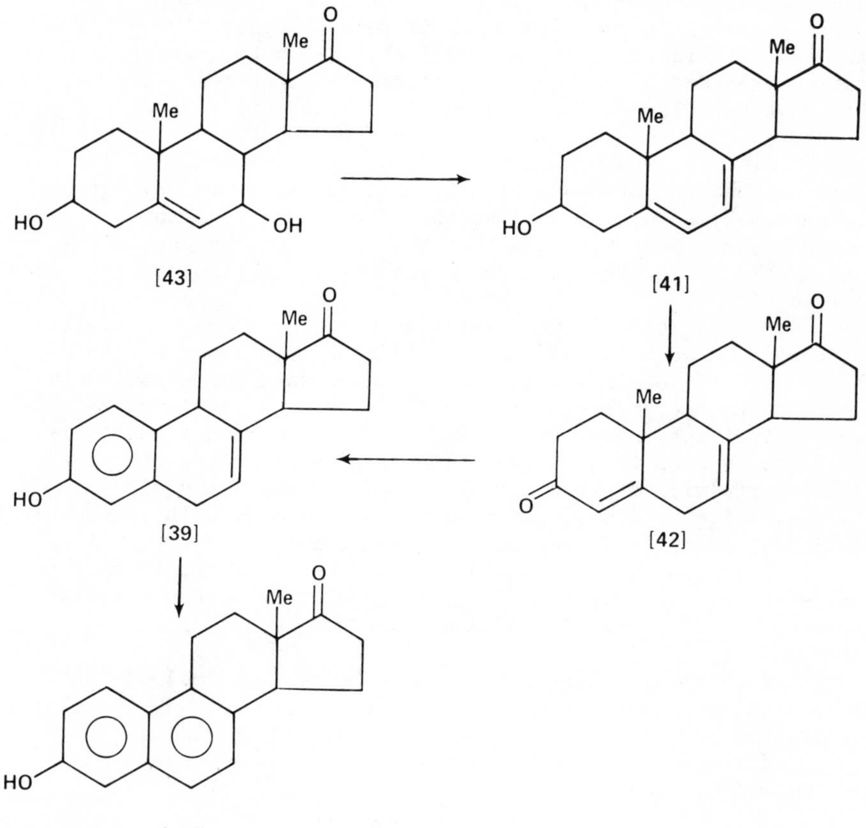

[43]

[41]

[42]

[39]

[40]

these compounds has also been demonstrated in perfused human
placenta (120); [41] and [42], as well as 3β,7α-dihydroxyandrost-
5-ene-17-one [43], were all converted into [39] and [40] and
their 17β-dihydro analogs. However, *in vivo* (122) and *in vitro*
(117) experiments have shown that [43] is not so metabolized
by the mare.

The aromatization of some Δ^7-C_{19}-steroids to Δ^7-oestrogens
has been studied *in vitro* by Givner et al. (121) using a
human placental preparation. Their results generally confirm
the earlier findings; it was found that Δ^7-oestrogens are not
formed from androstenedione, androstenolone, or testosterone,
but that steroids having an aromatic B ring were formed from
Δ^7-androstenolone and Δ^7-testosterone, that is, that the placental preparation was unable to introduce the Δ^7-bond. It is
evident that the presence of a Δ^7-bond (or a 7α-hydroxyl group)
is necessary before metabolism to a steroid having ring B aromatic becomes possible, and that this function must be introduced at a stage preceding those concerned with the aromatization
of ring A.

It is not clear what significance, if any, the conversion
of compounds such as [41] into equilin and equilenin by the
human placenta may have; the results discussed above obviously
demonstrate the presence of some interesting aromatizing enzymes,
but whether they have any relevance to the *in vivo* situation in
the mare is open to question. The experiments by Starka et al.
(120) certainly offer an attractive hypothesis: the allylic
hydroxylation of a Δ^5-C_{19}-steroid to yield [43], with subsequent
modification to [42], followed by aromatization of ring A in a
manner analogous to the sequence established for the formation
of oestrone. However, the early steps of this sequence are at
variance with the results of Ainsworth and Ryan (117), and
Bhavnani et al. (122).

Microbial aromatization of steroids to equilin [39] has
been studied by several groups; the results generally parallel
those obtained in the mammalian organism. Kluepfel and Vezina
(134), for example, found that several organisms could transform
androsta-1,4,7-triene-3,17-dione into both [39] and equilenin
[40], but that the same organisms could not convert androsta-1,4-
diene-3,17-dione into oestrone. However, 19-hydroxy-androsta-
4,7-diene-3,17-dione is transformed into [39] by several microorganisms (135); the presence of the Δ^7-bond in the substrate
is again seen to be necessary for the conversion to equilin and
equilenin.

24-3. NONOESTROGENIC AROMATIC STEROIDS

There are some other examples of partially aromatized steroids.

The alkaloid veratramine [44] from *Veratrum viride* is a modified
C-nor-D-homo-steroid; although nothing is known about the formation of its aromatic ring, it may be significant that jervine
[45], the chief nonaromatic alkaloid of *V. viride*, has the same
modified skeleton. Compounds with the same ring system as veratramine and jervine can be formed *in vitro* by base-catalyzed
transformations of normal steroids (123), and the changes involved
in these rearrangements may be analogous to the ones by which the
skeleton of the two alkaloids is formed.

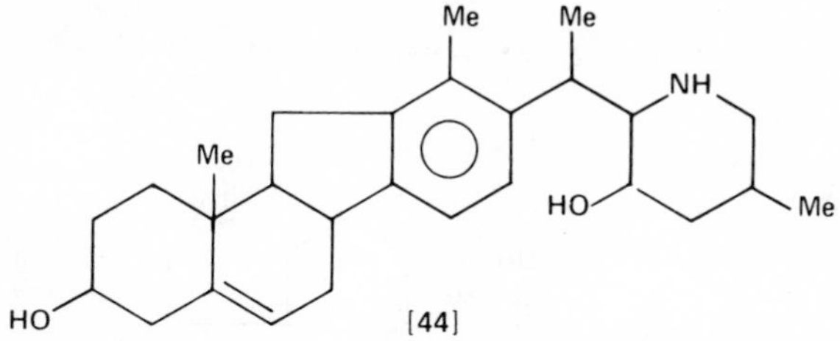

[44]

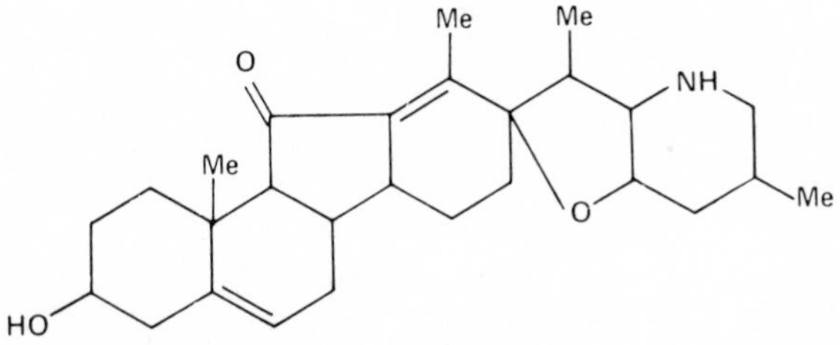

[45]

The biosynthesis of viridin [46], an antifungal substance
isolated from *Gliocladium virens*, has been elucidated by Grove et
al. (124). The structure of [46] would be compatible with its
being either a modified steroid, or a diterpenoid related to
cassaic acid. Decision in favor of the former possibility was
obtained by degradation of [46] biosynthesized from 2-14C-meva-
lonic acid; the distribution of label in the tetracarboxylic acid
formed during this degradation was found to be that expected for
a steroid resulting in the usual way from squalene, that is, by
tail-to-tail junction of two farnesyl residues. If [46] were an
aromatized diterpenoid, a different pattern would be expected
(see Scheme).

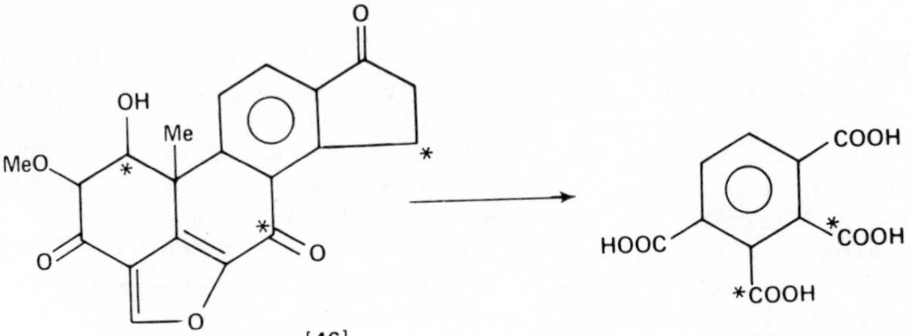

[46]

*C from 2 – ^{14}C – Mevalonic Acid

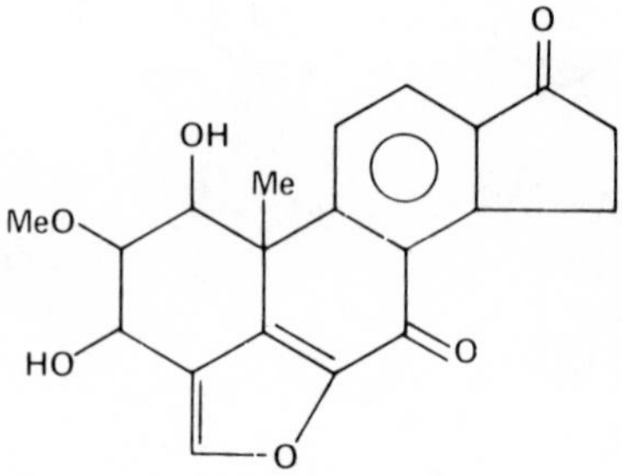

[47]

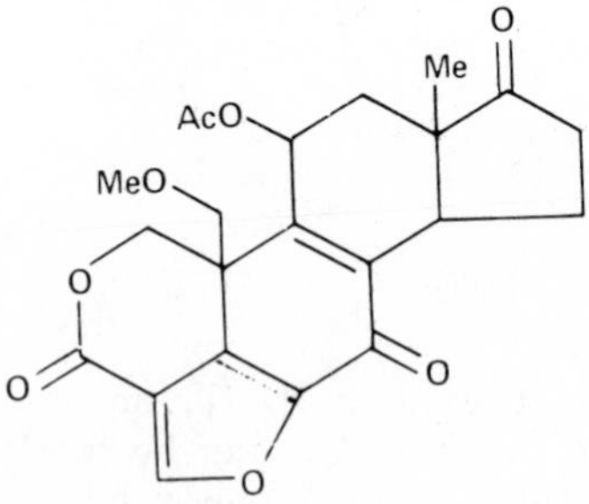

[48]

Trichoderma viride contains viridiol [47], the dihydro derivative of [46] (125); a related, though nonaromatic, metabolite,
wortmannin [48], has been isolated from *Penicillium wortmannii*
(126).   The structure of [48], with its vinylogous β-acetoxyketone
grouping, suggests that it may be related to some of the intermediates in the formation of the aromatic ring of viridin.
     Several interesting steroidal derivatives with an aromatic
D ring have been isolated from the insect-repellent plant

Nicandra physaloides (136, 137). The four metabolites, which differ only in the side-chain, are exemplified by nicandrenone [49].

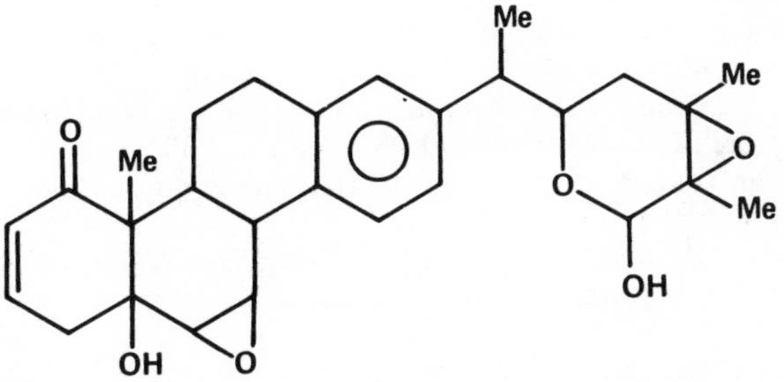

[49]

REFERENCES

1. R.I. Dorfman and F. Ungar, Metabolism of Steroid Hormones, Academic, New York, 1965.
2. El S. Amin, O. Awad, M. Abd El Samad, and M.N. Iskander, Phytochemistry 8, 295 (1969).
3. G.M. Jacobsohn, M.J. Frey, and R.B. Hochberg, Steroids 6, 93 (1965).
4. E. Staple, in Biogenesis of Natural Compounds, P. Bernfeld, Ed., 2nd ed., Pergamon, Oxford, 1967, p. 207.
5. P. Morand and J. Lyall, Chem. Rev. 68, 85 (1968).
6. N.T. Werthessen, E. Schwenk, and C. Baker, Science 117, 380 (1953).
7. J.L. Rabinowitz, Arch. Biochem. Biophys. 64, 285 (1956).
8. B.F. Rice and A. Segaloff, Steroids 7, 367 (1966).
9. M. Levitz, G.P. Condon, and J. Dancis, Fed. Proc. 14, 245 (1955).
10. K. Savard, K. Andrec, B.W.L. Brooksbank, C. Reyneri, R.I. Dorfman, R.D.H. Heard, R. Jacob, and S.S. Solomon, J. Biol. Chem. 231, 765 (1958).
11. D. Gospodarowicz, Acta Endocrinol. 47, 293 (1964).
12. K.J. Ryan and O.W. Smith, Rec. Prog. Hormone Res. 21, 367 (1965).
13. J.L. Rabinowitz and J.B. Rayland, Fed. Proc. 17, 293 (1958).
14. H. Werbin, J. Plotz, G.V. LeRoy, and E.M. Davis, J. Amer. Chem. Soc. 79, 1012 (1957).
15. M. Suzuki, K. Takahashi, M. Hirano, and K. Shindo, J. Exp. Med. 76, 89 (1962); Chem. Abstr. 57, 9139i (1962).
16. G. Telegdy, J.W. Weeks, N. Wiqvist, and E. Diczfalusy, Acta Endocrinol. 63, 105, 119 (1970).
17. K.J. Ryan and O.W. Smith, J. Biol. Chem. 236, 2204 (1961).

18. K.J. Ryan and O.W. Smith, J. Biol. Chem. 236, 705 (1961).
19. W.S. Lynn, Fed. Proc. 15, 305 (1956).
20. W.R. Slaunwhite and L.T. Samuels, J. Biol. Chem. 220, 341 (1956).
21. S. Solomon, R. Vande Wiele, and S. Lieberman, J. Amer. Chem. Soc. 78. 5453 (1956).
22. W.R. Miller and C.W. Turner, Steroids 2, 657 (1963).
23. K.J. Ryan, Acta Endocrinol. 44, 81 (1963).
24. K.J. Ryan and Z. Petro, J. Clin. Endocrinol. Metab. 26, 46 (1966).
25. K. Bloch, J. Biol. Chem. 157, 661 (1945).
26. N. Saba, O. Hechter, and D. Stone, J. Amer. Chem. Soc. 76, 3862 (1954).
27. K.D. Roberts, L. Bandy, and S. Lieberman, Biochemistry 8, 1259 (1969).
28. G. Constantopoulos, P.H. Satoh, and T.T. Tchen, Biochem. Biophys. Res. Commun. 8, 50 (1962).
29. K. Shimizu, M. Gut, and R.I. Dorfman, J. Biol. Chem. 237, 699 (1962).
30. K.J. Ryan and O.W. Smith, J. Biol. Chem. 236, 710 (1961).
31. P. Preumont, K.J. Ryan, E.V. Younglai, and S. Solomon, J. Clin. Endocrinol. Metab. 29, 1394 (1969).
32. M.L. Sweat, D.L. Berliner, M.J. Bryson, C. Nabors, J. Haskell, and E.G. Holstrom, Biochem. Biophys. Acta 40, 289 (1960).
33. J. Weisz and C.W. Lloyd, Endocrinology 77, 735 (1965).
34. P. Preumont, I.D. Cooke, and K.J. Ryan, Acta Endocrinol. 62, 449 (1969).
35. G.P. Vinson, J.K. Norymberski, and I.C. Jones, J. Endocrinol. 25, 557 (1963).
36. G.S. Rai, H. Brewer, and E. Witschi, Gen. Comp. Endocrinol. 12, 119 (1969) and refs. there.
37. S. Lieberman, L. Bandy, V. Lippman, and K.D. Roberts, Biochem. Biophys. Res. Commun. 34, 367 (1969).
38. R.E. Oakey and S.R. Stitch, J. Endocrinol. 39, 189 (1967); Biochem. J. 89, 57P (1963).
39. J. Kaiser, Acta Endocrinol. 47, 676 (1964).
40. O.W. Smith and K.J. Ryan, Endocrinology 69, 869 (1961).
41. E. Bolté, S. Mancuso, G. Eriksson, N. Wiqvist, and E. Diczfalusy, Acta Endocrinol. 45, 535 (1964).
42. L. Cédard and R. Knupper, Steroids 6, 307 (1965).
43. R.B. Wilcox and L.L. Engel, Steroids Supp. II, 249 (1965).
44. H. Brewer and P. Grill, Z. Physiol. Chem. 324, 254 (1961).
45. R.B. Arceo and K.J. Ryan, Acta Endocrinol. 56, 225 (1967).
46. B. Baggett, L.L. Engel, K. Savard, and R.I. Dorfman, J. Biol. Chem. 221, 931 (1956).
47. H.H. Wotiz, J.W. Davis, H.M. Lemon and M. Gut, J. Biol. Chem. 222, 487 (1956).

48. J. Ryan, J. Biol. Chem. $\underline{234}$, 268 (1959).
49. S. Sybulski, Amer. J. Obstetr. Gynecol. $\underline{105}$, 1055 (1969).
50. L. Stárka, J. Breuer, and H. Breuer, Naturwissenschaften $\underline{52}$, 540 (1965).
51. S.R. Stitch, R.E. Oakey, and S.S. Eccles, Biochem. J. $\underline{88}$, 70 (1968).
52. R. Ozon and H. Brewer, Z. Physiol. Chem. $\underline{337}$, 61 (1964).
53. L.R. Axelrod and J.W. Goldzieher, J. Clin. Endocrinol. Metab. $\underline{22}$, 431 (1962).
54. J. Ryan and O.W. Smith, J. Biol. Chem. $\underline{236}$, 2207 (1961).
55. V.B. Mahesh and R.B. Greenblatt, J. Clin. Endocrinol. Metab. $\underline{22}$, 441 (1962).
56. M.W. Noall, F. Alexander, and M.W. Allen, Biochem. Biophys. Acta $\underline{59}$, 520 (1962).
57. R.V. Short, J. Endocrinol. $\underline{23}$, 277 (1961).
58. M. Kalvert and E. Bloch, Endocrinology $\underline{82}$, 1021 (1968).
59. T. Nakao, M. Inaba, and Y. Omori, in Biogenesis and Action of Steroid Hormones, R.I. Dorfman, K. Yamasaki, and M. Dorfman, Eds., Geron-X, Inc., California, 1968, p. 366.
60. C.D. West and A.H. Naville, Biochemistry $\underline{1}$, 645 (1962).
61. A. Aakvaag and K.B. Eik-Nes, Biochim. Biophys. Acta $\underline{111}$, 273 (1965).
62. D. Fahmy, K. Griffiths, and A.C. Turnbull, Biochem. J. $\underline{107}$, 725 (1968).
63. E.E. Baulieu and F. Dray, J. Clin. Endocrinol. Metab. $\underline{23}$, 1298 (1963).
64. P.K. Siiteri and P.C. MacDonald, Steroids $\underline{2}$, 713 (1963).
65. V.B. Mahesh and R.B. Greenblatt, Acta Endocrinol. $\underline{41}$, 400 (1962).
66. D.L. Loriaux and M.W. Noall, Steroids $\underline{13}$, 143 (1969).
67. L. Ainsworth, M. Daenen, and K.J. Ryan, Endocrinology $\underline{84}$, 1421 (1969).
68. L. Ainsworth and K.J. Ryan, Steroids $\underline{14}$, 301 (1969).
69. M.M. Shahwan, R.E. Oakey, and S.R. Stitch, Acta Endocrinol. $\underline{60}$, 491 (1969); Proc. Biochem. Soc. $\underline{110}$, 30P (1968).
70. M.A. Kirschner, N. Wiqvist, and E. Diczfalusy, Acta Endocrinol. $\underline{53}$, 584 (1966).
71. S. Dell'Acqua, S. Mancuso, G. Eriksson, J.L. Ruse, S. Solomon, and E. Diczfalusy, Acta Endocrinol. $\underline{55}$, 401 (1967).
72. T. Nakayama, K. Arai, K. Satoh, T. Yanaihara, and K. Nagatomi, Endocrinol. Jap. $\underline{15}$, 255 (1968).
73. E.V. Younglai and S. Solomon, Biochemistry $\underline{7}$, 1881 (1968).
74. S. Mancuso, G. Benagiano, S. Dell'Acqua, M. Shapiro, N. Wiqvist, and E. Diczfalusy, Acta Endocrinol. $\underline{57}$, 208 (1968).
75. A.S. Meyer, Biochem. Biophys. Acta $\underline{17}$, 441 (1955).
76. A.S. Meyer, Experientia $\underline{11}$, 99 (1955).

77. M. Nishikawa and H. Hagiwara, Chem. Pharm. Bull. 6, 226 (1958); Bull. Agr. Chem. Soc. Japan 22, 212 (1958).
78. J.E. Longchampt, C. Gual, M. Ehrenstein, and R.I. Dorfman, Endocrinology 66, 416 (1960).
79. E. Charreau, W. Jung, L. Loring, and C. Villee, Steroids 12, 29 (1968).
80. D. Fahmy, K. Griffiths, and A.C. Turnbull, J. Endocrinol. 44, 133 (1969).
81. W.G. Kelly and S.R. Ranucci, J. Clin. Endocrinol. Metab. 28, 1401 (1958).
82. M. Hayano, J. Longchampt, W. Kelly, C. Gual, and R.I. Dorfman, Acta Endocrinol. Supp. 51, 699 (1960).
83. L.R. Axelrod and J.W. Goldzieher, J. Clin. Endocrinol. Metab. 22, 431 (1962).
84. N. Hollander, Endocrinology 71, 723 (1962).
85. M. Akhtar and S.J.M. Skinner, Biochem. J. 109, 318 (1968).
86. L.R. Axelrod, C. Matthijssen, P.N. Rao, and J.W. Goldzieher, Acta Endocrinol. 48, 383 (1965).
87. S.J.M. Skinner and A. Akhtar, Biochem. J. 114, 75 (1969).
88. T. Morato, M. Hayano, R.I. Dorfman and L.R. Axelrod, Biochem. Biophys, Res. Commun. 6, 334 (1961).
89. R.P.A. Sneedon and R.B. Turner, J. Amer. Chem. Soc. 77, 130 (1955).
90. H. Hagiwara, J. Pharm. Soc. Japan 80, 1671 (1960).
91. M. Ehrenstein and K. Otto, J. Org. Chem. 24, 2006 (1959).
92. C.J. Sih and A.M. Rahim, J. Pharm. Sci. 52, 1075 (1963).
93. R.M. Dodson and R.D. Muir, J. Amer. Chem. Soc. 83, 4627, 4631 (1961).
94. H.J. Brodie, G. Possanza, and J.D. Townsley, Biochim. Biophys. Acta 152, 770 (1968).
95. T. Morato, K. Raab, H.J. Brodie, M. Hayano, and R.I. Dorfman, J. Amer. Chem. Soc. 84, 3764 (1962).
96. S.M. Kupchan, C.J. Sih, N. Katsui, and O.E. Tayeb, J. Amer. Chem. Soc. 84, 1752 (1962).
97. H.R. Levy and P. Talalay, J. Amer. Chem. Soc. 79, 2658 (1957).
98. S. Kushinsky, J. Biol. Chem. 230, 31 (1958); C. Gaul, R.I. Dorfman, and S.R. Stitch. Biochim. Biophys. Acta 49, 387 (1961).
99. J.A. Zderic, A. Bowers, H. Carpio, and C. Djerassi, J. Amer. Chem. Soc. 80, 2596 (1958).
100. S.C. Pan, J. Semar, B. Junta, and P.A. Principe, Biotechnol. Bioeng. 11, 1183 (1969).
101. C.J. Sih, Biochem. Biophys. Res. Commun. 7, 87 (1962).
102. C.J. Sih, and R.E. Bennett, Biochim. Biophys. Acta 56, 584 (1962); C.J. Sih and K.C. Wang, J. Amer. Chem. Soc. 85, 2135 (1963). See also ref. 101.

103. K. Schubert, K.H. Böhme, and C. Hörhold, Steroids 4, 581
     (1964).
104. M. Nagasawa, M. Bae, G. Tamura, and K. Arima, Agr. Biol.
     Chem. 33, 1644 (1969).
105. J.D. Townsley and H.J. Brodie, Biochemistry 7, 33 (1968).
106. J.D. Townsley and H.J. Brodie, Biochem. J. 101, 25c (1966).
107. H.J. Brodie, K.J. Kripalani, and G. Possanza, J. Amer.
     Chem. Soc. 91, 1241 (1969).
108. J. Fishman and H. Guzik, J. Amer. Chem. Soc. 91, 2805 (1969);
     J. Fishman, H. Guzik, and D. Dixon, Biochemistry 8, 4304
     (1969).
109. R.A. Meigs and K.J. Ryan, Fed. Proc. Abstr. No. 2275, 28.
     660 (1969); J. Biol. Chem. 246, 83 (1971).
110. H.I. Bulker, R.D.H. Heard, and V.J. O'Donnell, Endocrinology
     60, 214 (1957).
111. L. Ainsworth and K.J. Ryan, Endocrinology 79, 875 (1966).
112. R.D.H. Heard, P.H. Jellinck, and V.J. O'Donnell, Endocrino-
     logy 57, 200 (1955).
113. H. Breuer and Ch. Mittermayer, Biochem. J. 86, 12P (1963).
114. R.D.H. Heard and V.J. O'Donnell, Endocrinology 54, 209
     (1954).
115. R.D.H. Heard, H.R. Jacobs, V.J. O'Donnell and F.G. Péron,
     J.C. Saffran, S. Solomon, L.M. Thompson, H. Willoughby,
     and C.H. Yates, Rec. Progr. Hormone Res. 9, 383 (1954).
116. P.D.G. Dean, D. Exley, and T.W. Goodwin, Phytochemistry 10,
     2215 (1971).
117. L. Ainsworth and K.J. Ryan, Endocrinology 79, 875 (1966).
118. K.L. Cheo and K.L. Loke, Steroids 11, 603 (1968).
119. L. Stárka and H. Breuer, Hoppe-Seiler's Z. Physiol. Chem.
     344, 124 (1966).
120. L. Stárka, H. Breuer, and L. Cedard, J. Endocrinol. 34,
     477 (1966).
121. M.L. Givner, G. Schilling, and D. Dvornik, Endocrinology
     83, 984 (1968),
122. B.R. Bhavnani, R.V. Short, and S. Solomon, Endocrinology
     85, 1172 (1969).
123. R. Hirschmann, C.S. Snoddy, C.F. Hiskey, and N.L. Wendler,
     J. Amer. Chem. Soc. 76, 4013 (1954).
124. M.M. Blight, J.J.W. Coppen, and J.F. Grove, Chem. Commun.
     1117 (1968); J. Chem. Soc. (C) 549 (1969).
125. J.S. Moffatt, J.D. Bu'Lock, and T.H. Yuen, Chem. Commun. 839
     (1969).
126. J. MacMillan, A.E. Vanstone, and S.K. Yeboah, Chem. Commun.
     613 (1968); J. MacMillan, T.J. Simpson, and S.K. Yeboah,
     ibid 1063 (1972).
127. R.Oh and B. Tamaoki, Acta Endocrinol. 67, 665 (1971).

128. H.J. Ringold, M. Hayano, and V. Stefanovic, J. Biol. Chem.
     238, 1960 (1963); Y.J.A. Hajj, ibid 247, 686 (1972).
129. T. Anjyo, M. Ito, H. Hosoda, and T. Nambara, Chem. Ind.
     (London) 385, (1972), and refs. there; Chem. Pharm.
     Bull. 20, 853 (1972).
130. L. Milewich and L.R. Axelrod, J. Endocrinol. 52, 137 (1972);
     56, 227 (1973).
131. Y. Osawa and D.G. Spaeth, Biochemistry 10, 66 (1971);
     J. Clin. Endocrinol. Metab. 38, 783 (1974).
132. L. Milewich and L.R. Axelrod, Endocrinology, 91, 1101
     (1972).
133. B.R. Bhavnani, R.V. Short, and S. Solomon, Endocrinology
     89, 1152 (1971).
134. D. Kleupfel and C. Vézina, Appl. Microbiol. 20, 515 (1970).
135. S.N. Sehgal and C. Vézina, Appl. Microbiol. 20, 875
     (1970).
136. R.B. Bates and D.J. Eckert, J. Amer. Chem. Soc. 94, 8258
     (1972).
137. M.J. Begley, L. Crombie, P.J. Ham, and D.A. Whiting,
     Chem. Commun. 1250 (1972).
138. B.R. Bhavnani and R.V. Short, Endocrinology 92, 657 (1973).
139. E. Lacroix, W. Eechaute, and I. Leusen, Steroids 23, 337
     (1974).
140. E.A. Thompson and P.K. Siiteri, Ann. N.Y. Acad. Sci. 212.
     378 (1973).

CHAPTER 25

Additional Biosynthetic Sequences Leading to Aromatic Rings

In the preceeding chapters an attempt has been made to summarize the extent of our present knowledge of the formation of aromatic structures by pathways which are of fairly general validity, or which have been the object of detailed investigation, such as the one leading to riboflavine. However, these cases do not by any means exhaust the possibilities for the biosynthesis of benzenoid structures; it would be surprising indeed if the immense diversity of synthetic activities of living organisms did not include examples of other pathways, and if there were no biochemical counterparts for some at least of the many remaining ways by which aromatic structures can be formed *in vitro* from nonbenzenoid precursors.

That such additional reaction types are biochemically feasible was shown as early as 1863 by Lautemann (1), who demonstrated the conversion of quinic to hippuric acid in man. Further extensive research into the aromatization of a variety of hydroaromatic compounds by the mammalian organism was carried out by Bernhard and by Dickens and his co-workers; the work has been reviewed by Dickens (2). The conversion of cyclohexane carboxylic acid and several of its derivatives into benzoic acid (literature in ref. 2), and of $\Delta^{4,6}$-dihydro-*o*-toluic acid into toluic acid (3) may be quoted as examples. The aromatization of hexahydrobenzoic acid has been studied by Mitoma et al. (4) and by Babior and Bloch (5). The reaction takes place in liver mitochondria in the presence of a variety of co-factors. The enzyme system has been purified (5), and the soluble enzymes have been shown to convert cyclohexane-carboxyl-CoA to benzoyl-CoA by way of 1-cyclohexenecarboxyl-CoA; the CoA- derivatives of 1- and 3-cyclohexenecarboxylic acids are aromatized by the enzyme system, but there is evidence against their being intermediates. Oxygen is not necessary and the reaction therefore proceeds without a hydroxylation step. *Corynebacterium cyclohexanicum* grows on cyclohexane carboxylic acid as carbon source and forms large amounts of 4-ketocyclohexane carboxylic acid as a major metabolite. Incubation of the metabolite with a crude extract of the organism under aerobic conditions produces *p*-hydroxy-benzoic acid (15).

While these investigations serve to point out biogenetic possibilities, such aromatizations of unnatural substances properly belong to the field of detoxification rather than that of biosynthesis.

Into the same category belongs the interesting conversion of *trans*-3,5-cyclohexadienediol into pyrocatechol (6); the reaction is carried out by an enzyme system (diol-dehydrogenase) isolated from a cell-free extract of rat liver. 1,2-Dihydroxy-1,2-dihydronaphthalene was similarly aromatized. By contrast, an extract from an *Aerobacter* sp. aromatized only the monocyclic compound, requiring DPN rather than TPN (cf. also Chapter 8).

Presumably related is the formation of tetrahydroxybenzo-quinone from *meso*- or *i*- inositol by *Pseudomonas beijerinckii*, which has been studied by Kluyver et al. (7). Actually, hexa-hydroxybenzene seems to be the primary product of this reaction, the quinone being formed by subsequent autoxidation. Tetrahy-droxybenzoquinone can not be formed from several compounds related to the inositols; according to Weygand et al. (8), activity is incorporated from *meso*-inositol but not from acetate or serine. The quinone thus seems to arise through a direct dehydrogenation rather than by fragmentation of the parent molecule; the dehydrogenase enzyme from *P. beijerinckii* has been characterized by Dworsky and Hoffmann–Ostenhof (9). The conversion of quercitol (1,2,3,4,5-pentahydroxycyclohexane) to pyrogallol by *Pseudomonas aromatica* (10) may be related to the transformation discussed above.

The mitomycin antibiotics elaborated by several *Streptomyces* spp. contain a benzoquinone grouping which seems to be formed more or less directly from glucose or ribose. (11); neither acetate nor the aromatic amino acids were incorporated, but D-glucosamine seemed to be an efficient precursor. That this aromatization may be related to the ones described above is suggested by the observation (12) that glucose can be converted into C-methyl inositol via a heptose intermediate by the alga *Chlorella fusa*. Further work on the biosynthesis of mitomycin B [1] by *S. verticillatus* (13) has confirmed the ready incor-poration of D-glucosamine; D-[1-^{14}C,^{15}N]-glucosamine was incor-porated into the antibiotic in a way which suggested that the

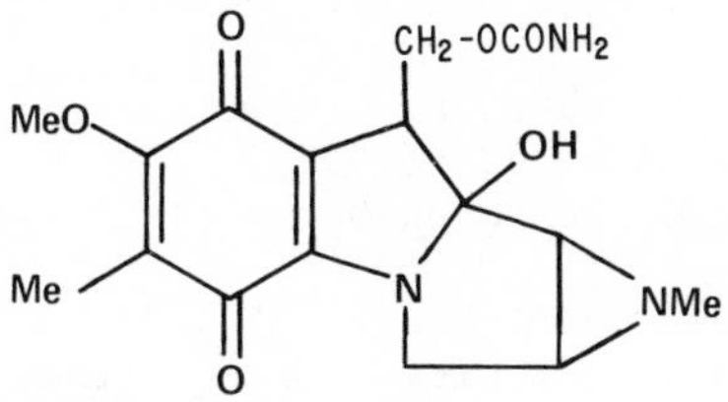

MeO

O

CH₂-OCONH₂

OH

Me

N

NMe

O

[1]

hexosamine can provide the nitrogen atom of the aziridine ring
in a specific manner, although a small proportion of the ^{15}N
label is randomized and incorporated into the two other nitrogen
atoms. Feeding [1-^{14}C, 6-^{3}H]-glucosamine (14) has demonstrated
the specific incorporation of the amino sugar. Shikimic acid
does not contribute to the aromatic ring, but pyruvic acid-1-^{14}C
provides the methyl group of the quinone; it is suggested that
dehydroquinic acid may be an immediate precursor.

The preceding examples of aromatization are supported.
by experimental evidence; there are, however many instances of
presumed aromatization, by pathways other than those of general
applicability, for which there is no proof and which rest on
inference alone. We have mentioned examples of such presumed
secondary aromatization in those sections dealing with shikimate-
and mevalonate-derived metabolites, where most of the examples
cited were based upon the co-occurrence of a fully aromatic
metabolite with its presumed nonaromatic precursor (see Chapter
16 and 22, respectively). In Section 17-2 the formation of the
central aromatic rings of perylene and *meso*-naphthobianthrone
derivatives was mentioned, as was the relationship between actiphenol and its alicyclic relatives, cycloheximide and the streptovitacins.

REFERENCES

1. E. Lautemann, Ann. 125, 9 (1863).
2. F. Dickens, Biochem. Soc. Symp. No. 5, 66 (1950).
3. C. Grundmann and I. Löw, Z. Physiol. Chem. 256, 141 (1938).
4. C. Mitoma, H.S. Posner, and F. Leonard, Biochim. Biophys.
    Acta 27, 156 (1958).
5. B.M. Babior and K. Bloch, J. Biol. Chem. 241, 3643 (1966).
6. P.K. Ayengar, O. Hayaishi, M. Nakajima, and I. Tomida,
    Biochim. Biophys. Acta 33, 111 (1959).

7. A.J. Kluyver, T. Hof, and A.G.J. Boezaardt, Enzymologia 7, 257 (1939).
8. F. Weygand, W. Brucker, H. Grisebach, and E. Schulze, Z. Naturforsch. 12b, 222 (1957).
9. P. Dworsky and O. Hoffmann-Ostenhof, Monatsh. 100, 1327 (1969).
10. M.W. Beijerinck, Chem. Zentr. 82 1232 (1911).
11. U. Hornemann and J.C. Cloyd, Chem. Commun. 301 (1971); G.S. Bezanson and L.C. Vining, Can. J. Biochem. 49, 911 (1971).
12. G. Wöber and O. Hoffmann-Ostenhof, Monatsh. 100, 369 (1969).
13. U. Hornemann and M.J. Aikman, Chem. Commun. 88 (1973).
14. U. Hornemann and M.J. Aikman, Lloydia 35, 470 (1972); U. Hornemann, J.P. Kehrer, and J. H. Eggert, Chem. Commun. 1045 (1974).
15. T. Kaneda, Biochem. Biophys. Res. Commun. 58, 140 (1974).

Index